Continuum Methods
of Physical Modeling

Springer-Verlag Berlin Heidelberg GmbH

Kolumban Hutter
Klaus Jöhnk

Continuum Methods of Physical Modeling

Continuum Mechanics, Dimensional Analysis, Turbulence

With 61 Figures, 14 Tables, 113 Exercises and Solutions

Springer

Professor Kolumban Hutter, Ph.D.
Technische Universität Darmstadt
Institut für Mechanik
Hochschulstrasse 1
64289 Darmstadt, Germany

Dr. Klaus D. Jöhnk
FNWI/IBED
University of Amsterdam
Nieuwe Achtergracht 127
1018WS Amsterdam, The Netherlands

The cover pictures: Laboratory avalanche simulation with a mixture of sand and gravel, at the Department of Mechanics, Darmstadt University of Technology, Germany and a powder snow avalanche in the Nepalese Himalaya, (Photo F. Tschirky, courtesy of Swiss Federal Institute of Snow and Avalanche Research, Davos, Switzerland).

Cataloging-in-Publication Data applied for

Bibliographic information published by Die Deutsche Bibliothek
Die Deutsche Bibliothek lists this publication in the Deutsche Nationalbibliografie; detailed bibliographic data is available in the Internet at <http://dnb.ddb.de>.

DOI 10.1007/978-3-662-06402-3

springeronline.com

Originally published by Springer-Verlag Berlin Heidelberg New York in 2004
MyCopy version of the original edition 2004

Typesetting by the authors using a Springer TEX macro package
Cover design: Erich Kirchner, Heidelberg

Printed on acid-free paper 55/3141/tr 5 4 3 2 1 0
www.springer.com/mycopy

Preface

This book is a considerable outgrowth of lecture notes on *Mechanics of environmentally related systems I*, which I hold since more than ten years in the Department of Mechanics at the Darmstadt University of Technology for upper level students majoring in mechanics, mathematics, physics and the classical engineering sciences. These lectures form a canon of courses over three semesters in which I present the foundations of continuum physics (first semester), those of physical oceanography and limnology (second semester) and those of soil, snow and ice physics in the geophysical context (third semester). The intention is to build an understanding of the mathematical foundations of the mentioned geophysical research fields combined with a corresponding understanding of the regional, but equally also the global, processes that govern the climate dynamics of our globe. The present book contains the material (and extensions of it) of the first semester; it gives an introduction into *continuum thermomechanics*, the methods of *dimensional analysis* and *turbulence modeling*. All these themes belong today to the everyday working methods of not only environmental physicists but equally also those engineers, who are confronted with continuous systems of solid and fluid mechanics, soil mechanics and generally the mechanics and thermodynamics of heterogeneous systems. The book addresses a broad spectrum of researchers, both at Universities and Research Laboratories who wish to familiarize themselves with the methods of "rational" continuum physics, and students from engineering and classical continuum physics.

Why, however, the threefold division in continuum thermodynamics, dimensional analysis and turbulence modeling? There are several reasons to this end. First, turbulence theory today is part of the working methods of every fluid dynamicist, especially in the geophysical context, such as meteorology, oceanography, limnology, not to mention all the technical applications in environmental and mechanical engineering. Second, turbulence research has, in the last twenty years perfected its theoretical formulation to such an extent, that one may well try to present some of its aspects from a viewpoint of general continuum mechanics. Third, it has become apparent in the past few years that for those aspects we are interested in, continuum thermodynamics possesses the right underlying structure to treat turbulence modeling in a particularly systematic fashion. In other words, one may base the constitutive theory of continuous materials on essentially the same, or at least very similar, concepts as the formulation of closure conditions in turbulence modeling. To my knowledge, such an approach has not been presented so far in book form. The advantage of such an approach is, however, a considerable increase of the transparency of the material to be learned, a clearer enlightenment into the concepts (which are indeed not very far apart) and, probably most effectively, a reduction of the amount of the topics that are new and must be absorbed for the first time. I emphasize here my opinion that turbulence modeling can profit from an access by continuum mechanics.

Which role is now played by dimensional analysis in this context? For one, modeling in turbulence theory, especially when closure conditions must be postulated, depends to a large degree upon simple concepts of dimensional analysis. Admittedly, this could well be presented without the explicit development of the BUCKINGHAM theorem. However, a clear and relatively rigorous presentation of the methods of dimensional analysis surely facilitates the basic understanding. Moreover, it is a simple fact that, because of the dimensional homogeneity of all equations in mathematical physics, a first bold understanding of a physical problem is gained with the aid of methods of dimensional analysis. At last this same statement also holds for rational continuum mechanics and has always been emphasized by its founder C.A. TRUESDELL. I concur, and this is why we give here a brief and incomplete introduction into this fascinating field of mathematical physics. Dimensional analysis precedes turbulence modeling in this book, because the former is used for the latter much more than vice versa.

A word about the role of thermodynamics seems equally to be in order. Today's researchers in rational continuum thermodynamics largely use the CLAUSIUS–DUHEM inequality and the COLEMAN–NOLL approach in deducing results in the particular research they are pursuing. In this book this approach towards the second law of thermodynamics will also be explained[1], but in a number of applications the more general entropy principle of MÜLLER will be used[2]. In so doing it will become apparent that the CLAUSIUS–DUHEM inequality, paired with the COLEMAN–NOLL approach of its exploitation would have been too restrictive in those cases and erroneous results would have been obtained. In this regard this book goes beyond most of the classical treatments of rational thermodynamics of the last two to three decades.

The intention of this book is, apart from presenting its treated subjects, a clear and (somewhat) rigorous mathematical presentation of them on the basis of limited knowledge as a prerequisite. Calculus or analysis of functions of a single and several variables, linear algebra and (only) the basics of ordinary and partial differential equations are assumed to be known (or having been learned once). Those subjects roughly form the mathematical tool which engineers in Germany learn during the first two years of their

[1] See e.g. C. A. TRUESDELL [243].

[2] I. MÜLLER [165].

university education. In the American system, senior undergraduate or first year graduate-education level is about the background needed to follow the material in this book. On the side of physics knowledge of strength of materials and dynamics or analytical dynamics courses or a basic course in classical physics should suffice to be able to follow the presented concepts.

Even though it is hoped that the book will also be used as a source book by researchers in the broad field of continuum physics, its intention is essentially to form a basis for teaching (and even more so learning). Great care has therefore been devoted in each chapter to formulate a number of exercises, and solutions are given in detail to most of these. The latter is justified for the following reasons: Often, the problems formulated in the exercises constitute complementary material to that presented in the main text of the respective chapters. Occasionally, a thought in a derivation of a certain fact is only briefly touched upon in the main text and the reader is asked to work out the details by himself/herself. At an other time a fact that is needed in the development of the material is only mentioned in the main text, and the reader is asked to corroborate the statement as an exercise. And, finally, additional material that could also be treated in the main text is explained in the exercises as an individual problem. In all these cases knowledge of the material dealt with in the exercises is assumed to be known in later chapters. This is also the reason why solutions to the stated problems are nearly completely outlined. A natural fringe benefit for the reader is obviously the fact of a self-control in his attempts to solve the problems. Most problems were stated for and solved in recitation hours with the students; considerable input has thereby been given by the students for which we express our sincere thanks.

The book has been drafted (first) in the German language jointly by both authors. First versions of Chaps. 1 to 6 were written from lecture notes of K. HUTTER by K. JÖHNK, when the second author was a postdoctoral assistant of the former. Chap. 7 and the two chapters of Part II: Dimensional Analysis were exclusively written by K. HUTTER. Rough drafts of Part III: Turbulence, i.e., Chaps. 10, 11 and 13 were written by K. JÖHNK, whilst Chap. 12 is due to K. HUTTER.

Most problem formulations and their solutions are due to K. HUTTER, but there is a number which are due to K. JÖHNK. Because of different professional assignments of K. JÖHNK since November 1997, the two authors were locally separated. This, together with K. JÖHNK more industrious professional involvement, made a collaboration as a consequence virtually impossible. For this reason, the homogenisation of the entire manuscript, the careful testing, reading and again reading, the dotting of all the i's, the incorporation of the References and the Index are all due to K. HUTTER. Consequently, even though we are both joint authors of the book, K. HUTTER is the sole author responsible for all the errors which still remain. He is particularly thankful to all the readers who would point out to him where they arise. A simple note by e-mail: hutter@mechanik.tu-darmstadt.de will suffice.

The English version now presented was drafted by K. HUTTER with the help of Dr. D. RAJ BARAL. Beside this invaluable help we also were assisted by my secretaries Mrs. R. DANNER and R. RUTSCHER and assistants H. HÜTTEMANN, A. DIENG, S. KTITAREVA, Y.-CH. TAI, E. VASSILIEVA, Y. WANG and A. WILLUWEIT, to all of whom we express our sincere thanks.

Before I finish this Preface let me state that writing a book can never be finished, a book has to be abandoned! This I am now going to do, well knowing that it bears its weaknesses, that I would now know how to do it better and being well aware that while pushing this project through all its stages needed isolation and separation from the beloved family members, who all deserve my deepest gratitude.

Darmstadt, Autumn 2003 *Kolumban Hutter*

Contents

Part II. Dimensional Analysis

Part III. Turbulence

Introduction

As already briefly outlined in the preface, this book is designed to give an introduction to the analytical methods of classical environmental physics in a form how it is needed by solid and fluid dynamicists in the context of geophysical applications. The original selection of the material and the completion of the first text has in the sequel been subject to several significant changes. The intention thereby was to select the content of the topics such that it could equally serve as a basis to any engineer who was interested in continuum mechanical material theory, be he involved with technical flows in complex geometries under conditions requiring advanced methods of turbulent closure, or a chemical engineer, soil mechanician, physicist, etc., to become acquainted with the *mathematical methods* of *physical modeling.* Indeed, irrespective of whether technical applications are in focus or systems of our environment are studied, an impressive work in an attempt to understand the dynamics of technical or environmental systems first lies in the deduction of the mathematical models which describe them.

In classical continuum physics an essential part of this step is the application of the fundamental laws of physics such as the conservation laws of mass, linear and angular momentum and energy as well as the second law of thermodynamics (and possibly the balance laws of electrodynamics in the form of the MAXWELL equations which shall not be dealt with in this book). These fundamental laws represent those mathematical statements which are valid for all bodies irrespective of their material behaviour; within the context of classical GALILEIan physics they are common to all physical phenomena and possess for this reason the notion of absolute truth.

However, the individual materials differ from one another, and so it is necessary to complement the mentioned balance laws that are valid for all media by separation statements which express that the individual materials react differently under the applied external driving mechanisms. These statements concern on the one hand the so-called constitutive relations of the material theory, on the other hand the *parameterization of the processes* which may take place on the level of the *substructure*, i.e., the behaviour of the processes on length and time scales which are not resolved in a model under consideration. The fields of science in which these questions are dealt with are continuum thermodynamics and turbulence theory.

Continuum mechanics and *continuum thermodynamics*[1] are the sciences which, in the past, have exclusively been dealing with the postulation of constitutive relations that concern the material properties of bodies subject to mechanical and thermal loads. Conceptually both are today essentially closed theories, in which rules and methods are formulated with the aid of which one may, for a particular material under certain processes, formulate and possibly reduce materially dependent equations in a physically correct, i.e., objective form. As a result so-called constitutive relations or material equations will emerge which are in conformity with the irreversibility requirements of physics (the second law of thermodynamics), but which are nevertheless sufficiently general that the material equations proposed in this form characterize a certain constitutive class. In other words, continuum thermodynamics is a theory which provides rules with the help of which postulated constitutive relations of a certain class may be so reduced that all those dependences, which would result in violation with the rules of physical objectivity, irreversibility (and other requirements to be formulated), are eliminated. At last, general constitutive laws will emerge for a certain class of material, which contain free parameter functions, which for a particular material in the postulated class of behaviour must be determined by the experimentalist.

Typical classes are elastic, thermoelastic, viscous and thermo-viscoelastic materials, but equally also the so-called rate independent plastic behaviour that is independent of the speeds by which processes are traversed. Even though the above mentioned classes of materials are typical for many ap-

[1] There are numerous textbooks dealing with continuum mechanics at different levels and with different intentions so that it is nearly impossible to present a balanced, let alone a complete, account. Introductory works written at times when continuum mechanics was in its early stages, or texts specifically aiming at an introduction are e.g. ERINGEN [66], [67]; JAUNZEMIS [113], LEIGH [132], MALVERN [143], BECKER & BÜRGER [40], CHADWICK [46], FUNG [81], SPENCER [218], GURTIN [91], ALTENBACH & ALTENBACH [12] and GREVE [88]. Specialized and more advanced texts are e.g. by WANG & TRUESDELL [252], MARSDEN & HUGHES [145], OGDEN [180], MÜLLER [165], GIESEKUS [83], WILMANSKI [256], ALBER [11], HAUPT [98] and LIU [138].

The basic monographs recommended for anyone having acquired some elementary knowledge of continuum mechanics, are however still the two seminal works of TRUESDELL & TOUPIN [238] and TRUESDELL & NOLL [240] on the classical and non-linear field theories of mechanics. These works provide a comprehensive representation of continuum mechanics with a wealth of quotations from the literature in a balanced historical context up to the mid sixties of the last century, which – with the exception of thermodynamics – is still representing the current understanding of continuum mechanics. The state of knowledge of rational thermodynamics, in the mid sixties just at its rapid development, is treated by TRUESDELL [243] in his monograph "Rational Thermodynamics" (2nd ed), but it is limited to the CLAUSIUS–DUHEM inequality and the COLEMAN–NOLL approach as a basis to the entropy imbalance. A review of more general approaches to the second law of thermodynamics can by found in HUTTER [104] and HUTTER & WANG [111]. Treatises on one such alternative are given by MÜLLER [163], [165] and MÜLLER & RUGGERI [167].

plications and their knowledge builds part of the tools of today's structural engineers, it is our experience that specialists in environmental physics are generally not so well familiar with them. There are, however, a number of applications in which the utmost demands of material complexity are required. For instance, sea ice under the driving atmosphere and ocean may be treated as a two-dimensional granular continuum with viscoplastic constitutive response; and the slow creeping deformation of snow in its deposit subject to the varying meteorological conditions is modeled thermomechanically as a mixture of snow, air, water and vapour with various phase change processes being possible and with a thermomechanically coupled viscoelastic response, the most complex behaviour one may think of.

Physical oceanography, limnology and meteorology are, however, special fields of fluid mechanics of ideal, viscous compressible or incompressible fluids and conceptionally therefore relatively simple. The most simple form of a material equation for the creeping deformation of ice in glaciers and large ice sheets, such as Greenland and Antarctica, is a thermomechanically coupled fluid, in which the fluidity (inverse viscosity) depends upon the applied stress and the temperature and the stress dependence is usually assumed to follow a power law, or more generally a polynomial law. If the natural impurities (dust, salts) and the induced anisotropy are accounted for, one is confronted with a material which exhibits the most complex constitutive behaviour that one may encounter today in material modeling. These examples may suffice to make clear to the reader that the environmental or climate physicist ought to know modern continuum mechanics as much as an engineer or material scientist does.

Snow, soil, but also temperate ice are *heterogeneous media*, i.e., materials composed of several constituents[2]. Temperate ice is a mixture of ice with water inclusions, snow consists of a granular "skeleton" made of ice crystals, air, water and water vapour, soil is a mingling of sand, water and air. In problems of dispersion or diffusion of a pollutant in the air or the water of our rivers, lakes or the ocean, the considered "fluid" is equally regarded as

---

[2] Heterogeneous media belong to the class of mixtures of immiscible constituents. Granular media, such as sand and soil, debris, suspension, mud and snow belong to them, but equally also porous media, such as sandstone, foams, sponges etc. Books that treat such media are scattered over a wide area of scientific expertise and include also accounts on the classical theories of mixtures of miscible constituents. Continuum thermodynamics of classical theories of mixtures are those of TRUESDELL [243] and MÜLLER [165]. Mixtures of heterogeneous bodies are treated in TRUESDELL [243], WILMANSKI [256] and GRAY [86]. Soil mechanical aspects are dealt with e.g. in BEAR [23], LEWIS & SCHREFLER [135], and COUSSY [52] and overviews (in the German language) can be found in EHLERS [62] and BLUHM [30]. Basic to the subject are the seminal articles by BIOT [26], [27] and those by GRAY & HASSANIAZADEH [95], [96], [97]. While this list is certainly not complete, a review, emphasising the historical development of soil mechanics is given by DE BOER [33], which gives a large number of references which, however, is incomplete in the very recent developments.

a mixture of the "bearing" fluid and a number of tracers. If these tracers are bio-chemically reacting substances (as for instance phyto- or zooplankton and phosphate and nitrate in the water of a lake or the ocean), then these concentrations are variables of a nutrient model. Likewise, the balance of the chemical elements, which describe the dynamics of ozone in the atmosphere may be interpreted as tracer balances, in which chemical reactions and the action of the sun monitor the mass productions of the individual tracer components. All the above examples are interpretable as *mixtures*; this is why mixtures are important concepts for the environmental physicist. For this reason it may now become understandable why in this book a relatively large portion is devoted to the continuum mechanical development of mixtures. In so doing not only the physical balance laws of mixtures for constituents with spin are dealt with, but equally also the material (constitutive) theory based on the thermomechanical irreversibility requirements. This procedure is regarded as optimal, because it allows us to build a quite natural understanding of the diffusion models of the so-called BOUSSINESQ fluids in turbulence theory; apart from this, prerequisites are thereby established which make the complicated formulations for mixtures such as soils, temperate ice, sea ice, etc. relatively easily understood.

Most technical fluid mechanical problems and likewise fluid flows in the geophysical and environmental context are not *laminar* but *turbulent*. For instance, if one observes the smoke of a quietly held cigarette, then in the first few centimeters its streakline forms a thin, coherent filament of soot particles; at a certain distance this filament is torn apart, and a chaotically looking whirling motion involving eddies is observed, which does not at all resemble the filament structure before. Usually, the air is not at rest, and the *smoky filament* is torn apart earlier and spread. Analogous behaviour can be observed in many fluid mechanical problems, however, in a much more intensified form. No well ordered motion is observed in these cases, but rather a motion which can be subdivided into an ordered mean motion and a superimposed fluctuating part. The latter consists of a number of large, smaller and very small eddies. Their form and size seems to depend only upon the flow of the fluid but not its material properties. Often it is so that one is not interested in the exact formation and structure of the eddies but only on the smooth mean motion. It seems plausible to assume that this mean motion is influenced by the kind, form and intensity of the "suspended" vortices and gyres; in other words, there must exist an interaction between the mean motion and the eddies. The description of this interaction between the micro- and macrostructure forms in the theory of turbulence the problem of closure of the equations and corresponds in material theory to the formulation of constitutive relations.

Therefore, *two separate steps are significant in the derivation of the equations of motion for turbulent processes.* First, this is the derivation of the balance laws for the mean motion from the physical balance laws of the entire motion[3]. This step involves an averaging procedure that traces back to OSBORN REYNOLDS [194]. The averaged equations contain terms, which a priori are not directly expressible as functions of the mean variables. These so-called correlation products are exactly those quantities, with the aid of which the second step can be completed, that is the formulation of the *turbulent closure conditions.* The correlation products must be expressed in terms of the mean field variables; the mathematical steps that are performed or the mathematical relations that are established correspond nearly identically to those performed in the constitutive theory of continuum mechanics. It is for this reason not only attractive but even more so very meaningful, to use this parallelism and to search for analogies. Indeed, it is so that establishing the closure conditions in turbulence theory is completely analogous to the formulation of the constitutive relations for gases in extended thermodynamics. The analogy with continuum thermodynamics is indeed evident and the common description on the basis of an unified concept is therefore not only attractive but compelling. This appears to be new in this book.

Both, the thermomechanical constitutive theory and the theory of turbulent closure are phenomenological theories, which leave a relatively large room for ad hoc assumptions involving free parameters. The finding of functional relations for these parameters is substantially facilitated if one employs to this end the methods of *dimensional analysis*[4]. In the postulation of turbulent closure conditions such methods are heavily used. If one therefore attempts to present a description of closure, the concepts of dimensional analysis and model theory appears to be indispensable. Dimensional analysis is, however, also of interest in its own right. It often offers to the engineer or physicist a method to find for a certain physical problem the significant dependences on the basis of a list of variables that might influence the process under consideration. Furthermore it may be advantageous in a study of a physical process to build a small size reproduction of the original set-up and to perform experiments at this smaller scale. The rules of transformation of measured physical quantities to the size of the prototype form the so-called *model theory*; to know them, and more generally to learn if and how down-

[3] Today's literature on turbulence is abundant and contains a large number of excellent treatises not all of which can be mentioned here. Mention might be made here of FRISCH [80], HINZE [101], MCCOMB [153], PIQUET [182], ROTTA [198], TENNEKES & LUMLEY [235]. All these works concentrate on the mechanics and heat flow of turbulence rather than its thermodynamics. A systematic treatment of turbulence from the viewpoint of irreversibility does not yet exist to our knowledge. First steps towards a turbulence theory incorporating the second law of thermodynamics were undertaken by AHMADI [2], [3], [5], [6], [7], [8], MARSHALL & NAGHDI [147], [148] and are based on the CLAUSIUS–DUHEM inequality and the COLEMAN–NOLL [50], [51] approach. The more flexible entropy principle of MÜLLER was generalised to turbulence closure by SADIKI & HUTTER [204], [205] and is explained in detail in this book.

[4] There are a number of books which are exclusively devoted to dimensional analysis; LANGHAAR [125], GOERTLER [84], SPURK [227] and BARENBLATT [19].

scaling of physical processes is possible, is a useful tool to all those who are confronted with the modeling of physical phenomena. In this book, dimensional analysis is dealt with in its middle part, however without an account of the modern group theoretic formulation. Nevertheless the theory is explained with the aid of a large number of introductory and explanatory examples, so that the beauty of dimensional analysis becomes understandable: to provide a deep and insightful understanding of physical processes with a comparably small investment of theoretical concepts.

Judging its content, the present book contains about material that can be taught in a two semester course for advanced undergraduate or first year graduate students. Our own preference is a selective choice of the material which concentrates in both its continuum mechanical as well as in the turbulence theory on the most significant principles and to let the students work through the chapters on dimensional analysis and model theory by themselves in seminars. With the worked-out exercises such a procedure is defendable. In this way a representative cross section through the treated topics can be taught and learned, respectively, in one semester, however not without hard work on the part of the student.

Part I

Continuum Mechanics

1. Basic Kinematics

1.1 Notation

List of the applied symbols:[1]

Symbol	Meaning
$\boldsymbol{a}$, $\boldsymbol{b}$, $\boldsymbol{c}$	General vectors
$\boldsymbol{A}$	General tensor of second rank, ALMANSI strain tensor
$\boldsymbol{A}^T$	Transpose of $\boldsymbol{A}$
$\boldsymbol{A}^{-1}$	Inverse of $\boldsymbol{A}$
$\boldsymbol{A}^{-T}$	Transpose of inverse of $\boldsymbol{A}$
$\mathcal{B}$	Body
$\mathcal{B}_R$	Reference configuration of the Body $\mathcal{B}$
$\mathcal{B}_t$	Present configuration of the Body $\mathcal{B}$
$\boldsymbol{B}$	Left–CAUCHY–GREEN Tensor (B_{ij})
$\boldsymbol{c}$	Translation
$\boldsymbol{C}$	Right–CAUCHY–GREEN Tensor ($C_{\alpha\beta}$)
$\mathfrak{c}^{\alpha}$	Volume production (rate) of mass of constituent α
dA, da	Surface elements, respectively, in the reference and present configuration
$d\boldsymbol{A}$, $d\boldsymbol{a}$	Surface elements in LAGRANGE and EULER representation
dV, dv	Volume elements, respectively, in the reference and present configuration; volume elements in LAGRANGE and EULER representation
Div, div	Divergence operators in LAGRANGE and EULER representation
$\boldsymbol{D}$	Strain rate tensor (D_{ij}), stretching tensor
$D^{\alpha}_{\xi\alpha}$	Mass diffusivity of constituent α
$\boldsymbol{D}^{\alpha}$	Relative stretching tensor of constituent α
$\boldsymbol{e}_i$	Basis vector ($i = 1, 2, 3$)
$\mathfrak{e}^{\alpha}$	Volume production (rate) of energy of constituent α
$\mathfrak{e}^{\alpha}_{Euclid}$	Non-convective volume production (rate) of energy of constituent α

[1] For vectors and tensors of higher ranks the component relations are given in brackets.

$\boldsymbol{E}$	FINGER strain tensor
$\mathcal{F}$	Flux of the physical variable $\mathcal{G}$ through the boundary $\partial\Omega$
$\boldsymbol{F}$, $\boldsymbol{f}$	General vector fields in LAGRANGE and EULER representation
$\boldsymbol{F}$	Deformation gradient with the components $F_{i\alpha}$
$\boldsymbol{g}$	Outer volume force (g_i) (normally the acceleration due to gravity)
Grad, grad	Gradient operators in LAGRANGE and EULER representation
$\mathcal{G}, \mathfrak{G}, \mathfrak{g}$ ...	Physical variable and its LAGRANGE and EULER representation
$\boldsymbol{G}$	GREEN strain tensor $(G_{\alpha\beta})$
$\mathbb{G}$	Symmetry group
h	Enthalpy
$\boldsymbol{H}$	Displacement gradient (H_{ij}), symmetry transformation
$I_{\boldsymbol{A}}$, $II_{\boldsymbol{A}}$, $III_{\boldsymbol{A}}$	Invariants of a three–dimensional tensor of second rank $\boldsymbol{A}$
J	Determinant of the deformation gradient $\boldsymbol{F}$
$\boldsymbol{j}^\alpha$	Diffusive mass flux of constituent α
k	Compression modulus
$\boldsymbol{k}$	Extra entropy flux vector
$\mathfrak{k}^\alpha$	Volume production (rate) of spin of constituent α
$\mathfrak{k}^\alpha_{Euclid}$ ...	Intrinsic or non-convective production density of momentum of constituent α
$\boldsymbol{K}$	PIOLA strain tensor
l^α	Specific spin supply of constituent α
L	Latent heat
$\boldsymbol{L}$	Spatial gradient of the velocity vector (L_{ij})
$m(\mathcal{B})$	Mass of the body $\mathcal{B}$
$\boldsymbol{m}^\alpha$	Couple stress tensor of constituent α (a third order tensor)
$\mathfrak{m}^\alpha$	Volume production (rate) of momentum of constituent α
$\mathfrak{m}^\alpha_{Euclid}$ ..	Intrinsic or non-convective production (rate)density of momentum of constituent α
$\boldsymbol{N}$, $\boldsymbol{n}$	Normal unit vectors in LAGRANGE and EULER representation (N_α, n_i),
$\boldsymbol{N}_{\mathfrak{S}}$, $\boldsymbol{n}_{\mathfrak{s}}$.	Normal vectors on the singular surface (reference, present configuration)
$\mathfrak{n} = \mathfrak{c}^\alpha/\hat{\rho}^\alpha$	Production rate density of volume fraction of constituent α
$\mathcal{O}$	Origin
$\boldsymbol{O}$	Orthogonal transformation (O_{ij})
$\mathbb{O}$	Group of orthogonal transformations
p	Pressure

$\mathcal{P}$ .... Production of the physical variable $\mathcal{G}$
$\mathfrak{P}$, $\mathfrak{p}$ ..... Surface production on singular surfaces
$\mathfrak{p}^\alpha$ .... Surface production (rate) of entropy of constituent α
$\boldsymbol{P}$ .... Gradient of transformation of the reference configuration
$\boldsymbol{Q}$, $\boldsymbol{q}$ ..... Heat flux vectors in LAGRANGE and EULER representation (Q_α, q_i)
$\mathfrak{r}$ .... Radiation
$\boldsymbol{R}$ .... Rotation(al) tensor ($R_{i\alpha}$)
$\mathbb{R}, \mathbb{R}^+$ ... Set of real numbers, set of non-negative real numbers
rot $\boldsymbol{a}$ .... Rotation of a vector field, $\operatorname{rot} \boldsymbol{a} = \nabla \times \boldsymbol{a}$
s .... Specific entropy
$\boldsymbol{s}^\alpha$ .... Specific spin density of constituent α (an axial vector)
$\mathfrak{S}$, $\mathfrak{s}$ ..... Singular surfaces (reference, present configuration)
$\mathfrak{S}^\alpha$ ...... Micromorphic spin production of constituent α
$\mathcal{S}$ .... Supply of the variable $\mathcal{G}$ in the region Ω
t, t_0 ..... Time, reference time
$\mathfrak{t}$ .... Stress vector
$\boldsymbol{t}$ .... CAUCHY stress tensor (t_{ij})
T .... Absolute (KELVIN) temperature
$\boldsymbol{T}$ .... First PIOLA–KIRCHHOFF stress tensor ($T_{i\alpha}$)
$\tilde{\boldsymbol{T}}$ .... Second PIOLA–KIRCHHOFF stress tensor ($\tilde{T}_{\alpha\beta}$)
tr $\boldsymbol{A}$ ..... Trace of the tensor $\boldsymbol{A}$
$\boldsymbol{u}$ .... Displacement (u_i)
$\boldsymbol{u}^\alpha = \boldsymbol{v}^\alpha - \boldsymbol{v}$ Diffusion velocity of constituent α
$\boldsymbol{U}$ .... Right stretch tensor ($U_{\alpha\beta}$)
$\boldsymbol{U}^\alpha$ ...... Relative deformation rate tensor of constituent α
$\mathbb{U}$ .... Group of unimodular transformations
$\boldsymbol{v}$ .... Velocity vector (v_i)
$\boldsymbol{v}^\alpha$ .... Velocity vector of constituent α
$\dot{\boldsymbol{v}}$ .... Acceleration
$V_\mathfrak{s}$ .... Propagation velocity of singular surfaces
$\boldsymbol{V}$ .... Left stretch tensor (V_{ij})
$\boldsymbol{V}^3_R, \boldsymbol{V}^3_t$. Three–dimensional vector space
$\boldsymbol{W}$, $\boldsymbol{w}$ ... Velocities of singular surfaces in reference (W_α) and present configuration (w_i)
$\boldsymbol{W}$ ...... Tensor of rotational velocity, vorticity or spin tensor (W_{ij}), of barycentric velocity
$\boldsymbol{W}^\alpha$ ..... Relative vorticity tensor of constituent α
$\mathfrak{X}$ .... Particle
$\dot{\boldsymbol{x}}$ .... Velocity vector
$\boldsymbol{X}$, $\boldsymbol{x}$ .... Position vector (X_α, x_i) for reference and present configuration
$\boldsymbol{Y}$, $\boldsymbol{y}$ .... Position vectors (Y_α, y_i)
$\boldsymbol{Z}$ .... Constraint stress

γ	Specific entropy production
Γ, γ	Specific densities of the physical variable $\mathcal{G}$ in LAGRANGE and EULER representation
$\Delta\boldsymbol{\omega}^\alpha$	Diffusive angular velocity of constituent α
δ_{ij}	KRONECKER symbol
ε	Internal energy
ε_D	Diffusive internal energy
ε_I	Inner internal energy
ϵ^α	Surface production (rate) of energy of constituent α
ε_{ijk}	Components of the fully antisymmetric tensor of third rank
$\boldsymbol{\varepsilon}$	Deformation tensor (ε_{ij})
$\xi^\alpha = \rho^\alpha/\rho$	Specific mass concentration of constituent α
η	Specific entropy supply
Θ	Temperature (absolute)
$\boldsymbol{\Theta}$	Barycentric specific moment of inertia (second order tensor)
$\boldsymbol{\Theta}^\alpha$	Specific moment of inertia of constituent α (second order tensor)
κ	Volume viscosity
κ_T	Coefficient of heat conduction
$\boldsymbol{\kappa}$	Transformation of reference configuration
Ψ_I	Inner HELMHOLTZ free energy
λ	Thermal conductivity
Λ	LAGRANGE multiplier
Λ^ε	LAGRANGE parameter for the energy equation
$\Lambda^{\boldsymbol{v}}$, $\Lambda^{\boldsymbol{v}}_\alpha$ ...	LAGRANGE parameter for the momentum equation, – of constituent α
Λ^ρ	LAGRANGE parameter for the balance of mass of the mixture
$\Lambda^{\xi\alpha}$	LAGRANGE parameter for the constituent mass balances
$\lambda_i^{(\boldsymbol{V})}$	i-th eigenvalue for the tensor $\boldsymbol{V}$
μ	Chemical potential
μ^α	Surface production (rate) of mass of constituent α
$\mu^{\xi\alpha}$	Chemical potential for constituent α
$\lambda^{\boldsymbol{A}}$, λ ...	Eigenvalue, of $\boldsymbol{A}$
μ, λ	Lamé parameters
μ	(Dynamic) viscosity
ν_α	Volume fraction of constituent α
Π, π	Specific productions in LAGRANGE and EULER representation for the physical variable $\mathcal{G}$
ρ_R, ρ	Densities (LAGRANGE, EULER representation)
Σ, ς	Specific supplies (LAGRANGE, EULER representation)

$\boldsymbol{\sigma}^{\alpha}$	......	Surface production (rate) of angular momentum of constituent α
τ	....	Time, history of a described material $\tau \in [0, \infty)$
$\boldsymbol{\tau}^{\alpha}$	......	Surface production (rate) of momentum of constituent α
ϕ	....	Rotation angle
$\hat{\varphi}$, φ	.....	Dissipation powers LAGRANGE, EULER representation)
$\tilde{\Phi}$, $\tilde{\phi}$	.....	Surface densities related to $\partial\Omega$, $\partial\omega$ (reference, present configuration)
$\boldsymbol{\Phi}$, $\boldsymbol{\phi}$	....	Entropy fluxes (LAGRANGE, EULER representation)
$\boldsymbol{\Phi}$, $\boldsymbol{\phi}$	....	Fluxes through the surfaces $\partial\Omega$, $\partial\omega$ (reference, present configuration)
$\boldsymbol{\chi}$, $\boldsymbol{\chi}^{-1}$	..	Motion function, inverse of the motion function
$\boldsymbol{\chi}$	....	Motion of the body $\mathcal{B}$ (χ_i)
ψ_h	....	Free enthalpy
Ψ	....	Value of a material functional
$\hat{\Psi}$	....	Material functional
$\boldsymbol{\omega}$	....	Rotation vector (ω_i), angular velocity of the mixture
$\boldsymbol{\omega}^{\alpha}$	......	Angular (spin) velocity of constituent α
Ω, ω	....	Regions (reference, present configuration)
$\partial\Omega$, $\partial\omega$	..	Boundaries of the regions Ω and ω
$\boldsymbol{\Omega}$	....	Spin tensor (Ω_{ij})

In the upcoming chapters we will introduce a number of physical variables such as scalars, vectors and tensors. Although we shall discuss each of these variables separately, it is worth to declare these variables and standardize the notations here. In addition, we make some remarks on notations below which should be used as reference.

General Rules Texts on continuum mechanics often start with a general introduction to vector and tensor algebra. We shall not do so here as it will be assumed that the elements of the Cartesian tensor calculus are known. A deep knowledge is anyway not required, and a reader willing to go through the computational steps and Exercises can, with moderate effort, familiarise himself/herself without difficulties. The following rules will be observed:

- Scalar variables are represented by Greek, Latin (etc.) letters in mathematical script types.
- Vectors and tensors (second or higher rank) are symbolically represented by boldfaced letters (e.g. $\boldsymbol{T}$), or by Cartesian index notations (e.g. T_{ij}).
- Scalar components of vectors and tensors are indexed by Latin and Greek letters, respectively. Similarly, the Latin letters are chosen for the *present configuration* and Greek letters for the *reference configuration* (see §1.2).
- The index $()_t$ indicates changes in time.

- We always consider a three–dimensional vector space $\mathbb{R}^3$. The representation of vectorial or tensorial variables is given in Cartesian coordinates which simplifies the tensor calculations significantly.
- Indices run from 1 to 3 subjected to three space coordinates, as usual.
- We use the EINSTEIN *summation convention*, which means that the scalar product of two vectors $\boldsymbol{a}, \boldsymbol{b} \in \mathbb{R}^3$ is written as

$$\boldsymbol{a} \cdot \boldsymbol{b} = \sum_{i=1}^{3} a_i b_i =: a_i b_i \ , \tag{1.1.1}$$

stating that the double indices are automatically summed. This summation rule applies to all tensor variables in index notation and their combinations.

Different authors use different rules of notation regarding the multiplication between vectors and tensors; however, we follow the notation usually observed in the mathematical literature. For further conventions on notation that follows one should refer to a book on tensor calculus[2]. We have already defined the scalar product between two vectors. In order to establish the further rules of multiplication we introduce the orthonormal basis vectors (in Cartesian coordinates) $\boldsymbol{e}_1$, $\boldsymbol{e}_2$ and $\boldsymbol{e}_3$ along x, y and z directions, for which

$$\boldsymbol{e}_i \cdot \boldsymbol{e}_j = \delta_{ij} := \begin{cases} 1 & i = j, \\ 0 & \text{otherwise} \end{cases} \tag{1.1.2}$$

holds. The symbol δ_{ij} is called the KRONECKER symbol.

For the scalar product of two vectors $\boldsymbol{a}$ and $\boldsymbol{b}$ which are expressed as $\boldsymbol{a} = a_i \boldsymbol{e}_i$, $\boldsymbol{b} = b_i \boldsymbol{e}_i$ relative to the basis $\boldsymbol{e}_i$ $(i = 1, 2, 3)$ with a_i, b_i $(i = 1, 2, 3)$ as their *Cartesian coordinates* one obtains the summation formula as given in (1.1.1); indeed, with the introduction of the KRONECKER symbol one has

$$\begin{aligned} \boldsymbol{a} \cdot \boldsymbol{b} &= a_i \boldsymbol{e}_i \cdot b_k \boldsymbol{e}_k = a_i b_k \boldsymbol{e}_i \cdot \boldsymbol{e}_k \\ &= a_i b_k \delta_{ik} = a_i b_i \ . \end{aligned} \tag{1.1.3}$$

Exactly analogously to the vectors expressed in components relative to their respective basis vectors, we can also express tensors in component form relative to their basis vectors. Moreover, tensors and vectors can also be connected through the dyadic product defined as follows:

- The *dyadic* or *tensor product* $\boldsymbol{a} \otimes \boldsymbol{b}$ of two vectors $\boldsymbol{a}$ and $\boldsymbol{b}$ is the transformation, which through an application to a vector $\boldsymbol{c}$ obeys the relation

$$(\boldsymbol{a} \otimes \boldsymbol{b})\, \boldsymbol{c} = (\boldsymbol{b} \cdot \boldsymbol{c})\, \boldsymbol{a} \ . \tag{1.1.4}$$

[2] Formal texts on tensor calculus are for instance DE BOER [31], [32], BETTEN [25], KLINGBEIL [120] – all in German –, BLOCK [28], BOWEN & WANG [38], MARSDEN & HOFFMANN [146] and ERICKSEN [65].

This expression can be corroborated via the basis $\{\boldsymbol{e}_i\}$ as follows:

$$
\begin{aligned}
(a_i\boldsymbol{e}_i \otimes b_j\boldsymbol{e}_j)\, c_k\boldsymbol{e}_k &\overset{(1)}{=} a_i b_j c_k\,(\boldsymbol{e}_i \otimes \boldsymbol{e}_j)\,\boldsymbol{e}_k \\
&= a_i b_j c_k\,(\boldsymbol{e}_j \cdot \boldsymbol{e}_k)\,\boldsymbol{e}_i = a_i(b_j c_j)\,\boldsymbol{e}_i \qquad (1.1.5)\\
&= (\boldsymbol{b}\cdot\boldsymbol{c})\,\boldsymbol{a}\,.
\end{aligned}
$$

Step (1) is only permitted when the operation $\otimes$ is multilinear; here, however, since this result has led to the definition (1.1.4) the multi-linearity is automatically proven.

- A tensor of second rank $\boldsymbol{A}$ is a linear transformation which maps a vector $\boldsymbol{a}$ to another vector $\boldsymbol{b}$ as follows:

$$
\boldsymbol{b} = \boldsymbol{A}\boldsymbol{a}\,. \qquad (1.1.6)
$$

One can show, using the dyadic product (1.1.4), that the Cartesian representation of this tensor is given by

$$
\boldsymbol{A} = A_{ij}\,\boldsymbol{e}_i \otimes \boldsymbol{e}_j\,. \qquad (1.1.7)
$$

Indeed, in *component expression* (Cartesian) the above product reads

$$
\begin{aligned}
\boldsymbol{A}\boldsymbol{a} &= A_{ij}(\boldsymbol{e}_i \otimes \boldsymbol{e}_j)a_k\boldsymbol{e}_k \\
&= A_{ij}a_k(\boldsymbol{e}_j \cdot \boldsymbol{e}_k)\boldsymbol{e}_i = \underbrace{A_{ij}a_j}_{=b_i}\,\boldsymbol{e}_i \qquad (1.1.8)\\
&= b_i\boldsymbol{e}_i = \boldsymbol{b}\,.
\end{aligned}
$$

According to (1.1.7) the *dyadic product* $\boldsymbol{a}\otimes\boldsymbol{b}$ of two vectors can be understood as a special tensor of second rank, whose component representation $\boldsymbol{a}\otimes\boldsymbol{b} = a_ib_j\boldsymbol{e}_i\otimes\boldsymbol{e}_j$ corresponds to the matrix

$$
[\boldsymbol{a}\otimes\boldsymbol{b}] = \begin{pmatrix} a_1b_1 & a_1b_2 & a_1b_3 \\ a_2b_1 & a_2b_2 & a_2b_3 \\ a_3b_1 & a_3b_2 & a_3b_3 \end{pmatrix}\,.
$$

Introducing the product operators, $\bullet$ (scalar product) and $\otimes$ (tensor product, or dyadic product), we can also establish further operations between tensors and vectors. Let $\boldsymbol{A} = A_{ij}\boldsymbol{e}_i\otimes\boldsymbol{e}_j$, $\boldsymbol{B} = B_{ij}\boldsymbol{e}_i\otimes\boldsymbol{e}_j$ be tensors of rank two and $\boldsymbol{a} = a_i\boldsymbol{e}_i$, $\boldsymbol{b} = b_i\boldsymbol{e}_i$ vectors, which are expressed in their Cartesian components.

- The *contracted* product between vectors and tensors is simply expressed in the sequence it is written down, e.g. as[3]

[3] This follows from (1.1.4) and the definition of the KRONECKER symbol (1.1.2), namely

$$
(\boldsymbol{e}_i\otimes\boldsymbol{e}_j)\boldsymbol{e}_k = (\boldsymbol{e}_j\cdot\boldsymbol{e}_k)\boldsymbol{e}_i = \delta_{jk}\boldsymbol{e}_i\,.
$$

$$\begin{aligned} \boldsymbol{aB} &= a_i \boldsymbol{e}_i B_{jk} \boldsymbol{e}_j \otimes \boldsymbol{e}_k = a_i B_{jk} \boldsymbol{e}_i (\boldsymbol{e}_j \otimes \boldsymbol{e}_k) = a_i B_{jk} \delta_{ij} \boldsymbol{e}_k \\ &= a_i B_{ik} \boldsymbol{e}_k \ , \end{aligned} \tag{1.1.9}$$

and similarly

$$\boldsymbol{Ba} = B_{ij} a_k (\boldsymbol{e}_i \otimes \boldsymbol{e}_j) \boldsymbol{e}_k = \ldots = B_{ij} a_j \boldsymbol{e}_i \ . \tag{1.1.10}$$

The result is in each case a vector. The sequence is, however, not changeable since the two multiplications do not *commute*.

- Similarly, the sequential arrangement[4] $\boldsymbol{AB}$ of two tensors corresponds to the product

$$\begin{aligned} \boldsymbol{AB} &= A_{ij} (\boldsymbol{e}_i \otimes \boldsymbol{e}_j)\, B_{kl} (\boldsymbol{e}_k \otimes \boldsymbol{e}_l) \\ &= A_{ij} B_{kl} (\boldsymbol{e}_i \otimes \boldsymbol{e}_j)(\boldsymbol{e}_k \otimes \boldsymbol{e}_l) \\ &= A_{ij} B_{kl}\, \underbrace{\boldsymbol{e}_i (\boldsymbol{e}_j \cdot \boldsymbol{e}_k) \otimes \boldsymbol{e}_l}_{\delta_{jk} (\boldsymbol{e}_i \otimes \boldsymbol{e}_l)} \\ &= A_{ik} B_{kl} \boldsymbol{e}_i \otimes \boldsymbol{e}_l \ . \end{aligned} \tag{1.1.11}$$

The result is again a tensor of second rank. In (Cartesian) component form this corresponds straightforwardly to the matrix multiplication of the respective matrices.

- The scalar product of two tensors of rank two, $\boldsymbol{A}$ and $\boldsymbol{B}$, as in the case of the scalar product of two vectors, is characterized by a point positioned between the two,

$$\begin{aligned} \boldsymbol{A} \cdot \boldsymbol{B} &= A_{ij} (\boldsymbol{e}_i \otimes \boldsymbol{e}_j) \cdot B_{kl} (\boldsymbol{e}_k \otimes \boldsymbol{e}_l) = A_{ij} B_{kl} \underbrace{(\boldsymbol{e}_i \otimes \boldsymbol{e}_j \cdot \boldsymbol{e}_k \otimes \boldsymbol{e}_l)}_{=:\delta_{jk} \boldsymbol{e}_i \cdot \boldsymbol{e}_l = \delta_{jk} \delta_{il}} \\ &= A_{ij} B_{kl} \delta_{jk} \delta_{il} = A_{ij} B_{ji} =: \operatorname{tr}(\boldsymbol{AB}) \ , \end{aligned} \tag{1.1.12}$$

which corresponds to the trace of the product of both tensors. In contrast to this, many texts write double point ($\boldsymbol{A} : \boldsymbol{B}$) to demonstrate double multiplication (relative to their respective basis); however, we will not follow such notation.

- The cross product of two vectors results again in a vector; one obtains it with the help of the LEVI–CIVITÀ ε–tensor. In Cartesian coordinates it is written in component form ε_{ijk} having the meaning

---

[4] In this calculation one uses the identity

$$(\boldsymbol{e}_i \otimes \boldsymbol{e}_j)\,(\boldsymbol{e}_k \otimes \boldsymbol{e}_l) = \boldsymbol{e}_i (\boldsymbol{e}_j \cdot \boldsymbol{e}_k) \otimes \boldsymbol{e}_l = \delta_{jk} (\boldsymbol{e}_i \otimes \boldsymbol{e}_l) \ .$$

For the proof of this identity one can multiply the left and right sides by an arbitrary basis vector and show them to be identical;

left side: $(\boldsymbol{e}_i \otimes \boldsymbol{e}_j)\,(\boldsymbol{e}_k \otimes \boldsymbol{e}_l) \boldsymbol{e}_m = (\boldsymbol{e}_i \otimes \boldsymbol{e}_j) \delta_{lm} \boldsymbol{e}_k = \delta_{lm} \delta_{jk} \boldsymbol{e}_i \ ,$

right side: $(\boldsymbol{e}_i (\boldsymbol{e}_j \cdot \boldsymbol{e}_k) \otimes \boldsymbol{e}_l)\, \boldsymbol{e}_m = \delta_{jk} (\boldsymbol{e}_i \otimes \boldsymbol{e}_l) \boldsymbol{e}_m = \delta_{jk} \delta_{lm} \boldsymbol{e}_i \ .$

$$\varepsilon_{ijk} = \begin{cases} 1 & (ijk) \text{ is a cyclic permutation of } (123), \\ -1 & (ijk) \text{ is an anticyclic permutation of } (123), \\ 0 & \text{otherwise;} \end{cases} \tag{1.1.13}$$

these are the components of the *completely antisymmetric tensor* of third rank. This tensor results while forming a cross product of two unit vectors (Cartesian) $\boldsymbol{e}_i$, namely[5]

$$\boldsymbol{e}_i \times \boldsymbol{e}_j = \varepsilon_{ijk} \boldsymbol{e}_k \, . \tag{1.1.14}$$

Thus the cross product of two vectors $\boldsymbol{a}$ and $\boldsymbol{b}$ is a vector perpendicular to both $\boldsymbol{a}$ and $\boldsymbol{b}$, that is, $\boldsymbol{a} \cdot (\boldsymbol{a} \times \boldsymbol{b}) = 0$ and $\boldsymbol{b} \cdot (\boldsymbol{a} \times \boldsymbol{b}) = 0$ hold, which is easily proved using

$$\begin{aligned} \boldsymbol{a} \times \boldsymbol{b} &= a_i \boldsymbol{e}_i \times b_j \boldsymbol{e}_j = a_i b_j \boldsymbol{e}_i \times \boldsymbol{e}_j \\ &= a_i b_j \varepsilon_{ijk} \boldsymbol{e}_k \, . \end{aligned} \tag{1.1.15}$$

As we can not avoid to use the same symbols for different variables, we will present a separate list of symbols for each part of the book. This will facilitate the reader to grasp the overall idea on symbols and notation. Needless to say that we shall be careful in consistently using the same symbols throughout the book whenever possible.

1.2 Basic Concepts, Motion

If one considers the usual equations of physics that describe different systems[6], then one recognizes certain similarities among them. Quite often the same terms appear in different contexts. Many statements of physics have in general a common structure and can be written in the form of *balance equations*. Recognition of the structure which we are going to derive shortly is enlightening and facilitates our general understanding. Before doing this we should, however, introduce some important basic concepts of continuum mechanics.

[5] In this formula use is made of the fact that $\{\boldsymbol{e}_i, \boldsymbol{e}_j, \boldsymbol{e}_k\}$, in this order, form a positively oriented triad of unit vectors.

[6] In the following we exclusively consider the thermo-mechanical systems within the GALILEIan–NEWTONian mechanics, i.e., electromagnetic, relativistic effects etc. are not taken into account even though these effects can have strong influences on our environment. For example, the light field is one of the most essential sources of life. In *limnology* (science of inland waters and ecosystems) and oceanography the light plays an important role; for instance, the thermal stratification in a lake is built up by the absorption of light. Such influences are usually incorporated into the equations by prescribed quantities such as the above mentioned thermal stratification or by boundary conditions.

Material Body The starting point of continuum mechanics is a *material body* $\mathcal{B}(\mathfrak{X})$, which consists of an infinite number of *material elements* $\mathfrak{X}$,

$$\mathcal{B} = \{\mathfrak{X}\} \ . \tag{1.2.1}$$

We consider these as primitive, i.e., *given* elements. Furthermore, we use the symbol $\{\cdot\}$ to mean "collection" or "set".

Position Vector In order to describe the motion of a material body every particle must be allocated a position. To the particles $\mathfrak{X} \in \mathcal{B}$ there exists a vector space $\boldsymbol{V}_R^3$ such that to every particle there is assigned a vector $\boldsymbol{X}$:

$$\begin{aligned} \hat{\boldsymbol{X}} : \mathcal{B} &\to \boldsymbol{V}_R^3 \\ \mathfrak{X} &\mapsto \boldsymbol{X} = \hat{\boldsymbol{X}}(\mathfrak{X}) \ . \end{aligned} \tag{1.2.2}$$

The *position vector* $\boldsymbol{X}$ identifies the individual particles of the body.

Reference Configuration The set of position vectors defined in a body $\mathcal{B}$ represents its *reference configuration*[7]

$$\mathcal{B}_R := \left\{ \hat{\boldsymbol{X}}(\mathfrak{X}) \,|\, \mathfrak{X} \in \mathcal{B} \right\} \ . \tag{1.2.3}$$

Generally this is identified as the configuration of a body in physical space at a fixed (or initial) time.

Material Coordinates The representation of the position vector $\boldsymbol{X}$ in EUCLIDian space is accomplished by choice of a coordinate system such that the position vector of a particle corresponds to a combination of three numbers[8] (coordinates). Such numbers are the components of the position vector given by

$$\boldsymbol{X} \equiv (X_1, X_2, X_3) \ , \quad \boldsymbol{X} = X_\alpha \boldsymbol{e}_\alpha \ ; \tag{1.2.4}$$

we call them the *material coordinates*, and the components are designated by Greek indices (X_α, $\alpha = 1,2,3$).

Present Configuration When a body $\mathcal{B}$ moves or deforms with time then a particle $\mathfrak{X}$ of it takes a new position at time $t \in \mathbb{R}^+$. As was the case for the reference configuration, to every particle of the body corresponds at a given time t a vector characterizing its position at time t. This can be expressed as

$$\begin{aligned} \hat{\boldsymbol{x}} : \mathcal{B} &\to \boldsymbol{V}_t^3 \\ \mathfrak{X} &\mapsto \boldsymbol{x} = \hat{\boldsymbol{x}}(\mathfrak{X}, t) \ . \end{aligned} \tag{1.2.5}$$

---

[7] The notation $\{a|b\}$ means "$\{a\}$ with the condition b satisfied".

[8] In this text, with only a few exceptions, straight orthogonal Cartesian coordinates are used.

The set of positions

$$\mathcal{B}_t := \left\{ \hat{\boldsymbol{x}}(\mathfrak{X}, t) \,|\, \mathfrak{X} \in \mathcal{B},\, t \in \mathbb{R}^+ \right\} \tag{1.2.6}$$

now represents the configuration of the body $\mathcal{B}$ at time t and is called *actual configuration* or *present configuration*.

Spatial Coordinates The representation of positions $\boldsymbol{x}$ at time t is again accomplished by a choice of a (Cartesian) coordinate system as follows

$$\boldsymbol{x} \equiv (x_1, x_2, x_3)\;, \quad \boldsymbol{x} = x_i \boldsymbol{e}_i\;; \tag{1.2.7}$$

these are called the *spatial coordinates.* In the present configuration the components are indicated by Latin indices in the form x_i, $i = 1, 2, 3$.

The reference and present configurations of a body can be visualized as the positions of the body (or its elements) at a fixed (reference) time t_0 and at all the following (present) times, respectively.

Motion We are now in the position to define the *motion* of a body as the succession of positions, which a particle $\mathfrak{X}$ traverses with time. The particle labelled $\mathfrak{X}$ found at position $\boldsymbol{X}$ in the reference configuration occupies a new position $\boldsymbol{x}$ after a certain time $t \in \mathbb{R}^+$. Then the motion of the particle can be described mathematically as the mapping

$$\begin{aligned} \chi : \mathcal{B}_R \times \mathbb{R}^+ &\to \mathcal{B}_t \\ (\boldsymbol{X}, t) &\mapsto \boldsymbol{x} = \chi(\boldsymbol{X}, t)\;. \end{aligned} \tag{1.2.8}$$

We assume that the motion χ is *continuously differentiable* in finite regions of the body or in the entire body so that the mapping (1.2.8) is invertible such that

$$\boldsymbol{X} = \chi^{-1}(\boldsymbol{x}, t) \tag{1.2.9}$$

holds. This means when all positions $\boldsymbol{x}$ in $\mathcal{B}_t$ and the motion $\chi(\boldsymbol{X}, t)$ at a fixed time t are known then the positions of the particles in the reference configuration can be determined.

The relationship between the body $\mathcal{B}$, its reference configuration $\mathcal{B}_R$ and its present configuration $\mathcal{B}_t$ is shown in Fig. 1.1.

Physical Variables Let us consider now an arbitrary physical variable $\mathcal{G}$ (e.g. mass or temperature) and describe its evolution in time. This variable is defined *with respect to a particle* $\mathfrak{X}$ at a certain time t as

$$\mathcal{G} = \hat{\mathcal{G}}(\mathfrak{X}, t)\;. \tag{1.2.10}$$

Such a functional relation is not very useful as it is not practical. It will be more meaningful to regard a physical variable as a function of the time and the position of the particle in the reference or present configurations. Such

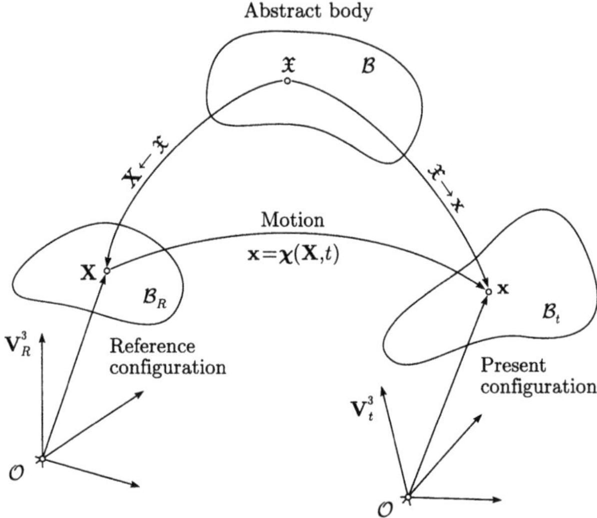

Fig. 1.1. Relation between the body $\mathcal{B}$, its reference configuration $\mathcal{B}_R$ and its present configuration $\mathcal{B}_t$.

representations can be derived from (1.2.10) so long as the functions $\hat{\boldsymbol{X}}(\mathfrak{X})$ and $\hat{\boldsymbol{x}}(\mathfrak{X}, t)$ are invertible, i.e., $\mathfrak{X} = \hat{\boldsymbol{X}}^{-1}(\boldsymbol{X})$ and $\mathfrak{X} = \hat{\boldsymbol{x}}^{-1}(\boldsymbol{x}, t)$ which we assume to hold. In this way one arrives at a description of a physical variable relative to the reference and present configuration, respectively. Only these will lead to a physically realizable field representation.

LAGRANGE **and** EULER **Representations** With the help of the mappings $\boldsymbol{X} = \hat{\boldsymbol{X}}(\mathfrak{X})$ or $\mathfrak{X} = \hat{\boldsymbol{X}}^{-1}(\boldsymbol{X})$ one obtains

$$\begin{aligned} \mathcal{G} &= \hat{\mathcal{G}}(\mathfrak{X}, t) = \hat{\mathcal{G}}(\hat{\boldsymbol{X}}^{-1}(\boldsymbol{X}), t) \\ &= \mathfrak{G}(\boldsymbol{X}, t) \, . \end{aligned} \tag{1.2.11}$$

This description is called the LAGRANGE*an description:* In this description a physical variable is expressed as a function of material coordinates and time.

If the description is with respect to the present configuration it is called EULER*ian representation.* In it, a physical variable is described via the spatial coordinates $\boldsymbol{x}$ and time t, whereby the inverse mapping of (1.2.5) $\mathfrak{X} = \hat{\boldsymbol{x}}^{-1}(\boldsymbol{x}, t)$ is used, so that

$$\begin{aligned} \mathcal{G} &= \hat{\mathcal{G}}(\hat{\boldsymbol{x}}^{-1}(\boldsymbol{x}, t), t) \\ &= \mathfrak{g}(\boldsymbol{x}, t) \, . \end{aligned} \tag{1.2.12}$$

Naturally, each of these representations can be transformed into the other. Even though both functions $\mathfrak{G}(\boldsymbol{X}, t)$ and $\mathfrak{g}(\boldsymbol{x}, t)$ have different forms, they obviously produce the same value,

$$\begin{aligned} \mathcal{G} &= \mathfrak{G}(\boldsymbol{X}, t) \quad \text{LAGRANGE representation,} \\ &= \mathfrak{g}(\boldsymbol{x}, t) \quad \text{EULER representation.} \end{aligned} \tag{1.2.13}$$

If one uses the definition of the motion (1.2.8) as well as its inverse function (1.2.9) so as to relate the reference configuration to the present configuration and vice versa, one has

$$\begin{aligned} \mathfrak{G}(\boldsymbol{X}, t) &= \mathfrak{G}(\chi^{-1}(\boldsymbol{x}, t), t) = \mathfrak{g}(\boldsymbol{x}, t) \\ \mathfrak{g}(\boldsymbol{x}, t) &= \mathfrak{g}(\chi(\boldsymbol{X}, t), t) = \mathfrak{G}(\boldsymbol{X}, t) . \end{aligned} \tag{1.2.14}$$

The derivatives of these functions with respect to their coordinates are not the same because the form of the functions, namely the coordinate–dependency is different, so

$$\frac{\partial \mathfrak{G}}{\partial X_\alpha} \neq \frac{\partial \mathfrak{g}}{\partial x_i} , \quad \alpha = 1, 2, 3 , \quad i = 1, 2, 3 . \tag{1.2.15}$$

Density and Mass Mass is an example of a physical variable. Since a material body consists of an infinite number of elements or particles, it is not so meaningful to assign to these particles a finite mass as in this case the mass of the whole body could become infinite. It is rather more appropriate to assign to every particle $\mathfrak{X}$ a positive *density* or *mass density*, which defines the mass per unit volume. Then the value of the density depends on the configuration used in question. If

$$\rho_R > 0, \qquad \rho > 0 \tag{1.2.16}$$

are the densities per unit volume in the reference and present configurations, respectively, then the mass of the body can be written as[9]

$$m(\mathcal{B}) = \iiint\limits_{\mathcal{B}_R} \rho_R \, \mathrm{d}V = \iiint\limits_{\mathcal{B}_t} \rho \, \mathrm{d}v . \tag{1.2.17}$$

The mass is therefore an *additive*, positive quantity, sometimes called a *measure.*

---

[9] Integration over the volume of the body in its reference and present configuration, respectively, is expressed here as a threefold integral e.g. $\iiint\limits_{\mathcal{B}_R}$. In ensuing developments we shall often simply use the abbreviated notation $\int\limits_{\mathcal{B}_R}$. Thus, instead of (1.2.17), we may also write

$$m(\mathcal{B}) = \int\limits_{\mathcal{B}_R} \rho_R \, \mathrm{d}V = \int\limits_{\mathcal{B}_t} \rho \, \mathrm{d}v .$$

Material Derivative Of special interest is the time rate of change of a physical variable relative to a fixed particle; the form of this derivative depends on the type of representation considered. The *material derivative* for a fixed material element may, in view of (1.2.11) and (1.2.12), be written as

$$\begin{aligned}\dot{\mathcal{G}} = \frac{\mathrm{d}\hat{\mathcal{G}}(\mathfrak{X},t)}{\mathrm{d}t} &= \frac{\partial \mathfrak{G}(\boldsymbol{X},t)}{\partial t} \\ &= \frac{\mathrm{d}\mathfrak{g}(\boldsymbol{x},t)}{\mathrm{d}t} = \frac{\partial \mathfrak{g}(\boldsymbol{x},t)}{\partial t} + \frac{\partial \mathfrak{g}(\boldsymbol{x},t)}{\partial x_i}\dot{x}_i \\ &= \mathfrak{g}_{,t} + \mathfrak{g}_{,i}\,\dot{x}_i \,.\end{aligned} \tag{1.2.18}$$

Here we have accounted for the fact that in the EULER representation the position vector is explicitly dependent on the time t.

In the above, we have introduced abbreviations for the notation of the various derivatives. Now we summarize them once again.

- The derivatives with respect to a coordinate are often written in index notation,

$$\frac{\partial \mathfrak{G}}{\partial X_\alpha} =: \mathfrak{G},_\alpha \quad \text{and} \quad \frac{\partial \mathfrak{g}}{\partial x_i} =: \mathfrak{g},_i \,, \quad i,\alpha = 1,2,3 \,. \tag{1.2.19}$$

- The gradients with respect to material and spatial coordinates are written symbolically as

$$\frac{\partial \mathfrak{G}}{\partial \boldsymbol{X}} \equiv \operatorname{Grad}\mathfrak{G} \,, \quad \frac{\partial \mathfrak{g}}{\partial \boldsymbol{x}} \equiv \operatorname{grad}\mathfrak{g} \,, \tag{1.2.20}$$

  where the definitions of these expressions depend on whether $\mathcal{G}$ is a scalar (e.g. temperature), a vector (e.g. momentum) or a tensor–valued variable. These are sometimes called material and spatial gradients. We write gradients relative to the reference configuration, i.e., the derivatives with respect to material coordinates with capital letters (Grad) to differentiate them from the spatial gradients (relative to the present configuration, grad). Analogous expressions will be introduced to exhibit the divergence of a vector or a tensor field in later sections.
- The partial or local derivative with respect to time is written as

$$\frac{\partial \mathfrak{g}}{\partial t} \equiv \mathfrak{g},_t \,, \tag{1.2.21}$$

  whilst the material or substantial derivative with respect to time is given by

$$\frac{\mathrm{d}\mathfrak{g}}{\mathrm{d}t} \equiv \dot{\mathfrak{g}} \,. \tag{1.2.22}$$

  In the LAGRANGEian representation, the partial derivative of a physical quantity with respect to time is written in the same fashion

$$\frac{\partial \mathfrak{G}}{\partial t} = \mathfrak{G},_t = \frac{\mathrm{d}\mathfrak{G}}{\mathrm{d}t} = \dot{\mathfrak{G}} \,. \tag{1.2.23}$$

1.3 Deformation Gradient

In the following developments further concepts on the description of the deformation of a body are needed. However, at this stage such concepts are presented only briefly, a deep understanding of these is not necessary in order to derive the balance equations. However, as we go ahead these concepts appear to be more and more essential, and slowly, further ideas will be introduced to broaden the framework presented here.

1.3.1 Definition of the Deformation Gradient

We consider a material body which is subjected to the motion $\boldsymbol{x} = \boldsymbol{\chi}(\boldsymbol{X}, t)$, Fig. 1.2. Suppose $\mathrm{d}\boldsymbol{X}$ is a material line element in the reference configuration; the same in the present configuration at time t is described by the relation

$$\mathrm{d}\boldsymbol{x} = \boldsymbol{F}\,\mathrm{d}\boldsymbol{X} \quad \text{or} \quad \mathrm{d}x_i = F_{i\alpha}\,\mathrm{d}X_\alpha\ . \tag{1.3.1}$$

The tensor $\boldsymbol{F}$ is known as the *deformation gradient*. Here we have $\boldsymbol{x} = \boldsymbol{\chi}(\boldsymbol{X}, t)$, so the deformation gradient is $\boldsymbol{F} = \operatorname{Grad} \boldsymbol{\chi}(\boldsymbol{X}, t)$, and the components of it are

$$F_{i\alpha} := \frac{\partial \chi_i(\boldsymbol{X}, t)}{\partial X_\alpha}\ . \tag{1.3.2}$$

It maps vectors from the reference configuration $\mathcal{B}_R \subset \boldsymbol{V}_R^3$ onto vectors in the present configuration $\mathcal{B}_t \subset \boldsymbol{V}_t^3$, and is therefore also known as a *two–point–tensor*. To distinguish between the coordinates of the reference and present configuration Latin and Greek indices are used so that the representation of the deformation gradient $\boldsymbol{F} = F_{i\alpha}(\boldsymbol{e}_i \otimes \boldsymbol{e}_\alpha)$ is expressed relative to both bases $\{\boldsymbol{e}_i\}$ and $\{\boldsymbol{e}_\alpha\}$ in the reference and present configuration, respectively.

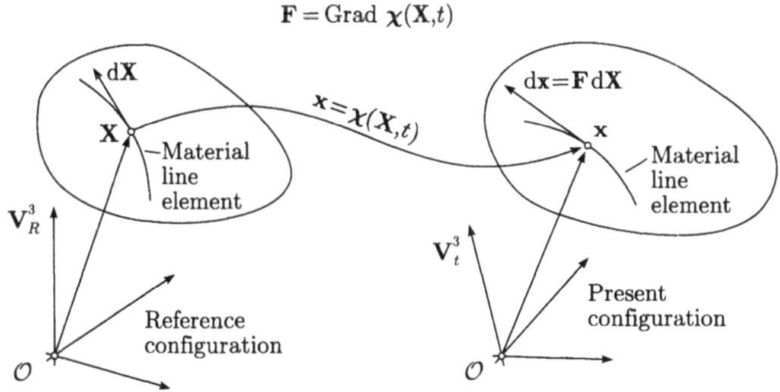

Fig. 1.2. Transformation of a vectorial material element.

The deformation gradient transforms a material line element from the reference to the present configuration. The inverse relation reads

$$\mathrm{d}\boldsymbol{X} = \boldsymbol{F}^{-1}\,\mathrm{d}\boldsymbol{x} \quad \text{or} \quad \mathrm{d}X_\alpha = F^{-1}_{\alpha i}\,\mathrm{d}x_i \tag{1.3.3}$$

with $\boldsymbol{F}^{-1} = F^{-1}_{\alpha i}(\boldsymbol{e}_\alpha \otimes \boldsymbol{e}_i)$. $\boldsymbol{F}$ is non-singular, since the motion $\boldsymbol{\chi}$ is invertible. The determinant of the deformation gradient[10] is therefore always non-zero, i.e.,

$$J := \det(\boldsymbol{F}) \neq 0\,. \tag{1.3.4}$$

The notation J for the determinant of $\boldsymbol{F}$ will consistently be used in the sequel.

1.3.2 Transformation of Surface and Volume Elements

With the help of the transformation of a line element which we have already demonstrated in the last subsection we can easily establish the transformation of a surface element $\mathrm{d}\boldsymbol{A}$. An (oriented) material surface element in the reference configuration can be expressed through two line elements, $\mathrm{d}\boldsymbol{X}_1$ and $\mathrm{d}\boldsymbol{X}_2$ as

$$\mathrm{d}\boldsymbol{A} = \mathrm{d}\boldsymbol{X}_1 \times \mathrm{d}\boldsymbol{X}_2\,. \tag{1.3.5}$$

With (1.3.1) this surface element in the present configuration is given by (see exercise)[11]

$$\begin{aligned} \mathrm{d}\boldsymbol{a} &= \mathrm{d}\boldsymbol{x}_1 \times \mathrm{d}\boldsymbol{x}_2 \\ &= \boldsymbol{F}\,\mathrm{d}\boldsymbol{X}_1 \times \boldsymbol{F}\,\mathrm{d}\boldsymbol{X}_2 = J\,\boldsymbol{F}^{-T}\,(\mathrm{d}\boldsymbol{X}_1 \times \mathrm{d}\boldsymbol{X}_2)\,. \end{aligned} \tag{1.3.6}$$

The transformation of a surface element from the reference to the present configuration is thus

$$\mathrm{d}\boldsymbol{a} = J\,\boldsymbol{F}^{-T}\,\mathrm{d}\boldsymbol{A}\,. \tag{1.3.7}$$

The transformation of a volume element dV from the reference to the present configuration follows directly from the above results,

---

[10] Caution is necessary in the calculation of the determinant of $\boldsymbol{F}$ to account for the bases of the coordinate systems of the reference and present configurations, since the deformation gradient $\boldsymbol{F}$, in general, is represented in two different coordinate systems at the same time; hence this must also be taken into account.

[11] The expression $\boldsymbol{A}^{-T}$ is

$$\boldsymbol{A}^{-T} := \left(\boldsymbol{A}^T\right)^{-1}\,,$$

and $(\bullet)^T$ denotes the transpose of $(\bullet)$. The statement follows from the identity ($\boldsymbol{a}, \boldsymbol{b}, \boldsymbol{c}$ are arbitrary vectors)

$$\det(\boldsymbol{F})(\boldsymbol{a} \times \boldsymbol{b}) \cdot \boldsymbol{c} \equiv (\boldsymbol{F}\boldsymbol{a} \times \boldsymbol{F}\boldsymbol{b}) \cdot \boldsymbol{F}\boldsymbol{c}\,.$$

One considers, here, the definition of the triple product via the determinant of the compound matrix, $\det(\boldsymbol{a}, \boldsymbol{b}, \boldsymbol{c})$, of individual vectors.

$$\mathrm{d}v = J\,\mathrm{d}V\;. \tag{1.3.8}$$

One can also verify this relation by using the definition of the triple product of the corresponding vectors (see the last footnote and Exercises).

1.4 Velocity, Acceleration and Velocity Gradient

The **velocity** of a material particle, which is the time rate of change of the position of the material particle, has already been introduced in (1.2.18). With the definition of the position vector (1.2.5) the velocity of a particle is defined as

$$\boldsymbol{v} := \dot{\boldsymbol{x}} = \frac{\mathrm{d}\hat{\boldsymbol{x}}(\mathfrak{X},t)}{\mathrm{d}t}\;. \tag{1.4.1}$$

In the LAGRANGE representation, with (1.2.8), this becomes

$$\boldsymbol{v} = \hat{\boldsymbol{v}}(\boldsymbol{X},t) = \frac{\partial\boldsymbol{\chi}(\boldsymbol{X},t)}{\partial t}\;. \tag{1.4.2}$$

The EULER representation of the velocity reads

$$\boldsymbol{v} = \hat{\boldsymbol{v}}(\boldsymbol{\chi}^{-1}(\boldsymbol{x},t),t) = \bar{\boldsymbol{v}}(\boldsymbol{x},t)\;, \tag{1.4.3}$$

where the particle, which at time t occupies the position $\boldsymbol{x}$, is held fixed.

Acceleration, Velocity Gradient, Stretching and Vorticity The time rate of change of the velocity of a material element is its *acceleration*. We express this in the EULER representation as

$$\dot{\boldsymbol{v}} = \frac{\mathrm{d}\bar{\boldsymbol{v}}(\boldsymbol{x},t)}{\mathrm{d}t} = \frac{\partial\bar{\boldsymbol{v}}}{\partial t} + \boldsymbol{L}\bar{\boldsymbol{v}}\;. \tag{1.4.4}$$

Here the spatial *velocity gradient* $\boldsymbol{L}$ is defined as

$$\boldsymbol{L} = L_{ij}\boldsymbol{e}_i \otimes \boldsymbol{e}_j := \operatorname{grad}\bar{\boldsymbol{v}} \quad \text{with} \quad L_{ij} = \bar{v}_{i\,,\,j}\;; \tag{1.4.5}$$

it is a second rank tensor. Since under almost all circumstances the velocity is differentiated with respect to $\boldsymbol{x}$ and not $\boldsymbol{X}$, the attribute "spatial" is not explicitly spelled out; so "velocity gradient" always means the spatial gradient.

The velocity gradient can be uniquely decomposed into symmetric and antisymmetric parts,

$$\boldsymbol{L} = \underbrace{\tfrac{1}{2}\left(\boldsymbol{L}+\boldsymbol{L}^T\right)}_{=:\,\boldsymbol{D}} + \underbrace{\tfrac{1}{2}\left(\boldsymbol{L}-\boldsymbol{L}^T\right)}_{=:\,\boldsymbol{W}}\;. \tag{1.4.6}$$

The antisymmetric part of the velocity gradient $\boldsymbol{W} = -\boldsymbol{W}^T$ is called the *vorticity* or the *spin tensor*, sometimes also called the *tensor of rotational velocity*. The symmetric part $\boldsymbol{D} = \boldsymbol{D}^T$ is called *strain rate tensor* or *stretching tensor*[12]; it will be encountered here quite often. The names of these quantities clearly reflect their meaning.

We will later quite often work with the stretching tensor, so we write it in Cartesian coordinates (x, y, z). In terms of the velocity field (u, v, w), it can be expressed in matrix form as

$$\boldsymbol{D} = \begin{pmatrix} \dfrac{\partial u}{\partial x} & \frac{1}{2}\left(\dfrac{\partial u}{\partial y} + \dfrac{\partial v}{\partial x}\right) & \frac{1}{2}\left(\dfrac{\partial u}{\partial z} + \dfrac{\partial w}{\partial x}\right) \\ \cdot & \dfrac{\partial v}{\partial y} & \frac{1}{2}\left(\dfrac{\partial v}{\partial z} + \dfrac{\partial w}{\partial y}\right) \\ \cdot & \cdot & \dfrac{\partial w}{\partial z} \end{pmatrix}, \tag{1.4.7}$$

where the symmetric elements of this matrix are indicated by dots.

Now, we like to derive the relation between the velocity gradient and the deformation gradient. Beginning with (1.4.2) the gradient of the velocity field in the reference configuration is obtained as

$$\operatorname{Grad} \hat{\boldsymbol{v}}(\boldsymbol{X}, t) = \operatorname{Grad} \frac{\partial \boldsymbol{\chi}(\boldsymbol{X}, t)}{\partial t} = \frac{\partial}{\partial t} \operatorname{Grad} \boldsymbol{\chi}(\boldsymbol{X}, t) = \dot{\boldsymbol{F}} . \tag{1.4.8}$$

On the other hand, because $\boldsymbol{x} = \boldsymbol{\chi}(\boldsymbol{X}, t)$ and $\boldsymbol{v} = \hat{\boldsymbol{v}}(\boldsymbol{X}, t) = \bar{\boldsymbol{v}}(\boldsymbol{x}, t) = \bar{\boldsymbol{v}}(\boldsymbol{\chi}(\boldsymbol{X}, t), t)$, the chain rule, applied on this gradient results in

$$\begin{aligned} \dot{\boldsymbol{F}} &= \operatorname{Grad} \hat{\boldsymbol{v}}(\boldsymbol{X}, t) = \operatorname{grad} \bar{\boldsymbol{v}}(\boldsymbol{x}, t) \operatorname{Grad} \boldsymbol{\chi}(\boldsymbol{X}, t) \\ &= \boldsymbol{L}(\boldsymbol{x}, t) \boldsymbol{F}(\boldsymbol{X}, t) \quad \text{or} \\ v_{i,\alpha} &= L_{ij} F_{j\alpha} . \end{aligned} \tag{1.4.9}$$

Solving this equation for the velocity gradient yields

$$\boldsymbol{L} = \dot{\boldsymbol{F}} \boldsymbol{F}^{-1} = \operatorname{grad} \boldsymbol{v} \quad \text{or} \quad L_{ij} = v_{i,j} . \tag{1.4.10}$$

A further important relation between the velocity and the deformation gradient, which is not derived here (see Exercises), is

$$\dot{J} = J \operatorname{div} \boldsymbol{v} = J\, v_{i,i} . \tag{1.4.11}$$

Here, $\operatorname{div} \boldsymbol{v}$ means the divergence of the velocity field relative to the present coordinates[13]. This expression describes the time rate of change of the relative volume change (1.3.8). In an *incompressible* or *density preserving medium*, in

---

[12] The appropriate name for this is *stretching tensor* even though it is not frequently found in the literature. The name *strain rate* suggests that the time integration of $\boldsymbol{D}$ gives rise to a strain tensor which, in general, is not the case. On the same ground it is equally not appropriate to write $\dot{\boldsymbol{\epsilon}}$ instead of $\boldsymbol{D}$.

[13] In Cartesian coordinates, the divergence of a vector field $\boldsymbol{f} = \boldsymbol{f}(\boldsymbol{x}) = \boldsymbol{F}(\boldsymbol{X})$ is written, relative to the present coordinates, as

which the volume does not change, $J = \text{const.}$ and therefore the velocity field must be source free (*solenoidal*), $\operatorname{div}\boldsymbol{v} = 0$. We can easily prove this statement by calculating the time rate of change of a volume element,

$$\begin{aligned}(\mathrm{d}v)^{\cdot} &= (J\,\mathrm{d}V)^{\cdot} = \dot{J}\,\mathrm{d}V = \operatorname{div}\boldsymbol{v}\,(J\,\mathrm{d}V)\\ &= \operatorname{div}\boldsymbol{v}\,\mathrm{d}v\,.\end{aligned} \tag{1.4.12}$$

This shows, then, that the time rate of change of the volume element is zero when the velocity field is solenoidal.

1.5 Deformation

1.5.1 Polar Decomposition of the Deformation Gradient

Every second rank tensor may be decomposed into two parts, one an orthogonal tensor and the other a positive definite symmetric tensor. In particular, this so-called polar decomposition can be carried out for the deformation gradient $\boldsymbol{F}$. The decomposition of $\boldsymbol{F}$ allows a closer interpretation of the deformation and will be of importance lateron. Note that, although we have a special tensor, the deformation gradient $\boldsymbol{F}$, in mind, the polar decomposition applies to every second rank tensor.

Polar Decomposition *Every second rank tensor $\boldsymbol{F}$ with* $\det\boldsymbol{F} \neq 0$ *permits two polar decompositions*, namely,

$$\boldsymbol{F} = \boldsymbol{R}\boldsymbol{U} = \boldsymbol{V}\boldsymbol{R}\,, \tag{1.5.1}$$

with the following properties:

- *$\boldsymbol{V}$ and $\boldsymbol{U}$ are symmetric,*

$$\boldsymbol{V} = \boldsymbol{V}^T\,, \quad \boldsymbol{U} = \boldsymbol{U}^T\,. \tag{1.5.2}$$

- *$\boldsymbol{V}$ and $\boldsymbol{U}$ are positive definite,*

$$\boldsymbol{x}\cdot\boldsymbol{V}\boldsymbol{x} \geq 0\,, \quad \boldsymbol{x}\cdot\boldsymbol{U}\boldsymbol{x} \geq 0\,, \quad \forall\boldsymbol{x} \in \mathbb{R}^3 \tag{1.5.3}$$

  and they have the same eigenvalues.
  This part of the deformation gradient corresponds to a pure strain; $\boldsymbol{U}$ and

---

$$\operatorname{div}\boldsymbol{f}(\boldsymbol{x}) = f_{i\,,\,i} = \frac{\partial f_1}{\partial x_1} + \frac{\partial f_2}{\partial x_2} + \frac{\partial f_3}{\partial x_3}\,,$$

whereas in reference coordinates it is written as

$$\operatorname{Div}\boldsymbol{F}(\boldsymbol{X}) = F_{\alpha\,,\,\alpha} = \frac{\partial F_1}{\partial X_1} + \frac{\partial F_2}{\partial X_2} + \frac{\partial F_3}{\partial X_3}\,.$$

Note again the different notation div and Div, respectively.

$\boldsymbol{V}$ are called the right and left stretch tensors, respectively[14]. *The denotation right and left implies that $\boldsymbol{U}$ stands to the right and $\boldsymbol{V}$ stands to the left of $\boldsymbol{R}$, no more!*

- *$\boldsymbol{R}$ is proper orthogonal,*

$$\boldsymbol{R}\boldsymbol{R}^T = \boldsymbol{R}^T\boldsymbol{R} = \boldsymbol{I}\,, \quad \det \boldsymbol{R} = +1\,. \tag{1.5.4}$$

  A proper orthogonal transformation corresponds to a rigid rotation, but no mirror reflection.

- *The polar decomposition is unique*[15]. ■

The above described polar decomposition enables us to interpret the deformation (Fig. 1.3) by stating that rotation follows stretch or stretch follows rotation.

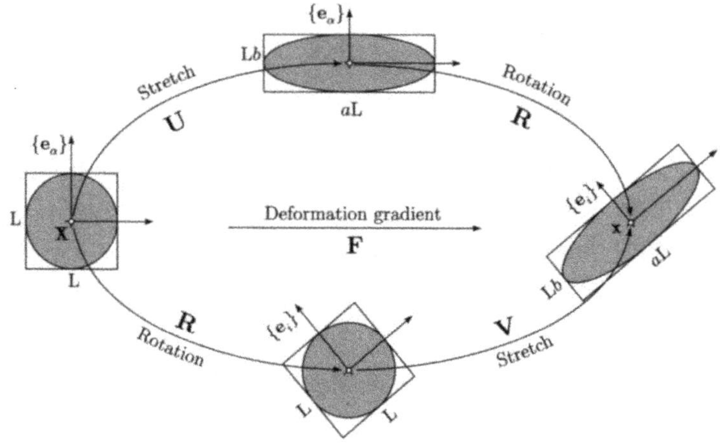

Fig. 1.3. Polar decomposition of the deformation gradient, interpreted as the compositional process of stretch followed by rotation or vice versa.

Proof of the Polar Decomposition For the proof of this theorem some basic knowledge of linear algebra is necessary which is assumed to be known. The proof starts with the statement that $\boldsymbol{C} := \boldsymbol{F}^T\boldsymbol{F}$ and $\boldsymbol{B} := \boldsymbol{F}\boldsymbol{F}^T$ are symmetric and positive definite transformations (see Exercises). Based on this symmetry property these tensors must possess spectral representations

[14] Note, we use "stretch" to denote deformation associated with "strain" and "stretching" to express "strain rate".

[15] The polar decomposition is unique for a tensor $\boldsymbol{F}$ which is non-singular. For details see TRUESDELL & NOLL [240], Sect. 23.

$$C = \sum_{\alpha} \lambda_{\alpha}^{(C)} \boldsymbol{e}_{\alpha} \otimes \boldsymbol{e}_{\alpha} \,, \quad B = \sum_{i} \lambda_{i}^{(B)} \boldsymbol{e}_{i} \otimes \boldsymbol{e}_{i} \tag{1.5.5}$$

with real *eigenvalues* $\lambda_\alpha^{(C)}$ and $\lambda_i^{(B)}$, which are positive since $\boldsymbol{B}$ and $\boldsymbol{C}$ are positive definite. The vectors $\boldsymbol{e}_\alpha$ and $\boldsymbol{e}_i$ are the eigenvectors in the reference and present configurations, respectively; these can be directly interpreted as the basis vectors of $\boldsymbol{V}_R^3$ and $\boldsymbol{V}_t^3$. On this ground, the right and left stretch tensors can be defined by

$$\begin{aligned} \boldsymbol{U} &:= \textstyle\sum_\alpha \lambda_\alpha^{(U)} \boldsymbol{e}_\alpha \otimes \boldsymbol{e}_\alpha \,, \quad \lambda_\alpha^{(U)} = \sqrt{\lambda_\alpha^{(C)}} \,, \\ \boldsymbol{V} &:= \textstyle\sum_i \lambda_i^{(V)} \boldsymbol{e}_i \otimes \boldsymbol{e}_i \,, \quad \lambda_i^{(V)} = \sqrt{\lambda_i^{(B)}} \,, \end{aligned} \tag{1.5.6}$$

where these are obviously symmetric and positive definite. This construction of the stretch tensors is unique and leads to $\boldsymbol{U}^2 = \boldsymbol{C}$ and $\boldsymbol{V}^2 = \boldsymbol{B}$. Particularly, the existence of inverse tensors $\boldsymbol{U}^{-1}$, $\boldsymbol{V}^{-1}$ is also guaranteed

$$\begin{aligned} \boldsymbol{U}^{-1} &= \textstyle\sum_\alpha \left(\lambda_\alpha^{(U)}\right)^{-1} \boldsymbol{e}_\alpha \otimes \boldsymbol{e}_\alpha \,, \\ \boldsymbol{V}^{-1} &= \textstyle\sum_i \left(\lambda_i^{(V)}\right)^{-1} \boldsymbol{e}_i \otimes \boldsymbol{e}_i \,. \end{aligned} \tag{1.5.7}$$

Thus, the first two points are verified.

With the right and left stretch tensors, we now can build orthogonal tensors $\boldsymbol{R} := \boldsymbol{F}\boldsymbol{U}^{-1}$ and $\bar{\boldsymbol{R}} := \boldsymbol{V}^{-1}\boldsymbol{F}$. The orthogonality results from

$$\begin{aligned} \boldsymbol{R}^T\boldsymbol{R} &= (\boldsymbol{F}\boldsymbol{U}^{-1})^T(\boldsymbol{F}\boldsymbol{U}^{-1}) = (\boldsymbol{U}^{-T}\boldsymbol{F}^T)(\boldsymbol{F}\boldsymbol{U}^{-1}) \\ &= \boldsymbol{U}^{-T}(\boldsymbol{F}^T\boldsymbol{F})\boldsymbol{U}^{-1} = \boldsymbol{U}^{-T}\boldsymbol{U}^2\boldsymbol{U}^{-1} = \boldsymbol{I} \,, \\ \bar{\boldsymbol{R}}\bar{\boldsymbol{R}}^T &= (\boldsymbol{V}^{-1}\boldsymbol{F})(\boldsymbol{V}^{-1}\boldsymbol{F})^T = (\boldsymbol{V}^{-1}\boldsymbol{F})(\boldsymbol{F}^T\boldsymbol{V}^{-T}) \\ &= \boldsymbol{V}^{-1}(\boldsymbol{F}\boldsymbol{F}^T)\boldsymbol{V}^{-T} = \boldsymbol{V}^{-1}\boldsymbol{V}^2\boldsymbol{V}^{-T} = \boldsymbol{I} \,. \end{aligned} \tag{1.5.8}$$

Therefore one has

$$\boldsymbol{F} = \boldsymbol{R}\boldsymbol{U} = \boldsymbol{V}\bar{\boldsymbol{R}} \,. \tag{1.5.9}$$

The uniqueness property of the polar decomposition follows from the fact that $\boldsymbol{C}$ and $\boldsymbol{B}$ are unique (via the spectral decomposition) according to their definition; thus $\boldsymbol{U}$ and $\boldsymbol{V}$ are also uniquely defined. Since the deformation gradient $\boldsymbol{F}$ is not singular, $\boldsymbol{R}$ and $\bar{\boldsymbol{R}}$ are also unique.

In order to show $\boldsymbol{R} = \bar{\boldsymbol{R}}$, we use the orthogonality of $\boldsymbol{R}$ and write

$$\begin{aligned} \boldsymbol{F} &= \boldsymbol{R}\boldsymbol{U} = \boldsymbol{R}\boldsymbol{U}(\boldsymbol{R}^T\boldsymbol{R}) = (\boldsymbol{R}\boldsymbol{U}\boldsymbol{R}^T)\boldsymbol{R} \\ &= \bar{\boldsymbol{V}}\boldsymbol{R} \\ &\overset{!}{=} \boldsymbol{V}\bar{\boldsymbol{R}} \,, \end{aligned} \tag{1.5.10}$$

where $\bar{\boldsymbol{V}} := (\boldsymbol{R}\boldsymbol{U}\boldsymbol{R}^T)$. As a result, one seemingly finds a further decomposition of $\boldsymbol{F}$, for which $\boldsymbol{B} = \boldsymbol{V}^2 = \bar{\boldsymbol{V}}^2$ is also valid. Because $\boldsymbol{B}$ is unique so must be also $\boldsymbol{V}$; both these transformations must therefore be identical, i.e., $\boldsymbol{V} = \bar{\boldsymbol{V}}$. Consequently, it immediately follows that $\boldsymbol{R} = \bar{\boldsymbol{R}}$. Exept for $\lambda^{(\boldsymbol{C})} = \lambda^{(\boldsymbol{B})}$ this completes the proof.

1.5.2 Strain Measures

We have learned so far about the right and left stretch tensors and their role in the polar decomposition; the squares of these tensors are called *right*–CAUCHY–GREEN *deformation tensor* and *left*–CAUCHY–GREEN *deformation tensor*, respectively,

$$\boldsymbol{C} := \boldsymbol{U}^2 = \boldsymbol{F}^T\boldsymbol{F}\,, \quad \boldsymbol{B} := \boldsymbol{V}^2 = \boldsymbol{F}\boldsymbol{F}^T\,. \tag{1.5.11}$$

There are still other deformation tensors found in the literature which constitute suitable measures of stretch or strain under certain conditions. These all can be derived from $\boldsymbol{U}$, $\boldsymbol{V}$ or $\boldsymbol{C}$, $\boldsymbol{B}$ and can be expressed as tensors either in the material configuration or in the spatial configuration. Examples of such tensors are, for the material representation and for the spatial description

- GREEN **strain tensor**

$$\boldsymbol{G} := \tfrac{1}{2}(\boldsymbol{C} - \boldsymbol{I}) = \tfrac{1}{2}(\boldsymbol{F}^T\boldsymbol{F} - \boldsymbol{I}), \tag{1.5.12}$$

- PIOLA **strain tensor**

$$\boldsymbol{K} := \tfrac{1}{2}(\boldsymbol{C}^{-1} - \boldsymbol{I}) = \tfrac{1}{2}(\boldsymbol{F}^{-1}\boldsymbol{F}^{-T} - \boldsymbol{I}), \tag{1.5.13}$$

- FINGER **strain tensor**

$$\boldsymbol{E} := \tfrac{1}{2}(\boldsymbol{I} - \boldsymbol{B}) = \tfrac{1}{2}(\boldsymbol{I} - \boldsymbol{F}\boldsymbol{F}^T), \tag{1.5.14}$$

- ALMANSI **strain tensor**

$$\boldsymbol{A} := \tfrac{1}{2}(\boldsymbol{I} - \boldsymbol{B}^{-1}) = \tfrac{1}{2}(\boldsymbol{I} - \boldsymbol{F}^{-T}\boldsymbol{F}^{-1})\,. \tag{1.5.15}$$

For a rigid body motion i.e., when $\boldsymbol{F} = \boldsymbol{R}$, all these strain tensors have the property to vanish. They are therefore non-zero only when a true deformation i.e., a strain or stretch arises. Some of the geometric interpretations will be treated in the Exercises[16].

[16] For as more detailed account on strain, see HAUPT (2000). There are still other strain measures which are functions of the strain measures introduced above. For instance, the logarithmic strain $\boldsymbol{G}^H = (1/2)\ln(\boldsymbol{I} + 2\boldsymbol{G})$ due to HENCKY is particularly useful (see XIAO et al, 2000).

1.5.3 Eigenvalues, Invariants and CAYLEY–HAMILTON Theorem of Tensors of the Second Rank

In this section we will introduce additional useful facts on second rank tensors, first, specially for the left and right stretch tensors.

- $\boldsymbol{U}$ and $\boldsymbol{V}$ have the same eigenvalues,

$$\lambda := \lambda^{(\boldsymbol{V})} = \lambda^{(\boldsymbol{U})} = \left(\lambda^{(\boldsymbol{B})}\right)^{\frac{1}{2}} = \left(\lambda^{(\boldsymbol{C})}\right)^{\frac{1}{2}} > 0 \, . \tag{1.5.16}$$

  Of these there are exactly three, which may have different values.
- Eigenvectors $\boldsymbol{e}^{(\boldsymbol{V})}$ of $\boldsymbol{V}$ and those of $\boldsymbol{U}$ ($\boldsymbol{e}^{(\boldsymbol{U})}$) are related to each other through

$$\boldsymbol{e}^{(\boldsymbol{V})} = \boldsymbol{R}\boldsymbol{e}^{(\boldsymbol{U})} \, . \tag{1.5.17}$$

- The eigenvalue equation or **characteristic equation** of $\boldsymbol{U}$ has the form

$$\lambda^3 - I_U\lambda^2 + II_U\lambda - III_U = 0 \, , \tag{1.5.18}$$

  where

$$I_U := \operatorname{tr}\boldsymbol{U} \, , \quad II_U := \tfrac{1}{2}\left(I_U^2 - I_{U^2}\right) \, , \quad III_U := \det\boldsymbol{U} \tag{1.5.19}$$

  are the **invariants** of the tensor $\boldsymbol{U}$.

In order to prove the first expression, we start with the eigenvalue equation

$$(\boldsymbol{U} - \lambda^{(\boldsymbol{U})}\boldsymbol{I})\boldsymbol{e}^{(\boldsymbol{U})} = \boldsymbol{0} \, , \tag{1.5.20}$$

which yields non-trivial solutions if $\det(\boldsymbol{U} - \lambda^{(\boldsymbol{U})}\boldsymbol{I}) = 0$. This is the *characteristic equation* of $\boldsymbol{U}$, with which one can calculate the corresponding eigenvalues. If we further use the relation

$$\boldsymbol{U} = \boldsymbol{R}^T\boldsymbol{V}\boldsymbol{R} \, , \tag{1.5.21}$$

which follows directly from the polar decomposition, it reads

$$\begin{aligned} 0 &= \det(\boldsymbol{U} - \lambda^{(\boldsymbol{U})}\boldsymbol{I}) = \det(\boldsymbol{R}^T\boldsymbol{V}\boldsymbol{R} - \lambda^{(\boldsymbol{U})}\boldsymbol{I}) \\ &= \det\left(\boldsymbol{R}^T(\boldsymbol{V} - \lambda^{(\boldsymbol{U})}\boldsymbol{I})\boldsymbol{R}\right) \\ &= \det\boldsymbol{R}^T \det(\boldsymbol{V} - \lambda^{(\boldsymbol{U})}\boldsymbol{I}) \det\boldsymbol{R} \\ \Rightarrow \quad & \det(\boldsymbol{V} - \lambda^{(\boldsymbol{U})}\boldsymbol{I}) = 0 \, , \end{aligned} \tag{1.5.22}$$

since $\det\boldsymbol{R} = 1$. It follows from here that the eigenvalues of $\boldsymbol{U}$ and $\boldsymbol{V}$ are identical, $\lambda := \lambda^{(\boldsymbol{V})} = \lambda^{(\boldsymbol{U})}$. Because $\boldsymbol{U}^2 = \boldsymbol{C}$ and $\boldsymbol{V}^2 = \boldsymbol{B}$ and in view of their spectral decomposition, one also has $\lambda = (\lambda^{(\boldsymbol{C})})^{1/2} = (\lambda^{(\boldsymbol{B})})^{1/2}$. Thus the first point is proved.

Similarly, from the eigenvalue equation for $\boldsymbol{U}$ one concludes

$$\begin{aligned} \mathbf{0} &= (\boldsymbol{U} - \lambda \boldsymbol{I})\, \boldsymbol{e}^{(U)} = \left(\boldsymbol{R}^T \boldsymbol{V} \boldsymbol{R} - \lambda \boldsymbol{I}\right) \boldsymbol{e}^{(U)} \\ &= \boldsymbol{R}^T (\boldsymbol{V} - \lambda \boldsymbol{I})\, \boldsymbol{R} \boldsymbol{e}^{(U)} , \end{aligned} \tag{1.5.23}$$

or, since $\boldsymbol{R}$ is non-singular, after multiplication from left with $\boldsymbol{R}$,

$$(\boldsymbol{V} - \lambda \boldsymbol{I})\, \boldsymbol{R} \boldsymbol{e}^{(U)} = \mathbf{0} . \tag{1.5.24}$$

Thus, $\boldsymbol{R}\boldsymbol{e}^{(U)}$ is the eigenvector corresponding to eigenvalue $\lambda = \lambda^{(V)}$; moreover, $\boldsymbol{e}^{(V)} = \boldsymbol{R}\boldsymbol{e}^{(U)}$, and thus the second point is likewise proved.

Finally, we can corroborate the characteristic equation (1.5.18) of a tensor $\boldsymbol{U}$, for example, by explicitly calculating $\det(\boldsymbol{U} - \lambda \boldsymbol{I}) = 0$. In case one chooses $\{\boldsymbol{e}^{(U)}\}$ as the basis, $\boldsymbol{U}$ has diagonal form

$$\boldsymbol{U} = \begin{pmatrix} \lambda_1 & 0 & 0 \\ 0 & \lambda_2 & 0 \\ 0 & 0 & \lambda_3 \end{pmatrix} , \tag{1.5.25}$$

with the eigenvalues λ_α, $\alpha = 1, 2, 3$. From this we can immediately write the characteristic equation with **invariants** I_U, II_U and III_U which are related to the eigenvalues as follows

$$\begin{aligned} I_U &:= \operatorname{tr} \boldsymbol{U} = \lambda_1 + \lambda_2 + \lambda_3 , \\ II_U &:= \tfrac{1}{2}\left(I_U^2 - I_{U^2}\right) = \lambda_1\lambda_2 + \lambda_2\lambda_3 + \lambda_3\lambda_1 , \\ III_U &:= \det \boldsymbol{U} = \lambda_1\lambda_2\lambda_3 , \end{aligned} \tag{1.5.26}$$

see Exercises for a general derivation.

One further statement in connection with the characteristic equation is the CAYLEY–HAMILTON **Theorem**. It states that the characteristic equation of a linear transformation $\boldsymbol{U}$ is not only fulfilled by its eigenvalues, but also by the transformation itself. This means,

$$\boldsymbol{U}^3 - I_U \boldsymbol{U}^2 + II_U \boldsymbol{U} - III_U \boldsymbol{I} = \mathbf{0} . \tag{1.5.27}$$

A proof can be found in any book on linear algebra.

1.5.4 Geometric Linearization

In dealing with the ensuing considerations we assume the same basis $\{\boldsymbol{e}_\alpha\} = \{\boldsymbol{e}_i \delta_{i\alpha}\}$ for the reference as well as the present configuration (see Fig 1.4).

Instead of considering the absolute positions in the present configuration we merely consider displacements relative to the reference configuration

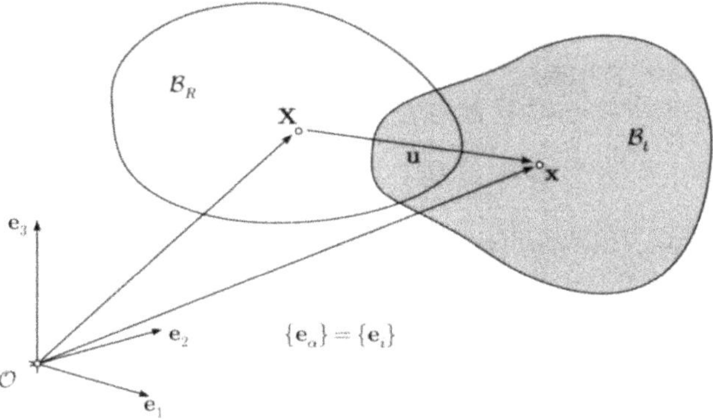

Fig. 1.4. Choice of the basis vectors for the reference and present configuration in geometric linearization.

$$\boldsymbol{u}(\boldsymbol{X},t) = \chi(\boldsymbol{X},t) - \boldsymbol{X} \; ; \tag{1.5.28}$$

the deformation gradient can hence be written as

$$\boldsymbol{F} = \boldsymbol{H} + \boldsymbol{I} \; . \tag{1.5.29}$$

The tensor $\boldsymbol{H}$ is called the *displacement gradient,*

$$\boldsymbol{H} := \frac{\partial \boldsymbol{u}}{\partial \boldsymbol{X}} \, , \quad H_{i\alpha} := \frac{\partial u_i}{\partial X_\alpha} \; . \tag{1.5.30}$$

For very small magnitude of the displacement gradient[17], $||\boldsymbol{H}|| \ll 1$, the quadratic or higher terms in the displacement gradient can be neglected. Such consideration is known as *geometric linearization.* In this linearization, the rotation and various strain tensors take the forms

[17] The norm of a vector or a tensor of rank two, $||\boldsymbol{v}||$ or $||\boldsymbol{T}||$, can be similarly introduced as the magnitude of a real number. It must satisfy the respective properties, such as positivity, homogeneity and triangular inequality. Commonly, the Euclidean distance, $||\boldsymbol{v}||_2 := \sqrt{\boldsymbol{v} \cdot \boldsymbol{v}}$ is chosen as the vector norm. The norm of a quadratic matrix $\boldsymbol{A}$ compatible with the Euclidean norm of a vector is given by the largest eigenvalue of the matrix formed by $\boldsymbol{A}^T\boldsymbol{A}$, $||\boldsymbol{A}|| := \sqrt{\lambda_{\max}(\boldsymbol{A}^T\boldsymbol{A})}$. This norm is always bigger or equal to the spectral radius $\rho(\boldsymbol{A})$ of the tensor $\boldsymbol{A}$, i.e., the largest eigenvalue (by magnitude) of $\boldsymbol{A}$.

$$\begin{aligned} \boldsymbol{R} &\approx \boldsymbol{I} + \tfrac{1}{2}(\boldsymbol{H} - \boldsymbol{H}^T) , \\ \boldsymbol{U} \approx \boldsymbol{V} &\approx \boldsymbol{I} + \tfrac{1}{2}(\boldsymbol{H} + \boldsymbol{H}^T) , \\ \boldsymbol{C} \approx \boldsymbol{B} &\approx \boldsymbol{I} + \boldsymbol{H} + \boldsymbol{H}^T , \\ \boldsymbol{G} &\approx \tfrac{1}{2}(\boldsymbol{H} + \boldsymbol{H}^T) . \end{aligned} \tag{1.5.31}$$

In the geometric linearization the GREEN strain tensor is the same as the symmetric part of the displacement gradient.

The geometric linearization implies in addition that derivatives with respect to the material coordinates can be approximated by the derivatives with respect to the present coordinates,

$$\frac{\partial f(\boldsymbol{x},t)}{\partial \boldsymbol{X}} = \frac{\partial f}{\partial \boldsymbol{x}} \frac{\partial \boldsymbol{x}}{\partial \boldsymbol{X}} = \frac{\partial f}{\partial \boldsymbol{x}} (\boldsymbol{I} + \boldsymbol{H}) \approx \frac{\partial f}{\partial \boldsymbol{x}} , \tag{1.5.32}$$

where $\boldsymbol{H}$ is negligible in comparison to $\boldsymbol{I}$. The *linearized strain tensor* is thus

$$\begin{aligned} \boldsymbol{\varepsilon} &:= \tfrac{1}{2}(\boldsymbol{H} + \boldsymbol{H}^T) = \tfrac{1}{2}(\operatorname{Grad} \boldsymbol{u} + (\operatorname{Grad} \boldsymbol{u})^T) \\ &\approx \tfrac{1}{2}(\operatorname{grad} \boldsymbol{u} + (\operatorname{grad} \boldsymbol{u})^T) . \end{aligned} \tag{1.5.33}$$

1.6 Exercises

In the following, let $f \in \mathbb{R}$ be a scalar field, $\boldsymbol{a}, \boldsymbol{b}, \boldsymbol{c} \in \mathbb{R}^3$ arbitrary vectors and $\boldsymbol{A} \in \mathbb{R}^{3\times3}$ an arbitrary tensor of rank two.

1. Express the following relations in component forms

$$\boldsymbol{a} \cdot (\boldsymbol{b} \times \boldsymbol{c}) , \quad \boldsymbol{A}\boldsymbol{a} \cdot (\boldsymbol{A}\boldsymbol{b} \times \boldsymbol{A}\boldsymbol{c}) , \quad \boldsymbol{b} \otimes \boldsymbol{A}$$

   and show that $\boldsymbol{a} \cdot (\boldsymbol{b} \times \boldsymbol{c}) = \boldsymbol{b} \cdot (\boldsymbol{c} \times \boldsymbol{a})$ and $\boldsymbol{a} \cdot (\boldsymbol{b} \times \boldsymbol{a}) = 0$.
2. The Nabla operator is defined as $\nabla = \boldsymbol{e}_i \dfrac{\partial}{\partial x_i}$. With this operator write

$$\operatorname{div} \boldsymbol{a} \equiv \nabla \cdot \boldsymbol{a} , \quad \operatorname{curl} \boldsymbol{a} \equiv \nabla \times \boldsymbol{a} \quad \text{and} \quad \operatorname{grad} f \equiv \nabla f$$

   in component form.
3. With the help of the component representation, prove the validity of the following identities:

$$\operatorname{curl} \operatorname{grad} f = \boldsymbol{0} \quad \text{and} \quad \operatorname{div} \operatorname{curl} \boldsymbol{a} = 0 .$$

4. Calculate the following expressions in Cartesian component forms:

$$\boldsymbol{a} \times (\boldsymbol{b} \times \boldsymbol{c}) , \quad \operatorname{curl} \operatorname{curl} \boldsymbol{a} , \quad \operatorname{grad} \boldsymbol{a} \quad \text{and} \quad \operatorname{div} \boldsymbol{A} .$$

5. Show that for a scalar, vector or a tensor field Φ the LAPLACE operator

$$\operatorname{lap}\Phi \equiv \operatorname{div}\operatorname{grad}\Phi$$

commutes with the operators curl, grad, and div, i.e.,

$$\operatorname{curl}(\operatorname{div}\operatorname{grad}\Phi) = \operatorname{div}\operatorname{grad}(\operatorname{curl}\Phi)\ .$$

(This of course is restricted to the cases for which the operations make sense)

6. With the fully antisymmetric tensor

$$\varepsilon_{ijk} := \begin{cases} 1 & ,\ (ijk)\text{ cyclic, e.g., }(123) \\ -1 & ,\ (ijk)\text{ anticyclic, e.g., }(321) \\ 0 & ,\ \text{otherwise} \end{cases} \tag{1.6.1}$$

show that the following relations hold:

a) $\varepsilon_{ijk}\varepsilon_{lmk} = \delta_{il}\delta_{jm} - \delta_{im}\delta_{jl}$,
b) $\varepsilon_{ijk}\varepsilon_{ljk} = 2\delta_{il}$,
c) $\varepsilon_{ijk}\varepsilon_{ijk} = 6$.

7. Prove with the previous results the following relations ($\boldsymbol{A}$ is a 3×3–matrix with components A_{ij}):

a) $\det \boldsymbol{A} = \frac{1}{6}\varepsilon_{ijk}\varepsilon_{lmn}A_{il}A_{jm}A_{kn}$
b) $\left(\boldsymbol{A}^{-1}\right)_{ij} = \dfrac{1}{2\det \boldsymbol{A}}\varepsilon_{ikl}\varepsilon_{jmn}A_{km}A_{ln}$.

8. The transformation of surface and volume elements from the reference to the present configuration is given by

$$\mathrm{d}v = J\,\mathrm{d}V \quad \text{and} \quad \mathrm{d}\boldsymbol{a} = J\boldsymbol{F}^{-T}\,\mathrm{d}\boldsymbol{A}\ . \tag{1.6.2}$$

Prove these transformation relations with the help of the identities

$$(\det \boldsymbol{F})\,(\boldsymbol{a}\times\boldsymbol{b})\cdot\boldsymbol{c} \equiv (\boldsymbol{F}\boldsymbol{a}\times\boldsymbol{F}\boldsymbol{b})\cdot\boldsymbol{F}\boldsymbol{c} \equiv \det(\boldsymbol{F}\boldsymbol{a},\boldsymbol{F}\boldsymbol{b},\boldsymbol{F}\boldsymbol{c})\ , \tag{1.6.3}$$

the triple product of the vectors $\boldsymbol{F}\boldsymbol{a}$, $\boldsymbol{F}\boldsymbol{b}$ and $\boldsymbol{F}\boldsymbol{c}$.

9. Let $\mathrm{d}\boldsymbol{x} = \boldsymbol{F}\,\mathrm{d}\boldsymbol{X}$ and $\mathrm{d}\boldsymbol{y} = \boldsymbol{F}\,\mathrm{d}\boldsymbol{Y}$ be two material line elements. Denote the unit vectors in the directions of these elements by

$$\begin{aligned} &\boldsymbol{e}_X, \boldsymbol{e}_Y && (\text{LAGRANGEan description})\ , \\ &\boldsymbol{e}_x, \boldsymbol{e}_y && (\text{EULERian description})\ . \end{aligned}$$

If $\mathrm{d}\boldsymbol{X}$ and $\mathrm{d}\boldsymbol{Y}$ are parallel, then two normal strains,

$$\varepsilon_L := \frac{|\mathrm{d}\boldsymbol{x}| - |\mathrm{d}\boldsymbol{X}|}{|\mathrm{d}\boldsymbol{X}|}, \quad \varepsilon_E := \frac{|\mathrm{d}\boldsymbol{x}| - |\mathrm{d}\boldsymbol{X}|}{|\mathrm{d}\boldsymbol{x}|}$$

can be defined. If $\mathrm{d}\boldsymbol{X}$ and $\mathrm{d}\boldsymbol{Y}$ are perpendicular to one another, likewise a shear strain $\gamma_{\boldsymbol{XY}}$ can be defined according to

$$\sin\gamma_{\boldsymbol{XY}} = \frac{\mathrm{d}\boldsymbol{x}}{\| \mathrm{d}\boldsymbol{x} \|} \cdot \frac{\mathrm{d}\boldsymbol{y}}{\| \mathrm{d}\boldsymbol{y} \|} \ .$$

This angle measures the deviation of the angle between $\mathrm{d}\boldsymbol{x}$ and $\mathrm{d}\boldsymbol{y}$ from 90°. Finally, prove that $\boldsymbol{G} = \boldsymbol{F}^T\boldsymbol{A}\boldsymbol{F}$.

10. Show that the left as well as the right CAUCHY–GREEN–deformation tensors ($\boldsymbol{C} = \boldsymbol{F}^T\boldsymbol{F}$ and $\boldsymbol{B} = \boldsymbol{F}\boldsymbol{F}^T$) are symmetric and positive definite.
11. Calculate the characteristic equation of a symmetric 3×3–matrix as well as the 3 invariants and show that these are really invariants under orthogonal transformations.
12. Using the CAYLEY–HAMILTON theorem prove the following identities

$$III_{\boldsymbol{A}} = \tfrac{1}{3}(I_{\boldsymbol{A}^3} - I_{\boldsymbol{A}}I_{\boldsymbol{A}^2} + II_{\boldsymbol{A}}I_{\boldsymbol{A}}) = \tfrac{1}{3}(I_{\boldsymbol{A}^3} - I_{\boldsymbol{A}}^3) + II_{\boldsymbol{A}}I_{\boldsymbol{A}} \ ,$$

$$\frac{\partial II_{\boldsymbol{A}}}{\partial \boldsymbol{A}} = I_{\boldsymbol{A}}\boldsymbol{I} - \boldsymbol{A}^T \ , \quad \frac{\partial III_{\boldsymbol{A}}}{\partial \boldsymbol{A}} = \boldsymbol{A}^{-T} III_{\boldsymbol{A}} \ .$$

13. Prove the identity of the expressions

$$\varphi := t_{ij}v_{j,i} = \mathrm{tr}(\boldsymbol{D}\boldsymbol{t}^T)$$

for the stress power. Here $\boldsymbol{t}$ is the symmetric CAUCHY stress tensor and $\boldsymbol{D}$ the symmetric part of the velocity gradient.

14. Calculate the total time derivative of the determinant of the deformation gradient,

$$\dot{J} = \frac{\mathrm{d}}{\mathrm{d}t}\det\boldsymbol{F} \ .$$

1.7 Solutions

1. With $\boldsymbol{a} = a_i\boldsymbol{e}_i$, $b_i\boldsymbol{e}_i$, $c_i\boldsymbol{e}_i$ and $\boldsymbol{A} = A_{ij}\,\boldsymbol{e}_i \otimes \boldsymbol{e}_j$ and using the rule cited in (1.1.14) one obtains the required expressions. The first expression (triple product) reads

$$\begin{aligned}\boldsymbol{a} \cdot (\boldsymbol{b} \times \boldsymbol{c}) &= a_i\boldsymbol{e}_i \cdot (b_j\boldsymbol{e}_j \times c_k\boldsymbol{e}_k) \\ &= a_ib_jc_k\boldsymbol{e}_i \cdot (\boldsymbol{e}_j \times \boldsymbol{e}_k) = a_ib_jc_k\boldsymbol{e}_i \cdot \varepsilon_{jkl}\boldsymbol{e}_l \\ &= a_ib_jc_k\varepsilon_{jkl}(\boldsymbol{e}_i \cdot \boldsymbol{e}_l) \ = a_ib_jc_k\varepsilon_{jkl}\delta_{il} = a_ib_jc_k\varepsilon_{ijk} \ .\end{aligned}$$

Through the exchange of indices and using the definition (1.6.1) for ε_{ijk}, this yields

$$\begin{aligned}a_ib_jc_k\varepsilon_{ijk} &= b_jc_ka_i\varepsilon_{jki} \quad \text{(cyclic interchanging)} \\ &= b_ic_ja_k\varepsilon_{ijk} \quad \text{(renaming the indices)} \ ,\end{aligned}$$

or in vector notation

$$\boldsymbol{a} \cdot (\boldsymbol{b} \times \boldsymbol{c}) = \boldsymbol{b} \cdot (\boldsymbol{c} \times \boldsymbol{a}) \ .$$

Substituting $\boldsymbol{c} = \boldsymbol{a}$, using $\boldsymbol{a} \times \boldsymbol{a} = \boldsymbol{0}$ leads to

$$\boldsymbol{a} \cdot (\boldsymbol{b} \times \boldsymbol{a}) = \boldsymbol{b} \cdot (\boldsymbol{a} \times \boldsymbol{a}) = 0 \ .$$

In the second expression of the problem formulation each of the vectors is multiplied by a tensor $\boldsymbol{A}$; so,

$$\begin{aligned} \boldsymbol{Aa} \cdot (\boldsymbol{Ab} \times \boldsymbol{Ac}) &= A_{il}a_l A_{jm}b_m A_{kn}c_n \varepsilon_{ijk} \\ &= A_{il}A_{jm}A_{kn}\varepsilon_{ijk}a_l b_m c_n \ . \end{aligned}$$

As shown in Exercise 8, the above relation is identical with $(\det \boldsymbol{A})\, \boldsymbol{a} \cdot (\boldsymbol{b} \times \boldsymbol{c})$, thus $\boldsymbol{Aa} \cdot (\boldsymbol{Ab} \times \boldsymbol{Ac}) = (\det \boldsymbol{A})\boldsymbol{a} \cdot (\boldsymbol{b} \times \boldsymbol{c})$.
The dyadic product of a vector (i.e., a tensor of first rank) with a tensor of rank two results in a tensor of third rank

$$\boldsymbol{b} \otimes \boldsymbol{A} = b_i \boldsymbol{e}_i \otimes A_{jk} \boldsymbol{e}_j \otimes \boldsymbol{e}_k = b_i A_{jk} \boldsymbol{e}_i \otimes \boldsymbol{e}_j \otimes \boldsymbol{e}_k \ ,$$

whose components $(b_i A_{jk})$ can be expressed as a $3 \times 3 \times 3$–scheme.

2. The Nabla operator ∇ is often used in the literature in order to express the gradient, divergence and curl of fields. With this operator, we write the following expressions in component representation as

$$\begin{aligned} \operatorname{grad} f &= \nabla f = \boldsymbol{e}_i \frac{\partial}{\partial x_i} f = \frac{\partial f}{\partial x_i} \boldsymbol{e}_i = f_{,i} \boldsymbol{e}_i \ , \\ \operatorname{div} \boldsymbol{a} &= \nabla \cdot \boldsymbol{a} = \frac{\partial}{\partial x_i} \boldsymbol{e}_i \cdot a_j \boldsymbol{e}_j = \frac{\partial a_j}{\partial x_i} \boldsymbol{e}_i \cdot \boldsymbol{e}_j \\ &= \frac{\partial a_j}{\partial x_i} \delta_{ij} = \frac{\partial a_i}{\partial x_i} = a_{i,i} \ , \\ \operatorname{curl} \boldsymbol{a} &= \nabla \times \boldsymbol{a} = \frac{\partial}{\partial x_i} \boldsymbol{e}_i \times a_j \boldsymbol{e}_j = \frac{\partial a_j}{\partial x_i} \boldsymbol{e}_i \times \boldsymbol{e}_j \\ &= \frac{\partial a_j}{\partial x_i} \varepsilon_{ijk} \boldsymbol{e}_k = a_{j,i} \varepsilon_{ijk} \boldsymbol{e}_k \ . \end{aligned}$$

The gradient is therefore a dyadic product with a field variable, which is assumed here as a scalar field f. The divergence is the contracted product of the Nabla operator with a field quantity; and the curl is the cross product of the Nabla operator and the field quantity.

3. It is to prove here that the curl of a gradient field $(\operatorname{grad} f)$ and the divergence of a vortex field $(\operatorname{curl} \boldsymbol{a})$ vanish identically. This also means that any gradient field is vortex free,

$$\begin{aligned}
\operatorname{curl}(\operatorname{grad} f) &= \frac{\partial}{\partial x_i}\boldsymbol{e}_i \times \left(\frac{\partial}{\partial x_j}\boldsymbol{e}_j f\right) = \frac{\partial^2 f}{\partial x_i\,\partial x_j}\boldsymbol{e}_i \times \boldsymbol{e}_j \\
&= \underline{f_{,ij}\;\varepsilon_{ijk}\boldsymbol{e}_k} \overset{(1)}{=} f_{,ji}\;\varepsilon_{ijk}\boldsymbol{e}_k \\
&\overset{(2)}{=} -f_{,ji}\;\varepsilon_{jik}\boldsymbol{e}_k \overset{(3)}{=} -\underline{f_{,ij}\;\varepsilon_{ijk}\boldsymbol{e}_k} = 0\,.
\end{aligned}$$

In step (1) the order of differentiation is interchanged, in step (2) and (3) the indices are interchanged and renamed. Comparing the underlined parts implies that curl (grad) must vanish.
A vortex field is source free,

$$\begin{aligned}
\operatorname{div}(\operatorname{curl}\boldsymbol{a}) &= \frac{\partial}{\partial x_i}\boldsymbol{e}_i \cdot \left(\frac{\partial}{\partial x_j}\boldsymbol{e}_j \times a_k\boldsymbol{e}_k\right) \\
&= \frac{\partial^2 a_k}{\partial x_i\,\partial x_j}\boldsymbol{e}_i \cdot (\boldsymbol{e}_j \times \boldsymbol{e}_k) = a_{k,ij}\varepsilon_{jkl}\boldsymbol{e}_i \cdot \boldsymbol{e}_l \\
&= a_{k,ji}\varepsilon_{jkl}\delta_{il} = a_{k,ji}\varepsilon_{kij} \\
&= -a_{k,ji}\varepsilon_{kji} = -a_{k,ij}\varepsilon_{kij} = 0\,;
\end{aligned}$$

the antisymmetry of ε_{ijk} and the interchangeability of the sequence of differentiation are exploited.

4. Twofold application of the cross products as well as the curl operators yields a product of two LEVI–CIVITÀ ε–tensors, which will be calculated in Exercise 6. Here, we sketch the result in advance

$$\begin{aligned}
\boldsymbol{a} \times (\boldsymbol{b} \times \boldsymbol{c}) &= a_i\boldsymbol{e}_i \times (b_j\boldsymbol{e}_j \times c_k\boldsymbol{e}_k) = a_ib_jc_k\boldsymbol{e}_i \times (\boldsymbol{e}_j \times \boldsymbol{e}_k) \\
&= a_ib_jc_k\boldsymbol{e}_i \times \boldsymbol{e}_l\varepsilon_{jkl} = a_ib_jc_k\varepsilon_{ilm}\varepsilon_{jkl}\boldsymbol{e}_m \\
&\overset{(1)}{=} a_ib_jc_k\varepsilon_{mil}\varepsilon_{jkl}\boldsymbol{e}_m \\
&\overset{(2)}{=} a_ib_jc_k(\delta_{mj}\delta_{ik} - \delta_{mk}\delta_{ij})\boldsymbol{e}_m \\
&= (a_ib_mc_i - a_ib_ic_m)\boldsymbol{e}_m = (\boldsymbol{a}\cdot\boldsymbol{c})\boldsymbol{b} - (\boldsymbol{a}\cdot\boldsymbol{b})\boldsymbol{c}\,,
\end{aligned}$$

where a cyclic interchange has been carried out in step (1). The result of Exercise 6 is used in step (2). Similarly the twofold application of the curl operator yields

$$\begin{aligned}
\operatorname{curl}\operatorname{curl}\boldsymbol{a} &= \frac{\partial}{\partial x_i}\boldsymbol{e}_i \times \left(\frac{\partial}{\partial x_j}\boldsymbol{e}_j \times a_k\boldsymbol{e}_k\right) = a_{k,ij}\boldsymbol{e}_i \times (\boldsymbol{e}_j \times \boldsymbol{e}_k) \\
&= a_{k,ij}(\delta_{mj}\delta_{ik} - \delta_{mk}\delta_{ij})\boldsymbol{e}_m = (a_{i,im} - a_{m,ii})\boldsymbol{e}_m \\
&= \frac{\partial}{\partial x_m}\boldsymbol{e}_m\left(\frac{\partial}{\partial x_i}\boldsymbol{e}_i \cdot a_j\boldsymbol{e}_j\right) - \frac{\partial}{\partial x_i}\boldsymbol{e}_i \cdot \left(\frac{\partial}{\partial x_j}\boldsymbol{e}_j(a_m\boldsymbol{e}_m)\right) \\
&= \operatorname{grad}(\operatorname{div}\boldsymbol{a}) - \operatorname{div}(\operatorname{grad}\boldsymbol{a})\,.
\end{aligned}$$

The gradient of a vector field occurs in the last result. As it was already made clear in Exercise 2 the gradient of a vector field corresponds to the dyadic product of the Nabla operator with that field,

$$(\operatorname{grad} \boldsymbol{a})^T = \nabla \otimes \boldsymbol{a} = \left(\frac{\partial}{\partial x_j} \boldsymbol{e}_j \otimes a_i \boldsymbol{e}_i \right) = a_{i,j} \boldsymbol{e}_j \otimes \boldsymbol{e}_i \ .$$

In this definition of the gradient please note the sequence of the indices, for the definition of the gradient of a vector field is

$$\operatorname{grad} \boldsymbol{a} = (\nabla \otimes \boldsymbol{a})^T = \left(\frac{\partial}{\partial x_j} \boldsymbol{e}_j \otimes a_i \boldsymbol{e}_i \right)^T = a_{i,j} \boldsymbol{e}_i \otimes \boldsymbol{e}_j \ .$$

In the same way, we build the divergence of a tensor field of higher order, we must take care on which index the derivative has to be performed; (the sequence is, however, insignificant in case of symmetric second rank tensors). The divergence of a tensor field of rank two is frequently symbolically written as

$$\operatorname{div} \boldsymbol{A} = \nabla \cdot \boldsymbol{A} \ ,$$

however, the dot does not imply the scalar product, but the contracted product and the formula is *not* correct as we shall see in a moment. According to our symbolic representation, the correct way of writing this is

$$\begin{aligned} \operatorname{div} \boldsymbol{A} &= \frac{\partial}{\partial x_k} \boldsymbol{e}_k \left(A_{ij} \boldsymbol{e}_j \otimes \boldsymbol{e}_i \right) \\ &= A_{ij,k} \boldsymbol{e}_k (\boldsymbol{e}_j \otimes \boldsymbol{e}_i) = A_{ij,k} \delta_{jk} \boldsymbol{e}_i = A_{ik,k} \boldsymbol{e}_i \ . \end{aligned}$$

This can also be written in terms of the Nabla operator as $\nabla \cdot \boldsymbol{A}^T$. Using the transpose of the operand $\boldsymbol{A}$ then yields

$$\begin{aligned} \operatorname{div} \boldsymbol{A}^T &= \frac{\partial}{\partial x_i} \boldsymbol{e}_i A_{jk} (\boldsymbol{e}_j \otimes \boldsymbol{e}_k) = A_{jk,i} \, \boldsymbol{e}_i (\boldsymbol{e}_j \otimes \boldsymbol{e}_k) \\ &= A_{jk,i} \, \delta_{ij} \boldsymbol{e}_k = A_{ik,i} \boldsymbol{e}_k \ , \end{aligned}$$

or with the Nabla operator $\operatorname{div} \boldsymbol{A}^T = \nabla \cdot \boldsymbol{A}$. It is clearly seen that one encounters difficulties in the symbolic representation with the Nabla operator: in calculating gradients and divergences of higher order tensors, care has to be taken with respect to which indices these should be differentiated.

5. A quite frequently appearing combination of derivatives is the LAPLACE operator

$$\operatorname{lap} \Phi = \operatorname{div}(\operatorname{grad} \Phi) \ ,$$

which is defined for an arbitrary scalar or tensor field. With the Nabla operator we write this as

$$\operatorname{lap} \Phi = \nabla \cdot \nabla \Phi = \nabla^2 \Phi \,.$$

Here, again, care must be taken with regard to the notation which one applies. The above given descriptions hold true for a scalar field Φ. For tensors of second (or higher) rank, $\boldsymbol{\Phi} = \Phi_{ij} \boldsymbol{e}_i \otimes \boldsymbol{e}_j$, one must be more careful and the operation is

$$\begin{aligned}
\operatorname{lap} \boldsymbol{\Phi} &= (\nabla \cdot \nabla) \boldsymbol{\Phi} \\
&= \Big(\frac{\partial}{\partial x_i} \boldsymbol{e}_i \cdot \frac{\partial}{\partial x_j} \boldsymbol{e}_j \Big) \Phi_{kl} (\boldsymbol{e}_k \otimes \boldsymbol{e}_l) \\
&= \Phi_{kl,ij} \delta_{ij} \boldsymbol{e}_k \otimes \boldsymbol{e}_l = \Phi_{kl,ii} \boldsymbol{e}_k \otimes \boldsymbol{e}_l \,.
\end{aligned}$$

In general, we have for an arbitrary tensor $\boldsymbol{\Phi}$

$$\operatorname{lap} \boldsymbol{\Phi} = \frac{\partial}{\partial x_i} \boldsymbol{e}_i \cdot \left(\frac{\partial}{\partial x_j} \boldsymbol{e}_j \boldsymbol{\Phi} \right) = \boldsymbol{\Phi}_{,ii} \,.$$

With the above results we can prove that the LAPLACE operator is interchangeable with the operators grad, div and curl. The first becomes

$$\operatorname{lap}(\operatorname{grad} \Phi) = (\Phi_{,j} \boldsymbol{e}_j)_{,ii} = \Phi_{,iij} \boldsymbol{e}_j = (\Phi_{,ii})_{,j} \boldsymbol{e}_j = \operatorname{grad}(\operatorname{lap} \Phi) \,,$$

and for the vector field $\boldsymbol{\Phi} = \Phi_k \boldsymbol{e}_k$, the LAPLACE operator is

$$\operatorname{lap}(\operatorname{div} \boldsymbol{\Phi}) = (\Phi_{k,k})_{,ii} = (\Phi_{k,ii})_{,k} = \operatorname{div}(\operatorname{lap} \boldsymbol{\Phi}) \,,$$

and similarly for the curl of a vector field

$$\operatorname{lap}(\operatorname{curl} \boldsymbol{\Phi}) = (\varepsilon_{ljk} \Phi_{k,j})_{,ii} \, \boldsymbol{e}_l = \varepsilon_{ljk} (\Phi_{k,ii})_{,j} \, \boldsymbol{e}_l = \operatorname{curl} (\operatorname{lap} \boldsymbol{\Phi}) \,.$$

6. To calculate the product of two LEVI–CIVITÀ ε–tensors, we introduce the component expression of a sixth rank tensor, the product is

$$\begin{aligned}
\varepsilon_{ijk} \varepsilon_{lmn} &= \begin{cases} 1 & \text{for } lmn \text{ and } ijk \text{ cyclically equal}\,, \\ -1 & \text{for } lmn \text{ and } ijk \text{ cyclically unequal}\,, \\ 0 & \text{otherwise}\,, \end{cases} \\
&= \det \begin{pmatrix} \delta_{il} & \delta_{jl} & \delta_{kl} \\ \delta_{im} & \delta_{jm} & \delta_{km} \\ \delta_{in} & \delta_{jn} & \delta_{kn} \end{pmatrix} .
\end{aligned}$$

This expression lets us calculate the contracted products in terms of their components

$$\varepsilon_{ijk}\varepsilon_{lmk} = \det \begin{pmatrix} \delta_{il} & \delta_{jl} & \delta_{kl} \\ \delta_{im} & \delta_{jm} & \delta_{km} \\ \delta_{ik} & \delta_{jk} & \delta_{kk} \end{pmatrix} .$$

We have $\delta_{kk} = \delta_{11} + \delta_{22} + \delta_{33} = 3$, and through multiplication, we obtain

$$\begin{aligned} \varepsilon_{ijk}\varepsilon_{lmk} = & \quad \delta_{il}(\delta_{jm} \cdot 3 - \delta_{km}\delta_{jk}) \\ & - \delta_{jl}(\delta_{im} \cdot 3 - \delta_{km}\delta_{ik}) \\ & + \delta_{kl}(\delta_{im}\delta_{jk} - \delta_{jm}\delta_{ik}) \\ = & \quad (3\delta_{il}\delta_{jm} - \delta_{il}\delta_{jm}) \\ & - (3\delta_{jl}\delta_{im} - \delta_{jl}\delta_{im}) \\ & + (\delta_{jl}\delta_{im} - \delta_{jm}\delta_{il}) \\ = & \quad \delta_{il}\delta_{jm} - \delta_{im}\delta_{jl} . \end{aligned}$$

With this result it follows that

$$\varepsilon_{ijk}\varepsilon_{ljk} = \delta_{il}\delta_{jj} - \delta_{ij}\delta_{jl} = 2\delta_{il} ,$$

and thus

$$\varepsilon_{ijk}\varepsilon_{ijk} = 2\delta_{ii} = 6 .$$

7. For a three dimensional matrix

$$\boldsymbol{A} = (\boldsymbol{A}_1, \boldsymbol{A}_2, \boldsymbol{A}_3) ,$$

which is decomposed into its column vectors, the determinant can be written as the triple product

$$\begin{aligned} \det \boldsymbol{A} &= \det(\boldsymbol{A}_1, \boldsymbol{A}_2, \boldsymbol{A}_3) \\ &= \boldsymbol{A}_1 \cdot (\boldsymbol{A}_2 \times \boldsymbol{A}_3) \\ &= A_{i1}A_{j2}A_{k3}\, \varepsilon_{ijk} . \end{aligned}$$

Through renaming

$$\det \boldsymbol{A} = A_{j1}A_{k2}A_{i3}\, \varepsilon_{jki}$$

and cyclically interchanging the indices we obtain

$$\det \boldsymbol{A} = A_{i3}A_{j1}A_{k2}\, \varepsilon_{ijk} .$$

Implementing all six permutations yields

$$\begin{aligned}
\det \boldsymbol{A} &= A_{i1}A_{j2}A_{k3}\ \varepsilon_{ijk} \\
&= A_{i2}A_{j3}A_{k1}\ \varepsilon_{ijk} \\
&= A_{i3}A_{j1}A_{k2}\ \varepsilon_{ijk} \\
&= -A_{i3}A_{j2}A_{k1}\ \varepsilon_{ijk} \\
&= -A_{i2}A_{j1}A_{k3}\ \varepsilon_{ijk} \\
&= -A_{i1}A_{j3}A_{k2}\ \varepsilon_{ijk}\ .
\end{aligned}$$

Adding these six identities and observing that we have a permutation of the indices (123) leads to

$$6\det \boldsymbol{A} = A_{il}A_{jm}A_{kn}\ \varepsilon_{ijk}\varepsilon_{lmn}\ ,$$

answering the first part of the problem. To prove the second, we exploit the relations

$$\boldsymbol{A}^{-1}\boldsymbol{A} = \boldsymbol{I} \quad \text{and} \quad (\boldsymbol{A}^{-1})_{ij}A_{jp} = \delta_{ip}\ ,$$

and show

$$\frac{1}{2\det \boldsymbol{A}}\varepsilon_{ikm}\varepsilon_{jln}A_{lk}A_{nm}A_{jp} \stackrel{!}{=} \delta_{ip}\ .$$

With the help of the expression derived above, this can be further written as

$$\varepsilon_{ikm}\varepsilon_{jln}A_{lk}A_{nm}A_{jp} \stackrel{!}{=} \tfrac{1}{3}\varepsilon_{qkm}\varepsilon_{jln}A_{lk}A_{nm}A_{jq}\delta_{ip}\ . \tag{1.7.1}$$

For $i = p$ we immediately obtain an identity.
For $i \neq p$, the right hand side of (1.7.1) vanishes. Therefore we must prove that the left hand side vanishes identically as well; indeed we realize that $\frac{1}{2}\varepsilon_{ikm}\varepsilon_{jln}A_{lk}A_{nm}$ is the subdeterminant[18] $\tilde{\boldsymbol{A}}$ associated with the element (j,i). Multiplying this with the element (A_{ji}) yields for $i = p$ the determinant. But if $i \neq p$, the expansion appears with the "false"

[18] The subdeterminant arises through cancellation of the i-th row and the j-th column, multiplied with $(-1)^{i+j}$. The determinant of $\boldsymbol{A}$ then gives

$$\det \boldsymbol{A} = \sum_j \tilde{A}_{ij}A_{ij} \quad (i \text{ arbitrary, but fixed}).$$

The expansion with respect to any other row always produces 0,

$$\sum_j \tilde{A}_{ij}A_{pj} = 0 \quad (p \neq i)\ .$$

column or one calculates the determinant of the matrix $\boldsymbol{A}$, in which the i-th column is substituted for the p-th column, producing straightforward zero.

8. The transformation of surface and volume elements from the reference to the present configuration can be carried out with the help of triple products. The line element in the present configuration $\mathrm{d}\boldsymbol{x}^i$ is connected to that in the reference configuration $\mathrm{d}\boldsymbol{X}^i$ through the deformation gradient via

$$\mathrm{d}\boldsymbol{x}^i = \boldsymbol{F}\mathrm{d}\boldsymbol{X}^i \,, \quad i = 1,2,3 \,.$$

a) The infinitesimal volume element is given by the triple product of three vectors

$$\mathrm{d}v = (\mathrm{d}\boldsymbol{x}^1 \times \mathrm{d}\boldsymbol{x}^2) \cdot \mathrm{d}\boldsymbol{x}^3 \,.$$

With the identity

$$(\det \boldsymbol{F})(\boldsymbol{a} \times \boldsymbol{b}) \cdot \boldsymbol{c} \equiv \det(\boldsymbol{F}\boldsymbol{a}, \boldsymbol{F}\boldsymbol{b}, \boldsymbol{F}\boldsymbol{c}) \equiv (\boldsymbol{F}\boldsymbol{a} \times \boldsymbol{F}\boldsymbol{b}) \cdot \boldsymbol{F}\boldsymbol{c} \,,$$

applied to $d\boldsymbol{x}^i = \boldsymbol{F}d\boldsymbol{X}^i$ the volume element becomes

$$\begin{aligned} \mathrm{d}v &= (\boldsymbol{F}\mathrm{d}\boldsymbol{X}^1 \times \boldsymbol{F}\mathrm{d}\boldsymbol{X}^2) \cdot \boldsymbol{F}\mathrm{d}\boldsymbol{X}^3 \\ &= (\det \boldsymbol{F})(\mathrm{d}\boldsymbol{X}^1 \times \mathrm{d}\boldsymbol{X}^2) \cdot \mathrm{d}\boldsymbol{X}^3 \\ &= J\mathrm{d}V \,, \end{aligned}$$

where $J = \det \boldsymbol{F}$ and $\mathrm{d}V = (\mathrm{d}\boldsymbol{X}^1 \times \mathrm{d}\boldsymbol{X}^2) \cdot \mathrm{d}\boldsymbol{X}^3$ is the volume element in the reference configuration.

b) The transformation of a surface element

$$\mathrm{d}\boldsymbol{a} = \mathrm{d}\boldsymbol{x}^1 \times \mathrm{d}\boldsymbol{x}^2 = \boldsymbol{F}\mathrm{d}\boldsymbol{X}^1 \times \boldsymbol{F}\mathrm{d}\boldsymbol{X}^2$$

is obtained from the above given identity when we replace $\boldsymbol{a} = \mathrm{d}\boldsymbol{X}^1, \boldsymbol{b} = \mathrm{d}\boldsymbol{X}^2$, and $\boldsymbol{c}$ is taken arbitrary. Then we get[19]

$$\begin{aligned} \det \boldsymbol{F} \underbrace{(\mathrm{d}\boldsymbol{X}^1 \times \mathrm{d}\boldsymbol{X}^2)}_{\mathrm{d}\boldsymbol{A}} \cdot \boldsymbol{c} &= (\boldsymbol{F}\mathrm{d}\boldsymbol{X}^1 \times \boldsymbol{F}\mathrm{d}\boldsymbol{X}^2) \cdot \boldsymbol{F}\boldsymbol{c} \\ &= (\mathrm{d}\boldsymbol{x}^1 \times \mathrm{d}\boldsymbol{x}^2) \cdot \boldsymbol{F}\boldsymbol{c} \\ &= \boldsymbol{F}^T \underbrace{(\mathrm{d}\boldsymbol{x}^1 \times \mathrm{d}\boldsymbol{x}^2)}_{\mathrm{d}\boldsymbol{a}} \cdot \boldsymbol{c} \quad \forall \boldsymbol{c} \in \mathbb{R}^3 \,. \end{aligned}$$

Because this must be valid for an arbitrary vector $\boldsymbol{c}$, multiplying from the left by $\boldsymbol{F}^{-T}$ finally gives

$$\mathrm{d}\boldsymbol{a} = (\det \boldsymbol{F})\boldsymbol{F}^{-T}\mathrm{d}\boldsymbol{A} \,.$$

[19] Here we have used

$$\boldsymbol{a} \cdot (\boldsymbol{A}\boldsymbol{b}) = a_i(A_{ij}b_j) = (A_{ij}a_i)b_j = (\boldsymbol{A}^T\boldsymbol{a}) \cdot \boldsymbol{b} \,.$$

9. Consider the following expression:

$$\begin{aligned}\tfrac{1}{2}\,(\mathrm{d}\boldsymbol{x}\cdot\mathrm{d}\boldsymbol{y}-\mathrm{d}\boldsymbol{X}\cdot\mathrm{d}\boldsymbol{Y}) &= \mathrm{d}\boldsymbol{X}\cdot\tfrac{1}{2}(\boldsymbol{F}^T\boldsymbol{F}-\boldsymbol{I})\,\mathrm{d}\boldsymbol{Y}\\ &= \mathrm{d}\boldsymbol{X}\cdot\boldsymbol{G}\mathrm{d}\boldsymbol{Y}\\ &= \mathrm{d}\boldsymbol{x}\cdot\tfrac{1}{2}(\boldsymbol{I}-\boldsymbol{F}^{-T}\boldsymbol{F}^{-1})\,\mathrm{d}\boldsymbol{y}\\ &= \mathrm{d}\boldsymbol{x}\cdot\boldsymbol{A}\mathrm{d}\boldsymbol{y}\,. \end{aligned}\tag{1.7.2}$$

Here $d\boldsymbol{x} = \boldsymbol{F}\mathrm{d}\boldsymbol{X}$ and $d\boldsymbol{y} = \boldsymbol{F}\mathrm{d}\boldsymbol{Y}$ was substituted and the definitions (1.5.12) and (1.5.15) were used. Formula (1.7.2) makes clear that $\boldsymbol{G}$ "lives" in the reference configuration whereas $\boldsymbol{A}$ "lives" in the present configuration.
Two natural definitions of the *normal (longitudinal)* strain are

$$\varepsilon_L := \frac{|\mathrm{d}\boldsymbol{x}|-|\mathrm{d}\boldsymbol{X}|}{|\mathrm{d}\boldsymbol{X}|}\,, \qquad \varepsilon_E := \frac{|\mathrm{d}\boldsymbol{x}|-|\mathrm{d}\boldsymbol{X}|}{|\mathrm{d}\boldsymbol{x}|}\,. \tag{1.7.3}$$

These may be called LAGRANGEan and EULERian longitudinal strains. $(1.7.3)_1$ is better known as engineering strain. $(1.7.3)_1$ implies

$$\begin{aligned}(\varepsilon_L+1)^2 = \frac{\mathrm{d}\boldsymbol{x}\cdot\mathrm{d}\boldsymbol{x}}{|\mathrm{d}\boldsymbol{X}||\mathrm{d}\boldsymbol{X}|} &= \boldsymbol{e_X}\cdot\boldsymbol{F}^T\boldsymbol{F}\,\boldsymbol{e_X}\\ &= 1+2\boldsymbol{e_X}\cdot\boldsymbol{G}\,\boldsymbol{e_X}\end{aligned}$$

or

$$\varepsilon_L = \sqrt{1+2\,\boldsymbol{e_X}\cdot\boldsymbol{G}\boldsymbol{e_X}}-1\,. \tag{1.7.4}$$

Similarly, with $(1.7.3)_2$

$$\varepsilon_E = \sqrt{1-2\,\boldsymbol{e_x}\cdot\boldsymbol{A}\boldsymbol{e_x}}+1\,. \tag{1.7.5}$$

Notice that the two strain measures are not the same.
There is also an angle between two non-collinear material line elements. Let $\mathrm{d}\boldsymbol{X}$ be orthogonal to $\mathrm{d}\boldsymbol{Y}$. Then

$$\mathrm{d}\boldsymbol{x}\cdot\mathrm{d}\boldsymbol{y} = |\mathrm{d}\boldsymbol{x}|\,|\mathrm{d}\boldsymbol{y}|\sin\gamma_{xy} = \mathrm{d}\boldsymbol{X}\cdot(\boldsymbol{I}+2\boldsymbol{G})\,\mathrm{d}\boldsymbol{Y} = 0+2\,\mathrm{d}\boldsymbol{X}\cdot\boldsymbol{G}\,\mathrm{d}\boldsymbol{Y}\,.$$

Therefore, since

$$\frac{|\mathrm{d}\boldsymbol{x}|}{|\mathrm{d}\boldsymbol{X}|} = \underbrace{(1+\varepsilon_L^{(X)})}_{\sqrt{1+2\,\boldsymbol{e_X}\cdot\boldsymbol{G}\boldsymbol{e_X}}}\,, \qquad \frac{|\mathrm{d}\boldsymbol{y}|}{|\mathrm{d}\boldsymbol{Y}|} = \underbrace{(1+\varepsilon_L^{(Y)})}_{\sqrt{1+2\,\boldsymbol{e_Y}\cdot\boldsymbol{G}\boldsymbol{e_Y}}}\,,$$

one gets

$$\sqrt{1+2\,\boldsymbol{e_X}\cdot\boldsymbol{G}\boldsymbol{e_X}}\sqrt{1+2\,\boldsymbol{e_Y}\cdot\boldsymbol{G}\boldsymbol{e_Y}}\;\sin\gamma_{xy} = 2\,\boldsymbol{e_X}\cdot\boldsymbol{G}\boldsymbol{e_Y}$$

or

$$\sin\gamma_{xy} = \frac{2\,\boldsymbol{e}_X\cdot\boldsymbol{G}\,\boldsymbol{e}_Y}{\sqrt{1+2\,\boldsymbol{e}_X\cdot\boldsymbol{G}\,\boldsymbol{e}_X}\,\sqrt{1+2\,\boldsymbol{e}_Y\cdot\boldsymbol{G}\,\boldsymbol{e}_Y}}\ . \tag{1.7.6}$$

Returning back to (1.7.2), one may prove the relation

$$\mathrm{d}\boldsymbol{X}\cdot\boldsymbol{G}\,\mathrm{d}\boldsymbol{Y} = \mathrm{d}\boldsymbol{x}\cdot\boldsymbol{A}\,\mathrm{d}\boldsymbol{y} = \mathrm{d}\boldsymbol{X}\cdot\boldsymbol{F}^T\boldsymbol{A}\boldsymbol{F}\,\mathrm{d}\boldsymbol{Y}\ ,$$

so that

$$\boldsymbol{G} = \boldsymbol{F}^T\boldsymbol{A}\boldsymbol{F} \quad\Longrightarrow\quad \boldsymbol{A} = \boldsymbol{F}^{-T}\boldsymbol{G}\boldsymbol{F}^{-1}\ .$$

10. Symmetry of the right and left CAUCHY–GREEN tensors follows from

$$\boldsymbol{C}^T = (\boldsymbol{F}^T\boldsymbol{F})^T = \boldsymbol{F}^T\left(\boldsymbol{F}^T\right)^T = \boldsymbol{F}^T\boldsymbol{F} = \boldsymbol{C}\ ,$$

$$\boldsymbol{B}^T = (\boldsymbol{F}\boldsymbol{F}^T)^T = \left(\boldsymbol{F}^T\right)^T\boldsymbol{F}^T = \boldsymbol{F}\boldsymbol{F}^T = \boldsymbol{B}\ .$$

Further the square of the line element gives

$$\begin{aligned} 0 \le \|\mathrm{d}\boldsymbol{x}\|^2 &= \mathrm{d}\boldsymbol{x}\cdot\mathrm{d}\boldsymbol{x} = \boldsymbol{F}\mathrm{d}\boldsymbol{X}\cdot\boldsymbol{F}\mathrm{d}\boldsymbol{X} = \mathrm{d}\boldsymbol{X}\cdot\boldsymbol{F}^T\boldsymbol{F}\mathrm{d}\boldsymbol{X} \\ &= \mathrm{d}\boldsymbol{X}\cdot\boldsymbol{C}\mathrm{d}\boldsymbol{X}\ , \end{aligned}$$

which demonstrates that $\boldsymbol{C}$ is positive definite. Analogously the positive definite property of $\boldsymbol{B}$ follows

$$\begin{aligned} 0 \le \|\mathrm{d}\boldsymbol{X}\|^2 &= \mathrm{d}\boldsymbol{X}\cdot\mathrm{d}\boldsymbol{X} = \boldsymbol{F}^{-1}\mathrm{d}\boldsymbol{x}\cdot\boldsymbol{F}^{-1}\mathrm{d}\boldsymbol{x} = \mathrm{d}\boldsymbol{x}\cdot\boldsymbol{F}^{-T}\boldsymbol{F}^{-1}\mathrm{d}\boldsymbol{x} \\ &= \mathrm{d}\boldsymbol{x}\cdot\boldsymbol{B}^{-1}\mathrm{d}\boldsymbol{x}\ . \end{aligned}$$

Because of the spectral decomposition, $\boldsymbol{B}^{-1}$ is positive definite if $\boldsymbol{B}$ is and vice versa.

11. The characteristic equation of the matrix $\boldsymbol{A}\in\mathbb{R}^{3\times 3}$ is given by

$$P(\lambda) = \det(\boldsymbol{A}-\lambda\boldsymbol{I}) = 0$$

or explicitly written as

$$\begin{vmatrix} A_{11}-\lambda & A_{12} & A_{13} \\ A_{21} & A_{22}-\lambda & A_{23} \\ A_{31} & A_{32} & A_{33}-\lambda \end{vmatrix} = 0\ .$$

Applying the CRAMER rule gives

$$\begin{aligned} &(A_{11}-\lambda)(A_{22}-\lambda)(A_{33}-\lambda) + A_{12}A_{23}A_{31} + A_{13}A_{21}A_{32} \\ &\quad - (A_{11}-\lambda)A_{23}A_{32} - A_{12}A_{21}(A_{33}-\lambda) - A_{13}(A_{22}-\lambda)A_{31} = 0 \end{aligned}$$

and thus results in

$$P(\lambda) = -\lambda^3 + \lambda^2 \underbrace{(A_{11} + A_{22} + A_{33})}_{=:I_A}$$

$$-\lambda \underbrace{(A_{11}A_{22} + A_{11}A_{33} + A_{22}A_{33} - A_{23}A_{32} - A_{12}A_{21} - A_{13}A_{31})}_{=:II_A}$$

$$\left.\begin{aligned} &+(A_{11}A_{22}A_{33} + A_{12}A_{23}A_{31} + A_{13}A_{21}A_{32} \\ &\quad - A_{11}A_{23}A_{32} - A_{12}A_{21}A_{33} - A_{13}A_{22}A_{31}) \end{aligned}\right\} =: III_A$$

$$= 0 \, .$$

In the above, the invariants are defined as

$$\begin{aligned} I_A &:= \mathrm{tr}(\boldsymbol{A}) = \boldsymbol{A} \cdot \boldsymbol{I} \, , \\ II_A &:= \tfrac{1}{2}\left[(\mathrm{tr}\,\boldsymbol{A})^2 - \mathrm{tr}(\boldsymbol{A}^2)\right] \, , \\ III_A &:= \det \boldsymbol{A} \, , \end{aligned}$$

where the representation for I_A is immediately apparent.
The second invariant is obtained by direct multiplication,

$$\begin{aligned} \mathrm{tr}(\boldsymbol{A}^2) &= \boldsymbol{A} \cdot \boldsymbol{A} \\ &= A_{11}^2 + A_{22}^2 + A_{33}^2 + 2A_{12}A_{21} + 2A_{13}A_{31} + 2A_{23}A_{32} \, , \end{aligned}$$

as well as

$$\begin{aligned} (\mathrm{tr}\,\boldsymbol{A})^2 &= (A_{11} + A_{22} + A_{33})^2 \\ &= A_{11}^2 + A_{22}^2 + A_{33}^2 + 2A_{11}A_{22} + 2A_{11}A_{33} + 2A_{22}A_{33} \end{aligned}$$

and finally subtracting the two.
In order to determine the third invariant we set λ to zero and equate the expressions

$$P(\lambda = 0) = \det \boldsymbol{A} \quad \Rightarrow \quad III_A = \det \boldsymbol{A} \, .$$

Now it is left to prove that I_A, II_A and III_A are invariant under arbitrary orthogonal transformations, i.e.,

$$\boldsymbol{A}^* = \boldsymbol{O}\boldsymbol{A}\boldsymbol{O}^T \quad \text{with} \quad \boldsymbol{O}\boldsymbol{O}^T = \boldsymbol{I} \, .$$

The first invariant is, in reality, invariant because

$$\begin{aligned} I_{A^*} &= \mathrm{tr}(\boldsymbol{A}^*) = \boldsymbol{A}^* \cdot \boldsymbol{I} = \boldsymbol{O}\boldsymbol{A}\boldsymbol{O}^T \cdot \boldsymbol{I} \\ &= \boldsymbol{O}\boldsymbol{A} \cdot \boldsymbol{O} = \boldsymbol{O}^T\boldsymbol{O}\boldsymbol{A} \cdot \boldsymbol{I} \\ &= \boldsymbol{A} \cdot \boldsymbol{I} = I_A \, . \end{aligned}$$

The second invariant can be expressed as

$$II_{\boldsymbol{A}} = \tfrac{1}{2}\left(I_{\boldsymbol{A}}^2 - I_{\boldsymbol{A}^2}\right) ,$$

which with the invariance of the first invariant and $(\boldsymbol{A}^*)^2 = \boldsymbol{O}\boldsymbol{A}\boldsymbol{O}^T\boldsymbol{O}\boldsymbol{A}\boldsymbol{O}^T = \boldsymbol{O}\boldsymbol{A}^2\boldsymbol{O}^T = (\boldsymbol{A}^2)^*$ immediately implies

$$\begin{aligned} II_{\boldsymbol{A}^*} &= \tfrac{1}{2}\left(I_{\boldsymbol{A}^*}^2 - I_{\boldsymbol{A}^{*2}}\right) \\ &= \tfrac{1}{2}\left(I_{\boldsymbol{A}}^2 - I_{\boldsymbol{A}^2}\right) = II_{\boldsymbol{A}} . \end{aligned}$$

The third invariant in any case satisfies the condition of invariance since

$$\begin{aligned} III_{\boldsymbol{A}^*} &= \det \boldsymbol{A}^* = \det(\boldsymbol{O}\boldsymbol{A}\boldsymbol{O}^T) \\ &= \det \boldsymbol{O} \det \boldsymbol{O}^T \det \boldsymbol{A} = \det(\boldsymbol{O}\boldsymbol{O}^T) \det \boldsymbol{A} \\ &= \det \boldsymbol{A} = III_{\boldsymbol{A}} . \end{aligned}$$

12. The invariants of a tensor $\boldsymbol{A}$ can be expressed in terms of the invariants of the powers of $\boldsymbol{A}$ with the help of CAYLEY–HAMILTON theorem. Applying this to the second invariant yields

$$II_{\boldsymbol{A}} = \tfrac{1}{2}\left[(\operatorname{tr} \boldsymbol{A})^2 - \operatorname{tr}(\boldsymbol{A}^2)\right] = \tfrac{1}{2}\left(I_{\boldsymbol{A}}^2 - I_{\boldsymbol{A}^2}\right) .$$

For the third invariant there follows by forming the trace of (1.5.27)

$$\begin{aligned} 0 &= \operatorname{tr}(\boldsymbol{A}^3 - I_{\boldsymbol{A}}\boldsymbol{A}^2 + II_{\boldsymbol{A}}\boldsymbol{A} - III_{\boldsymbol{A}}\boldsymbol{I}) \\ &= I_{\boldsymbol{A}^3} - I_{\boldsymbol{A}} I_{\boldsymbol{A}^2} + II_{\boldsymbol{A}} I_{\boldsymbol{A}} - 3 III_{\boldsymbol{A}} , \end{aligned}$$

or by solving for $III_{\boldsymbol{A}}$,

$$III_{\boldsymbol{A}} = \tfrac{1}{3}(I_{\boldsymbol{A}^3} - I_{\boldsymbol{A}} I_{\boldsymbol{A}^2} + II_{\boldsymbol{A}} I_{\boldsymbol{A}}) .$$

Substituting for the second invariant gives rise to the alternative form

$$III_{\boldsymbol{A}} = \tfrac{1}{3}\left(I_{\boldsymbol{A}^3} - \tfrac{3}{2} I_{\boldsymbol{A}} I_{\boldsymbol{A}^2} + \tfrac{1}{2} I_{\boldsymbol{A}}^3\right) ,$$

in which only the first invariants $I_{\boldsymbol{A}}$, $I_{\boldsymbol{A}^2}$ and $I_{\boldsymbol{A}^3}$ appear. Further, replacing $I_{\boldsymbol{A}^2}$ by $I_{\boldsymbol{A}}^2 - 2II_{\boldsymbol{A}}$ gives

$$III_{\boldsymbol{A}} = \tfrac{1}{3}\left(I_{\boldsymbol{A}^3} - I_{\boldsymbol{A}}^3\right) + II_{\boldsymbol{A}} I_{\boldsymbol{A}} .$$

The derivatives of invariants with respect to the tensor components can be given by use of the expressions which were derived just now. The derivative of the first invariant is

$$\frac{\partial I_{\boldsymbol{A}}}{\partial A_{ij}} = \frac{\partial A_{kk}}{\partial A_{ij}} = \delta_{ik}\delta_{jk} = \delta_{ij} \ ,$$

or symbolically,

$$\frac{\partial I_{\boldsymbol{A}}}{\partial \boldsymbol{A}} = \boldsymbol{I} \ .$$

This enables us to calculate the $\boldsymbol{A}$-derivative of the second invariant,

$$\begin{aligned}\frac{\partial II_{\boldsymbol{A}}}{\partial \boldsymbol{A}} &= \frac{1}{2}\frac{\partial}{\partial \boldsymbol{A}}\left(I_{\boldsymbol{A}}^2 - I_{\boldsymbol{A}^2}\right) = I_{\boldsymbol{A}}\frac{\partial I_{\boldsymbol{A}}}{\partial \boldsymbol{A}} - \frac{1}{2}\frac{\partial}{\partial \boldsymbol{A}}(\boldsymbol{A}\cdot\boldsymbol{A}) \\ &= I_{\boldsymbol{A}}\boldsymbol{I} - \boldsymbol{A}^T \ ,\end{aligned}$$

where in the last step

$$\frac{\partial}{\partial A_{ij}}(A_{lk}A_{kl}) = \delta_{il}\delta_{jk}A_{kl} + \delta_{ik}\delta_{jl}A_{lk} = 2A_{ji}$$

has been used so that

$$\frac{\partial I_{\boldsymbol{A}^2}}{\partial \boldsymbol{A}} = 2\boldsymbol{A}^T \ .$$

The $\boldsymbol{A}$-derivative of the third invariant is obtained by considering the terms used before

$$\begin{aligned}\frac{\partial III_{\boldsymbol{A}}}{\partial \boldsymbol{A}} &= \frac{\partial}{\partial \boldsymbol{A}}\left[\frac{1}{3}(I_{\boldsymbol{A}^3} - I_{\boldsymbol{A}}^3) + II_{\boldsymbol{A}}I_{\boldsymbol{A}}\right] \\ &= \frac{1}{3}\frac{\partial I_{\boldsymbol{A}^3}}{\partial \boldsymbol{A}} - I_{\boldsymbol{A}}^2\frac{\partial I_{\boldsymbol{A}}}{\partial \boldsymbol{A}} + I_{\boldsymbol{A}}\frac{\partial II_{\boldsymbol{A}}}{\partial \boldsymbol{A}} + II_{\boldsymbol{A}}\frac{\partial I_{\boldsymbol{A}}}{\partial \boldsymbol{A}} \\ &= (\boldsymbol{A}^2)^T - I_{\boldsymbol{A}}^2\boldsymbol{I} + I_{\boldsymbol{A}}(I_{\boldsymbol{A}}\boldsymbol{I} - \boldsymbol{A}^T) + II_{\boldsymbol{A}}\boldsymbol{I} \\ &= (\boldsymbol{A}^2 - I_{\boldsymbol{A}}\boldsymbol{A} + II_{\boldsymbol{A}}\boldsymbol{I})^T \ ,\end{aligned} \tag{1.7.7}$$

where in the second step

$$\begin{aligned}\frac{\partial I_{\boldsymbol{A}^3}}{\partial A_{ij}} &= \frac{\partial}{\partial A_{ij}}(A_{kl}A_{lm}A_{mk}) \\ &= \delta_{ik}\delta_{jl}A_{lm}A_{mk} + \delta_{il}\delta_{jm}A_{kl}A_{mk} + \delta_{im}\delta_{jk}A_{kl}A_{lm} \\ &= 3A_{jm}A_{mi}\end{aligned}$$

or

$$\frac{\partial I_{\boldsymbol{A}^3}}{\partial \boldsymbol{A}} = 3(\boldsymbol{A}^2)^T$$

is applied. We can use the CAYLEY-HAMILTON-theorem

$$\begin{aligned}&\boldsymbol{A}^3 - I_{\boldsymbol{A}}\boldsymbol{A}^2 + II_{\boldsymbol{A}}\boldsymbol{A} - III_{\boldsymbol{A}}\boldsymbol{I} = \boldsymbol{0} \\ \Longrightarrow\quad &\boldsymbol{A}^2 - I_{\boldsymbol{A}}\boldsymbol{A} + II_{\boldsymbol{A}}\boldsymbol{I} \quad = III_{\boldsymbol{A}}\boldsymbol{A}^{-1}\end{aligned}$$

to convert the last result (1.7.7) into

$$\frac{\partial III_{\boldsymbol{A}}}{\partial \boldsymbol{A}} = \boldsymbol{A}^{-T} III_{\boldsymbol{A}} \quad \text{or} \quad \frac{\partial III_{\boldsymbol{A}}}{\partial A_{ij}} = (\boldsymbol{A}^{-1})_{ji}\, III_{\boldsymbol{A}} \, .$$

13. The stress power φ is given by

$$\varphi := \operatorname{tr}(\operatorname{grad} \boldsymbol{v}\, \boldsymbol{t}^T) \, .$$

In component form this is

$$\begin{aligned}\varphi &= \operatorname{tr}(v_{i,j} \boldsymbol{e}_i \otimes \boldsymbol{e}_j \, t_{kl} \boldsymbol{e}_l \otimes \boldsymbol{e}_k) \\ &= \operatorname{tr}(v_{i,j} t_{kl} \delta_{jl} \boldsymbol{e}_i \otimes \boldsymbol{e}_k) = \operatorname{tr}(v_{i,j} t_{kj} \boldsymbol{e}_i \otimes \boldsymbol{e}_k) \\ &= v_{i,j} t_{ij} \, .\end{aligned}$$

Here we have assumed that the stress tensor is symmetric. Polar decomposition of the velocity gradient into symmetric and antisymmetric parts yields

$$\boldsymbol{L} = \boldsymbol{D} + \boldsymbol{W} \, ;$$

thus the stress power becomes

$$\begin{aligned}\varphi &= v_{i,j} t_{ij} = L_{ij} t_{ij} = D_{ij} t_{ij} + W_{ij} t_{ij} \\ &= D_{ij} t_{ij} = D_{ij} t_{ji} = \operatorname{tr}(\boldsymbol{D}\boldsymbol{t}) = \operatorname{tr}(\boldsymbol{D}\boldsymbol{t}^T) \, .\end{aligned}$$

In this derivation we have used that $\operatorname{tr}(\boldsymbol{W}\boldsymbol{t}) = 0$. [20]

14. The time derivative of the determinant of the deformation gradient can easily be computed if the results of the last exercise are used. We have

$$J = \det \boldsymbol{F} = III_{\boldsymbol{F}} \, .$$

Now, since the third invariant of $\boldsymbol{F}$ can be written as a function of $\boldsymbol{F}$ itself, the time derivative then follows as

$$\dot{J} = \frac{\mathrm{d} III_{\boldsymbol{F}}}{\mathrm{d}t} = \frac{\partial III_{\boldsymbol{F}}}{\partial F_{i\alpha}} \dot{F}_{i\alpha} \, .$$

The time derivative of the deformation gradient is related to the velocity gradient $\boldsymbol{L}$ through (1.4.9)

[20] In general this is valid for the trace of the product of a symmetric tensor $\boldsymbol{A}$ and an skewsymmetric tensor $\boldsymbol{S}$ since

$$\gamma = \operatorname{tr}(\boldsymbol{A}\boldsymbol{S}) = A_{ij} S_{ij} = -A_{ij} S_{ji} = -A_{ji} S_{ji} = -A_{ij} S_{ij} \, ,$$

where the dummy indices i and j are interchanged. Hence $\gamma = -\gamma$ and consequently $\gamma = 0$.

$$\dot{\boldsymbol{F}} = \boldsymbol{L}\boldsymbol{F} \quad \text{or} \quad \dot{F}_{i\alpha} = L_{ij} F_{j\alpha} \,,$$

from where the derivative of J is obtained as

$$\dot{J} = \frac{\partial III_{\boldsymbol{F}}}{\partial F_{i\alpha}} \dot{F}_{i\alpha} = \frac{\partial III_{\boldsymbol{F}}}{\partial F_{i\alpha}} L_{ij} F_{j\alpha} \,.$$

With the derivative of the invariant from the second-last Exercise this yields

$$\begin{aligned} \dot{J} &= F_{j\alpha} (\boldsymbol{F}^{-1})_{\alpha i} \, III_{\boldsymbol{F}} L_{ij} = III_{\boldsymbol{F}} \delta_{ji} L_{ij} \\ &= J L_{ii} = J \operatorname{tr}(\boldsymbol{L}) = J \operatorname{div} \boldsymbol{v} \,. \end{aligned}$$

In symbolic representation the derivative can now be written as

$$\begin{aligned} \dot{J} = \frac{\mathrm{d} III_{\boldsymbol{F}}}{\mathrm{d}t} &= \dot{\boldsymbol{F}} \cdot \frac{\partial III_{\boldsymbol{F}}}{\partial \boldsymbol{F}^T} = \boldsymbol{L}\boldsymbol{F} \cdot \boldsymbol{F}^{-1} III_{\boldsymbol{F}} \\ &= \boldsymbol{L} \cdot \boldsymbol{F}^{-1} \boldsymbol{F} \, III_{\boldsymbol{F}} = \boldsymbol{L} \cdot \boldsymbol{I} \, III_{\boldsymbol{F}} \\ &= I_{\boldsymbol{L}} \, III_{\boldsymbol{F}} = J \operatorname{div} \boldsymbol{v} \,. \end{aligned} \tag{1.7.8}$$

2. Balance Equations

2.1 General Balance Statements

2.1.1 Integral Form of the Balance Statements

In the preceding chapter, we focussed on (some of the) kinematic aspects related to the motion of a continuous body. In particular, the motion $\boldsymbol{\chi}(\boldsymbol{X}, t)$ was treated as a *given* function. However, it is in fact one of the main tasks of continuum mechanics to *calculate* the motion of the particles forming continuous bodies and, along with it, the evolution of the associated fields such as e.g. density and temperature. This can be done, once the relevant equations - usually functional differential equations - will have been established together with sufficient initial and/or boundary conditions. These equations comprise two sets of statements, the so-called *balance equations* of mass, momenta, energy and entropy and the *constitutive relations* describing the material behaviour of the body for which the spatial and temporal evolution of the field quantities, such as motion, density and temperature are sought. The balance equations have fairly general character and, in particular, contain no material specific information. The present section is devoted to the derivation of the global forms of the balance laws.

Material bodies are equipped with physical properties such as mass, momentum, energy, etc., and physical laws emerge by formulating relationships among these physical quantities. Experience has shown that a part of a body cut from the original body again enjoys the properties of a body, if the interactions of the two parts are properly accounted for; i.e., if the action of one body part is exerted on the second part of the original body and the reverse action is similarly exerted by the second body part on the first part. This *cutting principle*[1] is assumed to hold for any body part independent of its size, and may thus also be applied to an infinitesimal body element. Now, because a continuous material body consists of an infinite number of elements it is customary to assign the physical properties to these material points through the postulation of *densities of the physical variables* which are

[1] In engineering mechanics this principle is referred to as the *intersection principle* or the *free body principle.* In the literature it is primarily used in a mechanics context. However, it applies to any balance law if the additivity postulate is made.

defined per unit mass or unit volume. This process corresponds to the imposition of the *additivity assumption*, i.e., the value of a physical variable for a body is given by the summation (here the integration over the volume) of its values over the parts of the body (here the infinitesimal volume elements). The additivity assumption together with the cutting principle imply that a body may be thought to be decomposable into many parts, and each part of the body, complemented by the interactions on the cutting surfaces by the neighbouring subbodies, obeys the same principles as the whole body does. A more general approach starts with the assignment of physical quantities to the body $\mathcal{B}$ as a whole without the imposition of an additivity postulate; however, we shall not make use of this in this book. Thus one describes the physical properties by means of densities whose volume integrals give the corresponding variables for the body. These densities can be specified, as usual, with respect to both the reference and the present configuration.

Let $\mathcal{G}$ be a physical variable which characterizes a partial aspect of the state of the body at time t. Let Γ be the density of $\mathcal{G}$ assigned to every material element in the reference configuration and let γ denote the corresponding density in the present configuration. Further, we choose an open set Ω of a body with its bounding surface $\partial\Omega$ in the reference configuration (see Fig. 2.1) with respect to which the physical variable is evaluated. Obviously, the above set changes in the present configuration; the volume and the boundary surface in this configuration are $\omega(t)$ and $\partial\omega(t)$, respectively, which are explicitly dependent on time.

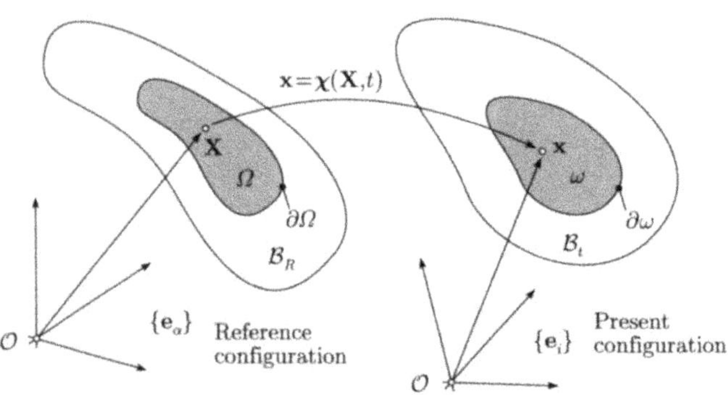

Fig. 2.1. Material domains in the reference and present configurations . Ω, ω and $\partial\Omega, \partial\omega$ denote the volumes and bounding surfaces of the material subsets of the body ($\mathcal{B}_R$ and $\mathcal{B}_t$ in the reference and present configurations, respectively).

The value of the physical variable $\mathcal{G}$ at time t is obtained by the additivity principle, i.e., by volume integration. In the reference configuration the variable $\mathcal{G}$ is given by

$$\mathcal{G}(t) = \int_{\Omega} \Gamma(\boldsymbol{X}, t)\, dV \,, \tag{2.1.1}$$

and in the present configuration it is

$$\mathcal{G}(t) = \int_{\omega} \gamma(\boldsymbol{x}, t)\, dv \,.$$

Using the definition of the motion (1.2.8) and the transformation of a volume element (1.3.8) we can relate these expressions as follows (see (1.2.14)),

$$\begin{aligned} \mathcal{G}(t) &= \int_{\omega} \gamma(\boldsymbol{x}, t)\, dv = \int_{\Omega} \gamma(\chi(\boldsymbol{X}, t), t)\, J\, dV \\ &= \int_{\Omega} \Gamma(\boldsymbol{X}, t)\, dV \,. \end{aligned} \tag{2.1.2}$$

Thus the volume densities of the physical variable $\mathcal{G}$ are related by

$$\Gamma = J \gamma \,. \tag{2.1.3}$$

Similar relations for surface densities and their transformation will be exhibited later.

Now we shall examine how the variable $\mathcal{G}$ in $\mathcal{B}$ may change with time. Such changes in general occur due to *external actions* as well as *internal processes* in the body itself. This can, in general, happen from two mechanisms. The first of these takes place over the volume of the subbody (e.g. gravitational force acts on each element of the body), and the other can take place over the surface (e.g. the surface force acting on or the heat flux conducted through the surface). Besides these two mechanisms the physical quantity can also change internally. We express all these effects together in terms of three different state or process variables.

The time rate of change of a physical variable $\mathcal{G}$ in a certain domain Ω or ω of a body under consideration is accomplished by the following contributions (in addition to these quantities the corresponding densities relative to individual material elements are also specified):

a) **Production** $\mathcal{P}$ — Densities: $\Pi(\boldsymbol{X}, t)$ or $\pi(\boldsymbol{x}, t)$
This quantity is produced within the domain Ω or ω, respectively.
Example: Production of heat in a body due to radioactive decay.

b) **Supply** $\mathcal{S}$ — Densities: $\Sigma(\boldsymbol{X}, t)$ or $\varsigma(\boldsymbol{x}, t)$
The supply or source density is exclusively supplied by action at a distance from outside of the body, where the whole domain – the volume – becomes directly influenced.
Examples: Radiation produces heat within a body. The gravitation produces changes in momentum in a body.

c) **Flux** $\mathcal{F}$ — Densities: $\tilde{\Phi}(\boldsymbol{X}, t, \boldsymbol{N})$ or $\tilde{\phi}(\boldsymbol{x}, t, \boldsymbol{n})$; here the densities are surface densities. These quantities flow from the outside into the body through the surface $\partial\Omega$ or $\partial\omega$.
Examples: Stress on the surface of a body. Heat flux through a surface.

Thus the time rate of change of the variable $\mathcal{G}$ of the body per unit time may be written as

$$\frac{\mathrm{d}\mathcal{G}}{\mathrm{d}t} = \mathcal{P} + \mathcal{S} + \mathcal{F} \,. \tag{2.1.4}$$

The quantities arising in this equation express the corresponding entities for the whole domain; they are equivalent to a sum of individual contributions for all particles contained in the domain Ω (or on the boundary $\partial\Omega$). Hereby Ω is a *material domain* of the body. The quantity $\mathcal{G}$ for the whole domain Ω is thus equal to the sum of all parts of it for the individual material points with position $\boldsymbol{X}$ at time t, i.e., the integration over the volume. The same explanation equally applies to the domain ω and the boundary $\partial\omega$ in the present configuration. Thus we have

$$\begin{aligned}
\mathcal{G} &= \int_{\Omega} \Gamma(\boldsymbol{X},t)\, dV &&= \int_{\omega} \gamma(\boldsymbol{x},t)\, dv \,, \\
\mathcal{P} &= \int_{\Omega} \Pi(\boldsymbol{X},t)\, dV &&= \int_{\omega} \pi(\boldsymbol{x},t)\, dv \,, \\
\mathcal{S} &= \int_{\Omega} \Sigma(\boldsymbol{X},t)\, dV &&= \int_{\omega} \varsigma(\boldsymbol{x},t)\, dv \,, \\
\mathcal{F} &= \int_{\partial\Omega} \tilde{\Phi}(\boldsymbol{X},t)\, dA &&= \int_{\partial\omega} \tilde{\phi}(\boldsymbol{x},t)\, da \,.
\end{aligned} \tag{2.1.5}$$

Here the flux (surface density) is integrated over the surface since it is only effective for a surface. With (2.1.5) the balance equation (2.1.4) can now be written as

$$\begin{aligned}
\frac{\mathrm{d}}{\mathrm{d}t}\int_{\Omega} \Gamma(\boldsymbol{X},t)\, dV &= \int_{\Omega} \big(\Pi(\boldsymbol{X},t) + \Sigma(\boldsymbol{X},t)\big)\, dV + \int_{\partial\Omega} \tilde{\Phi}(\boldsymbol{X},t,\boldsymbol{N})\, dA \,, \\
\frac{\mathrm{d}}{\mathrm{d}t}\int_{\omega} \gamma(\boldsymbol{x},t)\, dv &= \int_{\omega} \big(\pi(\boldsymbol{x},t) + \varsigma(\boldsymbol{x},t)\big)\, dv + \int_{\partial\omega} \tilde{\phi}(\boldsymbol{x},t,\boldsymbol{n})\, da \,,
\end{aligned} \tag{2.1.6}$$

in the LAGRANGE and EULER representations, respectively. $\boldsymbol{N}$ and $\boldsymbol{n}$ denote the unit normal vectors in the LAGRANGE and EULER representation. The dependence of $\boldsymbol{\Phi}$ and ϕ on $\boldsymbol{N}$ and $\boldsymbol{n}$, respectively, will be commented in the next Subsect. 2.1.2. Notice that the integration domain ω in the EULER representation changes with time whereas its counterpart Ω in the reference configuration remains constant.

The above statements (2.1.6) are the global balance statements for a physical variable in the LAGRANGE and EULER representations incorporating the additivity principle for an arbitrary material domain Ω or ω. These balance statements can be made more concrete and illustrated with the help of the following examples.

Example 1 Let $\mathcal{G}$ be the sum of money existing as cash, deposites, stocks, shares in a particular office of a bank. Its time rate of change is governed by the deposition to or withdrawal from the customers' accounts. The cash carried out or in through the door of the bank corresponds to the flux of money.

If the transaction is done by filling out a payment order from another office of another bank, it is interpreted as the supply because that sum increases by the payment order from an account of a different bank. The bank can still print notes or press coins; this corresponds to a production.

Example 2 As an alternative example, consider a bowl of vegetable which should be warmed in a microwave oven. Let $\mathcal{G}$ be the internal energy (heat) of the vegetable. The increase in energy of the vegetable in the microwave oven is primarily due to the radiation that stimulates the vibrations of the water molecules in the vegetable. Such a process is governed by the electromagnetic distant-action on each water molecule in the body: a supply. Some amount of heat is also transported by conduction through the surface of the vegetable; this represents a heat flux. Heat production does not exist in this particular example unless the vegetable is radioactively contaminated and producing heat by itself.

2.1.2 CAUCHY Lemma

In the foregoing description, see equation (2.1.6), we have assumed that the flux depends on the orientation of the surface or on the normal vector on the surface with respect to which the flux is taken. In order to clarify this statement, $\boldsymbol{N}$ or $\boldsymbol{n}$ are considered as additional arguments of the fluxes $\boldsymbol{\Phi}$ and $\boldsymbol{\phi}$, respectively, whose dependence on position and time is thus complemented by a direction. Because of this dependence on surface properties, namely $\boldsymbol{N}$ or $\boldsymbol{n}$, the quantities $\tilde{\Phi}$, $\tilde{\phi}$ are not just simple fields whose values would be determined if $\boldsymbol{x}$ and t are known[2]. These quantities do not represent any scalar or vectorial fields. For example, it makes no sense if we speak of the heat flux with respect to a point. Heat flux is always affiliated with a surface (e.g. heat flux at the Earth surface). This concept leads to an important consequence which shall be described in the following lemma.

CAUCHY **Lemma** *If the surface densities $\tilde{\phi}$ (or $\tilde{\Phi}$) depend on the normals $\boldsymbol{n}$ (or $\boldsymbol{N}$) at the surface, this dependency is linear:*

$$\tilde{\Phi}(\boldsymbol{X}, t, \boldsymbol{N}) = -\boldsymbol{\Phi}(\boldsymbol{X}, t)\boldsymbol{N} \quad \textit{or} \quad \tilde{\phi}(\boldsymbol{x}, t, \boldsymbol{n}) = -\boldsymbol{\phi}(\boldsymbol{x}, t)\boldsymbol{n} \,. \tag{2.1.7}$$

The multiplication on the right hand side is a contraction. ■

The proof of this lemma results from the balance laws applied to an infinitesimal tetrahedron, whose corners are located along lines parallel to the coordinate axes (Fig 2.2). The tetrahedron is reduced and shrunk to the origin at $(\boldsymbol{X})$ by preserving its geometry. Let its typical edge length be Δh, then using the mean value theorem of integral calculus the volume integral in

[2] It is tacitly assumed that the flux depends only on the outer normal vector at the surface points and not on differential geometric properties of the surface such as mean or GAUSSian curvature. This assumption has first been spelled out by CAUCHY and is referred to as the CAUCHY assumption.

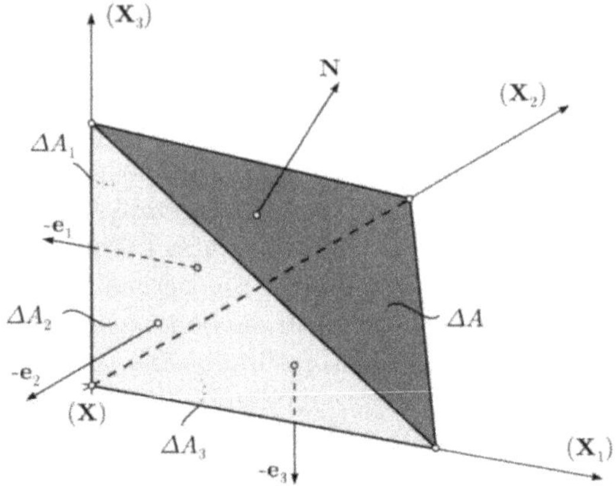

Fig. 2.2. Infinitesimal tetrahedron with normal vectors $\boldsymbol{N}$ on the abutting face and $-\boldsymbol{e}_\alpha$, $\alpha = 1, 2, 3$ on the side faces. The characteristic edge length of the tetrahedron is Δh, its surfaces are ΔA and ΔA_α for $\alpha = 1, 2, 3$ (LAGRANGEan description).

(2.1.6) becomes of order $O(\Delta h^3)$, whilst the surface integral remains of order $O(\Delta h^2)$. Here, the notation $f = O(a)$ or "f is $O(a)$" means that in the limit as $a \to 0$, f is proportional to a with a constant of proportionality different from 0 and ∞. Furthermore, we assume that all densities appearing in (2.1.6) are finite and therefore integrable. Thus, in the limiting case as $\Delta h \to 0$ the balance expressions (2.1.6) can be written[3] as

$$\begin{aligned} &\int_{\partial\Omega} \tilde{\Phi}(\boldsymbol{X}, t, \boldsymbol{N})\, dA + O(\Delta h^3) = \boldsymbol{0} \quad \text{or} \\ &\int_{\partial\omega} \tilde{\phi}(\boldsymbol{x}, t, \boldsymbol{n})\, da + O(\Delta h^3) = \boldsymbol{0} \,. \end{aligned} \tag{2.1.8}$$

Let the coordinate axis X_α be perpendicular to the surface element ΔA_α and let $\boldsymbol{N}$ be perpendicular to ΔA; then, geometrically, we can write

$$\Delta A_\alpha = N_\alpha \Delta A \,, \quad \alpha = 1, 2, 3 \,. \tag{2.1.9}$$

By dividing the surface integral of (2.1.8) into four parts corresponding to the four triangular segments of the tetrahedron and applying the mean value theorem for each part we obtain the following expression:

$$\sum_\alpha \tilde{\Phi}(\boldsymbol{X}_\alpha, t, -\boldsymbol{e}_\alpha) N_\alpha \Delta A + \tilde{\Phi}(\bar{\boldsymbol{X}}, t, \boldsymbol{N}) \Delta A + O(\Delta h^3) = \boldsymbol{0} \,, \tag{2.1.10}$$

[3] The following deductions are valid for both the LAGRANGE as well as the EULER representations. In this regard it should be sufficient to demonstrate the derivation for one of the representations, say for the LAGRANGE representation.

where we have also used (2.1.9) for the derivation. In the above equation $\boldsymbol{X}_\alpha$ is a convenient point within the triangle with the surface normal $-\boldsymbol{e}_\alpha$ and $\bar{\boldsymbol{X}}$ is a corresponding point within the triangle with the surface normal $\boldsymbol{N}$. Since the surface $\Delta A = O(\Delta h^2)$, in the limiting case as $\Delta h \to 0$, equation (2.1.10) reduces to

$$\sum_\alpha \tilde{\Phi}(\boldsymbol{X}, t, -\boldsymbol{e}_\alpha) N_\alpha + \tilde{\Phi}(\boldsymbol{X}, t, \boldsymbol{N}) = \boldsymbol{0} \, . \tag{2.1.11}$$

Further, by making a special choice of the normal vector $\boldsymbol{N} = \boldsymbol{e}_\beta$ one obtains (in this case the summation falls out)

$$-\tilde{\Phi}(\boldsymbol{X}, t, -\boldsymbol{e}_\beta) = \tilde{\Phi}(\boldsymbol{X}, t, \boldsymbol{e}_\beta) \, . \tag{2.1.12}$$

The flux through a surface with the unit normal $\boldsymbol{e}$ is equal to the negative flux through the corresponding surface with the unit normal $-\boldsymbol{e}$. In NEWTONian mechanics this statement is known as NEWTON's *third axiom*; the calculation shows that this statement is valid for every flux. All the more, NEWTON's axiom is not a real axiom since it directly results from the balance laws and the dependency of the flux on the surface normals[4] . If we now introduce the relation

$$\Phi_\alpha(\boldsymbol{X}, t) := \tilde{\Phi}(\boldsymbol{X}, t, -\boldsymbol{e}_\alpha) \tag{2.1.13}$$

in (2.1.11), we obtain

$$\tilde{\Phi}(\boldsymbol{X}, t, \boldsymbol{N}) = -\Phi_\alpha(\boldsymbol{X}, t) N_\alpha \tag{2.1.14}$$

in which EINSTEIN's summation convention is employed. Symbolically it is written as

$$\tilde{\Phi}(\boldsymbol{X}, t, \boldsymbol{N}) = -\boldsymbol{\Phi}(\boldsymbol{X}, t)\boldsymbol{N} \, , \quad \tilde{\phi}(\boldsymbol{x}, t, \boldsymbol{n}) = -\boldsymbol{\phi}(\boldsymbol{x}, t)\boldsymbol{n} \, , \tag{2.1.15}$$

where the second relation of the corresponding result is retained in the EULER representation. The choice of the sign is such that the flow from outside into the body is taken positive. With these relations the proof of (2.1.7) is achieved[5].

---

[4] One may also argue in the reverse order; namely for stresses and surface forces NEWTON's third law implies that the stress vector is linear in the surface normal. This backward interpretation of CAUCHY's hypothesis is, however, less stringent simply because it applies to surface forces and stresses only, and not the flux terms of a general balance law.

[5] The derivation of the result (2.1.15) requires a number of clarifying remarks to put it into the proper perspectives. The result was obtained by applying the balance law for a physical quantity to a material tetrahedronal volume with sharp edges and points for which unit normal vectors cannot uniquely be defined. Thus to make the above argument mathematically "clean" the tetrahedron must be smoothed out such that the edges and vertices become "diffuse" with uniquely defined normal vectors. This smoothing can formally be done and the limit to

2.1.3 Synopsis of General Balance Statements

CAUCHY's Lemma helps one to write the balance expressions (2.1.6) in abstract form. In doing so, one must however carefully distinguish between scalar and vector fields in the flux term. For example, when $\boldsymbol{f}$ is a vector field, such as the heat flux vector, the term $\boldsymbol{f}\boldsymbol{n}$ exhibits a scalar multiplication and should hence be better written as $\boldsymbol{f}\cdot\boldsymbol{n}$. The corresponding scalar valued surface density $\tilde{f}$ is

$$\tilde{f} = \boldsymbol{f}\cdot\boldsymbol{n} = f_i n_i \ .$$

If $\boldsymbol{f}$ is a tensorial field (e.g. stress tensor), then $\tilde{\boldsymbol{f}}$ is a vector valued density (stress vector), i.e.,

$$\tilde{\boldsymbol{f}} = \boldsymbol{f}\boldsymbol{n} = f_{ik}\boldsymbol{e}_i \otimes \boldsymbol{e}_k n_j \boldsymbol{e}_j = f_{ij} n_j \boldsymbol{e}_i \ .$$

where $\boldsymbol{e}_i$ is the unit vector in the i-th coordinate direction.

As a result, the general balance equations in their global forms are

$$\begin{aligned} \frac{\mathrm{d}\mathcal{G}}{\mathrm{d}t} &= \frac{\partial}{\partial t}\int_{\Omega} \Gamma(\boldsymbol{X},t)\,dV \\ &\quad + \int_{\Omega} \big(\Pi(\boldsymbol{X},t) + \Sigma(\boldsymbol{X},t)\big)\,dV - \int_{\partial\Omega} \boldsymbol{\Phi}(\boldsymbol{X},t)\boldsymbol{N}\,dA \ , \\ \frac{\mathrm{d}\mathcal{G}}{\mathrm{d}t} &= \frac{\mathrm{d}}{\mathrm{d}t}\int_{\omega} \gamma(\boldsymbol{x},t)\,dv \\ &\quad + \int_{\omega} \big(\pi(\boldsymbol{x},t) + \varsigma(\boldsymbol{x},t)\big)\,dv - \int_{\partial\omega} \boldsymbol{\phi}(\boldsymbol{x},t)\boldsymbol{n}\,da \ . \end{aligned} \tag{2.1.16}$$

In case of the LAGRANGE representation the total time derivative in the first term on the right–hand side of $(2.1.16)_1$ has been written here as partial derivative, as motivated in (1.2.18).

The relations between the variables in the LAGRANGE and EULER representations are obtained analogously to the transformation illustrated in (2.1.2). These are

$$\Gamma = J\gamma\ , \quad \Pi = J\pi\ , \quad \Sigma = J\varsigma\ , \quad \boldsymbol{\Phi} = J\boldsymbol{\phi}\boldsymbol{F}^{-T}\ , \tag{2.1.17}$$

expressions to be proved in the Exercises.

---

a tetrahedron with sharp edges and vertices can be performed such that in the limit (2.1.15) is obtained. The assumption that is needed is that the flux $\boldsymbol{\Phi}(\boldsymbol{\phi})$ is non-trivially defined with respect to surface measure, but that there is no specific flux quantity defined along the edges or at the vertices.

A historical account on CAUCHY's Lemma is given by TRUESDELL & TOUPIN [238], section 203, more general treatments, in which non-trivial edge fluxes and vertex fluxes are allowed for, are given by NOLL [177], NOLL & VIRGA [178] and DELL'ISOLA & SEPPECHER [56].

2.2 Local Balance Equation

In order to describe the motions of the material points in a body the above derived balance expressions are not sufficient because these expressions appear for various applications in an inadequate form. If we require for all fields *sufficient differentiability* in the description of the motion we can accomplish this by means of local balance statements. For the derivation of the local formulations the equations must be expressed in the form

$$\int_{\Omega} (Balance) \mathrm{d}V = 0 \,, \quad \forall \, \Omega \in \mathcal{B}_R \,. \tag{2.2.1}$$

Since this balance statement must be true for all volume elements Ω (satisfying the additivity principle), it results in the local balance expression

$$Balance = 0 \,.$$

This conclusion is obvious because the integrand "balance" is a continuous function.

From the general balance statement in the global form (2.1.16) one can also obtain the formula (2.2.1) when one performs the operations (i) interchange in the sequence of integration and time differentiation on the left-hand side, and (ii) transformation of the surface integral to the volume integral on the right–hand side.

The first step is the interchange of integration and differentiation. For the LAGRANGE representation this operation causes no difficulty. Here the integration and differentiation are interchangeable – provided Γ is differentiable – because the integration domain is temporally constant; so we simply have

$$\frac{\mathrm{d}\mathcal{G}}{\mathrm{d}t} = \frac{\partial}{\partial t} \int_{\Omega} \Gamma(\mathbf{X}, t) \, dV = \int_{\Omega} \frac{\partial \Gamma(\mathbf{X}, t)}{\partial t} \, dV \,. \tag{2.2.2}$$

In the EULER representation this is not the case as the integration domain $\omega(t)$ changes with time. The integration domain Ω, chosen in the reference configuration, becomes, in the present configuration, a different domain $\omega(t)$ at every time instance t.

2.2.1 REYNOLDS Transport Theorem

To carry out the differentiation for the term on the left-hand side of the balance equation $(2.1.16)_2$ we transform the respective variables from the present to the reference coordinates. The transformation of a volume element given in equation (1.3.8) implies (see (2.1.2)),

$$\mathcal{G} = \int_{\omega} \gamma \, dv = \int_{\Omega} \gamma J \, dV \,, \tag{2.2.3}$$

where γ is interpreted in the second integral as a function of the reference coordinates and the time, $\gamma = \hat{\gamma}(\boldsymbol{X}, t)$. A similar notation is used for the determinant of the deformation gradient, $J = \hat{J}(\boldsymbol{X}, t)$. The time differentiation of $\mathcal{G}$ is thus

$$\begin{aligned} \frac{\mathrm{d}\mathcal{G}}{\mathrm{d}t} &= \frac{\mathrm{d}}{\mathrm{d}t} \int_{\omega(t)} \gamma(\boldsymbol{x}, t)\, dv \\ &= \int_{\Omega} \frac{\partial}{\partial t} \Big(\hat{\gamma}(\boldsymbol{X}, t) \hat{J}(\boldsymbol{X}, t) \Big)\, dV \\ &= \int_{\Omega} \left(\frac{\partial \hat{\gamma}}{\partial t} J + \hat{\gamma} \frac{\partial \hat{J}}{\partial t} \right) dV \,. \end{aligned} \tag{2.2.4}$$

The time rate of change of the determinant of the deformation gradient has already been demonstrated in (1.4.11) to be $\partial \hat{J}/\partial t = (\mathrm{div}\boldsymbol{v})J$. Further, it should be noted that the material time derivative of a physical quantity in the LAGRANGE and EULER representation (1.2.18) must be the same,

$$\frac{\partial \hat{\gamma}(\boldsymbol{X}, t)}{\partial t} = \frac{\mathrm{d}\gamma(\boldsymbol{x}, t)}{\mathrm{d}t} \,; \tag{2.2.5}$$

this in turn enables us to re-transform the above integral to the present coordinates. Thus we obtain

$$\begin{aligned} \frac{\mathrm{d}\mathcal{G}}{\mathrm{d}t} &= \int_{\Omega} \left(\frac{\partial \hat{\gamma}}{\partial t} + \hat{\gamma}\, \mathrm{div}\boldsymbol{v} \right) J\, dV \\ &= \int_{\omega} \left(\frac{\mathrm{d}\gamma}{\mathrm{d}t} + \gamma\, \mathrm{div}\boldsymbol{v} \right) dv \,. \end{aligned} \tag{2.2.6}$$

This equation can also be written in another form; this is achieved by decomposing the total time derivative into a local term and a convective term, (1.2.18), to obtain[6]

$$\begin{aligned} \frac{\mathrm{d}\mathcal{G}}{\mathrm{d}t} &= \int_{\omega} \left(\frac{\partial \gamma}{\partial t} + (\mathrm{grad}\, \gamma)\, \boldsymbol{v} + \gamma\, \mathrm{div}\boldsymbol{v} \right) dv \\ &= \int_{\omega} \left(\frac{\partial \gamma}{\partial t} + \mathrm{div}(\gamma\, \boldsymbol{v}) \right) dv \,. \end{aligned} \tag{2.2.7}$$

[6] If the quantity γ under consideration is a scalar field, $\mathrm{div}(\gamma\, \boldsymbol{v})$ is given by the (customary) expression $(\gamma\, v_i)_{,\, i}$. If γ is a vector field, $\boldsymbol{\gamma} = \gamma_i \boldsymbol{e}_i$, the expression within the divergence indicates the *dyadic product* of the respective vectors, $\boldsymbol{\gamma} \otimes \boldsymbol{v}$. The divergence of this product is then

$$(\mathrm{grad}\, \boldsymbol{\gamma})\, \boldsymbol{v} + \boldsymbol{\gamma}\, \mathrm{div}\boldsymbol{v} = \mathrm{div}(\boldsymbol{\gamma} \otimes \boldsymbol{v}) \,,$$

or in component form

$$\gamma_{i\,,\,j} v_j + \gamma_i\, v_{j\,,\,j} = (\gamma_i\, v_j)_{,\,j} \,.$$

A more plausible physical interpretation of this expression is obtained if, on the basis of the divergence theorem, the second term in the integrand is expressed as a surface integral,

$$\begin{aligned}\frac{\mathrm{d}\mathcal{G}}{\mathrm{d}t} &= \int_{\omega}\left(\frac{\partial\gamma}{\partial t} + \operatorname{div}(\gamma\,\boldsymbol{v})\right) dv \\ &= \int_{\omega}\frac{\partial\gamma}{\partial t}\,dv + \int_{\partial\omega}\gamma\,(\boldsymbol{v}\cdot\boldsymbol{n})\,da\,.\end{aligned} \tag{2.2.8}$$

The first integral gives the temporal change of the density of the physical quantity within the volume ω; and the second integral considers the change of the physical quantity due to the change of the integration domain; this is the flux of the quantity through the surface.

The result thus obtained and its equivalent forms (2.2.6)–(2.2.7) are called the REYNOLDS *transport theorem.* Another, perhaps more succinct derivation of the transport theorem, can be obtained if the relation for the time rate of change of a volume element, as shown in (1.4.12), is used instead of the two-fold transformation from the reference to the present configuration and back, just discussed above. Thus the alternative procedure follows as

$$\begin{aligned}\frac{\mathrm{d}\mathcal{G}}{\mathrm{d}t} &= \frac{\mathrm{d}}{\mathrm{d}t}\int_{\omega}\gamma\,dv = \int_{\omega}\left(\dot{\gamma}\,dv + \gamma\,(dv)^{\cdot}\right) \\ &= \int_{\omega}\left(\dot{\gamma}\,dv + \gamma\operatorname{div}\boldsymbol{v}\,dv\right) \\ &= \int_{\omega}\left(\dot{\gamma} + \gamma\operatorname{div}\boldsymbol{v}\right)\,dv\,.\end{aligned} \tag{2.2.9}$$

The one dimensional analogue of the REYNOLDS transport theorem is known as the LEIBNIZ *integration rule.* The following formula exhibits an example of this rule:

$$\frac{\mathrm{d}}{\mathrm{d}t}\int_{a(t)}^{b(t)} f(x,t)dx = \int_{a(t)}^{b(t)}\frac{\partial f(x,t)}{\partial t}dx + \dot{b}\,f(b,t) - \dot{a}\,f(a,t) \tag{2.2.10}$$

in which the variable, t here, appears as an argument of the integration–limits as well as of the integrand function of the integral.

2.2.2 Local Balance Equations in the LAGRANGE Representation

If we desire the balance equations in local form, as sketched at the beginning of this section, we must change the surface integral to the volume integral. With the Divergence Theorem the surface integral can be written as [7]

[7] At this point, we have to distinguish between a vector and a tensor field. If $\boldsymbol{\Phi}$ is a vector field, $\boldsymbol{\Phi N}$ denotes the scalar product, $\tilde{\Phi}_{\alpha}N_{\alpha}$. In case $\boldsymbol{\Phi}$ is a second rank tensor field the operation $\tilde{\Phi}_{i\beta}N_{\beta}$ results in a vector.

$$\int_{\partial\Omega} \boldsymbol{\Phi}\boldsymbol{N}\,\mathrm{d}A = \int_{\Omega} \mathrm{Div}\,\boldsymbol{\Phi}\,dV \tag{2.2.11}$$

in the LAGRANGE representation.

Along with the rule (2.2.2) for interchanging the integration–differentiation sequence the balance expression in the LAGRANGE representation, $(2.1.16)_1$ gives

$$\int_{\Omega} \frac{\partial \Gamma}{\partial t}\,dV = \int_{\Omega} (\Pi + \Sigma - \mathrm{Div}\,\boldsymbol{\Phi})\,dV \tag{2.2.12}$$

or

$$\int_{\Omega} \left(\frac{\partial \Gamma}{\partial t} - \Pi - \Sigma + \mathrm{Div}\,\boldsymbol{\Phi} \right) dV = 0\,.$$

Since the above expression must hold true for an arbitrary volume element, this means the integrand must identically vanish. Thus the *local balance equation in the* LAGRANGE *representation* appears in its final form as

$$\frac{\partial \Gamma}{\partial t} = -\,\mathrm{Div}\,\boldsymbol{\Phi} + \Pi + \Sigma\,. \tag{2.2.13}$$

2.2.3 Local Balance Equations in the EULER Representation

The divergence theorem can also be used to convert the surface integral into the volume integral in the EULER representation, thus

$$\int_{\partial\omega} \boldsymbol{\phi}\boldsymbol{n}\,\mathrm{da} = \int_{\omega} \mathrm{div}\boldsymbol{\phi}\,dv\,. \tag{2.2.14}$$

By using the REYNOLDS transport theorem (2.2.6) the balance statement $(2.1.16)_2$ can be written as

$$\begin{aligned} &\int_{\omega} \left(\frac{\mathrm{d}\gamma}{\mathrm{d}t} + \gamma\,\mathrm{div}\boldsymbol{v} \right) dv = \int_{\omega} (\pi + \varsigma - \mathrm{div}\boldsymbol{\phi})\,dv \\ \text{or} \quad &\int_{\omega} \left(\frac{\mathrm{d}\gamma}{\mathrm{d}t} + \gamma\,\mathrm{div}\boldsymbol{v} - \pi - \varsigma + \mathrm{div}\boldsymbol{\phi} \right) dv = 0\,. \end{aligned} \tag{2.2.15}$$

As in the case for the LAGRANGE representation the above equation must also be valid for any arbitrary volume element; therefore the integrand must vanish, and the resulting equation for the *local balance equation in the* EULER *representation* is obtained as

$$\frac{\mathrm{d}\gamma}{\mathrm{d}t} + \gamma\,\mathrm{div}\boldsymbol{v} = -\mathrm{div}\boldsymbol{\phi} + \pi + \varsigma\,. \tag{2.2.16}$$

By decomposing the total time derivative into a partial derivative plus a convective term, the above equation can be written in one of the two forms

$$
\begin{aligned}
&\frac{\partial \gamma}{\partial t} + \operatorname{div}(\gamma \boldsymbol{v}) = -\operatorname{div}\boldsymbol{\phi} + \pi + \varsigma && (\gamma \text{ scalar field}), \\
&\frac{\partial \boldsymbol{\gamma}}{\partial t} + \operatorname{div}(\boldsymbol{\gamma} \otimes \boldsymbol{v}) = -\operatorname{div}\boldsymbol{\phi} + \boldsymbol{\pi} + \boldsymbol{\varsigma} && (\boldsymbol{\gamma} \text{ vector field}),
\end{aligned}
\tag{2.2.17}
$$

valid for a scalar and vector valued field, respectively.

2.3 Special Balance Equations

As already mentioned, a thermomechanical system is characterized by the specification of mass, momentum, angular momentum, energy and entropy. We shall establish balance equations for these five quantities. In so doing we shall assume that the reader has encountered these laws already in introductory physics courses and thus possesses some basic knowledge about them. Nevertheless when introducing them care will be observed in establishing a sufficient understanding even without reference to these laws. Because of the additivity assumption introduced earlier, densities of mass, momentum, angular momentum, energy and entropy are needed. These are defined as follows:

- *mass density* (or simply density), ρ,
- *momentum density*, $\rho \boldsymbol{v}$,
- *angular momentum density*, $\boldsymbol{x} \times \rho \boldsymbol{v}$,
- *energy density*, $\frac{1}{2}\rho v^2 + \rho\varepsilon$, ($\varepsilon$ is the specific internal energy),
- *entropy density*, ρs, (s is the specific entropy).

These are the expressions in the spatial description; in the material description ρ must simply be replaced by ρ_R. In ensuing discussions we shall frequently speak of mass, energy or entropy, etc. even though we actually mean the densities of these quantities.

The definitions in the above list are rather obvious. Once the mass density is specifically the corresponding momentum density simply follows from the definition "momentum equals mass times velocity". Somewhat special is the definition of the angular momentum density. Its formula reads "moment of momentum", $\boldsymbol{x} \times \rho \boldsymbol{v}$. This is not the most general situation one may encounter. The angular momentum of a moving rigid body with respect to a fixed point (the origin of the coordinate system) is given by the sum of its angular momentum with respect to its center of mass (the "self–angular momentum") plus the moment of its momentum acting at the center of mass. Analogously, the angular momentum of an infinitesimal material volume element at position $\boldsymbol{x}$ may consist of the "self–angular momentum" also called *spin* plus the moment of momentum. The form chosen by us in the above list indicates that the spin is not taken into account. This will be so until Chap.7, and so for the time being we restrict ourselves to continua without intrinsic spin. These are also called *non-polar* continua. The energy density constitutes of a kinetic and an internal contribution. The first is of dynamic nature, the second may be due to deformation, temperature variations and other non-mechanical causes. The fact that these two quantities together are used as the relevant energy density is in anticipation of the first law of thermodynamics which, roughly, states that all energies together form a conserved quantity. The entropy density does not yet disclose any physical appeal, which will become apparent later on.

2.3.1 Mass Balance

Let us consider a body in the reference configuration. Let its density distribution (to every particle $\mathfrak{X}$ a density is allocated) be indicated by $\rho_R(\boldsymbol{X}, t)$. The corresponding density at position $\boldsymbol{x}$ in the present configuration is indicated by $\rho(\boldsymbol{x}, t)$.

LAGRANGE **Representation** It is assumed that mass is a physical quantity which can not flow through a surface. Further, we assume that mass can neither be produced nor be supplied. Then to obtain the balance equation for mass we make the following substitutions

$$\Gamma^{(\rho)} = \rho_R(\boldsymbol{X}, t)\,, \quad \Pi^{(\rho)} = 0\,, \quad \Sigma^{(\rho)} = 0\,, \quad \boldsymbol{\Phi}^{(\rho)} = \boldsymbol{0}\,, \tag{2.3.1}$$

for the respective terms of mass density and its production, supply, flux–densities in the local balance equation (2.2.13). Consequently, the *mass balance equation* follows directly in the LAGRANGE representation as

$$\frac{\partial \rho_R(\boldsymbol{X}, t)}{\partial t} = 0 \quad \Longrightarrow \quad \rho_R = \rho_R(\boldsymbol{X})\,. \tag{2.3.2}$$

The density of an element is only a function of its reference position. This clearly verifies the initial claim that every element $\mathfrak{X}$ of the body is associated with a density.

EULER **Representation** The balance equation for mass in the EULER representation can be obtained, as before, with the substitutions

$$\gamma^{(\rho)} = \rho(\boldsymbol{x}, t)\,, \quad \pi^{(\rho)} = 0\,, \quad \varsigma^{(\rho)} = 0\,, \quad \boldsymbol{\phi}^{(\rho)} = \boldsymbol{0} \tag{2.3.3}$$

in (2.2.17) yielding the final forms of the local mass balance as

$$\frac{\partial \rho}{\partial t} + \mathrm{div}(\rho \boldsymbol{v}) = 0 \quad \text{or} \quad \frac{\mathrm{d}\rho}{\mathrm{d}t} + \rho\, \mathrm{div} \boldsymbol{v} = 0\,. \tag{2.3.4}$$

One often calls this equation the *continuity equation.*

A material is called *density preserving* when the density of a particle does not change with time. A consequence of the density preserving condition is the source free condition of the velocity field,

$$\text{density preserving} \quad \Longleftrightarrow \quad \frac{\mathrm{d}\rho}{\mathrm{d}t} = 0 \quad \Rightarrow \quad \operatorname{div}\boldsymbol{v} = 0\,. \tag{2.3.5}$$

However, in general, the reverse of the above chain of arguments does not hold true. Even though a source free (solenoidal) velocity field implies that the density does not change, this does by no means also imply that the density preserving of the material would follow. What can be inferred is that if in a process the velocity field of a body happens to be solenoidal, then the conditions are such that for a possibly compressible fluid the density happens to remain unchanged. From the view of the present knowledge in this context it is neither evident that the concepts "constant density" and "constant volume" do correspond to each other. But, that these seemingly distinct material properties are indeed the same and the terms "density preserving" and "volume preserving" are exchangeable follows from the following chain of arguments:

$$\dot{\rho} = 0 \iff \operatorname{div}\boldsymbol{v} = 0 \iff \dot{J} = J\operatorname{div}\boldsymbol{v} = 0 \iff \det \boldsymbol{F} = \text{const.} = 1$$

Volume conserving motions are called *isochoric* and transformations $\boldsymbol{F}$ whose determinant remains constant and equals unity are called *unimodular*. So isochoric motions give rise to an unimodular deformation field and a solenoidal velocity field and vice versa.

By exploiting the REYNOLDS transport theorem for *specific quantities per unit mass*, the mass balance equation (2.3.4) can be used in an elegant form; the fields must, however, be differentiable. To achieve this special form we substitute the density of a physical quantity, $\gamma = \rho\Psi$, in (2.2.9) to obtain the following result:

$$\begin{aligned}
\frac{\mathrm{d}}{\mathrm{d}t}\int_\omega \rho\Psi\, dv &= \int_\omega \Big((\rho\Psi)^{\cdot} + \rho\Psi\operatorname{div}\boldsymbol{v}\Big)\, dv \\
&= \int_\omega \Big(\rho\dot{\Psi} + \dot{\rho}\Psi + \rho\Psi\operatorname{div}\boldsymbol{v}\Big)\, dv \\
&= \int_\omega \Big(\rho\dot{\Psi} + \Psi\underbrace{[\dot{\rho} + \rho\operatorname{div}\boldsymbol{v}]}_{=0}\Big)\, dv \\
&= \int_\omega \Big(\rho\dot{\Psi}\Big)\, dv\,.
\end{aligned} \tag{2.3.6}$$

The material derivative is therefore simply transferred from outside the integral to the specific quantity Ψ. This is a very useful rule, however, the reader is cautioned to apply it only to fields that are differentiable.

2.3.2 Momentum Balance

Which quantities can change the momentum of a body? For 'normal' physical systems the three terms, namely production, supply and flux can be specified as

- **production of momentum**
  does not exist (this is a physical postulate, one says, momentum is a "conserved variable"),
- **supply of momentum**
  is governed by external volume forces or by densities of the volume forces (e.g. the gravitational force or weight),
- a **flux of momentum** through the surface
  is the result of the surface–force densities; these are the stress vectors acting on the surfaces. The corresponding tensor fields or momentum fluxes are called the stress tensors.

These assignments result in the following alternative statement

$$\frac{\mathrm{d}}{\mathrm{d}t}\{\text{momentum in } \mathcal{B}\} = \{\text{sum of all forces on } \partial\mathcal{B}\} + \{\text{sum of all forces in } \mathcal{B}\} .$$

It is evident that this assignment of production, supply and flux of momentum is reminiscent of NEWTON's *second law* according to which the time rate of change of the momentum of a body equals the sum of all forces acting on this body. We assume that the reader is familiar with this fundamental law of physics and also suppose that he or she has encountered it in the context of rigid body dynamics. Within that context it then is becoming quite clear that the volume forces and surface forces exerted on the body play the role of the supply and (negative) flux of momentum in the terminology of the balance laws. NEWTON [171] in his Principia never specified whether he addressed mass points or bodies of finite extent. A precise formulation was first given by L. EULER [70], and the local form of the momentum equation dates back to CAUCHY[8].

LAGRANGE **Representation** To derive the local form of the momentum balance equation in the LAGRANGE representation the following identifications are made,

$$\begin{aligned}
\Gamma^{(\rho v)} &= \rho_R \boldsymbol{v} \ , && \text{momentum (density)},\\
\Pi^{(\rho v)} &= \mathbf{0} \ , && \text{no production},\\
\Sigma^{(\rho v)} &= \rho_R \boldsymbol{g} \ , && \text{volume force (density)},\\
\boldsymbol{\Phi}^{(\rho v)} &= -\boldsymbol{T} \ , && \text{momentum flux (density)} \quad (\tilde{\boldsymbol{\Phi}}^{(\rho v)} = \boldsymbol{T}\boldsymbol{N}) .
\end{aligned} \tag{2.3.7}$$

The quantity $\boldsymbol{g}$, has the meaning of a force density per unit mass, and when considering it as the gravitation force it corresponds to the *acceleration due to gravity*. The momentum flux density is equal to the negative of the *stress tensor* $\boldsymbol{T}$.

[8] For a historical account see TRUESDELL & TOUPIN [238], Section 196, TRUESDELL [241] and SZABO [233].

In the LAGRANGE representation $\boldsymbol{T}$ is called the *first* PIOLA–KIRCHHOFF *stress tensor*. The *momentum balance* then reads

$$\frac{\partial}{\partial t}(\rho_R \boldsymbol{v}) = \operatorname{Div} \boldsymbol{T} + \rho_R \boldsymbol{g} , \tag{2.3.8}$$

which, with the mass balance equation (2.3.2), simplifies to yield[9]

$$\rho_R \frac{\mathrm{d}\boldsymbol{v}}{\mathrm{d}t} = \operatorname{Div} \boldsymbol{T} + \rho_R \boldsymbol{g} . \tag{2.3.9}$$

It is also meaningful to write this equation in component form. We let

$$\boldsymbol{v} = v_i \boldsymbol{e}_i , \quad \boldsymbol{T} = T_{i\alpha} \boldsymbol{e}_i \otimes \boldsymbol{e}_\alpha , \quad \boldsymbol{g} = g_i \boldsymbol{e}_i , \tag{2.3.10}$$

and for the Gradient and Divergence operators of $\boldsymbol{T}$ we write

$$\begin{aligned} \operatorname{Grad} \boldsymbol{T} &= T_{i\alpha,\beta}(\boldsymbol{e}_i \otimes \boldsymbol{e}_\alpha) \otimes \boldsymbol{e}_\beta , \\ \operatorname{Div} \boldsymbol{T} &= T_{i\alpha,\beta}(\boldsymbol{e}_i \otimes \boldsymbol{e}_\alpha) \cdot \boldsymbol{e}_\beta = T_{i\alpha,\alpha} \boldsymbol{e}_i . \end{aligned} \tag{2.3.11}$$

Together with the above component representations one obtains

$$\left(\rho_R \frac{\mathrm{d}v_i}{\mathrm{d}t} - T_{i\alpha,\alpha} - \rho_R g_i\right) \boldsymbol{e}_i = \boldsymbol{0} , \tag{2.3.12}$$

or if one restricts oneself to the component i,

$$\rho_R \frac{\mathrm{d}v_i}{\mathrm{d}t} = T_{i\alpha,\alpha} + \rho_R g_i . \tag{2.3.13}$$

EULER **Representation** In the EULER representation we have the following specifications in contrast to (2.3.7)

$$\gamma^{(\rho v)} = \rho \boldsymbol{v} , \quad \pi^{(\rho v)} = \boldsymbol{0} , \quad \varsigma^{(\rho v)} = \rho \boldsymbol{g} , \quad \phi^{(\rho v)} = -\boldsymbol{t} . \tag{2.3.14}$$

The tensor $\boldsymbol{t}$ is called CAUCHY *stress tensor*[10]. Substituting these expressions into the balance statement (2.2.16) we obtain

$$\frac{\mathrm{d}}{\mathrm{d}t}(\rho \boldsymbol{v}) + \rho \boldsymbol{v} \operatorname{div} \boldsymbol{v} = \operatorname{div} \boldsymbol{t} + \rho \boldsymbol{g} , \tag{2.3.15}$$

---

[9] It is not necessarily so that (2.3.9) is simpler than (2.3.8), but it is in a more direct form in which NEWTON's law is commonly known: "Mass times acceleration equals sum of the forces". The form (2.3.8) is the more general statement that is with advantage used for numerical analysis. Of course, the difference between the two versions, (2.3.8) and (2.3.9) is elementary in the LAGRANGE description. Its significance will become apparent when we deal with the EULER representation.

[10] Notice that we use a small $\boldsymbol{t}$ to denote this stress tensor. This is different from most other authors, who use $\boldsymbol{T}$ instead (which is the symbol we refer to as the first PIOLA KIRCHHOFF stress tensor).

or with (1.2.18),[11]

$$\frac{\partial}{\partial t}(\rho \boldsymbol{v}) + \operatorname{div}(\rho \boldsymbol{v} \otimes \boldsymbol{v}) = \operatorname{div}\boldsymbol{t} + \rho \boldsymbol{g} \ . \tag{2.3.16}$$

With the help of the continuity equation (2.3.4) the above relation becomes

$$\rho \frac{\mathrm{d}\boldsymbol{v}}{\mathrm{d}t} = \operatorname{div}\boldsymbol{t} + \rho \boldsymbol{g} \ , \tag{2.3.17}$$

where

$$\frac{\mathrm{d}\boldsymbol{v}}{\mathrm{d}t} = \frac{\partial \boldsymbol{v}}{\partial t} + (\operatorname{grad} \boldsymbol{v})\boldsymbol{v} = \frac{\partial \boldsymbol{v}}{\partial t} + \operatorname{grad}\left(\frac{|\boldsymbol{v}|^2}{2}\right) - \boldsymbol{v} \times \operatorname{curl} \boldsymbol{v} \ . \tag{2.3.18}$$

The identity of the last two expressions is best proved in Cartesian component form. In component form, (2.3.17) reads

$$\rho \frac{\mathrm{d}v_i}{\mathrm{d}t} = t_{ij},_j + \rho g_i \ , \tag{2.3.19}$$

in which

$$\frac{\mathrm{d}v_i}{\mathrm{d}t} = \frac{\partial v_i}{\partial t} + v_{i,j} v_j \ .$$

The *momentum balance equation* or the *equation of motion* is a general law valid for all materials. However, depending on the type of material considered, the CAUCHY stress tensor can have a different functional form. For example, for an isotropic and elastic rigid body one obtains the equations of motion, well known in seismology or elasticity theory, and for a NEWTONian fluid one obtains the NAVIER–STOKES equations.

Equation (2.3.17), in this general form, has first been derived by CAUCHY and was then applied by him to an elastic body. Therefore, these equations are known as CAUCHY *equations of motion.* The corresponding equations, for an ideal fluid in which $\boldsymbol{t}$ is only given by the pressure tensor $\boldsymbol{t} = -p\boldsymbol{I}$, were first given by LEONHARD EULER about 70 years earlier; therefore these equations are the EULER *equations of motion.* Their special form for linear viscous fluids, alternately, lead to the so called NAVIER–STOKES *equations.* We will focus our attentions to such formulations in Chap.5 "Material equations".

[11] The *dyadic product* $\boldsymbol{v} \otimes \boldsymbol{v}$ is defined as the second rank tensor $\boldsymbol{v} \otimes \boldsymbol{v} = v_i v_j \boldsymbol{e}_i \otimes \boldsymbol{e}_j$; in Cartesian coordinates and in matrix form it is expressed as

$$[\boldsymbol{v} \otimes \boldsymbol{v}] \equiv \begin{pmatrix} v_1 v_1 & v_1 v_2 & v_1 v_3 \\ v_2 v_1 & v_2 v_2 & v_2 v_3 \\ v_3 v_1 & v_3 v_2 & v_3 v_3 \end{pmatrix} \ .$$

2.3.3 Angular Momentum Balance

As with the balance of linear momentum we assume the reader to be familiar with the balance law of angular momentum as expressed e.g. for rigid body dynamics. In that context this balance law is expressed as the statement that the "time rate of change of the angular momentum of a body with respect to a point fixed in space (or the centre of gravity) equals the resulting moment of the forces acting on the body with respect to the same point". This law is one of the basic axioms of GALILEan physics and has first been formulated by L. EULER [70] and the corresponding equations for rigid bodies are known as the EULER equations[12]. Its application to deformable bodies with the inference that the CAUCHY stress tensor is symmetric is due to CAUCHY himself.

In general, angular momentum is "moment of momentum plus spin" and this composition (in its true semantic meaning) is verified by the fact that the angular momentum of a rigid body with respect to an arbitrary fixed point is the sum of the body's angular momentum relative to its centre of mass ($\hat{=}$ spin) plus the moment of the momentum at the centre of mass with respect to the arbitrary fixed point. This same concept can also be applied to volume elements of a continuous body.

For the angular momentum balance we make the suppositions that the particles do not have internal angular momentum (spin), and further, that they are not associated with quantities like volume moments (e.g. magnetic polarization) or surface couple stresses[13]. Let in the ensuing developments the angular momentum be defined with respect to the origin of the coordinate system.

LAGRANGE **Representation** Based on the above requirements we specify the angular momentum density, the production, the supply and the flux of the angular momentum as follows:

$$
\begin{aligned}
\Gamma^{(\rho x\times v)} &= \boldsymbol{x}\times\rho_R\boldsymbol{v}\,, && \text{angular momentum (density),}\\
&&& \text{density of moment of momentum,}\\
\Pi^{(\rho x\times v)} &= \mathbf{0}\,, && \text{no production of angular momentum,}\\
\Sigma^{(\rho x\times v)} &= \boldsymbol{x}\times\rho_R\boldsymbol{g}\,, && \text{density of moment of volume forces,}\\
\boldsymbol{\Phi}^{(\rho x\times v)} &= -\boldsymbol{x}\times\boldsymbol{T}\,, && \text{density of moment of surface forces.}
\end{aligned}
\qquad (2.3.20)
$$

All these quantities are moments of corresponding momentum densities[14]. With the above specifications, the angular momentum balance, resulting from

[12] For critical historical remarks on this see TRUESDELL & TOUPIN [238], TRUESDELL [241], [245], SZABO [233].

[13] The more general case will be dealt with in Chap. 7

[14] This very fact could be the reason why some – even prominent – scientists concluded that the balance law of angular momentum in the form of moment of momentum does not form an independent law. The result alone proves the contrary.

the general balance equation (2.2.13), takes the form

$$\frac{\mathrm{d}}{\mathrm{d}t}\left(\rho_R(\boldsymbol{x}\times\boldsymbol{v})\right) = \mathrm{Div}(\boldsymbol{x}\times\boldsymbol{T}) + \rho_R(\boldsymbol{x}\times\boldsymbol{g})\ . \tag{2.3.21}$$

From this, one finds (this computation should be repeated in the component expression for better understanding)[15]

$$\underbrace{\rho_R(\boldsymbol{v}\times\boldsymbol{v})}_{=\mathbf{0}} + \boldsymbol{x}\times\underbrace{\left(\rho_R\frac{\mathrm{d}\boldsymbol{v}}{\mathrm{d}t} - \mathrm{Div}\,\boldsymbol{T} - \rho_R\boldsymbol{g}\right)}_{=\mathbf{0}\ \text{momentum balance}} = \boldsymbol{T}^*\ , \tag{2.3.22}$$

where the vector $\boldsymbol{T}^*$ is in Cartesian components defined as

$$T_i^* := \varepsilon_{ijk}\frac{\partial x_j}{\partial X_\alpha}T_{k\alpha}\ . \tag{2.3.23}$$

Thus the *angular momentum balance* in the LAGRANGE representation is reduced to

$$\varepsilon_{ijk}\frac{\partial x_j}{\partial X_\alpha}T_{k\alpha} = 0\ ; \tag{2.3.24}$$

because of the antisymmetry of the ε-tensor[16] this implies

$$\frac{\partial x_k}{\partial X_\alpha}T_{j\alpha} = \frac{\partial x_j}{\partial X_\alpha}T_{k\alpha} \quad \text{or} \quad \boldsymbol{T}\boldsymbol{F}^T = (\boldsymbol{T}\boldsymbol{F}^T)^T\ , \tag{2.3.25}$$

where the definition of the deformation gradient has been used. This expression of the angular momentum balance states that the first PIOLA–KIRCHHOFF stress tensor $\boldsymbol{T}$ is not symmetric, only its product from the right with the transposed deformation gradient is symmetric.

EULER **Representation** In the EULER representation we specify, similarly to (2.3.20), the following angular momentum densities

$$\begin{aligned} \gamma^{(\rho x\times v)} &= \boldsymbol{x}\times\rho\boldsymbol{v}\ ,\\ \pi^{(\rho x\times v)} &= \mathbf{0}\ ,\\ \varsigma^{(\rho x\times v)} &= \boldsymbol{x}\times\rho\boldsymbol{g}\ ,\\ \boldsymbol{\phi}^{(\rho x\times v)} &= -\boldsymbol{x}\times\boldsymbol{t}\ . \end{aligned} \tag{2.3.26}$$

---

[15] The cross product of a vector $\boldsymbol{a}$ with a tensor $\boldsymbol{T}$ is defined (Cartesian coordinates) as

$$\boldsymbol{a}\times\boldsymbol{T} = a_i\boldsymbol{e}_i\times T_{j\alpha}\boldsymbol{e}_j\otimes\boldsymbol{e}_\alpha = a_iT_{j\alpha}\varepsilon_{ijl}\boldsymbol{e}_l\otimes\boldsymbol{e}_\alpha\ .$$

This is a matrix with the components $a_iT_{j\alpha}\varepsilon_{ijl}$.

[16] The trace of the product of a symmetric and an antisymmetric tensor is always equal to zero:

$$\begin{aligned} &A_{ij} = A_{ji}\ ,\quad B_{ij} = -B_{ji}\\ \Rightarrow\ &A_{ij}B_{ji} = -A_{ji}B_{ij} \qquad &&(\boldsymbol{B}\text{ is antisymmetric, }\boldsymbol{A}\text{ is symmetric})\\ &\phantom{A_{ij}B_{ji}} = -A_{ij}B_{ji} &&(\text{renaming the indices})\\ \Rightarrow\ &A_{ij}B_{ji} = -A_{ij}B_{ji} = 0\ . \end{aligned}$$

As their counterparts in the LAGRANGE representation, these quantities are the moments of the corresponding momentum densities. By substituting these specifications in the balance equation (2.2.16) we get[17]

$$\frac{\mathrm{d}}{\mathrm{d}t}\left(\rho(\boldsymbol{x}\times\boldsymbol{v})\right)+\rho(\boldsymbol{x}\times\boldsymbol{v})\mathrm{div}\boldsymbol{v}=\mathrm{div}(\boldsymbol{x}\times\boldsymbol{t})+\rho(\boldsymbol{x}\times\boldsymbol{g})\,, \tag{2.3.27}$$

or

$$(\boldsymbol{x}\times\boldsymbol{v})\underbrace{\left(\frac{\mathrm{d}\rho}{\mathrm{d}t}+\rho\mathrm{div}\boldsymbol{v}\right)}_{=0\text{ mass balance}}+\rho\underbrace{\boldsymbol{v}\times\boldsymbol{v}}_{=0}$$
$$+\boldsymbol{x}\times\underbrace{\left(\rho\frac{\mathrm{d}\boldsymbol{v}}{\mathrm{d}t}-\mathrm{div}\boldsymbol{t}-\rho\boldsymbol{g}\right)}_{=0\text{ momentum balance}}=\boldsymbol{t}^*\,, \tag{2.3.28}$$

where $\boldsymbol{t}^*$ can be written analogously to (2.3.23) in component form – the derivatives must now be taken with respect to the present coordinates –

$$t_i{}^* := \varepsilon_{ijk}\frac{\partial x_j}{\partial x_l}t_{kl}=\varepsilon_{ijk}t_{kj}\,. \tag{2.3.29}$$

The *angular momentum balance* in the EULER representation therefore implies that

$$\varepsilon_{ijk}t_{kj}=0\,, \tag{2.3.30}$$

or, because of the antisymmetry of the ε–tensor,

$$t_{kj}=t_{jk}\quad\text{rsp.}\quad \boldsymbol{t}=\boldsymbol{t}^T\,. \tag{2.3.31}$$

That is, *the* CAUCHY *stress tensor is symmetric.* The angular momentum balance, in contrast to the other balance equations, does not yield differential equations, but it implies symmetry conditions for the stress tensors.

One can prove (see Exercises) that in a system of a finite number of mass points, which, apart from the external forces, are exposed to only central forces, the law of angular momentum balance is identically satisfied if the momentum balance is satisfied. This means that for such a system the balance of angular momentum does not express an independent physical basic law. This fact occasionally manifests misunderstanding and confusion by claims that the law of angular momentum would not constitute an independent basic law. But this is not so. LEONHARD EULER was the first who pointed this out. In fact, the law of angular momentum is a physical axiom just like NEWTON's second law, and it provides new information; for the above analysis evidently shows that without this law the symmetry of the CAUCHY stress tensor would never have been concluded.

[17] In component form, $\mathrm{div}(\boldsymbol{x}\times\boldsymbol{t})$ can be written as

$$(\varepsilon_{ijk}x_it_{jl})_{,l}=\varepsilon_{ijk}(x_{i,l}\,t_{jl}+x_it_{jl,l})=\varepsilon_{ijk}(t_{ji}+x_it_{jl,l})\,.$$

2.3.4 Energy Balance

The energy balance corresponds physically to the *first law of thermodynamics*, which states that the mechanical and thermal energies (and all additional energies present) are not conserved individually but they are conserved *together*. The two energy quantities describing the energetic *state* of the body are the kinetic energy $\mathcal{T}$ and the internal energy $\mathcal{U}$ and the time rate of change of their sum $(\dot{\mathcal{T}} + \dot{\mathcal{U}})$ must be equal to the mechanical and non-mechanical energies supplied to the body from outside per unit time. These contributions are the *process quantities* $\mathcal{L}$ and $\mathcal{Q}$, the former being the power of working of the external forces the latter the energy supplied from outside other than mechanical working. Both have volume and surface contributions, so that

$$\frac{\mathrm{d}}{\mathrm{d}t}(T + U) = \mathcal{L}_v + \mathcal{L}_{\partial v} + \mathcal{Q}_v + \mathcal{Q}_{\partial v} \,. \tag{2.3.32}$$

If the non-mechanical energy supplies to the body are only of thermal origin (which will here be assumed), then $\mathcal{Q}$ is called the heat supplied to the body, and it constitutes *heat flow*, $\mathcal{Q}_{\partial v}$ through the boundary of V and heat supply, $\mathcal{Q}_v$ or *radiation*. Rearranging (2.3.31) according to

$$\frac{\mathrm{d}}{\mathrm{d}t}(\mathcal{T} + \mathcal{U}) = (\mathcal{L}_{\partial v} + \mathcal{Q}_{\partial v}) + (\mathcal{L}_v + \mathcal{Q}_v) \tag{2.3.33}$$

shows that the energy supplied per unit time to the body by external agents has a surface and a volume contribution. There is, however no production term, because by axiom the total energy is a conserved quantity.

LAGRANGE **Representation** For the derivation of the total energy balance equation the following quantities are specified:

$$\begin{aligned}
\Gamma^{(e)} &= \tfrac{1}{2}\rho_R \boldsymbol{v}\cdot\boldsymbol{v} + \rho_R\varepsilon, && \text{energy density (kinetic + internal)},\\
\Pi^{(e)} &= 0, && \text{physical postulate: no energy production},\\
\Sigma^{(e)} &= \rho_R \boldsymbol{g}\cdot\boldsymbol{v} + \rho_R\mathfrak{r}, && \text{supply of energy (power of the external volume forces + radiation)},\\
\boldsymbol{\Phi}^{(e)} &= -\boldsymbol{v}\boldsymbol{T} + \boldsymbol{Q}, && \text{energy flux density (negative power of surface force + heat flux)}.
\end{aligned} \tag{2.3.34}$$

The energy supply is given by the power of the external forces and the radiation $\rho\mathfrak{r}$. Likewise, the flux of the energy constitutes two parts, one is the energy flux density $\boldsymbol{Q}$, which, in pure thermomechanical processes, is equal to the heat flux density, and the other arises from the stresses due to surface forces.

The first law of thermodynamics in the LAGRANGE representation is thus given as

$$\frac{\mathrm{d}}{\mathrm{d}t}\left(\tfrac{1}{2}\rho_R \boldsymbol{v}\cdot\boldsymbol{v} + \rho_R\varepsilon\right) = -\operatorname{Div}\boldsymbol{Q} + \operatorname{Div}(\boldsymbol{v}\boldsymbol{T}) + \rho_R(\boldsymbol{g}\cdot\boldsymbol{v} + \mathfrak{r}) \,. \tag{2.3.35}$$

Simplifying this with the help of the other balance equations, (2.3.2) and (2.3.9), leads to[18]

$$\underbrace{\frac{\mathrm{d}\rho_R}{\mathrm{d}t}}_{=0}\left(\tfrac{1}{2}\boldsymbol{v}\cdot\boldsymbol{v} + \varepsilon\right) + \rho_R\frac{\mathrm{d}\varepsilon}{\mathrm{d}t} + \boldsymbol{v}\cdot\underbrace{\left(\rho_R\frac{\mathrm{d}\boldsymbol{v}}{\mathrm{d}t} - \operatorname{Div}\boldsymbol{T} - \rho_R\boldsymbol{g}\right)}_{=0\ \text{momentum balance}}$$

$$= -\operatorname{Div}\boldsymbol{Q} + \operatorname{Tr}\left((\operatorname{Grad}\boldsymbol{v})\boldsymbol{T}^T\right) + \rho_R\mathfrak{r} \,. \tag{2.3.36}$$

The principle of *conservation of energy* in the LAGRANGE representation thus yields the *balance of the internal energy*

$$\rho_R\frac{\mathrm{d}\varepsilon}{\mathrm{d}t} = -\operatorname{Div}\boldsymbol{Q} + \hat{\varphi} + \rho_R\mathfrak{r} \,, \quad \hat{\varphi} := \operatorname{Tr}\left((\operatorname{Grad}\boldsymbol{v})\boldsymbol{T}^T\right) . \tag{2.3.37}$$

Accordingly, the time rate of change of the internal energy is balanced by the flow of heat through the surface, the power $\hat{\varphi}$ of the stresses associated with the velocity gradients and, if present, radiation. In this connection $\hat{\varphi}$ can not be thought to be only due to dissipative processes alone although this happens to be so quite often. It may contain reversible and irreversible parts. Locally, the internal energy satisfies a balance expression, namely (2.3.37), in which $\boldsymbol{Q}$ is interpreted as flux, $\rho\mathfrak{r}$ as supply and $\hat{\varphi}$ as production. Because the production does not vanish here, the internal energy is thus not a conserved quantity.

EULER **Representation** The energy density, its production, supply and flux terms are given here by

$$\begin{aligned}
\gamma^{(e)} &= \tfrac{1}{2}\rho\boldsymbol{v}\cdot\boldsymbol{v} + \rho\varepsilon \,,\\
\pi^{(e)} &= 0 \,,\\
\varsigma^{(e)} &= \rho\boldsymbol{g}\cdot\boldsymbol{v} + \rho\mathfrak{r} \,,\\
\boldsymbol{\phi}^{(e)} &= -\boldsymbol{v}\boldsymbol{t} + \boldsymbol{q} \,,
\end{aligned} \tag{2.3.38}$$

[18] In index notation we write

$$\operatorname{Div}(\boldsymbol{v}\boldsymbol{T}) = \frac{\partial}{\partial X_\alpha}(v_i T_{i\alpha}) = v_{i,\alpha}T_{i\alpha} + v_i T_{i\alpha,\alpha} \,.$$

The second term is equal to $\boldsymbol{v}\cdot\operatorname{Div}\boldsymbol{T}$, and the first term is written as

$$v_{i,\alpha}T_{i\alpha} = \operatorname{Grad}\boldsymbol{v}\cdot\boldsymbol{T}^T = \operatorname{Tr}\left((\operatorname{Grad}\boldsymbol{v})\boldsymbol{T}^T\right) ,$$

where $\operatorname{Tr}(\boldsymbol{A}) = A_{\alpha\alpha}$ means the trace of the tensor $\boldsymbol{A}$. It is to be noted that the trace operators for the reference and present coordinates are designated as Tr or tr, analogously to the gradient and divergence operators Grad, grad and Div, div in the two different coordinate representations.

where $\boldsymbol{q}$ is the spatial energy flux density. Upon substitution in (2.2.16) one obtains

$$\frac{\mathrm{d}}{\mathrm{d}t}\left(\tfrac{1}{2}\rho\boldsymbol{v}\cdot\boldsymbol{v}+\rho\varepsilon\right)+\rho\left(\tfrac{1}{2}\boldsymbol{v}\cdot\boldsymbol{v}+\varepsilon\right)\mathrm{div}\boldsymbol{v}$$

$$=-\mathrm{div}\boldsymbol{q}+\mathrm{div}(\boldsymbol{v}\boldsymbol{t})+\rho(\boldsymbol{v}\cdot\boldsymbol{g}+\mathfrak{r})\,. \qquad (2.3.39)$$

The balance expressions (2.3.4), (2.3.17) and (2.3.31) then lead to[19]

$$\left(\tfrac{1}{2}\boldsymbol{v}\cdot\boldsymbol{v}+\varepsilon\right)\underbrace{\left[\frac{\mathrm{d}\rho}{\mathrm{d}t}+\rho\mathrm{div}\boldsymbol{v}\right]}_{=0\text{ mass balance}}+\boldsymbol{v}\cdot\underbrace{\left[\rho\frac{\mathrm{d}\boldsymbol{v}}{\mathrm{d}t}-\mathrm{div}\boldsymbol{t}-\rho\boldsymbol{g}\right]}_{=0\text{ momentum balance}}+\rho\frac{\mathrm{d}\varepsilon}{\mathrm{d}t}$$

$$=-\mathrm{div}\boldsymbol{q}+\mathrm{tr}(\boldsymbol{D}\boldsymbol{t})+\rho\mathfrak{r}\,. \qquad (2.3.40)$$

The local form of the *energy balance* in the EULER representation is thus given by

$$\rho\frac{\mathrm{d}\varepsilon}{\mathrm{d}t}=-\mathrm{div}\boldsymbol{q}+\varphi+\rho\mathfrak{r}\,,\quad \varphi:=\mathrm{tr}(\boldsymbol{D}\boldsymbol{t}) \qquad (2.3.41)$$

and is exactly of the same form as (2.3.37). The scalar φ denotes the specific power which the CAUCHY stress executes on the velocity gradients and is to be interpreted as a production term.

In addition to the local balance statements for the internal energy (2.3.37) and (2.3.41) one can also formulate a *balance equation for the kinetic energy.* This can be achieved through scalar multiplication of the momentum balance equation with the velocity vector, i.e.,

$$\begin{aligned}\boldsymbol{v}\cdot\left(\rho_R\frac{\partial\boldsymbol{v}}{\partial t}-\mathrm{Div}\,\boldsymbol{T}-\rho_R\boldsymbol{g}\right)&=0\,,\\ \boldsymbol{v}\cdot\left(\rho\frac{\mathrm{d}\boldsymbol{v}}{\mathrm{d}t}-\mathrm{div}\boldsymbol{t}-\rho\boldsymbol{g}\right)&=0\,;\end{aligned} \qquad (2.3.42)$$

leading with a few simplifications to

$$\begin{aligned}\rho_R\frac{\partial}{\partial t}\left(\frac{\boldsymbol{v}\cdot\boldsymbol{v}}{2}\right)&=\mathrm{Div}(\boldsymbol{v}\boldsymbol{T})-\hat{\varphi}+\boldsymbol{v}\cdot\rho_R\boldsymbol{g}\,,\\ \rho\frac{\mathrm{d}}{\mathrm{d}t}\left(\frac{\boldsymbol{v}\cdot\boldsymbol{v}}{2}\right)&=\mathrm{div}(\boldsymbol{v}\boldsymbol{t})-\varphi+\boldsymbol{v}\cdot\rho\boldsymbol{g}\,.\end{aligned} \qquad (2.3.43)$$

[19] $\mathrm{div}(\boldsymbol{v}\boldsymbol{t})=\ldots=\mathrm{tr}\left((\mathrm{grad}\,\boldsymbol{v})\boldsymbol{t}^T\right)+\boldsymbol{v}\cdot\mathrm{div}\boldsymbol{t}$.

Because the trace of the product of an antisymmetric and a symmetric tensor vanishes, and as the CAUCHY stress tensor is symmetric, we need to consider only the symmetric part of the velocity gradient,

$$\mathrm{sym}(\mathrm{grad}\,\boldsymbol{v})=\mathrm{sym}(\boldsymbol{L})=\boldsymbol{D}:=\tfrac{1}{2}\left(\boldsymbol{L}+\boldsymbol{L}^T\right)\ \text{ or }\ D_{ij}:=\frac{1}{2}\left(\frac{\partial v_i}{\partial x_j}+\frac{\partial v_j}{\partial x_i}\right).$$

Thus $\mathrm{tr}\left((\mathrm{grad}\,\boldsymbol{v})\boldsymbol{t}\right)=\mathrm{tr}(\boldsymbol{L}\boldsymbol{t})=\mathrm{tr}(\boldsymbol{D}\boldsymbol{t})$.

In these equations, the velocity vector in the LAGRANGE representation is expressed as $\boldsymbol{v} = \hat{\boldsymbol{v}}(\boldsymbol{X}, t)$, and in the EULER representation it is expressed as $\boldsymbol{v} = \bar{\boldsymbol{v}}(\boldsymbol{x}, t)$. These equations are the balance statements for the kinetic energy with the production terms given by $-\hat{\varphi}$ or $-\varphi$; thus, these terms emerge here as annihilations with opposite signs in contrast to the balance equation for the internal energy. This is quite natural, annihilated kinetic energy here appears as production of heat there. Finally, addition of the local balance equations for the internal energy (2.3.41) and for the kinetic energy (2.3.43) results again in the originally formulated conservation law (2.3.39), in which no production term is present.

2.3.5 Entropy Balance

Whilst mass, momentum, angular momentum and energy (with combined mechanical and thermal contributions) are conserved quantities so that their productions in the balance equations are set to zero, the entropy is not a conserved quantity. Regarding its inclusion in the thermomechanical system, it seems at first glance unclear why we require to consider an additional variable – the entropy – for a complete description of thermomechanical phenomena. Experience tells us, however, that the real physical processes are directional, i.e., they can proceed only in a certain chronology but not in the reverse of this. This principle of *irreversibility* can be accounted for by the introduction of the balance statement for entropy, in which one requires that its specific production can always have only one sign for all realistic thermomechanical processes. At the present state of understanding, this phenomenological idea may not provide much insight, but our understanding of this concept will soon become clearer. More precise specifications will be given in due course.

We consider the entropy (or its density) and temperature as primitive variables, that is, we take their existence as unquestioned and postulate a balance statement.

LAGRANGE **Representation** The following quantities are introduced to derive the entropy balance equation in the LAGRANGE representation:

$$\begin{aligned}
\Gamma^{(s)} &= \rho_R s \;, && \text{entropy density (s is the specific entropy),} \\
\Pi^{(s)} &= \rho_R \gamma \;, && \text{entropy production,} \\
\Sigma^{(s)} &= \rho_R \eta \;, && \text{entropy supply,} \\
\boldsymbol{\Phi}^{(s)} &= \boldsymbol{\Phi} \;, && \text{entropy flux.}
\end{aligned} \tag{2.3.44}$$

Substituting these quantities in the LAGRANGE form of the balance equation (2.2.13) and using the mass balance equation ($\mathrm{d}\rho_R/\mathrm{d}t = 0$) we obtain the *balance equation of entropy*

$$\rho_R \frac{\mathrm{d}s}{\mathrm{d}t} = -\operatorname{Div} \boldsymbol{\Phi} + \rho_R \gamma + \rho_R \eta \,. \tag{2.3.45}$$

EULER **Representation** Analogously to the LAGRANGE representation we make the following substitutions:

$$\begin{aligned} \gamma^{(s)} &= \rho s \ , \quad \text{entropy density,} \\ \pi^{(s)} &= \rho\gamma \ , \quad \text{entropy production,} \\ \varsigma^{(s)} &= \rho\eta \ , \quad \text{entropy supply,} \\ \boldsymbol{\phi}^{(s)} &= \boldsymbol{\phi} \ , \quad \text{entropy flux .} \end{aligned} \tag{2.3.46}$$

Using the mass balance (2.3.4) we may deduce the *entropy balance equation* in the EULER representation, i.e.,

$$\rho\frac{\mathrm{d}s}{\mathrm{d}t} = -\mathrm{div}\boldsymbol{\phi} + \rho\gamma + \rho\eta \ . \tag{2.3.47}$$

2.3.6 Second Law of Thermodynamics

Now we intend to scrutinize the entropy balance equation somewhat further and introduce the *second law of thermodynamics*. Besides, we shall prescribe the entropy flux and entropy supply in the same form as given in classical thermodynamics, namely

$$\eta = \frac{\mathfrak{r}}{\Theta} \ , \quad \boldsymbol{\phi} = \frac{\boldsymbol{q}}{\Theta} \quad \text{and} \quad \boldsymbol{\Phi} = \frac{\boldsymbol{Q}}{\Theta} \ . \tag{2.3.48}$$

The entropy supply is thus given by the energy supply (here radiation) divided by the *absolute temperature* Θ, the entropy flux is defined as the heat flux divided by Θ. These assignments can be motivated by classical thermostatics, see any book treating the theory of heat, and are taken over here from these earlier studies as simple axioms. They are reasonable for bodies consisting of only one component. In mixtures one must postulate a more general form of the entropy flux. In many cases, a further vectorial term is added in expressions $(2.3.48)_{2,3}$ such that entropy flux and heat flux are no longer collinear to each other.

Furthermore, it is assumed that for all permissible *thermodynamic processes* – these are the solutions of the balance equations (mass, momentum, energy) and the *material equations* (stress tensor, heat flux, etc.; these will be explained in later chapters) – *the second law of thermodynamics* holds in the form

$$\gamma \geq 0 \ , \tag{2.3.49}$$

i.e., entropy can only be produced, but can never be annihilated.

All thermodynamic processes must thus satisfy the following inequality

$$\begin{aligned} &\text{LAGRANGE:} \qquad \rho_R \frac{\mathrm{d}s}{\mathrm{d}t} + \mathrm{Div}\left(\frac{\boldsymbol{Q}}{\Theta}\right) - \rho_R \frac{\mathfrak{r}}{\Theta} \geq 0 \\ &\text{EULER:} \qquad \rho \frac{\mathrm{d}s}{\mathrm{d}t} + \mathrm{div}\left(\frac{\boldsymbol{q}}{\Theta}\right) - \rho \frac{\mathfrak{r}}{\Theta} \geq 0 \,, \\ &\qquad \forall \text{ thermodynamic processes} \end{aligned} \tag{2.3.50}$$

in the LAGRANGE and EULER description, respectively. This inequality or its corresponding global statement in integral form (2.1.4) is called the CLAUSIUS–DUHEM *inequality.*

The CLAUSIUS–DUHEM inequality is the most popular form of the second law of thermodynamics, but not the only one. According to the entropy principle of MÜLLER, the entropy flux is not a priori related to the heat flux and the absolute temperature, as in (2.3.48), but is rather considered as a material variable of general type and is determined by reduction. Similarly, the entropy supply η is not governed by the radiation and the absolute temperature, but rather determined as a combination of the momentum supply, energy supply and possibly additional supply terms. In many cases relations (2.3.48) are consequences of this general entropy principle. Thus, this principle then at least partly justifies the choice made by the CLAUSIUS–DUHEM relation.

Of greater consequence is, however, the tacit assumption implied by using the CLAUSIUS–DUHEM inequality that *there exists a scalar variable, the absolute temperature Θ, which takes non-negative values, $\Theta \geq 0$, and vanishes only at absolute zero.* All the more, we presuppose here knowledge of the concepts of classical thermostatics for a simple adiabatic system. In such a system, it was shown by CARATHEODORY [42] on the basis of very weak assumptions that a function Θ of the empirical temperature Θ, $\Theta(\theta)$, exists which is independent of the material for which it is defined and furthermore changes monotonically with the degree of coldness, i.e., the empirical temperature. Furthermore, this function changes monotonically with the degree of coldness, Θ. It is evident that $\Theta(\theta)$ possesses some notion of universality, and was therefore called *absolute temperature.* It can be identified with the temperature of an ideal gas, which was shown by LORD KELVIN to equally enjoy universal properties. It possesses zero value at the lowest possible temperature, and for water, it takes the value 273.15 K at the melting point at normal pressure.

The readers are referred to consult HUTTER [104], [109] and MÜLLER [163], [165] for more detailed descriptions of these issues.

2.4 Exercises

1. Let a density preserving fluid flow down an inclined plane (see Fig. 2.3). Its free surface geometry is given by $y = h(x,t)$. The coordinate plane

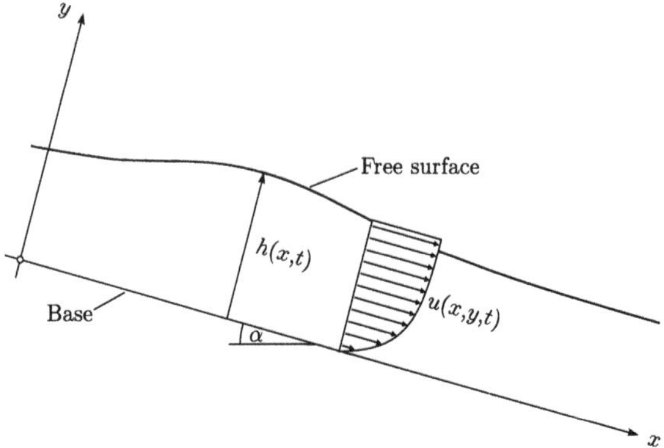

Fig. 2.3. Two dimensional flow of a fluid on an inclined plane. $h(x,t)$ denotes the free surface, (u,v) the velocity components.

$y = 0$ forms the lower boundary. The flow represents plane motion in which the velocity components in the x and y directions are given by u and v, respectively. Prove that the kinematic equation of the free surface $y = h(x,t)$ is given by

$$\frac{\partial h}{\partial t} + \frac{\partial h}{\partial x} u - v = 0 \,. \tag{2.4.1}$$

Then, integrate the continuity equation from $y = 0$ to $y = h(x,t)$ and show that the integration yields

$$\frac{\partial h}{\partial t} + \frac{\partial Q}{\partial x} = 0 \,, \quad Q := \int_0^{h(x,t)} u(x,y,t) dy \,. \tag{2.4.2}$$

In case of $Q = \hat{Q}(h)$, equation (2.4.2) turns into

$$\frac{\partial h}{\partial t} + \hat{C}(h) \frac{\partial h}{\partial x} = 0 \,, \quad \hat{C} := \frac{\mathrm{d}\hat{Q}}{\mathrm{d}h}(h) \,. \tag{2.4.3}$$

Prove that every differentiable F with

$$h = F\Big(x - \hat{C}(h)t\Big) \tag{2.4.4}$$

solves the differential equation (2.4.3). For $t = 0$, $h_0 = F(x)$ and the geometry of the surface is $y = h(x)$.

2. Consider a density preserving fluid in infinite, two dimensional space. The fluid emerges from a nozzle and spreads like a jet or plume (see Fig. 2.4). Let the flow be unsteady and let the spreading of the jet be symmetric to

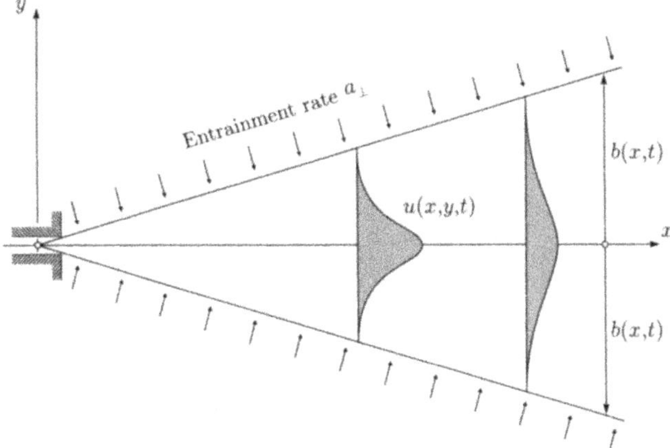

Fig. 2.4. Free jet with semi-width $b(x,t)$ which moves with velocity $\boldsymbol{v}_b$; $a_\perp$ denotes the entrainment flux through the surface.

the x axis. Show that the continuity equation and the kinematic equation for the jet boundary can be written in the form

$$\begin{aligned} \frac{\partial b}{\partial t} + \frac{\partial Q}{\partial x} &= a_\perp \sqrt{1 + \left(\frac{\partial b}{\partial x}\right)^2} = a \,, \\ Q = \hat{Q}(b) &= \int_0^b u(x,y,t)dy \end{aligned} \qquad (2.4.5)$$

through the integration over the half-width of the jet, where $a_\perp$ is the amount of the fluid entering through the jet boundary per unit time and length.

3. Multiply the local momentum balance dyadically with the velocity vector and show that the local balance equation in the EULER representation appears as

$$\begin{aligned} \frac{\partial}{\partial t}\left(\frac{\rho v_i v_j}{2}\right) + \left(\frac{\rho v_i v_j}{2} v_k\right)_{,k} &= \operatorname{sym}\left[(v_i t_{jk})_{,k}\right] \\ &+ \operatorname{sym}(\rho v_i g_j) + \operatorname{sym}(v_{i,k} t_{jk}) \,, \end{aligned} \qquad (2.4.6)$$

where the symmetrization concerns the indices i and j. As a result, the following identifications are valid

$$\begin{aligned}\gamma &= \tfrac{1}{2}\rho v_i v_j \ , \\ \pi &= \operatorname{sym}\left(v_{i,k} t_{jk}\right) \ , \\ \varsigma &= \operatorname{sym}\left(v_i \rho g_j\right) \ , \\ \phi &= -\operatorname{sym}\left(v_i t_{jk}\right) = -\tfrac{1}{2}\left(v_i t_{jk} + v_j t_{ik}\right) \ .\end{aligned} \tag{2.4.7}$$

4. Suppose that the angular momentum density comprises moment of momentum density plus spin, the supply of angular momentum is given by the moment of volume forces plus an intrinsic body couple density and the flux of the angular momentum is built from the moments of the stresses plus couple stresses. The production, however, vanishes. Derive a spin balance equation under these suppositions.
5. Prove that for a system of mass points subject to central forces the angular momentum balance is a consequence of the momentum balance.
6. Analogously to the REYNOLDS transport theorem derive an equivalent expression for the time rate of change of surface and line integrals

$$\Psi = \int_a \boldsymbol{\omega} \cdot \mathrm{d}\boldsymbol{a} \quad \text{and} \quad \psi = \int_l \boldsymbol{b} \cdot \mathrm{d}\boldsymbol{l} \ . \tag{2.4.8}$$

7. Derive a balance equation for the vorticity vector $\boldsymbol{\omega} := \operatorname{curl} \boldsymbol{v}$, valid for a density preserving, inviscid fluid ($\boldsymbol{t} = -p\boldsymbol{I}$).
8. For an elastic or ideal fluid, i.e., a barotropic fluid, $\rho = \rho(p)$, subject to conservative forces derive the HELMHOLTZ vorticity theorem and show that it has the form

$$\dot{\boldsymbol{\xi}} = (\operatorname{grad} \boldsymbol{v})\boldsymbol{\xi} \ , \qquad \boldsymbol{\xi} = \frac{\operatorname{curl} \boldsymbol{v}}{\rho} \ . \tag{2.4.9}$$

   Further show that with

$$\boldsymbol{\xi} = \boldsymbol{F}\, \boldsymbol{\xi}_R \ , \tag{2.4.10}$$

   (2.4.9) is integrated; $\boldsymbol{\xi}_R$ is the vector of the specific vorticity in the reference configuration. Using the polar decomposition of $\boldsymbol{F}$, prove that a material vortex filament is stretched and rotated during its motion.
9. Let $\boldsymbol{\Omega}$ be the constant angular velocity of a rotating coordinate system in an inertial system. Show that for a barotropic ideal fluid the HELMHOLTZ vorticity law has the form

$$\dot{\boldsymbol{\xi}}_a = \boldsymbol{L}\, \boldsymbol{\xi}_a \tag{2.4.11}$$

   where

$$\dot{\boldsymbol{\xi}}_a = \boldsymbol{\xi} + \frac{2\boldsymbol{\Omega}}{\rho} \tag{2.4.12}$$

   is the *absolute* vorticity vector.

10. Let λ and ψ_λ be scalar physical variables, which are assumed differentiable and let the evolution equation

$$\frac{\mathrm{d}\lambda}{\mathrm{d}t} = \psi_\lambda \tag{2.4.13}$$

be given. Further, let $\boldsymbol{\xi} = (\operatorname{curl} \boldsymbol{v} + 2\boldsymbol{\Omega})/\rho$ be the absolute vorticity vector per unit mass, see Exercise 9. Call

$$\pi_\lambda := \boldsymbol{\xi}_a \cdot \operatorname{grad} \lambda \,, \tag{2.4.14}$$

the *absolute potential vorticity for parameter* λ. Then show that for a barotropic ideal fluid the evolution equation

$$\frac{\mathrm{d}\pi_\lambda}{\mathrm{d}t} = \boldsymbol{\xi}_a \cdot \operatorname{grad} \psi_\lambda \tag{2.4.15}$$

holds. With $\operatorname{grad} \psi_\lambda = 0$, this equation yields

$$\frac{\mathrm{d}\pi_\lambda}{\mathrm{d}t} = 0 \,. \tag{2.4.16}$$

The potential vorticity is a materially conserved variable as long as ψ_λ is only a function of time.

11. How does the momentum balance equation change when we allow a production term for mass density ? This must be particularly considered when treating mixtures, in which one component can interact with any other.

2.5 Solutions

1. Since $F_s \equiv h(x,t) - y = 0$ represents the equation for the free surface for all times $t \in [0, \infty)$, $dF_s/dt \equiv 0$ must also hold, which directly leads to (2.4.1),

$$\begin{aligned} 0 = \frac{\mathrm{d}F_s}{\mathrm{d}t} &= \frac{\mathrm{d}h(x,t)}{\mathrm{d}t} - \frac{\mathrm{d}y}{\mathrm{d}t} \\ &= \frac{\partial h}{\partial t} + \frac{\partial h}{\partial x}\frac{\mathrm{d}x}{\mathrm{d}t} - \frac{\mathrm{d}y}{\mathrm{d}t} = \frac{\partial h}{\partial t} + \frac{\partial h}{\partial x} u - v \,, \end{aligned} \tag{2.5.1}$$

with $u = \dot{x}$ and $v = \dot{y}$ the velocities in the x and y directions, respectively. Integrating the continuity equation, $\operatorname{div} \boldsymbol{v} = 0$, from $y = 0$ to $y = h(x,t)$ and using the Leibniz rule (2.2.10), one gets

$$\begin{aligned} &\int_0^{h(x,t)} \left(\frac{\partial u}{\partial x} + \frac{\partial v}{\partial y} \right) \mathrm{d}y = 0 \qquad \Longrightarrow \\ &\frac{\partial}{\partial x} \int_0^{h(x,t)} u \, \mathrm{d}y - u(h) \frac{\partial h}{\partial x} + v(h) - v(0) = 0 \,. \end{aligned} \tag{2.5.2}$$

At the base the velocity must be tangential; thus $v(0) = 0$. On the other hand, for $y = h(x,t)$ (2.5.1) holds, so that one obtains from (2.5.2)

$$\frac{\partial h}{\partial t} + \frac{\partial}{\partial x} \int_0^{h(x,t)} u(x,y,t) \mathrm{d}y = 0 \, . \tag{2.5.3}$$

Using the definition of Q, the statement (2.4.2) follows from (2.5.3). With these relations one immediately obtains (2.4.3). Substituting these results into equation (2.4.4), equation (2.5.3) is satisfied, then with (2.4.4) follows the result

$$\begin{aligned} \frac{\partial h}{\partial t} + \hat{C}\frac{\partial h}{\partial x} &= \frac{\partial F(x-\hat{C}t)}{\partial t} + \hat{C}\frac{\partial F(x-\hat{C}t)}{\partial x} \\ &= F'\frac{\partial (x-\hat{C}t)}{\partial t} + \hat{C}F'\frac{\partial (x-\hat{C}t)}{\partial x} \\ &= -F'\hat{C} + F'\hat{C} \\ &= 0 \, . \end{aligned} \tag{2.5.4}$$

Equation (2.4.4) is known as the *kinematic wave equation.*

2. We imagine that the fluid jet is coloured so that its boundary $F_b = b(x,t) - y \equiv 0$ is clearly visible. $F_b \equiv 0$ holds for all time, and so we also have $\mathrm{d}F_b/\mathrm{d}t = 0$, where $\mathrm{d}/\mathrm{d}t$ represents the time rate of change following the motion of the jet boundary. Thus we have

$$\frac{\mathrm{d}F_b}{\mathrm{d}t} = \frac{\partial b}{\partial t} + \frac{\partial b}{\partial x}u_b - v_b = 0 \, . \tag{2.5.5}$$

A simple experiment shows that the velocity $\boldsymbol{v}_b = (u_b, v_b)$ of the boundary does not resemble the velocity v of a fluid particle at the time when this particle occupies the same position. The jet is diluted because of the fluid entering from the surrounding medium.

One can also write equation (2.5.5) as

$$\frac{\partial b}{\partial t} + \frac{\partial b}{\partial x}u - v = \frac{\partial b}{\partial x}(u - u_b) - (v - v_b) \, , \tag{2.5.6}$$

where u, v represent the components of the material-velocity vector. With the exterior unit normal vector $\boldsymbol{n}$,

$$\boldsymbol{n} = \left(-\frac{\partial b}{\partial x}, 1\right) \Big/ \sqrt{1 + \left(\frac{\partial b}{\partial x}\right)^2} \, , \tag{2.5.7}$$

(2.5.6) turns into

$$\frac{\partial b}{\partial t} + \frac{\partial b}{\partial x}u - v = \underbrace{\left(-(\boldsymbol{v} - \boldsymbol{v}_b) \cdot \boldsymbol{n}\right)}_{=:a_\perp} \sqrt{1 + \left(\frac{\partial b}{\partial x}\right)^2} \, , \tag{2.5.8}$$

which is valid for the jet boundary, $y = b(x,t)$. The variable $a_\perp$ is the areal flux of the surrounding fluid into the jet per unit jet-boundary length (dimensionally a velocity !). $a_\perp$ is known in fluid mechanics as *entrainment rate*. Next we consider the continuity equation $\partial u/\partial x + \partial v/\partial y = 0$ within the jet and integrate this from $y = 0$ to $y = b(x,t)$; then using the LEIBNIZ rule we obtain

$$\frac{\partial}{\partial x}\int_0^b u(x,y,t)\mathrm{d}y - \frac{\partial b(x,y,t)}{\partial x}u(x,b,t) + v(x,b,t) - v(x,0,t) = 0\,. \tag{2.5.9}$$

From symmetry reasons we have $v(x,0,t) = 0$, so $v(x,b,t)$ can be eliminated from (2.5.9) with the help of (2.5.8). One obtains

$$\frac{\partial b}{\partial t} + \frac{\partial Q}{\partial x} = a_\perp\sqrt{1+\left(\frac{\partial b}{\partial x}\right)^2}\,, \quad Q := \int_0^{b(x,t)} u(x,y,t)dy\,, \tag{2.5.10}$$

and hence formula (2.4.5).

3. We exploit the component expressions for the solution of this exercise. We multiply the i-th component of the momentum balance (2.3.19) by v_j; next we repeat this statement but with indices i and j interchanged and then add the two equations: $v_i \cdot (\text{momentum balance})_j + v_j \cdot (\text{momentum balance})_i$[20],

$$v_i\Big(\frac{\partial \rho v_j}{\partial t} + (\rho v_j v_k)_{,k} - t_{jk,k} - \rho g_j\Big) + v_j\Big(\frac{\partial \rho v_i}{\partial t} + (\rho v_i v_k)_{,k} - t_{ik,k} - \rho g_i\Big) = 0\,. \tag{2.5.11}$$

After a short calculation we obtain

$$\frac{\partial}{\partial t}(\rho v_i v_j) + (\rho v_i v_j v_k)_{,k} - 2\,\mathrm{sym}\left[(v_i t_{jk})_{,k}\right] - 2\,\mathrm{sym}\left(v_{i,k}t_{jk}\right) - 2\,\mathrm{sym}\left(\rho v_i g_j\right) + v_i v_j \underbrace{\left(\frac{\partial \rho}{\partial t} + (\rho v_k)_{,k}\right)}_{=0 \text{ mass balance}} = 0\,, \tag{2.5.12}$$

or

$$\frac{\partial}{\partial t}\left(\rho\frac{v_i v_j}{2}\right) + \left(\rho\frac{v_i v_j}{2}v_k\right)_{,k} = \mathrm{sym}\left[(v_i t_{jk})_{,k}\right] + \mathrm{sym}\left(v_{i,k}t_{jk}\right) + \mathrm{sym}\left(\rho v_i g_j\right)\,. \tag{2.5.13}$$

[20] Here $\mathrm{sym}\,\boldsymbol{A} := \frac{1}{2}\left(\boldsymbol{A} + \boldsymbol{A}^T\right)$ is the symmetric part of the tensor $\boldsymbol{A}$; in component form this reads $(\mathrm{sym}\,\boldsymbol{A})_{ij} = \frac{1}{2}\left(A_{ij} + A_{ji}\right)$.

Through the contraction of the indices i and j this becomes

$$\frac{\partial}{\partial t}\left(\rho\frac{v_i v_i}{2}\right)+\left(\rho\frac{v_i v_i}{2}v_k\right)_{,k}=\left(v_i t_{ik}\right)_{,k}+v_{i,k}t_{ik}+\rho v_i g_i\ , \tag{2.5.14}$$

or in symbolic form

$$\frac{\partial}{\partial t}\left(\rho\frac{\boldsymbol{v}\cdot\boldsymbol{v}}{2}\right)+\operatorname{div}\left(\frac{\rho\boldsymbol{v}\cdot\boldsymbol{v}}{2}\boldsymbol{v}\right)$$

$$=\operatorname{div}(\boldsymbol{v}\boldsymbol{t})-\operatorname{tr}(\operatorname{grad}\boldsymbol{v}\,\boldsymbol{t})+\boldsymbol{v}\cdot\rho\boldsymbol{g}\ , \tag{2.5.15}$$

which is the local balance of the kinetic energy (2.3.43). Here $\gamma=(\frac{1}{2}\rho)\boldsymbol{v}\cdot\boldsymbol{v}$ is the physical quantity (the kinetic energy), whose time variation is balanced by the production $\pi=-\operatorname{tr}(\operatorname{grad}\boldsymbol{v}\ \boldsymbol{t})$, the supply $\varsigma=\rho\boldsymbol{v}\cdot\boldsymbol{g}$ and the flux $\phi=-\boldsymbol{v}\boldsymbol{t}$ of the energy density.

4. Let the following identities hold for the EULER representation

$$\gamma^{\hat{D}}=\boldsymbol{x}\times\rho\boldsymbol{v}+\rho\tilde{\boldsymbol{s}}\ ,\qquad \tilde{\boldsymbol{s}}=\text{spin density},$$

$$\pi^{\hat{D}}=0\ ,$$

$$\varsigma^{\hat{D}}=\boldsymbol{x}\times\rho\boldsymbol{g}+\rho\boldsymbol{l}\ ,\qquad \boldsymbol{l}=\text{intrinsic body couple density},$$

$$\phi^{\hat{D}}=-\boldsymbol{x}\times\boldsymbol{t}-\boldsymbol{m}\ ,\qquad \boldsymbol{m}=\text{couple stresses}.$$

Substituting these expressions in the general balance equation (2.2.16) yields

$$(\boldsymbol{x}\times\boldsymbol{v})\underbrace{\left(\frac{\mathrm{d}\rho}{\mathrm{d}t}+\rho\operatorname{div}\boldsymbol{v}\right)}_{=0\text{ (mass balance)}}+\rho\underbrace{\boldsymbol{v}\times\boldsymbol{v}}_{=0}+\rho\frac{\mathrm{d}\tilde{\boldsymbol{s}}}{\mathrm{d}t}+\tilde{\boldsymbol{s}}\underbrace{\left(\frac{\mathrm{d}\rho}{\mathrm{d}t}+\rho\operatorname{div}\boldsymbol{v}\right)}_{=0\text{ (mass balance)}}$$

$$+\boldsymbol{x}\times\underbrace{\left(\rho\frac{\mathrm{d}\boldsymbol{v}}{\mathrm{d}t}-\operatorname{div}\boldsymbol{t}-\rho\boldsymbol{g}\right)}_{=0\text{ (momentum balance)}}=\operatorname{div}\boldsymbol{m}+\rho\boldsymbol{l}+\boldsymbol{t}^*\ , \tag{2.5.16}$$

from which one obtains the spin balance

$$\rho\frac{\mathrm{d}\tilde{\boldsymbol{s}}}{\mathrm{d}t}=\operatorname{div}\boldsymbol{m}+\rho\boldsymbol{l}+\boldsymbol{t}^*\ . \tag{2.5.17}$$

Here, $\boldsymbol{t}^*=\operatorname{dual}(\operatorname{skw}\,\boldsymbol{t})$ — see (2.3.29) — the dual-vector of the skew symmetric part of the stress tensor. Equation (2.5.17) is the required spin balance. We recognize that the CAUCHY stress tensor is symmetric only when $\tilde{\boldsymbol{s}}=\boldsymbol{0},\boldsymbol{m}=\boldsymbol{0}$ and $\boldsymbol{l}=\boldsymbol{0}$. In all other cases we expect skew symmetric parts.

5. Let m_i be the mass of a mass point i and $\boldsymbol{x}_i$ its position vector. Further, $\boldsymbol{K}_i$ is the force experienced from outside and $\boldsymbol{K}_{ij}$ is that experienced by the mass point i due to the mass point j, where the reaction principle $\boldsymbol{K}_{ij} = -\boldsymbol{K}_{ji}$ holds. Recalling NEWTON's law for a mass point i we have

$$m_i \ddot{\boldsymbol{x}}_i = \boldsymbol{K}_i + \sum_{\substack{j=1 \\ j \neq i}}^{N} \boldsymbol{K}_{ij} \,, \quad (i = 1, \cdots, N) \,. \tag{2.5.18}$$

N is the number of mass points in the system. Summation over all mass points results in

$$\sum_{i=1}^{N} m_i \ddot{\boldsymbol{x}}_i = \sum_{i=1}^{N} \boldsymbol{K}_i + \underbrace{\sum_{i=1}^{N} \sum_{\substack{j=1 \\ j \neq i}}^{N} \boldsymbol{K}_{ij}}_{=0} = \sum_{i=1}^{N} \boldsymbol{K}_i \tag{2.5.19}$$

because of the reaction principle.
The angular momentum of the system of mass points with respect to the point $\mathcal{O}$ is the angular momentum of all the mass points, namely

$$\hat{\boldsymbol{D}}^0 = \sum_{i=1}^{N} \boldsymbol{x}_i \times m_i \dot{\boldsymbol{x}}_i \,.$$

Its time rate of change is given by

$$\left(\hat{\boldsymbol{D}}^0\right)^{\cdot} = \sum_{i=1}^{N} (\boldsymbol{x}_i \times m_i \ddot{\boldsymbol{x}}_i + \dot{\boldsymbol{x}}_i \times m_i \dot{\boldsymbol{x}}_i) = \sum_{i=1}^{N} \boldsymbol{x}_i \times m_i \ddot{\boldsymbol{x}}_i \,. \tag{2.5.20}$$

Applying NEWTON's law, (2.5.18), we obtain from this

$$\left(\hat{\boldsymbol{D}}^0\right)^{\cdot} = \sum_{i=1}^{N} \boldsymbol{x}_i \times \big(\boldsymbol{K}_i + \sum_{\substack{j=1 \\ j \neq i}}^{N} \boldsymbol{K}_{ij}\big) = \sum_{i=1}^{N} \boldsymbol{x}_i \times \boldsymbol{K}_i \tag{2.5.21}$$

or if $\boldsymbol{M}_i^0 = \boldsymbol{x}_i \times \boldsymbol{K}_i$ is the moment of the exterior force acting at the mass point i,

$$\left(\hat{\boldsymbol{D}}^0\right)^{\cdot} = \sum_{i=1}^{N} \boldsymbol{M}_i^0 = \boldsymbol{M}^0 \,,$$

which is exactly the expression of the balance of angular momentum of the system. Finally, the moment of the internal forces $\boldsymbol{K}_{ij}$ vanishes, because for every mass point the forces $\boldsymbol{K}_{ij} = -\boldsymbol{K}_{ji}$ appear in pairs with opposite signs. If this were not the case then the angular momentum balance would not have resulted from the momentum balance of the system.

6. The REYNOLDS transport theorem (2.2.9) deals with the time rate of change of a physical variable which can be expressed as a volume integral of a density (e.g. mass density ρ, momentum density $\rho\boldsymbol{v}$). Similarly, one can also derive the time rate of change of a variable which is associated with surface or line densities.

   a) Let Ψ be a variable, which is given by the surface integral of the vector field $\boldsymbol{\omega}$

$$\Psi := \int\limits_{\partial\omega(t)} \boldsymbol{\omega}\cdot \mathrm{d}\boldsymbol{a}\,.$$

The time rate of change $\mathrm{d}\Psi/\mathrm{d}t$ can be worked out as in REYNOLDS transport theorem by interchanging the differentiation and integration sequence, however, for this one must transform the temporally changing integration domain $\partial\omega(t)$ from the present to the reference configuration. Thus, using the transformation rule for the surface elements $\mathrm{d}\boldsymbol{a} = J\boldsymbol{F}^{-T}\mathrm{d}\boldsymbol{A}$ we obtain

$$\begin{aligned}\frac{\mathrm{d}\Psi}{\mathrm{d}t} &= \frac{\mathrm{d}}{\mathrm{d}t}\int\limits_{\partial\omega}\boldsymbol{\omega}\cdot\mathrm{d}\boldsymbol{a} = \frac{\mathrm{d}}{\mathrm{d}t}\int\limits_{\partial\Omega}\hat{\boldsymbol{\omega}}\cdot J\boldsymbol{F}^{-T}\mathrm{d}\boldsymbol{A}\\ &= \int\limits_{\partial\Omega}\Big(\frac{\partial\hat{\boldsymbol{\omega}}}{\partial t}\cdot J\boldsymbol{F}^{-T} + \hat{\boldsymbol{\omega}}\cdot\frac{\partial\hat{J}}{\partial t}\boldsymbol{F}^{-T} + \hat{\boldsymbol{\omega}}\cdot J\frac{\partial\boldsymbol{F}^{-T}}{\partial t}\Big)\mathrm{d}\boldsymbol{A}\,,\end{aligned} \tag{2.5.22}$$

wherein $\boldsymbol{\omega} = \hat{\boldsymbol{\omega}}(\boldsymbol{X},\,t)$ as well as $J = \hat{J}(\boldsymbol{X},\,t)$. The time derivative of the determinant of deformation gradient is $\dot{J} = J\mathrm{div}\boldsymbol{v}$ – see (1.4.11). From the identity $\boldsymbol{F}\boldsymbol{F}^{-1} = \boldsymbol{I}$, the time derivative yields

$$\dot{\boldsymbol{F}}\boldsymbol{F}^{-1} = -\boldsymbol{F}(\boldsymbol{F}^{-1})^{\boldsymbol{\cdot}}$$

and with $\dot{\boldsymbol{F}} = \boldsymbol{L}\boldsymbol{F}$, see (1.4.10), this implies

$$\left(\boldsymbol{F}^{-T}\right)^{\boldsymbol{\cdot}} = -(\boldsymbol{F}^{-1}\boldsymbol{L})^{T} = -\boldsymbol{L}^{T}\boldsymbol{F}^{-T}\,.$$

Substituting this in (2.5.22) and transforming again to the present configuration one obtains

$$\begin{aligned}\frac{\mathrm{d}\Psi}{\mathrm{d}t} &= \int\limits_{\partial\Omega}\Big[(\dot{\hat{\boldsymbol{\omega}}} + \hat{\boldsymbol{\omega}}\mathrm{div}\boldsymbol{v})\cdot J\boldsymbol{F}^{-T} - \underbrace{\hat{\boldsymbol{\omega}}\cdot\boldsymbol{L}^{T}J\boldsymbol{F}^{-T}}_{\boldsymbol{L}\hat{\boldsymbol{\omega}}\cdot J\boldsymbol{F}^{-T}}\Big]\mathrm{d}\boldsymbol{A}\\ &= \int\limits_{\partial\omega}\big[\dot{\boldsymbol{\omega}} + (\mathrm{div}\boldsymbol{v}\,\boldsymbol{I} - \mathrm{grad}\,\boldsymbol{v})\boldsymbol{\omega}\big]\cdot\mathrm{d}\boldsymbol{a}\,.\end{aligned} \tag{2.5.23}$$

With the help of the identity

$$\mathrm{curl}\,(\boldsymbol{\omega}\times\boldsymbol{v}) = (\mathrm{div}\boldsymbol{v}\,\boldsymbol{I} - \mathrm{grad}\,\boldsymbol{v})\boldsymbol{\omega} - (\mathrm{div}\boldsymbol{\omega}\,\boldsymbol{I} - \mathrm{grad}\,\boldsymbol{\omega})\boldsymbol{v}$$

the above equation can be rewritten as

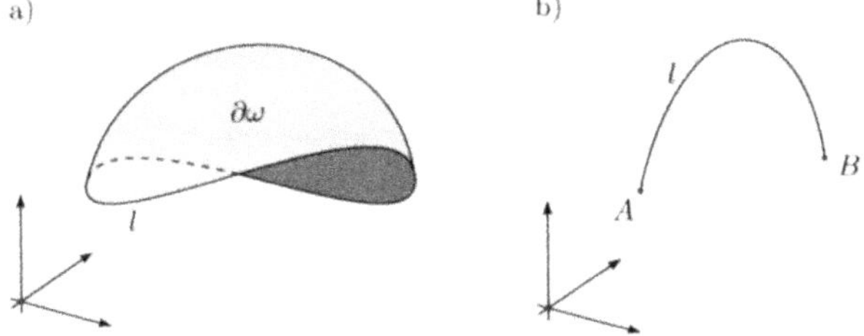

Fig. 2.5. a) Open surface $\partial\omega$ in three dimensional space with closed boundary, l. b) Line segment l between A and B.

$$\frac{\mathrm{d}\Psi}{\mathrm{d}t} = \frac{\mathrm{d}}{\mathrm{d}t}\int\limits_{\partial\omega} \boldsymbol{\omega}\cdot\mathrm{d}\boldsymbol{a}$$

$$= \int\limits_{\partial\omega}\Big[\frac{\partial\boldsymbol{\omega}}{\partial t} + \boldsymbol{v}\mathrm{div}\boldsymbol{\omega} + \mathrm{curl}\,(\boldsymbol{\omega}\times\boldsymbol{v})\Big]\cdot\mathrm{d}\boldsymbol{a}\,.$$

Now, consider an open surface $\partial\omega = a$ with a closed boundary l (see Fig. 2.5a), then using STOKES integral law this equation can be further modified to yield

$$\frac{\mathrm{d}\Psi}{\mathrm{d}t} = \frac{\mathrm{d}}{\mathrm{d}t}\int\limits_{\partial\omega} \boldsymbol{\omega}\cdot\mathrm{d}\boldsymbol{a} \tag{2.5.24}$$

$$= \int\limits_{\partial\omega}\left(\frac{\partial\boldsymbol{\omega}}{\partial t} + \boldsymbol{v}\mathrm{div}\boldsymbol{\omega}\right)\cdot\mathrm{d}\boldsymbol{a} + \oint_l(\boldsymbol{\omega}\times\boldsymbol{v})\cdot\mathrm{d}\boldsymbol{l}\,.$$

b) The time derivative of a line integral over the closed curve l (in the present configuration) is also derived in a similar manner. Let the line integral be given by

$$\psi = \int\limits_l \boldsymbol{b}\cdot\mathrm{d}\boldsymbol{x}\,,$$

where the curve l must not necessarily be a closed curve – it has end points, which shall be indicated by the indices $()_A$ and $()_B$.
The time rate of change of ψ can be calculated by interchanging differentiation and integration once the transformation into the reference configuration (l_o is the curve in the reference configuration) has been performed,

$$\frac{\mathrm{d}\psi}{\mathrm{d}t} = \frac{\mathrm{d}}{\mathrm{d}t}\int_{l} \boldsymbol{b}\cdot \mathrm{d}\boldsymbol{x} = \frac{\mathrm{d}}{\mathrm{d}t}\int_{l_o} \boldsymbol{b}\cdot \boldsymbol{F}\mathrm{d}\boldsymbol{X} = \int_{l_o} \frac{\partial}{\partial t}(\boldsymbol{b}\cdot\boldsymbol{F})\mathrm{d}\boldsymbol{X}$$

$$= \int_{l_o} (\dot{\boldsymbol{b}}\cdot\boldsymbol{F} + \boldsymbol{b}\cdot\dot{\boldsymbol{F}})\mathrm{d}\boldsymbol{X} = \int_{l} (\dot{\boldsymbol{b}} + \boldsymbol{b}\boldsymbol{L})\cdot \mathrm{d}\boldsymbol{x}$$

$$= \int_{l} \dot{\boldsymbol{b}}\cdot \mathrm{d}\boldsymbol{x} + \int_{l} \boldsymbol{b}\cdot \mathrm{d}\boldsymbol{v} ,$$

where $\boldsymbol{L}\mathrm{d}\boldsymbol{x} = \mathrm{d}\boldsymbol{v}$ is the velocity increment.
Replacing the vector field by the velocity, $\boldsymbol{b} = \boldsymbol{v}$, and choosing a closed curve for l, we obtain

$$\frac{\mathrm{d}\psi}{\mathrm{d}t} = \frac{\mathrm{d}}{\mathrm{d}t}\oint_{l} \boldsymbol{v}\cdot \mathrm{d}\boldsymbol{x} = \oint_{l} \dot{\boldsymbol{v}}\cdot \mathrm{d}\boldsymbol{x} , \tag{2.5.25}$$

since the second integral vanishes. One often defines the variable $\psi = \oint_{l} \boldsymbol{v}\cdot\mathrm{d}\boldsymbol{x}$ as the *circulation* of the velocity field along the closed curve l.
With the help of the integrals (2.5.24) and (2.5.25) one can derive the so called *vorticity law*. For example, equation (2.5.25) leads to

KELVIN's *circulation theorem:* If the material acceleration can be written as a gradient of a potential P, $\dot{\boldsymbol{v}} = \operatorname{grad} P$, then the circulation around any closed material line is conserved:

$$\frac{\mathrm{d}\psi}{\mathrm{d}t} = \int_{l} \dot{\boldsymbol{v}}\cdot \mathrm{d}\boldsymbol{x} = \oint_{l} \operatorname{grad} P\cdot \mathrm{d}\boldsymbol{x} = \oint_{l} \mathrm{d}P = 0 . \tag{2.5.26}$$

This is valid, e.g. in case of a density preserving, ideal fluid ($\boldsymbol{t} = -p\boldsymbol{I}$), $\mathrm{d}\boldsymbol{v}/\mathrm{d}t = -\operatorname{grad}(p/\rho + U)$ with $\boldsymbol{g} = -g\operatorname{grad} U$ as external potential field.
Under the same prerequisites HELMHOLTZ's vorticity balance can be derived, which is generally done in hydrodynamics.

7. The solution of this problem is carried out through the application of the curl operator in the desired equation or with the aid of the result from the last exercise. As in the previous case, under the requirement of a closed curve l, the circulation $\psi = \oint_{l} \boldsymbol{v}\cdot\mathrm{d}\boldsymbol{x}$ remains conserved. Thus

$$\frac{\mathrm{d}}{\mathrm{d}t}\int_{a} \boldsymbol{\omega}\cdot \mathrm{d}\boldsymbol{a} = \frac{\mathrm{d}}{\mathrm{d}t}\int_{a} \operatorname{curl}\boldsymbol{v}\cdot \mathrm{d}\boldsymbol{a} = \frac{\mathrm{d}}{\mathrm{d}t}\oint_{l} \boldsymbol{v}\cdot \mathrm{d}\boldsymbol{x} = 0 , \tag{2.5.27}$$

where a is the surface enclosed by the closed curve l. The surface integral over the vorticity vector is also a conserved variable. With equation (2.5.23) one can write (since $\operatorname{div}\boldsymbol{v} = 0$, here)

$$\int_a \left(\dot{\boldsymbol{\omega}} - (\operatorname{grad}\boldsymbol{v})\boldsymbol{\omega}\right) \cdot \mathrm{d}\boldsymbol{a} = 0$$

as a global balance equation, or with the requirement of the continuity of the integrand, this holds for an arbitrary domain a, implying

$$\dot{\boldsymbol{\omega}} = (\operatorname{grad}\boldsymbol{v})\boldsymbol{\omega} .$$

This is the HELMHOLTZ *vorticity transport equation* for a density preserving fluid.

8. We start with the momentum balance in the form

$$\frac{\mathrm{d}\boldsymbol{v}}{\mathrm{d}t} = \frac{\partial \boldsymbol{v}}{\partial t} + \operatorname{grad}|\frac{\boldsymbol{v}^2}{2}| - \boldsymbol{v} \times \operatorname{curl}\boldsymbol{v} = -\frac{1}{\rho}\operatorname{grad}p - \operatorname{grad}U \tag{2.5.28}$$

and build the curl of this to obtain

$$\frac{\partial \boldsymbol{\omega}}{\partial t} - \underbrace{\operatorname{curl}(\boldsymbol{v}\times\boldsymbol{\omega})}_{(i)} = \underbrace{-\operatorname{curl}\left(\frac{1}{\rho}\operatorname{grad}p\right)}_{(ii)} , \tag{2.5.29}$$

where the vorticity vector $\boldsymbol{\omega} = \operatorname{curl}\boldsymbol{v}$ is introduced. The terms indicated by (i) and (ii) are easily figured out in Cartesian index notation:

$$\begin{aligned}
(i) &= \epsilon_{ijk}(\epsilon_{klm}v_l\omega_m)_{,j} \\
&= [(\delta_{il}\delta_{jm} - \delta_{im}\delta_{jl})v_l\omega_m]_{,j} \\
&= (v_i\omega_j - v_j\omega_i)_{,j} \\
&= v_{i,j}\omega_j + v_i\omega_{j,j} - v_{j,j}\omega_i - v_j\omega_{i,j} \\
&= v_{i,j}\omega_j + \frac{1}{\rho}\frac{\mathrm{d}\rho}{\mathrm{d}t}\omega_i - \omega_{i,j}v_j ,
\end{aligned}$$

where $\omega_{j,j} \equiv 0$ and the mass balance equation has been used,

$$(ii) = \epsilon_{ijk}\left(\frac{1}{\rho}p_{,k}\right)_{,j} = \epsilon_{ijk}\left(\frac{1}{\rho}\right)_{,j} p_{,k} .$$

Substituting these results above and reverting again to symbolic notation yields

$$\frac{\partial \boldsymbol{\omega}}{\partial t} + (\operatorname{grad}\boldsymbol{\omega})\boldsymbol{v} - \frac{1}{\rho}\frac{\mathrm{d}\rho}{\mathrm{d}t}\boldsymbol{\omega} - (\operatorname{grad}\boldsymbol{v})\boldsymbol{\omega} = -\operatorname{grad}\left(\frac{1}{\rho}\right) \times \operatorname{grad}p ,$$

or dividing by ρ gives

$$\frac{\mathrm{d}}{\mathrm{d}t}\left(\frac{\boldsymbol{\omega}}{\rho}\right) = (\operatorname{grad}\boldsymbol{v})\frac{\boldsymbol{\omega}}{\rho} + \frac{\operatorname{grad}\rho \times \operatorname{grad}p}{\rho^3} . \tag{2.5.30}$$

Now let us define

$$\boldsymbol{\xi} := \frac{\boldsymbol{\omega}}{\rho} \quad \text{(vorticity per unit mass)} \tag{2.5.31}$$

and consider a barotropic fluid $p = p(\rho)$, then $\operatorname{grad}\rho$ and $\operatorname{grad}p$ are parallel and (2.5.30) takes the form

$$\frac{\mathrm{d}\boldsymbol{\xi}}{\mathrm{d}t} = (\operatorname{grad}\boldsymbol{v})\boldsymbol{\xi} = \boldsymbol{L}\boldsymbol{\xi} . \tag{2.5.32}$$

This is nothing but the HELMHOLTZ vorticity transport equation for a compressible, barotropic fluid. Let $\boldsymbol{\xi}_R$ be the specific vorticity in the reference configuration, then

$$\boldsymbol{\xi} = \boldsymbol{F}\boldsymbol{\xi}_R \tag{2.5.33}$$

necessarily solves (2.5.32) since

$$\dot{\boldsymbol{\xi}} = \dot{\boldsymbol{F}}\boldsymbol{\xi}_R = (\underbrace{\dot{\boldsymbol{F}}\boldsymbol{F}^{-1}}_{\boldsymbol{L}})\boldsymbol{F}\boldsymbol{\xi}_R = \boldsymbol{L}\boldsymbol{\xi} , \tag{2.5.34}$$

with which equation (2.5.32) is proved. Using the polar decomposition $\boldsymbol{F} = \boldsymbol{R}\boldsymbol{U}$ and the decomposition $\boldsymbol{L} = \boldsymbol{D} + \boldsymbol{W}$ with the usual definitions of $\boldsymbol{R}$, $\boldsymbol{U}$, $\boldsymbol{D}$, $\boldsymbol{W}$, we obtain

$$\dot{\boldsymbol{\xi}} = \boldsymbol{D}\boldsymbol{\xi} + \boldsymbol{W}\boldsymbol{\xi} \quad \text{or} \quad \boldsymbol{\xi} = \boldsymbol{R}(\boldsymbol{U}\boldsymbol{\xi}_R) . \tag{2.5.35}$$

Thus the vortex filaments are stretched as well as rotated. The left side of (2.5.35) demonstrates this in differential form and is additive whereas the right side expresses it in the integrated form and appears in product form, thus exhibiting stretching and rotation one after the other. Since $\boldsymbol{R}\boldsymbol{U} = \boldsymbol{V}\boldsymbol{R}$ their sequence can also be interchanged. This behaviour is called vortex stretching and vortex tilting .
In plane motion, the velocity vector is always perpendicular to the vorticity vector, thus $\dot{\boldsymbol{\xi}} = \boldsymbol{0}$. The vortex filaments in plane motion are therefore neither rotated nor stretched.

9. The momentum balance of a barotropic ideal fluid in a coordinate system rotating with constant angular velocity has the form

$$\begin{aligned} &\frac{\partial \boldsymbol{v}}{\partial t} + \operatorname{grad}\left(|\frac{\boldsymbol{v}\cdot\boldsymbol{v}}{2}|\right) - \boldsymbol{v} \times (\operatorname{curl}\boldsymbol{v} + 2\boldsymbol{\Omega}) \\ &\quad + \boldsymbol{\Omega} \times (\boldsymbol{\Omega} \times \boldsymbol{x}) = -\frac{1}{\rho}\operatorname{grad}p . \end{aligned} \tag{2.5.36}$$

Since

$$\boldsymbol{\Omega} \times (\boldsymbol{\Omega} \times \boldsymbol{x}) = (\boldsymbol{\Omega}(\boldsymbol{\Omega} \cdot \boldsymbol{x}) - |\boldsymbol{\Omega}|^2 \boldsymbol{x}) = -\operatorname{grad}\left(\frac{\Phi_\Omega}{2}\right) \tag{2.5.37}$$

with

$$\Phi_\Omega := (\boldsymbol{\Omega} \cdot \boldsymbol{\Omega})(\boldsymbol{x} \cdot \boldsymbol{x}) - (\boldsymbol{\Omega} \cdot \boldsymbol{x})^2 , \tag{2.5.38}$$

there follows from (2.5.36)

$$\frac{\partial \boldsymbol{v}}{\partial t} + \operatorname{grad}[\tfrac{1}{2}(\boldsymbol{v} \cdot \boldsymbol{v} - \Phi_\Omega)] - \boldsymbol{v} \times (\operatorname{curl} \boldsymbol{v} + 2\boldsymbol{\Omega}) = -\frac{1}{\rho} \operatorname{grad} p \,. \tag{2.5.39}$$

If the curl of this equation is considered and $\boldsymbol{\omega} = \operatorname{curl} \boldsymbol{v}$, $\boldsymbol{\omega}_a = \boldsymbol{\omega} + 2\boldsymbol{\Omega}$ is substituted, equation (2.5.39) gives

$$\frac{\partial \boldsymbol{\omega}_a}{\partial t} - \operatorname{curl}(\boldsymbol{x} \times \boldsymbol{\omega}_a) = -\operatorname{curl}\left(\frac{1}{\rho} \operatorname{grad} p\right) ,$$

an equation, which is identical with (2.5.29) of Exercise 8, where only $\boldsymbol{\omega}$ is to be replaced by $\boldsymbol{\omega}_a$. Following the same procedure, we obtain

$$\dot{\boldsymbol{\xi}}_a = \boldsymbol{L}\boldsymbol{\xi}_a \,, \qquad \boldsymbol{\xi}_a := \frac{\boldsymbol{\omega}_a}{\rho} = \frac{\boldsymbol{\omega} + 2\boldsymbol{\Omega}}{\rho} \,. \tag{2.5.40}$$

Here $\boldsymbol{\xi}_a$ is called the absolute vorticity per unit mass.

10. Scalar multiplication of equation (2.5.40) of Exercise 9 with $\operatorname{grad} \lambda$ gives

$$\dot{\boldsymbol{\xi}}_a \cdot \operatorname{grad} \lambda = \big((\operatorname{grad} \boldsymbol{v})\boldsymbol{\xi}_a\big) \cdot \operatorname{grad} \lambda \,. \tag{2.5.41}$$

Adding the identity

$$\boldsymbol{\xi}_a \cdot (\operatorname{grad} \lambda)^{\cdot} = \boldsymbol{\xi}_a \cdot \operatorname{grad} \dot{\lambda} - \big((\operatorname{grad} \boldsymbol{v})\boldsymbol{\xi}_a\big) \cdot \operatorname{grad} \lambda \tag{2.5.42}$$

thus yields

$$(\boldsymbol{\xi}_a \cdot \operatorname{grad} \lambda)^{\cdot} = \boldsymbol{\xi}_a \cdot \operatorname{grad} \dot{\lambda} \,. \tag{2.5.43}$$

Introducing the definition of the *absolute potential vorticity for parameter* λ and using the evolution equation $\dot{\lambda} = \Psi_\lambda$ implies

$$\dot{\pi}_\lambda = \boldsymbol{\xi}_a \cdot \operatorname{grad} \Psi_\lambda ,$$

from where, with $\operatorname{grad} \Psi_\lambda = \boldsymbol{0}$, we deduce

$$\dot{\pi}_\lambda = 0 \,. \tag{2.5.44}$$

This is the *conservation law of potential vorticity* due to ERTEL [68][21].

[21] There is a rather substantial literature on ERTEL's law of conservation of potential vorticity, some of which is given in TRUESDELL & TOUPIN [238]; a series of lectures and a series of reprints of ERTEL's papers is given in SCHRÖDER [212] and SCHRÖDER & TREDER [211], [212], respectively.

11. Let R be the mass produced per unit time and unit volume; then the mass balance equation is given by

$$\frac{\mathrm{d}\rho}{\mathrm{d}t} + \rho \mathrm{div} \boldsymbol{v} = R \, .$$

The local momentum balance

$$\frac{\partial}{\partial t}(\rho \boldsymbol{v}) + \mathrm{div}(\rho \boldsymbol{v} \otimes \boldsymbol{v}) = \mathrm{div} \boldsymbol{t} + \rho \boldsymbol{g} \tag{2.5.45}$$

or

$$\underbrace{\left(\frac{\partial \rho}{\partial t} + \mathrm{div}(\rho \boldsymbol{v})\right)}_{\text{left side of mass balance}} \boldsymbol{v} + \rho \left(\frac{\partial \boldsymbol{v}}{\partial t} + \mathrm{div}(\boldsymbol{v} \otimes \boldsymbol{v})\right) = \mathrm{div} \boldsymbol{t} + \rho \boldsymbol{g}$$

can thus be transformed into

$$\rho \left(\frac{\mathrm{d}\boldsymbol{v}}{\mathrm{d}t} + \boldsymbol{v} \, \mathrm{grad} \, \boldsymbol{v}\right) = \mathrm{div} \boldsymbol{t} + \rho \boldsymbol{g} - R \boldsymbol{v} \, .$$

Thus an extra term appears in the resulting momentum balance equation, which is the consequence of the production term, R. Notice, however, that the original momentum balance (2.5.45) does not contain the production term R.

3. Jump Conditions

In the preceding chapter we derived the differential equations for the time evolution of certain physical variables. For those derivations we assumed that all field quantities are continuously differentiable within the body, i.e., within $\mathcal{B}_R$ and $\mathcal{B}_t$, (the reference and present configurations), respectively. These assumptions were also implemented for the application of the REYNOLDS transport theorem and the Divergence Theorem. The derivation of the local balance equations, as demonstrated in the last chapter, is no longer possible in those forms if the associated field variables are not continuously differentiable in the whole domain. When the variables do not satisfy the continuity conditions at a surface of a body then the global balance laws imply the so called *jump conditions* that must hold on surfaces across which certain field variables are not continuous. These jump conditions can be interpreted as boundary or transition conditions at boundary surfaces. Particular surfaces are:

Material Surface A *material surface* (or also a *material line*), analogous to a material body, is defined as a surface (line) within a body which is formed by the same material elements or particles at all times.

Singular Surface A surface within a material body across which a physical quantity experiences a discontinuity is called a *singular surface.*

The discontinuities can have different degrees. For example, a variable can experience a *finite jump* across a singular surface; such a variable is the density when we consider e.g. the ocean and the atmosphere as a body and the ocean surface as a singular surface. It can also be imaginable that a variable grows rapidly as the surface is approached and eventually becomes infinite; the so called VAN DER WAALS intermolecular forces in fluids may approximately be considered as an example for this situation, in which these forces ultimately govern the surface tension and are thus responsible for forming fluid droplets and capillary rises. Here, we shall restrict ourselves to singular surfaces across which a physical quantity suffers a finite jump or becomes infinite only in such a form that certain integrals can exist. The discontinuity which the motion $\boldsymbol{x} = \boldsymbol{\chi}(\boldsymbol{X}, t)$ can experience may be such that $\boldsymbol{\chi}$ itself remains continuous across the singular surface – otherwise cracks are formed (such discontinuities are called VOLTERRA-*dislocations*) – but higher derivatives of $\boldsymbol{\chi}$ may be discontinuous (e.g. velocity, deformation gradient).

Examples of singular surfaces are (compare Fig. 3.1):

- A surface separating two immiscible fluids is a material surface. The surface of a lake or the ocean in the system of atmosphere–ocean is such a material surface. The density distribution across this singular surface is discontinuous. Notice, however, that the sea surface can only be considered to be material if no evaporation of water into the atmosphere occurs. Since evaporation of water or precipitation of rain nearly always occurs, the sea surface cannot really be taken as material, only nearly so.
- Consider the surface separating ice and water in a frozen lake, the surface in this case is non-material, as the material of one side (ice) can be converted into the material of the other side (water) – and vice versa. The singular surface thus does not consist of the same material elements at all times.
- A sliding surface between two different bodies is a material singular surface, provided no material is abraded. An example is the sliding surface at the base of a temperate glacier if no melting processes occur.
  The example of a sliding surface requires specification. The sole of the sliding "shoe" is obviously a material surface because this is formed by the same material particles at all times. This is also true for the sliding track, but the sliding bed consists of different material points. Finally one may also define the geometric interface at which the sole and the bed touch one another and thus locally agrees with the sole and the bed as a singular surface, but in this case its points neither share the points of the sole nor of the bed during their motion. If material is abraded by the frictional forces established by the sliding process then one must be careful; if only material from the sole is abraded but not from the bed, then the surface of the sole is non-material but the bed is material. Of course, abrasion generally occurs on both sides, and then neither of the surfaces is material.
- The shock front in supersonic flows is a non-material singular surface.
- Every phase–boundary surface separating two aggregate states is a singular surface; it is a material surface if no phase–change occurs between them, but is non-material if one phase evolves at the costs of the other.

3.1 General Formulation of Jump Conditions

3.1.1 Jump Through a Surface

Let a body be given in which there exists an orientable, smooth interface, not necessarily material, across which a physical variable may be discontinuous, but in the remaining part of the body the variable is supposed to be continuously differentiable. Let such a surface be designated by $\mathfrak{S}$ and $\mathfrak{s}$ in the LAGRANGE and EULER representations, respectively, and let both these representations of the surface be assumed smooth and orientable. They are described by differentiable implicit functions

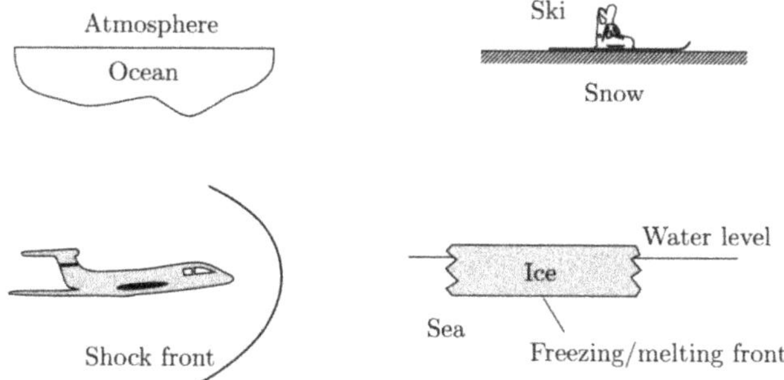

Fig. 3.1. Examples of singular surfaces.

$$F_{\mathfrak{S}}(\boldsymbol{X},t) = 0 \quad \text{and} \quad f_{\mathfrak{s}}(\boldsymbol{x},t) = 0 \tag{3.1.1}$$

in the reference and present configurations, respectively. Surfaces for which $f = f_{\mathfrak{s}}(\boldsymbol{x})$ holds are called *stationary surfaces* and those for which $F = F_{\mathfrak{S}}(\boldsymbol{X})$ holds are known as *material surfaces.* Physically, one observes these surfaces in the spatial representation. The equation $F_{\mathfrak{S}}(\boldsymbol{X},t) = 0$ represents the "pull–back" of $f_{\mathfrak{s}}(\boldsymbol{x},t) = 0$ to the reference configuration, i.e., it is the pre-image of the singular surface in the present configuration. Since the motion $\chi(\boldsymbol{X},t)$ is continuous, this referential surface is well defined.

With (3.1.1) the unit normal vectors at these surfaces can be written as

$$\boldsymbol{N}_{\mathfrak{S}} = \frac{\operatorname{Grad} F_{\mathfrak{S}}}{\|\operatorname{Grad} F_{\mathfrak{S}}\|}\,, \quad \boldsymbol{n}_{\mathfrak{s}} = \frac{\operatorname{grad} f_{\mathfrak{s}}}{\|\operatorname{grad} f_{\mathfrak{s}}\|}\,, \tag{3.1.2}$$

and the corresponding kinematic equations are

$$\frac{\partial F_{\mathfrak{S}}}{\partial t} + \left(\operatorname{Grad} F_{\mathfrak{S}}\right) \cdot \boldsymbol{W} = 0\,, \quad \frac{\partial f_{\mathfrak{s}}}{\partial t} + \left(\operatorname{grad} f_{\mathfrak{s}}\right) \cdot \boldsymbol{w} = 0\,, \tag{3.1.3}$$

where $\boldsymbol{W}$ and $\boldsymbol{w}$ denote the velocities of the singular surfaces in the reference and present configurations, respectively. $\boldsymbol{w}$ is usually known as the *displacement velocity* and $\boldsymbol{W}$ as the *propagation velocity.* It is evident that only the normal components of these velocities, $W_{\mathfrak{S}} = \boldsymbol{W} \cdot \boldsymbol{N}_{\mathfrak{S}}$ and $w_{\mathfrak{s}} = \boldsymbol{w} \cdot \boldsymbol{n}_{\mathfrak{s}}$, are physically relevant.

The singular surface divides the domain of the material body under consideration in two parts. We designate the variables on each side of this surface either with a $(\cdot)^+$ or with a $(\cdot)^-$ sign. The index $(\cdot)^+$ hereby is so chosen that it identifies the side into which the normal vector at the singular surface is directed. Figure 3.2 elucidates these concepts and the rule of the sign conventions.

By combining (3.1.2) and (3.1.3), the propagation and displacement speeds can be written as

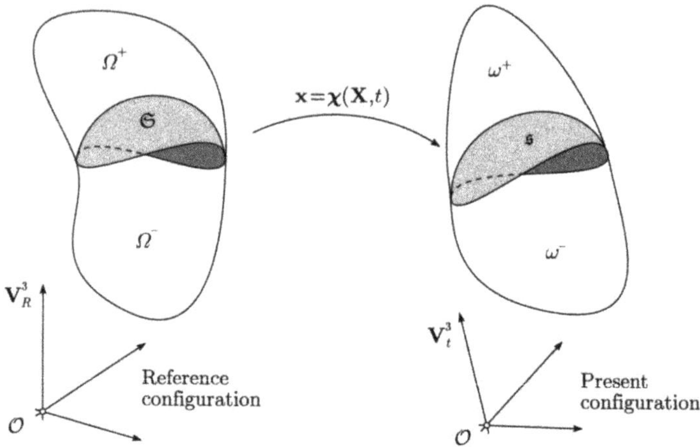

Fig. 3.2. A singular surface $\mathfrak{S}$ divides the body Ω (in the reference configuration) in two parts, whose volume segments are denoted by Ω^+, Ω^-. The same principle also applies to the body in the present configuration.

$$W_{\mathfrak{S}} = -\frac{\dfrac{\partial F_{\mathfrak{S}}}{\partial t}}{\|\operatorname{Grad} F_{\mathfrak{S}}\|}\,, \quad w_{\mathfrak{s}} = -\frac{\dfrac{\partial f_{\mathfrak{s}}}{\partial t}}{\|\operatorname{grad} f_{\mathfrak{s}}\|}\,. \tag{3.1.4}$$

In addition the time evolution of the singular surface leads to the kinematic equation in the form

$$\begin{aligned}\frac{\partial F_{\mathfrak{S}}}{\partial t} &= -W_{\mathfrak{S}}\|\operatorname{Grad} F_{\mathfrak{S}}\| \\ &= \frac{\partial f_{\mathfrak{s}}}{\partial t} + \operatorname{grad} f_{\mathfrak{s}} \cdot \dot{\boldsymbol{x}} = \frac{\mathrm{d} f_{\mathfrak{s}}}{\mathrm{d}t} \\ &= -(w_{\mathfrak{s}} - \dot{\boldsymbol{x}} \cdot \boldsymbol{n}_{\mathfrak{s}})\,\|\operatorname{grad} f_{\mathfrak{s}}\|\,.\end{aligned} \tag{3.1.5}$$

One calls the expression

$$W_{(\mathrm{i})} = w_{\mathfrak{s}} - \dot{\boldsymbol{x}} \cdot \boldsymbol{n}_{\mathfrak{s}} \tag{3.1.6}$$

the *instantaneous propagation velocity.* This quantity agrees with $W_{\mathfrak{S}}$ when one identifies the reference and the present configuration with each other; it represents the normal velocity of the singular surface relative to the particles instantaneously sitting upon the surface. Here the particle velocity, in general, can take different values on the positive and negative sides of the singular surface; this implies that one must differentiate between $W_{(\mathrm{i})}^+$ and $W_{(\mathrm{i})}^-$ and consequently also between $W_{\mathfrak{S}}^+$ and $W_{\mathfrak{S}}^-$. On the other hand $w_{\mathfrak{s}}$ is unique. This is the reason why the position of the singular surface in the reference configuration can not be computed in its evolution by using the propagation velocity, but rather by computing at each time a "pull–back" from the present configuration to the reference configuration.

Jump Let us consider a physical quantity $\mathcal{G}$ which experiences a finite discontinuity across the singular surface, but remains continuously differentiable in the rest of the domain. The quantity possesses different values on both sides of the surface. The limiting values of this quantity at a point of the singular surface are denoted by $\mathcal{G}^+$ and $\mathcal{G}^-$. The difference in these values is called the *jump* of the quantity $\mathcal{G}$ over the surface. The jump is assumed finite and is written as

$$[\![\mathcal{G}]\!] := \mathcal{G}^+ - \mathcal{G}^- \,. \tag{3.1.7}$$

Occasionally one also requires the arithmetic mean of $\mathcal{G}^+$ and $\mathcal{G}^-$, defined by

$$\langle\!\langle \mathcal{G} \rangle\!\rangle := \tfrac{1}{2}\left(\mathcal{G}^+ + \mathcal{G}^-\right) \,. \tag{3.1.8}$$

The following arithmetic rules hold for any arbitrary scalar, vectorial and tensorial variable a, b, c:

$$\begin{aligned}
&\bullet\ [\![a]\!] = 0 \quad \Rightarrow \quad a^+ = a^- = a \,, \\
&\bullet\ [\![a + b]\!] = [\![a]\!] + [\![b]\!] \,, \\
&\bullet\ [\![a - b]\!] = [\![a]\!] - [\![b]\!] \,, \\
&\bullet\ [\![c\, a]\!] = c[\![a]\!] \quad \text{only when } [\![c]\!]{=}0, \\
&\bullet\ [\![a\, b]\!] = a^+b^+ - a^-b^- \neq [\![a]\!]\,[\![b]\!], \\
&\bullet\ [\![a \cdot b]\!] = [\![a]\!] \cdot \langle\!\langle b \rangle\!\rangle + \langle\!\langle a \rangle\!\rangle \cdot [\![b]\!] \quad \text{for products of scalars and} \\
&\quad \text{scalar products with vectors.}
\end{aligned} \tag{3.1.9}$$

Kinematically important classes of singular surfaces are given as follows:

- **Material surfaces** For such surfaces the relation $W_{(\mathrm{i})} = 0$ holds and with this $W_{\mathfrak{S}} = 0$ or also $w_{\mathfrak{s}} = \dot{\boldsymbol{x}} \cdot \boldsymbol{n}_{\mathfrak{s}}$. According to this definition sliding surfaces are included within the class of material surfaces. The normal velocity of a particle on both sides of a singular surface is given by the unique displacement velocity $w_{\mathfrak{s}}$.
- **Vortex surfaces** or vortex sheets are defined by those singular surfaces through which the (material) velocity field suffers a jump only along the tangent to the surface:

$$[\![\dot{\boldsymbol{x}}]\!] \neq \boldsymbol{0} \quad \text{but} \quad [\![\dot{\boldsymbol{x}} \cdot \boldsymbol{n}_{\mathfrak{s}}]\!] = 0 \,. \tag{3.1.10}$$

  Sliding surfaces are material vortex–surfaces. In that case $\dot{\boldsymbol{x}} \cdot \boldsymbol{n}_{\mathfrak{s}} = 0$. However, when $\dot{\boldsymbol{x}} \cdot \boldsymbol{n}_{\mathfrak{s}} \neq 0$, then these vortex sheets are non–material.
- **Shock surface** or **Shocks** are those singular surfaces through which the normal component of the velocity experiences a jump,

$$[\![\dot{\boldsymbol{x}} \cdot \boldsymbol{n}_{\mathfrak{s}}]\!] \neq 0 \,. \tag{3.1.11}$$

From these definitions, we can draw the following conclusions: (*i*) The propagation velocity $W_{(\mathrm{i})}$ or $W_{\mathfrak{S}}$ in shocks can not be continuous through singular surfaces, (*ii*) material shock surfaces do not exist and (*iii*) $W_{(\mathrm{i})}$ and $W_{\mathfrak{S}}$ are continuous across vortex surfaces.

3.1.2 Modified Transport Theorem

The general balance equations described in the preceding chapter must now be modified because the derivations were accomplished by assuming the field variables as continuously differentiable. Analogous to the corresponding derivation we must consider the time rate of change of a volume integral and the transformation of a surface integral to a volume integral.

LAGRANGE **representation** When we interchange the differentiation and integration in the reference configuration we must now realize that the singular surface moves within the body. We assume that the singular surface divides the integration domain in two parts. The respective volume parts that are separated by the surface are indicated by Ω^+ and Ω^-, and the corresponding surfaces are denoted by $\partial\Omega^+$ and $\partial\Omega^-$. The whole mantle surface consists of $\partial\Omega^+ \setminus \mathfrak{S}$ and $\partial\Omega^- \setminus \mathfrak{S}$ plus the singular surface $\mathfrak{S}$ as described[1] in Fig. 3.3.

If one considers both volumes Ω^+ and Ω^- separately, one must account for the fact that they change with time despite the fact that they are referential. Under such circumstances both terms in line 2 of (3.1.12) must be subjected to the REYNOLDS transport theorem. The outer surface $\partial\Omega^\pm \setminus \mathfrak{S}$ is material and has no contribution. However, the interface $\mathfrak{S}$ moves itself with velocity $\boldsymbol{W}$. Thus one obtains

$$
\begin{aligned}
\frac{\mathrm{d}\mathcal{G}}{\mathrm{d}t} = \frac{\mathrm{d}}{\mathrm{d}t}\int_{\Omega} \Gamma(\boldsymbol{X},t)\,dV &= \frac{\mathrm{d}}{\mathrm{d}t}\int_{\Omega^+} \Gamma\,dV + \frac{\mathrm{d}}{\mathrm{d}t}\int_{\Omega^-} \Gamma\,dV \\
&= \int_{\Omega^+} \frac{\partial \Gamma}{\partial t}\,dV - \int_{\mathfrak{S}} \Gamma^+ \boldsymbol{W}^+ \cdot \boldsymbol{N}_{\mathfrak{S}}\,dA \\
&\quad + \int_{\Omega^-} \frac{\partial \Gamma}{\partial t}\,dV + \int_{\mathfrak{S}} \Gamma^- \boldsymbol{W}^- \cdot \boldsymbol{N}_{\mathfrak{S}}\,dA \\
&= \int_{\Omega} \frac{\partial \Gamma}{\partial t}\,dV - \int_{\mathfrak{S}} [\![\Gamma \boldsymbol{W} \cdot \boldsymbol{N}_{\mathfrak{S}}]\!]\,dA\,.
\end{aligned}
\tag{3.1.12}
$$

Here a minus sign appears in the surface integral over $\mathfrak{S}$ for the positive side of the body since the normal vector at the singular surface is directed to the positive side.

[1] We emphasize that $\partial\Omega^+$ is not only that part of the surface $\partial\Omega$ that is separated by the surface $\mathfrak{S}$ but the whole surface of Ω^+ comprising also the singular surface. This description of the domain boundary deviates partly from those of other authors and has the consequence that certain formulas appear different in case of the derivation of the transport theorem.

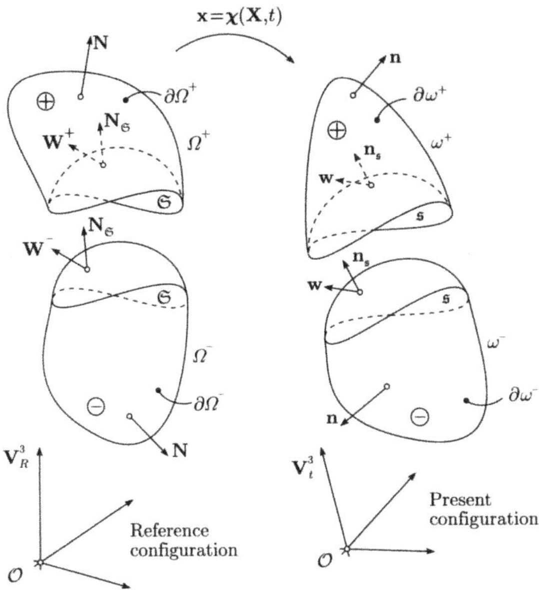

Fig. 3.3. Material body with singular surface in the present and reference configuration. The body is separated in two parts by the singular surface which are indicated by + and −. The boundary surfaces $\partial\Omega^{\pm}$, $\partial\omega^{\pm}$ are the outer boundaries governed by the mantle surfaces plus the singular surfaces of the body in the two configurations. $\mathfrak{S}$ and $\mathfrak{s}$ indicate the singular surfaces, $\boldsymbol{N}_{\mathfrak{S}}$ and $\boldsymbol{n}_{\mathfrak{s}}$ represent the unit normal vectors, and finally $\boldsymbol{W}^{\pm}$ or $\boldsymbol{w}$ the velocities of these surfaces in the two different configurations.

EULER **Representation** In the present configuration, we are concerned with the time rate of change of the domain area $\omega(t)$. Our aim in this regard is to examine the integral

$$\frac{\mathrm{d}}{\mathrm{d}t}\int_{\omega}\gamma\,dv=\frac{\mathrm{d}}{\mathrm{d}t}\int_{\omega^{+}}\gamma\,dv+\frac{\mathrm{d}}{\mathrm{d}t}\int_{\omega^{-}}\gamma\,dv\,, \tag{3.1.13}$$

whereby both domains of the body are considered non-material. Figure 3.3 demonstrates the partitioning of the body.

Here, since the singular surface moves with the velocity $\boldsymbol{w}$, one can exploit the transport theorem for both parts on the right-hand side of (3.1.13); the direction of the normal vector $\boldsymbol{n}_{\mathfrak{s}}$, pointing into the positive side in our case must again be noticed with care:

$$\frac{\mathrm{d}}{\mathrm{d}t}\int_{\omega^{\pm}}\gamma\,dv=\int_{\omega^{\pm}}\frac{\partial\gamma}{\partial t}\,dv+\int_{\partial\omega^{\pm}\setminus\mathfrak{s}}\gamma\,(\boldsymbol{v}\cdot\boldsymbol{n})\,da\mp\int_{\mathfrak{s}}\gamma^{\pm}(\boldsymbol{w}\cdot\boldsymbol{n}_{\mathfrak{s}})\,da\,, \tag{3.1.14}$$

thus, by adding the two contributions, one obtains (Exercise)

$$\frac{\mathrm{d}}{\mathrm{d}t}\int_{\omega}\gamma\,dv = \int_{\omega}\frac{\partial\gamma}{\partial t}\,dv + \int_{\partial\omega}\gamma\,(\boldsymbol{v}\cdot\boldsymbol{n})\,da - \int_{\mathfrak{s}}[\![\gamma(\boldsymbol{w}\cdot\boldsymbol{n}_{\mathfrak{s}})]\!]\,da\,. \tag{3.1.15}$$

In the second term on the right of this expression, however, the surface integral can not be converted into a volume integral as the Divergence Theorem can not be directly applied in case of a discontinuity of a field variable; nevertheless this can be achieved for the two subdomains separately. For this reason one complements the surface integral in (3.1.14), which is performed over the open domains $\partial\omega^- \setminus \mathfrak{s}$ or $\partial\omega^+ \setminus \mathfrak{s}$, by a surface integral over the singular surface, whereby the corresponding fields for the positive and negative sides must be substituted, viz.,

$$\begin{aligned}\frac{\mathrm{d}}{\mathrm{d}t}\int_{\omega^\pm}\gamma\,dv = &\int_{\omega^\pm}\frac{\partial\gamma}{\partial t}\,dv \mp \int_{\mathfrak{s}}\gamma^\pm(\boldsymbol{w}\cdot\boldsymbol{n}_{\mathfrak{s}})\,da\\ &\underbrace{+\int_{\partial\omega^\pm\setminus\mathfrak{s}}\gamma\,(\boldsymbol{v}\cdot\boldsymbol{n})\,da}_{(1)}\\ &\underbrace{\mp\int_{\mathfrak{s}}\gamma^\pm\,(\boldsymbol{v}^\pm\cdot\boldsymbol{n}_{\mathfrak{s}})\,da}_{(2)}\underbrace{\pm\int_{\mathfrak{s}}\gamma^\pm\,(\boldsymbol{v}^\pm\cdot\boldsymbol{n}_{\mathfrak{s}})\,da}_{(3)}\,.\end{aligned} \tag{3.1.16}$$

The terms (1) and (2) can now be combined to give surface integrals over the bounding surfaces of both partial volume regions. Since the fields have been assumed ab initio to be continuously differentiable within the partial volumes, one can apply the Divergence Theorem for these subregions independently. Thus the result corresponding to expression (3.1.15) becomes

$$\begin{aligned}\frac{\mathrm{d}}{\mathrm{d}t}\int_{\omega}\gamma\,dv = &\int_{\omega^+}\frac{\partial\gamma}{\partial t}\,dv + \int_{\partial\omega^+}\gamma\,(\boldsymbol{v}\cdot\boldsymbol{n})\,da\\ &+\int_{\omega^-}\frac{\partial\gamma}{\partial t}\,dv + \int_{\partial\omega^-}\gamma\,(\boldsymbol{v}\cdot\boldsymbol{n})\,da\\ &-\int_{\mathfrak{s}}[\![\gamma(\boldsymbol{w}\cdot\boldsymbol{n}_{\mathfrak{s}})]\!]\,da + \int_{\mathfrak{s}}[\![\gamma(\boldsymbol{v}\cdot\boldsymbol{n}_{\mathfrak{s}})]\!]\,da\\ = &\int_{\omega^+}\left(\frac{\partial\gamma}{\partial t}+\operatorname{div}(\gamma\boldsymbol{v})\right)dv + \int_{\omega^-}\left(\frac{\partial\gamma}{\partial t}+\operatorname{div}(\gamma\boldsymbol{v})\right)dv\\ &-\int_{\mathfrak{s}}[\![\gamma\,((\boldsymbol{w}-\boldsymbol{v})\cdot\boldsymbol{n}_{\mathfrak{s}})]\!]\,da\,,\end{aligned} \tag{3.1.17}$$

where the term (3) has been combined to give a jump term, and the volume integrals over ω^+ and ω^- could also be combined to a single integral.

Comparing (3.1.15) with (3.1.17), there follows

$$\int_{\partial\omega}\gamma(\boldsymbol{v}\cdot\boldsymbol{n})\,da = \int_{\omega}\operatorname{div}(\gamma\boldsymbol{v})\,dv + \int_{\mathfrak{s}}[\![\gamma\boldsymbol{v}\cdot\boldsymbol{n}_{\mathfrak{s}}]\!]\,da\,. \tag{3.1.18}$$

This may be interpreted as the Divergence Theorem applied to a domain ω for a field γ and a material velocity $\boldsymbol{v}$, which can experience a discontinuity across the surface $\mathfrak{s}$ within the volume ω.

3.1.3 General Balance Equations and Jump Conditions

In order to establish the balance equations one must in addition compute the surface integral of the flux of a physical variable. This can principally be accomplished analogously to the above procedure by dividing the volume into its positive and negative parts. We shall impose this procedure for the flux term in the reference configuration as an example. To start with, let us divide the surface $\partial\Omega$ into the positive and negative mantle surfaces as

$$\int_{\partial\Omega} \boldsymbol{\Phi N}\,\mathrm{d}A = \int_{\partial\Omega^+\setminus\mathfrak{S}} \boldsymbol{\Phi N}\,\mathrm{d}A + \int_{\partial\Omega^-\setminus\mathfrak{S}} \boldsymbol{\Phi N}\,\mathrm{d}A\,, \tag{3.1.19}$$

where the open surface-integrals can be closed by the corresponding extension procedure as before. This yields

$$\begin{aligned}\int_{\partial\Omega} \boldsymbol{\Phi N}\,\mathrm{d}A &= \int_{\partial\Omega^+\setminus\mathfrak{S}} \boldsymbol{\Phi N}\,\mathrm{d}A - \int_{\mathfrak{S}} \boldsymbol{\Phi}^+\boldsymbol{N}_{\mathfrak{S}}\,\mathrm{d}A \\ &\quad + \int_{\partial\Omega^-\setminus\mathfrak{S}} \boldsymbol{\Phi N}\,\mathrm{d}A + \int_{\mathfrak{S}} \boldsymbol{\Phi}^-\boldsymbol{N}_{\mathfrak{S}}\,\mathrm{d}A \\ &\quad + \int_{\mathfrak{S}} \boldsymbol{\Phi}^+\boldsymbol{N}_{\mathfrak{S}}\,\mathrm{d}A - \int_{\mathfrak{S}} \boldsymbol{\Phi}^-\boldsymbol{N}_{\mathfrak{S}}\,\mathrm{d}A \\ &= \int_{\partial\Omega^+} \boldsymbol{\Phi N}\,\mathrm{d}A + \int_{\partial\Omega^-} \boldsymbol{\Phi N}\,\mathrm{d}A \\ &\quad + \int_{\mathfrak{S}} [\![\boldsymbol{\Phi N}_{\mathfrak{S}}]\!]\,\mathrm{d}A\,.\end{aligned} \tag{3.1.20}$$

Now the Divergence Theorem can be applied to the integrals over the closed surfaces, since all the existing variables are continuously differentiable within the partial volume regions under consideration. We may thus write

$$\begin{aligned}\int_{\partial\Omega} \boldsymbol{\Phi N}\,\mathrm{d}A &= \int_{\Omega^+} \mathrm{Div}\,\boldsymbol{\Phi}\,\mathrm{d}V + \int_{\Omega^-} \mathrm{Div}\,\boldsymbol{\Phi}\,\mathrm{d}V + \int_{\mathfrak{S}} [\![\boldsymbol{\Phi N}_{\mathfrak{S}}]\!]\,\mathrm{d}A\,, \\ \int_{\partial\omega} \boldsymbol{\phi n}\,\mathrm{d}a &= \int_{\omega^+} \mathrm{div}\boldsymbol{\phi}\,\mathrm{d}v + \int_{\omega^-} \mathrm{div}\boldsymbol{\phi}\,\mathrm{d}v + \int_{\mathfrak{s}} [\![\boldsymbol{\phi n}_{\mathfrak{s}}]\!]\,\mathrm{d}a\,,\end{aligned} \tag{3.1.21}$$

where the corresponding result for the EULERian formulation has been added in the second line; its derivation is analogous.

The volume integral for the supply (from outside) of the variable $\mathcal{G}$, by principle, is not influenced by the presence of the internal singular surface, so that the division into positive and negative parts has only formal character

$$\begin{aligned} \int_{\Omega} \Sigma\, dV &= \int_{\Omega^+} \Sigma\, dV + \int_{\Omega^-} \Sigma\, dV\,, \\ \int_{\omega} \varsigma\, dV &= \int_{\omega^+} \varsigma\, dv + \int_{\omega^-} \varsigma\, dv\,. \end{aligned} \tag{3.1.22}$$

Somewhat different appears the situation for the (internal) production term. When splitting the volume integrals of the production density of the physical quantity $\mathcal{G}$ into integrals over the domains Ω^+, Ω^- or ω^+, ω^-, one must, in general, also allow for the existence of an additional production term of the variable $\mathcal{G}$ on the singular surface. For this reason, the *surface density of production* $\mathfrak{P}$ or $\mathfrak{p}$ are introduced. Thus one obtains

$$\begin{aligned} \int_{\Omega} \Pi\, dV &= \int_{\Omega^+} \Pi\, dV + \int_{\Omega^-} \Pi\, dV + \int_{\mathfrak{S}} \mathfrak{P}\, dA\,, \\ \int_{\omega} \pi\, dv &= \int_{\omega^+} \pi\, dv + \int_{\omega^-} \pi\, dv + \int_{\mathfrak{s}} \mathfrak{p}\, da\,. \end{aligned} \tag{3.1.23}$$

Naturally, one may also suppose that quantities other than the production may, on the singular surface, generate additional contributions. For example, the flux of the variable at the cutting line of the surfaces Ω^+, Ω^- with the singular surfaces ($\partial\mathfrak{S} = \Omega \cap \mathfrak{S}$) can give rise to a flux term. This term is a directional contribution, a flux in the direction $\boldsymbol{\nu}_{\partial\mathfrak{S}}$, where $\boldsymbol{\nu}_{\partial\mathfrak{S}}$ is the unit normal vector in the singular surface but equally perpendicular to the line $\partial\mathfrak{S}$; after the application of the Divergence Theorem for a surface integral, this yields

$$\int_{\partial\mathfrak{S}} \boldsymbol{\Phi}^{(\mathfrak{P})} \boldsymbol{\nu}_{\mathfrak{S}}\, dS = \int_{\mathfrak{S}} \mathrm{Div}_{\mathfrak{S}}\, \boldsymbol{\Phi}^{(\mathfrak{P})}\, dA\,,$$

so that this term must be added to the jump conditions ($\mathrm{Div}_{\mathfrak{S}}$ indicates the surface divergence in the material representation). Such an extension is for example necessary when surface stresses are acting on singular surfaces. We shall not be dealing with these terms in this book.

Combining (3.1.12) or (3.1.17), (3.1.21), (3.1.22) and (3.1.23), the balance equations (2.1.16) in the Lagrange and Euler representations, respectively, read

$$\begin{aligned} &\int_{\Omega^+} \left(\frac{\partial \Gamma}{\partial t} + \mathrm{Div}\, \boldsymbol{\Phi} - \Pi - \Sigma \right) dV \\ &\quad + \int_{\Omega^-} \left(\frac{\partial \Gamma}{\partial t} + \mathrm{Div}\, \boldsymbol{\Phi} - \Pi - \Sigma \right) dV \\ &\quad = \int_{\mathfrak{S}} \left([\![\Gamma \boldsymbol{W} \cdot \boldsymbol{N}_{\mathfrak{S}} - \boldsymbol{\Phi} \boldsymbol{N}_{\mathfrak{S}}]\!] + \mathfrak{P} \right) dA \end{aligned} \tag{3.1.24}$$

and

$$\begin{aligned}
&\int_{\omega^+}\left(\frac{\partial\gamma}{\partial t}+\operatorname{div}(\gamma\boldsymbol{v})+\operatorname{div}\boldsymbol{\phi}-\pi-\varsigma\right)dv\\
&\quad+\int_{\omega^-}\left(\frac{\partial\gamma}{\partial t}+\operatorname{div}(\gamma\boldsymbol{v})+\operatorname{div}\boldsymbol{\phi}-\pi-\varsigma\right)dv\\
&\quad=\int_{\mathfrak{s}}\left([\![\gamma(\boldsymbol{w}-\boldsymbol{v})\cdot\boldsymbol{n}_{\mathfrak{s}}-\boldsymbol{\phi}\boldsymbol{n}_{\mathfrak{s}}]\!]+\mathfrak{p}\right)da\,.
\end{aligned}\tag{3.1.25}$$

In the above equations the integration domains Ω or ω as well as the respective volume regions are arbitrary; thus, the integrands of the volume integrals in (3.1.24) and (3.1.25) must vanish identically. This yields, since both integrals lead to the same results, the so-called local balance equations (2.2.13) and (2.2.16) as derived in the previous chapter, which hold true at all points of the body except at a singular surface. Likewise, the surface integrals must vanish for an arbitrary area of the singular surface; as a result, the corresponding integrand must equally vanish identically. Hence follow the *jump conditions* in the LAGRANGE and EULER representations which are valid at all points of the singular surface,

$$[\![\boldsymbol{\Phi}\boldsymbol{N}_{\mathfrak{S}}-\Gamma\boldsymbol{W}\cdot\boldsymbol{N}_{\mathfrak{S}}]\!]=\mathfrak{P}\,,\quad[\![\boldsymbol{\phi}\boldsymbol{n}_{\mathfrak{s}}-\gamma(\boldsymbol{w}-\boldsymbol{v})\cdot\boldsymbol{n}_{\mathfrak{s}}]\!]=\mathfrak{p}\,.\tag{3.1.26}$$

These formulas of the jump conditions neglect not only the above mentioned surface fluxes but equally also surface contributions of the physical quantities themselves. There are formulations of interfacial thermomechanics in which all such contributions are accounted for. What emerges are thermodynamic theories of continuous interfaces. These will, however, not be part of our analysis. For relatively recent works see KOSINSKI [123], ALTS and HUTTER [14], [15], [16], [17], KOSINSKI & MURDOCH [124] and GURTIN [92].

3.2 Special Jump Conditions

Now we can easily furnish the corresponding jump conditions for the physical balance equations as given in Sect. 2.3. One must simply substitute the specified variables of (2.3.1), (2.3.7) etc. in the general jump conditions derived above. The results will also be depicted later in tabular form together with the local balance equations. In case of the jump condition for entropy in the LAGRANGE and EULER representation, it is assumed that there always appears a positive surface production, but such a term is not present in all other cases.

Mass With (2.3.1) there follows from (3.1.26),

$$[\![\rho_R\boldsymbol{W}]\!]\cdot\boldsymbol{N}_{\mathfrak{S}}=0\,,\quad[\![\rho(\boldsymbol{w}-\boldsymbol{v})]\!]\cdot\boldsymbol{n}_{\mathfrak{s}}=0\,.\tag{3.2.1}$$

Let us suppose that an observer moves with a displacement velocity $\boldsymbol{w}$, and the material particle with the velocity $\boldsymbol{v}_*=\boldsymbol{v}-\boldsymbol{w}$ relative to the observer.

Then the jump condition $(3.2.1)_2$ becomes $[\![\rho \boldsymbol{v}_*]\!] \cdot \boldsymbol{n}_{\mathfrak{s}} = 0$. The flow of mass per unit time from one side onto a unit surface of $\mathfrak{s}$ must therefore be equal to the same flow of mass into the other side, i.e., $\rho^+ \boldsymbol{v}_*^+ \cdot \boldsymbol{n}_{\mathfrak{s}} = \rho^- \boldsymbol{v}_*^- \cdot \boldsymbol{n}_{\mathfrak{s}}$. The variable

$$\mathcal{M} := \rho^{\pm}(\boldsymbol{v}^{\pm} - \boldsymbol{w}) \cdot \boldsymbol{n}_{\mathfrak{s}} \tag{3.2.2}$$

is known as *mass flux*; because of its continuity it assumes the same value on both sides of the singular surface, $\mathcal{M}^+ = \mathcal{M}^- = \mathcal{M}$. So, the reader can easily convince himself that ρ can only have a jump across a material or a shock surface.

Momentum With (2.3.7) there follows

$$[\![\rho_R \boldsymbol{v}(\boldsymbol{W} \cdot \boldsymbol{N}_{\mathfrak{S}}) + \boldsymbol{T}\boldsymbol{N}_{\mathfrak{S}}]\!] = \boldsymbol{0}\,, \quad [\![\rho \boldsymbol{v}(\boldsymbol{w} - \boldsymbol{v}) \cdot \boldsymbol{n}_{\mathfrak{s}} + \boldsymbol{t}\boldsymbol{n}_{\mathfrak{s}}]\!] = \boldsymbol{0}\,. \tag{3.2.3}$$

Consideration of the continuity of the mass flux leads to a frequently used form of the momentum jump condition in the EULER representation: With (3.2.1) together with the definition of the mass flux, $(3.2.3)_2$ yields

$$[\![\boldsymbol{v}]\!]\mathcal{M} - [\![\boldsymbol{t}\boldsymbol{n}_{\mathfrak{s}}]\!] = \boldsymbol{0}\,. \tag{3.2.4}$$

Therefore, if the traction is not continuous across a non-material surface, the momentum jump of the mass flux is responsible for this discontinuity.

On a material surface the conditions $\boldsymbol{W} \cdot \boldsymbol{N}_{\mathfrak{S}} = 0$ and $\boldsymbol{w} \cdot \boldsymbol{n}_{\mathfrak{s}} = \boldsymbol{v} \cdot \boldsymbol{n}_{\mathfrak{s}}$ imply the continuity of the normal part of the stress tensor (stress vector or traction on the singular surface),

$$\begin{aligned} [\![\boldsymbol{T}\boldsymbol{N}_{\mathfrak{S}}]\!] &= \boldsymbol{0}\,, \qquad \text{for } \boldsymbol{W} \cdot \boldsymbol{N}_{\mathfrak{S}} = 0\,, \\ [\![\boldsymbol{t}\boldsymbol{n}_{\mathfrak{s}}]\!] &= \boldsymbol{0}\,, \qquad \text{for } \boldsymbol{w} \cdot \boldsymbol{n}_{\mathfrak{s}} = \boldsymbol{v} \cdot \boldsymbol{n}_{\mathfrak{s}}\,. \end{aligned} \tag{3.2.5}$$

Angular Momentum With (2.3.20) the jump conditions take here the form

$$\begin{aligned} &[\![\boldsymbol{x} \times \rho_R \boldsymbol{v}(\boldsymbol{W} \cdot \boldsymbol{N}_{\mathfrak{S}}) + \boldsymbol{x} \times \boldsymbol{T}\boldsymbol{N}_{\mathfrak{S}}]\!] \\ &\qquad = \boldsymbol{x} \times [\![\rho_R \boldsymbol{v}(\boldsymbol{W} \cdot \boldsymbol{N}_{\mathfrak{S}}) + \boldsymbol{T}\boldsymbol{N}_{\mathfrak{S}}]\!] = \boldsymbol{0}\,, \\ &\boldsymbol{x} \times [\![\rho \boldsymbol{v}\Big((\boldsymbol{w} - \boldsymbol{v}) \cdot \boldsymbol{n}_{\mathfrak{s}}\Big) + \boldsymbol{t}\boldsymbol{n}_{\mathfrak{s}}]\!] = \boldsymbol{0}\,. \end{aligned} \tag{3.2.6}$$

This equation, as is straightforwardly seen, is identically satisfied with the satisfaction of the jump condition of momentum and is therefore redundant.

The reason why the jump of angular momentum does not produce any new expression lies solely in the postulates for the identities Γ, $\tilde{\Phi}$, Σ and Π (or γ, $\tilde{\phi}$, ς, π), which are the moments of the corresponding momentum densities. If specifications like spin density $\rho \boldsymbol{s}$, couple stress $\boldsymbol{m}$ and specific body couple $\rho \boldsymbol{l}$ would be made, the jump relation for the angular momentum would be given by an essential non-redundant equation (Exercise 4).

Energy With (2.3.34) the jump condition of energy in the LAGRANGE and EULER representations, respectively, takes the form

$$
\begin{aligned}
&[\![\rho_R\left(\tfrac{1}{2}\boldsymbol{v}^2+\varepsilon\right)\boldsymbol{W}+(\boldsymbol{v}\boldsymbol{T}-\boldsymbol{Q})]\!]\cdot\boldsymbol{N}_{\mathfrak{S}}=0\,,\\
&[\![\rho\left(\tfrac{1}{2}\boldsymbol{v}^2+\varepsilon\right)(\boldsymbol{w}-\boldsymbol{v})+(\boldsymbol{v}\boldsymbol{t}-\boldsymbol{q})]\!]\cdot\boldsymbol{n}_{\mathfrak{s}}=0\,.
\end{aligned}
\tag{3.2.7}
$$

These equations can further be simplified with the help of the mass jump condition. In the Euler representation and with the aid of (3.2.2) one obtains

$$
[\![\tfrac{1}{2}\boldsymbol{v}^2+\varepsilon]\!]\mathcal{M}-[\![(\boldsymbol{v}\boldsymbol{t}-\boldsymbol{q})\cdot\boldsymbol{n}_{\mathfrak{s}}]\!]=0\,. \tag{3.2.8}
$$

Of significance, in this regard, is the jump condition for material singular surfaces $(\boldsymbol{w}-\boldsymbol{v})\cdot\boldsymbol{n}_{\mathfrak{s}}=0$ or $\boldsymbol{W}\cdot\boldsymbol{N}_{\mathfrak{S}}=0$. In this case, as already explained, the stress traction vector remains continuous; thus

$$
[\![\boldsymbol{Q}-\boldsymbol{v}\boldsymbol{T}]\!]\cdot\boldsymbol{N}_{\mathfrak{S}}=0\,,\quad [\![\boldsymbol{q}-\boldsymbol{v}\boldsymbol{t}]\!]\cdot\boldsymbol{n}_{\mathfrak{s}}=0\,,
$$

or with (3.2.5)

$$
[\![\boldsymbol{Q}\cdot\boldsymbol{N}_{\mathfrak{S}}]\!]-[\![\boldsymbol{v}]\!]\cdot\boldsymbol{T}\boldsymbol{N}_{\mathfrak{S}}=0\,,\qquad [\![\boldsymbol{q}\cdot\boldsymbol{n}_{\mathfrak{s}}]\!]-[\![\boldsymbol{v}]\!]\cdot\boldsymbol{t}\boldsymbol{n}_{\mathfrak{s}}=0\,. \tag{3.2.9}
$$

These relations hold without any further reduction if the two body parts slide upon one another. Physically, these equations express the fact that the heat lost by conduction must be generated within the surface by frictional dissipation due to sliding. If the velocity is continuous then the two body parts stick together, resulting in the continuity of the normal components of the heat flux vectors.

Entropy The entropy jump equation is obtained by inserting the identifications of (2.3.44) in the general jump condition, thus yielding

$$
\begin{aligned}
&[\![\rho_R s\boldsymbol{W}-\boldsymbol{\Phi}]\!]\cdot\boldsymbol{N}_{\mathfrak{S}}=-\mathfrak{P}_s\,,\\
&[\![\rho s(\boldsymbol{w}-\boldsymbol{v})-\boldsymbol{\phi}]\!]\cdot\boldsymbol{n}_{\mathfrak{s}}=-\mathfrak{p}_s\,,
\end{aligned}
\tag{3.2.10}
$$

where a positive surface production is permitted. With the implementations made in §2.3.6 (second law) one obtains

$$
\begin{aligned}
&[\![\rho_R s\boldsymbol{W}-\frac{\boldsymbol{Q}}{\Theta}]\!]\cdot\boldsymbol{N}_{\mathfrak{S}}=-\mathfrak{P}_s\le 0\,,\\
&[\![\rho s(\boldsymbol{w}-\boldsymbol{v})-\frac{\boldsymbol{q}}{\Theta}]\!]\cdot\boldsymbol{n}_{\mathfrak{s}}=-\mathfrak{p}_s\le 0\,,
\end{aligned}
\tag{3.2.11}
$$

in which the entropy flux is now represented by its CLAUSIUS –DUHEM expression.

3.3 Balance Statements and Jump Conditions for a 1–Component–System (Summary)

We shall now summarize the balance equations and jump conditions in the LAGRANGE and EULER representations for the five physical variables: mass, momentum, angular momentum, energy and entropy. Table 3.1 summarizes the local balance equations in their general form. Tables 3.2 and 3.3 list the physical variables under consideration and their production, supply and flux terms. Tables 3.4 and 3.5 exhibit the balance equations and the corresponding jump conditions for these physical variables.

Table 3.1. General form of local balance equations in the LAGRANGE and EULER representation.

Local balance equations
LAGRANGE **representation**
$\dfrac{\partial \Gamma}{\partial t} = -\operatorname{Div} \boldsymbol{\Phi} + \Pi + \Sigma$
$\left[\!\left[\boldsymbol{\Phi} \boldsymbol{N}_{\mathfrak{S}} - \Gamma \boldsymbol{W} \cdot \boldsymbol{N}_{\mathfrak{S}} \right]\!\right] = \mathfrak{P}$
EULER **representation**
$\dfrac{\mathrm{d}\gamma}{\mathrm{d}t} + \gamma \operatorname{div} \boldsymbol{v} = -\operatorname{div} \boldsymbol{\phi} + \pi + \varsigma$
$\left[\!\left[\boldsymbol{\phi} \boldsymbol{n}_{\mathfrak{s}} - \gamma (\boldsymbol{w} - \boldsymbol{v}) \cdot \boldsymbol{n}_{\mathfrak{s}} \right]\!\right] = \mathfrak{p}$

The notations represent the following definitions:

Γ, γ	Physical variable
$\boldsymbol{\Phi}, \boldsymbol{\phi}$	Flux of the physical variable
Π, π	Production
Σ, ς	Supply over the volume
$\mathfrak{P}, \mathfrak{p}$	Surface production on the singular surface
$\boldsymbol{N}_{\mathfrak{S}}, \boldsymbol{n}_{\mathfrak{s}}$	Unit normal vector at the singular surface
$\boldsymbol{W}, \boldsymbol{w}$	Velocity of the singular surface
$\boldsymbol{v}$	Velocity of a material particle

These quantities transform between the reference and present configuration according to

$$\Gamma = J\gamma\,, \quad \Pi = J\pi\,, \quad \Sigma = J\varsigma\,, \quad \boldsymbol{\Phi} = J\boldsymbol{\phi}\boldsymbol{F}^{-T}\,,$$

in which

$$\begin{aligned} \boldsymbol{F} &= \operatorname{Grad} \boldsymbol{\chi}(\boldsymbol{X}, t), \\ J &= \det \boldsymbol{F} \end{aligned}$$

are the deformation gradient and its JACOBIan determinant.

Table 3.2. Physical quantities for the derivation of balance statements in the LAGRANGE representation.

LAGRANGE **representation**				
Variables $\mathcal{G}$	Corresponding densities Γ	Production Π	Supply Σ	Flux $\boldsymbol{\Phi}$
Mass	ρ_R	0	0	0
Momentum	$\rho_R \boldsymbol{v}$	0	$\rho_R \boldsymbol{g}$	$-\boldsymbol{T}$
Angular momentum	$\rho_R \boldsymbol{x} \times \boldsymbol{v}$	0	$\rho_R \boldsymbol{x} \times \boldsymbol{g}$	$-\boldsymbol{x} \times \boldsymbol{T}$
Energy	$\frac{1}{2}\rho_R \boldsymbol{v}^2 + \rho_R \varepsilon$	0	$\rho_R \boldsymbol{g} \cdot \boldsymbol{v} + \rho_R \mathfrak{r}$	$\boldsymbol{Q} - \boldsymbol{v}\boldsymbol{T}$
Entropy	$\rho_R s$	$\rho_R \gamma \geq 0$	$\rho_R \dfrac{\mathfrak{r}}{\Theta}$	$\dfrac{\boldsymbol{Q}}{\Theta}$

The notations have the following definitions:

ρ_R Density (in the reference configuration),
$\boldsymbol{x}, \boldsymbol{v}$ Position vector, velocity,
$\boldsymbol{g}$ External force-field per unit mass (acceleration due to gravity),
$\boldsymbol{T}$ First PIOLA–KIRCHHOFF–stress tensor,
$\varepsilon, \mathfrak{r}$ Specific internal energy and radiation energy,
$\boldsymbol{Q}$ Material heat flux vector,
s, γ Specific entropy, entropy production,
Θ Absolute temperature .

Table 3.3. Physical quantities for the derivation of balance statements in the EULER representation.

EULER **representation**				
Variables $\mathcal{G}$	Corresponding densities γ	Production π	Supply ς	Flux $\boldsymbol{\phi}$
Mass	ρ	0	0	0
Momentum	$\rho \boldsymbol{v}$	0	$\rho \boldsymbol{g}$	$-\boldsymbol{t}$
Angular momentum	$\rho \boldsymbol{x} \times \boldsymbol{v}$	0	$\rho \boldsymbol{x} \times \boldsymbol{g}$	$-\boldsymbol{x} \times \boldsymbol{t}$
Energy	$\frac{1}{2}\rho \boldsymbol{v}^2 + \rho \varepsilon$	0	$\rho \boldsymbol{g} \cdot \boldsymbol{v} + \rho \mathfrak{r}$	$\boldsymbol{q} - \boldsymbol{v}\boldsymbol{t}$
Entropy	ρs	$\rho \gamma \geq 0$	$\rho \dfrac{\mathfrak{r}}{\Theta}$	$\dfrac{\boldsymbol{q}}{\Theta}$

The quantities appearing in addition to the Table 3.2 have the following definitions:

ρ Density (in the present configuration),
$\boldsymbol{t}$ CAUCHY stress tensor,
$\boldsymbol{q}$ Spatial heat flux vector.

The remaining definitions are as in Table 3.2.

Table 3.4. Local balance statements and jump conditions in the LAGRANGE representation for mass, momentum, angular momentum, energy and entropy.

LAGRANGE **balance statements**	
Mass balance	$\frac{\mathrm{d}\rho_R}{\mathrm{d}t} = 0 \quad \Rightarrow \quad \rho_R = \rho_R(\boldsymbol{X})$
Momentum balance	$\rho_R \frac{\mathrm{d}\boldsymbol{v}}{\mathrm{d}t} = \operatorname{Div} \boldsymbol{T} + \rho_R \boldsymbol{g}$
Angular momentum balance	$\boldsymbol{T}\boldsymbol{F}^T = (\boldsymbol{T}\boldsymbol{F}^T)^T$
Energy balance	$\rho_R \frac{\mathrm{d}\varepsilon}{\mathrm{d}t} = -\operatorname{Div} \boldsymbol{Q} + \operatorname{tr}(\operatorname{Grad} \boldsymbol{v}\boldsymbol{T}^T) + \rho_R \mathfrak{r}$
Entropy balance	$\rho_R \frac{\mathrm{d}s}{\mathrm{d}t} + \operatorname{Div}\left(\frac{\boldsymbol{Q}}{\Theta}\right) - \rho_R \frac{\mathfrak{r}}{\Theta} = \rho_R \gamma \geq 0$
Jump conditions	
Density	$[\![\rho_R \boldsymbol{W}]\!] \cdot \boldsymbol{N}_{\mathfrak{S}} = 0$
Momentum	$[\![\rho_R \boldsymbol{v}(\boldsymbol{W} \cdot \boldsymbol{N}_{\mathfrak{S}}) + \boldsymbol{T}\boldsymbol{N}_{\mathfrak{S}}]\!] = \boldsymbol{0}$
Angular momentum	—
Energy	$[\![\rho_R \left(\frac{1}{2}\boldsymbol{v}^2 + \varepsilon\right) \boldsymbol{W} + (\boldsymbol{v}\boldsymbol{T} - \boldsymbol{Q})]\!] \cdot \boldsymbol{N}_{\mathfrak{S}} = 0$
Entropy	$[\![\rho_R s \boldsymbol{W} - \frac{\boldsymbol{Q}}{\Theta}]\!] \cdot \boldsymbol{N}_{\mathfrak{S}} = -\mathfrak{P}_s \leq 0$

The balance equation of entropy specifies the second law in the special form of the CLAUSIUS–DUHEM–inequality (see §2.3.6). In the corresponding jump conditions a (positive) surface production of entropy, $\mathfrak{P}_s$, is introduced. The notations of these quantities are shown in Tables 3.2 and 3.3. In some of the later applications (Chaps. 5, 7 and 12) we shall use more general entropy balance statements, in which the entropy flux is not specified as heat flux divided by absolute temperature.

Table 3.5. Local balance statements and corresponding jump conditions for mass, momentum, angular momentum, energy and entropy in the EULER representation.

EULER **balance statements**	
Mass balance	$\frac{\mathrm{d}\rho}{\mathrm{d}t} + \rho \operatorname{div} \boldsymbol{v} = 0$
Momentum balance	$\rho \frac{\mathrm{d}\boldsymbol{v}}{\mathrm{d}t} = \operatorname{div} \boldsymbol{t} + \rho \boldsymbol{g}$
Angular momentum balance	$\boldsymbol{t} = \boldsymbol{t}^T$
Energy balance	$\rho \frac{\mathrm{d}\varepsilon}{\mathrm{d}t} = -\operatorname{div} \boldsymbol{q} + \operatorname{tr}(\operatorname{grad} \boldsymbol{v t}) + \rho \mathfrak{r}$
Entropy balance	$\rho \frac{\mathrm{d}s}{\mathrm{d}t} + \operatorname{div}\left(\frac{\boldsymbol{q}}{\Theta}\right) - \rho \frac{\mathfrak{r}}{\Theta} = \rho\gamma \geq 0$
Jump conditions	
Density	$[\![\rho(\boldsymbol{v} - \boldsymbol{w})]\!] \cdot \boldsymbol{n}_s = 0$ or $[\![\mathcal{M}]\!] = 0$
Momentum	$[\![\rho\boldsymbol{v}\left((\boldsymbol{w} - \boldsymbol{v}) \cdot \boldsymbol{n}_s\right) + \boldsymbol{t}\boldsymbol{n}_s]\!] = \boldsymbol{0}$ or $[\![\boldsymbol{v}]\!]\mathcal{M} - [\![\boldsymbol{t}\boldsymbol{n}_s]\!] = \boldsymbol{0}$
Angular momentum	—
Energy	$[\![\rho\left(\frac{1}{2}\boldsymbol{v}^2 + \varepsilon\right)(\boldsymbol{w} - \boldsymbol{v}) + (\boldsymbol{v t} - \boldsymbol{q})]\!] \cdot \boldsymbol{n}_s = 0$
Entropy	$[\![\rho s(\boldsymbol{w} - \boldsymbol{v}) - \frac{\boldsymbol{q}}{\Theta}]\!] \cdot \boldsymbol{n}_s = -\mathfrak{p}_s \leq 0$

The balance equation of entropy implies the second law in the special form of the CLAUSIUS–DUHEM inequality (see §2.3.6). In case of the corresponding jump conditions a (positive) surface production of entropy, $\mathfrak{p}_s$, is introduced. For notation see Tables 3.2 and 3.3. In some of the later applications (Chaps. 5, 7 and 12) we shall use more general entropy balance statements, in which the entropy flux is not specified as heat flux divided by absolute temperature.

3.4 Exercises

1. Suppose that the following jump conditions hold for a singular surface

$$[\![a]\!] \neq 0\,, \quad [\![b]\!] \neq 0\,, \quad [\![c]\!] = 0\,.$$

Calculate the expressions (or show):

$$[\![a+b]\!]\,, \quad [\![a+c]\!]\,, \quad [\![a\cdot c]\!]\,, \quad [\![a\cdot(b+c)]\!]\,,$$

$$[\![a\cdot b]\!] \neq [\![a]\!]\cdot[\![b]\!]\,, \quad [\![a]\!]\langle\!\langle b\rangle\!\rangle + \langle\!\langle a\rangle\!\rangle[\![b]\!] = [\![ab]\!]\,.$$

2. a) Show that on a shock surface, $[\![W_{(\mathrm{i})}]\!] \neq 0$.
   b) Show for a non–material vortex surface that the density ρ is continuous, for accelerating waves that the propagation velocity $W_{(\mathrm{i})}$ is continuous.
   c) If the flux and the production of a quantity γ are continuous, $[\![\boldsymbol{\phi}\boldsymbol{n}_s]\!] = \mathbf{0}$, then the quantity γ is also continuous on a non–material vortex surface.
3. Let $f(x,t) = 0$ be the free-surface equation of an ice sheet. Show that

$$\left(\frac{\mathrm{d}f}{\mathrm{d}t}\right)^{-} = -\|\,\mathrm{grad}\,f\|\,\frac{\mathcal{M}}{\rho^{-}}\,, \tag{3.4.1}$$

where $\mathcal{M}$ is the mass flux through the surface ($\mathcal{M} > 0$ for mass addition),

$$\mathcal{M} = \rho^{-}(\boldsymbol{w} - \boldsymbol{v}^{-})\cdot\boldsymbol{n}_s\,, \tag{3.4.2}$$

and $()^{-}$ indicates the side of the ice sheet.
4. Derive the jump condition for the angular momentum of a spin continuum (without production term).
5. Assume that the base of an ice sheet, given by the equation $z = f_b(x,\,y,\,t)$, is temperate (with temperature exactly at the pressure melting point). Derive the kinematic boundary condition for this surface and formulate the thermal jump condition for the calculation of the melting rate.

3.5 Solutions

1. Computational rules for jump quantities: If the quantities a and b experience a jump, $[\![a]\!] \neq 0$, $[\![b]\!] \neq 0$, but the quantity c is continuous, $[\![c]\!] = 0$, then

$$
\begin{aligned}
[\![a+b]\!] &= (a^+ + b^+) - (a^- + b^-) = a^+ - a^- + b^+ - b^- \\
&= [\![a]\!] + [\![b]\!] \,, \\
[\![a+c]\!] &= (a^+ + c) - (a^- + c) = a^+ - a^- \\
&= [\![a]\!] \,, \\
[\![ac]\!] &= (a^+ c) - (a^- c) = (a^+ - a^-)c \\
&= [\![a]\!]c \,, \\
[\![a(b+c)]\!] &= a^+(b^+ + c) - a^-(b^- + c) \\
&= a^+b^+ - a^-b^- + (a^+ - a^-)c = [\![ab]\!] + [\![a]\!]c \,, \\
[\![ab]\!] &= (a^+b^+) - (a^-b^-) \,, \quad \text{but} \\
[\![a]\!][\![b]\!] &= (a^+ - a^-)(b^+ - b^-) \\
&= a^+b^+ + a^-b^- - a^-b^+ - a^+b^- \neq [\![ab]\!] \,, \\
[\![a]\!]\langle\!\langle b\rangle\!\rangle + \langle\!\langle a\rangle\!\rangle[\![b]\!] &= \tfrac{1}{2}(a^+ - a^-)(b^+ + b^-) + \tfrac{1}{2}(a^+ + a^-)(b^+ - b^-) \\
&= \tfrac{1}{2}(a^+b^+ + a^+b^- - a^-b^+ - a^-b^- \\
&\quad + a^+b^+ - a^+b^- + a^-b^+ - a^-b^-) \\
&= a^+b^+ - a^-b^- = [\![ab]\!] \,.
\end{aligned}
$$

2. a) According to the definition of the jump condition on a shock surface one has $[\![\boldsymbol{v}\cdot\boldsymbol{n}_\mathfrak{s}]\!] \neq 0$, and therefore

$$
[\![W_{(\mathrm{i})}]\!] = [\![(\boldsymbol{w}-\boldsymbol{v})\cdot\boldsymbol{n}_\mathfrak{s}]\!] = -[\![\boldsymbol{v}\cdot\boldsymbol{n}_\mathfrak{s}]\!] \neq 0 \,.
$$

b) On a vortex surface, on the other hand, $[\![\boldsymbol{v}\cdot\boldsymbol{n}_\mathfrak{s}]\!] = 0$; an acceleration wave is defined such that $\boldsymbol{v}$ and $\boldsymbol{F}$ are continuous across the singular surface, but not the derivatives $\dot{\boldsymbol{v}}, \dot{\boldsymbol{F}}$ and $\operatorname{grad}\boldsymbol{F}$. Then, it follows from (3.2.1)

$$
[\![\rho(\boldsymbol{w}-\boldsymbol{v})\cdot\boldsymbol{n}_\mathfrak{s}]\!] = [\![\rho]\!]\underbrace{(\boldsymbol{w}-\boldsymbol{v})\cdot\boldsymbol{n}_\mathfrak{s}}_{\neq 0} = 0 \,,
$$

and here, the second factor does not vanish; as a result, the density must be continuous i.e. $[\![\rho]\!] = 0$. On the other hand, from

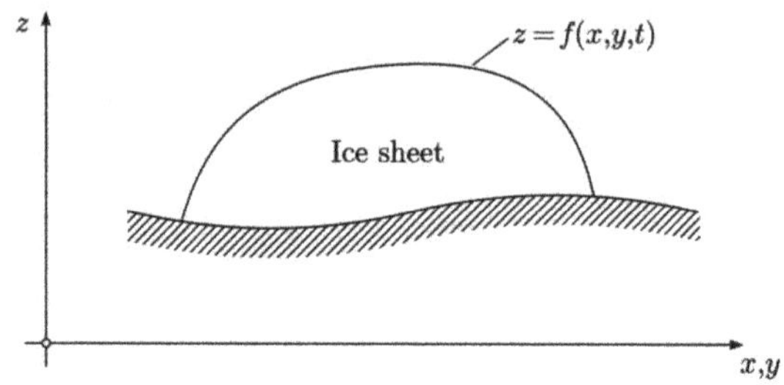

Fig. 3.4. Geometry of a spatially three-dimensional ice-sheet with free surface $z = f(x, y, t)$.

the definition of the instantaneous propagation velocity we have $W_{(\mathrm{i})} = w_s - \boldsymbol{v} \cdot \boldsymbol{n}_s$; with the continuity of $\boldsymbol{v} \cdot \boldsymbol{n}_s$, the continuity of $W_{(\mathrm{i})}$ holds.

c) The general jump balance (3.1.26) for a quantity γ is given by

$$[\![\phi \boldsymbol{n}_s]\!] - [\![\gamma(\boldsymbol{w} - \boldsymbol{v}) \cdot \boldsymbol{n}_s]\!] = \mathfrak{p} \,.$$

Since according to the statement of the problem both terms, $[\![\phi \boldsymbol{n}_s]\!]$ and $\mathfrak{p}$ vanish, this implies $[\![\gamma(\boldsymbol{w} - \boldsymbol{v}) \cdot \boldsymbol{n}_s]\!] = 0$, which in turn yields, as in the case of the density, $[\![\gamma]\!] = 0$.

3. The kinematic equation of a surface $f(\boldsymbol{x}, t) = 0$ reads

$$\frac{\mathrm{d}f}{\mathrm{d}t} = \frac{\partial f}{\partial t} + (\operatorname{grad} f) \cdot \boldsymbol{w} = 0 \,, \tag{3.5.1}$$

where $\boldsymbol{w}$ is the velocity of the singular surface. Taking $()^-$ as the quantities representing the ice side, we get the following expression from (3.5.1)

$$\begin{aligned} \frac{\partial f}{\partial t} + (\operatorname{grad} f) \cdot \boldsymbol{v}^- &= \|\operatorname{grad} f\| \, (\boldsymbol{v}^- - \boldsymbol{w}) \cdot \frac{\operatorname{grad} f}{\|\operatorname{grad} f\|} \\ &= \|\operatorname{grad} f\| \, (\boldsymbol{v}^- - \boldsymbol{w}) \cdot \boldsymbol{n}_s \,, \end{aligned} \tag{3.5.2}$$

where $\boldsymbol{n}_s := \operatorname{grad} f / \| \operatorname{grad} f \|$ is defined such that it represents the outer unit normal vector at the surface. The jump condition for the mass balance is given by

$$[\![\rho(\boldsymbol{w} - \boldsymbol{v}) \cdot \boldsymbol{n}_s]\!] = 0 \Rightarrow \rho^-(\boldsymbol{w} - \boldsymbol{v})^- \cdot \boldsymbol{n}_s = \mathcal{M} \,. \tag{3.5.3}$$

Hereby, $\mathcal{M} > 0$ corresponds to addition of mass through snow fall and $\mathcal{M} < 0$ represents the ablation through melting of ice. Substituting (3.5.3) in (3.5.2) one obtains

$$\frac{\partial f}{\partial t} + (\operatorname{grad} f) \cdot \boldsymbol{v}^- = -\|\operatorname{grad} f\| \frac{\mathcal{M}}{\rho^-} . \tag{3.5.4}$$

Notice that $\mathcal{M}/\rho^- = a_\perp$ corresponds to a velocity: it is the ice volume flux perpendicular to the surface per unit surface element. The left-hand side of (3.5.4) represents the material derivative $(df/dt)^-$ following the motion of the ice particle; thus

$$\left(\frac{\mathrm{d}f}{\mathrm{d}t}\right)^- = -\|\operatorname{grad} f\| \frac{\mathcal{M}}{\rho^-} . \tag{3.5.5}$$

4. With the following specifications (see Exercise 3 of Chap. 2) in the EULER and LAGRANGE representation

$$\gamma = \boldsymbol{x} \times \rho \boldsymbol{v} + \rho \tilde{\boldsymbol{s}} \quad , \quad \Gamma = \boldsymbol{x} \times \rho_R \dot{\boldsymbol{x}} + \rho_R \tilde{\boldsymbol{s}}$$

$$\boldsymbol{\phi} = -\boldsymbol{x} \times \boldsymbol{t} - \boldsymbol{m} \quad , \quad \boldsymbol{\Phi} = -\boldsymbol{x} \times \boldsymbol{T} - \boldsymbol{M} ,$$

(3.1.26) in the LAGRANGE representation gives

$$\begin{aligned} \mathbf{0} &= [\![(\boldsymbol{x} \times \rho_R \dot{\boldsymbol{x}} + \rho_R \tilde{\boldsymbol{s}})(\boldsymbol{W} \cdot \boldsymbol{N}_{\mathfrak{S}})]\!] + [\![\boldsymbol{x} \times \boldsymbol{T}\boldsymbol{N}_{\mathfrak{S}} + \boldsymbol{M}\boldsymbol{N}_{\mathfrak{S}}]\!] \\ &= \boldsymbol{x} \times \underbrace{[\![\rho_R \dot{\boldsymbol{x}} (\boldsymbol{W} \cdot \boldsymbol{N}_{\mathfrak{S}}) + \boldsymbol{T}\boldsymbol{N}_{\mathfrak{S}}]\!]}_{=0 \text{ (momentum jump condition)}} + [\![\rho_R \tilde{\boldsymbol{s}} (\boldsymbol{W} \cdot \boldsymbol{N}_{\mathfrak{S}}) + \boldsymbol{M}\boldsymbol{N}_{\mathfrak{S}}]\!] \\ &= [\![\rho_R \tilde{\boldsymbol{s}} (\boldsymbol{W} \cdot \boldsymbol{N}_{\mathfrak{S}})]\!] + [\![\boldsymbol{M}\boldsymbol{N}_{\mathfrak{S}}]\!] , \end{aligned}$$

and similarly in the EULER representation

$$\begin{aligned} \mathbf{0} &= [\![(\boldsymbol{x} \times \rho \boldsymbol{v} + \rho \tilde{\boldsymbol{s}})(\boldsymbol{w} - \boldsymbol{v}) \cdot \boldsymbol{n}_s]\!] + [\![\boldsymbol{x} \times \boldsymbol{t}\boldsymbol{n}_s + \boldsymbol{m}\boldsymbol{n}_s]\!] \\ &= \boldsymbol{x} \times \underbrace{[\![\rho \boldsymbol{v} (\boldsymbol{w} - \boldsymbol{v}) \cdot \boldsymbol{n}_s + \boldsymbol{t}\boldsymbol{n}_s]\!]}_{=0 \text{ (momentum jump condition)}} + [\![\rho \tilde{\boldsymbol{s}} (\boldsymbol{w} - \boldsymbol{v}) \cdot \boldsymbol{n}_s + \boldsymbol{m}\boldsymbol{n}_s]\!] \\ &= [\![\rho \tilde{\boldsymbol{s}} (\boldsymbol{w} - \boldsymbol{v}) \cdot \boldsymbol{n}_s]\!] + [\![\boldsymbol{m}\boldsymbol{n}_s]\!] . \end{aligned}$$

These are the required jump conditions for the angular momentum balance of a spin continuum.

5. Let us represent the ice side with "−" and the ice-base side with "+"; then the kinematic equation following from the equation of the base, $f_b(x, y, t) - z = 0$, is given by

$$\begin{aligned} &\frac{\partial f_b}{\partial t} + \frac{\partial f_b}{\partial x} u^- + \frac{\partial f_b}{\partial y} v^- - w^- \\ &\qquad = -\frac{\partial f_b}{\partial x}(u_b - u^-) - \frac{\partial f_b}{\partial y}(v_b - v^-) + w_b - w^- . \end{aligned} \tag{3.5.6}$$

With

$$\boldsymbol{n}_b = \frac{(\partial f_b/\partial x,\ \partial f_b/\partial y,\ -1)}{\sqrt{1+(\partial f_b/\partial x)^2+(\partial f_b/\partial y)^2}} \tag{3.5.7}$$

the coordinate-invariant representation is

$$\frac{\partial f_b/\partial t}{\sqrt{1+(\partial f_b/\partial x)^2+(\partial f_b/\partial y)^2}} + \boldsymbol{v}^- \cdot \boldsymbol{n}_b = \underbrace{(\boldsymbol{v}^- - \boldsymbol{w}_b)\cdot \boldsymbol{n}_b}_{a_b^\perp} . \tag{3.5.8}$$

The outer normal $\boldsymbol{n}_b$ is directed to the ice base, and $a_b^\perp$ is the so-called melting rate (positive for melting, negative for freezing).
The equation determining $a_b^\perp$ is obtained from the energy jump condition $(3.2.7)_2$, which appears in the following form under the application of the mass jump condition

$$[\![\boldsymbol{v}\boldsymbol{t}\cdot\boldsymbol{n}_b]\!] - [\![\boldsymbol{q}\cdot\boldsymbol{n}_b]\!] = -[\![\frac{\boldsymbol{v}\cdot\boldsymbol{v}}{2}+\varepsilon]\!]\,\underbrace{\rho^-(\boldsymbol{w}_b-\boldsymbol{v}^-)\cdot\boldsymbol{n}_b}_{-\rho^- a_b^\perp} . \tag{3.5.9}$$

The individual terms can be specified or re-formulated as follows:

$$\begin{aligned}
[\![\boldsymbol{v}\boldsymbol{t}\cdot\boldsymbol{n}_b]\!] &= [\![\boldsymbol{v}\cdot\boldsymbol{t}\boldsymbol{n}_b]\!] \\
&= [\![\boldsymbol{v}_\perp(-p^\perp)]\!] + [\![\boldsymbol{v}_\parallel\cdot\boldsymbol{\tau}]\!] \\
&= -[\![\boldsymbol{v}_\perp]\!]\,p^\perp + [\![\boldsymbol{v}_\parallel\cdot\boldsymbol{\tau}]\!] ,
\end{aligned} \tag{3.5.10}$$

$$[\![\boldsymbol{q}\cdot\boldsymbol{n}_b]\!] = \boldsymbol{q}^+\cdot\boldsymbol{n}_b - \boldsymbol{q}^-\cdot\boldsymbol{n}_b = -Q^\perp_{Geoth} - \boldsymbol{q}^-\cdot\boldsymbol{n}_b , \tag{3.5.11}$$

$$[\![\varepsilon]\!] = L - p_\perp[\![1/\rho]\!] . \tag{3.5.12}$$

The above simplifications and formulations are accomplished on the basis of thermodynamic results of Chap. 6, namely, that $p_\perp$ is continuous across the phase change surface, $[\![p^\perp]\!]=0$, that $\boldsymbol{q}^+\cdot\boldsymbol{n}_b$ is replaced by the negative geothermal heat flow, that $[\![\varepsilon]\!]=\varepsilon_{water}-\varepsilon_{ice}$ is the jump of the internal energy for H_2O and is given by the latent heat L, the jump of the specific volume $[\![1/\rho]\!]$ and the pressure as stated in (3.5.12). Equations (3.5.10), (3.5.11),(3.5.12) are substituted in (3.5.9) to yield

$$\begin{aligned}
-[\![\boldsymbol{v}_\perp]\!]\,p_\perp + [\![\boldsymbol{v}_\parallel\cdot\boldsymbol{\tau}]\!] + Q^\perp_{Geoth} + \boldsymbol{q}^-\cdot\boldsymbol{n}_b \\
= a_b^\perp\,\rho^-\left([\![\frac{\boldsymbol{v}\cdot\boldsymbol{v}}{2}]\!] + L - p_\perp[\![\tfrac{1}{\rho}]\!]\right) ,
\end{aligned} \tag{3.5.13}$$

which can be simplified to yield

$$[\![\boldsymbol{v}_\parallel]\!]\cdot\boldsymbol{\tau} + Q^\perp_{Geoth} + \boldsymbol{q}^-\cdot\boldsymbol{n}_b = \rho^-\,L\,a_b^\perp . \tag{3.5.14}$$

In arriving at this equation it was assumed that on the right-hand side the latent heat contribution overrides the contributions from the kinetic energy and the pressure. Furthermore, $[\![\boldsymbol{v}_\perp]\!]\,p_\perp$ was ignored and $[\![\tau]\!]=\boldsymbol{0}$ was assumed.

4. Moving Reference Systems

4.1 Transformation of Position Vectors

In Chap. 2 it was assumed that balance equations were referred to an observer at rest in an *inertial system*. In this chapter our intention is not to concentrate on the issues like existence of such a reference system or on the ideas of constructing such a system, rather we presume that there exists such a reference system relative to which the quantities like momentum density, momentum supply and flux of momentum, as shown in Chap. 2, are uniquely defined. In the following, we shall demonstrate that the balance equations can also be formulated relative to an observer, which moves himself relative to a fixed observer, i.e., the observer performs a translatory and/or rotatory motion. In geophysical applications this situation arises quite frequently, because for a number of processes the Earth can not be identified with an inertial system. This implies that one must take the Earth's motion in the respective equations into account and modify them accordingly. Theoretically, this can be done by introducing additional terms of the relative motion – CORIOLIS, centripetal and EULER accelerations. A change of the reference system, however, will also play an important role for the material equations as will be seen in Chap. 5.

We consider two different *reference systems*, one fixed and the other in motion[1]. To every reference system belongs a reference point, the so-called origin, from which one measures distances or defines position vectors in space relative to the reference point. Figure 4.1 shows two such reference systems and the relationship between the position vectors to the same point measured in both reference systems. Notice that this representation is perfectly free of the choice of the *coordinate system* which one may choose in each reference system and with which one may define the components of a position vector (or velocity, etc.). The respective coordinate system can be chosen arbitrarily or matched physically to the real facts. The components of the position vector can have different values depending upon the coordinate system chosen in the analysis. In NEWTONian mechanics the physical space is EUCLIDian, and the

[1] The reader should not confuse the term "reference system" with "reference configuration". All transformations in this chapter "live" in the *present configuration*.

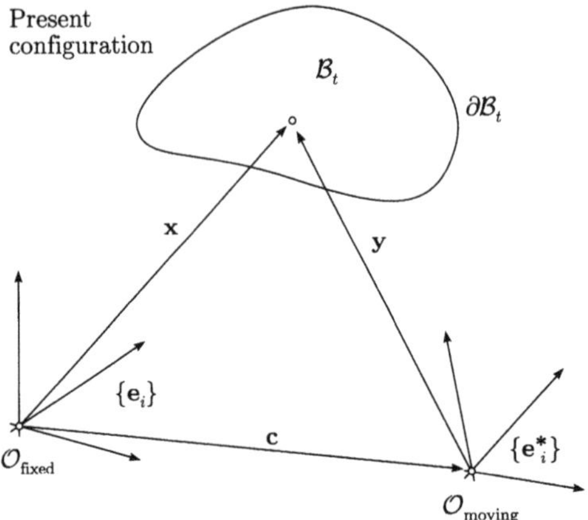

Fig. 4.1. Fixed and moving reference systems with the origin $\mathcal{O}_{\text{fixed}}$ and basis $\{\boldsymbol{e}_i\}$ or $\mathcal{O}_{\text{moving}}$ and $\{\boldsymbol{e}_i{}^*\}$, respectively. Vectors $\boldsymbol{x}$ and $\boldsymbol{y}$ are the position vectors of a given point with respect to the fixed and the moving system, measured relative to the fixed basis. The vector $\mathbf{c}$ gives the displacement between both reference points.

observer in one reference system chooses his own basis $\{\boldsymbol{e}_i\}$ or $\{\boldsymbol{e}_i{}^*\}$ to form the different coordinate systems. The following representation is essentially independent of the choice of the coordinate system. Therefore, when one must perform calculations, the first task is to choose a coordinate system.

Let $\boldsymbol{x}$ be a position vector of a material point in space relative to the fixed reference system and $\boldsymbol{y}$ the position vector of the same point in the moving reference system. Obviously, there exists a relation between the position vectors given by

$$\boldsymbol{x} = \boldsymbol{y} + \boldsymbol{c} \,, \tag{4.1.1}$$

where $\boldsymbol{c}$ is the displacement (translation) of both reference systems or, better, of their origins, and $\boldsymbol{x}$, $\boldsymbol{y}$ and $\boldsymbol{c}$ are referred to the basis of the fixed observer. Let us begin by assuming that each observer refers not only the position vectors to his own chosen origin but also chooses his own basis, which is attached to the origin. In Fig 4.1 these bases are Cartesian. The bases can now be related to each other, however, the moving basis performs a translation as well as a rotation. Finally, there are also possibilities of *mirror reflections* when one changes from right to left oriented bases; however, these will be excluded here. The transformation, i.e., the rotation is described by the *rotation matrix* $\boldsymbol{O}$ with the properties

$$\boldsymbol{O}\boldsymbol{O}^T = \boldsymbol{O}^T\boldsymbol{O} = \boldsymbol{I} \quad \text{and} \quad \det \boldsymbol{O} = +1 \,, \tag{4.1.2}$$

i.e., it is *orthogonal* [2] and, in principle, time dependent.

[2] The orthogonality condition merely implies

In Cartesian coordinates, a rotation about the z-axis is represented by

$$[O] = \begin{pmatrix} \cos\phi & \sin\phi & 0 \\ -\sin\phi & \cos\phi & 0 \\ 0 & 0 & 1 \end{pmatrix} , \tag{4.1.3}$$

where ϕ is the rotation angle.

With this rotation of the coordinate systems relative to one another the position vector $\boldsymbol{y}$ in the moving coordinate system may be represented relative to the fixed coordinate system on the one hand, and relative to the moving and rotated system on the other hand. Quantities in the rotating coordinate system are indicated with stars. If $\{\boldsymbol{e}_i\}$ and $\{\boldsymbol{e}_i{}^*\}$ represent the basis vectors in the fixed and moving coordinate systems, respectively, then the position vector $\boldsymbol{y}$ can be written in its component form as

$$\begin{aligned} \boldsymbol{y} &= y_1\boldsymbol{e}_1 + y_2\boldsymbol{e}_2 + y_3\boldsymbol{e}_3 , \\ \boldsymbol{y}^* &= y_1{}^*\boldsymbol{e}_1{}^* + y_2{}^*\boldsymbol{e}_2{}^* + y_3{}^*\boldsymbol{e}_3{}^* . \end{aligned} \tag{4.1.4}$$

These representations are related to each other according to

$$\begin{aligned} \boldsymbol{y}^* &= \boldsymbol{O}\boldsymbol{y} \qquad \text{or} \quad \boldsymbol{y} = \boldsymbol{O}^T\boldsymbol{y}^* , \\ y_i{}^* &= O_{ij}y_j \qquad \text{or} \quad y_i = O_{ji}y_j{}^* . \end{aligned} \tag{4.1.5}$$

A calculation of such a rotation with $\boldsymbol{c} = \boldsymbol{0}$, for example, and a rotation of 45° in two dimensions is demonstrated in Fig. 4.2. Thus, the position vector in the fixed reference system is represented as

$$\begin{aligned} &\boldsymbol{x} = \boldsymbol{y} + \boldsymbol{c} = \boldsymbol{O}^T\boldsymbol{y}^* + \boldsymbol{c} , \\ &\text{or} \quad \boldsymbol{y}^* = \boldsymbol{O}\boldsymbol{x} - \boldsymbol{c}^* \quad \text{with} \quad \boldsymbol{c}^* = \boldsymbol{O}\boldsymbol{c} . \end{aligned} \tag{4.1.6}$$

On the one hand, this equation expresses the position of a particle in a fixed reference system with the components of the fixed coordinate system ($\boldsymbol{x}$) and, on the other hand, in a moving reference frame with the components of the fixed coordinate ($\boldsymbol{y}$), which, in turn, may be represented relative to the moving coordinate system ($\boldsymbol{O}^T\boldsymbol{y}^*$). Thus, the same point can be represented by its components y_i^* in the moving coordinate system as well by x_i in the fixed coordinate system.

Transformations (4.1.6) are called EUCLIDian transformations.

$$\det(\boldsymbol{O}\boldsymbol{O}^T) = \det\boldsymbol{O}\,\det\boldsymbol{O}^T = \det\boldsymbol{I} = 1 \quad \Longrightarrow \quad \det\boldsymbol{O} = \pm 1 .$$

In orthogonal transformations, the distance between any two points remains the same; indeed, if $\boldsymbol{d}^* = \boldsymbol{O}\boldsymbol{d}$, the following identity holds:

$$\boldsymbol{d}^* \cdot \boldsymbol{d}^* = \boldsymbol{O}\boldsymbol{d} \cdot \boldsymbol{O}\boldsymbol{d} = \boldsymbol{d} \cdot \boldsymbol{O}^T\boldsymbol{O}\boldsymbol{d} = \boldsymbol{d} \cdot \boldsymbol{d} .$$

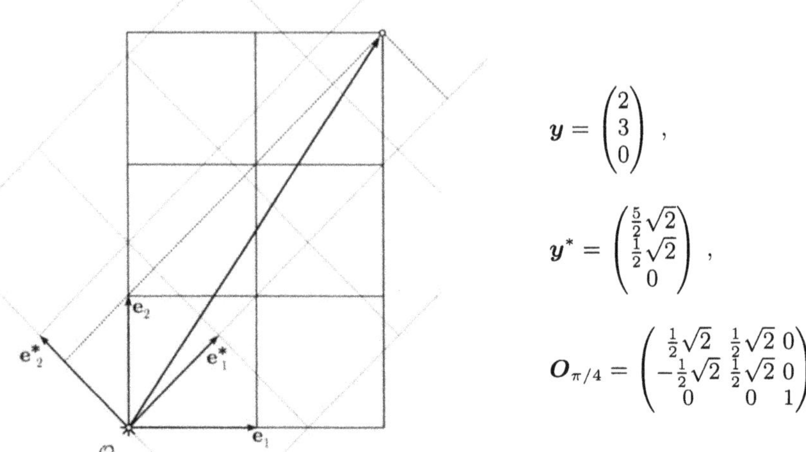

Fig. 4.2. Demonstration of a position vector in two coordinate systems that are rotated relative to one another.

4.2 Velocity and Acceleration

Since we wish to trace the motion of material points we differentiate the position vector $\boldsymbol{y}$ with respect to time to derive the velocity. From (4.1.5),

$$\dot{\boldsymbol{y}} = \left(\boldsymbol{O}^T\boldsymbol{y}^*\right)^{\boldsymbol{\cdot}} = \left(\boldsymbol{O}^T\right)^{\boldsymbol{\cdot}}\boldsymbol{y}^* + \boldsymbol{O}^T\left(\boldsymbol{y}^*\right)^{\boldsymbol{\cdot}} . \tag{4.2.1}$$

Note here that $(\boldsymbol{y}^{\boldsymbol{\cdot}})^* \neq (\boldsymbol{y}^*)^{\boldsymbol{\cdot}}$! The first represents the (absolute) time rate of change of the quantity $\boldsymbol{y}$ in the moving system, whereas the second represents the change of the quantity $\boldsymbol{y}$ in the moving system. Since this representation can change with respect to time in both systems both expressions are different from each other.

Using the transformation (4.2.1), (4.1.5) gives, in the moving system,

$$\dot{\boldsymbol{y}} = \left(\dot{\boldsymbol{O}}^T\boldsymbol{O}\right)\boldsymbol{y} + \boldsymbol{O}^T\left(\boldsymbol{y}^*\right)^{\boldsymbol{\cdot}} , \tag{4.2.2}$$

and in the fixed reference system we have

$$\dot{\boldsymbol{x}} = \dot{\boldsymbol{c}} + \left(\dot{\boldsymbol{O}}^T\boldsymbol{O}\right)\boldsymbol{y} + \boldsymbol{O}^T\left(\boldsymbol{y}^*\right)^{\boldsymbol{\cdot}} . \tag{4.2.3}$$

This equation describes the time rate of change of the position vector in the fixed coordinate system in terms of the so-called *frozen velocity* $\boldsymbol{v}_f := \dot{\boldsymbol{c}} +$

$\dot{O}^T O y$ and the *relative velocity* $v_r := O^T (y^*)^{\bullet}$. The frozen velocity[3] describes the motion of the material point under consideration in the fixed coordinate system if one identifies its motion with that of the moving coordinate system; it is given by the sum of the *translation velocity* $\dot{c}$ and the *rotation velocity* $\dot{O}^T O y$. It is obtained by momentarily *freezing* the particle with the moving frame. These velocity contributions are actually governed by the motion of the coordinate system. The relative velocity, in contrast, arises due to the motion of a particle relative to the moving coordinate system.

The rotational part can now be written in terms of the *spin matrix*[4], or in short, the *spin*, through

$$\bar{\Omega} := \dot{O}^T O \,. \tag{4.2.4}$$

The spin matrix is skew-symmetric, $\Omega = -\Omega^T$; indeed, from (4.1.2) we have

$$\left(O^T O\right)^{\bullet} = \dot{I} = 0 \quad \text{or} \quad \dot{O}^T O + O^T \dot{O} = 0 \,, \tag{4.2.5}$$

so that

$$\bar{\Omega} = \dot{O}^T O = -O^T \dot{O} = -\left(\dot{O}^T O\right)^T = -\bar{\Omega}^T \,. \tag{4.2.6}$$

A skew-symmetric matrix has, in three dimensions, exactly three different components – the diagonal elements must all vanish, and the three elements below the diagonal appear with the mirror reflection and negation of the elements above the diagonal. These three components of the spin matrix are indicated by the letters ω_1, ω_2, ω_3 and are arranged in matrix form as

$$[\bar{\Omega}] = \begin{pmatrix} 0 & -\omega_3 & \omega_2 \\ \omega_3 & 0 & -\omega_1 \\ -\omega_2 & \omega_1 & 0 \end{pmatrix} \,. \tag{4.2.7}$$

Every skew-symmetric matrix can be assigned to its *dual vector* with three components and vice versa. The skew-symmetric spin matrix corresponds to the vector of *angular velocity* ω defined by

$$\bar{\Omega}_{ij} = -\varepsilon_{ijk}\omega_k \,, \quad \omega_i = -\tfrac{1}{2}\varepsilon_{ijk}\,\bar{\Omega}_{jk} \,. \tag{4.2.8}$$

In vector form we can write this definition as

$$\bar{\Omega} a =: \omega \times a = -a \times \omega \qquad \forall\, a \in \mathbb{R}^3 \,, \tag{4.2.9}$$

---

[3] Searching for the English expression of the velocity v_f and, later, the acceleration a_f, was difficult as in no texts we could find one. So, we chose to call "Führungsgeschwindigkeit" "frozen velocity" and "Führungsbeschleunigung" "frozen acceleration". "Velocity of transport" and "acceleration of transport" are suggestions one could equally use.

[4] Notice that this is not the usual definition of the spin tensor, which is given later by (see e.g. (5.2.2)) $\Omega = \dot{O}O^T$. With this we have $\Omega_{ij} = \varepsilon_{ijk}\omega_k$, $\omega_i = \frac{1}{2}\varepsilon_{ijk}\Omega_{jk}$, since $\Omega = -\bar{\Omega}$, as can be shown.

or in component form

$$\bar{\Omega}_{ij} a_j =: -\varepsilon_{ijk} a_j \omega_k \, , \tag{4.2.10}$$

where $\boldsymbol{a}$ is an arbitrary vector.

The velocity of a particle can now be expressed as

$$\dot{\boldsymbol{x}} = \dot{\boldsymbol{c}} + \bar{\boldsymbol{\Omega}} \boldsymbol{y} + \boldsymbol{O}^T \left(\boldsymbol{y}^* \right)^{\boldsymbol{\cdot}} \, . \tag{4.2.11}$$

The acceleration is given by the second derivative of $\boldsymbol{x}$ with respect to time, holding the particle fixed. Substituting the velocity $\dot{\boldsymbol{y}}$ from (4.2.1) and, further, using the transformation (4.1.2), one obtains from (4.2.11)

$$\begin{aligned} \ddot{\boldsymbol{x}} &= \ddot{\boldsymbol{c}} + \dot{\bar{\boldsymbol{\Omega}}} \boldsymbol{y} + \bar{\boldsymbol{\Omega}} \dot{\boldsymbol{y}} + \dot{\boldsymbol{O}}^T \left(\boldsymbol{y}^* \right)^{\boldsymbol{\cdot}} + \boldsymbol{O}^T \left(\boldsymbol{y}^* \right)^{\boldsymbol{\cdot\cdot}} \\ &= \ddot{\boldsymbol{c}} + \dot{\bar{\boldsymbol{\Omega}}} \boldsymbol{y} + \bar{\boldsymbol{\Omega}} \left(\boldsymbol{\Omega} \boldsymbol{y} + \boldsymbol{O}^T \left(\boldsymbol{y}^* \right)^{\boldsymbol{\cdot}} \right) \\ &\quad + \dot{\boldsymbol{O}}^T \boldsymbol{O} \boldsymbol{O}^T \left(\boldsymbol{y}^* \right)^{\boldsymbol{\cdot}} + \boldsymbol{O}^T \left(\boldsymbol{y}^* \right)^{\boldsymbol{\cdot\cdot}} \, . \end{aligned} \tag{4.2.12}$$

Thus,

$$\ddot{\boldsymbol{x}} = \underbrace{\ddot{\boldsymbol{c}} + \dot{\bar{\boldsymbol{\Omega}}} \boldsymbol{y} + \bar{\boldsymbol{\Omega}} \bar{\boldsymbol{\Omega}} \boldsymbol{y}}_{=: \, \boldsymbol{a}_f} + \underbrace{2 \bar{\boldsymbol{\Omega}} \boldsymbol{O}^T \left(\boldsymbol{y}^* \right)^{\boldsymbol{\cdot}}}_{=: \, \boldsymbol{a}_c \, = \, 2 \boldsymbol{\Omega} \boldsymbol{v}_r} + \underbrace{\boldsymbol{O}^T \left(\boldsymbol{y}^* \right)^{\boldsymbol{\cdot\cdot}}}_{=: \, \boldsymbol{a}_r} \, . \tag{4.2.13}$$

In the above expression, $\boldsymbol{a}_f$ denotes the *acceleration frozen to the moving system*, $\boldsymbol{a}_c$ the CORIOLIS *acceleration* and $\boldsymbol{a}_r$ the *relative acceleration.* Regarding a physical interpretation of these various acceleration terms the following explanation may be helpful: The acceleration frozen to the moving system is the acceleration of the point instantaneously coinciding with the position of the particle but frozen to the moving coordinate system. The CORIOLIS acceleration indicates that the velocity vector in the moving system is rotating relative to the fixed system, and similarly, the relative acceleration results from the acceleration of a particle in the moving system (measured in the fixed coordinate system – therefore the factor $\boldsymbol{O}^T$).

With the definition of the vector of the angular velocity (4.2.9) we write (4.2.13) as

$$\ddot{\boldsymbol{x}} = \ddot{\boldsymbol{c}} + \dot{\boldsymbol{\omega}} \times \boldsymbol{y} + \boldsymbol{\omega} \times (\boldsymbol{\omega} \times \boldsymbol{y}) + 2 \boldsymbol{\omega} \times \boldsymbol{v}_r + \boldsymbol{a}_r \, . \tag{4.2.14}$$

The various terms of this equation have the following definitions:

- $\dot{\boldsymbol{\omega}} \times \boldsymbol{y}$ — EULER *acceleration,*
- $\boldsymbol{\omega} \times (\boldsymbol{\omega} \times \boldsymbol{y})$ — *centripetal acceleration,* (4.2.15)
- $2 \boldsymbol{\omega} \times \boldsymbol{v}_r$ — CORIOLIS *acceleration.*

The expression (4.2.14) governs the absolute acceleration in a moving reference system, where the position vector is related to the fixed coordinate system $\boldsymbol{y} = \boldsymbol{O}^T \boldsymbol{y}^*$ as are the relative velocity $\boldsymbol{v}_r$ and the relative acceleration $\boldsymbol{a}_r$.

Let us consider a moving system and modify the momentum balance (2.3.17) as suggested by (4.2.14). For a system with translational and rotational motion relative to the fixed system, the acceleration must be written as

$$\left.\frac{\mathrm{d}\boldsymbol{v}}{\mathrm{d}t}\right|_{\text{fixed}} = \underbrace{\frac{\partial \boldsymbol{v}_r}{\partial t} + (\operatorname{grad} \boldsymbol{v}_r)\boldsymbol{v}_r}_{=\boldsymbol{a}_r} + \ddot{\boldsymbol{c}} + 2\boldsymbol{\omega} \times \boldsymbol{v}_r + \boldsymbol{\omega} \times (\boldsymbol{\omega} \times \boldsymbol{y}) + \dot{\boldsymbol{\omega}} \times \boldsymbol{y} \,. \tag{4.2.16}$$

In principle, one can eliminate the index from the relative motion and write $\boldsymbol{v}$ instead of $\boldsymbol{v}_r$, if simultaneously the derivative on the left-hand side is that of an observer in the fixed system. The EULER acceleration only plays a role in geophysics when one intends to examine the rotational variations of the Earth (e.g. changes in the length of days). These effects are very small for the Earth's rotation and play a negligible role in the usual processes that are relevant in Geophysics. Furthermore, the centripetal acceleration should not be considered any further; this term can be included in the body force via a potential and is usually part of the definition of the acceleration due to gravity (consult Exercises). And, finally, the translational acceleration of the Earth on its orbit, $\ddot{\boldsymbol{c}}$ is equally negligible. So, in geophysical applications (4.2.16) is replaced by

$$\left.\frac{\mathrm{d}\boldsymbol{v}}{\mathrm{d}t}\right|_{\text{fixed}} = \frac{\partial \boldsymbol{v}}{\partial t} + (\operatorname{grad} \boldsymbol{v})\boldsymbol{v} + 2\boldsymbol{\omega} \times \boldsymbol{v} \,. \tag{4.2.17}$$

4.3 Transformation Properties of Balance Equations

For a moving reference system, as explained already in the last section, one must consider the additional terms in the momentum balance equation. The equation changes while going from an inertial frame to a moving system, *the momentum balance is therefore system dependent.* At this stage our intention is to examine the system dependency of the balance equations listed in Chap. 2. The transformation properties of the physical balance laws under EUCLIDian transformations are controversely discussed in the literature. In fact this is an issue that seems to be difficult to understand even though its concepts are actually easy to understand.

4.3.1 Invariance and Indifference of Equations

The following statements of the indifference and invariance of equations[5], respectively, concern not only the balance equations to be examined here, but

[5] The terms "invariance" and "indifference" are not unanimously used to have the same meaning from author to author. It seems that continuum mechanicians

also the material equations to be treated in the next chapter. While considering the transformation properties of equations we will focus on two different aspects. One of them refers to the transformation *invariance* if an equation does not change its form under a transformation. This concept implies, if applied, that the various terms are differently interpreted. In the momentum balance one interprets the additional terms that appear due to the transformation to a moving system as virtual forces. The CORIOLIS force belongs to them. These terms act physically as if they were external forces affecting the system (without doing work). Thus, these additional terms, with external forces, can be combined together to a single transformed force; if $\boldsymbol{f}$ is the external force in a system then

$$\boldsymbol{f}^* = \boldsymbol{O}(\boldsymbol{f} - \boldsymbol{i}) \tag{4.3.1}$$

is the transformation rule for a force or force per unit mass. The vector $\boldsymbol{i}$ represents the action of the forces due to the relative motion and will be specified later (it comprises essentially the terms in the second line of (4.2.16)). Clearly, this expression is system dependent because it explicitly involves the translational acceleration $\ddot{\boldsymbol{c}}$ and the angular velocity $\boldsymbol{\Omega}(\boldsymbol{\omega})$ of the moving system. These are the reasons why the physical variables which do not explicitly involve this system dependency are special and called *objective* whilst all those which involve it are called *non-objective.*

With the definition (4.3.1) of the transformation of the force the momentum balance remains (form–) invariant, i.e., the equation of the transformed variables looks exactly like the original equation. However, this equation can be system dependent. Further, and complementing this expression, an equation is called *indifferent* when no system dependence appears in case of a change in the reference system.

We remark that the above definitions are connected with the kind of transformation to which they are applied. More specifically, an equation may be indifferent for a class of transformations, but not for another class. A transformation of the form

$$\boldsymbol{y}^* = \boldsymbol{A}\boldsymbol{x} + \boldsymbol{V}t + \boldsymbol{C}\ , \quad \boldsymbol{A}, (\boldsymbol{A}\boldsymbol{A}^T = \boldsymbol{I}), \boldsymbol{V}, \boldsymbol{C} \quad \text{truly constant}\ , \tag{4.3.2}$$

is called a GALILEI*an transformation*. It means that two observers move with constant velocity relative to each other. The starred system moves uniformly relative to the non-starred system. The most general form of transformation between two observers which are referred to two reference frames is the EUCLID*ian transformation*

$$\boldsymbol{y}^* = \boldsymbol{O}(t)\boldsymbol{x} - \boldsymbol{c}^*(t)\ , \quad \boldsymbol{O}\boldsymbol{O}^T = \boldsymbol{I}\ , \tag{4.3.3}$$

have not yet agreed on a unique terminology. So the reader is cautioned to accept the definitions used in this text and then to translate to his terminology, if it is different from the one used here.

as it was defined by (4.1.6). This is the general transformation group whose distances are not changed. It is easily seen that the momentum equations are indifferent with respect to GALILEIan transformations, but not EUCLIDian transformations, but they are invariant with respect to both.

From now on, all discussions concerning transformation properties like invariance or indifference are based on EUCLIDian transformations. In this regard, any scalar a, vector $\boldsymbol{a}$, second rank tensor $\boldsymbol{A}$ and tensor of rank n, $\mathbb{A}$ are termed, respectively, *objective scalar*, *objective vector* and *objective tensor* quantities when they satisfy the following transformation rules under EUCLIDian transformations:

$$a^* = a\,, \quad \boldsymbol{a}^* = \boldsymbol{O}\boldsymbol{a}\,, \quad \boldsymbol{A}^* = \boldsymbol{O}\boldsymbol{A}\boldsymbol{O}^T\,,$$
$$\mathbb{A}^*_{i_1 i_2 \cdots i_n} = \boldsymbol{O}_{i_1 j_1}\boldsymbol{O}_{i_2 j_2}\cdots\boldsymbol{O}_{i_n j_n}\mathbb{A}_{j_1 j_2 \cdots j_n}\,. \tag{4.3.4}$$

With these definitions the idea of objectivity is linked to a certain transformation group. For our purposes this group is only the EUCLIDian transformation group.

4.3.2 Important Transformation Properties

In order to examine the transformation properties of the balance equations we must, at first, specify those of the physical variables arising in those equations. Let (see (4.1.6))

$$\boldsymbol{y}^* = \boldsymbol{O}(\boldsymbol{x} - \boldsymbol{c}) \quad \text{or} \quad \boldsymbol{x} = \boldsymbol{O}^T\boldsymbol{y}^* + \boldsymbol{c} \tag{4.3.5}$$

be the EUCLIDian transformation from a starred to an unstarred system[6] or vice versa. Scalar quantities, density ρ, internal energy ε, radiation power $\mathfrak{r}$, entropy s and temperature Θ are objective scalars, i.e., the values of these quantities do not change when changing the reference system. Among the other quantities, e.g. those later derived from material equations such as heat flux vector $\boldsymbol{q}$ and the CAUCHY stress tensor $\boldsymbol{t}$ are obviously objective variables (an objective vector and an objective second rank tensor, respectively); this means, their basis vectors must only be changed relative to the rotating system – the corresponding orthogonal transformation must be used. The velocity, on the other hand, is not an objective vector since the transformation is system dependent through the occurrence of $\dot{\boldsymbol{c}}$ and $\dot{\boldsymbol{O}}^T\boldsymbol{O}$, etc. The transformation properties of the external force $\boldsymbol{f} = \rho\boldsymbol{g}$ is left open for the time being (see, however, the preceding subsection). In summary, we can compile the needed transformation rules as follows:

[6] In the following we often write $\boldsymbol{x}^*$ in place of $\boldsymbol{y}^*$, then

$$\boldsymbol{x}^* = \boldsymbol{O}(\boldsymbol{x} - \boldsymbol{c})\,, \quad \boldsymbol{x} = \boldsymbol{O}^T\boldsymbol{x}^* + \boldsymbol{c}\,.$$

$$\begin{aligned} \rho^* &= \rho\,, \quad \varepsilon^* = \varepsilon\,, \quad s^* = s \\ \mathfrak{r}^* &= \mathfrak{r}\,, \quad \Theta^* = \Theta \end{aligned} \tag{4.3.6}$$

for the scalar quantities,

$$\boldsymbol{q}^* = \boldsymbol{O}\boldsymbol{q}\,, \quad \boldsymbol{t}^* = \boldsymbol{O}\boldsymbol{t}\boldsymbol{O}^T \tag{4.3.7}$$

for the heat–flux vector and the CAUCHY–stress tensor and

$$\boldsymbol{v}^* = \boldsymbol{O}\boldsymbol{v} + \dot{\boldsymbol{O}}\boldsymbol{x} - (\boldsymbol{O}\boldsymbol{c})^{\cdot} \quad \text{or} \quad \boldsymbol{v} = \boldsymbol{O}^T\boldsymbol{v}^* + \dot{\boldsymbol{O}}^T\boldsymbol{y}^* + \dot{\boldsymbol{c}} \tag{4.3.8}$$

(from (4.3.5)) for the velocity vector. In the transformation formulas for the velocity vector two additional system dependent terms appear. This indicates that the velocity vector does not transform as an objective vector under EUCLIDian transformations (but it is objective under GALILEIan transformations).

Further, we must also transform the terms involving derivatives in the balance equations. The Exercises show that the spatial gradient and the spatial divergence of objective quantities are objective, implying the following transformation rules

$$\begin{aligned} \operatorname{div}^*\boldsymbol{q}^* &= \operatorname{div}\boldsymbol{q} && \text{(objective scalar)}\,, \\ \operatorname{div}^*\boldsymbol{t}^* &= \boldsymbol{O}(\operatorname{div}\boldsymbol{t}) && \text{(objective vector)}\,. \end{aligned} \tag{4.3.9}$$

Moreover, the material time derivative of an objective scalar is equally an objective scalar,

$$\frac{\mathrm{d}\rho^*}{\mathrm{d}t} = \frac{\mathrm{d}\rho}{\mathrm{d}t}\,, \quad \frac{\mathrm{d}\varepsilon^*}{\mathrm{d}t} = \frac{\mathrm{d}\varepsilon}{\mathrm{d}t}\,, \quad \frac{\mathrm{d}s^*}{\mathrm{d}t} = \frac{\mathrm{d}s}{\mathrm{d}t}\,, \tag{4.3.10}$$

and the divergence of the velocity field is an objective scalar, i.e.,

$$\begin{aligned} \operatorname{div}^*\boldsymbol{v}^* &= \frac{\partial v_i^*}{\partial x_i^*} = \frac{\partial}{\partial x_i^*}\Big(O_{ij}v_j + \dot{O}_{ij}x_j - (O_{ij}c_j)^{\cdot}\Big) \\ &= O_{ij}\frac{\partial v_j}{\partial x_k}\frac{\partial x_k}{\partial x_i^*} + \dot{O}_{ij}O_{ij} = \frac{\partial v_j}{\partial x_k}\underbrace{O_{ij}O_{ik}}_{\delta_{jk}} + \dot{O}_{ij}O_{ij}\,. \end{aligned} \tag{4.3.11}$$

In this equation, the last term in the first line vanishes, because $(O_{ij}\boldsymbol{c}_j)^{\cdot}$ is not space dependent, and in the second line the spin matrix $\boldsymbol{\Omega} = \dot{\boldsymbol{O}}^T\boldsymbol{O}$ is antisymmetric so its trace vanishes, $\dot{O}_{ij}O_{ij} = 0$. With these results (4.3.11) gives

$$\operatorname{div}^*\boldsymbol{v}^* = \operatorname{div}\boldsymbol{v}\,. \tag{4.3.12}$$

These objectivity relations will be used next to examine the transformation properties of the balance equations.

4.3.3 Invariance of Balance Equations

Mass Balance Referring to the objectivity of the mass density, $\rho = \rho^*$, along with relations (4.3.10) and (4.3.12), the balance equation of mass can be given in both systems, viz.,

$$\frac{\mathrm{d}\rho}{\mathrm{d}t} + \rho \operatorname{div} \boldsymbol{v} = 0 \quad \Rightarrow \quad \frac{\mathrm{d}\rho^*}{\mathrm{d}t} + \rho^* \operatorname{div}^* \boldsymbol{v}^* = 0 \,. \tag{4.3.13}$$

Thus the transformed equation has the same form as the original equation and is thus invariant with respect to a change of the reference system. Notice that $\boldsymbol{v}^*$ is a non-objective (system dependent) quantity, but $\operatorname{div} \boldsymbol{v}^*$ is not.

Momentum Balance Because of the transformation of the velocity (4.3.8) the acceleration transforms as

$$\frac{\mathrm{d}\boldsymbol{v}}{\mathrm{d}t} = \boldsymbol{O}^T \frac{\mathrm{d}\boldsymbol{v}^*}{\mathrm{d}t} + 2\dot{\boldsymbol{O}}^T \boldsymbol{v}^* + \ddot{\boldsymbol{O}}^T \boldsymbol{y}^* + \ddot{\boldsymbol{c}} \,. \tag{4.3.14}$$

Suppose that the external force $\boldsymbol{f} = \rho \boldsymbol{g}$ is an objective vector,

$$\rho^* \boldsymbol{g}^* = \boldsymbol{O}(\rho \boldsymbol{g}) \,, \tag{4.3.15}$$

then with $(4.3.9)_2$, (4.3.14) and (4.3.15) the transformation of the momentum balance

$$\rho \frac{\mathrm{d}\boldsymbol{v}}{\mathrm{d}t} = \operatorname{div} \boldsymbol{t} + \rho \boldsymbol{g} \tag{4.3.16}$$

upon multiplication by $\boldsymbol{O}$ from the left turns into the equation

$$\rho^* \frac{\mathrm{d}\boldsymbol{v}^*}{\mathrm{d}t} = \operatorname{div}^* \boldsymbol{t}^* + \rho^* \boldsymbol{g}^* - \rho^* \left(2\boldsymbol{O}\dot{\boldsymbol{O}}^T \boldsymbol{v}^* + \boldsymbol{O}\ddot{\boldsymbol{O}}^T \boldsymbol{y}^* + \boldsymbol{O}\ddot{\boldsymbol{c}} \right) \,. \tag{4.3.17}$$

With regard to (4.2.13) we can identify the following as additional terms:

$$\begin{aligned}
&\bullet \quad \text{Frozen acceleration} \quad \boldsymbol{a}_f^* = \boldsymbol{O}\boldsymbol{a}_f \quad \text{with} \quad \boldsymbol{a}_f = \ddot{\boldsymbol{O}}^T \boldsymbol{y}^* + \ddot{\boldsymbol{c}} \\
&\bullet \quad \text{Coriolis acceleration} \quad \boldsymbol{a}_c^* = \boldsymbol{O}\boldsymbol{a}_c \quad \text{with} \quad \boldsymbol{a}_c = 2\dot{\boldsymbol{O}}^T \boldsymbol{v}^* \,.
\end{aligned} \tag{4.3.18}$$

Clearly (4.3.17) is system dependent. Considering the transformation of the external forces, (4.3.15) and (4.3.18) yield

$$\boldsymbol{g}^* = \boldsymbol{O}(\boldsymbol{g} - \boldsymbol{a}_f - \boldsymbol{a}_c) = \boldsymbol{O}(\boldsymbol{g} - \boldsymbol{i}) \,. \tag{4.3.19}$$

It follows that equations (4.3.16) and (4.3.17) have the same form. $\boldsymbol{i} = \boldsymbol{a}_f + \boldsymbol{a}_c$ is the sum of the frozen and CORIOLIS acceleration as measured in the moving frame but referred to the reference frame. It constitutes the frame dependent term.

Let us repeat the result again: *Under* EUCLID*ian transformations the momentum balance equation is invariant – see equation (4.3.17) – but it is frame dependent through the inertial terms (4.3.18).*

Energy Balance The expression for the stress power is found to be an objective scalar. Indeed, we have

$$\varphi = (\operatorname{grad} \boldsymbol{v}) \cdot \boldsymbol{t} = \boldsymbol{D} \cdot \boldsymbol{t} \,, \tag{4.3.20}$$

where $\boldsymbol{t}$ is symmetric. Since the stretching tensor $\boldsymbol{D}$ and the Cauchy stress tensor $\boldsymbol{t}$ are both objective tensors, we have

$$\begin{aligned} \varphi &= \boldsymbol{D} \cdot \boldsymbol{t} = \operatorname{tr}(\boldsymbol{O}^T \boldsymbol{D}^* \boldsymbol{O} \boldsymbol{O}^T \boldsymbol{t}^* \boldsymbol{O}) = \operatorname{tr}(\boldsymbol{O}^T \boldsymbol{D}^* \boldsymbol{t}^* \boldsymbol{O}) \\ &= \boldsymbol{D}^* \cdot \boldsymbol{t}^* = \varphi^* \,. \end{aligned} \tag{4.3.21}$$

Using the objectivity conditions of (4.3.6), (4.3.7) and $(4.3.9)_1$ the energy balance equation

$$\rho \frac{\mathrm{d}\varepsilon}{\mathrm{d}t} = -\operatorname{div} \boldsymbol{q} + (\operatorname{grad} \boldsymbol{v}) \cdot \boldsymbol{t} + \rho \mathfrak{r} \tag{4.3.22}$$

transforms into

$$\rho^* \frac{\mathrm{d}\varepsilon^*}{\mathrm{d}t} = -\operatorname{div}^* \boldsymbol{q}^* + (\operatorname{grad}^* \boldsymbol{v}^*) \cdot \boldsymbol{t}^* + \rho^* \mathfrak{r}^* \,, \tag{4.3.23}$$

which has the same form as (4.3.22) and thus, this equation is invariant under EUCLIDian transformation.

Entropy Balance Exploiting the same arguments as in the case of the energy balance equation yields the transformed entropy balance equation as

$$\rho^* \frac{\mathrm{d}s^*}{\mathrm{d}t} + \operatorname{div}^* \left(\frac{\boldsymbol{q}^*}{\Theta^*} \right) - \rho^* \frac{\mathfrak{r}^*}{\Theta^*} = \rho^* \gamma^* > 0 \,, \tag{4.3.24}$$

where the entropy production γ was considered an objective scalar: $\gamma^* = \gamma$. Hence the entropy balance is also invariant.

4.3.4 Invariance of Jump Conditions

On a singular surface the jump conditions of the general balance equations hold, see Chap. 3. These jump equations must naturally satisfy the invariance requirements with respect to the EUCLIDian transformations, i.e., they must be independent of the choice of the reference system. This will now be demonstrated.

Let us first consider the transformation of the velocity difference, which appears in the jump conditions of mass (see Table 3.5),

$$\mathcal{M} := \rho(\boldsymbol{v} - \boldsymbol{w}) \cdot \boldsymbol{n}_s \,, \tag{4.3.25}$$

and also in the terms of the type $[\![\boldsymbol{v}]\!]$. The transformation of the velocity (4.3.8) yields

$$\boldsymbol{v}_1^* = \boldsymbol{O}\boldsymbol{v}_1 + \dot{\boldsymbol{O}}\boldsymbol{x} - (\boldsymbol{O}\boldsymbol{c})^{\boldsymbol{\cdot}} \,, \quad \boldsymbol{v}_2^* = \boldsymbol{O}\boldsymbol{v}_2 + \dot{\boldsymbol{O}}\boldsymbol{x} - (\boldsymbol{O}\boldsymbol{c})^{\boldsymbol{\cdot}} \,, \tag{4.3.26}$$

where the "action point" $\boldsymbol{x}$ is the same for both velocities. The difference is now given by

$$\boldsymbol{v}_1^* - \boldsymbol{v}_2^* = \boldsymbol{O}(\boldsymbol{v}_1 - \boldsymbol{v}_2) \, . \tag{4.3.27}$$

A difference of velocity vectors at the same point of action is always an objective vector.

Mass The transformation of the mass flux (4.3.25) can be given as

$$\begin{aligned} \mathcal{M}^* &= \rho^*(\boldsymbol{v}^* - \boldsymbol{w}^*) \cdot \boldsymbol{n}_{\mathfrak{s}}^* \\ &= \rho \boldsymbol{O}(\boldsymbol{v} - \boldsymbol{w}) \cdot \boldsymbol{O}\boldsymbol{n}_{\mathfrak{s}} = \rho(\boldsymbol{v} - \boldsymbol{w}) \cdot \boldsymbol{O}^T \boldsymbol{O}\boldsymbol{n}_{\mathfrak{s}} \\ &= \rho(\boldsymbol{v} - \boldsymbol{w}) \cdot \boldsymbol{n}_{\mathfrak{s}} = \mathcal{M} \, , \end{aligned} \tag{4.3.28}$$

where we used the fact that the density is an objective scalar and the normal vector is an objective vector. The mass flux is transformed as an objective scalar under EUCLIDian transformations. There follows

$$[\![\mathcal{M}]\!] := [\![\rho(\boldsymbol{v} - \boldsymbol{w}) \cdot \boldsymbol{n}_{\mathfrak{s}}]\!] = 0 \quad \Rightarrow \quad [\![\mathcal{M}^*]\!] = 0 \tag{4.3.29}$$

is invariant; with (4.3.28) $[\![\mathcal{M}]\!] = 0$ immediately implies $[\![\mathcal{M}^*]\!] = 0$.

Momentum With the aid of (4.3.29) the jump condition of momentum can be written as

$$-\mathcal{M}[\![\boldsymbol{v}]\!] + [\![\boldsymbol{t}\boldsymbol{n}_{\mathfrak{s}}]\!] = \boldsymbol{0} \, . \tag{4.3.30}$$

The jump of the velocity is objective according to (4.3.27), and here $\boldsymbol{t}$ and $\boldsymbol{n}_{\mathfrak{s}}$ are likewise objective; thus, is an objective vector. Hence the transformed momentum jump condition (after multiplying it by $\boldsymbol{O}$),

$$-\mathcal{M}^*[\![\boldsymbol{v}^*]\!] + [\![\boldsymbol{t}^*\boldsymbol{n}_{\mathfrak{s}}^*]\!] = \boldsymbol{0} \, , \tag{4.3.31}$$

is invariant under EUCLIDian transformations.

Energy The proof of the invariance of the energy jump condition is somewhat tedious to conduct. The kinetic energy $\boldsymbol{v}^2/2$ and the friction term $\boldsymbol{v}\boldsymbol{t}\cdot\boldsymbol{n}_{\mathfrak{s}}$ are clearly not objective (and neither is the jump of these quantities). First, the combination of both terms is recognized as an objective term. In order to prove this recall the definition of the mean value (3.1.8) and the rule (3.1.9) for the jump condition, specially recall the identities

$$[\![\boldsymbol{a} \cdot \boldsymbol{b}]\!] = [\![\boldsymbol{a}]\!] \cdot \langle\!\langle \boldsymbol{b} \rangle\!\rangle + \langle\!\langle \boldsymbol{a} \rangle\!\rangle \cdot [\![\boldsymbol{b}]\!] \, . \tag{4.3.32}$$

In the energy jump condition

$$-\mathcal{M}[\![\tfrac{1}{2}\boldsymbol{v}^2 + \varepsilon]\!] + [\![\boldsymbol{v}\boldsymbol{t} - \boldsymbol{q}]\!] \cdot \boldsymbol{n}_{\mathfrak{s}} = 0 \tag{4.3.33}$$

we first consider only the expressions for the kinetic and the frictional energy. By use of (4.3.32) on both terms we deduce

$$
\begin{aligned}
-\mathcal{M}\left[\!\left[\tfrac{1}{2}v^2\right]\!\right] + \left[\!\left[\boldsymbol{v}\boldsymbol{t}\cdot\boldsymbol{n}_s\right]\!\right] &= -\mathcal{M}\left[\!\left[\boldsymbol{v}\right]\!\right]\cdot\langle\!\langle\boldsymbol{v}\rangle\!\rangle + \left[\!\left[\boldsymbol{v}\right]\!\right]\cdot\langle\!\langle\boldsymbol{t}\boldsymbol{n}_s\rangle\!\rangle + \langle\!\langle\boldsymbol{v}\rangle\!\rangle\cdot\left[\!\left[\boldsymbol{t}\boldsymbol{n}_s\right]\!\right] \\
&= -\underbrace{\left(\mathcal{M}\left[\!\left[\boldsymbol{v}\right]\!\right] + \left[\!\left[\boldsymbol{t}\boldsymbol{n}_s\right]\!\right]\right)}_{=0 \quad (4.3.30)}\cdot\langle\!\langle\boldsymbol{v}\rangle\!\rangle + \left[\!\left[\boldsymbol{v}\right]\!\right]\cdot\langle\!\langle\boldsymbol{t}\boldsymbol{n}_s\rangle\!\rangle \\
&= \left[\!\left[\boldsymbol{v}\right]\!\right]\cdot\langle\!\langle\boldsymbol{t}\boldsymbol{n}_s\rangle\!\rangle\,. \qquad (4.3.34)
\end{aligned}
$$

The energy jump condition can therefore also be expressed as

$$
-\mathcal{M}\left[\!\left[\varepsilon\right]\!\right] + \left[\!\left[\boldsymbol{v}\right]\!\right]\cdot\langle\!\langle\boldsymbol{t}\boldsymbol{n}_s\rangle\!\rangle - \left[\!\left[\boldsymbol{q}\right]\!\right]\cdot\boldsymbol{n}_s = 0\,. \qquad (4.3.35)
$$

In this equation, $\mathcal{M}, \varepsilon$ are objective scalars, $\boldsymbol{n}_s, \boldsymbol{q}$ are objective vectors and $\boldsymbol{t}$ is an objective tensor. Since, here, with (4.3.27) the velocity jump is an objective vector, (4.3.35) is objective, i.e., under EUCLIDian transformations the equation appears invariant, viz.,

$$
\mathcal{M}^*\left[\!\left[\varepsilon^*\right]\!\right] + \left[\!\left[\boldsymbol{v}^*\right]\!\right]\cdot\langle\!\langle\boldsymbol{t}^*\boldsymbol{n}_s^*\rangle\!\rangle - \left[\!\left[\boldsymbol{q}^*\right]\!\right]\cdot\boldsymbol{n}_s^* = 0\,. \qquad (4.3.36)
$$

Entropy The jump condition for the entropy reads

$$
\mathcal{M}\left[\!\left[s\right]\!\right] - \left[\!\left[\frac{\boldsymbol{q}}{\Theta}\right]\!\right]\cdot\boldsymbol{n}_s = -\mathfrak{p}_s \le 0\,. \qquad (4.3.37)
$$

Under the requirement that temperature as well as the surface production of entropy are objective scalars, this leads to the invariance of the entropy jump condition

$$
\mathcal{M}^*\left[\!\left[s^*\right]\!\right] - \left[\!\left[\frac{\boldsymbol{q}^*}{\Theta^*}\right]\!\right]\cdot\boldsymbol{n}_s^* = -\mathfrak{p}_s^* \le 0\,, \qquad (4.3.38)
$$

using the special presumption that $\mathfrak{p}_s^* = \mathfrak{p}_s$.

Let us summarize: Density, internal energy, entropy, radiation, temperature, heat flux and CAUCHY stress tensor, have all been found to be objective variables, i.e., these variables transform under EUCLIDian transformations like objective scalars, vectors and tensors. This then has shown that the *balance equations of mass, momentum, energy and entropy* are invariant, i.e., they have the same form in all EUCLIDian systems. However, the balance equations can exhibit system dependent terms like the additional forces due to the relative motion in the momentum balance. These system dependent terms contain the acceleration of the system via the translational acceleration $\ddot{\boldsymbol{c}}$ and the angular velocity $\boldsymbol{\Omega}$ of the reference system.

4.4 Exercises

1. Considering $-\boldsymbol{\omega}\times(\boldsymbol{\omega}\times\boldsymbol{x})$ as a specific force per unit mass; construct the corresponding potential.

2. How large is the deviation of the acceleration due to gravity from the pure gravitational acceleration considering the spherical, rotational Earth and how does the angular deviation from the plumb line vary with the geographical latitude ?
3. Consider a cylindrical container filled partly by a volume preserving fluid with free surface, and rotated with constant angular velocity about its vertical axis. In the stationary case, when the fluid is at rest in the rotating system, the hydrostatic pressure equation is given by

$$\operatorname{grad} p = \rho(\boldsymbol{g} - \boldsymbol{\omega} \times (\boldsymbol{\omega} \times \boldsymbol{x})) .$$

Show that the surfaces of constant pressure (isobares) are rotational paraboloids described by the equation

$$z = -\frac{\omega^2}{2} r^2 + \text{const} ,$$

where r is the distance of a point from the axis of rotation.
4. Consider a fluid on the rotating Earth, whose density is only a function of temperature and assume this fluid to be stratified (i.e., there exist non-vanishing spatial density gradients). Under the same condition as in Exercise 3, the hydrostatic pressure equation is again given by (4.5.2). Prove that for such a fluid the axial density variation, $\partial\rho/\partial z$ and the radial density variation, $\partial\rho/\partial r$ are related by

$$\frac{\partial \rho}{\partial r} = -\frac{\omega^2}{g} r \frac{\partial \rho}{\partial z} .$$

Thus there follows: *A purely axial density variation is not possible.* If there exists a linear relationship between the density ρ and temperature Θ, then the relation

$$\frac{\partial \Theta}{\partial r} = -\frac{\omega^2}{g} r \frac{\partial \Theta}{\partial z}$$

holds. Show further that this equation and the heat conduction equation $\operatorname{div} \operatorname{grad} \Theta = 0$, which here reads

$$\frac{1}{r}\frac{\partial}{\partial r}\left(r\frac{\partial \Theta}{\partial r}\right) + \frac{\partial^2 \Theta}{\partial z^2} = 0 ,$$

can not simultaneously be identically satisfied. This implies: *A stratified rotationally symmetric fluid at rest can not exist in a rotational system.* Finally, show that a state of rest can exist if the centrifugal terms ($\omega^2 = 0$) are ignored. It possesses linear stratification in the z–direction.
5. For a rotating body (e.g. the fluids of the Earth's outer core) verify the statement proved in Exercise 4 that no basic rest state with stratification can exist. There must be a convective flow. This time assume that the

gravitational acceleration is not constant, rather, it is to be computed from the associated gravitational potential G via a POISSON equation

$$\operatorname{lap} G = -4\pi \Gamma \rho \,,$$

where Γ is the gravitation constant and lap is the LAPLACE operator.

6. In the decomposition of the velocity gradients into symmetric and antisymmetric parts, $\boldsymbol{L} = \boldsymbol{D} + \boldsymbol{W}$, one can also express $\boldsymbol{W} = -\boldsymbol{W}^T$ through its dual vectors; how does this condition look like?
7. Prove that under EUCLIDian transformations $\boldsymbol{y}^* = \boldsymbol{O}\boldsymbol{x} + \boldsymbol{c}^*$ the following transformation rules hold

$$\left.\begin{array}{ll} \bullet & \text{Deformation gradient} \quad \boldsymbol{F}^* = \boldsymbol{O}\boldsymbol{F} \\ \bullet & \text{Rotation tensor} \quad \boldsymbol{R}^* = \boldsymbol{O}\boldsymbol{R} \end{array}\right\} \tag{4.4.1}$$

$$\left.\begin{array}{ll} \bullet & \text{Right CAUCHY–GREEN tensor} \quad \boldsymbol{C}^* = \boldsymbol{C} \\ \bullet & \text{GREEN strain tensor} \quad \boldsymbol{G}^* = \boldsymbol{G} \\ \bullet & \text{KARNI–REINER strain tensor} \quad \boldsymbol{K}^* = \boldsymbol{K} \\ \bullet & \text{Right stretch tensor} \quad \boldsymbol{U}^* = \boldsymbol{U} \end{array}\right\} \tag{4.4.2}$$

$$\left.\begin{array}{ll} \bullet & \text{Left CAUCHY-GREEN tensor} \quad \boldsymbol{B}^* = \boldsymbol{O}\boldsymbol{B}\boldsymbol{O}^T \\ \bullet & \text{FINGER strain tensor} \quad \boldsymbol{E}^* = \boldsymbol{O}\boldsymbol{E}\boldsymbol{O}^T \\ \bullet & \text{ALMANSI strain tensor} \quad \boldsymbol{A} = \boldsymbol{O}\boldsymbol{A}\boldsymbol{O}^T \\ \bullet & \text{Left stretch tensor} \quad \boldsymbol{V}^* = \boldsymbol{O}\boldsymbol{V}\boldsymbol{O}^T \end{array}\right\} \tag{4.4.3}$$

   The columns of the matrix of the deformation gradient and spin tensor transform like three objective vectors under EUCLIDian transformations; the elements of all right strain tensors ($\boldsymbol{C}$, $\boldsymbol{G}$, $\boldsymbol{K}$, $\boldsymbol{U}$) transform like scalars; however, the elements of all left strain tensors ($\boldsymbol{B}$, $\boldsymbol{E}$, $\boldsymbol{A}$, $\boldsymbol{V}$) transform like objective tensors of rank two.
8. Considering the definition $\boldsymbol{T} = (\det \boldsymbol{F})\boldsymbol{t}\boldsymbol{F}^{-T}$ of the first PIOLA–KIRCHHOFF stress tensor, which depends on the CAUCHY stress tensor and the deformation gradient, prove that under EUCLIDian transformations this tensor transforms like

$$\boldsymbol{T}^* = \boldsymbol{O}\boldsymbol{T}$$

   i.e., as three objective vectors. Further prove also that

$$\operatorname{Div}^* \boldsymbol{T}^* = \boldsymbol{O} \operatorname{Div} \boldsymbol{T}$$

   holds. Finally prove the invariance of the momentum balance in the present configuration.
9. Let $\boldsymbol{a}$ and $\boldsymbol{T}$ be an objective vector and an objective, symmetric tensor of rank two, respectively. Prove that neither $\dot{\boldsymbol{a}}$ nor $\dot{\boldsymbol{T}}$ are objective. Prove, however, that the following quantities can be interpreted as objective time derivatives of $\boldsymbol{a}$ und $\boldsymbol{T}$:

$$\overset{\triangledown}{\boldsymbol{a}} = (\dot{\boldsymbol{a}} - \boldsymbol{W}\boldsymbol{a}) - \boldsymbol{D}\boldsymbol{a} \,, \quad \overset{\triangledown}{\boldsymbol{T}} = \dot{\boldsymbol{T}} - \boldsymbol{L}\boldsymbol{T} - \boldsymbol{T}\boldsymbol{L}^T \,,$$

$$\overset{\vartriangle}{\boldsymbol{a}} = (\dot{\boldsymbol{a}} - \boldsymbol{W}\boldsymbol{a}) + \boldsymbol{D}\boldsymbol{a} \,, \quad \overset{\vartriangle}{\boldsymbol{T}} = \dot{\boldsymbol{T}} + \boldsymbol{L}^T\boldsymbol{T} + \boldsymbol{T}\boldsymbol{L} \,,$$

$$\overset{\circ}{\boldsymbol{a}} = (\dot{\boldsymbol{a}} - \boldsymbol{W}\boldsymbol{a}) \,, \qquad \overset{\circ}{\boldsymbol{T}} = \dot{\boldsymbol{T}} - \boldsymbol{W}\boldsymbol{T} + \boldsymbol{T}\boldsymbol{W} \,.$$

4.5 Solutions

1. The centrifugal potential can be calculated from

$$f := \operatorname{grad} Z = -\boldsymbol{\omega} \times (\boldsymbol{\omega} \times \boldsymbol{x}) \,.$$

Choose a coordinate system so that

$$\boldsymbol{\omega} = \omega \boldsymbol{e}_z \quad \text{and} \quad \boldsymbol{x} = r\boldsymbol{e}_r + z\boldsymbol{e}_z$$

(compare Fig. 4.3), then follows

$$\operatorname{grad} Z = -\omega^2 r \boldsymbol{e}_r \quad \text{or} \quad \frac{\partial Z}{\partial r} = -\omega^2 r \,.$$

This can be easily integrated to

$$Z(r) = -\tfrac{1}{2}\omega^2 r^2 + \text{const} \quad \text{or} \quad \frac{\partial Z}{\partial r} = -\omega^2 r + \text{const} \,.$$

The centrifugal potential can be combined with the gravity potential to form the total gravitational potential. Assuming the Earth as a sphere of mass M yields this potential to be

$$W = \frac{\Gamma M}{R} - \frac{1}{2}\omega^2 (R^2 - z^2) \,,$$

where $R = \sqrt{r^2 + z^2}$ is the radial distance to the center. In general, one can also formulate the centrifugal potential for an arbitrary coordinate system, i.e.,

$$Z = -\frac{1}{2}(\boldsymbol{\omega} \times \boldsymbol{x})^2 \,,$$

which can be immediately verified by forming the gradient

$$\operatorname{grad} Z = -\tfrac{1}{2} \operatorname{grad}\left[(\boldsymbol{\omega} \times \boldsymbol{x}) \cdot (\boldsymbol{\omega} \times \boldsymbol{x})\right] = -\boldsymbol{\omega} \times (\boldsymbol{I}\boldsymbol{\omega} \times \boldsymbol{x}) = \boldsymbol{\omega} \times (\boldsymbol{\omega} \times \boldsymbol{x}) \,.$$

2. The gravitational acceleration on the surface of a sphere having the same mass and the mean radius R as the Earth is

$$\mid \boldsymbol{g} \mid = g = 9.81 \text{ms}^{-2} \,.$$

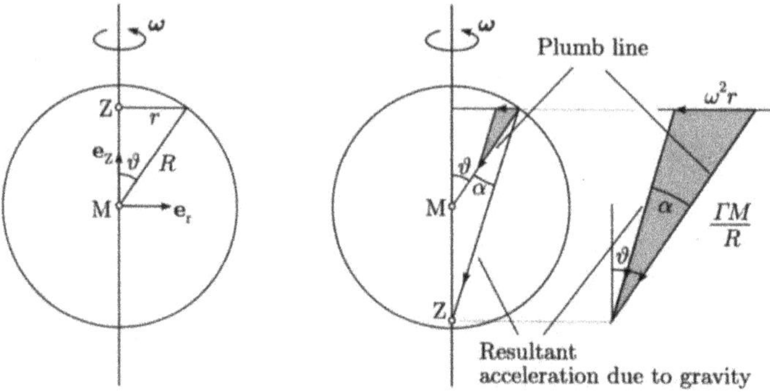

Fig. 4.3. Polar coordinate for a section through the Earth. The gravity acceleration is directed radially to the center, the direction of the gravitational acceleration (due to gravity plus centripetal acceleration) deviates an angle α from this direction, which itself depends on the geographical latitude, from the gravitational acceleration (see Exercise 2). The force triangle shows how the resultant acceleration is composed of pure gravity and centripetal acceleration.

Choosing a Cartesian coordinate system with the rotation axis as z axis, one has (for a special choice $y = 0$)

$$\boldsymbol{\omega} = (0,0,1) \quad \text{and} \quad \boldsymbol{x} = (R\sin\vartheta, 0, R\cos\vartheta) ,$$

and one obtains the magnitude of the centripetal acceleration as

$$|\boldsymbol{g}^{(z)}| = |-\boldsymbol{\omega} \times (\boldsymbol{\omega} \times \boldsymbol{x})| = |\omega^2 R \sin\vartheta \boldsymbol{e}_x| = \omega^2 R \sin\vartheta .$$

At the equator ($\vartheta = 90°$) this gives

$$|\boldsymbol{g}^{(z)}| = 0.034\,\mathrm{m\,s^{-2}} ,$$

where use has been made of the mean radius of the Earth $R = 6371\,\mathrm{km}$ and the angular velocity $\omega = 2\pi/24\,\mathrm{h}$. The above value corresponds to about 0.35% of the gravitational acceleration. In the present application, therefore, the centripetal acceleration becomes negligible as compared to the gravitational acceleration; moreover, the value of $9.81\,\mathrm{m\,s^{-2}}$ is only a mean value which varies from pole ($g = 9.83\,\mathrm{m\,s^{-2}}$) to the equator ($g = 9.78\,\mathrm{m\,s^{-2}}$) due to the almost elliptical shape of the Earth. Also the angular deviation from the plumb line (for a spherical surface) is small. Let us consider the triangle formed by the acceleration vectors; thus one obtains

$$\tan\alpha = \frac{g^{(z)}\cos\vartheta}{g + g^{(z)}\sin\vartheta} = \frac{\cos\vartheta\sin\vartheta}{\dfrac{g}{\omega^2 R} + \sin^2\vartheta} . \tag{4.5.1}$$

Therefore at the pole ($\vartheta = 0°$) as well as the equator ($\vartheta = 90°$) the angular deviation is zero, as expected. The maximum deviation is obtained from (4.5.1) at the geographical latitude of

$$\sin\vartheta_0 = \sqrt{\frac{g}{2g+\omega^2 R}} = 0.7065 \quad \text{or} \quad \vartheta_0 = 44^\circ \ ,$$

where the value $\alpha_{max} = 0.1^\circ$.
This very small deviation of the plumb line from the normal to the center of the Earth is basically due to the centripetal acceleration. It allows us to assume that the acceleration due to gravity is in the direction of the Earth's center or perpendicular to the Earth's surface. Such an assumption shall be used frequently, later.

3. Consider the experimental set up as shown in Fig. 4.4 with the rotation axis in the z–direction. Let us choose polar coordinates so that the angular velocity and the position vector can be written as

$$\boldsymbol{\omega} = \omega \boldsymbol{e}_z \ , \quad \boldsymbol{x} = r\hat{\boldsymbol{e}}_r + z\hat{\boldsymbol{e}}_z \ .$$

Then, the hydrostatic equation is given by

$$\operatorname{grad} p = \rho\big[-g\hat{\boldsymbol{e}}_z - \omega^2 r \underbrace{\hat{\boldsymbol{e}}_z \times (\hat{\boldsymbol{e}}_z \times \hat{\boldsymbol{e}}_r)}_{-\hat{\boldsymbol{e}}_r}\big] \ ,$$

the solution to this equation is

$$p - p_0 = -\rho g z + \rho\frac{\omega^2 r^2}{2} \ . \tag{4.5.2}$$

Here, p_0 is the ambient pressure. Surfaces of constant pressure are given by

$$-\rho g z + \rho\frac{\omega^2 r^2}{2} = \text{const}$$

or

$$z = \frac{\omega^2 r^2}{2g} + \text{const} \ .$$

This is an equation of a rotational paraboloid.

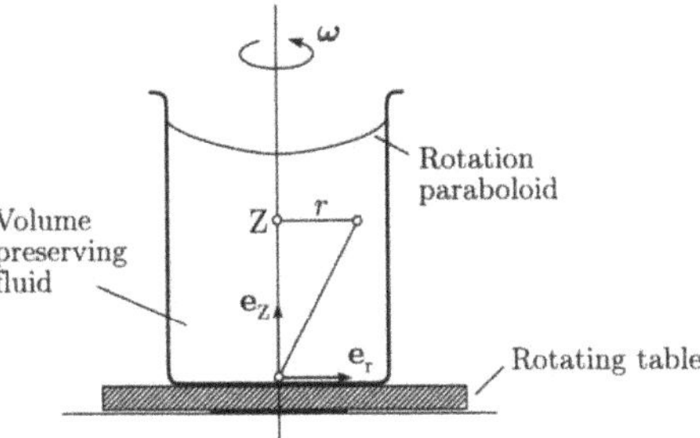

Fig. 4.4. The surface of a fluid in a rotating cylinder forms a rotational paraboloid.

4. Starting from the hydrostatic pressure equation we obtain a differential equation for the density when we apply the curl operator to this equation

$$\operatorname{curl}\operatorname{grad} p = \operatorname{curl}\left(-\rho g\hat{\boldsymbol{e}}_z + \rho\omega^2 r\hat{\boldsymbol{e}}_r\right) = \boldsymbol{0}$$
$$\Rightarrow \qquad \frac{\partial\rho}{\partial r} + \frac{\omega^2 r}{g}\frac{\partial\rho}{\partial z} = 0 \,. \tag{4.5.3}$$

From this one may conclude: A pure axial density variation is not possible, the fluid is driven to build a corresponding radial density gradient. Postulating a relation between the density and temperature

$$\rho = \hat{\rho}(\Theta) \,,$$

equation (4.5.3) can be written as

$$\frac{\partial\Theta}{\partial r} + \frac{\omega^2 r}{g}\frac{\partial\Theta}{\partial z} = 0 \,. \tag{4.5.4}$$

The temperature should, in addition, satisfy the (stationary) heat transport equation

$$\operatorname{div}\operatorname{grad}\Theta = 0 \quad \text{or} \quad \frac{1}{r}\frac{\partial}{\partial r}\left(r\frac{\partial\Theta}{\partial r}\right) + \frac{\partial^2\Theta}{\partial z^2} = 0 \,. \tag{4.5.5}$$

Solving the differential equations (4.5.4) and (4.5.5), we obtain only one possible solution of a homogeneous temperature distribution, thus there can be no stratification in a rotating fluid at rest. If we neglect the centrifugal term, there follows $\mathrm{d}\Theta/\mathrm{d}r = 0$ from (4.5.4) and therefore $\Theta = \Theta_0 + z\,\delta\Theta$, with the LAPLACE equation (4.5.5), a linear temperature distribution in the z direction. Therefore, in a slowly rotating system a linear stratification can exist for a fluid at rest in a moving system.

5. We shall now show in a somewhat different fashion, that in a rotating system no rest with nontrivial stratification can exist. We write the gravitational potential as

$$W := G + \tfrac{1}{2}\left(\boldsymbol{\Omega}\times\boldsymbol{x}\right)^2 \quad \text{with} \quad \boldsymbol{g} = \operatorname{grad} G \,.$$

The hydrostatic pressure is now given by

$$\operatorname{grad} p = \rho \operatorname{grad} W \,.$$

Forming the curl of this equation yields

$$0 = \operatorname{curl}\left(\rho \operatorname{grad} W\right) = \operatorname{grad}\rho \times \operatorname{grad} W \,,$$

corresponding to (4.5.3). Physically, the interpretation of this equation is that surfaces of constant density are parallel to surfaces of constant gravitational potential. Expressed alternatively, the density must be a function of the potential W

$$\rho = \hat{\rho}(W) .$$

Since we start with the equation of state for the density as a function of temperature

$$\rho = \tilde{\rho}(\Theta) ,$$

the above statement also holds for the temperature:

$$\Theta = \hat{\Theta}(W) .$$

Now this temperature must identically satisfy the heat transport equation

$$\operatorname{div} \operatorname{grad} \Theta = \operatorname{div} \operatorname{grad} \hat{\Theta}(W) = 0 ,$$

or

$$\Theta' \operatorname{div} \operatorname{grad} W + \Theta''(\operatorname{grad} W)^2 = 0 ,$$

where $\Theta' = \mathrm{d}\Theta/\mathrm{d}W = f(W)$ again must be a function of the potential. This potential identically satisfies the POISSON equation

$$\operatorname{div} \operatorname{grad} W = -4\pi\Gamma\rho + 2\omega^2 ,$$

with Γ the gravitational constant. From this then follows,

$$(4\pi\Gamma\rho - 2\omega^2) f(W) - (\operatorname{grad} W)^2 f'(W) = 0 . \tag{4.5.6}$$

Since here the density ρ is a function of W, the gradient of W must be constant on the potential surfaces $W = \text{const}$ in case of $f'(W) \neq 0$. This implies, however, that two equipotential surfaces (surfaces of constant potentials) must always maintain the same distances with each other. In a rotating system this is, however, not possible (because of the centripetal acceleration). Therefore we write

$$f'(W) \equiv 0 \quad \text{or} \quad f(W) = \text{const} ,$$

and specially $f(W) = 0$ because of (4.5.6). With this we can show that $\Theta' = f(W) = 0$ holds, i.e., in a rotating fluid the temperature can only be a constant. A stratification is only possible if additional flows are present, which are excluded in a hydrostatic pressure balance. This statement is called in geophysics the VON ZEIPEL theorem.

6. The antisymmetric part of the velocity gradient $\boldsymbol{W} = -\boldsymbol{W}^T$ is defined as

$$\boldsymbol{W} = \tfrac{1}{2}\left(\operatorname{grad} \boldsymbol{v} - (\operatorname{grad} \boldsymbol{v})^T\right) \quad \text{or} \quad W_{ij} = \tfrac{1}{2}\left(v_{i,j} - v_{j,i}\right) .$$

For this, the dual vector $\frac{1}{2}\boldsymbol{\omega}$ is given by the condition

$$\boldsymbol{W}\boldsymbol{a} \quad = \tfrac{1}{2}\boldsymbol{\omega} \times \boldsymbol{a} \quad \text{or}$$

$$W_{ij}a_j = -\tfrac{1}{2}\varepsilon_{ijk}\omega_k a_j \quad \forall \boldsymbol{a} \in \mathbb{R}^3 \, .$$

Multiplying the equation $W_{ij} = -\frac{1}{2}\varepsilon_{ijk}\omega_k$ with ε_{ijm}, there follows, because $\varepsilon_{ijm}\varepsilon_{ijk} = 2\delta_{mk}$,

$$\omega_m = -\varepsilon_{ijm}W_{ij} = -\tfrac{1}{2}\varepsilon_{ijm}(v_{i,j} - v_{j,i}) \, .$$

With the antisymmetry of $\varepsilon_{ijk} = -\varepsilon_{jik}$ we obtain

$$\omega_m = \varepsilon_{ijm}v_{j,i} \quad \text{or} \quad \boldsymbol{\omega} = \operatorname{curl} \boldsymbol{v} \, .$$

The dual vector associated with $\boldsymbol{W}$ is thus equal to $\frac{1}{2}$ times the vorticity vector.

7. Starting from $\boldsymbol{y}^* = \boldsymbol{O}\boldsymbol{x} - \boldsymbol{c}^*$, with $\boldsymbol{y}^* = \hat{\chi}^*(\boldsymbol{X}, t)$ and $\boldsymbol{x}^* = \hat{\chi}(\boldsymbol{X}, t)$, and through differentiation with respect to $\boldsymbol{X}$, one may write

$$\boldsymbol{F}^* = \boldsymbol{O}\boldsymbol{F} \, , \quad \text{with} \quad \boldsymbol{F}^* = \operatorname{Grad} \hat{\chi}^* \, , \quad \boldsymbol{F} = \operatorname{Grad} \hat{\chi} \, .$$

With these results one simply has via the definition of $\boldsymbol{C}$

$$\boldsymbol{C}^* = \boldsymbol{F}^{*T}\boldsymbol{F}^* = (\boldsymbol{O}\boldsymbol{F})^T(\boldsymbol{O}\boldsymbol{F}) = \boldsymbol{F}^T\boldsymbol{O}^T\boldsymbol{O}\boldsymbol{F} = \boldsymbol{F}^T\boldsymbol{F} = \boldsymbol{C} \, .$$

The transformation rule for the right strain tensors follows from their definitions (1.5.11)–(1.5.13), which all can be expressed in terms of $\boldsymbol{C}$. Similarly,

$$\boldsymbol{B}^* = \boldsymbol{F}^*\boldsymbol{F}^{*T} = (\boldsymbol{O}\boldsymbol{F})(\boldsymbol{O}\boldsymbol{F})^T = \boldsymbol{O}\boldsymbol{B}\boldsymbol{O}^T \, ,$$

and again, $\boldsymbol{V}$, $\boldsymbol{E}$ and $\boldsymbol{A}$ are defined in (1.5.11), (1.5.14) and (1.5.15), and are expressible exclusively as functions of $\boldsymbol{B}$, e.g.

$$\boldsymbol{B}^{*-1} = \boldsymbol{O}\boldsymbol{B}^{-1}\boldsymbol{O}^T \, ,$$

in which calculating $\boldsymbol{B}^*\boldsymbol{B}^{*-1}$ yields

$$\boldsymbol{B}^*\boldsymbol{B}^{*-1} = \boldsymbol{O}\boldsymbol{B}^{-1}\boldsymbol{O}^T\boldsymbol{O}\boldsymbol{B}\boldsymbol{O}^T = \boldsymbol{O}\boldsymbol{B}^{-1}\boldsymbol{B}\boldsymbol{O}^T = \boldsymbol{O}\boldsymbol{O}^T = \boldsymbol{I}$$

as expected.

Finally, the transformation rule for $\boldsymbol{R}$ follows from the polar decomposition

$$\boldsymbol{F}^* = \boldsymbol{R}^*\boldsymbol{U}^* = \boldsymbol{O}\boldsymbol{F} = \boldsymbol{O}\boldsymbol{R}\boldsymbol{U} \, ,$$

with $\boldsymbol{U} = \boldsymbol{U}^*$ one obtains $\boldsymbol{R}^* = \boldsymbol{O}\boldsymbol{R}$. The property that $\boldsymbol{F}$ and $\boldsymbol{R}$ are transformed like three objective vectors and the right strain tensors like 9 objective scalar elements whereas the left strain tensors transform like objective tensors can be easily understood when we recall that the right stain tensors are defined in the reference configuration, the left strain tensors in the present configuration, whereas $\boldsymbol{F}$ and $\boldsymbol{R}$ are so-called two-point variables defined in both configurations. Because EUCLIDian transformations are effective in the present configuration they act only on the basis vectors of the associated vector space.

There are also authors[7] who claim, on the basis of the transformation properties (4.4.1) and (4.4.2), that the deformation gradient, the rotational and the right strain tensors characterize non-objective quantities. However, we do not follow these terminologies and interpret $\boldsymbol{F}$, $\boldsymbol{R}$, $\boldsymbol{C}$, $\boldsymbol{G}$, $\boldsymbol{K}$, $\boldsymbol{U}$ as objective quantities.

8. We are now in the position to derive the transformation properties of the determinant of $\boldsymbol{F}$, viz.,

$$J^* = \det \boldsymbol{F}^* = \det(\boldsymbol{OF}) = (\det \boldsymbol{O})(\det \boldsymbol{F}) = J \,,$$

here $\det \boldsymbol{O} = 1$; this result holds for orthogonal transformations. With these we obtain

$$\begin{aligned} \boldsymbol{T}^* &= (\det \boldsymbol{F}^*)\boldsymbol{t}^*\boldsymbol{F}^{*-T} = (\det \boldsymbol{F})\boldsymbol{OtO}^T\left((\boldsymbol{OF})^{-1}\right)^T \\ &= (\det \boldsymbol{F})\boldsymbol{OtO}^T\left(\boldsymbol{F}^{-1}\boldsymbol{O}^T\right)^T = (\det \boldsymbol{F})\boldsymbol{OtO}^T\boldsymbol{OF}^{-T} \\ &= \boldsymbol{O}\left[(\det \boldsymbol{F})\boldsymbol{tF}^{-T}\right] = \boldsymbol{OT} \,. \end{aligned} \tag{4.5.7}$$

The first PIOLA–KIRCHHOFF stress tensor transforms as three objective vectors under EUCLIDian transformations. Here the Div–operator is regarded in the reference configuration which is transformed as $\mathrm{Div}^* = \mathrm{Div}$ so

$$\mathrm{Div}^* \boldsymbol{T}^* = \boldsymbol{O}\,\mathrm{Div}\,\boldsymbol{T} \,. \tag{4.5.8}$$

In order to find the transformation properties of the momentum balance equation in the LAGRANGE representation,

$$\rho_R \frac{\mathrm{d}\boldsymbol{v}}{\mathrm{d}t} = \mathrm{Div}\,\boldsymbol{T} + \rho_R \boldsymbol{g} \,, \tag{4.5.9}$$

we exploit (4.3.14) to express $\mathrm{d}\boldsymbol{v}/\mathrm{d}t$ in terms of $\mathrm{d}\boldsymbol{v}^*/\mathrm{d}t$. One obtains from (4.5.9) that

$$\rho_R \frac{\mathrm{d}\boldsymbol{v}^*}{\mathrm{d}t} = \underbrace{\boldsymbol{O}\,\mathrm{Div}\,\boldsymbol{T}}_{\mathrm{Div}^*\boldsymbol{T}^*} + \rho_R \boldsymbol{Og} - \rho_R \boldsymbol{O}(\underbrace{2\dot{\boldsymbol{O}}^T\boldsymbol{v}^*}_{\boldsymbol{a}_c} + \underbrace{\ddot{\boldsymbol{O}}^T\boldsymbol{y}^* + \ddot{\boldsymbol{c}}}_{\boldsymbol{a}_f})$$

or

$$\rho_R \frac{\mathrm{d}\boldsymbol{v}^*}{\mathrm{d}t} = \mathrm{Div}^*\boldsymbol{T}^* + \rho_R \boldsymbol{f}^* \,, \tag{4.5.10}$$

where we have substituted

$$\boldsymbol{f}^* = \boldsymbol{O}(\boldsymbol{f} - \boldsymbol{a}_f - \boldsymbol{a}_c) \,, \quad \boldsymbol{f} = \boldsymbol{g}$$

as in (4.3.19). The equation (4.5.10) demonstrates the invariance of the momentum balance with respect to the EUCLIDian transformations.

[7] See e.g. ALTENBACH & ALTENBACH [12].

9. We start with the recognition that under EUCLIDian transformations the relations

$$\begin{aligned} \boldsymbol{L}^* &= \boldsymbol{O}\boldsymbol{L}\boldsymbol{O}^T + \dot{\boldsymbol{O}}\boldsymbol{O}^T , \\ \boldsymbol{W}^* &= \boldsymbol{O}\boldsymbol{W}\boldsymbol{O}^T + \dot{\boldsymbol{O}}\boldsymbol{O}^T \end{aligned} \tag{4.5.11}$$

hold, in which starred variables are formulated in the rotating system. Since $\boldsymbol{a}$ is presumed objective, $\boldsymbol{a}^* = \boldsymbol{O}\boldsymbol{a}$, we may deduce

$$\begin{aligned} (\boldsymbol{a}^*)^\circ &= (\boldsymbol{a}^*)^\cdot - \boldsymbol{W}^*\boldsymbol{a}^* \\ &= (\boldsymbol{O}\boldsymbol{a})^\cdot - \boldsymbol{O}\boldsymbol{W}\boldsymbol{a} - \dot{\boldsymbol{O}}\boldsymbol{a} \\ &= \dot{\boldsymbol{O}}\boldsymbol{a} + \boldsymbol{O}\dot{\boldsymbol{a}} - \boldsymbol{O}\boldsymbol{W}\boldsymbol{a} - \dot{\boldsymbol{O}}\boldsymbol{a} \\ &= \boldsymbol{O}(\dot{\boldsymbol{a}} - \boldsymbol{W}\boldsymbol{a}) = \boldsymbol{O}\overset{\circ}{\boldsymbol{a}} , \end{aligned} \tag{4.5.12}$$

which proves the objectivity of $\overset{\circ}{\boldsymbol{a}}$. Since $\overset{\triangle}{\boldsymbol{a}}$ and $\overset{\triangledown}{\boldsymbol{a}}$ differ from $\overset{\circ}{\boldsymbol{a}}$ only through an additional term, and since $\boldsymbol{D}\boldsymbol{a}$ is itself an objective quantity i.e. $\boldsymbol{D}^*\boldsymbol{a}^* = \boldsymbol{O}\boldsymbol{D}\boldsymbol{a}$, it follows that $\overset{\triangle}{\boldsymbol{a}}$ and $\overset{\triangledown}{\boldsymbol{a}}$ are objective vectors.

With regard to stresses one can, upon using $\boldsymbol{L} = \boldsymbol{D} + \boldsymbol{W}$, show that

$$\begin{aligned} \overset{\triangledown}{\boldsymbol{T}} &= \overset{\circ}{\boldsymbol{T}} - (\boldsymbol{D}\boldsymbol{T} + \boldsymbol{T}\boldsymbol{D}) , \\ \overset{\triangle}{\boldsymbol{T}} &= \overset{\circ}{\boldsymbol{T}} + (\boldsymbol{D}\boldsymbol{T} + \boldsymbol{T}\boldsymbol{D}) . \end{aligned} \tag{4.5.13}$$

Since $(\boldsymbol{D}\boldsymbol{T} + \boldsymbol{T}\boldsymbol{D})$ is by itself an objective tensor, then objectivity of $\overset{\circ}{\boldsymbol{T}}$ may be corroborated as follows:

$$\begin{aligned} (\boldsymbol{T}^*)^\circ &= (\boldsymbol{T}^*)^\cdot - \boldsymbol{W}^*\boldsymbol{T}^* + \boldsymbol{T}^*\boldsymbol{W}^* \\ &= \boldsymbol{O}\dot{\boldsymbol{T}}\boldsymbol{O}^T + \dot{\boldsymbol{O}}\boldsymbol{T}\boldsymbol{O}^T + \boldsymbol{O}\boldsymbol{T}\dot{\boldsymbol{O}} - \boldsymbol{O}\boldsymbol{W}\boldsymbol{T}\boldsymbol{O}^T - \dot{\boldsymbol{O}}\boldsymbol{T}\boldsymbol{O}^T \\ &\quad + \boldsymbol{O}\boldsymbol{T}\boldsymbol{W}\boldsymbol{O}^T + \boldsymbol{O}\boldsymbol{T}\underbrace{\boldsymbol{O}^T\dot{\boldsymbol{O}}\boldsymbol{O}^T}_{-\boldsymbol{O}^T\boldsymbol{O}\dot{\boldsymbol{O}}^T = -\dot{\boldsymbol{O}}^T} \\ &= \boldsymbol{O}\{\dot{\boldsymbol{T}} - \boldsymbol{W}\boldsymbol{T} + \boldsymbol{T}\boldsymbol{W}\}\boldsymbol{O}^T = \boldsymbol{O}\overset{\circ}{\boldsymbol{T}}\boldsymbol{O}^T . \end{aligned} \tag{4.5.14}$$

Therefore $\overset{\circ}{\boldsymbol{T}}$ is an objective tensor, qed.

5. Material Equations

Until now we were devoting our attention to the formulation of the basic principles. In Chaps. 2 and 3 we concentrated on the formulation of physical laws representing the conservation laws for mass, momentum, angular momentum and energy as well as the balance statement of entropy. These laws represent the basis of classical physics. They were formulated for one-component continua which are material bodies formed by one kind of material. Later, these laws will be extended to multi-component systems in order to treat also mixtures.

One of the important characteristics of the balance equations is that these – either written in their local forms in both the LAGRANGE and EULER representations or as jump conditions – hold for material bodies, but the material properties are not yet defined in their formulations. The balance statements hold for solids, fluids, gases irrespectively upon how the individual solids, fluids, gases may differ from each other; in other words *the balance equations for mass, linear momentum, angular momentum, energy and entropy represent laws that embrace all material behaviour*. They form, say, the superordinate principle.

Now it is obvious that individual materials such as water and oil, or wood and steel, or oxygen and nitrogen react differently to the same external physical driving conditions. There must therefore also exist laws which can describe different material properties that apparently separate the various materials from one another. These laws are the *material* or *constitutive equations.* That the balance equations are not sufficient for determining the field variables, follows from the fact that, provided the external forcings, the fields $\boldsymbol{g}$ and $\mathfrak{r}$, are known, there are five balance equations (mass balance [1], momentum balance [3], energy balance [1]) apart from the entropy imbalance (see Table 3.5 – we consider the EULER representation, for example) and three equations for the symmetry of the stress tensor for the determination of 18 variables (density [1], velocity [3], stress tensor [9], energy [1], heat flux vector [3] and temperature [1])[1]. For given boundary and initial conditions these 18 vari-

[1] There are several ways to count the number of variables. Above we have regarded the conservation law of angular momentum as a generic equation; thus, the fact that the skew symmetric part of the CAUCHY stress tensor must vanish was counted as three equations and the CAUCHY stress was assumed to possess nine independent components. Equally justified is to assume this symmetry *ab initio*

ables are clearly not uniquely determinable. The objective of the postulation of the constitutive equations is to attain a closed system of equations with which the field variables can (at least, in principle) be uniquely determined. In this regard one recognizes the material equations as *closure conditions*. The closed system, consisting of the balance equations of mass, momentum, angular momentum, energy and the material equations is called the system of *field equations*; these equations are in general complicated equations – partial differential equations, functional differential equations etc. – and, as a rule, the existence of solutions for the field variables is assumed but can seldom be proved.

What role does the entropy inequality (second law of thermodynamics) play in this connection? The most obvious would possibly be to exclude for a closed system of balance equations of mass, momentum, angular momentum, energy and material equations, all these solutions which contradict the entropy inequality. That means, in this interpretation the second law would restrict processes and a material scientist would formulate the constitutive laws for a material according to his please and desire. The more appropriate and also physically more reasonable concept is that the entropy inequality should be applied in such a way *that it restricts the material equations* such that for any process which obeys the field equations, the second law is identically satisfied. We will transform these views in the following sections into mathematical expressions.

5.1 Formulation of the General Material Equations

5.1.1 Thermodynamic Processes

Material equations are based on experimental findings and motivated by observations and leave enough space for a variety of assumptions. Therefore it may be possible that two researchers studying identical materials, e.g. water or lubricating oil, in the same environment and under identical driving conditions, can write different material equations. From this, one may conclude that an absolute material law can not be assigned to a material even though particular aspects of its behaviour with regard to the driving processes can be described by equations of a certain form. In other words, certain *constitutive equations define a certain material behaviour*, an absolute material description of a physical material cannot be expected.

Despite the fact that there exists a large degree of freedom in postulating the material laws, one must pay attention to certain rules in formulating them. Some of them are self-evident. As an example one should actually unquestionably assume that the remembrance of the future is not possible;

and to state that the CAUCHY stress tensor is symmetric and has six independent components.

the material behaviour at the present time can depend only on the past processes from infinitely far away to present. This is occasionally called the *principle of determinism.* Apart from such natural philosophical evidences material laws make specifications about the possible dependency of certain variables (stress tensor, heat flux vector, internal energy) on the independent fields like density, temperature and the state of motion. In so doing, the influence of these variables can well embrace their past history. In such a case, one speaks of material with *memory.* Moreover, the state of a certain material point may also be influenced not only by the processes arising at its immediate neighbours but by those of all points of the body; this behaviour accounts for non-local properties.

The material theory is founded upon three rules:

- *Constitutive equations* (=material equations) must be independent of the observers.
- *Symmetry properties* of the body shall be portrayed through the material equations.
- Material equations must satisfy the second law of thermodynamics (entropy principle).

Additional restrictions e.g. on the kinematics may further influence the form of the material equations. This can for example be the assumption of density preserving or incompressibility, or the restriction of the motion to that of a rigid body.

Let us now regard the motion χ, the temperature Θ and the density ρ as the independent variables and suppose that the external fields, the radiation and the external volume forces are prescribed, known functions. It then suffices to establish material equations for the stress tensor (with 6 independent components – assuming the symmetry of the stress tensor), the heat flux vector and the internal energy in order to arrive at closed field equations.

Field Equations and Thermodynamic Processes The balance equations (Tables 3.4 and 3.5) together with the material equations form an integrable system of equations. This combined system is called the *field equations.* Every solution of these field equation will be called a *thermodynamic process*[2]. It is now absolutely plausible to assume, that through the alteration of the material laws the extent of the manifold of the thermodynamical processes will likewise alter. If the material laws are so chosen that for no corresponding

[2] Even though no proof is usually given at this stage of the analysis, it is always tacitly assumed that the balance laws and constitutive equations together – i.e., the field equations – possess a unique solution, at least for some finite time. Otherwise one could not proceed in the development of the theory. Mathematicians may prove existence of solutions once the theory is developed and thus give an a posteriori justification of the model in these cases. Thus, the assumption of the existence of solutions of the field equations is a matter of "être de raison", without the reliance on which any further analysis would be meaningless.

thermodynamic process the second law of thermodynamics is violated, then such a process is called *admissible.* In other words, only those thermodynamic processes are admissible which satisfy the second law of thermodynamics (entropy inequality). The solutions of the field equations defining such processes are particularized through the given initial and boundary conditions. In the modern literature, particularly that on turbulence, a process satisfying the second law of thermodynamics is also called *physically realizable.* So, the thermodynamically admissible processes are those that are physically realizable.

5.1.2 Non-Local Material Equations with Memory

In the following considerations the formulation will be treated in the EULER representation. The stress, heat flux and internal energy at a material point $\boldsymbol{X}$ of a body $\mathcal{B}$ depend on the history, i.e., the temporal succession of the independent variables, density ρ, temperature Θ and motion χ, and the state of the material at all other points of the body, $\boldsymbol{Y} \in \mathcal{B}$. One recognizes these dependencies as the *memory of the material* as well as its *non-locality.* Existence of an explicit dependency on the position is called *inhomogeneity.*

Let Ψ be one of the three quantities $\boldsymbol{t}$, $\boldsymbol{q}$ and ε, then the general form of a non-local material with memory is given by the functional $\hat{\Psi}(\cdot)$ as follows,

$$\begin{gathered}\Psi(t) = \mathop{\hat{\Psi}}_{\substack{\tau=0\\ \boldsymbol{Y}\in\mathcal{B}}}^{\infty} \Big(\chi(\boldsymbol{Y}, t-\tau), \Theta(\boldsymbol{Y}, t-\tau), \rho(\boldsymbol{Y}, t-\tau), \boldsymbol{X}\Big)\,, \\ \Psi \in \{\boldsymbol{t},\, \boldsymbol{q},\, \varepsilon\}\,. \end{gathered} \tag{5.1.1}$$

The argument $t-\tau$ runs through all temporal points in the past, from $t-\infty$ to the present time $t-0$, describing a *memory* or *capacity of remembering.* The dependency on every point of the body $\boldsymbol{Y} \in \mathcal{B}$ is the expression of *non-locality.* The explicit dependency of the considered position $\boldsymbol{X}$ is the expression of the *inhomogeneity* of the material. The symbols $\tau = 0$, $\boldsymbol{Y} \in \mathcal{B}$ and ∞ standing below and above the functional symbol $\hat{\Psi}$ are nothing else than a short–hand reminder that the argument τ varies from 0 to ∞ and $\boldsymbol{Y}$ may vary to be any point in $\mathcal{B}$.

5.1.3 Simple Body and Finite Memory

The general formulation of a material equation through the functional equation (5.1.1) is still too complex. No real material has been described at this level of complexity. We restrict it and only consider special materials.

Simple Body If the material properties of the particle $\boldsymbol{X}$ do not depend upon the histories of the motion, the temperature and density of all particles $\boldsymbol{Y}$ in the entire body, but if they depend on the histories of these variables only in the immediate neighbourhood of $\boldsymbol{X}$, then the body is called a *simple material.*

The restriction to the immediate neighbourhood of $\boldsymbol{X}$ can then be described by the dependency of the values and their gradients (the first derivatives) – of $\boldsymbol{\chi}$, Θ and ρ. This corresponds to a Taylor series expansion of these variables about the position $\boldsymbol{X}$ in which the second and higher order terms are ignored. In this way one considers only local effects.[3] With this restriction the material equations become

$$\begin{aligned}\Psi(t) = \overset{\infty}{\underset{\tau=0}{\hat{\Psi}}}\Big(&\boldsymbol{\chi}(\boldsymbol{X},t-\tau), \boldsymbol{F}(\boldsymbol{X},t-\tau),\\ &\Theta(\boldsymbol{X},t-\tau), \operatorname{Grad}\Theta(\boldsymbol{X},t-\tau),\\ &\rho(\boldsymbol{X},t-\tau), \operatorname{Grad}\rho(\boldsymbol{X},t-\tau), \boldsymbol{X}\Big)\,.\end{aligned} \tag{5.1.2}$$

Here the density itself is a function of the deformation gradient, viz.,

$$J = \frac{dv}{dV} = \frac{\rho_R}{\rho} \quad \Rightarrow \quad \rho = \frac{\rho_R}{|\det \boldsymbol{F}|}\,,$$

and since this dependency is already contained in the functional $\hat{\Psi}(\cdot)$ the density will be omitted as a variable from the argument list of (5.1.2). Moreover, the density gradient can be expressed in terms of the first derivative of the deformation gradient and consequently it involves the second derivative of the motion. However, in a simple material only the first derivatives should be considered; the material equation for a simple body thus takes the form

$$\begin{aligned}&\Psi(t) = \overset{\infty}{\underset{\tau=0}{\hat{\Psi}}}\Big(\boldsymbol{\chi}(\boldsymbol{X},t-\tau), \boldsymbol{F}(\boldsymbol{X},t-\tau), \Theta(\boldsymbol{X},t-\tau), \operatorname{Grad}\Theta(\boldsymbol{X},t-\tau), \boldsymbol{X}\Big)\,,\\ &\Psi \in \{\boldsymbol{t},\, \boldsymbol{q},\, \varepsilon\}\,.\end{aligned} \tag{5.1.3}$$

Many material laws with relaxation effects belong to this class. As a simple example, let us consider the following Gedanken model. Let a rod be subjected to tensile forces. Let the tension at a point be denoted by σ, and let the strain along the direction of its length be given by ε. Close observation may have led us to conclude that an increment of the strain $d\varepsilon(s)$ at time $s = t - \tau$ in the past may invoke a tension increment $d\sigma(t)$ at the present time t that is proportional to the strain increment at time s, $d\sigma(t) = \kappa\, d\varepsilon(s)$. Further, observation may allow us to infer that the longer the time difference between t and s or the larger τ is, the smaller will be the influence; stating in

[3] The assumption of existence of a TAYLOR series requires differentiability of the motion: The series converges within the convergence radius and, as a rule, does not incorporate the whole body. With this expansion alone one already neglects a certain degree of non-locality. Through the truncation of the TAYLOR series after the second term one restricts oneself necessarily to a minimum that affects the material properties within the neighbourhood of the point $\boldsymbol{X}$. In this restricted sense one calls the theories non-local theories that permit higher gradients of the motion $\boldsymbol{\chi}(\boldsymbol{X},\, t)$ as arguments in the functional $\hat{\Psi}(\cdot)$.

other words, κ is not constant but a monotonically decreasing function with growing argument τ, $\kappa(\tau_1) > \kappa(\tau_2)$ for $\tau_1 < \tau_2$. Thus,

$$d\sigma(t) = \kappa(t-s)\, d\varepsilon(s) \; . \tag{5.1.4}$$

This equation is expression of a *fading memory*. By integrating (5.1.4) over all strain increments in the past there follows

$$\begin{aligned} \sigma(t) &= \int_{\varepsilon(-\infty)}^{\varepsilon(t)} \kappa(t-s)\, d\varepsilon(s) = \int_{-\infty}^{t} \kappa(t-s) \frac{\mathrm{d}\varepsilon}{\mathrm{d}s}\, ds \\ &= \int_0^{\infty} \kappa(\tau)\dot{\varepsilon}(t-\tau)\, d\tau \; . \end{aligned} \tag{5.1.5}$$

Quite often $\kappa(\tau)$ is set equal to $\kappa(\tau) = \kappa_0 \exp(-\tau/\tau_0)$, corresponding to an exponentially fading memory with relaxation time τ_0. Despite its exponential decay the memory time is infinitely large, however.

Naturally, such integral representations are somewhat complex. Simplifications are sought in the next steps.

Restricted (Bounded) Memory It is extremely rare that a material can remember its whole history through which it is affected. All the more, the memory of the material is *bounded*, $0 \leq \tau \leq \tau_0 < \infty$. To express this, we employ a Taylor series expansion for one of our variables – say $\boldsymbol{\chi}(t-\tau)$ – about the point $\tau = 0$, viz.,

$$\boldsymbol{\chi}(t-\tau) = \sum_{n=0}^{\infty} \frac{(-1)^n}{n!} \boldsymbol{\chi}^{(n)}(t)\, \tau^n \quad \text{with} \quad \boldsymbol{\chi}^{(n)} = \left. \frac{\partial^n \boldsymbol{\chi}}{\partial t^n} \right|_{\tau=0} \; . \tag{5.1.6}$$

The other independent variables in (5.1.3) can be treated analogously. In the material equation (5.1.3) then, instead of using $\boldsymbol{\chi}(t-\tau)$, $\tau \in \mathbb{R}^+$ as independent variable, the expansion coefficients can be employed, where this is considered to a certain degree, $n = 0, ..., N$. The constitutive equation for a material with memory, then, takes the form

$$\begin{aligned} \Psi(t) = \hat{\Psi}\Big(&\boldsymbol{\chi}(t), \boldsymbol{\chi}^{(1)}(t), \ldots, \boldsymbol{\chi}^{(N)}(t), \boldsymbol{F}(t), \ldots, \\ &\Theta(t), \ldots, \operatorname{Grad}\Theta(t), \ldots, \boldsymbol{X}\Big) \; , \end{aligned} \tag{5.1.7}$$

$$\Psi \in \{\mathbf{t},\, \boldsymbol{q},\, \varepsilon\} \; .$$

Such a material is called *rate dependent of degree* N with respect to $\boldsymbol{\chi}$. The expansion of the other variables can be accomplished in a similar manner, but the number of derivatives that are taken into account may be different in general. Thus the material is of rate type with different degrees in each variable.

The example of the rod considered in connection with equation (5.1.3) would lead here for $N = 1$ to the relation $\sigma(t) = \hat{\sigma}(\varepsilon,\, \dot{\varepsilon})$, which in the linear case becomes

$$\sigma(t) = E\varepsilon(t) + \eta\dot{\varepsilon}(t) \tag{5.1.8}$$

with two constants E and η. These are a modulus of elasticity and a viscosity.

We will, in the forthcoming examples, treat the simplest cases, such as a material without memory or at most of degree 1 in the deformation gradient. First of all it should be instructive to illustrate the above derivation by some examples.

5.1.4 Examples of Simple Material Equations

Starting with the above considered general formulations of the material equations we go through frequently appearing special cases (simple bodies with bounded memory). These material laws will later be further analyzed as they are only briefly introduced here.

Thermoelastic Body There are materials in which the CAUCHY stress tensor, the heat flux vector and the internal energy depend upon the deformation gradient, the temperature and its gradients at a position under consideration but not on their history. Such a body is described by the functions

$$\begin{aligned} \boldsymbol{t}(t) &= \hat{\boldsymbol{t}}\big(\boldsymbol{F}, \Theta, \operatorname{Grad}\Theta, \boldsymbol{X}\big)\ , \\ \boldsymbol{q}(t) &= \hat{\boldsymbol{q}}\big(\boldsymbol{F}, \Theta, \operatorname{Grad}\Theta, \boldsymbol{X}\big)\ , \\ \varepsilon(t) &= \hat{\varepsilon}\big(\boldsymbol{F}, \Theta, \operatorname{Grad}\Theta, \boldsymbol{X}\big) \end{aligned} \tag{5.1.9}$$

and is called a *thermoelastic* body. Further simplifications of these equations may be made by introducing the following

Definitions

- When the temperature gradient does not appear in the material equation then the material is called *elastic with temperature dependence.*
- If the material properties are independent of the temperature itself in addition to its gradient, then such a material is called *purely elastic.* The nature of the material equations in this case differs from the above one as the constitutive relations are

$$\boldsymbol{t}(t) = \hat{\boldsymbol{t}}\big(\boldsymbol{F}, \boldsymbol{X}\big)\ , \quad \boldsymbol{q}(t) = \boldsymbol{0}\ , \quad \varepsilon(t) = \hat{\varepsilon}\big(\boldsymbol{F}, \boldsymbol{X}\big)\ . \tag{5.1.10}$$

■

Here the statement $\boldsymbol{q} = \boldsymbol{0}$ is not yet fully understandable as it is the unique consequence of thermodynamics. Nevertheless, it should be, at least, plausible to assume that there is no heat flow in a purely elastic material[4].

If one desires to write these material properties in the reference configuration then one must consider the transformation rules between the two

[4] The alert reader will realize that, since $\boldsymbol{q}$ is a vector and $\boldsymbol{F}$ is a two–point tensor, the only solution of $\boldsymbol{q} = \boldsymbol{q}(\boldsymbol{F})$ is the zero function $\boldsymbol{q} = \boldsymbol{0}$.

configurations (Table 3.1). Since the transformation rules only depend on the deformation gradient, which is already contained in the argument list, the material equations have the same dependency as shown above: thus

$$\begin{aligned} \boldsymbol{T}(t) &= \hat{\boldsymbol{T}}(\boldsymbol{F}, \Theta, \operatorname{Grad}\Theta, \boldsymbol{X}) , \\ \boldsymbol{Q}(t) &= \hat{\boldsymbol{Q}}(\boldsymbol{F}, \Theta, \operatorname{Grad}\Theta, \boldsymbol{X}) , \\ \varepsilon(t) &= \hat{\varepsilon}(\boldsymbol{F}, \Theta, \operatorname{Grad}\Theta, \boldsymbol{X}) \end{aligned} \tag{5.1.11}$$

for a *thermoelastic material* and

$$\begin{aligned} \boldsymbol{T}(t) &= \hat{\boldsymbol{T}}(\boldsymbol{F}, \boldsymbol{X}) , \\ \boldsymbol{Q}(t) &= \boldsymbol{0} , \\ \varepsilon(t) &= \hat{\varepsilon}(\boldsymbol{F}, \boldsymbol{X}) \end{aligned} \tag{5.1.12}$$

for a purely *elastic body.*

Ideal, Compressible Fluids If the material property of an elastic material depends only on the determinant of the deformation gradient, i.e., on the density, then the material laws are given by

$$\boldsymbol{t}(t) = -p(\rho)\boldsymbol{I} , \quad \varepsilon(t) = \hat{\varepsilon}(\rho) . \tag{5.1.13}$$

Here a possibility of inhomogeneity, i.e., a dependence on $\boldsymbol{X}$ is ignored. One calls the material law (5.1.13) also an *ideal (perfect) compressible* or a *barotropic fluid.* In English the denotation *elastic fluid* is more frequent. The fact that the CAUCHY stress tensor is proportional to the unit tensor with a scalar prefactor $p(\rho)$ must naturally be proved which will follow later. The function $p(\rho)$ is called the *elastic pressure.*

Viscous Thermoelastic Body Let us consider a body of which the material behaviour does not only depend on the deformation gradient, the temperature and its gradient, but also on the time rate of change of the deformation gradient; such a material with bounded memory is given by the constitutive relation

$$\begin{aligned} \boldsymbol{t}(t) &= \hat{\boldsymbol{t}}(\boldsymbol{F}, \boldsymbol{L}, \Theta, \operatorname{Grad}\Theta, \boldsymbol{X}) , \\ \boldsymbol{q}(t) &= \hat{\boldsymbol{q}}(\boldsymbol{F}, \boldsymbol{L}, \Theta, \operatorname{Grad}\Theta, \boldsymbol{X}) , \\ \varepsilon(t) &= \hat{\varepsilon}(\boldsymbol{F}, \boldsymbol{L}, \Theta, \operatorname{Grad}\Theta, \boldsymbol{X}) . \end{aligned} \tag{5.1.14}$$

Here we have used the fact that the time derivative of the deformation gradient can be written as $\dot{\boldsymbol{F}} = \boldsymbol{L}\boldsymbol{F}$, so that in place of a dependence on $\dot{\boldsymbol{F}}$ we are allowed to write $\boldsymbol{L}$. Further assumptions can be introduced with the following specifications:

- If the material behaviour is additionally independent of the temperature and temperature gradient the material is called *viscoelastic.*

- If the material property depends on the determinant of the deformation gradient, i.e., the density ρ, in addition to the velocity gradient, then such a material is called a *heat conducting, viscous, elastic fluid.*
- The stress tensor of a viscous, compressible fluid is given by the law

$$\boldsymbol{t}(t) = -p(\rho)\boldsymbol{I} + \hat{\boldsymbol{t}}^{E}(\boldsymbol{D}) . \tag{5.1.15}$$

  Here $\hat{\boldsymbol{t}}^{E}(\boldsymbol{D})$ is called the *extra stress tensor* and $\boldsymbol{D}$ is the symmetric part of the velocity gradient $\boldsymbol{L}$.
- If the dependency on $\boldsymbol{D}$ is non-linear, then one speaks of a *non-*NEWTON*ian* material.
- If the dependency $\hat{\boldsymbol{t}}^{E}(\boldsymbol{D})$ is linear then this material is called a NEWTON*ian* material law; the field equations (specially the momentum balance) corresponding to such a material law are then called the NAVIER–STOKES *equations.*

We close the listing of these examples by making the following remarks:

- The material behaviour just explained above is described as being NEWTONian because NEWTON developed such a model in the third edition of his famous work, Principia [171], which states that the stress is linearly connected to the strain rate.
- In example (5.1.8) the stress is written as a linear function of the strain and the strain rate. A material, whose constitutive law is given by (5.1.8) and obeys a linear relationship, is termed a VOIGT or KELVIN material. More interesting would be to introduce as a material equation an inverse relation $\varepsilon = \hat{\varepsilon}(\sigma, \dot{\sigma})$ instead of $\sigma = \hat{\sigma}(\varepsilon, \dot{\varepsilon})$, or more generally $f(\varepsilon, \dot{\varepsilon}, \sigma, \dot{\sigma}) = 0$. In this regard, a law of the form

$$\dot{\varepsilon} = \frac{\dot{\sigma}}{E} + \frac{\sigma}{\eta} \tag{5.1.16}$$

  is also possible in place of (5.1.8). A material of this type is called a MAXWELL fluid. At the moment, however, it is not insightful how one can generalize (5.1.8) or (5.1.16) into general tensor equations or how they may be derived from (5.1.1).
- In (5.1.9) and partly also in later formulas, $\boldsymbol{F}$ and $\operatorname{Grad}\Theta$ are used as independent variables. These are gradients with respect to the material coordinates. One can also substitute $\operatorname{grad}\Theta$ instead of $\operatorname{Grad}\Theta$, and because of the relation $\operatorname{Grad}\Theta = \operatorname{grad}\Theta\,\boldsymbol{F}$ the type of the material does not change by this alteration.
- Constitutive relations of the form $\boldsymbol{t}^{E} = \hat{\boldsymbol{t}}^{E}(\boldsymbol{D})$ arise in a variety of forms; two classes are *rate dependent* and *rate independent* stress-stretching relations. The former are also called *viscous* the latter are *plastic*. For a rate independent material $\hat{\boldsymbol{t}}^{E}(\boldsymbol{D}) = \hat{\boldsymbol{t}}^{E}(\lambda\boldsymbol{D})$ for any $\lambda \in (0, \infty)$.

5.2 Material Objectivity

With the assumption of locality and bounded memory the material equations obtained so far are considerably simplified. The physical requirement of the *observer invariance or material objectivity*, however, allows us to further reduce the material equations. This postulate says that *material laws should be independent of the choice of the reference frame or the observer*, i.e., changing a reference system does not change the material behaviour. The effects of this postulate will be examined in the following subsections.

5.2.1 Transformations of the Reference System

The change of a reference system to another reference system was studied in Chap. 4 where the quantities such as velocity and acceleration were transformed from an inertial system to another system by performing a rigid body motion. In (4.1.5) we considered a general change of the reference frame. Let us start with that form in which the distance of two points does not change.

The most general form of a transformation between two reference systems moving against each other is the EUCLID*ian transformation*, as discussed in Chap. 4, see also Fig. 4.1 for the choice of relations of position vectors in a fixed and in a moving reference system,

$$\boldsymbol{y}^* = \boldsymbol{O}(t)\boldsymbol{x} - \boldsymbol{c}^*(t) \; . \tag{5.2.1}$$

In this equation $\boldsymbol{y}^*$ is the position vector of a material point in the moving reference system. The tensor $\boldsymbol{O}(t)$ is a time dependent orthogonal transformation such that $\boldsymbol{O}\boldsymbol{O}^T = \boldsymbol{I}$, which describes the rotation of both systems against each other; $\boldsymbol{c}^*(t)$ is a time dependent translation between the origins of the two systems. We also distinguish the systems by writing the variables of one with a star and those of the other without an asterisk.

5.2.2 Transformation of Physical Quantities

The motion of a particle in an inertial system (in the following often called the non-starred system) is given by $\boldsymbol{x} = \boldsymbol{\chi}(\boldsymbol{X}, t)$, and in the moving (starred) system it is given by $\boldsymbol{y}^* = \boldsymbol{\chi}^*(\boldsymbol{X}, t)$; one can derive all kinematic quantities from these functions and calculate and relate them to each other. These relations describe then the transformation properties of the respective variables under EUCLIDian transformations. For example, with the help of (5.2.1) the following transformation rules may easily be verified:

$$
\begin{aligned}
\boldsymbol{v}^* &= \boldsymbol{O}\boldsymbol{v} + \dot{\boldsymbol{O}}\boldsymbol{x} - \dot{\boldsymbol{c}}^* , \\
\boldsymbol{v}_1^* - \boldsymbol{v}_2^* &= \boldsymbol{O}(\boldsymbol{v}_1 - \boldsymbol{v}_2) , \quad \text{provided} \quad \boldsymbol{x}_1 = \boldsymbol{x}_2 = \boldsymbol{x} \\
\boldsymbol{F}^* &= \boldsymbol{O}\boldsymbol{F} , \\
\boldsymbol{L}^* &= \boldsymbol{O}\boldsymbol{L}\boldsymbol{O}^T + \boldsymbol{\Omega} , \\
\boldsymbol{D}^* &= \boldsymbol{O}\boldsymbol{D}\boldsymbol{O}^T , \\
\boldsymbol{W}^* &= \boldsymbol{O}\boldsymbol{W}\boldsymbol{O}^T + \boldsymbol{\Omega} , \\
\boldsymbol{B}^* &= \boldsymbol{F}^*\boldsymbol{F}^{*T} = \boldsymbol{O}\boldsymbol{B}\boldsymbol{O}^T , \\
\mathrm{d}v^* &= \mathrm{d}v ,
\end{aligned}
\tag{5.2.2}
$$

where $\boldsymbol{\Omega} = \dot{\boldsymbol{O}}\boldsymbol{O}^T$, for the velocity $\boldsymbol{v}$, the velocity difference $\boldsymbol{v}_1 - \boldsymbol{v}_2$[5], the deformation gradient $\boldsymbol{F}$, the spatial velocity gradient $\boldsymbol{L}$, the stretching tensor $\boldsymbol{D} = \frac{1}{2}(\boldsymbol{L} + \boldsymbol{L}^T)$, the spin tensor $\boldsymbol{W} = \frac{1}{2}(\boldsymbol{L} - \boldsymbol{L}^T)$, the left Cauchy–Green tensor $\boldsymbol{B}$ as well as the volume increment $\mathrm{d}v$. It is apparent that the volume element (a scalar), the velocity difference (a vector) and the stretching tensor like the left Cauchy–Green strain tensor (of rank 2) transform as

$$
a^* = a , \quad \boldsymbol{b}^* = \boldsymbol{O}\boldsymbol{b} , \quad \boldsymbol{A}^* = \boldsymbol{O}\boldsymbol{A}\boldsymbol{O}^T , \tag{5.2.3}
$$

where a, $\boldsymbol{b}$, $\boldsymbol{A}$ are a scalar, vector and a tensor of second rank, whereas the transformation laws for the velocity, its spatial gradient and the spin tensor contain besides $\boldsymbol{O}$ also $\dot{\boldsymbol{O}}$ (or/and $\boldsymbol{\Omega}$) as well as $\dot{\boldsymbol{c}}^*$ as additional transformation variables. There are therefore physical quantities, which are elements of mathematical vector spaces and transform particularly simply under EUCLIDian transformations, namely according to (5.2.3), and others which contain system dependent terms via $\dot{\boldsymbol{O}}$, $\boldsymbol{\Omega}$ and $\dot{\boldsymbol{c}}^*$. This suggests the following.

Definition *Physical quantities (scalars a, vectors $\boldsymbol{b}$ and tensors $\boldsymbol{A}$) are called objective scalars, objective vectors and objective tensors when they are transformed under* EUCLID*ian transformations according to (5.2.3).* ∎

According to this the volume increment is an *objective* scalar, the velocity difference is an *objective* vector and the stretching tensor as well as the left CAUCHY–GREEN tensor are *objective* tensors of rank two, but the velocity is not an objective vector. Of special interest are the transformation rules for the deformation gradient: $\boldsymbol{F}^* = \boldsymbol{O}\boldsymbol{F}$; every column of $\boldsymbol{F}$ transforms as an objective vector. The deformation gradient $\boldsymbol{F}$ transforms as three objective

[5] The velocities $\boldsymbol{v}_1$ and $\boldsymbol{v}_2$ are thought to be applied at the same location; this is what is meant by $\boldsymbol{x}_1 = \boldsymbol{x}_2 = \boldsymbol{x}$. Otherwise the above rule does not apply.

vectors. We will also agree to call such quantities objective quantities. In this regard the right CAUCHY–GREEN tensor $\boldsymbol{C} = \boldsymbol{F}^T\boldsymbol{F}$ is also objective, since here

$$\boldsymbol{C}^* = \boldsymbol{F}^{*T}\boldsymbol{F}^* = \boldsymbol{F}^T\boldsymbol{O}^T\boldsymbol{O}\boldsymbol{F} = \boldsymbol{F}^T\boldsymbol{F} = \boldsymbol{C} \tag{5.2.4}$$

holds. Every component of $\boldsymbol{C}$ transforms as an objective scalar under EUCLIDian transformations[6].

It can also be easily shown (Exercise) that the material time derivatives of an objective vector $\boldsymbol{b}$ or an objective tensor $\boldsymbol{A}$ are not objective; however, the combinations

$$\begin{aligned} \overset{\circ}{\boldsymbol{b}} &:= \dot{\boldsymbol{b}} - \tfrac{1}{2}(\boldsymbol{L} - \boldsymbol{L}^T)\boldsymbol{b}\,, \\ \overset{\diamond}{\boldsymbol{A}} &:= \dot{\boldsymbol{A}} + \boldsymbol{A}\mathrm{div}\boldsymbol{v} - \boldsymbol{A}\boldsymbol{L}^T - \boldsymbol{L}\boldsymbol{A}^T \end{aligned} \tag{5.2.5}$$

represent an objective vector or an objective second rank tensor. Here the symbols $\circ$ and $\diamond$ indicate objective derivatives.

Next, we shall examine the transformation properties of the internal energy, the heat flux and the stress tensor. The internal energy is a scalar quantity. We know that it does not change its value when we change the reference system. The transformation of such objective scalars is therefore

$$\varepsilon^* = \varepsilon\,. \tag{5.2.6}$$

The spatial heat flux vector $\boldsymbol{q}$ has different components in the moving coordinate system than in the non-moving system (relative to a rotating system the vector appears rotated). The value must naturally be conserved but the direction can be determined through the rotation of the observer. With this, the spatial heat flux vector transforms according to

$$\boldsymbol{q}^* = \boldsymbol{O}\boldsymbol{q}\,, \tag{5.2.7}$$

which is the transformation rule of an objective vector.

To identify the transformation property of the CAUCHY stress tensor is somewhat complicated. Let us consider the stress vector (traction),

$$\mathsf{t} = \boldsymbol{t}\boldsymbol{n}\,, \tag{5.2.8}$$

on a surface with surface normal $\boldsymbol{n}$. In a rotating system the unit normal vector rotates accordingly; its value must always remain unity, so it transforms as

$$\boldsymbol{n}^* = \boldsymbol{O}\boldsymbol{n}\,. \tag{5.2.9}$$

[6] Some authors call $\boldsymbol{F}$ and $\boldsymbol{C}$ non-objective because both are tensors (defined in their respective spaces) but do not transform as objective tensors when subject to EUCLIDian transformations. We do not follow them and retain the objectivity property for $\boldsymbol{F}$ and $\boldsymbol{C}$. Such variants of the notion of objectivity are e.g. employed by ALTENBACH & ALTENBACH [12].

The surface force (the traction vector) transforms in the rotating reference system analogously as

$$\mathrm{t}^* = \boldsymbol{O}\mathrm{t} \,. \tag{5.2.10}$$

With the assumption that in the moving system the traction vector and the stress tensor are related to each other according to

$$\mathrm{t}^* = \boldsymbol{t}^*\boldsymbol{n}^* \tag{5.2.11}$$

there follows

$$\mathrm{t}^* = \boldsymbol{O}\mathrm{t} = \boldsymbol{O}\boldsymbol{t}\boldsymbol{n} = \boldsymbol{t}^*\boldsymbol{n}^* = \boldsymbol{t}^*\boldsymbol{O}\boldsymbol{n} \,, \tag{5.2.12}$$

from which, with the orthogonality of $\boldsymbol{O}$, the transformation rule of the CAUCHY stress tensor takes the form

$$\boldsymbol{t}^* = \boldsymbol{O}\boldsymbol{t}\boldsymbol{O}^T \,. \tag{5.2.13}$$

If we consider the material instead of the spatial representation, the material heat flux $\boldsymbol{Q}$ and the first PIOLA–KIRCHHOFF stress tensor $\boldsymbol{T}$ undergo the transformations (see also the transformation of fluxes of a physical variable (2.1.17) from the spatial to the material representation)

$$\boldsymbol{Q} = J\boldsymbol{q}\boldsymbol{F}^{-T} \,, \quad \boldsymbol{T} = J\boldsymbol{t}\boldsymbol{F}^{-T} , \tag{5.2.14}$$

with $\boldsymbol{q}$ and $\boldsymbol{t}$ in the spatial representation. It follows through the application of the known transformation rules (5.2.6), (5.2.7) and (5.2.13) for $\boldsymbol{q}$, $\boldsymbol{t}$ and $\boldsymbol{F}$, that the quantities $\boldsymbol{Q}$ and $\boldsymbol{T}$ transform as

$$\boldsymbol{Q}^* = \boldsymbol{Q} \,, \quad \boldsymbol{T}^* = \boldsymbol{O}\boldsymbol{T} \,. \tag{5.2.15}$$

We summarize: *Internal energy, temperature and density are objective scalars, the spatial heat flux vector is an objective vector, the* CAUCHY *stress tensor is an objective tensor. In contrast, the material heat flux vector is transformed as three objective scalars and the first* PIOLA–KIRCHHOFF *stress tensor transforms as three objective vectors.* On the other hand, the velocity and acceleration are not objective quantities because their transformations (see also Chap. 4) do not obey these rules, since further terms such as the CORIOLIS acceleration arise which depend on the transformation of the moving system.

Finally, we like to point out that the balance equations are invariant under EUCLIDian transformations as shown in Chap. 4, they exhibit the same form in every reference system; however, the momentum balance is explicitly system dependent because of the additional terms due to the relative motion. We will use the term *indifference* for transformations which are system independent and further emphasize that the balance laws are invariant under EUCLIDian transformations but not indifferent. The reader is suggested to consult the Exercises (also in Chap. 4) to become more familiar with these concepts.

5.2.3 Indifference of the Material Equations

Let us consider a material law. It can not be ruled out from the outset that this law is independent of the system variables $\dot{\boldsymbol{O}}$, $\boldsymbol{\Omega}$ and $\boldsymbol{c}$. In a non-inertial system these equations could, analogously to the momentum balance, depend on those variables even though this dependency must dissolve in an inertial system. Thus, the system independency of the material equations can not be forcefully claimed. Nevertheless, it has become practical evidence that there are (virtually) no known cases in which constitutive equations are actually system dependent[7]. The postulate of the indifference of the material equations against observer transformations will be called the

Rule or **principle of material objectivity:**

A material law must be indifferent against a change of a reference system. ■

That is, with the requirement of the rule of material objectivity the balance equations are invariant against EUCLIDian transformations, but not indifferent. The material equations however are indifferent. The rules of material objectivity demand that the material laws, such as those of the stress tensor, are independent of the reference system. Both observers must secure exactly the same (stress) state (of which the representation may be different in different reference systems). Indeed both observers relate their measurements to their respective systems. The condition of material objectivity demands that both observers measure the same state of stress with the *same material law*, independent of the corresponding reference system. In both systems under consideration, therefore, the material laws (5.1.3) read, respectively,

$$
\begin{aligned}
&\Psi(t) = \overset{\infty}{\underset{\tau=0}{\hat{\Psi}}}\Big(\boldsymbol{\chi}(t-\tau), \boldsymbol{F}(t-\tau), \Theta(t-\tau), \operatorname{Grad}\Theta(t-\tau), \boldsymbol{X}\Big)\,,\\
&\Psi^*(t) = \overset{\infty}{\underset{\tau=0}{\hat{\Psi}}}\Big(\boldsymbol{\chi}^*(t-\tau), \boldsymbol{F}^*(t-\tau), \Theta^*(t-\tau), \operatorname{Grad}\Theta^*(t-\tau), \boldsymbol{X}\Big)\,,\\
&\Psi \in \{\boldsymbol{t}, \boldsymbol{q}, \varepsilon\}\,.
\end{aligned}
\tag{5.2.16}
$$

For simplicity, the argument $\boldsymbol{X}$ will henceforth be omitted from the argument list. It is very important to remark that the functional in the starred system

[7] The formulation is not verbalized any stronger because a system dependence, however weak, can not be ruled out. This is the reason why *rule* of material frame indifference is the preferred denotation.

There have been studies which illustrate that the material equation for the constitutive relations need not be frame independent. MÜLLER [162] studies the frame dependence of the stress tensor and heat flux vector as well as the electric current (MÜLLER [164]). Modern theories such as extended thermodynamics (MÜLLER & RUGGERI [167]) and turbulence theory (LUMLEY & TENNEKES [235]; AHMADI [9], [4]; SADIKI & HUTTER [204], [205] make it clear that the concept of material frame indifference cannot be maintained under all circumstances. However, as a rule, it is convenient and should be maintained as long as possible.

(i.e., the material law) carries no star, i.e., we have here the same material law, independent of the reference system. The absence of the star from the functional, i.e., the choice $\hat{\Psi}^*\big((\cdot)^* \big) = \hat{\Psi}\big((\cdot)^* \big)$ in the starred system is exactly the mathematical implementation of the rule of material objectivity.

If the observer (*) knows the transformation of his system to the non-starred system, then he is in the position to transform the values of variables measured by him. For this he must transform his deformation gradient etc. (the argument list), as well as his stress tensor etc. (the field variable for which the material law yields) from the starred to the unstarred system. With the above given implementations the transformations are

$$
\begin{aligned}
\boldsymbol{\chi}^* &= \boldsymbol{O}\boldsymbol{\chi} - \boldsymbol{c}^* , & \varepsilon^* &= \varepsilon , \\
\boldsymbol{F}^* &= \boldsymbol{O}\boldsymbol{F} , & \boldsymbol{q}^* &= \boldsymbol{O}\boldsymbol{q} , \\
\Theta^* &= \Theta , & \boldsymbol{t}^* &= \boldsymbol{O}\boldsymbol{t}\boldsymbol{O}^T , \\
\operatorname{Grad}\Theta^* &= \operatorname{Grad}\Theta .
\end{aligned}
\tag{5.2.17}
$$

The material laws (5.2.16) thus must satisfy the conditions

$$
\begin{aligned}
\varepsilon^*(t) = \varepsilon(t) \quad &: \quad \overset{\infty}{\underset{\tau=0}{\hat{\varepsilon}}}\Big(\boldsymbol{O}\boldsymbol{\chi} - \boldsymbol{c}^*, \boldsymbol{O}\boldsymbol{F}, \Theta, \operatorname{Grad}\Theta, \boldsymbol{X}\Big) \\
&\quad = \overset{\infty}{\underset{\tau=0}{\hat{\varepsilon}}}\Big(\boldsymbol{\chi}, \boldsymbol{F}, \Theta, \operatorname{Grad}\Theta, \boldsymbol{X}\Big) , \\
\boldsymbol{q}^*(t) = \boldsymbol{O}\boldsymbol{q}(t) \quad &: \quad \overset{\infty}{\underset{\tau=0}{\hat{\boldsymbol{q}}}}\Big(\boldsymbol{O}\boldsymbol{\chi} - \boldsymbol{c}^*, \boldsymbol{O}\boldsymbol{F}, \Theta, \operatorname{Grad}\Theta, \boldsymbol{X}\Big) \\
&\quad = \boldsymbol{O}\, \overset{\infty}{\underset{\tau=0}{\hat{\boldsymbol{q}}}}\Big(\boldsymbol{\chi}, \boldsymbol{F}, \Theta, \operatorname{Grad}\Theta, \boldsymbol{X}\Big) , \\
\boldsymbol{t}^*(t) = \boldsymbol{O}\boldsymbol{t}(t)\boldsymbol{O}^T : &\quad \overset{\infty}{\underset{\tau=0}{\hat{\boldsymbol{t}}}}\Big(\boldsymbol{O}\boldsymbol{\chi} - \boldsymbol{c}^*, \boldsymbol{O}\boldsymbol{F}, \Theta, \operatorname{Grad}\Theta, \boldsymbol{X}\Big) \\
&\quad = \boldsymbol{O}\, \overset{\infty}{\underset{\tau=0}{\hat{\boldsymbol{t}}}}\Big(\boldsymbol{\chi}, \boldsymbol{F}, \Theta, \operatorname{Grad}\Theta, \boldsymbol{X}\Big)\boldsymbol{O}^T ,
\end{aligned}
\tag{5.2.18}
$$

identically for all orthogonal transformations $\boldsymbol{O}$, translations $\boldsymbol{c}^*$, all processes and all histories $0 \leq \tau < \infty$. $\boldsymbol{O}$ and $\boldsymbol{c}^*$ are time dependent, and all other arguments are functions of $\boldsymbol{X}$ and $t - \tau$, which in (5.2.18) have been omitted in order not to overload the formulas. In the following subsections we will draw the consequences implied by these identities.

5.2.4 Observer–Invariant Material Equations

Because the equations (5.2.18) must be valid with respect to an arbitrary orthogonal transformation and translation, one can choose special transformations $\boldsymbol{O}$ and $\boldsymbol{c}^*$.

Transformation $\boldsymbol{O}(t) = \boldsymbol{I}$; $\boldsymbol{c}^*(t) = \boldsymbol{\chi}(t)$ Substituting this transformation in $(5.2.18)_3$ gives

$$\overset{\infty}{\underset{\tau=0}{\hat{\boldsymbol{t}}}}\left(\boldsymbol{0}, \boldsymbol{F}, \Theta, \operatorname{Grad}\Theta, \boldsymbol{X}\right) = \overset{\infty}{\underset{\tau=0}{\hat{\boldsymbol{t}}}}\left(\boldsymbol{\chi}, \boldsymbol{F}, \Theta, \operatorname{Grad}\Theta, \boldsymbol{X}\right), \tag{5.2.19}$$

with analogous results for the spatial heat flux vector $\boldsymbol{q}$ and the internal energy ε. This implies that *the material laws are not allowed to explicitly depend on the motion* $\boldsymbol{\chi}(t-\tau)$. This is valid for the whole history of the motion. Further, it implies that *velocity and acceleration and all other higher time derivatives of the motion can have no influence on the material laws.* In the ensuing development we will omit the motion $\boldsymbol{\chi}$ as a variable from the argument list. The translation of the systems, $\boldsymbol{c}^*$, is thus irrelevant.

Transformation $\boldsymbol{O}(t) = \boldsymbol{R}^T(t)$ Starting from the polar decomposition of the deformation gradient (1.5.1), $\boldsymbol{F} = \boldsymbol{R}\boldsymbol{U}$, one may choose the transformation of the inverse of the rotational part of the deformation gradient. Substituting in (5.2.18) $\boldsymbol{F} = \boldsymbol{R}\boldsymbol{U}$ and using $\boldsymbol{O} = \boldsymbol{R}^T$ yields

$$\begin{aligned}
\boldsymbol{t}(t) &= \overset{\infty}{\underset{\tau=0}{\hat{\boldsymbol{t}}}}\left(\boldsymbol{F}, \Theta, \operatorname{Grad}\Theta, \boldsymbol{X}\right) = \boldsymbol{R}\,\overset{\infty}{\underset{\tau=0}{\hat{\boldsymbol{t}}}}\left(\boldsymbol{U}, \Theta, \operatorname{Grad}\Theta, \boldsymbol{X}\right)\boldsymbol{R}^T, \\
\boldsymbol{q}(t) &= \overset{\infty}{\underset{\tau=0}{\hat{\boldsymbol{q}}}}\left(\boldsymbol{F}, \Theta, \operatorname{Grad}\Theta, \boldsymbol{X}\right) = \boldsymbol{R}\,\overset{\infty}{\underset{\tau=0}{\hat{\boldsymbol{q}}}}\left(\boldsymbol{U}, \Theta, \operatorname{Grad}\Theta, \boldsymbol{X}\right), \\
\varepsilon(t) &= \overset{\infty}{\underset{\tau=0}{\hat{\varepsilon}}}\left(\boldsymbol{F}, \Theta, \operatorname{Grad}\Theta, \boldsymbol{X}\right) = \overset{\infty}{\underset{\tau=0}{\hat{\varepsilon}}}\left(\boldsymbol{U}, \Theta, \operatorname{Grad}\Theta, \boldsymbol{X}\right).
\end{aligned} \tag{5.2.20}$$

Every objective material law must obey these conditions and every material law satisfying these conditions is objective. The last statement is however, not yet proved; it says that the representations for the CAUCHY stress tensor, the spatial heat flux vector and the internal energy in (5.2.20) are sufficient for the objectivity requirement (5.2.18) to be satisfied identically for all orthogonal $\boldsymbol{O}(t)$ (Exercises). One recognizes further that the stress tensor and the heat flux vector are only *explicitly* dependent on the rotational part of the deformation gradient, and the internal energy does not exhibit such a dependency at all. The intrinsic material functional includes only the stretch part of $\boldsymbol{F}$ via the right stretch tensor $\boldsymbol{U}$. Instead of $\boldsymbol{U}$ any other dependence of $\boldsymbol{U}$-dependent deformation can arise, e.g. $\boldsymbol{C} = \boldsymbol{U}^2$ or the GREEN's strain tensor $\boldsymbol{G} = \frac{1}{2}(\boldsymbol{C} - \boldsymbol{I})$ or the HENCKY strain $\boldsymbol{G}_H = \ln \boldsymbol{G}$, and other right strain measures $\boldsymbol{f}(\boldsymbol{U})$.

Second PIOLA–KIRCHHOFF **Stress Tensor** If the *second* PIOLA–KIRCHHOFF *stress tensor* is used instead of the CAUCHY stress tensor and if the heat flux vector in the reference configuration is employed, viz.,

$$\tilde{\boldsymbol{T}} := J\boldsymbol{F}^{-1}\boldsymbol{t}\boldsymbol{F}^{-T}, \quad \boldsymbol{Q} = J\boldsymbol{F}^{-1}\boldsymbol{q}, \tag{5.2.21}$$

then it is easy to show that the functional will depend only on $\boldsymbol{U}$ (or the right CAUCHY–GREEN tensor $\boldsymbol{C}$ or the GREEN strain tensor $\boldsymbol{G}$, respectively),

$$
\begin{aligned}
\tilde{\boldsymbol{T}}(t) &= \mathop{\hat{\tilde{\boldsymbol{T}}}}_{\tau=0}^{\infty}\Big(\boldsymbol{F}, \Theta, \operatorname{Grad}\Theta, \boldsymbol{X}\Big) = \mathop{\hat{\tilde{\boldsymbol{T}}}}_{\tau=0}^{\infty}\Big(\boldsymbol{C}, \Theta, \operatorname{Grad}\Theta, \boldsymbol{X}\Big), \\
\boldsymbol{Q}(t) &= \mathop{\hat{\boldsymbol{Q}}}_{\tau=0}^{\infty}\Big(\boldsymbol{F}, \Theta, \operatorname{Grad}\Theta, \boldsymbol{X}\Big) = \mathop{\hat{\boldsymbol{Q}}}_{\tau=0}^{\infty}\Big(\boldsymbol{C}, \Theta, \operatorname{Grad}\Theta, \boldsymbol{X}\Big), \\
\varepsilon(t) &= \mathop{\hat{\varepsilon}}_{\tau=0}^{\infty}\Big(\boldsymbol{F}, \Theta, \operatorname{Grad}\Theta, \boldsymbol{X}\Big) = \mathop{\hat{\varepsilon}}_{\tau=0}^{\infty}\Big(\boldsymbol{C}, \Theta, \operatorname{Grad}\Theta, \boldsymbol{X}\Big).
\end{aligned} \tag{5.2.22}
$$

There are still other forms of the objective material equations; we shall recall some of them at later stages.

Dependency on the Spatial Velocity Gradient In a material which depends on the time derivative of the deformation gradient (viscous material), then, besides the deformation gradient, the velocity gradient, $\boldsymbol{L}$ appears in the spatial argument list, because $\dot{\boldsymbol{F}} = \boldsymbol{L}\boldsymbol{F}$. The transformation rule for the velocity gradient has been given in (5.2.2) as

$$
\boldsymbol{L}^* = \boldsymbol{O}\boldsymbol{L}\boldsymbol{O}^T + \boldsymbol{\Omega}\,. \tag{5.2.23}
$$

The spatial velocity gradient is not an objective quantity because of the second term in (5.2.23). One can, however, not conclude that the material law (which must be objective) is independent of the velocity gradient. This dependency is given by the symmetric part, $\boldsymbol{D}$ of $\boldsymbol{L} = \boldsymbol{D} + \boldsymbol{W}$ alone, which namely is objective, $\boldsymbol{D}^* = \boldsymbol{O}\boldsymbol{D}\boldsymbol{O}^T$. Material equations for which a dependence on the time rate of change of the deformation gradient is assumed, may, owing to material objectivity, solely depend on the stretching or strain rate, the symmetric part $\boldsymbol{D}$ of the velocity gradient,

$$
\begin{aligned}
&\Psi(t) = \hat{\Psi}\Big(\boldsymbol{F}, \boldsymbol{L}, \Theta, \operatorname{Grad}\Theta, \boldsymbol{X}\Big) = \hat{\Psi}\Big(\boldsymbol{F}, \boldsymbol{D}, \Theta, \operatorname{Grad}\Theta, \boldsymbol{X}\Big), \\
&\Psi \in \{\boldsymbol{t}, \boldsymbol{q}, \varepsilon\}\,.
\end{aligned} \tag{5.2.24}
$$

For this reason the material equation of a visco-elastic fluid, (5.1.15) was written only in terms of $\boldsymbol{D}$ and not of $\boldsymbol{L}$.

The reader is also encouraged to consult the Exercises to deepen his/her knowledge in establishing constitutive relations.

5.3 Material Symmetry

5.3.1 Change of the Reference Configuration

The rule of material objectivity describes the indifference of the material equations under a change of frame. Constitutive equations must also reflect the symmetry properties of a body. Thus, for example, the material properties of a body with cubic symmetry must be invariant against the rotation of the reference configuration by 90 degrees. The *material symmetry* describes such invariance properties of bodies due to changes in the reference system. Hereby the material functionals depend now on the chosen reference configuration, and therefore are different from each other if they are referred to different reference configurations. In the following we indicate the change of a reference system by the symbol $(\cdot)^\circ$.

A change between two reference systems is defined by the one-to-one correspondence of the respective position vectors (the *names* of the particles),

$$\boldsymbol{X}^\circ = \boldsymbol{\kappa}(\boldsymbol{X}) \quad \text{and} \quad \boldsymbol{X} = \boldsymbol{\kappa}^{-1}(\boldsymbol{X}^\circ) \,. \tag{5.3.1}$$

A material particle $\mathfrak{X}$ with name $\boldsymbol{X}$ is named $\boldsymbol{X}^\circ$ in the new reference configuration. The motion of the particle or its position in the present configuration is then given with respect to both reference systems (see Fig. 5.1)

$$\boldsymbol{x} = \begin{cases} \chi(\boldsymbol{X},t) = \chi(\boldsymbol{\kappa}^{-1}(\boldsymbol{X}^\circ),t) = \chi^\circ(\boldsymbol{X}^\circ,t) \,, \\ \chi^\circ(\boldsymbol{X}^\circ,t) = \chi^\circ(\boldsymbol{\kappa}(\boldsymbol{X}),t) = \chi(\boldsymbol{X},t) \,. \end{cases} \tag{5.3.2}$$

The deformation gradient relative to the $^\circ$-reference system is therefore

$$\boldsymbol{F}^\circ = \frac{\partial \chi^\circ(\boldsymbol{X}^\circ,t)}{\partial \boldsymbol{X}^\circ} \,. \tag{5.3.3}$$

The transformation between the two deformation gradients is obtained via the gradient of the transformation (5.3.1) or

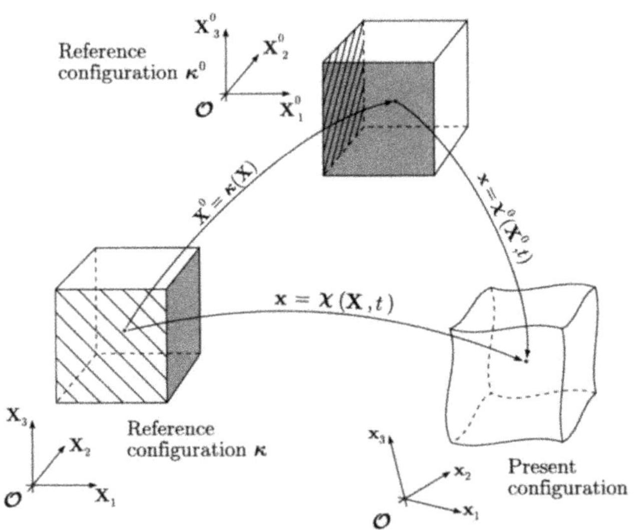

Fig. 5.1. Motion of a body with respect to two different reference configurations.

$$P := \operatorname{Grad} \boldsymbol{\kappa}(\boldsymbol{X}) \tag{5.3.4}$$

from (5.3.2) through differentiation using the chain rule of differentiation

$$\boldsymbol{F} = \boldsymbol{F}^{\circ}\boldsymbol{P} \quad \text{or} \quad \boldsymbol{F}^{\circ} = \boldsymbol{F}\boldsymbol{P}^{-1} \,. \tag{5.3.5}$$

It is evident that a function which describes the temporal progression of a physical quantity of a particle from the reference configuration depends on the choice of this reference configuration, however, the value of the quantity must be independent of this choice; i.e., one can write the density, temperature and temperature gradient as follows

$$\begin{aligned} \rho &= \rho^{\circ}(\boldsymbol{X}^{\circ}, t) = \rho(\boldsymbol{X}, t) \,, \\ \Theta &= \Theta^{\circ}(\boldsymbol{X}^{\circ}, t) = \Theta(\boldsymbol{X}, t) \,, \\ \operatorname{Grad}^{\circ} \Theta &= \operatorname{Grad} \Theta \boldsymbol{P}^{-1} \,. \end{aligned} \tag{5.3.6}$$

Analogous rules hold for the dependent constitutive variables. Let Ψ be the value of such a variable for a particle at time t. Then this value, on the one hand, can be calculated with the help of a functional $\hat{\Psi}(\cdot, \boldsymbol{X})$ of the history of deformation gradient, temperature and temperature gradient in the reference configuration, in which the particle is indicated by $\boldsymbol{X}$, viz.,

$$\Psi(t) = \underset{\tau=0}{\overset{\infty}{\hat{\Psi}}} \Big(\boldsymbol{F}(\boldsymbol{X}, t-\tau), \Theta(\boldsymbol{X}, t-\tau), \operatorname{Grad} \Theta(\boldsymbol{X}, t-\tau), \boldsymbol{X} \Big) \,. \tag{5.3.7}$$

On the other hand, the value of Ψ is also calculable with the functional $\hat{\Psi}^{\circ}(\cdot, \boldsymbol{X}^{\circ})$, whose independent variables represent the corresponding process histories from the reference configuration in which the same particle is denoted by $\boldsymbol{X}^{\circ}$, viz.,

$$\Psi(t) = \underset{\tau=0}{\overset{\infty}{\hat{\Psi}^{\circ}}} \Big(\boldsymbol{F}^{\circ}(\boldsymbol{X}^{\circ}, t-\tau), \Theta^{\circ}(\boldsymbol{X}^{\circ}, t-\tau), \operatorname{Grad}^{\circ} \Theta(\boldsymbol{X}^{\circ}, t-\tau), \boldsymbol{X}^{\circ} \Big) \,. \tag{5.3.8}$$

In the following discussions we consider only those materials for which there exists a global reference configuration of the type such that the same material equations hold at all material points of the body. Such a material is called *homogeneous* and the reference configuration in which all material points have common material laws is called *natural*. For these materials the explicit dependency on $\boldsymbol{X}$ and $\boldsymbol{X}^{\circ}$ is removed in (5.3.7), (5.3.8), and the value of the quantity Ψ for the constant process histories $\boldsymbol{F} = \boldsymbol{I}$, $\Theta = \Theta_0$, $\operatorname{Grad} \Theta_0 = \boldsymbol{0}$ in the whole body for all particles is the same constant – for the stress tensor and the heat flux vector e.g. zero. Now, we presume that all materials considered here possess such a *relaxed state* and call the corresponding configuration *natural*. It is to be expected that symmetry properties are specially simple with respect to the natural reference configurations. The material equations relative to both reference systems read in this case

$$\begin{aligned} \Psi(t) &= \overset{\infty}{\underset{\tau=0}{\hat{\Psi}}}\left(\boldsymbol{F}, \Theta, \operatorname{Grad}\Theta\right) = \overset{\infty}{\underset{\tau=0}{\hat{\Psi}^\circ}}\left(\boldsymbol{F}\boldsymbol{P}^{-1}, \Theta, \operatorname{Grad}\Theta\boldsymbol{P}^{-1}\right), \\ \Psi &\in \{\boldsymbol{t}, \boldsymbol{q}, \varepsilon\}\,. \end{aligned} \tag{5.3.9}$$

Hereby a dependency of the quantities $\boldsymbol{F}$, Θ, $\operatorname{Grad}\Theta$ on $(t-\tau)$ is tacitly understood. An equivalent representation is obviously also

$$\Psi(t) = \overset{\infty}{\underset{\tau=0}{\hat{\Psi}}}\left(\boldsymbol{F}^\circ\boldsymbol{P}, \Theta^\circ, \operatorname{Grad}^\circ\Theta\boldsymbol{P}\right) = \overset{\infty}{\underset{\tau=0}{\hat{\Psi}^\circ}}\left(\boldsymbol{F}^\circ, \Theta^\circ, \operatorname{Grad}^\circ\Theta\right), \tag{5.3.10}$$

where it is irrelevant whether the independent variables are characterized by $(\cdot)^\circ$ or $(\cdot)$; obviously (5.3.9) and (5.3.10) hold for all processes.

5.3.2 Symmetry Condition

If a material exhibits a certain symmetry, the material equations must be able to describe this property. With the help of the just introduced concept of the changes of the reference–system the conditions expressing the material symmetry can be mathematically formulated. To this end it is remarked that the transformation $\boldsymbol{X}^\circ = \boldsymbol{\kappa}(\boldsymbol{X})$ or its gradient $\boldsymbol{P}$ can not be fully arbitrary. It is physically easily understandable (and will be explained below) that only such transformations are permitted for changes of reference configurations which preserve the volume of the body. A transformation with this property is called *unimodular*, and it must satisfy the condition

$$\det\boldsymbol{P} = 1; \tag{5.3.11}$$

orthogonal transformations – rotations, mirror reflections – are examples of unimodular transformations.

In order to introduce the symmetry conditions, let us imagine the following experiment: A material particle $\mathfrak{X}$ is subjected to a certain process history for $\boldsymbol{F}$, Θ and $\operatorname{Grad}\Theta$ in the configuration $\boldsymbol{X}$. For this process let the value of Ψ at time t be given by $\Psi_{\boldsymbol{X}}$. In a second experiment, let the particle be subjected to the same process history, but in the configuration $\boldsymbol{X}^\circ$. The value of Ψ at time t for the process history from this configuration is given by $\Psi_{\boldsymbol{X}^\circ}$, and is in general different from $\Psi_{\boldsymbol{X}}$. Should $\Psi_{\boldsymbol{X}} = \Psi_{\boldsymbol{X}^\circ}$, however, be valid, then both reference configurations stand in a special relation to one another; they are called *symmetric* to each other, and

$$\begin{aligned} &\overset{\infty}{\underset{\tau=0}{\hat{\Psi}}}\left(\boldsymbol{F}(t-\tau), \Theta(t-\tau), \operatorname{Grad}\Theta(t-\tau)\right) \\ &= \overset{\infty}{\underset{\tau=0}{\hat{\Psi}^\circ}}\left(\boldsymbol{F}(t-\tau), \Theta(t-\tau), \operatorname{Grad}\Theta(t-\tau)\right) \end{aligned} \tag{5.3.12}$$

holds. If we indicate the gradients of this special transformation of the reference configuration with $\boldsymbol{H} = \boldsymbol{P}$, then it follows from (5.3.12) that

$$\overset{\infty}{\underset{\tau=0}{\hat{\Psi}}}\Big(\boldsymbol{F}, \Theta, \operatorname{Grad}\Theta\Big) = \overset{\infty}{\underset{\tau=0}{\hat{\Psi}}}\Big(\boldsymbol{FH}, \Theta, \operatorname{Grad}\Theta\boldsymbol{H}\Big) , \tag{5.3.13}$$

using (5.3.10) for $\Psi \in \{\boldsymbol{t},\, \boldsymbol{q},\, \varepsilon\}$. Replacing $\boldsymbol{F}$ by $\boldsymbol{FH}^{-1}$ and the corresponding $\operatorname{Grad}\Theta$ by $\operatorname{Grad}\Theta\boldsymbol{H}^{-1}$ – which is permitted, since they simply represent other possible deformation gradient and temperature gradient fields –, then there follows the equivalent statement

$$\overset{\infty}{\underset{\tau=0}{\hat{\Psi}}}\Big(\boldsymbol{FH}^{-1}, \Theta, \operatorname{Grad}\Theta\boldsymbol{H}^{-1}\Big) = \overset{\infty}{\underset{\tau=0}{\hat{\Psi}}}\Big(\boldsymbol{F}, \Theta, \operatorname{Grad}\Theta\Big) . \tag{5.3.14}$$

Here $\boldsymbol{F}$, Θ and $\operatorname{Grad}\Theta$ are functions of $(t - \tau)$. The two reference configurations are not to be distinguished in this case. Every transformation which satisfies this equation is called *symmetry or isotropy transformation.*

Symmetry Group The set of all unimodular transformations $\boldsymbol{H}$, which leaves the material equations (5.3.13) invariant relative to the reference configuration, is called the *symmetry group* relative to that configuration.[8] The symmetry group $\mathbb{G}$ is dependent on the choice of the reference system. It transforms with a change of the reference system (with $\boldsymbol{F} = \boldsymbol{F}^\circ\boldsymbol{P}$) according to

$$\mathbb{G}^\circ = \boldsymbol{P}\mathbb{G}\boldsymbol{P}^{-1} . \tag{5.3.15}$$

We can convince ourselves of the correctness of this as follows. Using on every side of (5.3.13) the transformation rule (5.3.9), one obtains

$$\begin{aligned} &\overset{\infty}{\underset{\tau=0}{\hat{\Psi}^\circ}}\Big(\boldsymbol{FP}^{-1}, \Theta, \operatorname{Grad}\Theta\boldsymbol{P}^{-1}\Big) \\ &\quad = \overset{\infty}{\underset{\tau=0}{\hat{\Psi}^\circ}}\Big(\boldsymbol{FHP}^{-1}, \Theta, \operatorname{Grad}\Theta\boldsymbol{HP}^{-1}\Big) . \end{aligned} \tag{5.3.16}$$

Substituting $\boldsymbol{F} = \boldsymbol{F}^\circ\boldsymbol{P}$ and $\operatorname{Grad}\Theta = \operatorname{Grad}^\circ\Theta\boldsymbol{P}$, there follows

$$\begin{aligned} &\overset{\infty}{\underset{\tau=0}{\hat{\Psi}^\circ}}\Big(\boldsymbol{F}^\circ, \Theta, \operatorname{Grad}^\circ\Theta\Big) \\ &\quad = \overset{\infty}{\underset{\tau=0}{\hat{\Psi}^\circ}}\Big(\boldsymbol{F}^\circ(\boldsymbol{PHP}^{-1}), \Theta, \operatorname{Grad}^\circ\Theta(\boldsymbol{PHP}^{-1})\Big) . \end{aligned} \tag{5.3.17}$$

[8] The transformations $\boldsymbol{H} \in \mathbb{G}$ satisfy the *group axioms* (Exercise):

- The composition of two symmetry transformations is again a symmetry transformation.
- The symmetry transformations are associative.
- The identical transformation $\boldsymbol{I}$ is an element of the group.
- The inverse transformation $\boldsymbol{H}^{-1}$ is an element of the group.

This is now the isotropy condition referred to the °-configuration. In other words, the transformations, which let the material equations invariant relative to this configuration, are exactly $\boldsymbol{PHP}^{-1}$, and thus transform as the symmetry group according to (5.3.15). This was first proved by NOLL [176]; (5.3.15) is therefore called *NOLL's rule of symmetry transformations.*

5.3.3 Group of Unimodular Transformations

From NOLL's rule it also follows that no more general transformation must be allowed as an element of $\mathbb{G}$ than the unimodular transformations. In case $\boldsymbol{P}$ is a pure dilatation, $\boldsymbol{P} = \lambda \boldsymbol{I}$, then one necessarily has $\boldsymbol{PHP}^{-1} = \boldsymbol{H}$. In other words, *a material functional cannot distinguish between reference configurations which are transformed to each other by pure dilatations.* Now, every arbitrary non-singular transformation may be decomposed into a *spherical* and an unimodular part according to

$$\boldsymbol{P} = \underbrace{(\det \boldsymbol{P})^{\frac{1}{3}} \boldsymbol{I}}_{\text{spherical tensor}} \ \underbrace{(\det \boldsymbol{P})^{-\frac{1}{3}} \boldsymbol{P}}_{\text{unimodular}} \ . \tag{5.3.18}$$

An isotropic transformation (with such a spherical tensor) eliminates itself in the transformation of the group (5.3.15). This means that the symmetry group of a body does not change under a pure dilatation of the configuration; this is perfectly understandable: the molecular structure of a body does not alter when it is subject to a volume change.

We call the symmetry groups, which embrace all unimodular transformations, $\mathbb{G} = \mathbb{U}$. Since the most general symmetry group of a body is the unimodular group, the symmetry group of an arbitrary body must always be a subgroup $\mathbb{G} \subseteq \mathbb{U}$. One can show that the number of these sub-groups is finite.

5.3.4 Classification of the Symmetry Properties

Isotropic Body Now, we will classify a body according to the symmetry group it belongs to. We call a body *isotropic* when the *symmetry conditions (*5.3.13*) are fulfilled at least for all orthogonal transformations.* That is, an arbitrary rotation or a mirror reflection does not alter the material behaviour. Because the orthogonal transformations are included in the unimodular transformations, all isotropic bodies are described by the groups

$$\mathbb{O} \subseteq \mathbb{G} \subseteq \mathbb{U} \, , \tag{5.3.19}$$

where $\mathbb{O}$ stands for the group of all orthogonal transformations. All groups of isotropic bodies are thus bounded by the orthogonal and unimodular groups. The answer of the question, how many groups bounded by $\mathbb{O}$ and $\mathbb{U}$ may exist, is significant because it clears how many types of isotropic bodies can in principle be defined. NOLL [176] has proved that in reality there are only two such isolated groups of symmetry transformations, namely

$$\mathbb{G} = \mathbb{O} \quad \text{and} \quad \mathbb{G} = \mathbb{U} \,. \tag{5.3.20}$$

An isotropic body is therefore a body whose symmetry group is either equal to the unimodular group or else equal to the orthogonal group. These bodies possess a very high symmetry. All other bodies are anisotropic and possess a lower degree of symmetry.

Fluids A body whose symmetry group is the unimodular group and thus possesses a very high degree of symmetry is called a *fluid,*

$$\mathbb{G} = \mathbb{U} \,, \quad \text{fluid.} \tag{5.3.21}$$

This definition of a fluid given by NOLL is evident, when one visualizes a container with water: Initially we suppose the fluid is at rest. When the fluid in the container is strongly stirred then no one molecule will probably occupy its original position when the container is brought again to a state of rest. But it is highly improbable that one recognizes a difference in the physical behaviour of the water; it is the same material as before. From this it follows that every configuration – also the present configuration – can be a reference configuration. Since the water is so defined that it is density preserving, all motions are isochoric and the respective deformation gradient $\boldsymbol{F}$ is unimodular ($\det \boldsymbol{F} = 1$).

Isotropic Solid A body whose symmetry group is equal to the orthogonal group, and whose material equations are only invariant against rotation, is called an *isotropic solid,*

$$\mathbb{G} = \mathbb{O} \,, \quad \text{isotropic solid.} \tag{5.3.22}$$

(Anisotropic) Solid, Crystal Classes A body whose symmetry group is a proper subgroup of the orthogonal group is called an *anisotropic solid,*

$$\mathbb{G} \subset \mathbb{O} \,, \quad \text{anisotropic solid.} \tag{5.3.23}$$

There is only a finite number of symmetry groups with this property, and it forms the *32 crystal classes*[9]. The lowest (trivial) symmetry group, which is possible, are those transformations which merely allow the identity as a transformation, i.e., the material is fully asymmetric.

---

[9] See e.g. A.E.H. LOVE [139], A treatise on the mathematical theory of elasticity, or any book on solid state physics, in which crystallography is treated, e.g., KITTEL [119], EHRENTRAUT [63].

5.4 Material Equations for Isotropic Bodies

Every material belonging to a certain symmetry group $\mathbb{G}$ must also be objective. At the same time the objectivity conditions (5.2.18) as well as the conditions of the material symmetry (5.3.14) must be fulfilled. Combining these statements shows that the material functionals of simple bodies must fulfill the conditions

$$\begin{aligned}
&\mathop{\hat{\varepsilon}}_{\tau=0}^{\infty}\Big(\boldsymbol{F}(\boldsymbol{X},t-\tau),\Theta(\boldsymbol{X},t-\tau),\mathrm{Grad}\,\Theta(\boldsymbol{X},t-\tau)\Big)\\
&\qquad=\mathop{\hat{\varepsilon}}_{\tau=0}^{\infty}\Big(\boldsymbol{O}\boldsymbol{F}(t-\tau)\boldsymbol{H}^{-1},\Theta(t-\tau),\mathrm{Grad}\,\Theta(t-\tau)\boldsymbol{H}^{-1}\Big)\,,\\
&\boldsymbol{O}\mathop{\hat{\boldsymbol{q}}}_{\tau=0}^{\infty}\Big(\boldsymbol{F}(t-\tau),\Theta(t-\tau),\mathrm{Grad}\,\Theta(t-\tau)\Big)\\
&\qquad=\mathop{\hat{\boldsymbol{q}}}_{\tau=0}^{\infty}\Big(\boldsymbol{O}\boldsymbol{F}(t-\tau)\boldsymbol{H}^{-1},\Theta(t-\tau),\mathrm{Grad}\,\Theta(t-\tau)\boldsymbol{H}^{-1}\Big)\,,\\
&\boldsymbol{O}\mathop{\hat{\boldsymbol{t}}}_{\tau=0}^{\infty}\Big(\boldsymbol{F}(t-\tau),\Theta(t-\tau),\mathrm{Grad}\,\Theta(t-\tau)\Big)\boldsymbol{O}^{T}\\
&\qquad=\mathop{\hat{\boldsymbol{t}}}_{\tau=0}^{\infty}\Big(\boldsymbol{O}\boldsymbol{F}(t-\tau)\boldsymbol{H}^{-1},\Theta(t-\tau),\mathrm{Grad}\,\Theta(t-\tau)\boldsymbol{H}^{-1}\Big)\,,
\end{aligned}\tag{5.4.1}$$

for all orthogonal transformations of the reference system, $\boldsymbol{O}(t)$ and all symmetry transformations $\boldsymbol{H} \in \mathbb{G}$. For isotropic bodies $\boldsymbol{H}$ is any element of the unimodular or orthogonal groups, thus at least the full orthogonal group; this is temporally constant in this restriction and shall be designated by $\boldsymbol{\mathcal{Q}}$. In contrast, $\boldsymbol{O}$ is time dependent. Restricting the last transformations likewise to the temporally constant element $\boldsymbol{\mathcal{Q}}$ of the original group, (5.4.1) becomes

$$\begin{aligned}
&\mathop{\hat{\varepsilon}}_{\tau=0}^{\infty}\Big(\boldsymbol{F}(\boldsymbol{X},t-\tau),\Theta(\boldsymbol{X},t-\tau),\mathrm{Grad}\,\Theta(\boldsymbol{X},t-\tau)\Big)\\
&\qquad=\mathop{\hat{\varepsilon}}_{\tau=0}^{\infty}\Big(\boldsymbol{\mathcal{Q}}\boldsymbol{F}(t-\tau)\boldsymbol{\mathcal{Q}}^{T},\Theta(t-\tau),\boldsymbol{\mathcal{Q}}\,\mathrm{Grad}\,\Theta(t-\tau)\Big)\,,\\
&\boldsymbol{\mathcal{Q}}\mathop{\hat{\boldsymbol{q}}}_{\tau=0}^{\infty}\Big(\boldsymbol{F}(t-\tau),\Theta(t-\tau),\mathrm{Grad}\,\Theta(t-\tau)\Big)\\
&\qquad=\mathop{\hat{\boldsymbol{q}}}_{\tau=0}^{\infty}\Big(\boldsymbol{\mathcal{Q}}\boldsymbol{F}(t-\tau)\boldsymbol{\mathcal{Q}}^{T},\Theta(t-\tau),\boldsymbol{\mathcal{Q}}\,\mathrm{Grad}\,\Theta(t-\tau)\Big)\,,\\
&\boldsymbol{\mathcal{Q}}\mathop{\hat{\boldsymbol{t}}}_{\tau=0}^{\infty}\Big(\boldsymbol{F}(t-\tau),\Theta(t-\tau),\mathrm{Grad}\,\Theta(t-\tau)\Big)\boldsymbol{\mathcal{Q}}^{T}\\
&\qquad=\mathop{\hat{\boldsymbol{t}}}_{\tau=0}^{\infty}\Big(\boldsymbol{\mathcal{Q}}\boldsymbol{F}(t-\tau)\boldsymbol{\mathcal{Q}}^{T},\Theta(t-\tau),\boldsymbol{\mathcal{Q}}\,\mathrm{Grad}\,\Theta(t-\tau)\Big)\,,\\
&\forall\ \text{orthogonal transformations, } \boldsymbol{\mathcal{Q}}.
\end{aligned}\tag{5.4.2}$$

Functionals, which satisfy these conditions, are called *scalar, vectorial and tensorial isotropic functionals* with respect to the orthogonal transformations. All functionals, which satisfy these conditions, represent constitutive equations for an isotropic body. It is self-evident that the identities (5.4.2) imply drastic reductions in the material functionals. In the following subsections, we will list them for the specific cases, that the material behaviour does not depend on the whole history of the deformation gradient, the temperature and the temperature gradient, but only on the values of these variables in a small neighbourhood of the present time.

It is also plausible that the conditions (5.4.2) are necessary for objective, isotropic functionals, but not automatically sufficient, because the rule of material objectivity is fulfilled only for temporally constant $\boldsymbol{\mathcal{Q}} \in \mathbb{O}$, but not for general $\boldsymbol{O}(t) \in \mathbb{O}$. In many cases the resulting functionals are also sufficient in this regard, i.e., the rule of material objectivity for all $\boldsymbol{O}(t) \in \mathbb{O}$ is fulfilled in reality. But this must be scrutinized in every individual case.

The conditions (5.4.2) will be examined individually for different materials, specially for fluids and isotropic solids. Hereby we restrict ourselves to the examination of the stress tensor. To this end certain mathematical properties of isotropic functions are needed which we now proceed to derive.

5.4.1 Isotropic Functions for a Symmetric Tensor, a Vector and a Scalar

In general the derivation of the reduced form of the isotropic functionals which satisfy (5.4.2) is complicated. In fact such reductions have almost exclusively only been constructed for functions. We are also doing this here but will restrict ourselves to simple examples. Complicated cases can be taken from the literature (SMITH, 1965; SPENCER, 1986; WANG, 1970).

First we provide some hints for notation; in the following we shall speak of a (second rank) tensor $\boldsymbol{K}$ and a vector $\boldsymbol{k}$, if these quantities transform under time independent orthogonal isotropy transformations (rotations, mirror reflections) as

$$\boldsymbol{K}^* = \boldsymbol{\mathcal{Q}}\boldsymbol{K}\boldsymbol{\mathcal{Q}}^T \,, \quad \boldsymbol{k}^* = \boldsymbol{\mathcal{Q}}\boldsymbol{k} \qquad \forall\, \boldsymbol{\mathcal{Q}} \in \mathbb{O} \tag{5.4.3}$$

where $\det \boldsymbol{\mathcal{Q}} = \pm 1$. With these relations all objective tensors and vectors are simply tensors and vectors, and $\boldsymbol{F}$, $\boldsymbol{B}$ and $\boldsymbol{C}$ are all tensors of rank two.

Let $(\sigma,\ \boldsymbol{s},\ \boldsymbol{S})$ and $(\mu,\ \boldsymbol{a},\ \boldsymbol{A})$ be two triples of physical quantities, which can be scalars $(\sigma,\ \mu)$, vectors $(\boldsymbol{s},\ \boldsymbol{a})$ and symmetric second rank tensors $(\boldsymbol{S},\ \boldsymbol{A})$ according to the above declaration. Let us assume that

$$\sigma = \hat{\sigma}(\mu, \boldsymbol{a}, \boldsymbol{A}) \,, \quad \boldsymbol{s} = \hat{\boldsymbol{s}}(\mu, \boldsymbol{a}, \boldsymbol{A}) \,, \quad \boldsymbol{S} = \hat{\boldsymbol{S}}(\mu, \boldsymbol{a}, \boldsymbol{A}) \tag{5.4.4}$$

are scalar, vector and tensor valued isotropic functions of their arguments. Then necessarily the identities

$$\begin{aligned} \hat{\sigma}(\mu, \boldsymbol{\mathcal{Q}}\boldsymbol{a}, \boldsymbol{\mathcal{Q}}\boldsymbol{A}\boldsymbol{\mathcal{Q}}^T) &= \hat{\sigma}(\mu, \boldsymbol{a}, \boldsymbol{A}) \,, \\ \hat{\boldsymbol{s}}(\mu, \boldsymbol{\mathcal{Q}}\boldsymbol{a}, \boldsymbol{\mathcal{Q}}\boldsymbol{A}\boldsymbol{\mathcal{Q}}^T) &= \boldsymbol{\mathcal{Q}}\hat{\boldsymbol{s}}(\mu, \boldsymbol{a}, \boldsymbol{A}) \,, \\ \widehat{\boldsymbol{S}}(\mu, \boldsymbol{\mathcal{Q}}\boldsymbol{a}, \boldsymbol{\mathcal{Q}}\boldsymbol{A}\boldsymbol{\mathcal{Q}}^T) &= \boldsymbol{\mathcal{Q}}\hat{\boldsymbol{S}}(\mu, \boldsymbol{a}, \boldsymbol{A})\boldsymbol{\mathcal{Q}}^T \end{aligned} \tag{5.4.5}$$

must hold for all time–independent $\boldsymbol{\mathcal{Q}} \in \mathbb{O}$.

Scalar Valued Isotropic Functions Let us first consider the scalar valued function $\hat{\sigma}(\cdot)$. Based on the above condition $(5.4.5)_1$, one can simplify the dependency of the function $\hat{\sigma}(\cdot)$ on the triple $(\mu,\ \boldsymbol{a},\ \boldsymbol{A})$. It is easy to see that the following scalar quantities can be constructed from μ, $\boldsymbol{a}$ and $\boldsymbol{A}$

$$\mathcal{A} := \left\{ I_{\boldsymbol{A}}, II_{\boldsymbol{A}}, III_{\boldsymbol{A}}, \boldsymbol{a}\cdot\boldsymbol{a}, \boldsymbol{a}\cdot\boldsymbol{A}\boldsymbol{a}, \boldsymbol{a}\cdot\boldsymbol{A}^2\boldsymbol{a}, \mu \right\} , \tag{5.4.6}$$

which are independent of each other and all of which are invariant under orthogonal transformations, e.g.

$$\begin{aligned} \boldsymbol{a}^*\cdot(\boldsymbol{A}^*)^2\boldsymbol{a}^* &= (\boldsymbol{\mathcal{Q}}\boldsymbol{a})\cdot\left(\boldsymbol{\mathcal{Q}}\boldsymbol{A}^2\boldsymbol{\mathcal{Q}}^T\right)\boldsymbol{\mathcal{Q}}\boldsymbol{a} = (\boldsymbol{\mathcal{Q}}\boldsymbol{a})\cdot\boldsymbol{\mathcal{Q}}\boldsymbol{A}^2\boldsymbol{a} \\ &= \boldsymbol{a}\cdot\boldsymbol{\mathcal{Q}}^T\boldsymbol{\mathcal{Q}}\boldsymbol{A}^2\boldsymbol{a} = \boldsymbol{a}\cdot\boldsymbol{A}^2\boldsymbol{a}\ . \quad \text{etc.} \end{aligned} \tag{5.4.7}$$

It is also evident that there are no further independent scalar combinations of $\boldsymbol{a}$ and $\boldsymbol{A}$, because these lead with the help of the CAYLEY–HAMILTON theorem to the above introduced combinations. Thus $\mathcal{A}$ contains all independent isotropic scalars of $\boldsymbol{a}$ and $\boldsymbol{A}$. It follows that the most general scalar valued function of $(\mu,\ \boldsymbol{a},\ \boldsymbol{A})$ has the form

$$\sigma = \hat{\sigma}(\mathcal{A})\ . \tag{5.4.8}$$

Parenthetically, it may also be mentioned that the above argument is not restricted to only one scalar μ. One can extend the dependency of $\hat{\sigma}(\cdot)$ to an arbitrary number of scalars.

Vector Valued Isotropic Functions In order to analyze the functional dependency of a vector valued isotropic function of μ, $\boldsymbol{a}$, $\boldsymbol{A}$, we choose an arbitrary fixed vector $\boldsymbol{k}$ and examine the dependency of the associated scalar product $\Phi := \boldsymbol{s}\cdot\boldsymbol{k}$. One then finds for $\boldsymbol{s}$ an isotropic function $(5.4.5)_2$ of the variables μ, $\boldsymbol{a}$, $\boldsymbol{A}$, if one has found for Φ a scalar isotropic function of the variables μ, $\boldsymbol{a}$, $\boldsymbol{A}$, $\boldsymbol{k}$, that is linear in $\boldsymbol{k}$ (corresponding to the produced scalar product $\hat{\boldsymbol{s}}(\cdot)\cdot\boldsymbol{k}$). To the invariant quantities $\mathcal{A}$ just found above and independent of $\boldsymbol{k}$ the quantities

$$\boldsymbol{a}\cdot\boldsymbol{k}\ ,\quad \boldsymbol{A}\boldsymbol{a}\cdot\boldsymbol{k}\ ,\quad \boldsymbol{A}^2\boldsymbol{a}\cdot\boldsymbol{k}\ , \tag{5.4.9}$$

are added which are linear in $\boldsymbol{k}$. The product $\boldsymbol{k}\cdot\boldsymbol{k}$ must not be considered as a further invariant, because it is quadratic; likewise $\boldsymbol{A}^3\boldsymbol{a}\cdot\boldsymbol{k}$ falls out of consideration, because $\boldsymbol{A}^3$ can be expressed in terms of $\boldsymbol{I}$, $\boldsymbol{A}$ and $\boldsymbol{A}^2$ via the CAYLEY–HAMILTON theorem, so that there emerges a linear combination of the quantities (5.4.9). The most general isotropic function for Φ must thus result in a linear combination of elements in (5.4.9)

$$\Phi = \boldsymbol{s}\cdot\boldsymbol{k} = \left(\alpha_1\boldsymbol{a} + \alpha_2\boldsymbol{A}\boldsymbol{a} + \alpha_3\boldsymbol{A}^2\boldsymbol{a}\right)\cdot\boldsymbol{k} \tag{5.4.10}$$

with $\alpha_i = \widehat{\alpha}_i(\mathcal{A})$, $i = 1, 2, 3$. This must hold for an arbitrary vector $\boldsymbol{k}$, which implies

$$\begin{aligned} \boldsymbol{s} &= \hat{\boldsymbol{s}}(\mu, \boldsymbol{a}, \boldsymbol{A}) = \alpha_1 \boldsymbol{a} + \alpha_2 \boldsymbol{A}\boldsymbol{a} + \alpha_3 \boldsymbol{A}^2 \boldsymbol{a} \, , \\ \alpha_i &= \widehat{\alpha}_i(\mathcal{A}) \, . \end{aligned} \tag{5.4.11}$$

This is the required form of the dependency of a vector valued isotropic function, which can be represented as a linear combination of the vectors $\boldsymbol{a}$, $\boldsymbol{A}\boldsymbol{a}$ and $\boldsymbol{A}^2\boldsymbol{a}$. Corresponding to the three vector components of $\boldsymbol{s}$, three scalar valued functions α_i, $i = 1, 2, 3$ appear.

Tensor Valued Isotropic Functions Let $\boldsymbol{K}$ be an arbitrary symmetric tensor of rank 2, then $\Phi = \boldsymbol{S} \cdot \boldsymbol{K}$ is a scalar. Analogously to the above procedure for a vector, one can find for $\boldsymbol{S}$ an isotropic function of the variables μ, $\boldsymbol{a}$, $\boldsymbol{A}$ if one has found for Φ an isotropic function of the variables μ, $\boldsymbol{a}$, $\boldsymbol{A}$, $\boldsymbol{K}$ which is linear in $\boldsymbol{K}$. Scalar isotropic quantities for μ, $\boldsymbol{a}$, $\boldsymbol{A}$ are $\mathcal{A}$, and corresponding invariant scalar combination of $\boldsymbol{a}$, $\boldsymbol{A}$ and $\boldsymbol{K}$ that are linear in $\boldsymbol{K}$ are

$$\begin{aligned} &\boldsymbol{I} \cdot \boldsymbol{K} \, , \quad \boldsymbol{A} \cdot \boldsymbol{K} \, , \quad \boldsymbol{A}^2 \cdot \boldsymbol{K} \, , \quad \boldsymbol{a} \otimes \boldsymbol{a} \cdot \boldsymbol{K} \, , \\ &\operatorname{sym}((\boldsymbol{a}\boldsymbol{A}) \otimes \boldsymbol{a}) \cdot \boldsymbol{K} \, , \quad \operatorname{sym}\left((\boldsymbol{a}\boldsymbol{A}^2) \otimes \boldsymbol{a}\right) \cdot \boldsymbol{K} \, . \end{aligned} \tag{5.4.12}$$

Here we have restricted the last two expressions to the symmetric parts of $(\boldsymbol{a}\boldsymbol{A}) \otimes \boldsymbol{a}$ or $(\boldsymbol{a}\boldsymbol{A}^2) \otimes \boldsymbol{a}$, because $\boldsymbol{S}$ is also symmetric. It is easy to demonstrate the objectivity of these quantities (Exercise), and it is also possible to show – but not easy – that every other scalar combination of $\boldsymbol{a}$, $\boldsymbol{A}$ and $\boldsymbol{K}$, linear in $\boldsymbol{K}$, must be a linear combination of the quantities (5.4.12). Thus

$$\begin{aligned} \Phi &= \boldsymbol{S} \cdot \boldsymbol{K} \\ &= \Big\{ \beta_1 \boldsymbol{I} + \beta_2 \boldsymbol{A} + \beta_3 \boldsymbol{A}^2 + \beta_4 \boldsymbol{a} \otimes \boldsymbol{a} \\ &\quad + 2\beta_5 \operatorname{sym}((\boldsymbol{a}\boldsymbol{A}) \otimes \boldsymbol{a}) + 2\beta_6 \operatorname{sym}\left((\boldsymbol{a}\boldsymbol{A}^2) \otimes \boldsymbol{a}\right) \Big\} \cdot \boldsymbol{K} \end{aligned} \tag{5.4.13}$$

with $\mathcal{A}$-dependent coefficients β_i, $i = 1, 2, ..., 6$. Since this holds for arbitrary tensors $\boldsymbol{K}$, there follows for the dependency of a tensor valued isotropic function

$$\begin{aligned} \boldsymbol{S} &= \hat{\boldsymbol{S}}(\mu, \boldsymbol{a}, \boldsymbol{A}) \\ &= \beta_1 \boldsymbol{I} + \beta_2 \boldsymbol{A} + \beta_3 \boldsymbol{A}^2 + \beta_4 \boldsymbol{a} \otimes \boldsymbol{a} \\ &\quad + 2\beta_5 \operatorname{sym}((\boldsymbol{a}\boldsymbol{A}) \otimes \boldsymbol{a}) + 2\beta_6 \operatorname{sym}\left((\boldsymbol{a}\boldsymbol{A}^2) \otimes \boldsymbol{a}\right) \end{aligned} \tag{5.4.14}$$

with $\beta_i = \widehat{\beta}_i(\mathcal{A})$, $i = 1, 2, ..., 6$.

Finally it should be mentioned that in the set $\mathcal{A}$ the scalar invariants (5.4.6) $I_{\boldsymbol{A}}$, $II_{\boldsymbol{A}}$, $III_{\boldsymbol{A}}$ are called the principal invariants of the tensor $\boldsymbol{A}$ and are given[10] by (1.5.19).

[10] Naturally there are also other invariants, which are obtained from the combinations of the principal invariants. Such a set is, for example,

5.4.2 Symmetric Tensor as an Isotropic Function of two Symmetric Tensors

Often it is necessary to deal with the dependency of an isotropic function of two symmetric tensors. Such a case appears e.g. when one is interested to examine the effect of induced anisotropy upon the constitutive behaviour of the stress tensor. In addition to the usual dependency of the stress tensor on a tensor valued variable (5.4.1) one must incorporate a further tensorial dependency, which describes the anisotropy of the material.

Let $\boldsymbol{S}$ be a symmetric tensor, which is supposed to be an isotropic tensor function of two symmetric tensors $\boldsymbol{A}_{(1)}$ and $\boldsymbol{A}_{(2)}$. The isotropy requirement then acclaims that

$$\hat{\boldsymbol{S}}(\boldsymbol{Q}\boldsymbol{A}_{(1)}\boldsymbol{Q}^T,\ \boldsymbol{Q}\boldsymbol{A}_{(2)}\boldsymbol{Q}^T) = \boldsymbol{Q}\hat{\boldsymbol{S}}(\boldsymbol{A}_{(1)},\ \boldsymbol{A}_{(2)})\boldsymbol{Q}^T \tag{5.4.15}$$

for all time independent orthogonal tensors $\boldsymbol{Q}$. With the methods of linear algebra, which surpass the mathematical prerequisites of this text, one can show that $\boldsymbol{S}$ must have the following isotropic representation:

$$\begin{aligned}\boldsymbol{S} = {} & \gamma_1\boldsymbol{I} + \gamma_2\boldsymbol{A}_{(1)} + \gamma_3\boldsymbol{A}_{(2)} + \gamma_4\boldsymbol{A}^2_{(1)} + \gamma_5\boldsymbol{A}^2_{(2)} \\ & + \gamma_6\left(\boldsymbol{A}_{(1)}\boldsymbol{A}_{(2)} + \boldsymbol{A}_{(2)}\boldsymbol{A}_{(1)}\right) + \gamma_7\left(\boldsymbol{A}^2_{(1)}\boldsymbol{A}_{(2)} + \boldsymbol{A}_{(2)}\boldsymbol{A}^2_{(1)}\right) \\ & + \gamma_8\left(\boldsymbol{A}_{(1)}\boldsymbol{A}^2_{(2)} + \boldsymbol{A}^2_{(2)}\boldsymbol{A}_{(1)}\right) + \gamma_9\left(\boldsymbol{A}^2_{(1)}\boldsymbol{A}^2_{(2)} + \boldsymbol{A}^2_{(2)}\boldsymbol{A}^2_{(1)}\right)\end{aligned} \tag{5.4.16}$$

with coefficients γ_i, $i = 1, \ldots, 9$, which are in general functions of the following scalar invariants

$$\begin{aligned}\mathcal{A} = \Big\{ & I_{\boldsymbol{A}_{(1)}}, II_{\boldsymbol{A}_{(1)}}, III_{\boldsymbol{A}_{(1)}}, I_{\boldsymbol{A}_{(2)}}, II_{\boldsymbol{A}_{(2)}}, III_{\boldsymbol{A}_{(2)}}, \\ & I_{\boldsymbol{A}_{(1)}\boldsymbol{A}_{(2)}}, I_{\boldsymbol{A}_{(1)}\boldsymbol{A}^2_{(2)}}, I_{\boldsymbol{A}^2_{(1)}\boldsymbol{A}_{(2)}}, I_{\boldsymbol{A}^2_{(1)}\boldsymbol{A}^2_{(2)}} \Big\} .\end{aligned} \tag{5.4.17}$$

The proof for this can been taken from the literature already mentioned. But it should be remarked that no higher powers of $\boldsymbol{A}_{(1)}$ or $\boldsymbol{A}_{(2)}$ can appear, because such terms can be transformed by means of the CAYLEY–HAMILTON theorem. That no dependency on the third invariants of the product of $\boldsymbol{A}_{(1)}$ and $\boldsymbol{A}_{(2)}$ can persist, is easily proved (Exercise).

$$I'_{\boldsymbol{A}} = I_{\boldsymbol{A}}\ , \quad II'_{\boldsymbol{A}} = -\tfrac{1}{2}I_{\boldsymbol{A}^2}\ , \quad III'_{\boldsymbol{A}} = -\tfrac{1}{3}I_{\boldsymbol{A}^3}\ ,$$

so that $\mathcal{A}$ may be replaced by

$$\mathcal{A}' = \left\{I'_{\boldsymbol{A}}, II'_{\boldsymbol{A}}, III'_{\boldsymbol{A}}, \boldsymbol{a}\cdot\boldsymbol{a}, \boldsymbol{a}\cdot\boldsymbol{A}\boldsymbol{a}, \boldsymbol{a}\cdot\boldsymbol{A}^2\boldsymbol{a}, \mu\right\}\ .$$

A Simple Example The isotropy representation of a tensor valued function, which solely depends on a scalar, is, according to (5.4.14), given as

$$\boldsymbol{S} = \hat{\boldsymbol{S}}(\mu) \quad \Rightarrow \quad \boldsymbol{S} = \gamma_1 \boldsymbol{I} \, , \tag{5.4.18}$$

where γ_1 now depends only on the scalar μ.

Let us prove this statement here in an alternative way: Let $\boldsymbol{S} = \boldsymbol{S}^T$ be an isotropic symmetric second rank tensor. Then from the isotropy condition (5.4.2),

$$\boldsymbol{S} = \boldsymbol{Q}\boldsymbol{S}\boldsymbol{Q}^T \, , \tag{5.4.19}$$

it follows that the tensor $\boldsymbol{S}$ can be a multiple of the unit tensor. For the proof, we start with the statement that every symmetric second rank tensor is diagonalizable and expressible in the coordinate system of the principal axes through the diagonal matrix

$$\boldsymbol{S}_{\text{diag}} = \begin{pmatrix} \lambda_1^{(\boldsymbol{S})} & 0 & 0 \\ 0 & \lambda_2^{(\boldsymbol{S})} & 0 \\ 0 & 0 & \lambda_3^{(\boldsymbol{S})} \end{pmatrix} . \tag{5.4.20}$$

With this, the isotropy condition can be written as

$$\boldsymbol{S}_{\text{diag}} = \boldsymbol{Q}\boldsymbol{S}_{\text{diag}}\boldsymbol{Q}^T \, . \tag{5.4.21}$$

Here this condition must hold for an arbitrary transformation $\boldsymbol{Q}$, for example one can choose the transformations

$$\boldsymbol{Q}_1 = \begin{pmatrix} \cos\phi & \sin\phi & 0 \\ -\sin\phi & \cos\phi & 0 \\ 0 & 0 & 1 \end{pmatrix} , \quad \boldsymbol{Q}_2 = \begin{pmatrix} 1 & 0 & 0 \\ 0 & \cos\phi & \sin\phi \\ 0 & -\sin\phi & \cos\phi \end{pmatrix} \tag{5.4.22}$$

with arbitrary ϕ. Substituting both of these transformations consecutively into (5.4.21) and performing explicitly the matrix products leads to

$$\lambda_1^{(\boldsymbol{S})} = \lambda_2^{(\boldsymbol{S})} \quad \text{and} \quad \lambda_2^{(\boldsymbol{S})} = \lambda_3^{(\boldsymbol{S})} \quad \Rightarrow \quad \lambda_1^{(\boldsymbol{S})} = \lambda_2^{(\boldsymbol{S})} = \lambda_3^{(\boldsymbol{S})} \, , \tag{5.4.23}$$

i.e., all three eigenvalues must have the same value, with which the above statement is proved.

Further applications are formulated in the Exercises.

5.4.3 Elastic or Barotropic Fluids

A fluid is defined as a material, whose material functionals are invariant under unimodular transformations of the reference configurations. The symmetry condition (5.3.13) for an elastic fluid reads

$$\hat{\Psi}(\boldsymbol{F}) = \hat{\Psi}(\boldsymbol{F}\boldsymbol{H}) \qquad \forall \boldsymbol{H} \in \mathbb{U} \,. \tag{5.4.24}$$

Because this symmetry condition must hold for all unimodular transformations, also time dependent ones, it is permissible to choose a special transformation such that

$$\boldsymbol{H} = (\det \boldsymbol{F})^{\frac{1}{3}} \boldsymbol{F}^{-1} \quad \Rightarrow \quad \det \boldsymbol{H} = 1 \,. \tag{5.4.25}$$

If this is substituted in (5.4.24), one obtains

$$\hat{\Psi}(\boldsymbol{F}) = \hat{\Psi}\big((\det \boldsymbol{F})^{\frac{1}{3}} \boldsymbol{I}\big) \,. \tag{5.4.26}$$

The material functionals of an elastic fluid can therefore only depend on the determinant of the deformation gradient and, consequently, on the density of the body, since $\rho_R = \rho \det \boldsymbol{F}$. Consequently one may also write

$$\varepsilon = \hat{\varepsilon}(\rho)\,, \quad \boldsymbol{q} = \hat{\boldsymbol{q}}(\rho)\,, \quad \boldsymbol{t} = \hat{\boldsymbol{t}}(\rho) \quad \text{elastic fluid} \,. \tag{5.4.27}$$

Until now we have only examined the symmetry properties; the material equations must, however, also satisfy the principle of objectivity (5.2.18), that is, we write (the density is an objective scalar)

$$\boldsymbol{O}\hat{\boldsymbol{q}}(\rho) = \hat{\boldsymbol{q}}(\rho)\,, \quad \boldsymbol{O}\hat{\boldsymbol{t}}(\rho)\boldsymbol{O}^T = \hat{\boldsymbol{t}}(\rho) \,. \tag{5.4.28}$$

According to the statement (5.4.21) of the last subsection and the result to be treated in the Exercises one can only fulfill this postulate if the stress tensor is a multiple of the unit tensor and the heat flux vector vanishes. The final form of the material equations of an elastic fluid is therefore

$$\varepsilon = \hat{\varepsilon}(\rho)\,, \quad \boldsymbol{q} = \boldsymbol{0}\,, \quad \boldsymbol{t} = -p(\rho)\boldsymbol{I}\,, \tag{5.4.29}$$

where the scalar quantity p is called the pressure. It is a matter of the second law of thermodynamics to show that there exists a relation between the internal energy and the pressure function.

5.4.4 Viscous Fluids

Nonlinear Viscous Fluid The material equations of a viscous fluid depend, via the deformation gradient $\boldsymbol{F}$, on the first time derivative of the deformation gradient $\dot{\boldsymbol{F}} = \boldsymbol{L}\boldsymbol{F}$. In case of a viscoelastic material thus $\Psi = \hat{\Psi}(\boldsymbol{F}, \boldsymbol{L})$ holds, or here, since we are concerned with a fluid, $\Psi = \hat{\Psi}(\rho, \boldsymbol{L})$; the dependency on $\boldsymbol{F}$ is reduced to a dependency on the density. We have already seen that a dependency on the whole velocity gradient $\boldsymbol{L}$ is not permissible as it is not objective but, clearly, it may depend on its objective symmetric part $\boldsymbol{D} = \frac{1}{2}(\boldsymbol{L}+\boldsymbol{L}^T)$. The stress tensor of a viscous fluid must therefore be of the form

$$\boldsymbol{t} = \hat{\boldsymbol{t}}(\rho, \boldsymbol{D}) \,. \tag{5.4.30}$$

Material objectivity now requires

$$\boldsymbol{O}\hat{\boldsymbol{t}}(\rho, \boldsymbol{D})\boldsymbol{O}^T = \hat{\boldsymbol{t}}(\rho, \boldsymbol{O}\boldsymbol{D}\boldsymbol{O}^T) \,. \tag{5.4.31}$$

The functional for the stress tensor is an isotropic tensor function of a scalar and a tensor and can, with the aid of (5.4.14), therefore be written as

$$\begin{aligned} &\boldsymbol{t} = a_0\boldsymbol{I} + \nu_1\boldsymbol{D} + \nu_2\boldsymbol{D}^2 \,, \\ &a_0 = -p + \nu_0 \,, \quad \nu_0 \xrightarrow{\boldsymbol{D}\to\boldsymbol{0}} 0 \\ &p = \hat{p}(\rho) \,, \quad \nu_i = \hat{\nu}_i(\rho, I_{\boldsymbol{D}}, II_{\boldsymbol{D}}, III_{\boldsymbol{D}}) \,, \quad i = 0,1,2. \end{aligned} \tag{5.4.32}$$

This is the form of the constitutive relation of the stress for a viscous fluid, as it is obtained by employing the principle of material objectivity and the symmetry requirements. Hereby the division of a_0 into $-p = -\hat{p}(\rho)$ and ν_0 such that with $\boldsymbol{D} \to \boldsymbol{0}$ also $\nu_0 \to 0$ holds, is a result, that can only be explained by means of thermodynamic arguments (2nd law). One calls such a material as REINER–RIVLIN *fluid.*

Newtonian Fluid If one supposes that the stress tensor depends only linearly on the strain rate tensor $\boldsymbol{D}$, then the quadratic term drops out in the material equation for the REINER–RIVLIN fluid, and the material coefficient ν_0 depends, at most, on the first (linear) invariant of $\boldsymbol{D}$ whilst ν_1 is independent of $\boldsymbol{D}$. Moreover, ν_0 must vanish for a vanishing stretching tensor $\boldsymbol{D}$; so for the coefficients the relations

$$\nu_0 = \nu_0'(\rho)\,\mathrm{tr}(\boldsymbol{D}) \,, \quad \nu_1 = \nu_1(\rho) \,, \quad \nu_2 = 0 \tag{5.4.33}$$

must hold. If one introduces the *volume viscosity* κ and the *shear viscosity* μ (or simply viscosity), given by

$$\nu_0' = \kappa - \tfrac{2}{3}\mu \,, \quad \nu_1 = 2\mu \,, \tag{5.4.34}$$

of which both can depend on density, the stress tensor for a *compressible,* NEWTON*ian fluid* becomes

$$\begin{aligned} \boldsymbol{t} &= \Big(-p + \big(\kappa - \tfrac{2}{3}\mu\big)\,\mathrm{tr}\,\boldsymbol{D}\Big)\boldsymbol{I} + 2\mu\boldsymbol{D} \\ &= (-p + \kappa\,\mathrm{tr}\,\boldsymbol{D})\boldsymbol{I} + 2\mu\boldsymbol{D}' \,. \end{aligned} \tag{5.4.35}$$

Here we have introduced in the second alternative form the *deviator* of the stretching tensor

$$\boldsymbol{D}' := \boldsymbol{D} - \tfrac{1}{3}(\operatorname{tr}\boldsymbol{D})\boldsymbol{I} \quad \text{with} \quad \operatorname{tr}\boldsymbol{D}' = 0 \,. \tag{5.4.36}$$

One also occasionally calls $\boldsymbol{D}'$ *distorsion tensor.*

Many gases and fluids occurring in nature are well described by this material equation, whereby, because of thermodynamic considerations, one further admits a temperature dependency of the pressure and viscosity. This is not considered here, but the temperature can be carried along with the whole derivation without changing it; the coefficients simply become temperature dependent.

Often the volume viscosity is neglected or set to zero, $\kappa = 0$; one speaks in such compressible cases of the STOKES *assumption.*

5.4.5 Isotropic Elastic Solid

Isotropic Elastic Solid The derivation of the reduced form of the material equations for the stress tensor for an isotropic, elastic solid strongly resembles the above procedure. Only the symmetry group of the isotropic solid is smaller than that of the fluid. Material equations must be invariant with regard to arbitrary rotations of the reference configuration; the group of the symmetry transformation is then equal to the group of the orthogonal transformations, $\mathbb{G} = \mathbb{O}$. The symmetry condition is thus

$$\hat{\Psi}(\boldsymbol{F}) = \hat{\Psi}(\boldsymbol{F}\boldsymbol{H}) \qquad \forall\, \boldsymbol{H} \in \mathbb{O} \,. \tag{5.4.37}$$

One can modify, to some extent, the dependency of the material function by choosing a special symmetry transformation. To this end one needs the polar decomposition of the deformation gradient, $\boldsymbol{F} = \boldsymbol{V}\boldsymbol{R}$, and the definition of the left CAUCHY–GREEN tensor, $\boldsymbol{B} = \boldsymbol{F}\boldsymbol{F}^T$.

The rotational part of the polar decomposition of the deformation gradient is orthogonal, $\boldsymbol{R}\boldsymbol{R}^T = \boldsymbol{I}$. If we choose the symmetry transformation as

$$\boldsymbol{H} = \boldsymbol{R}^T \quad \Rightarrow \quad \boldsymbol{H}\boldsymbol{H}^T = \boldsymbol{I} \,, \tag{5.4.38}$$

then by substituting this into the material equation of the CAUCHY stress tensor and simultaneously employing (5.4.37) there follows

$$\hat{\mathfrak{t}}(\boldsymbol{F}) = \hat{\mathfrak{t}}(\boldsymbol{F}\boldsymbol{R}^T) = \hat{\mathfrak{t}}(\boldsymbol{V}\boldsymbol{R}\boldsymbol{R}^T) = \hat{\mathfrak{t}}(\boldsymbol{V}) \,. \tag{5.4.39}$$

The stress tensor of an isotropic, elastic solid can thus depend only on the left stretch tensor $\boldsymbol{V}$, or on quantities which themselves are representable as tensor functions $\boldsymbol{f}(\boldsymbol{V})$. These are e.g. the left CAUCHY–GREEN deformation tensor $\boldsymbol{B}$, the FINGER strain tensor $\boldsymbol{E}$ and the ALMANSI strain tensor $\boldsymbol{A}$ (see Subsect. 1.5.2), but equally also other strain measures that are expressible in terms of $\boldsymbol{V}$. Choosing the left CAUCHY–GREEN deformation tensor, one obtains

$$\boldsymbol{t} = \hat{\boldsymbol{t}}(\boldsymbol{B}) , \quad \boldsymbol{B} = \boldsymbol{F}\boldsymbol{F}^T = \boldsymbol{V}^2 . \tag{5.4.40}$$

Material objectivity now requires

$$\boldsymbol{O}\hat{\boldsymbol{t}}(\boldsymbol{B})\boldsymbol{O}^T = \hat{\boldsymbol{t}}(\boldsymbol{O}\boldsymbol{B}\boldsymbol{O}^T) . \tag{5.4.41}$$

The functional of the stress tensor is thus given by an isotropic tensor function of the tensor $\boldsymbol{B}$; therefore, (5.4.14) implies (there is no dependency on a vector)

$$\begin{aligned} \boldsymbol{t} &= a_0\boldsymbol{I} + a_1\boldsymbol{B} + a_2\boldsymbol{B}^2 , \\ a_i &= \hat{a}_i(I_{\boldsymbol{B}}, II_{\boldsymbol{B}}, III_{\boldsymbol{B}}) , \quad i = 0,1,2 . \end{aligned} \tag{5.4.42}$$

For the heat flux vector one obtains

$$\boldsymbol{O}\hat{\boldsymbol{q}}(\boldsymbol{B}) = \hat{\boldsymbol{q}}(\boldsymbol{O}\boldsymbol{B}\boldsymbol{O}^T) , \quad \forall \boldsymbol{O} \in \mathbb{O} . \tag{5.4.43}$$

This is the isotropy condition of a vector function, which depends only on a symmetric tensor. In view of (5.4.11) only $\boldsymbol{q} = \boldsymbol{0}$ can satisfy this condition.

To summarize, an isotropic elastic solid can be described by the equations

$$\begin{aligned} \varepsilon = \hat{\varepsilon}(I_{\boldsymbol{B}}, II_{\boldsymbol{B}}, III_{\boldsymbol{B}}) , \quad \boldsymbol{q} = \boldsymbol{0} , \quad \boldsymbol{t} = a_0\boldsymbol{I} + a_1\boldsymbol{B} + a_2\boldsymbol{B}^2 , \\ a_i = \hat{a}_i(I_{\boldsymbol{B}}, II_{\boldsymbol{B}}, III_{\boldsymbol{B}}) , \quad i = 0,1,2 . \end{aligned} \tag{5.4.44}$$

In this representation $\boldsymbol{B}$ can be replaced by $\boldsymbol{V}$, $\boldsymbol{E}$, $\boldsymbol{A}$ or any other $\boldsymbol{V}$-dependent tensor valued strain measure. Further, one can, with the help of the CAYLEY–HAMILTON theorem, for example replace $\boldsymbol{B}^2$ and obtain a representation of the form

$$\begin{aligned} \boldsymbol{t} &= b_0\boldsymbol{I} + b_1\boldsymbol{B} + b_2\boldsymbol{B}^{-1} , \\ b_i &= \hat{b}_i(I_{\boldsymbol{B}}, II_{\boldsymbol{B}}, III_{\boldsymbol{B}}) , \quad i = 0,1,2 . \end{aligned} \tag{5.4.45}$$

There still remains the task to show by thermodynamic arguments how the internal energy ε (or any other thermodynamic potential, as such functions are called) is connected to the stress tensor. This is a task of thermodynamics. Nonlinear elasticity is a popular and extensively treated field of continuum mechanics. Two specialized books on the subject are WANG & TRUESDELL [250], [252] and OGDEN [180].

Hooke Law Let us consider small deformations from a natural state in which the stress tensor vanishes, as in Sect. 1.5.4. Then the left CAUCHY–GREEN deformation tensor is approximately given by[11]

[11] The step $\operatorname{Grad}\boldsymbol{u} \simeq \operatorname{grad}\boldsymbol{u}$ is essential, because $\boldsymbol{B}$ "lives" in the present configuration:

$$\operatorname{Grad}\boldsymbol{u} = \frac{\partial\boldsymbol{u}}{\partial\boldsymbol{\chi}} = \frac{\partial\boldsymbol{u}}{\partial\boldsymbol{x}}\frac{\partial\boldsymbol{x}}{\partial\boldsymbol{\chi}} = \frac{\partial\boldsymbol{u}}{\partial\boldsymbol{x}}\left(1 + \frac{\partial\boldsymbol{u}}{\partial\boldsymbol{\chi}}\right) \simeq \frac{\partial\boldsymbol{u}}{\partial\boldsymbol{x}} .$$

$$\boldsymbol{B} \approx \boldsymbol{I} + 2\boldsymbol{\varepsilon} \,, \quad \boldsymbol{\varepsilon} = \tfrac{1}{2}(\operatorname{Grad}\boldsymbol{u} + (\operatorname{Grad}\boldsymbol{u})^T) \simeq \tfrac{1}{2}(\operatorname{grad}\boldsymbol{u} + (\operatorname{grad}\boldsymbol{u})^T) \,, \tag{5.4.46}$$

where $\boldsymbol{\varepsilon}$ is the linearized strain tensor and $\boldsymbol{u} := \boldsymbol{x} - \boldsymbol{X}$ is the displacement field. The material equation for the stress tensor in the geometric linearization is then given by

$$\boldsymbol{t} = \phi_0 \boldsymbol{I} + \phi_1 \boldsymbol{\varepsilon} \,, \quad \phi_i = \hat{\phi}_i(I_{\boldsymbol{\varepsilon}}, II_{\boldsymbol{\varepsilon}}, III_{\boldsymbol{\varepsilon}}) \,, \quad i = 0, 1 \,. \tag{5.4.47}$$

If one now requests a linear relationship between the stress and strain tensor then the coefficient ϕ_0 can at most depend linearly on the first invariant of the strain tensor, and ϕ_1 must be a constant, viz.,

$$\phi_0 = \lambda I_{\boldsymbol{\varepsilon}} \,, \quad \phi_1 = 2\mu \,. \tag{5.4.48}$$

The linearized stress tensor then reads

$$\boldsymbol{t} = \lambda I_{\boldsymbol{\varepsilon}} \boldsymbol{I} + 2\mu\boldsymbol{\varepsilon} = k I_{\boldsymbol{\varepsilon}} \boldsymbol{I} + 2\mu\boldsymbol{\varepsilon}' \,, \tag{5.4.49}$$

where $\boldsymbol{\varepsilon}' = \boldsymbol{\varepsilon} - \frac{1}{3} I_{\boldsymbol{\varepsilon}} \boldsymbol{I}$ is the deviator of the linearized strain tensor. Equation (5.4.49) is called HOOKE*'s law* in the linear elasticity theory. The coefficients μ, λ appearing in this equation are called the LAMÉ *parameters*, which according to $k = \lambda + \frac{2}{3}\mu$ determine the bulk modulus. Hence, two parameters characterize the stress–strain relationship of a linear elastic solid, λ and μ or k and μ. With these parameters further quantities such as the modulus of elasticity can be derived.

In the formulation of HOOKE's law we have presumed that the displacement gradient $\operatorname{Grad}\boldsymbol{u}$ of the displacement vector $\boldsymbol{u}$ is small and can be replaced by $\operatorname{grad}\boldsymbol{u}$ with sufficient accuracy. Only with such approximations is a stress–strain relation as introduced in (5.4.49) truly linearized. Often in modern elasticity theory the two steps, *physical* and *geometric linearizations* are decoupled from each other. In this spirit we shall define a *physically linearized* isotropic elastic body by a stress–strain relation given by, for example,

$$\boldsymbol{t} = \bar{k} I_{\boldsymbol{E}} \boldsymbol{I} + 2\bar{\mu}\boldsymbol{E}' \,, \tag{5.4.50}$$

where $\boldsymbol{E}$ denotes the FINGER strain tensor and $\bar{k}$ and $\bar{\mu}$ are two constants. Even though linear in form, this law is not linear in the displacement vector and can therefore be found useful in geometric non-linear theories. Another physically linearized law is

$$\tilde{\boldsymbol{T}} = \tilde{k} I_{\boldsymbol{G}} \boldsymbol{I} + 2\tilde{\mu}\boldsymbol{G}' \,, \tag{5.4.51}$$

in which the second PIOLA–KIRCHHOFF stress tensor is related linearly to the GREEN strain tensor; $\tilde{k}$ and $\tilde{\mu}$ are again constants. Naturally, the relation between (5.4.50) and (5.4.51) is non-linear and especially, $\bar{k}$ and $\tilde{k}$ as well as $\bar{\mu}$ and $\tilde{\mu}$ can not be transformed into each other because these are all assumed to be constants.

5.4.6 Isotropic Viscoelastic Solid

General Non-linear Laws The material law of an isotropic viscoelastic solid depends on the deformation gradient $\boldsymbol{F}$ and its time derivative $\dot{\boldsymbol{F}}$, $\Psi = \hat{\Psi}(\boldsymbol{F},\dot{\boldsymbol{F}})$. The CAUCHY stress tensor $\boldsymbol{t}$ is also regarded as a function of these quantities, $\boldsymbol{t} = \hat{\boldsymbol{t}}(\boldsymbol{F},\dot{\boldsymbol{F}})$; then from Subsects. 5.4.4 and 5.4.5, it can be concluded that $\boldsymbol{t}$ can depend on $\boldsymbol{F}$ through $\boldsymbol{B}$ and that its dependence on $\dot{\boldsymbol{F}}$ is only through $\boldsymbol{D} = \frac{1}{2}(\boldsymbol{L} + \boldsymbol{L}^T)$. Thus the CAUCHY stress may be written as $\boldsymbol{t} = \hat{\boldsymbol{t}}\,(\boldsymbol{B}, \boldsymbol{D})$ or, since the FINGER strain tensor $\boldsymbol{E}$ is connected to $\boldsymbol{B}$ through $\boldsymbol{E} = \frac{1}{2}\,(\boldsymbol{B} - \boldsymbol{I})$, $\boldsymbol{t} = \hat{\boldsymbol{t}}\,(\boldsymbol{E}, \boldsymbol{D})$. Now the transformation properties under EUCLIDian transformations for these quantities are given by

$$\boldsymbol{t}^* = \boldsymbol{O}\boldsymbol{t}\boldsymbol{O}^T\,, \quad \boldsymbol{E}^* = \boldsymbol{O}\boldsymbol{E}\boldsymbol{O}^T\,, \quad \boldsymbol{D}^* = \boldsymbol{O}\boldsymbol{D}\boldsymbol{O}^T\,,$$

so that for the above material law one obtains

$$\boldsymbol{O}\hat{\boldsymbol{t}}\,(\boldsymbol{E}, \boldsymbol{D})\,\boldsymbol{O}^T = \hat{\boldsymbol{t}}\left(\boldsymbol{O}\boldsymbol{E}\boldsymbol{O}^T, \boldsymbol{O}\boldsymbol{D}\boldsymbol{O}^T\right)\,, \quad \forall \boldsymbol{O} \in \mathbb{O}\,.$$

This is precisely the requirement of an isotropic tensor function of two symmetric tensor arguments. One can therefore apply equations (5.4.16) and (5.4.17) with $\boldsymbol{S} = \boldsymbol{t}$, $\boldsymbol{A}_{(1)} = \boldsymbol{E}$ and $\boldsymbol{A}_{(2)} = \boldsymbol{D}$ so that the most general stress law of an isotropic viscoelastic material with bounded memory takes the form

$$\begin{aligned}\boldsymbol{t} = {} & \gamma_1\boldsymbol{I} + \gamma_2\boldsymbol{E} + \gamma_3\boldsymbol{D} + \gamma_4\boldsymbol{E}^2 + \gamma_5\boldsymbol{D}^2 \\ & +\gamma_6\left(\boldsymbol{E}\boldsymbol{D} + \boldsymbol{D}\boldsymbol{E}\right) + \gamma_7\left(\boldsymbol{E}^2\boldsymbol{D} + \boldsymbol{D}\boldsymbol{E}^2\right) \\ & +\gamma_8\left(\boldsymbol{E}\boldsymbol{D}^2 + \boldsymbol{D}^2\boldsymbol{E}\right) + \gamma_9\left(\boldsymbol{E}^2\boldsymbol{D}^2 + \boldsymbol{D}^2\boldsymbol{E}^2\right)\,,\end{aligned} \tag{5.4.52}$$

with coefficients $\gamma_i, i = 1, \ldots, 9$ that may be functions of the following scalar invariants:

$$\mathcal{A} = \left\{I_{\boldsymbol{E}}, I\!I_{\boldsymbol{E}}, I\!I\!I_{\boldsymbol{E}}, I_{\boldsymbol{D}}, I\!I_{\boldsymbol{D}}, I\!I\!I_{\boldsymbol{D}}, I_{\boldsymbol{E}\boldsymbol{D}}, I_{\boldsymbol{E}\boldsymbol{D}^2}, I_{\boldsymbol{E}^2\boldsymbol{D}}, I_{\boldsymbol{E}^2\boldsymbol{D}^2}\right\}\,. \tag{5.4.53}$$

For the heat flux vector one has $\boldsymbol{q} = \hat{\boldsymbol{q}}\,(\boldsymbol{E}, \boldsymbol{D})$, but there exists no vector valued isotropic function of two symmetric tensors, so $\boldsymbol{q} = \boldsymbol{0}$, and for the internal energy, we also have $\varepsilon = \hat{\varepsilon}\,(\mathcal{A})$. A possible isotropic viscoelastic solid with bounded memory is therefore given by relation (5.4.52) for the CAUCHY stress tensor, $\boldsymbol{q} = \boldsymbol{0}$ and $\varepsilon = \hat{\varepsilon}\,(\mathcal{A})$. There are also others.

If one considers the LAGRANGE representation instead of the EULER description, then the second PIOLA–KIRCHHOFF stress tensor $\tilde{\boldsymbol{T}}$ will be represented as a function of the GREEN strain tensor $\boldsymbol{G}$ and its material derivative $\dot{\boldsymbol{G}}$, thus $\tilde{\boldsymbol{T}} = \tilde{\boldsymbol{T}}(\boldsymbol{G}, \dot{\boldsymbol{G}})$. Under orthogonal transformations of the reference configuration (i.e., transformations guaranteeing symmetric behaviour) the transformation rules

$$\tilde{\boldsymbol{T}}^* = \boldsymbol{O}\tilde{\boldsymbol{T}}\boldsymbol{O}^T\,, \quad \boldsymbol{G}^* = \boldsymbol{O}\boldsymbol{G}\boldsymbol{O}^T\,, \quad \boldsymbol{G}^* = \boldsymbol{O}\dot{\boldsymbol{G}}\boldsymbol{O}^T\,,$$

apply, and therefore,

$$O\hat{\tilde{T}}\left(G,\dot{G}\right)O^T=\hat{\tilde{T}}\left(OGO^T,O\dot{G}O^T\right) .$$

Admitting all orthogonal transformations in this identity implies that the conditions of an isotropic tensor function of two tensor valued variables prevail so that one can write the most general viscoelastic law for an isotropic solid also as

$$\begin{aligned}\tilde{T} &= \Gamma_1 I+\Gamma_2 G+\Gamma_3\dot{G}+\Gamma_4 G^2+\Gamma_5\dot{G}^2\\ &+\Gamma_6\left(G\dot{G}+\dot{G}G\right)+\Gamma_7\left(G^2\dot{G}+\dot{G}G^2\right)\\ &+\Gamma_8\left(G\dot{G}^2+\dot{G}^2G\right)+\Gamma_9\left(G^2\dot{G}^2+\dot{G}^2G^2\right)\end{aligned} \tag{5.4.54}$$

with $\Gamma_\alpha=\Gamma_\alpha\left(\bar{\mathcal{A}}\right)$, $\alpha=1,\ldots,9$ and

$$\bar{\mathcal{A}}=\left\{I_G,II_G,III_G,I_{\dot{G}},II_{\dot{G}},III_{\dot{G}},I_{G\dot{G}},I_{G\dot{G}^2},I_{G^2\dot{G}},I_{G^2\dot{G}^2}\right\} .$$

The laws (5.4.52) and (5.4.54) are equivalent. For the material heat flux vectors one obtains $Q=0$ and for the internal energy $\varepsilon=\hat{\varepsilon}\left(\bar{\mathcal{A}}\right)$.

The laws (5.4.52) and (5.4.54) form the most general constitutive relations for the stress tensor of a viscoelastic material with bounded memory for an isotropic solid. In general, these representations can be further simplified by use of the second law of thermodynamics; and practically, certain terms will be neglected (simply to reduce the complexity of the formulas). The farthest reaching simplifications are the linear laws.

Physically Linear Viscoelastic Laws Often one considers the special cases, for which t and $\tilde{T}$ depend linearly on the strain tensors and the strain–rate tensors. Then one has in (5.4.52) $\gamma_4=\cdots=\gamma_9=0$ and

$$\gamma_1=\lambda_E I_E+\lambda_D I_D\ ,\quad \gamma_2=2\mu_E\ ,\quad \gamma_3=2\mu_D\ , \tag{5.4.55}$$

with constants $\lambda_E,\lambda_D,\mu_E,\mu_D$, so that

$$t=(\lambda_E I_E I+2\mu_E E)+(\lambda_D I_D I+2\mu_D D) \tag{5.4.56}$$

or with $\lambda_E+\frac{2}{3}\mu_E=\kappa_E$, $\lambda_D+\frac{2}{3}\mu_D=\kappa_D$,

$$t=(\kappa_E I_E I+2\mu_E E')+(\kappa_D I_D I+2\mu_D D')\ , \tag{5.4.57}$$

where $E'=E-\frac{1}{3}I_E I$ and $D'=D-\frac{1}{3}I_D I$ denote the deviators of the FINGER strain tensor and the stretching tensor. The coefficients κ_E and μ_E are the bulk and the shear modulus, and κ_D and μ_D are called bulk viscosity and shear viscosity, respectively.

Denoting the spherical part of the tensor in (5.4.57) with $I_t I$ and the deviator with $t'=t-\frac{1}{3}I_t I$, it is easily shown that (5.4.57) implies

$$I_t 3\kappa_E I_E + 3\kappa_D I_D\ , \qquad t' 2\mu_E E' + 2\mu_D D'\ . \tag{5.4.58}$$

With this decomposition, we achieved to assign the spherical deformation to a spherical tensor and the deviatoric deformations to a stress deviator.

Likewise, one can also restrict oneself to linearity in $\boldsymbol{G}$ and $\dot{\boldsymbol{G}}$ in the law (5.4.54) and thus obtains with $\Gamma_4 = \cdots = \Gamma_9 = 0$ and

$$\Gamma_1 = \lambda_G I_G + \lambda_{\dot{G}} I_{\dot{G}}\ , \quad \Gamma_2 = 2\mu_G\ , \quad \Gamma_3 = 2\mu_{\dot{G}} \tag{5.4.59}$$

the relations

$$\begin{aligned} \tilde{\boldsymbol{T}} &= (\lambda_G I_G \boldsymbol{I} + 2\mu_G \boldsymbol{G}) + \left(\lambda_{\dot{G}} I_{\dot{G}} \boldsymbol{I} + 2\mu_{\dot{G}} \dot{\boldsymbol{G}}\right) \quad \text{or} \\ \tilde{\boldsymbol{T}} &= (\kappa_G I_G \boldsymbol{I} + 2\mu_G \boldsymbol{G}') + \left(\kappa_{\dot{G}} I_{\dot{G}} \boldsymbol{I} + 2\mu_{\dot{G}} \dot{\boldsymbol{G}}'\right)\ . \end{aligned} \tag{5.4.60}$$

Here too, the first invariant and the deviator become

$$I_{\tilde{T}} = 3\kappa_G I_G + 3\kappa_{\dot{G}} I_{\dot{G}}\ , \quad \tilde{\boldsymbol{T}}' = 2\mu_G \boldsymbol{G}' + 2\mu_{\dot{G}} \dot{\boldsymbol{G}}'\ . \tag{5.4.61}$$

The representations (5.4.58) and (5.4.61) are apparently the three–dimensionally transformed postulate (5.1.8) of a KELVIN or VOIGT body.

Finally, we can also restrict ourselves in the laws (5.4.58) and (5.4.61) to small displacements from the reference configuration. Then, according to the geometric linearization (see Subsect. 1.5.4)

$$\boldsymbol{G} \approx \boldsymbol{E} \approx \boldsymbol{\varepsilon}\ , \quad \dot{\boldsymbol{G}} \approx \boldsymbol{D} \approx \dot{\boldsymbol{\varepsilon}}\ , \quad \tilde{\boldsymbol{T}} \approx \boldsymbol{t}\ , \tag{5.4.62}$$

so that one obtains

$$I_t \approx I_{\tilde{T}} = 3\kappa_\varepsilon I_\varepsilon + 3\kappa_{\dot{\varepsilon}} I_{\dot{\varepsilon}}, \quad \boldsymbol{t}' \approx \tilde{\boldsymbol{T}}' = 2\mu_\varepsilon \boldsymbol{\varepsilon}' + 2\mu_{\dot{\varepsilon}} \dot{\boldsymbol{\varepsilon}}' \tag{5.4.63}$$

as the linearized stress laws for isotropic viscoelastic solid bodies. Finally, we emphasize that in the above linear representations the various coefficients are to be expected to be numerically distinct for the parameterizations.

5.5 Anisotropic Solids

In this section we will not present a complete account on anisotropy of crystals – for this, there exist well known books – but merely a short overview on how one can treat the constitutive equations of anisotropic materials[12]. Applications are encountered in the theory of elasticity, in the theory of heat conduction, in general thermoviscoelasticity as well as in many other branches of physics, such as the theory of turbulence (Chap. 12).

[12] Books in which anisotropic properties are dealt with generally restrict considerations to linear material behaviour. Such books are by LOVE [139], but HAUPT [98] and EHRENTRAUT [63] also treat the nonlinear case.

5.5.1 Linear Stress Strain Relation, Linear Heat Conduction

In the foregoing section the theory of *isotropic* constitutive equations was treated with rather broad generality. A similar extensive treatment of the representation–theory for anisotropic material equations is incomparably more complex and demands the background of group theory with corresponding representation theorems. Here we proceed with a somewhat more moderate procedure and postulate that the anisotropic material laws, that we are concerned with, are derivable from "potentials" and the derived relations are *linear*.

Let $\boldsymbol{S}$ and $\boldsymbol{s}$ be a symmetric second rank tensor and a vector, respectively, which both depend on a symmetric tensor $\boldsymbol{A}$ and a vector $\boldsymbol{a}$ as follows

$$\boldsymbol{S} = \hat{\boldsymbol{S}}(\boldsymbol{A},\ \boldsymbol{a})\ , \quad \boldsymbol{s} = \hat{\boldsymbol{s}}(\boldsymbol{A},\ \boldsymbol{a})\ . \tag{5.5.1}$$

Assume, moreover that a potential $\Xi = \hat{\Xi}(\boldsymbol{A},\ \boldsymbol{a})$ exists with the properties that

$$\boldsymbol{S} = \frac{\partial \hat{\Xi}(\boldsymbol{A},\ \boldsymbol{a})}{\partial \boldsymbol{A}}\ , \quad \boldsymbol{s} = \frac{\partial \hat{\Xi}(\boldsymbol{A},\ \boldsymbol{a})}{\partial \boldsymbol{a}}\ . \tag{5.5.2}$$

Often, the dependencies of the potential Ξ on $\boldsymbol{A}$ and $\boldsymbol{a}$ are separated and describe no interaction. Then the relation

$$\hat{\Xi}(\boldsymbol{A},\ \boldsymbol{a}) = \hat{\Phi}(\boldsymbol{A}) + \hat{\Psi}(\boldsymbol{a})\ , \tag{5.5.3}$$

holds such that (5.5.2) becomes

$$\boldsymbol{S} = \frac{\partial \hat{\Phi}(\boldsymbol{A})}{\partial \boldsymbol{A}}\ , \quad \boldsymbol{s} = \frac{\partial \hat{\Psi}(\boldsymbol{a})}{\partial \boldsymbol{a}}\ . \tag{5.5.4}$$

In this section Φ and Ψ shall also be called potentials.

We emphasize here that the relations (5.5.2) or (5.5.4) are insofar restrictive, as they exclude in the case of elasticity for example the so-called *hypo-elastic* materials. The *hyper-elastic* behaviour[13] of crystals can, however, be treated with the help of the potentials (5.5.2), (5.5.4). If $\boldsymbol{S}$ and $\boldsymbol{s}$ are restricted to be linearly dependent on $\boldsymbol{A}$ and $\boldsymbol{a}$, then the potentials Φ and Ψ are quadratic functions of $\boldsymbol{A}$ and $\boldsymbol{a}$. In most situations one argues as follows: $\boldsymbol{A}$ and $\boldsymbol{a}$ are interpreted as generalized measures of deformation, and the processes are assumed to be so natured that $||\boldsymbol{A}||$ and $|\boldsymbol{a}|$ can be presumed small. Then one can expand the potentials in TAYLOR series of $\boldsymbol{A}$ and $\boldsymbol{a}$. Doing this and omitting the cubic and higher order terms one obtains

$$\begin{aligned} \Phi &= \alpha + \boldsymbol{\beta} \cdot \boldsymbol{A} + \tfrac{1}{2}\boldsymbol{\eta} \cdot (\boldsymbol{A} \otimes \boldsymbol{A}) + \cdots \quad , \\ \Psi &= \alpha' + \boldsymbol{\beta}' \cdot \boldsymbol{a} + \tfrac{1}{2}\boldsymbol{\kappa} \cdot \boldsymbol{a} \otimes \boldsymbol{a} + \cdots\ , \end{aligned} \tag{5.5.5}$$

or in Cartesian tensor form

[13] The definition will shortly be given.

$$\Phi = \alpha + \beta_{\alpha\beta} A_{\alpha\beta} + \tfrac{1}{2}\eta_{\alpha\beta\gamma\delta} A_{\alpha\beta} A_{\gamma\delta} + \cdots \quad , \qquad \Psi = \alpha' + \beta'_{\alpha} a_{\alpha} + \tfrac{1}{2}\kappa_{\alpha\beta} a_{\alpha} a_{\beta} + \cdots \, . \tag{5.5.6}$$

The parameters $\alpha, \alpha', \boldsymbol{\beta}, \boldsymbol{\beta}', \boldsymbol{\kappa}$ and $\boldsymbol{\eta}$ are two scalars, a vector, two second and fourth rank tensors, respectively. The constant terms in (5.5.5) and (5.5.6) are not required to be specified, because of the differentiations (5.5.4) that must be performed to calculate $\boldsymbol{S}$ and $\boldsymbol{s}$; so they can be set to zero: $\alpha = 0,\ \alpha' = 0$. Under most situations the linear terms $\boldsymbol{\beta} = \boldsymbol{0},\ \boldsymbol{\beta}' = \boldsymbol{0}$ also vanish, resulting in $\boldsymbol{S}(\boldsymbol{A} = \boldsymbol{0}) = \boldsymbol{0}$ and $\boldsymbol{s}(\boldsymbol{a} = \boldsymbol{0}) = \boldsymbol{0}$. Such behaviour prevails e.g. when the equilibrium leads to stress free conditions. In such cases it may be sufficient to write Φ and Ψ in the form

$$\Phi = \tfrac{1}{2}\eta_{\alpha\beta\gamma\delta} A_{\alpha\beta} A_{\gamma\delta} \, , \quad \Psi = \tfrac{1}{2}\kappa_{\alpha\beta} a_{\alpha} a_{\beta} \, . \tag{5.5.7}$$

The components of the tensor $\boldsymbol{S}$ and the vector $\boldsymbol{s}$ are then given by

$$\boldsymbol{S} = \boldsymbol{\eta} \boldsymbol{A} \, , \quad S_{\alpha\beta} = \eta_{\alpha\beta\gamma\delta} A_{\gamma\delta} \quad , \qquad \boldsymbol{s} = \boldsymbol{\kappa} \boldsymbol{a} \, , \quad s_{\alpha} = \kappa_{\alpha\beta} a_{\beta} \, . \tag{5.5.8}$$

In the following, as usual, we demand that under symmetry transformations, i.e., when changing the reference system, then $\boldsymbol{S}, \boldsymbol{s}, \boldsymbol{A}, \boldsymbol{a}$ transform as symmetric second rank tensors and vectors, respectively. Such transformations form a subgroup $\mathbb{G}$ of the orthogonal group $\mathbb{O}$. With $\boldsymbol{H} \in \mathbb{G}$ one has

$$\boldsymbol{S}^{+} = \boldsymbol{H}\boldsymbol{S}\boldsymbol{H}^{T} \, , \quad \boldsymbol{A}^{+} = \boldsymbol{H}\boldsymbol{A}\boldsymbol{H}^{T} \, , \qquad \boldsymbol{s}^{+} = \boldsymbol{H}\boldsymbol{s} \, , \quad \boldsymbol{a}^{+} = \boldsymbol{H}\boldsymbol{a} \, , \tag{5.5.9}$$

where $(\cdot)^{+}$ denotes the variables $(\cdot)$ referred to the reference system $(\ )^{+}$.

It is a relatively simple exercise, which is based on the properties (5.5.9), to demonstrate that $\boldsymbol{\eta}$, $\boldsymbol{\kappa}$, Φ and Ψ in changing the reference system, i.e., under the transformation $\boldsymbol{H} \in \mathbb{O}$, transform as symmetric tensors of fourth and second rank as well as two scalars. For $\boldsymbol{H} \in \mathbb{O}$ we thus have

$$\eta^{+}_{\alpha\beta\gamma\delta} = H_{\alpha\mu} H_{\beta\nu} H_{\gamma\rho} H_{\delta\zeta} \eta_{\mu\nu\rho\zeta} \quad , \qquad \kappa^{+}_{\alpha\beta} = H_{\alpha\mu} H_{\beta\nu} \kappa_{\mu\nu} \quad , \qquad \Phi^{+} = \Phi \, , \quad \Psi^{+} = \Psi \, . \tag{5.5.10}$$

We will use these transformation rules in order to analyze certain symmetry properties of linear anisotropic solids.

5.5.2 Symmetry Properties of the Coefficients

The coefficient "matrix" $\eta_{\alpha\beta\gamma\delta}$ $(\kappa_{\alpha\beta})$ is composed of $3 \times 3 \times 3 \times 3 = 81$ $(3 \times 3 = 9)$ elements. The number of the independent elements is further reduced on the grounds of the symmetry assumptions on $\boldsymbol{S}, \boldsymbol{A}$ as well as the properties that $\boldsymbol{S}$ and $\boldsymbol{s}$ are derivable from potentials.

Because of the symmetry of $\boldsymbol{S}$ and $\boldsymbol{A}$, (5.5.8) implies

$$S_{\alpha\beta} = \eta_{\alpha\beta\gamma\delta} A_{\gamma\delta} = S_{\beta\alpha} = \eta_{\beta\alpha\gamma\delta} A_{\gamma\delta} = \eta_{\alpha\beta\gamma\delta} A_{\delta\gamma} = \eta_{\alpha\beta\delta\gamma} A_{\gamma\delta} \,, \tag{5.5.11}$$

from which one concludes

$$\eta_{\alpha\beta\gamma\delta} = \eta_{\beta\alpha\gamma\delta} = \eta_{\alpha\beta\delta\gamma} \,, \tag{5.5.12}$$

so that of the original 81 independent components of $\boldsymbol{\eta}$ only 36 are independent. Regarding the potentials Φ and Ψ the relations

$$\frac{\partial^2 \Phi}{\partial A_{\alpha\beta}\, \partial A_{\gamma\delta}} = \eta_{\alpha\beta\gamma\delta} \,, \qquad \frac{\partial^2 \Psi}{\partial a_\alpha\, \partial a_\beta} = \kappa_{\alpha\beta} \tag{5.5.13}$$

hold. If Φ and Ψ should be uniquely determined from $\boldsymbol{A}$ and $\boldsymbol{a}$, then the integrability conditions

$$\frac{\partial^2 \Phi}{\partial A_{\alpha\beta}\, \partial A_{\gamma\delta}} = \frac{\partial^2 \Phi}{\partial A_{\gamma\delta}\, \partial A_{\alpha\beta}} \,, \qquad \frac{\partial^2 \Psi}{\partial a_\alpha\, \partial a_\beta} = \frac{\partial^2 \Psi}{\partial a_\beta\, \partial a_\alpha} \tag{5.5.14}$$

must hold which, because of (5.5.13) yield the further symmetry conditions

$$\eta_{\alpha\beta\gamma\delta} = \eta_{\gamma\delta\alpha\beta} \,, \qquad \kappa_{\alpha\beta} = \kappa_{\beta\alpha} \,. \tag{5.5.15}$$

These reduce the number of independent components of $\boldsymbol{\eta}$ to at most 21 and those for $\boldsymbol{\kappa}$ to 6[14].

Because of the reduction of the number of independent components of the fourth rank tensor $\boldsymbol{\eta}$ from 81 to 21 it is relevant that the symmetric tensors $\boldsymbol{S}$ and $\boldsymbol{A}$ can be interpreted as six dimensional vectors over the space $\mathbb{R}^6$; we may then introduce the 6-vectors

$$\begin{aligned} \mathfrak{S} &= (S_{11},\, S_{22},\, S_{33},\, 2S_{23},\, 2S_{31},\, 2S_{12}) \,, \\ \mathcal{A} &= (A_{11},\, A_{22},\, A_{33},\, A_{23},\, A_{31},\, A_{12}) \,, \end{aligned} \tag{5.5.16}$$

whose components $\mathfrak{S}_p$, $\mathcal{A}_p$, $p = 1, ..., 6$ as represented in (5.5.16) are given by the tensor components of $\boldsymbol{S}$ and $\boldsymbol{A}$. (Precaution should be taken for the

[14] In the elasticity theory this symmetry is characterized with the concept of *hyperelasticity* or GREEN *elasticity*. When this symmetry is not available and the stress tensor can not be derived from a potential, then the elasticity is called *hypoelasticity* or CAUCHY *elasticity*. It arises most often when small deformations are superimposed upon an intermediate configuration with prestress and the associated stress increments are related to the strain increments.

In the theory of heat conduction the assumption of the symmetry of $\boldsymbol{\kappa}$ is often called CURIE *principle* and is attributed to the ONSAGER *reciprocity relations.* The latter can, however, only be understood in connection with a thermodynamic theory.

factor 2 in the three last components of $\mathfrak{S}$). The relation $\boldsymbol{S} = \boldsymbol{\eta}\boldsymbol{A}$ can then be written in the form

$$\mathfrak{S} = \mathbb{C}\boldsymbol{\mathcal{A}}\,, \tag{5.5.17}$$

where $\mathbb{C}$ is a symmetric 6×6 matrix and exhibits 21 independent components if (5.5.17) is derivable from a potential. This notation has originally been introduced by VOIGT [249]. Now, Φ can be interpreted as a function of $\boldsymbol{\mathcal{A}}$, $\Phi = \hat{\Phi}(\boldsymbol{\mathcal{A}})$, and in the linear case[15] it is

$$\begin{aligned}
\Phi &= \tfrac{1}{2}\mathbb{C}\cdot\boldsymbol{\mathcal{A}}\otimes\boldsymbol{\mathcal{A}} = \tfrac{1}{2}C_{pq}\mathcal{A}_p\mathcal{A}_q \\
&= \tfrac{1}{2}C_{11}\mathcal{A}_1^2 + C_{12}\mathcal{A}_1\mathcal{A}_2 + C_{13}\mathcal{A}_1\mathcal{A}_3 + 2C_{14}\mathcal{A}_1\mathcal{A}_4 + 2C_{15}\mathcal{A}_1\mathcal{A}_5 + 2C_{16}\mathcal{A}_1\mathcal{A}_6 \\
&\quad + \tfrac{1}{2}C_{22}\mathcal{A}_2^2 + C_{23}\mathcal{A}_2\mathcal{A}_3 + 2C_{24}\mathcal{A}_2\mathcal{A}_4 + 2C_{25}\mathcal{A}_2\mathcal{A}_5 + 2C_{26}\mathcal{A}_2\mathcal{A}_6 \\
&\quad + \tfrac{1}{2}C_{33}\mathcal{A}_3^2 + 2C_{34}\mathcal{A}_3\mathcal{A}_4 + 2C_{35}\mathcal{A}_3\mathcal{A}_5 + 2C_{36}\mathcal{A}_3\mathcal{A}_6 \\
&\quad + 2C_{44}\mathcal{A}_4\mathcal{A}_4 + 4C_{45}\mathcal{A}_4\mathcal{A}_5 + 4C_{46}\mathcal{A}_4\mathcal{A}_6 \\
&\quad + 2C_{55}\mathcal{A}_5\mathcal{A}_5 + 4C_{56}\mathcal{A}_5\mathcal{A}_6 \\
&\quad + 2C_{66}\mathcal{A}_6^2\,,
\end{aligned} \tag{5.5.18}$$

where the symmetry conditions $C_{pq} = C_{qp}$ are applied and thus

$$\mathfrak{S} = \mathbb{C}\boldsymbol{\mathcal{A}}\,, \quad \mathfrak{S}_p = C_{pq}\mathcal{A}_q \tag{5.5.19}$$

holds. The correspondence of the fourth rank tensor $\boldsymbol{\eta}$ with all the listed symmetries to the components of the symmetric 6×6–matrix $\mathbb{C}$ can be easily obtained by writing the quadratic potential Φ with respect to both quantities $\boldsymbol{A}$ and $\boldsymbol{\mathcal{A}}$, respectively. In the original notation we have

$$\begin{aligned}
\Phi &= \tfrac{1}{2}\eta_{\alpha\beta\gamma\delta}A_{\alpha\beta}A_{\gamma\delta} = \\
&\tfrac{1}{2}\eta_{1111}A_{11}A_{11} + \eta_{1122}A_{11}A_{22} + \eta_{1133}A_{11}A_{33} + 2\eta_{1123}A_{11}A_{23} + 2\eta_{1113}A_{11}A_{13} + 2\eta_{1112}A_{11}A_{12} \\
&\quad + \tfrac{1}{2}\eta_{2222}A_{22}A_{22} + \eta_{2233}A_{22}A_{33} + 2\eta_{2223}A_{22}A_{23} + 2\eta_{2213}A_{22}A_{13} + 2\eta_{2212}A_{22}A_{12} \\
&\quad + \tfrac{1}{2}\eta_{3333}A_{33}A_{33} + 2\eta_{3323}A_{33}A_{23} + 2\eta_{3313}A_{33}A_{13} + 2\eta_{3312}A_{33}A_{12} \\
&\quad + 2\eta_{2323}A_{23}A_{23} + 4\eta_{2313}A_{23}A_{13} + 4\eta_{2312}A_{23}A_{12} \\
&\quad + 2\eta_{1313}A_{13}A_{13} + 4\eta_{1312}A_{13}A_{12} \\
&\quad + 2\eta_{1212}A_{12}A_{12}.
\end{aligned} \tag{5.5.20}$$

---

[15] In the component representations (5.5.18) and (5.5.20) the EINSTEIN summation convention is applied over six components.

Here care must be taken with the different prefactors in the summation, which reflects the symmetry: Terms such as η_{1111} appear only once, terms like η_{1122} appear twice ($\eta_{1122}+\eta_{2211}$), terms of the type η_{1123} or η_{2323} four times and terms η_{2313} eight times ! (One distinguishes between the different combination possibilities of the index pairs). Substituting the definition of $\mathcal{A}$ in (5.5.20), one obtains

$$
\begin{aligned}
\Phi= \\
\tfrac{1}{2}\eta_{1111}\mathcal{A}_1^2+\eta_{1122}\mathcal{A}_1\mathcal{A}_2+\eta_{1133}\mathcal{A}_1\mathcal{A}_3+2\eta_{1123}\mathcal{A}_1\mathcal{A}_4+2\eta_{1113}\mathcal{A}_1\mathcal{A}_5+2\eta_{1112}\mathcal{A}_1\mathcal{A}_6 \\
+\tfrac{1}{2}\eta_{2222}\mathcal{A}_2^2 +\eta_{2233}\mathcal{A}_2\mathcal{A}_3+2\eta_{2223}\mathcal{A}_2\mathcal{A}_4+2\eta_{2213}\mathcal{A}_2\mathcal{A}_5+2\eta_{2212}\mathcal{A}_2\mathcal{A}_6 \\
+\tfrac{1}{2}\eta_{3333}\mathcal{A}_3^2 +2\eta_{3323}\mathcal{A}_3\mathcal{A}_4 +2\eta_{3313}\mathcal{A}_3\mathcal{A}_5+2\eta_{3312}\mathcal{A}_3\mathcal{A}_6 \\
+2\eta_{2323}\mathcal{A}_4^2 \quad +4\eta_{2313}\mathcal{A}_4\mathcal{A}_5+4\eta_{2312}\mathcal{A}_4\mathcal{A}_6 \\
+2\eta_{1313}\mathcal{A}_5^2 \quad +4\eta_{1312}\mathcal{A}_5\mathcal{A}_6 \\
+2\eta_{1212}\mathcal{A}_6^2 \,.
\end{aligned}
\tag{5.5.21}
$$

Comparing this with (5.5.18), there follows

$$
(\mathbb{C})_{pq} = \begin{bmatrix}
\eta_{1111} & \eta_{1122} & \eta_{1133} & 2\eta_{1123} & 2\eta_{1113} & 2\eta_{1112} \\
 & \eta_{2222} & \eta_{2233} & 2\eta_{2223} & 2\eta_{2213} & 2\eta_{2212} \\
 & & \eta_{3333} & 2\eta_{3323} & 2\eta_{3313} & 2\eta_{3312} \\
 & & & 4\eta_{2323} & 4\eta_{2313} & 4\eta_{2312} \\
 & & & & 4\eta_{1313} & 4\eta_{1312} \\
 & & & & & 4\eta_{1212}
\end{bmatrix}
\tag{5.5.22}
$$

as the correspondence of the coefficients of $\mathbb{C}$ and $\boldsymbol{\eta}$ with the indices

$$
\begin{array}{ccc}
p & \to & \alpha\beta(p) \\
1 & \to & 11 \\
2 & \to & 22 \\
3 & \to & 33 \\
4 & \to & 23 \\
5 & \to & 13 \\
6 & \to & 12 \,,
\end{array}
\qquad p=1,\dots,6 \,,
\tag{5.5.23}
$$

which shall be consistently used. In the literature both representations are common. The matrix representation is the older, founded by VOIGT and is used by solid state physicists today, but sometimes with different definitions of $\mathfrak{S}$ and $\mathbb{C}$. It should be noticed that in the notation used here, because of the definition of $\mathfrak{S}$, the components $\mathfrak{S}_p$ ($p \geq 4$) are twice the corresponding components $S_{\alpha\beta}$. In the following, as usual, we shall use the tensor notation and deal with the components of $\boldsymbol{\eta}$.

Finally we write, to be complete, the potential $\Psi(\boldsymbol{a})$ in the form

$$\begin{aligned}\Psi = \tfrac{1}{2}\kappa_{\alpha\beta}a_\alpha a_\beta = \tfrac{1}{2}\kappa_{11}a_1^2 &+ \kappa_{12}a_1a_2 + \kappa_{13}a_1a_3 \\ &+ \tfrac{1}{2}\kappa_{22}a_2^2 \ + \kappa_{23}a_2a_3 \\ &\qquad\qquad + \tfrac{1}{2}\kappa_{33}a_3^2 \ .\end{aligned} \tag{5.5.24}$$

It determines $\boldsymbol{s}$ through differentiation in accordance with (5.5.8).

5.5.3 Symmetry Transformations for Anisotropic Bodies

The linear constitutive relations introduced in the last subsection for materials with unspecified symmetries have led to a fourth and second rank tensor with 21 and 6 independent components, respectively. In case of special symmetries the number of these independent components must be reduced. We will now show how one specifies this reduction for selected symmetries.

If a certain symmetry property prevails, then the material response does not change with respect to such a symmetry transformation $\boldsymbol{H}$. In the following we will not specifically identify the variables referred to a fixed reference system but we will indicate when they are referred to any other reference system and designate them by writing $(\cdot)^+$. If a material is symmetric relative to $\boldsymbol{H} \in \mathbb{G}$ then

$$\Phi = \Phi^+ \quad \text{and} \quad \Psi = \Psi^+ \ , \quad \forall\, \boldsymbol{H} \in \mathbb{G} \ . \tag{5.5.25}$$

Thus: *Every symmetry transformation leaves the potentials invariant.* In order to exactly examine the transformation properties of $\boldsymbol{A}$ and $\boldsymbol{a}$, the formulas

$$\boldsymbol{a}^+ = \boldsymbol{H}\boldsymbol{a} \ , \quad \boldsymbol{A}^+ = \boldsymbol{H}\boldsymbol{A}\boldsymbol{H}^T \tag{5.5.26}$$

must be substituted in the functional relations for Φ and Ψ and then the conditions (5.5.25) must be satisfied. In so doing the fact that the reference configurations related by $\boldsymbol{H}$ are symmetric, is then expressed as a functional relation between $\boldsymbol{\eta}^+$, $\boldsymbol{\kappa}^+$ and $\boldsymbol{\eta}$, $\boldsymbol{\kappa}$.

To examine the different symmetries we wish to choose the reference system in such a way that its Cartesian coordinates with selected symmetry axes agree with the assumed body symmetry. Starting from this, the transformations belonging to a certain symmetry group $\mathbb{G}$ (rotations, mirror reflections) are then performed, and with this the number of the independent components of $\boldsymbol{\eta}$ and $\boldsymbol{\kappa}$ is reduced. We will demonstrate this procedure for some special symmetry groups, because a general examination would be too involved. We shall deal with the following cases (see also the sketches in Table 5.5.4):

- **Orthotropy** Symmetry with respect to rotations of 180° (π). Material properties are independent of the direction (forth and back).

- **Orthotropy, Horizontally Regular** Symmetry with respect to 180°–rotations and additional symmetry with respect to 90°–rotations about one particular symmetry axis.

- **Orthotropy, Horizontal Isotropy** also called **Transverse Isotropy** Orthotropy and invariance with respect to arbitrary rotations about one given axis of orthotropic symmetry

- **Regular (Cubic)** Symmetric with respect to 90°–rotations about three fixed perpendicular axes (automatically orthotropic).

- **Isotropic:** Symmetric with respect to arbitrary rotations.

- **Hexagonal** Symmetry with respect to $n \times 30°$–rotations about a fixed axis ($\hat{=} c$-axis).

a) Orthotropic Solid Let a solid be given whose material properties differ from each other along three mutually perpendicular directions. The potentials Φ and Ψ must then remain invariant to every rotation of three mutually perpendicular axes about an angle of 180°. We choose the reference configuration such that the coordinate axes X, Y, Z lie in the symmetry axes of the body.

- **Rotation About Z–axis:** If $\boldsymbol{H}$ is such a rotation about the angle ϕ, then

$$\boldsymbol{H} = \begin{pmatrix} \cos\phi & \sin\phi & 0 \\ -\sin\phi & \cos\phi & 0 \\ 0 & 0 & 1 \end{pmatrix} \overset{(\phi=180°)}{=} \boldsymbol{H} = \begin{pmatrix} -1 & 0 & 0 \\ 0 & -1 & 0 \\ 0 & 0 & 1 \end{pmatrix} . \tag{5.5.27}$$

Thus

$$(\boldsymbol{A}^{+}) = (\boldsymbol{H}\boldsymbol{A}\boldsymbol{H}^{T}) = \begin{pmatrix} A_{11} & A_{12} & -A_{13} \\ A_{12} & A_{22} & -A_{23} \\ -A_{13} & -A_{23} & A_{33} \end{pmatrix} . \tag{5.5.28}$$

Now writing the quadratic form (5.5.20) for Φ^{+},

$$\Phi^{+} = \tfrac{1}{2}\eta_{\alpha\beta\gamma\delta} A^{+}_{\alpha\beta} A^{+}_{\gamma\delta} \tag{5.5.29}$$

(here the coefficient $\eta_{\alpha\beta\gamma\delta}$ does not carry $(\cdot)^{+}$) and applying the result (5.5.28), then one obtains again a representation of the form (5.5.20), which will not be repeated here but which should be characterized by the following triangular representation:

$$\begin{pmatrix} \cdot & \cdot & \cdot & - & - & \cdot \\ & \cdot & \cdot & - & - & \cdot \\ & & \cdot & - & - & \cdot \\ & & & \cdot & \cdot & - \\ & & & & \cdot & - \\ & & & & & \cdot \end{pmatrix} . \tag{5.5.30}$$

The minus signs in (5.5.30) should indicate that in (5.5.20) at the corresponding positions follows no addition, but a subtraction. The condition $\Phi^+ = \Phi$ is abbreviated as follows

$$\begin{pmatrix} \cdot & \cdot & \cdot & - & - & \cdot \\ & \cdot & \cdot & - & - & \cdot \\ & & \cdot & - & - & \cdot \\ & & & \cdot & \cdot & - \\ & & & & \cdot & - \\ & & & & & \cdot \end{pmatrix} = \begin{pmatrix} \cdot & \cdot & \cdot & \cdot & \cdot & \cdot \\ & \cdot & \cdot & \cdot & \cdot & \cdot \\ & & \cdot & \cdot & \cdot & \cdot \\ & & & \cdot & \cdot & \cdot \\ & & & & \cdot & \cdot \\ & & & & & \cdot \end{pmatrix} , \tag{5.5.31}$$

which (understood as an identity of two quadratic forms) is only identically satisfied when the prefactors of the terms with negative signs vanish i.e., when

$$\eta_{1123} = \eta_{2223} = \eta_{3323} = \eta_{2312} = \eta_{1113} = \eta_{2213} = \eta_{3313} = \eta_{1312} = 0 \, . \tag{5.5.32}$$

- **Rotation About the Y-axis** A rotation of 180° about the y-axis yields

$$\boldsymbol{H} = \begin{pmatrix} -1 & 0 & 0 \\ 0 & 1 & 0 \\ 0 & 0 & -1 \end{pmatrix} , \tag{5.5.33}$$

so that

$$(\boldsymbol{A}^+) = (\boldsymbol{H}\boldsymbol{A}\boldsymbol{H}^T) = \begin{pmatrix} A_{11} & -A_{12} & A_{13} \\ -A_{12} & A_{22} & -A_{23} \\ A_{13} & -A_{23} & A_{33} \end{pmatrix} . \tag{5.5.34}$$

The postulation $\Phi^+ = \Phi$ leads now to the identity

$$\begin{pmatrix} \cdot & \cdot & \cdot & - & \cdot & - \\ & \cdot & \cdot & - & \cdot & - \\ & & \cdot & - & \cdot & - \\ & & & \cdot & - & \cdot \\ & & & & \cdot & - \\ & & & & & \cdot \end{pmatrix} = \begin{pmatrix} \cdot & \cdot & \cdot & \cdot & \cdot & \cdot \\ & \cdot & \cdot & \cdot & \cdot & \cdot \\ & & \cdot & \cdot & \cdot & \cdot \\ & & & \cdot & \cdot & \cdot \\ & & & & \cdot & \cdot \\ & & & & & \cdot \end{pmatrix} , \tag{5.5.35}$$

which is satisfied provided that the coefficients

$$\eta_{1123} = \eta_{2223} = \eta_{3323} = \eta_{2313} = \eta_{1112} = \eta_{2212} = \eta_{3312} = \eta_{1312} \tag{5.5.36}$$

vanish. We leave it to the reader to prove that a further rotation of 180° about the X-axis results in no further reductions.

For orthotropic materials the coefficient matrix $\mathbb{C}$, when referred to principal axes, is therefore given by the scheme

$$(\mathbb{C})_{pq} = \begin{pmatrix} \eta_{1111} & \eta_{1122} & \eta_{1133} & 0 & 0 & 0 \\ & \eta_{2222} & \eta_{2233} & 0 & 0 & 0 \\ & & \eta_{3333} & 0 & 0 & 0 \\ & & & 4\eta_{2323} & 0 & 0 \\ & & & & 4\eta_{1313} & 0 \\ & & & & & 4\eta_{1212} \end{pmatrix} \qquad (5.5.37)$$

with nine independent coefficients $\eta_{\alpha\beta\gamma\delta}$. Likewise, one can also deal with the reduction for Φ. One considers here $\Phi^+ = \Phi$, uses $\boldsymbol{a}^+ = \boldsymbol{H}\boldsymbol{a}$, where $\boldsymbol{H}$ is given by (5.5.27) and (5.5.33), and verifies that the invariance of Φ under rotations by 180° about the Z- and Y-axis yields $\kappa_{\alpha\beta} = 0$ for $\alpha \neq \beta$, so that κ becomes a diagonal matrix

$$(\boldsymbol{\kappa})_{\alpha\beta} = \begin{pmatrix} \kappa_{11} & 0 & 0 \\ 0 & \kappa_{22} & 0 \\ 0 & 0 & \kappa_{33} \end{pmatrix} \qquad (5.5.38)$$

with three independent elements.

b) Orthotropic Solid, Regular with Respect to the XY-Plane A material which obeys this symmetry condition satisfies the conditions of an orthotropic material relative to the X-, Y- and Z-axes, but is simultaneously also distinguished by the symmetry properties that it behaves the same way in the X- and Y- directions in planes parallel to the XY plane. A rotation of 90 degrees about the Z- axis is thus a symmetry transformation. Because of $(5.5.27)_1$, one obtains with $\phi = 90°$

$$\boldsymbol{H} = \begin{pmatrix} 0 & 1 & 0 \\ -1 & 0 & 0 \\ 0 & 0 & 1 \end{pmatrix}, \quad \boldsymbol{H}^T = \begin{pmatrix} 0 & -1 & 0 \\ 1 & 0 & 0 \\ 0 & 0 & 1 \end{pmatrix} \qquad (5.5.39)$$

and thus

$$(\boldsymbol{H}\boldsymbol{A}\boldsymbol{H}^T) = \begin{pmatrix} A_{22} & -A_{21} & A_{32} \\ -A_{12} & A_{11} & -A_{13} \\ A_{32} & -A_{31} & A_{33} \end{pmatrix}, \qquad (5.5.40)$$

$$\boldsymbol{H}\boldsymbol{a} = \begin{pmatrix} a_2 \\ -a_1 \\ a_3 \end{pmatrix}. \qquad (5.5.41)$$

For the calculation of Φ^+ in (5.5.20) because of (5.5.40) only the interchanges

$$A_{11} \longleftrightarrow A_{22}, \quad A_{12} \longleftrightarrow -A_{21}, \quad A_{13} \longleftrightarrow \mp A_{32}$$

must be made. If $\Phi^+ = \Phi$ should be valid for a coefficient matrix according to (5.5.37) (the orthotropy condition may be assumed ab initio), then the identities

$$\eta_{1111} = \eta_{2222}, \quad \eta_{2323} = \eta_{1313}, \quad \eta_{1133} = \eta_{2233} \qquad (5.5.42)$$

must be valid. Further reductions are not possible, and one obtains

$$(\mathbb{C})_{pq} = \begin{pmatrix} \eta_{1111} & \eta_{1122} & \eta_{1133} & 0 & 0 & 0 \\ & \eta_{1111} & \eta_{1133} & 0 & 0 & 0 \\ & & \eta_{3333} & 0 & 0 & 0 \\ & & & 4\eta_{2323} & 0 & 0 \\ & & & & 4\eta_{2323} & 0 \\ & & & & & 4\eta_{1212} \end{pmatrix} , \tag{5.5.43}$$

which is a matrix with 6 independent coefficients. For $\boldsymbol{\kappa}$ one obtains correspondingly

$$(\boldsymbol{\kappa}_{\alpha\beta}) = \begin{pmatrix} \kappa_{11} & 0 & 0 \\ 0 & \kappa_{11} & 0 \\ 0 & 0 & \kappa_{33} \end{pmatrix} . \tag{5.5.44}$$

c) Orthotropic, Horizontally Isotropic (Transversally Isotropic) Solids If a material behaves isotropically in planes parallel to the XY-plane, then invariance of Φ against arbitrary rotations about the Z-axis must be required. Especially one requires invariance under an infinitesimal rotation $\delta\phi$ about the Z-axis. The matrix $\boldsymbol{H}$ in $(5.5.27)_1$ is then given by

$$\boldsymbol{H} = \begin{pmatrix} 1 & \delta\phi & 0 \\ -\delta\phi & 1 & 0 \\ 0 & 0 & 1 \end{pmatrix} + \mathcal{O}(\delta\phi^2) \tag{5.5.45}$$

and $\boldsymbol{HAH}^T$ can be calculated to yield

$$\boldsymbol{HAH}^T \approx \boldsymbol{A} + \delta\phi \begin{pmatrix} 2A_{12} & A_{22}-A_{11} & A_{23} \\ A_{22}-A_{11} & -2A_{12} & -A_{13} \\ A_{23} & -A_{12} & A_{33} \end{pmatrix} + \mathcal{O}(\delta\phi^2) . \tag{5.5.46}$$

If one, now, determines Φ^+, then the expansion

$$\Phi^+ = \Phi + \delta\,\phi\,\Phi_1 + \mathcal{O}(\delta\phi^2) \tag{5.5.47}$$

holds, where Φ_1 is given by

$$\Phi_1 = (2\eta_{1111} - 2\eta_{1122} - 4\eta_{1212})(A_{11} - A_{22}) , \tag{5.5.48}$$

so that from $\Phi^+ = \Phi$ the condition

$$2\eta_{1212} = \eta_{1111} - \eta_{1122} \tag{5.5.49}$$

follows. The matrix $(\mathbb{C})$ has now only 5 independent coefficients and is given by (5.5.43) in which (5.5.49) must be observed. It may also be shown that with general rotations about the Z-axis no further restrictions can be derived.

d) Regular (Cubic) Crystal This symmetry is characterized by orthotropy, which in all spatial directions shows the same behaviour. One can thus start from (5.5.37) and eliminate the differences which are observable in the three spatial directions; this can be achieved by the equalities

$$\begin{aligned} \eta_{1122} = \eta_{1133} = \eta_{2233} \,, \\ \eta_{1212} + \eta_{1313} = \eta_{2323} \,, \end{aligned} \tag{5.5.50}$$

so that the matrix $\mathbb{C}$ now possesses three different coefficients

$$(\mathbb{C})_{pq} = \begin{pmatrix} \eta_{1111} & \eta_{1122} & \eta_{1122} & 0 & 0 & 0 \\ & \eta_{1111} & \eta_{1122} & 0 & 0 & 0 \\ & & \eta_{1111} & 0 & 0 & 0 \\ & & & 4\eta_{2323} & 0 & 0 \\ & & & & 4\eta_{2323} & 0 \\ & & & & & 4\eta_{2323} \end{pmatrix} . \tag{5.5.51}$$

The $\boldsymbol{\kappa}$–matrix is reduced to $\boldsymbol{\kappa} = \kappa \boldsymbol{I}$.

e) Isotropic Solid These materials show the properties which are given by the cubic symmetry (5.5.51) with the additional restrictions (5.5.49), see parts c) and d). If we make the identifications

$$\eta_{1122} =: \lambda \,, \quad \eta_{1111} =: (\lambda + 2\mu) \,, \tag{5.5.52}$$

where λ and μ are commonly known as LAMÉ constants in the theory of elasticity, we conclude from (5.5.49) and (5.5.51) that $\eta_{2323} = \mu$. Thus, only two independent coefficients characterize linear isotropic bodies. The $(\mathbb{C})$ matrix is given by

$$(\mathbb{C})_{pq} = \begin{pmatrix} \lambda + 2\mu & \lambda & \lambda & 0 & 0 & 0 \\ & \lambda + 2\mu & \lambda & 0 & 0 & 0 \\ & & \lambda + 2\mu & 0 & 0 & 0 \\ & & & 4\mu & 0 & 0 \\ & & & & 4\mu & 0 \\ & & & & & 4\mu \end{pmatrix} , \tag{5.5.53}$$

from which the relation

$$\boldsymbol{S} = \lambda(\operatorname{tr} \boldsymbol{A})\boldsymbol{I} + 2\mu\boldsymbol{A} = (\lambda + \tfrac{2}{3}\mu)(\operatorname{tr} \boldsymbol{A})\boldsymbol{I} + 2\mu(\boldsymbol{A} - \tfrac{1}{3} \operatorname{tr} \boldsymbol{A}\boldsymbol{I}) \tag{5.5.54}$$

may be deduced.

f) Hexagonal Crystal Snow and ice are made of hexagonal ice crystals. Thus, this symmetry is also important. The derivations of the form of the $\mathbb{C}$ matrix will be treated as an Exercise.

5.5.4 Stokes Assumption

In fluid dynamics of linearly viscous media the assumption of a vanishing volume viscosity is denoted as the STOKES *assumption*. It presumes that the volume changes occur without dissipation. The second rank tensor $\boldsymbol{A}$ then agrees with the stretching tensor $\boldsymbol{D}$ in this case, and vanishing volume viscosity implies that the factor of the volume stretching $\operatorname{tr}\boldsymbol{D}$ should vanish. In accordance with the notation of (5.5.54) the condition

$$\lambda + \tfrac{2}{3}\mu \equiv 0 \quad \Rightarrow \quad \lambda \equiv -\tfrac{2}{3}\mu \tag{5.5.55}$$

must hold. In the linear isotropic elasticity theory this assumption corresponds to the assumption of density preserving. The $\mathbb{C}$ matrix of an isotropic body satisfying the STOKES assumption thus contains only one single independent coefficient, and $\boldsymbol{S}$ is a deviator. It is at least for the isotropic case so that relation (5.5.55) follows from the statement that $\boldsymbol{S}$ is independent of $\operatorname{tr}\boldsymbol{A}$ as well as from the condition that $\boldsymbol{S}$ is a deviator for all $\boldsymbol{A}$.

A general useful relation for these assumptions does not exist, but we will indicate the following statement as the STOKES assumption.

Stokes Assumption *If in the constitutive law* $\boldsymbol{S} = \boldsymbol{\eta}\boldsymbol{A}$ *the tensor* $\boldsymbol{S}$ *is independent of the trace* $\operatorname{tr}\boldsymbol{A}$, *then this assumption will be called Stokes assumption.*

We will now comply with these additional conditions in the law $\boldsymbol{S} = \boldsymbol{\eta}\boldsymbol{A}$ of the above considered symmetries. If the law $\boldsymbol{S} = \boldsymbol{\eta}\boldsymbol{A}$ is independent of the trace of $\boldsymbol{A}$, then one may write

$$\begin{aligned} S_{\alpha\beta} &= \eta_{\alpha\beta\gamma\delta}[(A_{\gamma\delta} - \tfrac{1}{3}\operatorname{tr}\boldsymbol{A}\,\delta_{\gamma\delta}) + \tfrac{1}{3}\operatorname{tr}\boldsymbol{A}\,\delta_{\gamma\delta}] \\ &= \tfrac{1}{3}(\operatorname{tr}\boldsymbol{A})\eta_{\alpha\beta\gamma\gamma} + \eta_{\alpha\beta\gamma\delta}(A_{\gamma\delta} - \tfrac{1}{3}\boldsymbol{A}\,\delta_{\gamma\delta}) \,, \end{aligned} \tag{5.5.56}$$

and consequently must conclude that

$$\eta_{\alpha\beta\gamma\gamma} = 0 \,, \quad (\alpha,\, \beta = 1,\, 2,\, 3) \,. \tag{5.5.57}$$

Conversely, from the condition $S_{\alpha\alpha} = 0$ follows that for all $\boldsymbol{A}$ one necessarily has $\eta_{\alpha\alpha\gamma\delta} = 0$, which implies the same because of the symmetry $\eta_{\alpha\beta\gamma\delta} = \eta_{\gamma\delta\alpha\beta}$. It is therefore so that equality of the results deduced from both requirements is tied to the existence of a potential Φ. In general, relations (5.5.57) represent 6 equations with the coefficients $\eta_{\alpha\beta\gamma\delta}$, so that under the assumption of the validity of the STOKES assumption not 21 but only 15 coefficients of $\boldsymbol{\eta}$ remain independent. In the presence of orthotropy, the relations (5.5.57) with $\alpha \neq \beta$ are trivially satisfied, and there remain the relations

$$\eta_{11\gamma\gamma} = \eta_{22\gamma\gamma} = \eta_{33\gamma\gamma} = 0 \,, \tag{5.5.58}$$

from which one deduces

$$\begin{aligned}
\eta_{1122} &= \tfrac{1}{2}(\eta_{3333} - \eta_{2222} - \eta_{1111}) \,, \\
\eta_{2233} &= \tfrac{1}{2}(\eta_{1111} - \eta_{3333} - \eta_{2222}) \,, \\
\eta_{3311} &= \tfrac{1}{2}(\eta_{2222} - \eta_{1111} - \eta_{3333}) \,.
\end{aligned} \tag{5.5.59}$$

From the 9 independent coefficients of $\boldsymbol{\eta}$ of an *orthotropic crystal* under the validity of the STOKES assumption only six coefficients remain independent, because three are given by (5.5.59).

In case of orthotropic, horizontally regular solids the relation $\eta_{1111} = \eta_{2222}$ holds additionally, so that (5.5.59) reduces to

$$\eta_{1122} = \tfrac{1}{2}(\eta_{3333} - 2\eta_{1111}) \,, \quad \eta_{2233} = \eta_{3311} = -\tfrac{1}{2}\eta_{3333} \,. \tag{5.5.60}$$

The matrix (5.5.43) contains thus only 4 independent coefficients.

In a *transversally isotropic solid* relation (5.5.49) must also be satisfied, a relation, which, with $(5.5.60)_1$, takes the following form

$$\eta_{1212} = \eta_{1111} - \tfrac{1}{4}\eta_{3333} \,, \tag{5.5.61}$$

so that only three coefficients of $\boldsymbol{\eta}$ are now independent.

In the *regular (cubic) case*, $\mathbb{C}$ has the form (5.5.51), and because of (5.5.59), (5.5.49) and (5.5.50) it yields

$$\begin{aligned}
\eta_{1122} &= \eta_{2233} = \eta_{3311} = -\tfrac{1}{2}\eta_{1111} \,, \\
\eta_{1212} &= \eta_{2323} = \eta_{3131} = \tfrac{3}{4}\eta_{1111} \,,
\end{aligned} \tag{5.5.62}$$

such that only one coefficient of $\mathbb{C}$ remains independent.

Finally, for the *isotropic case* relations (5.5.53) and (5.5.62), which must hold with the identification (5.5.52), yields $\lambda + \frac{2}{3}\mu = 0$, as concluded already in (5.5.55).

We summarize these results in Table 5.1.

Table 5.1. Symmetry conditions for $\mathbb{C}$ and $\boldsymbol{\kappa}$ in accordance with equation (5.5.8). Form of $(\mathbb{C})$ and $(\boldsymbol{\kappa})$ matrices.

Case	$\mathfrak{S} = \mathbb{C}\boldsymbol{\mathcal{A}}, \quad \boldsymbol{S} = \boldsymbol{\eta}\boldsymbol{\mathcal{A}}$ (Form of $\mathbb{C}$–matrix) without STOKES assumption	with STOKES assumption	$\boldsymbol{s} = \boldsymbol{\kappa}\boldsymbol{a}$ $(\boldsymbol{\kappa})$ matrix
General ortho-tropic	$\begin{array}{cccccc} \eta_{1111} & \eta_{1122} & \eta_{1133} & 0 & 0 & 0 \\ & \eta_{2222} & \eta_{2233} & 0 & 0 & 0 \\ & & \eta_{3333} & 0 & 0 & 0 \\ & & & 4\eta_{2323} & 0 & 0 \\ & & & & 4\eta_{1313} & 0 \\ & & & & & 4\eta_{1212} \end{array}$ 9 coefficients	$\begin{array}{cccccc} \eta_{1111} & \frac{1}{2}(\eta_{3333} - 2\eta_{1111}) & -\frac{1}{2}\eta_{3333} & 0 & 0 & 0 \\ & \eta_{2222} & -\frac{1}{2}\eta_{3333} & 0 & 0 & 0 \\ & & \eta_{3333} & 0 & 0 & 0 \\ & & & 4\eta_{2323} & 0 & 0 \\ & & & & 4\eta_{1313} & 0 \\ & & & & & 4\eta_{1111} - \eta_{3333} \end{array}$ 6 coefficients	$\begin{array}{ccc} \kappa_{11} & 0 & 0 \\ 0 & \kappa_{22} & 0 \\ 0 & 0 & \kappa_{33} \end{array}$ 3 coefficients
orthotropic horizon-tally regular	$\begin{array}{cccccc} \eta_{1111} & \eta_{1122} & \eta_{1133} & 0 & 0 & 0 \\ & \eta_{1111} & \eta_{1133} & 0 & 0 & 0 \\ & & \eta_{3333} & 0 & 0 & 0 \\ & & & 4\eta_{2323} & 0 & 0 \\ & & & & 4\eta_{2323} & 0 \\ & & & & & 4\eta_{1212} \end{array}$ 6 coefficients	$\begin{array}{cccccc} \eta_{1111} & \frac{1}{2}(\eta_{3333} - 2\eta_{1111}) & -\frac{1}{2}\eta_{3333} & 0 & 0 & 0 \\ & \eta_{1111} & -\frac{1}{2}\eta_{3333} & 0 & 0 & 0 \\ & & \eta_{3333} & 0 & 0 & 0 \\ & & & 4\eta_{2323} & 0 & 0 \\ & & & & 4\eta_{2323} & 0 \\ & & & & & 4\eta_{1212} \end{array}$ 4 coefficients	$\begin{array}{ccc} \kappa_{11} & 0 & 0 \\ 0 & \kappa_{11} & 0 \\ 0 & 0 & \kappa_{33} \end{array}$ 2 coefficients
transverse isotropic	$\begin{array}{cccccc} \eta_{1111} & \eta_{1122} & \eta_{1133} & 0 & 0 & 0 \\ & \eta_{1111} & \eta_{1133} & 0 & 0 & 0 \\ & & \eta_{3333} & 0 & 0 & 0 \\ & & & 4\eta_{2323} & 0 & 0 \\ & & & & 4\eta_{2323} & 0 \\ & & & & & \frac{1}{2}(\eta_{1111} - \eta_{1122}) \end{array}$ 5 coefficients	$\begin{array}{cccccc} \eta_{1111} & \frac{1}{2}(\eta_{3333} - 2\eta_{1111}) & -\frac{1}{2}\eta_{3333} & 0 & 0 & 0 \\ & \eta_{1111} & -\frac{1}{2}\eta_{3333} & 0 & 0 & 0 \\ & & \eta_{3333} & 0 & 0 & 0 \\ & & & 4\eta_{2323} & 0 & 0 \\ & & & & 4\eta_{2323} & 0 \\ & & & & & \eta_{1111} - \frac{1}{4}\eta_{3333} \end{array}$ 3 coefficients	$\begin{array}{ccc} \kappa_{11} & 0 & 0 \\ 0 & \kappa_{11} & 0 \\ 0 & 0 & \kappa_{33} \end{array}$ 2 coefficients

Table 5.1. (continued)

Case	without STOKES assumption	with STOKES assumption	$(\boldsymbol{\kappa})$ matrix
regular (cubic)	$\begin{array}{cccccc} \eta_{1111} & \eta_{1122} & \eta_{1122} & 0 & 0 & 0 \\ & \eta_{1111} & \eta_{1122} & 0 & 0 & 0 \\ & & \eta_{1111} & 0 & 0 & 0 \\ & & & 4\eta_{2323} & 0 & 0 \\ & & & & 4\eta_{2323} & 0 \\ & & & & & 4\eta_{2323} \end{array}$ 3 coefficients	$\begin{array}{cccccc} \eta_{1111} & -\frac{1}{2}\eta_{1111} & -\frac{1}{2}\eta_{1111} & 0 & 0 & 0 \\ & \eta_{1111} & -\frac{1}{2}\eta_{1111} & 0 & 0 & 0 \\ & & \eta_{1111} & 0 & 0 & 0 \\ & & & 3\eta_{1111} & 0 & 0 \\ & & & & 3\eta_{1111} & 0 \\ & & & & & 3\eta_{1111} \end{array}$ 1 coefficient	$\begin{array}{ccc} \kappa_{11} & 0 & 0 \\ 0 & \kappa_{11} & 0 \\ 0 & 0 & \kappa_{11} \end{array}$ 1 coefficient
isotropic	$\begin{array}{cccccc} \lambda+2\mu & \lambda & \lambda & 0 & 0 & 0 \\ & \lambda+2\mu & \lambda & 0 & 0 & 0 \\ & & \lambda+2\mu & 0 & 0 & 0 \\ & & & 4\mu & 0 & 0 \\ & & & & 4\mu & 0 \\ & & & & & 4\mu \end{array}$ 2 coefficients	$\begin{array}{cccccc} \frac{4}{3}\mu & \frac{2}{3}\mu & \frac{2}{3}\mu & 0 & 0 & 0 \\ & \frac{4}{3}\mu & -\frac{2}{3}\mu & 0 & 0 & 0 \\ & & \frac{4}{3}\mu & 0 & 0 & 0 \\ & & & 4\mu & 0 & 0 \\ & & & & 4\mu & 0 \\ & & & & & 4\mu \end{array}$ 1 coefficient	$\begin{array}{ccc} \kappa_{11} & 0 & 0 \\ 0 & \kappa_{22} & 0 \\ 0 & 0 & \kappa_{33} \end{array}$ 1 coefficient
hexagonal (Axis)	$\begin{array}{cccccc} \eta_{1111} & \eta_{1122} & \eta_{1133} & 0 & 0 & 0 \\ & \eta_{1111} & \eta_{1133} & 0 & 0 & 0 \\ & & \eta_{3333} & 0 & 0 & 0 \\ & & & 4\eta_{2323} & 0 & 0 \\ & & & & 4\eta_{2323} & 0 \\ & & & & & 2(\eta_{1111} - \eta_{1122}) \end{array}$ 5 coefficients	$\begin{array}{cccccc} \eta_{1111} & \frac{1}{2}(\eta_{3333} - 2\eta_{1111}) & -\frac{1}{2}\eta_{3333} & 0 & 0 & 0 \\ & \eta_{1111} & -\frac{1}{2}\eta_{3333} & 0 & 0 & 0 \\ & & \eta_{3333} & 0 & 0 & 0 \\ & & & 4\eta_{2323} & 0 & 0 \\ & & & & 4\eta_{2323} & 0 \\ & & & & & (4\eta_{1111} - \eta_{3333}) \end{array}$ 3 coefficients	$\begin{array}{ccc} \kappa_{11} & 0 & 0 \\ 0 & \kappa_{11} & 0 \\ 0 & 0 & \kappa_{33} \end{array}$ 2 coefficients

5.6 Internal Constraint Conditions

Until now we have assumed that the body under consideration can perform all (smooth) processes if it only is exposed to external sources (force, energy supply). However, there are also bodies which perform certain deformations or process changes only if extremely large external forces are applied. For instance, the volume of a fluid body such as water changes, in general, only due to the application of a very high pressure; or a fiber reinforced material hardly undergoes length changes in the direction of the fibers, because these are practically inextensible. For the sake of simplification of the theoretical formulation one can consider that the part of the deformation which hardly arises is completely suppressed. The assumption of *volume preserving* is an example of such a simplification and states that the deformations should be restricted to *isochoric motion.* Such a process restriction is known as an *internal constraint*; it can naturally be sustained if internal constraint forces are in effect so as to maintain the constraint conditions. As in maintaining the motion of a mass point along a curve, the reactive forces acting perpendicular to the curve do no work, *the internal constraint stresses satisfying the constraint conditions perform no work either.* Such a restriction, as we will see, permits us to determine the internal constraint stresses.

In Subsect. 2.3.1, while considering the mass balance, we showed that $\dot{\rho}/\rho = -\dot{J}/J$, where $J = \det \boldsymbol{F}$, stating that the volume and density preserving are identical concepts because of the correspondence

$$\underbrace{\dot{J} = 0}_{\text{Volume preserving}} \quad \Leftrightarrow \quad \underbrace{\dot{\rho} = 0}_{\text{Density preserving}} \quad , \tag{5.6.1}$$

however, these two relations do not in general agree with the "incompressibility" statement. The latter states that the equation of state for density $\rho = \hat{\rho}(\Theta, p, \ldots)$ is independent of pressure,

$$\frac{\partial \hat{\rho}}{\partial p} = 0 \, , \quad \text{incompressibility,} \tag{5.6.2}$$

and is a thermodynamic statement (see Sect. 5.7). However, for the convenience of language, in general, the concept *incompressibility* is ordinarily regarded as *volume* or *density preserving.* We shall not follow this custom.

In the discussion of the material equations it was demonstrated that the field equations can be interpreted as defining equations for the functions of density ρ, velocity $\boldsymbol{v}$ and temperature Θ. For a material which satisfies the constraint condition of density preserving, the density does no longer form an independent variable since it is constant. The loss of this variable as an unknown field must be compensated by a new variable; this variable acts as the constraint stress, which becomes the pressure in case of the internal constraint condition of density preserving.

Let us consider constraint conditions (written as implicit equations), which depend only on the deformation gradient, viz.,

$$\phi\big(\boldsymbol{F}(\boldsymbol{X},t)\big) = 0\,, \tag{5.6.3}$$

then these are maintained by a constraint force (i.e. a stress) $\boldsymbol{Z}$. The stress tensor of the material thus obtains an additional term,

$$\boldsymbol{t} = \hat{\boldsymbol{t}}(\boldsymbol{F},\ldots) + \boldsymbol{Z}\,, \tag{5.6.4}$$

where $\hat{\boldsymbol{t}}(\boldsymbol{F},\ldots)$ is assumed to be prescribed as a constitutive variable, whereas the constraint stress must be brought into connection with the restricted deformation expressed by (5.6.3). We will now determine this constraint stress.

Differentiating the equation of the constraint condition, $\phi = 0$, with respect to time, thereby using the chain rule, one obtains

$$\frac{\partial\phi}{\partial F_{i\alpha}}\dot{F}_{i\alpha} = \operatorname{tr}\left(\frac{\partial\phi}{\partial\boldsymbol{F}}\dot{\boldsymbol{F}}^T\right) = 0\,. \tag{5.6.5}$$

Because the constraint stress does not perform any work, the related dissipation term in the energy balance must vanish. This requires

$$\operatorname{tr}(\boldsymbol{L}\boldsymbol{Z}^T) = 0\,. \tag{5.6.6}$$

Using the relations $\dot{\boldsymbol{F}} = \boldsymbol{L}\boldsymbol{F}$, $\boldsymbol{L} = \operatorname{grad}\boldsymbol{v}$, the above equation can be written as

$$\operatorname{tr}(\boldsymbol{Z}\boldsymbol{L}^T) = \operatorname{tr}\left(\boldsymbol{Z}\boldsymbol{F}^{-T}\dot{\boldsymbol{F}}^T\right) = 0\,. \tag{5.6.7}$$

Both (5.6.5) and (5.6.7) can be interpreted geometrically when the quantities $\dot{\boldsymbol{F}}$, $\partial\phi/\partial\boldsymbol{F}$ and $\boldsymbol{Z}\boldsymbol{F}^{-T}$ are interpreted as vectors of a 9-dimensional space; the trace of the product of two tensors can then be viewed as a simple scalar product in this representation. The vector $\partial\phi/\partial\boldsymbol{F}$ is then normal to the surface $\phi(\boldsymbol{F}) = 0$ and orthogonal to all vectors $\dot{\boldsymbol{F}}$; this follows from the first of both equations. From the last equation it then follows that $\boldsymbol{Z}\boldsymbol{F}^{-T}$ must also be orthogonal to all the vectors $\dot{\boldsymbol{F}}$. With this, one can conclude that both vectors must be parallel and therefore proportional to each other. Thus it follows

$$\boldsymbol{Z}\boldsymbol{F}^{-T} \parallel \frac{\partial\phi}{\partial\boldsymbol{F}} \quad\Rightarrow\quad \boldsymbol{Z} = \Lambda\frac{\partial\phi}{\partial\boldsymbol{F}}\boldsymbol{F}^T\,. \tag{5.6.8}$$

The proportionality factor (LAGRANGE *multiplier*) Λ remains undetermined and must be considered as an independent field. Note further that the constraint condition implies a material law and thus must satisfy the principle of material objectivity, $\phi(\boldsymbol{F}) = \phi(\boldsymbol{O}\boldsymbol{F})$.

Finally, we like to mention that for a body subjected to N different constraint conditions $\phi_i(\boldsymbol{F}) = 0$, $i = 1,\ldots,N$, the constraint stress must be given by

$$\boldsymbol{Z} = \sum_{i=1}^{N}\Lambda_i\frac{\partial\phi_i}{\partial\boldsymbol{F}}\boldsymbol{F}^T\,. \tag{5.6.9}$$

This immediately follows from the fact that N constraint conditions can be considered as a single constraint condition of the form $\phi(\boldsymbol{F}) = \sum_{i=1}^{N} \alpha_i \phi_i(\boldsymbol{F}) = 0$ with arbitrary coefficients α_i, $i = 1, ..., N$[16].

5.6.1 Density Preserving as a Constraint Condition

A very important constraint condition which we will frequently encounter is the assumption of *density preserving* of a material. In order to describe this condition with the help of the method just discussed, these equations must be written in terms of the deformation gradient. From $\rho = \rho_R \det \boldsymbol{F}$ for a density preserving material it directly follows that $\det \boldsymbol{F} = 1$. The constraint condition for a density preserving material is therefore

$$\phi(\boldsymbol{F}) \equiv \det \boldsymbol{F} - 1 = 0\,, \qquad \text{constraint condition for density preserving.} \tag{5.6.10}$$

According to Sect. 1.4 (see also Exercise 1.12) the derivative of the determinant of the deformation gradient with respect to the deformation gradient is given by

$$\frac{\partial \det \boldsymbol{F}}{\partial \boldsymbol{F}} = \det \boldsymbol{F}\, \boldsymbol{F}^{-T}\,. \tag{5.6.11}$$

The constraint stress for a density preserving material is then given by

$$\boldsymbol{Z} = \Lambda \frac{\partial \phi}{\partial \boldsymbol{F}} \boldsymbol{F}^T = \Lambda \frac{\partial \det \boldsymbol{F}}{\partial \boldsymbol{F}} \boldsymbol{F}^T = \Lambda \underbrace{(\det \boldsymbol{F})}_{=1} \boldsymbol{I} = \Lambda \boldsymbol{I}\,. \tag{5.6.12}$$

Here the constraint condition has been used in the last step. The constraint condition for density preserving introduces an isotropic stress tensor or an all sided pressure, ensuring that a volume change is compensated by a corresponding value of the pressure tensor in the material. One writes the LAGRANGE multiplier as $\Lambda = -p$, to indicate a pressure here. The stress tensor of a density preserving material can now be written as a combination of the

[16] Another, perhaps more stringent argumentation, is as follows: The condition of density preserving (5.6.5) must be fulfilled subject to the constraint (5.6.7). If this constraint is incorporated in (5.6.5) with the method of LAGRANGE parameters, (5.6.7) is multiplied with Λ^{-1} and the resulting equation subtracted from (5.6.5). This process yields

$$\mathrm{tr}\left(\left(\frac{\partial \phi}{\partial \boldsymbol{F}} - \Lambda^{-1} \boldsymbol{Z} \boldsymbol{F}^{-T}\right) \dot{\boldsymbol{F}}\right) = 0\,,$$

which must hold true for all $\dot{\boldsymbol{F}}$. This necessarily requires that

$$\frac{\partial \phi}{\partial \boldsymbol{F}} - \Lambda^{-1} \boldsymbol{Z} \boldsymbol{F}^{-T} = 0\,,$$

which is equivalent to (5.6.8).

pressure (the isotropic stress tensor due to the constraint condition) and the stress tensor $\boldsymbol{t}^E$, which is known as the *extra stress tensor*,

$$\boldsymbol{t} = -p\boldsymbol{I} + \boldsymbol{t}^E \quad \text{with} \quad \operatorname{tr} \boldsymbol{t}^E = 0 \,. \tag{5.6.13}$$

Here, for $\boldsymbol{t}^E$ the trace free condition has been required (i.e., $\boldsymbol{t}^E$ is a deviator), because every isotropic stress contribution can be absorbed in the (free) pressure, and the latter belongs to the independent field quantities. The trace free condition of the extra stress tensor is often advantageous as this simplifies the form of the equations.

Thus, for a density preserving isotropic elastic material the material equation (5.4.44) is replaced by

$$\boldsymbol{t} = -p\boldsymbol{I} + a_1\boldsymbol{B} + a_2\boldsymbol{B}^2 \,, \tag{5.6.14}$$

where a_1 and a_2 are only functions of $I_{\boldsymbol{B}}$ and $II_{\boldsymbol{B}}$, since $III_{\boldsymbol{B}} = J^2 = 1$, and the isotropic (spherical) part $a_0\boldsymbol{I}$ is absorbed in the constraint pressure.

5.6.2 Other Constraint Conditions

Density preserving is a kinematic constraint that is expressible as a functional equation $\phi(\boldsymbol{F}) = 0$, or $\phi(\boldsymbol{C}) = 0$. Other constraints, not related to the deformation of a body, are equally thinkable. Consider for example a body which is incompressible. We have taken the position here to call a material density preserving if the density ρ for a particle is truly constant. Alternatively we wish to call a material *incompressible* if the thermal equation of state – this is the equation expressing the density as a function of temperature and pressure $\rho = \rho(T,p)$ – does not depend upon the pressure, i.e., $\rho = \rho(T)$.

We shall see in the next subsection that the thermal equation of state is functionally derivable from another scalar function, the caloric equation of state (the HELMHOLTZ free energy or another thermodynamic potential). So, strictly, incompressibility should be viewed as an asymptotic limit of a compressible material behaviour in which the compressibility is becoming vanishingly small.

There have been attempts to view the restriction of the thermal equation of state to an equation of the form $\rho = \rho(T)$ as a thermodynamic constraint. The associated constraint variable may then be a heat flux, entropy, internal energy or a combination of these and the postulate from which the form of this variable is derived is the postulate that *the constraint variables do not produce entropy* on any thermomechanical processes that are realizable with their presence.

Situations that one may encounter are e.g. as follows: (i) In a fibre reinforced material the fibres may be inextensible under an applied force but encounter an elongation under variation of the temperature. (ii) Similarly, fibres may be ideally heat conducting, whilst the matrix material may act as a thermal insulator. The unworkable hypothesis that was used to derive the form of the constraint stress for a density preserving material does no longer serve as a universal criterion in these more general situations because in an inextensible fibre the stress acting in the fibre may well perform work when the fibre experiences a thermal expansion.

These more general constraint conditions are necessarily tied to the entropy principle which we have not yet touched upon. We shall not go any deeper into them for reasons of space, the interested reader may, however, consult GREEN et al. [87], GURTIN & PODIO-GIUDUGLI [90] and ALTS [13].

5.7 Entropy Principle

The material equations experience further restrictions via the second law of thermodynamics. From the balance equation of entropy it follows that the inequality (2.3.50) must be satisfied by all admissible processes. This implies that every permissible choice of material equations specifies a system of field equations (balance for mass, momentum, energy plus material equations), of which the solutions must conform with the second law, i.e., this inequality must be fulfilled. We shall now demonstrate how the *entropy principle* is analyzed for special material equations. For this purpose we first use the entropy inequality in the form of the CLAUSIUS–DUHEM inequality and afterwards present the more general entropy principle due to MÜLLER. For the former we set

$$\eta = \frac{\mathfrak{r}}{\Theta}\,, \quad \phi = \frac{\boldsymbol{q}}{\Theta} \quad \text{and} \quad \boldsymbol{\Phi} = \frac{\boldsymbol{Q}}{\Theta}\,, \tag{5.7.1}$$

as specified in Tables 3.2 and 3.3.

5.7.1 Viscous Heat Conducting Compressible Fluid

The material equations possess the general form

$$\Psi = \hat{\Psi}(\rho, \boldsymbol{D}, \Theta, \operatorname{grad}\Theta)\,, \quad \Psi \in \{\varepsilon, s, \boldsymbol{q}, \boldsymbol{t}\}\,. \tag{5.7.2}$$

The balance equation of energy and the entropy inequality,

$$\begin{aligned} \rho\dot{\varepsilon} &= -\mathrm{div}\boldsymbol{q} + \mathrm{tr}(\boldsymbol{tD}) + \rho\mathfrak{r}\,, \\ \rho\dot{s} + \mathrm{div}\left(\frac{\boldsymbol{q}}{\Theta}\right) &- \rho\frac{\mathfrak{r}}{\Theta} \geq 0\,, \end{aligned} \tag{5.7.3}$$

can be combined by eliminating the radiation to yield

$$\rho\left(\Theta\dot{s} - \dot{\varepsilon}\right) + \mathrm{tr}(\boldsymbol{tD}) - \frac{\boldsymbol{q}\cdot\operatorname{grad}\Theta}{\Theta} \geq 0\,. \tag{5.7.4}$$

The expression on the left-hand side is the specific entropy production multiplied by the (positive) absolute temperature. Considering the dependency of the material equations on the density, the stretching, the temperature and the temperature gradient, the time derivative of Ψ, following the chain rule of differentiation, is given as

$$\dot{\Psi} = \frac{\partial \hat{\Psi}}{\partial \rho}\dot{\rho} + \operatorname{tr}\big(\frac{\partial \hat{\Psi}}{\partial \boldsymbol{D}}\dot{\boldsymbol{D}}^T\big) + \frac{\partial \hat{\Psi}}{\partial (\operatorname{grad}\Theta)} \cdot (\operatorname{grad}\Theta)^{\cdot} + \frac{\partial \hat{\Psi}}{\partial \Theta}\dot{\Theta} . \tag{5.7.5}$$

If the same operation is applied for $\dot{s}$ and $\dot{\varepsilon}$ in (5.7.4) and substituted in the resulting inequality by using the mass balance,

$$\dot{\rho} = -\rho \operatorname{div}\boldsymbol{v} = -\rho \operatorname{tr}\boldsymbol{D} = -\rho \boldsymbol{I} \cdot \boldsymbol{D} , \tag{5.7.6}$$

we obtain the inequality in the form

$$\begin{aligned} &\rho\Big(\Theta\frac{\partial s}{\partial \boldsymbol{D}} - \frac{\partial \varepsilon}{\partial \boldsymbol{D}}\Big)\cdot\dot{\boldsymbol{D}} + \rho\Big(\Theta\frac{\partial s}{\partial \Theta} - \frac{\partial \varepsilon}{\partial \Theta}\Big)\dot{\Theta} \\ &+\rho\Big(\Theta\frac{\partial s}{\partial(\operatorname{grad}\Theta)} - \frac{\partial \varepsilon}{\partial(\operatorname{grad}\Theta)}\Big)\cdot(\operatorname{grad}\Theta)^{\cdot} \\ &+\Big[-\rho^2\Big(\Theta\frac{\partial s}{\partial \rho} - \frac{\partial \varepsilon}{\partial \rho}\Big)\boldsymbol{I} + \boldsymbol{t}\Big]\cdot\boldsymbol{D} - \frac{\boldsymbol{q}\cdot\operatorname{grad}\Theta}{\Theta} \geq 0 , \end{aligned} \tag{5.7.7}$$

which must be satisfied by all admissible thermodynamic processes; these are those which satisfy the balance equations of mass, momentum and energy as well as the material equations. The mass balance has just been considered by substituting (5.7.6), and the momentum and the energy balances must not additionally be accounted for because in every process there exist external force and radiation fields which themselves can be selected such that these two equations are identically satisfied. It is also worthwhile to remark that a process in which for a particle with the independent constitutive variables ρ, $\boldsymbol{D}$, Θ, $\operatorname{grad}\Theta$, the time derivatives $\dot{\boldsymbol{D}}$, $\dot{\Theta}$ as well as $(\operatorname{grad}\Theta)^{\cdot}$ can be assigned to have arbitrary values, and the emerging process is an admissible thermodynamic process. Finally, it is pointed out that inequality (5.7.7) is linear in the variables $\dot{\boldsymbol{D}}$, $\dot{\Theta}$ and $(\operatorname{grad}\Theta)^{\cdot}$.

This form of exploiting the second law restricts considerations to the analysis of *open systems*, in which external, arbitrary supply terms can appear in the momentum and energy laws, and both of these balance equations do not affect the evaluation of the entropy principle i.e., the CLAUSIUS–DUHEM inequality. The assumption of arbitrariness of the external supply term, that is mathematically so convenient, may physically be questionable by the argument that the "physical world" may not be so general as to allow arbitrarily large or small external source terms. It is here, where the methods of exploitation of the entropy principle deviates among different authors. Nevertheless, the form of this analysis was first proposed by COLEMAN & NOLL [50] and since then dominates the scene in rational thermodynamics. As one easily recognizes from the above calculations, such a procedure presupposes knowledge of the absolute temperature. It is simply taken over from classical thermodynamics of simple systems. Likewise, the entropy flux and the entropy production are taken from the DUHEM and TRUESDELL relations

$$\text{entropy flux} = \frac{\text{heat flux}}{\text{absolute temperature}}\,, \tag{5.7.8}$$

$$\text{entropy supply} = \frac{\text{energy supply}}{\text{absolute temperature}}\,, \tag{5.7.9}$$

however, their validity is not automatically ascertained. We will see in the next section, how a new entropy principle is obtained through modification of the second law.

Let us collect the individual variables vectorially as follows:

$$\begin{aligned}
\boldsymbol{\alpha} &:= \left(\dot{\boldsymbol{D}}, \dot{\Theta}, (\operatorname{grad}\Theta)^{\bullet}\right)\,,\\
\boldsymbol{b} &:= \Bigg(\rho\left(\Theta\frac{\partial s}{\partial \boldsymbol{D}} - \frac{\partial \varepsilon}{\partial \boldsymbol{D}}\right), \rho\left(\Theta\frac{\partial s}{\partial \Theta} - \frac{\partial \varepsilon}{\partial \Theta}\right),\\
&\qquad \rho\left(\Theta\frac{\partial s}{\partial(\operatorname{grad}\Theta)} - \frac{\partial \varepsilon}{\partial(\operatorname{grad}\Theta)}\right)\Bigg)\,,\\
\Gamma &:= \left[-\rho^2\left(\Theta\frac{\partial s}{\partial \rho} - \frac{\partial \varepsilon}{\partial \rho}\right)\boldsymbol{I} + \boldsymbol{t}\right]\cdot\boldsymbol{D} - \frac{\boldsymbol{q}\cdot\operatorname{grad}\Theta}{\Theta}\,.
\end{aligned} \tag{5.7.10}$$

Then, equation (5.7.7) can also be written as

$$\boldsymbol{b}\cdot\boldsymbol{\alpha} + \Gamma \geq 0\,, \tag{5.7.11}$$

where $\boldsymbol{b}$ and Γ do not depend on $\boldsymbol{\alpha}$. Then we have the following

Theorem *Let $\boldsymbol{\alpha}$ and $\boldsymbol{b}$ be vectors in the n-dimensional space $\mathbb{R}^n$, and Γ a scalar in $\mathbb{R}$; and let $\boldsymbol{b}$ and Γ be independent of the components of $\boldsymbol{\alpha}$. Then*

$$\boldsymbol{b}\cdot\boldsymbol{\alpha} + \Gamma \geq 0\,, \quad \forall\, \boldsymbol{\alpha}\in\mathbb{R}^n$$

necessarily implies

$$\boldsymbol{b} = \boldsymbol{0}\,, \quad \Gamma \geq 0\,. \tag{5.7.12}$$

■

For the **proof**, note that the conditions (5.7.12) are sufficient to satisfy (5.7.11). They are, however, also necessary. Indeed, for $\boldsymbol{\alpha} = (0,\dots,\alpha_j,\dots,0)$, from (5.7.11) one has

$$\alpha_j b_j + \Gamma \geq 0 \quad (\text{no summation over } j). \tag{5.7.13}$$

Choosing $b_j \neq 0$ and $\alpha_j = -(\Gamma+\epsilon)/b_j$ with $\epsilon > 0$, (5.7.13) implies $-\epsilon > 0$, which is a contradiction; thus only $b_j = 0$ can satisfy the condition. Since this holds for all $j = 1, ..., n$, so (5.7.12) is proved.

With the above identification for $\boldsymbol{b}$ one obtains from the condition $\boldsymbol{b} = \boldsymbol{0}$

$$\begin{aligned} \frac{\partial \varepsilon}{\partial \boldsymbol{D}} &= \Theta \frac{\partial s}{\partial \boldsymbol{D}} , \\ \frac{\partial \varepsilon}{\partial \Theta} &= \Theta \frac{\partial s}{\partial \Theta} , \\ \frac{\partial \varepsilon}{\partial (\operatorname{grad} \Theta)} &= \Theta \frac{\partial s}{\partial (\operatorname{grad} \Theta)} . \end{aligned} \tag{5.7.14}$$

These equations must hold as identities. If one forms the mixed second derivatives of the internal energy and considers that the interchange of the sequence of differentiation of the functions ε and s must be immaterial for these functions to be unique, then

$$\frac{\partial^2 \varepsilon}{\partial \Theta\, \partial \boldsymbol{D}} \overset{(1)}{=} \frac{\partial s}{\partial \boldsymbol{D}} + \Theta \frac{\partial^2 s}{\partial \Theta\, \partial \boldsymbol{D}} = \frac{\partial^2 \varepsilon}{\partial \boldsymbol{D}\, \partial \Theta} \overset{(2)}{=} \Theta \frac{\partial^2 s}{\partial \boldsymbol{D}\, \partial \Theta} , \tag{5.7.15}$$

and one obtains

$$\frac{\partial s}{\partial \boldsymbol{D}} = \boldsymbol{0} . \tag{5.7.16}$$

In the above $\overset{(1)}{=}$ and $\overset{(2)}{=}$ mean that $(5.7.14)_1$ and $(5.7.14)_2$ are used, respectively. Analogously, forming mixed derivatives with respect to Θ and $\operatorname{grad} \Theta$ we deduce from (5.7.14)

$$\frac{\partial s}{\partial (\operatorname{grad} \Theta)} = \boldsymbol{0} . \tag{5.7.17}$$

In this way, the following relations are obtained

$$\begin{aligned} \frac{\partial s}{\partial \boldsymbol{D}} &= \boldsymbol{0} , & \frac{\partial s}{\partial (\operatorname{grad} \Theta)} &= \boldsymbol{0} , \\ \frac{\partial \varepsilon}{\partial \boldsymbol{D}} &= \boldsymbol{0} , & \frac{\partial \varepsilon}{\partial (\operatorname{grad} \Theta)} &= \boldsymbol{0} . \end{aligned} \tag{5.7.18}$$

For a viscous heat conducting fluid, the internal energy and the entropy can neither be functions of the stretching tensor nor the temperature gradient. Thus the entropy principle introduces the three conditions

$$\varepsilon = \hat{\varepsilon}(\rho, \Theta) , \quad s = \hat{s}(\rho, \Theta) , \quad \frac{\partial \varepsilon}{\partial \Theta} = \Theta \frac{\partial s}{\partial \Theta} . \tag{5.7.19}$$

Now there still remains the residual inequality $\Gamma \geq 0$ in (5.7.12), or

$$\Gamma = \left[-\rho^2 \left(\Theta \frac{\partial s}{\partial \rho} - \frac{\partial \varepsilon}{\partial \rho} \right) \boldsymbol{I} + \boldsymbol{t} \right] \cdot \boldsymbol{D} - \frac{\boldsymbol{q} \cdot \operatorname{grad} \Theta}{\Theta} \geq 0 . \tag{5.7.20}$$

It is to be emphasized that in addition to ρ and Θ, Γ depends on $\boldsymbol{D}$ and $\operatorname{grad} \Theta$. From the inequality $\Gamma \geq 0$, we then can derive further restrictions. For this we define thermodynamic equilibrium first.

Thermodynamic equilibrium *is a thermodynamic process for which the temperature and velocity are uniformly distributed, i.e.,*

$$\operatorname{grad}\Theta = \mathbf{0} \quad \textit{and} \quad \boldsymbol{D} = \mathbf{0} \quad \Longleftrightarrow \quad \textit{equilibrium}\,. \tag{5.7.21}$$

■

This implies that the inequality $\Gamma \geq 0$ yields

$$\Gamma_{|E} = 0 \quad \Rightarrow \quad \Gamma_{|E} = \text{minimum} \tag{5.7.22}$$

in equilibrium. The index $|_E$ indicates this equilibrium. In thermodynamic equilibrium the quantity Γ assumes its minimum and the value of this minimum is zero. This must naturally be so, for Γ is the entropy production (multiplied with the positive temperature) which, as expected, vanishes, when constant, uniform processes are in effect[17]. The minimum property can now be expressed mathematically; because $\Gamma = \Gamma(\operatorname{grad}\Theta, \boldsymbol{D}, \ldots)$ is a function of the independent variables 'temperature gradient and stretching tensor', the requirement of the existence of a minimum in equilibrium leads to the following necessary conditions (recall the definition of an extreme value of a function $f(x, y, \ldots)$ of several variables),

$$\left(\frac{\partial \Gamma}{\partial \boldsymbol{D}}\right)_{|E} = \mathbf{0}\,, \quad \left(\frac{\partial \Gamma}{\partial(\operatorname{grad}\Theta)}\right)_{|E} = \mathbf{0}\,,$$

$$\begin{pmatrix} \dfrac{\partial^2 \Gamma}{\partial \boldsymbol{D}\,\partial \boldsymbol{D}} & \dfrac{\partial^2 \Gamma}{\partial \boldsymbol{D}\,\partial(\operatorname{grad}\Theta)} \\ \dfrac{\partial^2 \Gamma}{\partial(\operatorname{grad}\Theta)\,\partial \boldsymbol{D}} & \dfrac{\partial^2 \Gamma}{\partial(\operatorname{grad}\Theta)\,\partial(\operatorname{grad}\Theta)} \end{pmatrix}_{|E} \quad \text{is positive semi-definite.} \tag{5.7.23}$$

It can be easily shown that the first two conditions yield

$$\boldsymbol{t}_{|E} = \rho^2\left(\Theta\frac{\partial s}{\partial \rho} - \frac{\partial \varepsilon}{\partial \rho}\right)\boldsymbol{I} =: -p\boldsymbol{I}\,, \qquad \boldsymbol{q}_{|E} = \mathbf{0}\,, \tag{5.7.24}$$

where the pressure p is used as an abbreviation for the scalar expression in front of the unit tensor. The equilibrium stress in a viscous, heat conducting fluid is therefore isotropic and determined by the entropy as well as the internal energy, and the equilibrium heat flux vector vanishes.

Combining these results with the condition (5.7.19) implies

$$\begin{gathered} s = \hat{s}(\rho, \Theta)\,, \quad \varepsilon = \hat{\varepsilon}(\rho, \Theta)\,, \\ \frac{\partial s}{\partial \rho} = \frac{1}{\Theta}\left(\frac{\partial \varepsilon}{\partial \rho} - \frac{p}{\rho^2}\right), \quad \frac{\partial s}{\partial \Theta} = \frac{1}{\Theta}\frac{\partial \varepsilon}{\partial \Theta}\,; \end{gathered} \tag{5.7.25}$$

[17] Conversely, the thermodynamic equilibrium can also be defined by requiring $\Gamma = 0$ which leads to (5.7.21) and (5.7.22).

so one obtains the total differential of the internal energy by suitably combining both derivatives of the entropy as follows

$$\mathrm{d}\varepsilon = \frac{\partial \varepsilon}{\partial \rho}\mathrm{d}\rho + \frac{\partial \varepsilon}{\partial \Theta}\mathrm{d}\Theta = \Theta\,\mathrm{d}s + \frac{p}{\rho^2}\,\mathrm{d}\rho \tag{5.7.26}$$

or

$$\mathrm{d}s = \frac{1}{\Theta}\left(\mathrm{d}\varepsilon + p\,\mathrm{d}\left(\frac{1}{\rho}\right)\right) . \tag{5.7.27}$$

This is called the GIBBS *equation.* Even though this relation is derived for thermostatic equilibrium it is equally valid for *all* thermodynamic processes; this follows from the fact that ε and s are, in general, i.e., for all admissible processes only functions of ρ and Θ. The GIBBS equation is thus a generally valid result[18]. Equation $(5.7.25)_1$ can be summarized as the defining equation for the pressure when the constitutive equations for the internal energy and the entropy are given. It is customary to call the material equation for the pressure, $p = \hat{p}(\rho, \Theta)$, the *thermal equation of state* , and those for the internal energy and the entropy as *caloric equations of state.* The second law in the form of the CLAUSIUS–DUHEM inequality therefore also implies that the material equations are not prescribed independently of each other; all the more, the caloric equations of state determine the thermal equation of state. This can be better understood by introducing the LEGENDRE *transformation*

$$\Psi_\varepsilon := \varepsilon - \Theta s \tag{5.7.28}$$

by which one introduces the HELMHOLTZ *free energy.* Because of (5.7.25) s and ε are functions of ρ and Θ, and so is the HELMHOLTZ free energy, $\Psi_\varepsilon = \hat{\Psi}_\varepsilon(\rho, \Theta)$. Eliminating ε from (5.7.25) and (5.7.26) then leads to

$$s = -\frac{\partial \hat{\Psi}_\varepsilon}{\partial \Theta} , \quad p = \rho^2 \frac{\partial \hat{\Psi}_\varepsilon}{\partial \rho} . \tag{5.7.29}$$

Defining the caloric equation of state for the HELMHOLTZ free energy thus suffices to determine the specific entropy and the pressure. Incidentally, with (5.7.29) the potential character of the HELMHOLTZ free energy is clear as both these variables $(-s,\, p/\rho^2)$ are determinable as the gradient of the free energy with respect to Θ and ρ:

$$(-s,\, p/\rho^2) = \nabla_{\Theta,\rho}\Psi_\varepsilon(\Theta,\, \rho) . \tag{5.7.30}$$

In this respect the variables $(-s,\, p/\rho^2)$ and $(\Theta,\, \rho)$ are denoted as *canonical.*

In this connection we can now deepen the concept of incompressibility. In Sect. 5.6 on internal constraints a material was called incompressible when

[18] This is a typical property of the entropy principle in the CLAUSIUS–DUHEM form. In other entropy principles different results may be obtained, see e.g. HUTTER [104].

the density is not a function of the pressure, $\partial\rho/\partial p = 0$; i.e., when $\partial p/\partial\rho$ becomes singular. The theory presented here can not describe a material which has a thermal equation of state given by $\rho = \hat{\rho}(\Theta)$. We will come back to this point later on.

Now, there still remain the conditions that describe *thermodynamic non-equilibrium conditions.* To this end we introduce with

$$\boldsymbol{t}^E = \boldsymbol{t} + p\boldsymbol{I} \ , \quad \boldsymbol{t}^E{}_{|E} = \boldsymbol{0} \ , \tag{5.7.31}$$

the extra stress tensor, which describes the deviation from the equilibrium stress, and with which the residual inequality (5.7.20) takes the form

$$\Gamma = \boldsymbol{t}^E \cdot \boldsymbol{D} - \frac{\boldsymbol{q} \cdot \operatorname{grad}\Theta}{\Theta} \geq 0 \ . \tag{5.7.32}$$

Let us assume – as a simplification – that $\boldsymbol{t}^E$ can not depend on the temperature gradient and $\boldsymbol{q}$ cannot depend on the stretching tensor:

$$\boldsymbol{t}^E = \hat{\boldsymbol{t}}^E(\rho, \Theta, \boldsymbol{D}) \ , \quad \boldsymbol{q} = \hat{\boldsymbol{q}}(\rho, \Theta, \operatorname{grad}\Theta) \ . \tag{5.7.33}$$

The most general isotropic representations for these functions – see (5.4.11) and (5.4.14) – are

$$\begin{aligned} &\boldsymbol{t}^E = \alpha_1 \boldsymbol{I} + \alpha_2 \boldsymbol{D} + \alpha_3 \boldsymbol{D}^2 \ , \quad \boldsymbol{q} = -\lambda \operatorname{grad}\Theta \ , \\ &\alpha_i = \hat{\alpha}_i(\rho, \Theta, I_{\boldsymbol{D}}, II_{\boldsymbol{D}}, III_{\boldsymbol{D}}) \ , \quad i = 1,2,3 \ , \\ &\alpha_1(\rho, \Theta, \boldsymbol{0}, \boldsymbol{0}, \boldsymbol{0}) = \boldsymbol{0} \ , \\ &\lambda = \hat{\lambda}(\rho, \Theta, \|\operatorname{grad}\Theta\|) \ . \end{aligned} \tag{5.7.34}$$

Apart from the restrictions (5.7.33) these relations represent the most general constitutive equations for the extra stress tensor and the heat flux vector of a viscous, heat conducting, compressible fluid. Naturally, the coefficient functions α_i are restricted through the statement $(5.7.23)_3$.

The satisfaction of (5.7.32) is complicated for the general non-linear material laws (5.7.34), so let us describe the case that $\boldsymbol{t}^E$ and $\boldsymbol{q}$ depend linearly on $\boldsymbol{D}$ and $\operatorname{grad}\Theta$. Then, see (5.4.35),

$$\boldsymbol{t}^E = \kappa I_{\boldsymbol{D}} \boldsymbol{I} + 2\mu \boldsymbol{D}' \ , \quad \boldsymbol{q} = -\lambda \operatorname{grad}\Theta \ , \tag{5.7.35}$$

hold, in which κ, μ and λ are the bulk and shear viscosities and the thermal conductivity which are functions of ρ and Θ. With (5.7.35) the residual inequality becomes

$$\Gamma = \kappa I_{\boldsymbol{D}}^2 + 2\mu \boldsymbol{D}' \cdot \boldsymbol{D}' + \lambda \frac{\|\operatorname{grad}\Theta\|^2}{\Theta} \geq 0 \ . \tag{5.7.36}$$

If one defines

$$x := \sqrt{2} I_{\boldsymbol{D}} \,, \quad y := \sqrt{4 \boldsymbol{D}' \cdot \boldsymbol{D}'} \,, \quad z := \sqrt{\frac{2}{\Theta}} \| \operatorname{grad} \Theta \| \,, \tag{5.7.37}$$

one obtains

$$\Gamma = \kappa \frac{x^2}{2} + \mu \frac{y^2}{2} + \lambda \frac{z^2}{2} \,. \tag{5.7.38}$$

The dissipation Γ is a quadratic function of three scalar variables, x, y and z with coefficients which themselves can depend on the density and the temperature. The thermodynamic equilibrium is indicated by $x = y = z = 0$, so that $\Gamma_{|E} = \min$ is given by the statement

$$\frac{\partial \Gamma}{\partial x}_{|E} = 0 \,, \quad \frac{\partial \Gamma}{\partial y}_{|E} = 0 \,, \quad \frac{\partial \Gamma}{\partial z}_{|E} = 0 \,,$$

$$\begin{pmatrix} \dfrac{\partial^2 \Gamma}{\partial x^2} & 0 & 0 \\ 0 & \dfrac{\partial^2 \Gamma}{\partial y^2} & 0 \\ 0 & 0 & \dfrac{\partial^2 \Gamma}{\partial z^2} \end{pmatrix} = \begin{pmatrix} \kappa & 0 & 0 \\ 0 & \mu & 0 \\ 0 & 0 & \lambda \end{pmatrix}_{|E} \text{ is positive semi-definite,} \tag{5.7.39}$$

which can only be fulfilled when

$$\kappa = \hat{\kappa}(\rho, \Theta) \geq 0 \,, \quad \mu = \hat{\mu}(\rho, \Theta) \geq 0 \,, \quad \lambda = \hat{\lambda}(\rho, \Theta) \geq 0 \,. \tag{5.7.40}$$

With this, in a linear, heat conducting fluid the bulk and shear viscosities as well as the coefficient of thermal conductivity are compatible with the second law, if these are non-negative functions of density and temperature.

5.7.2 Viscous, Heat Conducting and Density Preserving Fluids

An essential difference for the application of the entropy principle to a special material is realized when one demands additional constraint conditions. Here we shall briefly demonstrate the procedure for a viscous heat conducting density preserving fluid. The material equations now no longer depend on the density,

$$\Psi = \hat{\Psi}(\boldsymbol{D}, \Theta, \operatorname{grad} \Theta) \,, \quad \Psi \in \{\varepsilon, s, \boldsymbol{q}, \boldsymbol{t}\} \,. \tag{5.7.41}$$

Indeed the material must satisfy the constraint condition

$$\rho = \text{constant} \quad \Rightarrow \quad \operatorname{div} \boldsymbol{v} = 0 \quad \text{or} \quad \operatorname{tr} \boldsymbol{D} = 0 \,. \tag{5.7.42}$$

The additional assumption of density preserving as a constraint condition requires that this is to be considered as an auxiliary condition to satisfy the entropy inequality. This is accomplished via a LAGRANGE multiplier Λ, in which $\Lambda \operatorname{tr} \boldsymbol{D}$ is added in the entropy inequality. The entropy inequality (5.7.4) becomes, since the density dependency falls out,

$$\begin{aligned}
&\rho\Big(\Theta\frac{\partial s}{\partial \boldsymbol{D}}-\frac{\partial \varepsilon}{\partial \boldsymbol{D}}\Big)\cdot\dot{\boldsymbol{D}}+\rho\Big(\Theta\frac{\partial s}{\partial \Theta}-\frac{\partial \varepsilon}{\partial \Theta}\Big)\dot{\Theta}\\
&+\rho\Big(\Theta\frac{\partial s}{\partial(\operatorname{grad}\Theta)}-\frac{\partial \varepsilon}{\partial(\operatorname{grad}\Theta)}\Big)(\operatorname{grad}\Theta)^{\cdot}\\
&+\Big(\boldsymbol{Z}+\boldsymbol{t}^E\Big)\cdot\boldsymbol{D}-\frac{\boldsymbol{q}\cdot\operatorname{grad}\Theta}{\Theta}+\Lambda\operatorname{tr}\boldsymbol{D}\geq 0\,,
\end{aligned} \tag{5.7.43}$$

in which

$$\boldsymbol{t}=\Big(\boldsymbol{Z}+\boldsymbol{t}^E\Big)\,,\quad \operatorname{tr}\boldsymbol{t}^E=0 \tag{5.7.44}$$

and a free constraint stress $\boldsymbol{Z}$ is incorporated; the extra stress tensor $\boldsymbol{t}^E$ is taken as a deviator because its spherical part is absorbed in $\boldsymbol{Z}$.

The meaning of the incorporation of internal constraint conditions in the entropy inequality is expressed by the fact that the extended inequality (5.7.43), in contrast to the original inequality, should be satisfied for arbitrary deformations, also those, which do not obey the internal constraint conditions; thus if (5.7.43) is to hold for arbitrary stretching tensors, then it must do so also for those with $\operatorname{tr}\boldsymbol{D}\neq 0$. Since the inequality is explicitly linear in the variables $\dot{\boldsymbol{D}}$, $\dot{\Theta}$ and $(\operatorname{grad}\Theta)^{\cdot}$, which can take arbitrary values, we follow the same argumentation chain as in the last section and deduce

$$s=\hat{s}(\Theta)\,,\quad \varepsilon=\hat{\varepsilon}(\Theta)\,,\quad d\varepsilon=\Theta\,ds\,. \tag{5.7.45}$$

For fulfilling the residual inequality

$$\Big(\boldsymbol{Z}+\boldsymbol{t}^E\Big)\cdot\boldsymbol{D}-\frac{\boldsymbol{q}\cdot\operatorname{grad}\Theta}{\Theta}+\Lambda\operatorname{tr}\boldsymbol{D}\geq 0$$

it is meaningful to decompose the constraint stress in its deviatoric part $\boldsymbol{Z}'$ and its isotropic part as follows

$$\boldsymbol{Z}=\boldsymbol{Z}'+\tfrac{1}{3}(\operatorname{tr}\boldsymbol{Z})\,\boldsymbol{I}\,,\quad \tfrac{1}{3}\operatorname{tr}\boldsymbol{Z}=:-p\,. \tag{5.7.46}$$

With this the residual inequality reduces to

$$\Gamma=\boldsymbol{Z}'\cdot\boldsymbol{D}'+\boldsymbol{t}^E\cdot\boldsymbol{D}'+(-p+\Lambda)\operatorname{tr}\boldsymbol{D}-\frac{\boldsymbol{q}\cdot\operatorname{grad}\Theta}{\Theta}\geq 0\,. \tag{5.7.47}$$

This inequality can only be satisfied for arbitrary $\operatorname{tr}\boldsymbol{D}$ when the LAGRANGE multiplier equals the pressure

$$\Lambda=p\,. \tag{5.7.48}$$

When the thermodynamic equilibrium is again defined by $\boldsymbol{D}_{|E}=\boldsymbol{0}$ and $\operatorname{grad}\Theta_{|E}=\boldsymbol{0}$, then $(5.7.23)_{1,2}$ imply

$$\frac{\partial \Gamma}{\partial \boldsymbol{D}'}_{|E} = \boldsymbol{Z}' + \boldsymbol{t}^E{}_{|E} = \boldsymbol{0}\,, \qquad \frac{\partial \Gamma}{\partial\, \operatorname{grad} \Theta}_{|E} = \boldsymbol{q}_{|E} = \boldsymbol{0}\,. \tag{5.7.49}$$

Now $\boldsymbol{Z}'$ is arbitrarily assignable, $\boldsymbol{t}^E$, however, is a constitutive variable; i.e., the first of relations (5.7.49) can only be meaningful when

$$\boldsymbol{Z}' = \boldsymbol{0} \tag{5.7.50}$$

is chosen. As in Sect. 5.6 the result can be expressed in a different way such that the constraint stress associated with the density preserving is the pressure. With (5.7.50), (5.7.49) becomes

$$\boldsymbol{t}^E{}_{|E} = \boldsymbol{0} \quad \text{and} \quad \boldsymbol{q}_{|E} = \boldsymbol{0}\,. \tag{5.7.51}$$

Specifying the extra stress tensor and the heat flux vector as isotropic functions of the form $\boldsymbol{t}^E = \hat{\boldsymbol{t}}^E(\Theta, \boldsymbol{D})$ and $\boldsymbol{q} = \hat{\boldsymbol{q}}(\Theta, \operatorname{grad}\Theta)$ yields

$$\begin{aligned} &\boldsymbol{t}^E = \alpha_2 \boldsymbol{D} + \alpha_3 \left(\boldsymbol{D}^2 + \tfrac{2}{3} II_{\boldsymbol{D}} \boldsymbol{I}\right)\,, \quad \boldsymbol{q} = -\lambda \operatorname{grad}\Theta\,, \\ &\alpha_i = \hat{\alpha}_i(\Theta, II_{\boldsymbol{D}}, III_{\boldsymbol{D}})\,, \quad i = 2, 3\,, \quad \lambda = \hat{\lambda}(\Theta, \|\operatorname{grad}\Theta\|)\,, \end{aligned} \tag{5.7.52}$$

or, in the linear case with $\alpha_2 = 2\mu$ and $\boldsymbol{D} = \boldsymbol{D}'$ (here div $\boldsymbol{v} = 0$),

$$\boldsymbol{t}^E = 2\mu \boldsymbol{D}'\,, \quad \boldsymbol{q} = -\lambda \operatorname{grad}\Theta \tag{5.7.53}$$

with

$$\mu = \hat{\mu}(\Theta) \geq 0\,, \quad \lambda = \hat{\lambda}(\Theta) \geq 0\,. \tag{5.7.54}$$

5.7.3 Pressure and Extra Stress as Independent Variables

It is often reasonable in a heat conducting, viscous, compressible fluid to replace the pressure and the extra stress tensor as independent constitutive variables by the density and the stretching tensor and thus to choose

$$\Psi = \hat{\Psi}(p, \Theta, \boldsymbol{t}^E, \operatorname{grad}\Theta)\,, \quad \Psi \in \left\{\frac{1}{\rho},\ \varepsilon,\ s,\ \boldsymbol{D},\ \boldsymbol{q}\right\}\,. \tag{5.7.55}$$

Dependent material quantities are now the specific volume $1/\rho$, the internal energy ε, the entropy s, the stretching tensor $\boldsymbol{D}$ and the heat flux vector $\boldsymbol{q}$.

One way to derive the material equations in such cases is to start with the HELMHOLTZ free energy $\Psi_\varepsilon = \hat{\Psi}_\varepsilon(\rho, \Theta)$ and to obtain the result

$$s = -\frac{\partial \hat{\Psi}_\varepsilon}{\partial \Theta}\,, \quad p = \rho^2 \frac{\partial \hat{\Psi}_\varepsilon}{\partial \rho}\,. \tag{5.7.56}$$

Using the LEGENDRE transformation

$$\psi_h = \Psi_\varepsilon + \frac{p}{\rho} = \varepsilon - \Theta s + \frac{p}{\rho} \tag{5.7.57}$$

and thus introducing the *free enthalpy* or the GIBBS *free energy* ψ_h, yields the total differential of ψ_h as follows

$$\begin{aligned} d\psi_h &= d\Psi_\varepsilon + \frac{1}{\rho}dp - \frac{p}{\rho^2}d\rho \\ &= \frac{\partial\hat{\Psi}_\varepsilon}{\partial\rho}d\rho + \frac{\partial\hat{\Psi}_\varepsilon}{\partial\Theta}d\Theta + \frac{1}{\rho}dp - \frac{p}{\rho^2}d\rho \\ &\overset{(5.7.56)}{=} \frac{1}{\rho}dp - s d\Theta = \frac{\partial\hat{\psi}_h}{\partial p}dp + \frac{\partial\hat{\psi}_h}{\partial\Theta}d\Theta\,. \end{aligned} \tag{5.7.58}$$

The free enthalpy ψ_h is thus a function of p and Θ, $\psi_h = \hat{\psi}_h(p,\Theta)$. Furthermore, the relations

$$\frac{1}{\rho} = \frac{\partial\hat{\psi}_h}{\partial p}\,, \quad s = -\frac{\partial\hat{\psi}_h}{\partial\Theta} \quad \Longrightarrow \quad \left(\frac{1}{\rho}, -s\right) = \nabla_{p,\Theta}\,\hat{\psi}_h(p,\Theta) \tag{5.7.59}$$

must hold. The caloric equation of state is thus the equation for the GIBBS free energy, which lets the entropy s and the specific volume $1/\rho$ be determined by differentiating with respect to temperature and the pressure. It follows that $(1/\rho, -s)$ and (p,Θ) are canonical variables.

The entropy inequality (5.7.4), which with (5.7.57) and $\boldsymbol{t} = -p\boldsymbol{I} + \boldsymbol{t}^E$ takes the form

$$\begin{aligned} &\left(1 - \rho\frac{\partial\hat{\psi}_h}{\partial p}\right)\dot{p} - \rho\left(\frac{\partial\hat{\psi}_h}{\partial\Theta} + s\right)\dot{\Theta} - \rho\frac{\partial\hat{\psi}_h}{\partial\boldsymbol{t}^E}\cdot(\boldsymbol{t}^E)^{\cdot} \\ &\quad - \rho\frac{\partial\hat{\psi}_h}{\partial\,\mathrm{grad}\,\Theta}\cdot(\mathrm{grad}\,\Theta)^{\cdot} + \boldsymbol{t}^E\cdot\boldsymbol{D} - \frac{\boldsymbol{q}\cdot\mathrm{grad}\,\Theta}{\Theta} \geq 0\,, \end{aligned} \tag{5.7.60}$$

and reduces, in view of (5.7.59), to the statement

$$\Gamma = \boldsymbol{t}^E\cdot\boldsymbol{D} - \frac{\boldsymbol{q}\cdot\mathrm{grad}\,\Theta}{\Theta} \geq 0\,. \tag{5.7.61}$$

If one defines thermodynamic equilibrium as a process for which no entropy is produced, then $\boldsymbol{t}^E{}_{|E} = \boldsymbol{0}$ and $\mathrm{grad}\,\Theta_{|E} = \boldsymbol{0}$ must hold, such that

$$\Gamma_{|E} = 0 \quad \Rightarrow \quad \Gamma_{|E} = \text{minimum.} \tag{5.7.62}$$

Necessary conditions for this are

$$\left(\frac{\partial\Gamma}{\partial\boldsymbol{t}^E}\right)_{|E} = \boldsymbol{0}\,, \quad \left(\frac{\partial\Gamma}{\partial(\mathrm{grad}\,\Theta)}\right)_{|E} = \boldsymbol{0}\,,$$

$$\begin{pmatrix} \dfrac{\partial^2\Gamma}{\partial\boldsymbol{t}^E\,\partial\boldsymbol{t}^E} & \dfrac{\partial^2\Gamma}{\partial\boldsymbol{t}^E\,\partial(\mathrm{grad}\,\Theta)} \\ \dfrac{\partial^2\Gamma}{\partial(\mathrm{grad}\,\Theta)\,\partial\boldsymbol{t}^E} & \dfrac{\partial^2\Gamma}{\partial(\mathrm{grad}\,\Theta)\,\partial(\mathrm{grad}\,\Theta)} \end{pmatrix}_{|E} \quad \text{is positive semi-definite.} \tag{5.7.63}$$

The first two conditions imply the statements

$$\boldsymbol{D}(p,\Theta,\boldsymbol{0},\boldsymbol{0}) = \boldsymbol{D}_{|E} = \boldsymbol{0}\,, \quad \boldsymbol{q}(p,\Theta,\boldsymbol{0},\boldsymbol{0}) = \boldsymbol{q}_{|E} = \boldsymbol{0}\,, \tag{5.7.64}$$

as one would have expected. Let us now assume that the stretching tensor does not depend on the temperature gradient and the heat flux vector does not depend on the extra stress tensor; then from the representation of these quantities as isotropic tensor and vector functions, we conclude that

$$\begin{aligned}
&\boldsymbol{D} = \alpha_1 \boldsymbol{I} + \alpha_2 \boldsymbol{t}^E + \alpha_3 (\boldsymbol{t}^E)^2\,, \quad \boldsymbol{q} = -\lambda \operatorname{grad} \Theta\,, \\
&\alpha_i = \hat{\alpha}_i(p,\Theta, I_{\boldsymbol{t}^E}, II_{\boldsymbol{t}^E}, III_{\boldsymbol{t}^E})\,, \quad i=1,2,3\,, \quad \alpha_{1|E} = 0\,, \\
&\lambda = \hat{\lambda}(p,\Theta, \|\operatorname{grad}\Theta\|)\,,
\end{aligned} \tag{5.7.65}$$

in which the restrictions on the coefficients α_i and λ follow by evaluating the statement $(5.7.63)_3$. Starting from the representations, which are linear in $\boldsymbol{t}^E$ and $\operatorname{grad}\Theta$, we deduce

$$\boldsymbol{D} = g I_{\boldsymbol{t}^E} \boldsymbol{I} + \tfrac{1}{2} f \boldsymbol{t}^{E'}\,, \quad \operatorname{tr} \boldsymbol{t}^{E'} = 0\,, \quad \boldsymbol{q} = -\lambda \operatorname{grad}\Theta\,, \tag{5.7.66}$$

in which g, f, λ are functions of p and Θ. With (5.7.66) the entropy production inequality can be written as

$$\Gamma = \frac{g}{3}(I_{\boldsymbol{t}^E})^2 + \tfrac{1}{2} f \boldsymbol{t}^{E'} \cdot \boldsymbol{t}^{E'} + \frac{\lambda}{\Theta} \|\operatorname{grad}\Theta\|^2 \geq 0 \tag{5.7.67}$$

or

$$\begin{gathered}
\Gamma = g\frac{x^2}{2} + f\frac{y^2}{2} + \lambda\frac{z^2}{2}\,, \\
x := \sqrt{\frac{2}{3}} I_{\boldsymbol{t}^E}\,, \quad y := \sqrt{\boldsymbol{t}^{E'} \cdot \boldsymbol{t}^{E'}}\,, \quad z := \sqrt{\frac{2}{\Theta}} \|\operatorname{grad}\Theta\|\,.
\end{gathered} \tag{5.7.68}$$

Thus Γ is a positive semi-definite quadratic form in x, y and z, so that

$$g \geq 0\,, \quad f \geq 0\,, \quad \lambda \geq 0 \tag{5.7.69}$$

are necessary conditions for fulfilling the second law. One calls g the *bulk fluidity* and f the *shear fluidity*. These are the inverses of the viscosities κ and μ (Exercise).

5.8 Entropy Principle of MÜLLER

In the last section we formulated the second law in the form of the CLAUSIUS–DUHEM inequality and simplified its mathematical exploitation effectively by using two assumptions. These assumptions were

- $$\text{entropy supply} = \frac{\text{energy supply}}{\text{absolute temperature}}\,,$$
  $$\text{entropy flux} = \frac{\text{heat flux}}{\text{absolute temperature}}\,.$$
- The balance laws of linear momentum and energy accommodate non-vanishing supply terms, which can be prescribed arbitrarily, and when necessary, can take every value we please.

The first assumption is restricting, because it assumes the existence of the absolute temperature; in addition it fails for mixtures (see Chap. 7) and must be modified there. The second assumption is physically presumptuous, because it assumes that our "universe" is natured in such a way that, for a body, when necessary, there exists always a neighbourhood for which the external forces and the radiation take values as we please.

In the endeavour of softening these assumptions I. MÜLLER formulated a weaker form of entropy principle, which, nevertheless, satisfies all necessary requirements of an irreversibility statement and reads as follows:

Entropy Principle:

1) *In every material body there exists an additive quantity, the specific entropy s, which obeys a balance equation*

$$\rho\frac{ds}{dt} = -\operatorname{div}\boldsymbol{\phi} + \rho\eta + \rho\gamma\,, \tag{5.8.1}$$

in which $\boldsymbol{\phi}$ is the entropy flux, η the specific entropy supply and γ the specific entropy production.

2) *The specific entropy s and the entropy flux $\boldsymbol{\phi}$ are material quantities for which, according to the rule of equipresence, the same material laws hold as for the remaining constitutive quantities.*
3) *The entropy production must for all thermodynamic processes be a non-negative quantity,*

$$\gamma \geq 0 \quad \textit{for all thermodynamic processes,} \tag{5.8.2}$$

i.e., for all solutions of the field equations (these are the balance equations plus the constitutive relations together).

4) *The supply terms, which appear in the balance equations, can not influence the material behaviour.*
5) *There exist special material singular surfaces, the so-called* ideal walls, *between two continua, across which the (empirical) temperature and the tangential velocity are continuous.* ■

5.8.1 Heat Conducting Compressible Fluid

We shall apply this entropy principle for the simplest case, a heat conducting, compressible fluid; then the material equations are

$$\Psi = \hat{\Psi}(\rho,\, \theta,\, \operatorname{grad}\theta)\,, \quad \Psi \in \{\varepsilon,\, s,\, \boldsymbol{q},\, \boldsymbol{t},\, \boldsymbol{\phi}\}\,. \tag{5.8.3}$$

Mass, momentum, energy and entropy balances are given by

$$\begin{aligned}
&\frac{\partial \rho}{\partial t} + \operatorname{div}(\rho \boldsymbol{v}) = 0\,,\\
&\rho \frac{\mathrm{d}\boldsymbol{v}}{\mathrm{d}t} - \operatorname{div}\hat{\boldsymbol{t}} - \rho \boldsymbol{g} = 0\,,\\
&\rho \frac{\mathrm{d}\hat{\varepsilon}}{\mathrm{d}t} + \operatorname{div}\hat{\boldsymbol{q}} - \operatorname{tr}(\hat{\boldsymbol{t}}\boldsymbol{D}) - \rho \mathfrak{r} = 0\,,\\
&\rho \frac{\mathrm{d}\hat{s}}{\mathrm{d}t} + \operatorname{div}\hat{\boldsymbol{\phi}} - \rho \eta \geq 0\,,
\end{aligned} \tag{5.8.4}$$

in which ρ, θ (the empirical temperature) and $\boldsymbol{v}$ are to be considered as independent field quantities and the constitutive equations are thought to be substituted (which is indicated by the notation $(\hat{.})$). A thermodynamic process is a solution of the equations $(5.8.4)_{1,2,3}$, and the entropy principle demands that the entropy inequality $(5.8.4)_4$ must be fulfilled by all fields, which also satisfy the field equations $(5.8.4)_{1,2,3}$[19].

It is plausible to think that one can satisfy this statement by the following modification of the original entropy inequality:

$$\begin{aligned}
&\rho \frac{\mathrm{d}\hat{s}}{\mathrm{d}t} + \operatorname{div}\hat{\boldsymbol{\phi}} - \rho \eta\\
&- \Lambda^{\rho} \left\{ \frac{\partial \rho}{\partial t} + \operatorname{div}(\rho \boldsymbol{v}) \right\} - \boldsymbol{\Lambda}^{\boldsymbol{v}} \cdot \left\{ \rho \frac{\mathrm{d}\boldsymbol{v}}{\mathrm{d}t} - \operatorname{div}\hat{\boldsymbol{t}} - \rho \boldsymbol{g} \right\}\\
&- \Lambda^{\varepsilon} \left\{ \rho \frac{\mathrm{d}\hat{\varepsilon}}{\mathrm{d}t} + \operatorname{div}\hat{\boldsymbol{q}} - \operatorname{tr}(\hat{\boldsymbol{t}}\boldsymbol{D}) - \rho \mathfrak{r} \right\}\\
&\geq 0\,.
\end{aligned} \tag{5.8.5}$$

In this inequality the balance equations of mass, momentum and energy multiplied by the corresponding so-called LAGRANGE parameters are subtracted,

[19] This entropy principle is more general than the principle using the CLAUSIUS–DUHEM inequality and the COLEMAN–NOLL approach by the fact that the form of the constitutive relation for the entropy flux is kept free within the constitutive class under study and not a priori set in relation to heat flux and absolute temperature. It is different also by the fact that the concept of absolute temperature is a derived one, i.e., the measure of coldness of a body is the empirical temperature and the absolute temperature is functionally related to it (if it is meaningful at all). And third, external source terms are required not to affect the material behaviour of a body. This latter point is contrary to the COLEMAN–NOLL approach.

and it is immediately prudent that (5.8.4) imply (5.8.5). The inverse of this is also true, which was proved by LIU [136]. This proof will be given in the Appendix to this chapter. LIU's theorem states that both statements: (i) Satisfy the inequality (5.8.5) for *unrestricted fields* and (ii) satisfy the inequality $(5.8.4)_4$ by simultaneously satisfying the field equations $(5.8.4)_{1,2,3}$ are equivalent. It is easy to fulfill the extended inequality, but one must determine the unknown LAGRANGE multipliers, which is again a matter of tedious calculations.

If the constitutive equations (5.8.3) are substituted in (5.8.5), and differentiations with respect to time and space coordinates using the chain rule are executed, one obtains the resulting modified inequality in the form

$$
\begin{aligned}
&\rho\left(\frac{\partial \hat{s}}{\partial \theta}-\Lambda^{\varepsilon}\frac{\partial \hat{\varepsilon}}{\partial \theta}\right)\dot{\theta}+\rho\left(\frac{\partial \hat{s}}{\partial \rho}-\Lambda^{\varepsilon}\frac{\partial \hat{\varepsilon}}{\partial \rho}-\frac{\Lambda^{\rho}}{\rho}\right)\dot{\rho}\\
&+\rho\left(\frac{\partial \hat{s}}{\partial\,\mathrm{grad}\,\theta}-\Lambda^{\varepsilon}\frac{\partial \hat{\varepsilon}}{\partial\,\mathrm{grad}\,\theta}\right)(\mathrm{grad}\,\theta)^{\cdot}\\
&+\left\{\frac{\partial \hat{\boldsymbol{\phi}}}{\partial \rho}-\Lambda^{\varepsilon}\frac{\partial \hat{\boldsymbol{q}}}{\partial \rho}+\boldsymbol{\Lambda}^{\boldsymbol{v}}\frac{\partial \hat{\boldsymbol{t}}}{\partial \rho}\right\}\cdot\mathrm{grad}\,\rho\\
&+\left\{\frac{\partial \hat{\boldsymbol{\phi}}}{\partial\,\mathrm{grad}\,\theta}-\Lambda^{\varepsilon}\frac{\partial \hat{\boldsymbol{q}}}{\partial\,\mathrm{grad}\,\theta}+\boldsymbol{\Lambda}^{\boldsymbol{v}}\frac{\partial \hat{\boldsymbol{t}}}{\partial\,\mathrm{grad}\,\theta}\right\}\cdot\mathrm{grad}(\mathrm{grad}\,\theta)\\
&-\rho\,\boldsymbol{\Lambda}^{\boldsymbol{v}}\cdot\dot{\boldsymbol{v}}\\
&+\left\{\frac{\partial \hat{\boldsymbol{\phi}}}{\partial \theta}-\Lambda^{\varepsilon}\frac{\partial \hat{\boldsymbol{q}}}{\partial \theta}+\boldsymbol{\Lambda}^{\boldsymbol{v}}\frac{\partial \hat{\boldsymbol{t}}}{\partial \theta}\right\}\cdot\mathrm{grad}\,\theta\\
&+\Lambda^{\varepsilon}\,\mathrm{tr}\left[\left(\hat{\boldsymbol{t}}-\rho\frac{\Lambda^{\rho}}{\Lambda^{\varepsilon}}\boldsymbol{I}\right)\boldsymbol{D}\right]-\rho\eta+\rho\boldsymbol{g}\cdot\boldsymbol{\Lambda}^{\boldsymbol{v}}+\rho\mathfrak{r}\Lambda^{\varepsilon}\\
&\geq 0\,.
\end{aligned}
\tag{5.8.6}
$$

This inequality is simplified in a first step, in which point 4 of the entropy principle should be evaluated. It says that the material properties should not be influenced by the supply terms, and from this it follows that the LAGRANGE parameter $\Lambda^{\rho}, \boldsymbol{\Lambda}^{\boldsymbol{v}}$ and Λ^{ε} cannot depend on η, $\boldsymbol{g}$ and $\mathfrak{r}$, as well as that

$$
\eta=\Lambda^{\varepsilon}\mathfrak{r}+\boldsymbol{\Lambda}^{\boldsymbol{v}}\cdot\boldsymbol{g}\,. \tag{5.8.7}
$$

The entropy supply is a linear combination of the energy supply and the momentum supply, whereby the factors are simply the LAGRANGE multipliers of the corresponding equations. The reader may recall that the entropy supply was postulated in the CLAUSIUS–DUHEM inequality as $\eta=\mathfrak{r}/\theta$, where θ is the absolute temperature; this is obviously a special case of (5.8.7).

Substituting (5.8.7) in (5.8.6), one obtains an inequality which is linear in

$$\boldsymbol{\beta} = [\dot{\theta}\,,\, \dot{\rho}\,,\, (\operatorname{grad}\theta)^{\cdot}\,,\, \operatorname{grad}\rho\,,\, \operatorname{grad}(\operatorname{grad}\theta)\,,\, \boldsymbol{D}]^T \tag{5.8.8}$$

and expressable in the form

$$\boldsymbol{\alpha}\cdot\boldsymbol{\beta} + \Gamma \geq 0\,. \tag{5.8.9}$$

The vector $\boldsymbol{\alpha}$ is given[20] by the prefactors of $\boldsymbol{\beta}$ (in the first four lines of (5.8.6)); Γ embraces the last three lines of (5.8.6) without the supply terms.

It follows from LIU's theorem that $\boldsymbol{\beta}$ is arbitrarily choosable at a fixed material point – in other words, it is possible to construct an admissible thermodynamic process with arbitrary $\boldsymbol{\beta}$. Thus, necessary and sufficient condition for (5.8.9) to hold is $\boldsymbol{\alpha} = \boldsymbol{0}$ (and $\Gamma \geq 0$), or

$$\begin{aligned}
&\frac{\partial \hat{s}}{\partial \theta} - \Lambda^{\varepsilon}\frac{\partial \hat{\varepsilon}}{\partial \theta} = 0\,,\\
&\frac{\partial \hat{s}}{\partial \rho} - \Lambda^{\varepsilon}\frac{\partial \hat{\varepsilon}}{\partial \rho} - \frac{\Lambda^{\rho}}{\rho} = 0\,,\\
&\frac{\partial \hat{s}}{\partial \operatorname{grad}\theta} - \Lambda^{\varepsilon}\frac{\partial \hat{\varepsilon}}{\partial \operatorname{grad}\theta} = \boldsymbol{0}\,,\\
&\frac{\partial \hat{\boldsymbol{\phi}}}{\partial \rho} - \Lambda^{\varepsilon}\frac{\partial \hat{\boldsymbol{q}}}{\partial \rho} + \boldsymbol{\Lambda}^{\boldsymbol{v}}\frac{\partial \hat{\boldsymbol{t}}}{\partial \rho} = \boldsymbol{0}\,,\\
&\left\{\frac{\partial \hat{\boldsymbol{\phi}}}{\partial \operatorname{grad}\theta} - \Lambda^{\varepsilon}\frac{\partial \hat{\boldsymbol{q}}}{\partial \operatorname{grad}\theta} + \boldsymbol{\Lambda}^{\boldsymbol{v}}\frac{\partial \hat{\boldsymbol{t}}}{\partial \operatorname{grad}\theta}\right\}_{\mathrm{sym}} = \boldsymbol{0}\,,\\
&\hat{\boldsymbol{t}} = \rho\frac{\Lambda^{\rho}}{\Lambda^{\varepsilon}}\boldsymbol{I} = -p\boldsymbol{I}\,,
\end{aligned} \tag{5.8.10}$$

equations, which must be fulfilled as identities. These constrain the constitutive equations for $\hat{s}$, $\hat{\varepsilon}$, $\hat{\boldsymbol{q}}$, $\hat{\boldsymbol{t}}$ and $\hat{\boldsymbol{\phi}}$ but can also be viewed as determining equations for Λ^{ε}, Λ^{ρ} and $\boldsymbol{\Lambda}^{\boldsymbol{v}}$. This last interpretation can be applied to conclude, that the LAGRANGE multipliers, as these are determined alone by constitutive quantities, themselves, can only depend on the independent constitutive variables. This implies, specially, that these can not depend on $\dot{\boldsymbol{v}}$. (5.8.6) is therefore also linear in $\dot{\boldsymbol{v}}$, and from this it follows that

$$\boldsymbol{\Lambda}^{\boldsymbol{v}} = \boldsymbol{0}\,. \tag{5.8.11}$$

The LAGRANGE multiplier of the momentum equation vanishes, or the momentum equation does not modify the analysis of the entropy inequality – at least not in this restricted theory for a compressible heat conducting fluid.

[20] One could be tempted to regard the term $\boldsymbol{\Lambda}^{\boldsymbol{v}}\cdot\dot{\boldsymbol{v}}$ as linear in $\dot{\boldsymbol{v}}$, however, this is not so – at least not at this stage of the computations, because the LAGRANGE multipliers can depend on $\dot{\boldsymbol{v}}$ in addition to ρ, θ, $\operatorname{grad}\theta$.

This was assumed in the last section in which the entropy principle was employed as CLAUSIUS–DUHEM inequality and the COLEMAN–NOLL approach was used for its exploitation. Finally, the CAUCHY stress is isotropic and known as soon as Λ^ε and Λ^ρ are determined.

The next step in the evaluation of the identities (5.8.10) consists in the determination of the LAGRANGE multiplier Λ^ε. Here as well one starts from explicit representations of the entropy flux and heat flux as objective vector valued isotropic functions,

$$\begin{aligned} \boldsymbol{\phi} &= -\phi_1(\rho,\, \theta,\, \|\operatorname{grad}\theta\|^2)\, \operatorname{grad}\theta\,, \\ \boldsymbol{q} &= -\mathrm{q}_1(\rho,\, \theta,\, \|\operatorname{grad}\theta\|^2)\, \operatorname{grad}\theta\,. \end{aligned} \tag{5.8.12}$$

Substituting these assumptions in the second to the last of the relations (5.8.10) results, by considering (5.8.11), in

$$\begin{aligned} &\phi_1 \boldsymbol{I} + \frac{\partial \phi_1}{\partial \|\operatorname{grad}\theta\|^2} 2 \operatorname{grad}\theta \otimes \operatorname{grad}\theta \\ &\quad = \Lambda^\varepsilon \mathrm{q}_1 \boldsymbol{I} + \Lambda^\varepsilon \frac{\partial \mathrm{q}_1}{\partial \|\operatorname{grad}\theta\|^2} 2 \operatorname{grad}\theta \otimes \operatorname{grad}\theta\,, \end{aligned} \tag{5.8.13}$$

which must be satisfied for arbitrary values of $\operatorname{grad}\theta$. From this one obtains

$$\phi_1 = \Lambda^\varepsilon \mathrm{q}_1 \quad \text{and} \quad \frac{\partial \Lambda^\varepsilon}{\partial \|\operatorname{grad}\theta\|^2} = 0\,. \tag{5.8.14}$$

The entropy flux is thus collinear with the heat flux, whereby the factor is given by the LAGRANGE multiplier of the energy equation. In view of $(5.8.14)_2$ the latter is not a function of the temperature gradient.

Using (5.8.14) with (5.8.12) and (5.8.11) in $(5.8.10)_4$, one obtains

$$\Lambda^\varepsilon \frac{\partial \mathrm{q}_1}{\partial \rho} + \frac{\partial \Lambda^\varepsilon}{\partial \rho} \mathrm{q}_1 = \Lambda^\varepsilon \frac{\partial \mathrm{q}_1}{\partial \rho} \quad \Rightarrow \quad \frac{\partial \Lambda^\varepsilon}{\partial \rho} \mathrm{q}_1 = 0\,, \tag{5.8.15}$$

from which with $\mathrm{q}_1 \neq 0$ (which is to be required) follows that Λ^ε is no longer permitted to be a function of ρ. In summary, one obtains, from the relations $(5.8.10)_{4,5}$

$$\boldsymbol{\phi} = \Lambda^\varepsilon(\theta)\, \boldsymbol{q}\,, \tag{5.8.16}$$

a result, which approaches the CLAUSIUS–DUHEM assumption $\Lambda^\varepsilon(\theta) = 1/\theta$, very closely where θ indicates the absolute temperature. Presently, however, $\Lambda^\varepsilon(\theta)$ is still a materially dependent function of the empirical temperature θ.

In order to prove that the LAGRANGE multiplier of the energy, $\Lambda^\varepsilon(\theta)$, is independent of the material properties, (at least within the material class of the heat conducting compressible fluids) let us recall the last property of the entropy principle, namely, that between two such materials there exist material singular surfaces with the property of an ideal wall across which the temperature and tangential velocity experience no jumps. We now consider two heat conducting compressible fluids, which are separated by a material singular surface through which the empirical temperature is continuous. The jump conditions of entropy and energy read in this case,

$$[\![\boldsymbol{\phi}\cdot\boldsymbol{n}]\!] = [\![\Lambda^{\varepsilon}\boldsymbol{q}\cdot\boldsymbol{n}]\!] = 0 \quad \text{and} \quad [\![\boldsymbol{q}\cdot\boldsymbol{n}]\!] = 0\,,$$

or

$$[\![\Lambda^{\varepsilon}]\!]\,\boldsymbol{q}\cdot\boldsymbol{n} = 0 \quad \Longrightarrow \quad [\![\Lambda^{\varepsilon}]\!] = 0 \quad \text{provided that} \quad \boldsymbol{q}\cdot\boldsymbol{n} \neq 0\,. \tag{5.8.17}$$

In other words, $\Lambda^{\varepsilon}(\theta)^{+} = \Lambda^{\varepsilon}(\theta)^{-}$; or, the LAGRANGE multiplier is the same function of empirical temperature on both sides of the ideal wall. Since the fluids on both sides of the ideal wall can be arbitrary within their constitutive class, then follows the material independency of $\Lambda^{\varepsilon}(\theta)$ within this class. One refers to $\Lambda^{\varepsilon}(\theta)$ as the coldness function (or simply coldness) and its reciprocal value as the absolute temperature,

$$\Theta(\theta) = \frac{1}{\Lambda^{\varepsilon}(\theta)}\,. \tag{5.8.18}$$

Thus the relations $(5.8.10)_{4,5}$ are exploited[21].

We now turn our attention to the identities $(5.8.10)_{1,2,3}$, where we will simultaneously make use of the result $\Lambda^{\varepsilon} = \Lambda^{\varepsilon}(\theta)$. Differentiating $(5.8.10)_1$ with respect to $\operatorname{grad}\theta$ and $(5.8.10)_3$ with respect to θ, one can derive the following chain of equations

$$\begin{aligned}\frac{\partial^2 \hat{s}}{\partial \operatorname{grad}\theta\,\partial\theta} &= \Lambda^{\varepsilon}\frac{\partial^2 \hat{\varepsilon}}{\partial \operatorname{grad}\theta\,\partial\theta} = \frac{\partial^2 \hat{s}}{\partial\theta\,\partial \operatorname{grad}\theta}\\ &= \Lambda^{\varepsilon}\frac{\partial^2 \hat{\varepsilon}}{\partial\theta\,\partial \operatorname{grad}\theta} + \frac{\partial \Lambda^{\varepsilon}}{\partial\theta}\frac{\partial \hat{\varepsilon}}{\partial \operatorname{grad}\theta}\,.\end{aligned} \tag{5.8.19}$$

Since the sequence of differentiation of the functions $\hat{s}$ and $\hat{\varepsilon}$ with respect to θ and $\operatorname{grad}\theta$ must be irrelevant, (5.8.19) implies, since $\partial\Lambda^{\varepsilon}(\theta)/\partial\theta \neq 0$, that $\hat{\varepsilon}$ can not be a function of $\operatorname{grad}\theta$, $\partial\hat{\varepsilon}/\partial \operatorname{grad}\theta = 0$. Resubstituting this result in $(5.8.10)_3$ shows then that $\hat{s}$ does not depend on $\operatorname{grad}\theta$ either, and – after having this shown for $\hat{s}$ and $\hat{\varepsilon}$ – the same must also hold for Λ^{ρ}. Thus, one has the classical result

$$s = \hat{s}(\rho,\theta)\,, \quad \varepsilon = \hat{\varepsilon}(\rho,\theta)\,, \quad \Lambda^{\rho} = \hat{\Lambda}^{\rho}(\rho,\theta)\,. \tag{5.8.20}$$

It is still to be demonstrated from (5.8.10) that the first two identities, which can be combined together, yield

[21] Notice that the above result was obtained by exploiting only the relations (5.8.10) involving entropy flux and heat flux, but not entropy and internal energy. This is typical. Characteristic is equally the fact that isotropy relations had also to be used to achieve the result.

$$\mathrm{d}\hat{s} = \Lambda^{\varepsilon}\left[\frac{\partial\hat{\varepsilon}}{\partial\theta}\mathrm{d}\theta + \left(\frac{\partial\hat{\varepsilon}}{\partial\rho} + \frac{\Lambda^{\rho}}{\rho\Lambda^{\varepsilon}}\right)\mathrm{d}\rho\right] = \frac{1}{\Theta}\left[\mathrm{d}\hat{\varepsilon} + \hat{p}\,\mathrm{d}\left(\frac{1}{\rho}\right)\right], \tag{5.8.21}$$

where we have identified the ratio $\rho\Lambda^{\rho}/\Lambda^{\varepsilon}$ with the thermodynamic pressure via

$$\hat{p}(\rho,\theta) = -\frac{\Lambda^{\rho}(\rho,\theta)}{\Lambda^{\varepsilon}(\theta)}\rho\,. \tag{5.8.22}$$

This pressure is the same as that introduced in $(5.8.10)_6$. Equation (5.8.21) is known as GIBBS *equation*, which expresses the total differential of the entropy as the product of the inverse of the absolute temperature times the total differential of the internal energy plus the additional term $\hat{p}\,\mathrm{d}(1/\rho)$.

Naturally, the GIBBS equation must also satisfy an integrability condition, which is obtained from the cross differentiations of the coefficients of (5.8.21); the result is

$$\frac{\mathrm{d}\ln\Lambda^{\varepsilon}}{\mathrm{d}\theta} = \frac{1}{\Lambda^{\varepsilon}}\frac{\mathrm{d}\Lambda^{\varepsilon}}{\mathrm{d}\theta} = \frac{\partial\hat{p}/\partial\theta}{(\partial\hat{\varepsilon}/\partial\rho)\rho^2 - \hat{p}}\,. \tag{5.8.23}$$

Integrating this equation yields

$$\ln\frac{\Lambda^{\varepsilon}}{\Lambda^{\varepsilon}_0} = -\ln\frac{\Theta}{\Theta_0} = \int\limits_{\theta_0}^{\theta}\frac{\partial\hat{p}/\partial\theta}{(\partial\hat{\varepsilon}/\partial\rho)\rho^2 - \hat{p}(\rho,\bar{\theta})}\,d\bar{\theta}$$

or

$$\Theta(\theta) = \Theta_0\exp\left\{-\int_{\theta_0}^{\theta}\frac{\partial\hat{p}/\partial\theta}{(\partial\hat{\varepsilon}/\partial\rho)\rho^2 - \hat{p}(\rho,\bar{\theta})}\,d\bar{\theta}\right\}. \tag{5.8.24}$$

The absolute temperature Θ is thus known as a function of the empirical temperature, if one knows $\hat{p}(\rho,\theta)$ and $(\partial\hat{\varepsilon}/\partial\rho)(\rho,\theta)$ as functions of their variables *for any heat conducting compressible fluid.* The left-hand side of (5.8.24) is materially independent, and thus so must be its right-hand side. Conversely, when one knows the function $\Theta(\theta)$, the pressure and the internal energy can not be chosen arbitrarily from each other, since relation (5.8.24) must be obeyed.

Choosing an *ideal gas* as the special fluid for which the equations of state are

$$p = R\rho\,\Theta(\theta)\,, \quad \varepsilon = \hat{\varepsilon}(\theta)\,, \tag{5.8.25}$$

where R is the gas constant, then (5.8.24) exhibits the identity $\Theta(\theta) = \Theta(\theta)$. This can be taken as motivation to set

$$\Theta(\theta) =: T\,. \tag{5.8.26}$$

This was suggested by LORD KELVIN[22]. One calls T the *absolute* or KELVIN *temperature.* Using this relation one can replace in all relations the empirical temperature θ by the absolute temperature; this we shall now do.

[22] WILLIAM THOMPSON (1824–1907), since 1892 Lord KELVIN, Professor of natural philosophy and theoretical physics at Glasgow University.

If with

$$\Psi := \varepsilon - Ts = \hat{\Psi}(\rho, T) \tag{5.8.27}$$

the HELMHOLTZ free energy is introduced, then the GIBBS equation (5.8.21) takes the form

$$\left(\frac{\partial\hat{\Psi}}{\partial\rho} - \frac{\hat{p}}{\rho^2}\right) \mathrm{d}\rho + \left(\frac{\partial\hat{\Psi}}{\partial T} + \hat{s}\right) \mathrm{d}T = 0 \,, \tag{5.8.28}$$

which must be satisfied for arbitrary differentials $\mathrm{d}\rho$ and $\mathrm{d}T$. Consequently,

$$s = -\frac{\partial\hat{\Psi}}{\partial T} \,, \quad \hat{p} = \rho^2 \frac{\partial\hat{\Psi}}{\partial\rho} \,. \tag{5.8.29}$$

The entropy and the thermodynamic pressure are thus calculable from the prescribed thermodynamic potential $\hat{\Psi}(\rho, \theta)$, the HELMHOLTZ free energy. The restrictions on the constitutive functions, which are imposed by the second law, appear especially concise in this form.

Thus the identities (5.8.10) are exploited and there remains the analysis of the residual inequality $\Gamma \geq 0$, or

$$\Pi^s = T\Gamma = -\frac{\boldsymbol{q} \cdot \operatorname{grad} T}{T} \geq 0 \,. \tag{5.8.30}$$

Thermodynamic equilibrium is defined as a process, which produces no entropy, given by $\operatorname{grad} T = \mathbf{0}$. The necessary conditions for this are the statements

$$\left(\frac{\partial \Pi^s}{\partial \operatorname{grad} T}\right)_{|E} = \mathbf{0} \,, \tag{5.8.31}$$

$$\frac{\partial^2 \Pi^s}{(\partial \operatorname{grad} T)^2}_{|E} \quad \text{is positive semidefinite} \,, \tag{5.8.32}$$

where $|_E$ indicates evaluation at equilibrium. Performing the differentiation (5.8.31) in (5.8.30) results in

$$\boldsymbol{q}_{|E} = \mathbf{0} \,. \tag{5.8.33}$$

The equilibrium heat flux vector vanishes. With the isotropic representations

$$\begin{aligned} \boldsymbol{t} &= -\hat{p}(\rho, T)\boldsymbol{I} \\ \boldsymbol{q} &= -\mathrm{q}_1(\rho, T, \|\operatorname{grad} T\|^2) \operatorname{grad} T \end{aligned} \tag{5.8.34}$$

one can exploit (5.8.32). The only relation which results from criterion (5.8.32) is

$$\mathrm{q}_1(\rho, T, 0) \geq 0 \,. \tag{5.8.35}$$

The nonlinear material equations (5.8.34) for the stress tensor and the heat flux vector are thus compatible with the entropy principle, if $\hat{p}$ is derived via

(5.8.29) from the HELMHOLTZ free energy, and the thermal conductivity q_1 at grad $T = 0$ is non-negative.

The entropy principle of MÜLLER, in this example of a heat conducting compressible fluid, has lead to the same results, as would have been obtained with the application of the CLAUSIUS–DUHEM inequality carried out in accordance with the COLEMAN–NOLL approach. However, these results were obtained with the much weaker formulation. It was *proved* by the entropy principle of MÜLLER that the momentum balance does not influence the exploitation of the entropy principle; the absolute temperature was not assumed *a priori to exist*, but one has *proved* that it can be interpreted as the inverse of the LAGRANGE multiplier of the internal energy balance, and further one has *shown* that it represents a quantity independent of the material. Finally, *rewriting* the relations given at the beginning of this section,

$$\text{entropy supply} = \frac{\text{energy supply}}{\text{absolute temperature}} ,$$
$$\text{entropy flux} = \frac{\text{heat flux}}{\text{absolute temperature}}$$

– in connection with the heat conducting compressible fluids – we can say that these relations are now proved statements. These facts mediate to the model equations, which are derived from MÜLLER's entropy principle, strengthened credibility. But it is also likely that for general material laws both entropy principles – CLAUSIUS–DUHEM inequality with the exploitation of COLEMAN–NOLL on the one hand and the more general entropy principle of MÜLLER on the other – do not necessarily furnish the same results. This is so in general and must be scrutinized on a case by case basis. In this regard it is advisable to apply MÜLLER's entropy principle whenever possible.

5.8.2 Heat Conducting Density Preserving Fluid

In a heat conducting density preserving fluid the independent constitutive variables are θ and grad θ only; thus

$$\Psi = \hat{\Psi}(\theta, \operatorname{grad}\theta),\ \Psi \in \{\varepsilon, s, \boldsymbol{q}, \boldsymbol{t}^E, \phi\} \tag{5.8.36}$$

where

$$\boldsymbol{t} = -p\boldsymbol{I} + \boldsymbol{t}^E , \tag{5.8.37}$$

in which p is the constraint pressure and $\boldsymbol{t}^E$ the extra stress tensor which may be taken to be a deviator, $\operatorname{tr}\boldsymbol{t}^E = 0$. The balance of mass reduces to $\operatorname{div}\boldsymbol{v} = 0$, so that the entropy inequality, extended by the field-equation constraints, see (5.8.5) for comparison, becomes

$$
\begin{aligned}
&\rho\frac{\mathrm{d}\hat{s}}{\mathrm{d}t} + \operatorname{div}\hat{\boldsymbol{\phi}} - \rho\eta \\
&- \Lambda^{\rho}\operatorname{div}\boldsymbol{v} - \boldsymbol{\Lambda}^{\boldsymbol{v}}\cdot\left\{\rho\frac{\mathrm{d}\boldsymbol{v}}{\mathrm{d}t} + \operatorname{div}(p\boldsymbol{I}) - \operatorname{div}(\hat{\boldsymbol{t}}^{E}) - \rho\boldsymbol{g}\right\} \\
&- \Lambda^{\varepsilon}\left\{\rho\frac{\mathrm{d}\hat{\varepsilon}}{\mathrm{d}t} + \operatorname{div}\hat{\boldsymbol{q}} + p\operatorname{div}\boldsymbol{v} - \operatorname{tr}(\hat{\boldsymbol{t}}^{E}\boldsymbol{D}) - \rho\mathfrak{r}\right\} \\
&\geq 0\,.
\end{aligned}
\tag{5.8.38}
$$

This inequality must hold for arbitrary independent fields, i.e., also deformation fields which do not satisfy the condition $\operatorname{div}\boldsymbol{v} = 0$. The constitutive relations are thought to be substituted in (5.8.38); this is made visible by writing any constitutive variable f as $\hat{f}$.

Performing the differentiations, using the chain rule wherever needed yields the inequality

$$
\begin{aligned}
&\rho\left(\frac{\partial\hat{s}}{\partial\theta} - \Lambda^{\varepsilon}\frac{\partial\hat{\varepsilon}}{\partial\theta}\right)\dot{\theta} + \rho\left(\frac{\partial\hat{s}}{\partial\operatorname{grad}\theta} - \Lambda^{\varepsilon}\frac{\partial\hat{\varepsilon}}{\partial\operatorname{grad}\theta}\right)(\operatorname{grad}\theta)^{\cdot} \\
&+ \left\{\frac{\partial\hat{\boldsymbol{\phi}}}{\partial\operatorname{grad}\theta} - \Lambda^{\varepsilon}\frac{\partial\hat{\boldsymbol{q}}}{\partial\operatorname{grad}\theta} + \boldsymbol{\Lambda}^{\boldsymbol{v}}\frac{\partial\hat{\boldsymbol{t}}^{E}}{\partial\operatorname{grad}\theta}\right\}\cdot\operatorname{grad}(\operatorname{grad}\theta) \\
&- \rho\,\boldsymbol{\Lambda}^{\boldsymbol{v}}\cdot\dot{\boldsymbol{v}} - \boldsymbol{\Lambda}^{\boldsymbol{v}}\cdot\operatorname{grad}p \\
&+ \left\{\frac{\partial\hat{\boldsymbol{\phi}}}{\partial\theta} - \Lambda^{\varepsilon}\frac{\partial\hat{\boldsymbol{q}}}{\partial\theta} + \boldsymbol{\Lambda}^{\boldsymbol{v}}\frac{\partial\hat{\boldsymbol{t}}^{E}}{\partial\theta}\right\}\cdot\operatorname{grad}\theta - \Lambda^{\varepsilon}\left(p + \frac{\Lambda^{\rho}}{\Lambda^{\varepsilon}}\right)\operatorname{div}\boldsymbol{v} \\
&+ \Lambda^{\varepsilon}\operatorname{tr}\left[\hat{\boldsymbol{t}}^{E}\left(\boldsymbol{D} - \frac{1}{3}I_{\boldsymbol{D}}\boldsymbol{I}\right)\right] - \rho\eta + \rho\boldsymbol{g}\cdot\boldsymbol{\Lambda}^{\boldsymbol{v}} + \rho\mathfrak{r}\Lambda^{\varepsilon} \\
&\geq 0\,.
\end{aligned}
\tag{5.8.39}
$$

Because the material is assumed to be independent of the external sources (item 4 in the entropy principle) one necessarily has

$$
\eta = \Lambda^{\varepsilon}\mathfrak{r} + \boldsymbol{\Lambda}^{\boldsymbol{v}}\cdot\boldsymbol{g}\,. \tag{5.8.40}
$$

Moreover, since inequality (5.8.39) is linear in the variables $\dot{\theta}$, $(\operatorname{grad}\theta)^{\cdot}$, $\operatorname{grad}(\operatorname{grad}\theta)$, $\operatorname{div}\boldsymbol{v}$ and $\left(\boldsymbol{D} - \frac{1}{3}I_{\boldsymbol{D}}\boldsymbol{I}\right)$, which all may have any arbitrarily assigned values, we have

$$
\begin{aligned}
&\frac{\partial \hat{s}}{\partial \theta} - \Lambda^{\varepsilon} \frac{\partial \hat{\varepsilon}}{\partial \theta} = 0\,, \\
&\frac{\partial \hat{s}}{\partial \operatorname{grad} \theta} - \Lambda^{\varepsilon} \frac{\partial \hat{\varepsilon}}{\partial \operatorname{grad} \theta} = \mathbf{0}\,, \\
&\left\{ \frac{\partial \hat{\boldsymbol{\phi}}}{\partial \operatorname{grad} \theta} - \Lambda^{\varepsilon} \frac{\partial \hat{\boldsymbol{q}}}{\partial \operatorname{grad} \theta} - \boldsymbol{\Lambda}^{\boldsymbol{v}} \frac{\partial \hat{\boldsymbol{t}}^{E}}{\partial \operatorname{grad} \theta} \right\}_{\text{sym}} = \mathbf{0}\,, \\
&\Lambda^{\rho} = -\Lambda^{\varepsilon} p\,, \\
&\boldsymbol{t}^{E} = \mathbf{0}\,.
\end{aligned}
\qquad (5.8.41)
$$

These identities show that the LAGRANGE multipliers $\Lambda^{\varepsilon}, \boldsymbol{\Lambda}^{\boldsymbol{v}}$ may be viewed as constitutive quantities. Furthermore, Λ^{ρ} is determined by Λ^{ε} and the constraint pressure. These facts imply that in particular $\boldsymbol{\Lambda}^{\boldsymbol{v}}$ does not depend on $\dot{\boldsymbol{v}}$ so that the inequality (5.8.39) is also linear in $\dot{\boldsymbol{v}}$. As a consequence

$$
\boldsymbol{\Lambda}^{\boldsymbol{v}} = \mathbf{0}\,. \qquad (5.8.42)
$$

The momentum equation does not influence the thermodynamics.

Using the representations of $\boldsymbol{\phi}$ and $\boldsymbol{q}$ as isotropic functions of their variables, it is now straightforward to show that $(5.8.41)_3$ implies that Λ^{ε} is only a function of the empirical temperature, $\Lambda^{\varepsilon} = \Lambda^{\varepsilon}(\theta)$, and $\boldsymbol{q}$ and $\boldsymbol{\phi}$ are collinear such that

$$
\boldsymbol{\phi} = \Lambda^{\varepsilon}(\theta)\, \boldsymbol{q}\,, \qquad (5.8.43)
$$

for details see the paragraph from (5.8.12) - (5.8.16). Similarly, with the aid of item 5) of the entropy principle it may also be demonstrated that $\Lambda^{\varepsilon}(\theta)$ is a universal function of the empirical temperature, (see the arguments leading to (5.8.18)), so that

$$
\Theta(\theta) = \frac{1}{\Lambda^{\varepsilon}(\theta)} = T \qquad (5.8.44)
$$

may be identified with the absolute temperature.

Let us focus the attention now on the identities $(5.8.41)_{1,2}$. Differentiating $(5.8.41)_1$ with respect to $\operatorname{grad} \theta$ and $(5.8.41)_2$ with respect to θ and comparing the two emerging results shows that

$$
s = \hat{s}(\theta)\,, \quad \varepsilon = \hat{\varepsilon}(\theta)\,, \qquad (5.8.45)
$$

provided that $\Lambda^{\varepsilon}(\theta)$ is a nontrivial function of θ; thus the GIBBS relation of a density preserving heat conducting fluid takes the form

$$
\mathrm{d}s = \Lambda^{\varepsilon}(\theta) \mathrm{d}\varepsilon\,. \qquad (5.8.46)
$$

The results (5.8.45) and (5.8.46) are also interesting for the following fact: There is no relation like (5.8.23) or (5.8.24) in a density preserving fluid, in which a certain combination of derivatives of the pressure and internal energy would be related to the logarithmic derivative of Λ^{ε}.

With the identities (5.8.41)-(5.8.44) being satisfied inequality (5.8.39) reduces to

$$-\frac{\boldsymbol{q}\cdot \operatorname{grad}\theta}{\Theta(\theta)} \geq 0\,, \tag{5.8.47}$$

where $\Theta(\theta) > 0$. With $\boldsymbol{q} = -q_1(\theta, \operatorname{grad}\theta)\operatorname{grad}\theta$ this implies

$$q_1(\theta, 0) \geq 0\,. \tag{5.8.48}$$

The proof follows the same lines as that which led to (5.8.35).

This completes the thermodynamic analysis for a heat conducting density preserving fluid; the results that this more general entropy principle delivers are in this case the same as those obtained by the CLAUSIUS-DUHEM inequality.

Supplement: Proof of Liu's Theorem[23] We now return to the balance equations (5.8.4) and the material equations (5.8.3). Substituting the material equations (5.8.3) into the balance equations (5.8.4), the resulting balance equations of mass, momentum and energy – known as field equations – can be written in the form

$$\boldsymbol{A}\boldsymbol{x} + \boldsymbol{b} = \boldsymbol{0}\,, \tag{5.8.49}$$

in which $\boldsymbol{x}$, $\boldsymbol{A}$ and $\boldsymbol{b}$ are given by

$$\boldsymbol{x} = \{\,\dot{\theta}\,,\, \dot{\rho}\,,\, (\operatorname{grad}\theta)^{\cdot}\,,\, \operatorname{grad}\rho\,,\, \operatorname{grad}(\operatorname{grad}\theta)\,\}\,,$$

$$\boldsymbol{A} = \begin{pmatrix} 0 & 1 & 0 & 0 & 0 \\ 0 & 0 & 0 & -\dfrac{\partial\hat{\boldsymbol{t}}}{\partial\rho} & -\dfrac{\partial\hat{\boldsymbol{t}}}{\partial\operatorname{grad}\theta} \\ \rho\dfrac{\partial\hat{\varepsilon}}{\partial\theta} & \rho\dfrac{\partial\hat{\varepsilon}}{\partial\rho} & \rho\dfrac{\partial\hat{\varepsilon}}{\partial\operatorname{grad}\theta} & \dfrac{\partial\hat{\boldsymbol{q}}}{\partial\rho} & \dfrac{\partial\hat{\boldsymbol{q}}}{\partial\operatorname{grad}\theta} \end{pmatrix}, \tag{5.8.50}$$

$$\boldsymbol{b}^T = \left(\rho\operatorname{div}\boldsymbol{v}\,,\, \rho\dot{\boldsymbol{v}} - \frac{\partial\hat{\boldsymbol{t}}}{\partial\theta}\operatorname{grad}\theta - \rho\boldsymbol{g}\,,\, \rho\frac{\partial\hat{\varepsilon}}{\partial\theta} - \operatorname{tr}(\hat{\boldsymbol{t}}\boldsymbol{D}) - \rho\mathfrak{r}\right).$$

Likewise the entropy inequality takes the form

$$\boldsymbol{\alpha}\cdot\boldsymbol{x} + \beta \geq 0 \tag{5.8.51}$$

with

[23] See LIU [136], or MÜLLER [165]. Actually, LIU's theorem is a special case of a much broader theorem well known in operations research. A proof in that context can be found in SCHRIJVER [210], but the theorem dates back to FARKAS [71] and MINKOWSKI [155], see also HAUSER and KIRCHNER [99].

$$
\begin{aligned}
\boldsymbol{\alpha}^T &= \left(\rho\frac{\partial \hat{s}}{\partial \theta}\,,\ \rho\frac{\partial \hat{s}}{\partial \rho}\,,\ \rho\frac{\partial \hat{s}}{\partial\,\mathrm{grad}\,\theta}\,,\ \frac{\partial \phi}{\partial \rho}\,,\ \left(\frac{\partial \phi}{\partial\,\mathrm{grad}\,\theta}\right)_{\mathrm{sym}} \right)\,,\\
\beta &= \frac{\partial \phi}{\partial \theta}\cdot \mathrm{grad}\,\theta\,.
\end{aligned} \tag{5.8.52}
$$

Equations (5.8.49) and the inequality (5.8.51) are linear in the variables $\boldsymbol{x}$, since these variables are not contained in the set of constitutive variables. If one considers a solution of the balance equations at a position in space and time in the form $\boldsymbol{A}\boldsymbol{x}+\boldsymbol{b}=\boldsymbol{0}$, then from the dimension of $\boldsymbol{A}$ it is apparent that this equation allows a whole variety of higher dimensions from which $\boldsymbol{x}$ can originate, if $\boldsymbol{A}$ and $\boldsymbol{b}$ are fixed. Indeed, $\boldsymbol{A}$ possesses in any case more columns than rows. This being assured it is, however, still not clear whether to all these values of $\boldsymbol{x}$ at fixed $\boldsymbol{A}$ and $\boldsymbol{b}$ there belong in reality globally meaningful fields as solutions of the balance equations. Actually, it can be applied even for an empty solution set. This is shown in HAUSER & KIRCHNER (2002). In many cases the desired proof is carried out with the conditions of CAUCHY & KOVALEVSKAYA being fulfilled. This is the case here, but we mention it only without proof.

Theorem *Let a matrix $\boldsymbol{A}$ and vectors $\boldsymbol{x}$, $\boldsymbol{b}$ and $\boldsymbol{\alpha}$, as well as a scalar β be given. In view of their dimensions these are assumed compatible with the statements*

$$
\boldsymbol{A}\boldsymbol{x} + \boldsymbol{b} = \boldsymbol{0} \quad \textit{and} \quad \boldsymbol{\alpha}\cdot\boldsymbol{x} + \beta \geq 0\,. \tag{5.8.53}
$$

Assume, moreover, that the linear equation system $(5.8.53)_1$ has for $\boldsymbol{x}$ a non-empty solution set $\mathbb{S}$. Then the following statements are equivalent:

(a) For all $\boldsymbol{x} \in \mathbb{S}$ the inequality $\boldsymbol{\alpha}\cdot\boldsymbol{x} + \beta \geq 0$ holds.
(b) There exists a vector $\boldsymbol{\lambda} \neq \boldsymbol{0}$, such that

$$
\boldsymbol{\alpha}^T - \boldsymbol{\lambda}^T\boldsymbol{A} = \boldsymbol{0} \quad \textit{and} \quad \beta - \boldsymbol{\lambda}\cdot\boldsymbol{b} \geq 0\,. \tag{5.8.54}
$$

Proof

i) From *(b)* follows *(a)*: We multiply the first equation from *(b)* with an arbitrary $\boldsymbol{x}$ of dimension of $\boldsymbol{\alpha}$ and add this to the second inequality from *(b)*. This yields

$$
\begin{aligned}
&\beta - \boldsymbol{\lambda}\cdot\boldsymbol{b} + (\boldsymbol{\alpha}^T - \boldsymbol{\lambda}^T\boldsymbol{A})\boldsymbol{x} \geq 0\\
&\boldsymbol{\lambda}\cdot \underbrace{(\boldsymbol{A}\boldsymbol{x} + \boldsymbol{b})}_{=0 \text{ since } \boldsymbol{x}\in\mathbb{S}} + (\boldsymbol{\alpha}\cdot\boldsymbol{x} + \beta) \geq 0 \quad \longrightarrow \quad (\boldsymbol{\alpha}\cdot\boldsymbol{x} + \beta) \geq 0\,,
\end{aligned}
$$

which proves statement *(a)*.

ii) From *(a)* follows *(b)*: This is shown by contradicting the opposite assumption: Let us therefore assume, *(a)* holds, but there does not exist a suitable $\boldsymbol{\lambda}$ with $\boldsymbol{\alpha} - \boldsymbol{\lambda}^T\boldsymbol{A} = \boldsymbol{0}$. Then the vector $\boldsymbol{\alpha}^T$ is linearly independent of the rows of the matrix $\boldsymbol{A}$. This lets us to find a vector with the property

$\boldsymbol{A}\boldsymbol{x}_0 = \boldsymbol{0}$, but at the same time $\boldsymbol{\alpha}\cdot\boldsymbol{x}_0 \neq 0$. One now adds a suitable multiple $\delta\boldsymbol{x}_0$ of this vector to a solution of $\boldsymbol{A}\boldsymbol{x} + \boldsymbol{b} = \boldsymbol{0}$, one obtains again a solution vector because $\boldsymbol{A}(\boldsymbol{x} + \delta\boldsymbol{x}_0) = \boldsymbol{A}\boldsymbol{x}$. On the other hand, one is now able to violate the inequality $\boldsymbol{\alpha}\cdot(\boldsymbol{x} + \delta\boldsymbol{x}_0) + \beta \geq 0$ arbitrarily. Thus a contradiction to the assumption that the statement *(a)* holds has been obtained. Therefore, there exists really a $\boldsymbol{\lambda}$ with $\boldsymbol{\alpha}^T - \boldsymbol{\lambda}^T\boldsymbol{A} = \boldsymbol{0}$, and thus one necessarily has $\beta - \boldsymbol{\lambda}\cdot\boldsymbol{b} \geq 0$. ■

In a postscript to this law let us mention that the statement $\boldsymbol{\alpha}^T - \boldsymbol{\lambda}^T\boldsymbol{A} = \boldsymbol{0}$ in our example corresponds to the identities (5.8.10), and the second relation (5.8.54), $\beta - \boldsymbol{\lambda}\cdot\boldsymbol{b} \geq 0$ corresponds to the residual inequality (5.8.30) which has been obtained from the last two rows of (5.8.6).

5.9 Exercises

1. Determine the transformation properties (under EUCLIDian transformations) of the left and right stretch tensors, $\boldsymbol{V}$ and $\boldsymbol{U}$ as well as the rotation tensor $\boldsymbol{R}$.
2. What is the constraint condition for a rigid body? Which property possesses the deformation gradient in this case?
3. On the basis of the principle of material objectivity, which the constraint condition $\phi(\boldsymbol{F}) = 0$ must obey, show that the latter can also be written as $\phi(\boldsymbol{C}) = 0$, where $\boldsymbol{C} = \boldsymbol{F}^T\boldsymbol{F}$ is the right CAUCHY-GREEN tensor. With this result then the relation
$$\boldsymbol{Z} = \Lambda\frac{\partial\phi}{\partial\boldsymbol{F}}\boldsymbol{F}^T = 2\Lambda\boldsymbol{F}\frac{\partial\hat{\phi}}{\partial\boldsymbol{C}}\boldsymbol{F}^T$$
must hold.
4. Consider a fiber reinforced material composed of non-stretchable fibres, which possess the directional field $\boldsymbol{A}(\boldsymbol{x}), |\boldsymbol{A}| = 1$ in the reference configuration. Determine the internal constraint condition and the associated constraint stresses.
5. For an elastic material, $\boldsymbol{t} = \hat{\boldsymbol{t}}(\boldsymbol{F})$, show that a pure rigid body motion can not generate stresses (this follows from the material objectivity).
6. Show that the unimodular matrix $\boldsymbol{P}$ with $\det(\boldsymbol{P}) = 1$, form a group (with respect to matrix multiplication). Give an example of a non-orthogonal unimodular matrix.
7. Unimodular matrices are also called "volume preserving". Prove that a prism shows this behaviour of volume preservation under an unimodular transformation.
8. Show that the EUCLIDian transformations ($\boldsymbol{x}^* = \boldsymbol{O}(t)\boldsymbol{x} + \boldsymbol{c}(t)$) form a group with $\boldsymbol{O}\boldsymbol{O}^T = \boldsymbol{I}$.
9. Prove that the set of all symmetry transformations of a body form a group.

10. Show that (for a viscous heat conducting fluid) in thermodynamic equilibrium the stresses are isotropic and the heat flux vanishes:

$$\boldsymbol{t}_{|E} = \rho^2 \left(T \frac{\partial s}{\partial \rho} - \frac{\partial \epsilon}{\partial \rho} \right) \boldsymbol{I} = -p \boldsymbol{I} \,, \quad \boldsymbol{q}_{|E} = \boldsymbol{0} \,.$$

11. Consider the second PIOLA–KIRCHHOFF stress tensor $\tilde{\boldsymbol{T}} = \tilde{\boldsymbol{T}}(\boldsymbol{C})$ of an isotropic elastic solid. How can you visualise its functional dependency on the right CAUCHY–GREEN tensor $\boldsymbol{C}$ in the most general case? (Hint: An orthogonal symmetry transformation of the reference system is helpful.)
12. Show that the quantities $\overset{\circ}{\boldsymbol{b}}$ and $\overset{\diamond}{\boldsymbol{A}}$ defined in (5.2.5) are objective.
13. The stress-power in the EULER representation is given by $\varphi = \operatorname{tr}\left(\boldsymbol{L}\boldsymbol{t}^T\right)$. Assume that under EUCLIDian transformations, φ behaves like an objective scalar and show that the CAUCHY stress tensor must then necessarily be symmetric, $\boldsymbol{t} = \boldsymbol{t}^T$.
14. Prove that the representations

$$\begin{aligned}
\boldsymbol{t}(t) &= \overset{\infty}{\underset{\tau=0}{\hat{\boldsymbol{t}}}}\left(\boldsymbol{F}, \Theta, \operatorname{Grad}\Theta, \boldsymbol{X}\right) = \boldsymbol{R}\, \overset{\infty}{\underset{\tau=0}{\hat{\boldsymbol{t}}}}\left(\boldsymbol{U}, \Theta, \operatorname{Grad}\Theta, \boldsymbol{X}\right)\boldsymbol{R}^T, \\
\boldsymbol{q}(t) &= \overset{\infty}{\underset{\tau=0}{\hat{\boldsymbol{q}}}}\left(\boldsymbol{F}, \Theta, \operatorname{Grad}\Theta, \boldsymbol{X}\right) = \boldsymbol{R}\, \overset{\infty}{\underset{\tau=0}{\hat{\boldsymbol{q}}}}\left(\boldsymbol{U}, \Theta, \operatorname{Grad}\Theta, \boldsymbol{X}\right), \\
\varepsilon(t) &= \overset{\infty}{\underset{\tau=0}{\hat{\varepsilon}}}\left(\boldsymbol{F}, \Theta, \operatorname{Grad}\Theta, \boldsymbol{X}\right) = \overset{\infty}{\underset{\tau=0}{\hat{\varepsilon}}}\left(\boldsymbol{U}, \Theta, \operatorname{Grad}\Theta, \boldsymbol{X}\right),
\end{aligned}$$

corresponding to (5.2.20) are sufficient to fulfill the principle of material objectivity. In these equations the independent variables are functions of $\boldsymbol{X}$ and $(t - \tau)$.
15. Show that the objective material functionals for the second PIOLA–KIRCHHOFF stress tensor, the heat flux vector and the internal energy for a viscoelastic body can be written in the form

$$\begin{aligned}
\tilde{\boldsymbol{T}}(t) &= \hat{\boldsymbol{T}}\left(\boldsymbol{C}, \dot{\boldsymbol{C}}, \operatorname{Grad}\Theta, \boldsymbol{X}\right) \,, \\
\tilde{\boldsymbol{Q}}(t) &= \hat{\boldsymbol{Q}}\left(\boldsymbol{C}, \dot{\boldsymbol{C}}, \operatorname{Grad}\Theta, \boldsymbol{X}\right) \,, \\
\tilde{\varepsilon}(t) &= \hat{\varepsilon}\left(\boldsymbol{C}, \dot{\boldsymbol{C}}, \operatorname{Grad}\Theta, \boldsymbol{X}\right) \,,
\end{aligned} \tag{5.9.1}$$

and show further that one can substitute the GREEN strain tensor and its time derivative, $\boldsymbol{G}$,$\dot{\boldsymbol{G}}$ instead of $\boldsymbol{C}$,$\dot{\boldsymbol{C}}$.
16. In the LAGRANGE representation the stress power is given by $\hat{\varphi} = \operatorname{Tr}\left(\operatorname{Grad}\boldsymbol{v}\,\boldsymbol{T}^T\right)$. Show that the objective representation for $\hat{\varphi}$ has the form

$$\hat{\varphi} = \tfrac{1}{2}\operatorname{Tr}\left(\tilde{\boldsymbol{T}}\dot{\boldsymbol{C}}\right) = \operatorname{Tr}\left(\tilde{\boldsymbol{T}}\dot{\boldsymbol{G}}\right) \,.$$

17. The n-th RIVLIN–ERICKSEN tensor $\boldsymbol{A}_{(n)}$ is related to the $(n+1)$-th via the recurrence relation

$$\boldsymbol{A}_{(n+1)} = \dot{\boldsymbol{A}}_{(n)} + \boldsymbol{L}^T \boldsymbol{A}_{(n)} + \boldsymbol{A}_{(n)} \boldsymbol{L} \, , \quad \boldsymbol{A}_{(n)} = {\boldsymbol{A}_{(n)}}^T \, .$$

If $\boldsymbol{A}_{(0)} = \boldsymbol{I}$, $\boldsymbol{A}_{(1)} = 2\boldsymbol{D}$ becomes an objective tensor. Show, by induction, that all RIVLIN–ERICKSEN tensors are objective.

18. Let $\boldsymbol{f}(\boldsymbol{A})$ be an analytic isotropic symmetric second rank tensor function of a symmetric second rank tensor $\boldsymbol{A}$. Show by using the CAYLEY–HAMILTON theorem that $\boldsymbol{f}$ has the representation

$$\boldsymbol{f}(\boldsymbol{A}) = a_0 \boldsymbol{I} + a_1 \boldsymbol{A} + a_2 \boldsymbol{A}^2 \, , \quad a_i = \hat{a}_i(I_{\boldsymbol{A}}, II_{\boldsymbol{A}}, III_{\boldsymbol{A}}) \, , \quad i = 0, 1, 2 \, .$$

19. Let $\boldsymbol{q}$ be an isotropic vector function dependent only on a scalar μ: $\boldsymbol{q} = \hat{\boldsymbol{v}}(\mu)$. Then the relation $\boldsymbol{q} = \boldsymbol{0}$ must necessarily hold.

20. A thermoelastic solid is given by the material equations of the form $\Psi = \hat{\Psi}(\boldsymbol{C}, \Theta, \operatorname{Grad} \Theta)$, where $\boldsymbol{C} = \boldsymbol{F}^T \boldsymbol{F}$ denotes the right CAUCHY–GREEN deformation tensor and $\operatorname{Grad} \Theta$ the material temperature gradient, and $\Psi \in \{s, \Psi_\varepsilon, \tilde{\boldsymbol{T}}, \boldsymbol{Q}\}$ stand for the entropy s, the HELMHOLTZ free energy Ψ_ε, the second PIOLA–KIRCHHOFF tensor $\tilde{\boldsymbol{T}}$ and the material heat flux $\boldsymbol{Q}$. Show that

$$\tilde{\boldsymbol{T}} = \rho_R \frac{\partial \hat{\Psi}_\varepsilon}{\partial \boldsymbol{C}} \, , \quad s = -\frac{\partial \hat{\Psi}_\varepsilon}{\partial \Theta} \, .$$

21. Let the dual, linear-viscous material equations

$$\boldsymbol{t}^E = \kappa (\operatorname{tr} \boldsymbol{D}) \boldsymbol{I} + 2\mu \boldsymbol{D}' \, , \quad \boldsymbol{D} = g \left(\operatorname{tr} \boldsymbol{t}^E \right) \boldsymbol{I} + \tfrac{1}{2} f \boldsymbol{t}^{E'} \, ,$$

be given, in which $(\kappa, \mu) = (\kappa, \mu)(\rho, \Theta)$ and $(g, f) = (g, f)(p, \Theta))$. Assume, moreover, that for p a thermal equation of state $p = \hat{p}(\rho, \Theta)$ holds. $\boldsymbol{D}'$ and $\boldsymbol{t}^{E'}$ are the deviators of $\boldsymbol{D}$ und $\boldsymbol{t}^E$. Prove the following relations

$$g(p, \Theta) = \frac{1}{9\kappa(\rho, \Theta)} \, , \quad f(p, \Theta) = \frac{1}{\mu(\rho, \Theta)} \, .$$

22. The transformation between a fixed and moving reference system is given by the EUCLIDian transformation as

$$\boldsymbol{y}^* = \boldsymbol{O}(t) \boldsymbol{x} - \boldsymbol{c}^*(t) \, .$$

Verify with this the transformation rules given in (5.2.2), the transformations for the left CAUCHY–GREEN tensor $\boldsymbol{B} = \boldsymbol{F}\boldsymbol{F}^T$, the right CAUCHY–GREEN tensor $\boldsymbol{C} = \boldsymbol{F}^T \boldsymbol{F}$ and the first and the second PIOLA–KIRCHHOFF stress tensors $\boldsymbol{T} = J \boldsymbol{t} \boldsymbol{F}^{-T}$ and $\tilde{\boldsymbol{T}} = J \boldsymbol{F}^{-1} \boldsymbol{t} \boldsymbol{F}^{-T}$.

23. For a NEWTONian fluid, write the momentum balance equation. How does one simplify this under the incompressibility condition and the STOKES assumption? What is the stress power in this case? Assume that the viscosities (μ, λ) are prescribed.

24. For a hexagonal crystal determine the number of independent coefficients (eigenvalues) of the matrix $(\mathbb{C})$ for the law $\mathfrak{S} = \mathbb{C} \mathcal{A}$ as given in (5.5.22).

5.10 Solutions

1. Starting from the polar decomposition of the deformation gradient

$$\boldsymbol{F} = \boldsymbol{R}\boldsymbol{U} = \boldsymbol{V}\boldsymbol{R}$$

and its transformation

$$\boldsymbol{F}^* = \boldsymbol{O}\boldsymbol{F} \, ,$$

the transformations for the left and the right CAUCHY-GREEN tensors are given by

$$\boldsymbol{B}^* = \boldsymbol{O}\boldsymbol{B}\boldsymbol{O}^T \quad \text{with} \quad \boldsymbol{B} = \boldsymbol{F}\boldsymbol{F}^T = \boldsymbol{V}^2,$$

$$\boldsymbol{C}^* = \boldsymbol{C} \qquad \text{with} \quad \boldsymbol{C} = \boldsymbol{F}^T\boldsymbol{F} = \boldsymbol{U}^2 \, .$$

From here, one obtains the transformations of $\boldsymbol{V}$ and $\boldsymbol{U}$ as

$$\boldsymbol{V}^{*2} = \boldsymbol{O}\boldsymbol{V}^2\boldsymbol{O}^T \, , \quad \boldsymbol{U}^{*2} = \boldsymbol{U}^2 \, ,$$

which can only be fulfilled when

$$\boldsymbol{V}^* = \boldsymbol{O}\boldsymbol{V}\boldsymbol{O}^T \quad \text{and} \quad \boldsymbol{U}^* = \boldsymbol{U} \, .$$

Then this implies

$$\boldsymbol{R}^* = \boldsymbol{O}\boldsymbol{R} \, .$$

2. A rigid body is characterized by the fact that all distances (lengths and angles) between two or three points remain conserved,

$$d\boldsymbol{x}_1 \cdot d\boldsymbol{x}_2 = d\boldsymbol{X}_1 \cdot d\boldsymbol{X}_2 \, .$$

Using the transformation of the line elements $d\boldsymbol{x} = \boldsymbol{F}d\boldsymbol{X}$, we obtain

$$d\boldsymbol{x}_1 \cdot d\boldsymbol{x}_2 = \boldsymbol{F}d\boldsymbol{X}_1 \cdot \boldsymbol{F}d\boldsymbol{X}_2 = \boldsymbol{F}^T\boldsymbol{F}d\boldsymbol{X}_1 \cdot d\boldsymbol{X}_2 = d\boldsymbol{X}_1 \cdot d\boldsymbol{X}_2 \, ,$$

and thus $\boldsymbol{F}^T\boldsymbol{F} = \boldsymbol{I}$. The constraint condition of a rigid body is thus

$$\phi(\boldsymbol{F}) = \boldsymbol{F}^T\boldsymbol{F} - \boldsymbol{I} = \boldsymbol{0} \, .$$

The deformation gradient is an orthogonal tensor and thus describes a pure rotation.

3. From $\phi(\boldsymbol{F}) = \phi(\boldsymbol{O}\boldsymbol{F})$ for all $\boldsymbol{O} \in \mathbb{O}$ follows, with the polar decomposition $\boldsymbol{F} = \boldsymbol{R}\boldsymbol{U}$ and the special choice of the orthogonal transformation $\boldsymbol{O} = \boldsymbol{R}^T$, that $\phi = \tilde{\phi}(\boldsymbol{U}) = \hat{\phi}(\boldsymbol{C})$ holds. Thus the constraint condition can also be written as

$$\hat{\phi}(\boldsymbol{C}) = 0 \, .$$

From a differentiation of this with respect to $\boldsymbol{F}$, using the chain rule and with the relations $\boldsymbol{C} = \boldsymbol{F}^T\boldsymbol{F}$ or $C_{\alpha\beta} = F_{j\alpha}F_{j\beta}$, we deduce

$$\begin{aligned}\frac{\partial \phi}{\partial F_{i\gamma}} &= \frac{\partial \hat{\phi}(\boldsymbol{C})}{\partial C_{\alpha\beta}} \frac{\partial}{\partial F_{i\gamma}} (F_{j\alpha} F_{j\beta}) \\ &= \frac{\partial \hat{\phi}}{\partial C_{\alpha\beta}} (\delta_{ij}\delta_{\gamma\alpha} F_{j\beta} + F_{j\alpha}\delta_{ij}\delta_{\beta\gamma}) \\ &= \frac{\partial \hat{\phi}}{\partial C_{\alpha\beta}} (\delta_{\gamma\alpha} F_{i\beta} + \delta_{\beta\gamma} F_{i\alpha}) \ .\end{aligned}$$

Therefore, with (5.6.8), the constraint stress $\boldsymbol{Z}$ is given by

$$\begin{aligned}Z_{ij} = \Lambda \frac{\partial \hat{\phi}}{\partial F_{i\gamma}} F_{j\gamma} &= \Lambda \frac{\partial \hat{\phi}}{\partial C_{\alpha\beta}} (\delta_{\gamma\alpha} F_{i\beta} + \delta_{\beta\gamma} F_{i\alpha}) F_{j\gamma} \\ &= \Lambda F_{j\alpha} \frac{\partial \hat{\phi}}{\partial C_{\alpha\beta}} F_{i\beta} + \Lambda F_{j\beta} \frac{\partial \hat{\phi}}{\partial C_{\alpha\beta}} F_{i\alpha} \ .\end{aligned}$$

Since $\boldsymbol{C}$ is symmetric, we have

$$\boldsymbol{Z} = \Lambda \boldsymbol{F} \frac{\partial \hat{\phi}}{\partial \boldsymbol{C}} \boldsymbol{F}^T + \Lambda \boldsymbol{F} \frac{\partial \hat{\phi}}{\partial \boldsymbol{C}} \boldsymbol{F}^T = 2\Lambda \boldsymbol{F} \frac{\partial \hat{\phi}}{\partial \boldsymbol{C}} \boldsymbol{F}^T \ . \tag{5.10.1}$$

4. The direction field of fibres in the present configuration is given by

$$\boldsymbol{a} = \boldsymbol{F}\boldsymbol{A} \ ;$$

$\boldsymbol{a}$ must remain a unit vector if the fibres are not stretchable; from this the internal constraint condition follows as

$$\boldsymbol{a} \cdot \boldsymbol{a} = \boldsymbol{F}\boldsymbol{A} \cdot \boldsymbol{F}\boldsymbol{A} = 1 \quad \text{and} \quad \Phi(\boldsymbol{F}) = \boldsymbol{A} \cdot \boldsymbol{F}^T\boldsymbol{F}\boldsymbol{A} - 1 = 0$$

or with the result from Exercise 3,

$$\hat{\phi}(\boldsymbol{C}) = \boldsymbol{A} \cdot \boldsymbol{C}\boldsymbol{A} - 1 = 0 \ .$$

The derivative of the constraint condition with respect to $\boldsymbol{C}$ gives

$$\frac{\partial \hat{\phi}}{\partial \boldsymbol{C}} = \boldsymbol{A} \otimes \boldsymbol{A} \ ,$$

and thus from (5.10.1)

$$\begin{aligned}\boldsymbol{Z} &= 2\Lambda \boldsymbol{F}\boldsymbol{A} \otimes \boldsymbol{A}\boldsymbol{F}^T = 2\Lambda (\boldsymbol{F}\boldsymbol{A}) \otimes (\boldsymbol{F}\boldsymbol{A}) \\ &= 2\Lambda \boldsymbol{a} \otimes \boldsymbol{a} \ .\end{aligned}$$

The constraint stresses are given by the dyadic product of the effective directional field.

5. The principle of material objectivity demands that

$$t(F) = R\hat{g}(C)R^T , \tag{5.10.2}$$

where $F = RU$, $U^2 = C$. If the natural state is stress free, then for $F = I$ we have

$$t(I) = \hat{g}(I) = 0 . \tag{5.10.3}$$

In case of a rigid body motion $F_{\text{rigid}} = Q$ holds, where $Q \in \mathbb{O}$ is an orthogonal matrix. Then $C_{\text{rigid}} = I$ holds, so that from (5.10.2) and (5.10.3) one obtains

$$\hat{t}(Q) = Q\hat{t}(I)Q^T = 0 , \quad \text{qed.}$$

An arbitrary rigid body motion creates no stress in an elastic body.

6. Let P_1, $P_2 \in \mathbb{U}$ (unimodular matrices) with $\det(P_i) = 1$ $(i = 1, 2)$; then the validity of the group axioms for matrix multiplication follows immediately,
   a. $\det(P_1P_2) = \det(P_2P_1) = \det(P_1)\det(P_2) = 1$, i.e., the combination of two elements of the group is again an element of the group, $P = P_1P_2 \in \mathbb{U}$.
   b. There exists a unit element, the unit matrix I, with

   $$P = IP = PI$$

   and

   $$\det(I) = 1, \quad \text{i.e.} \quad I \in \mathbb{U} .$$

   c. To every unimodular matrix P there exists an inverse element, denoted by P^{-1}, the inverse matrix. Since $\det(P) = 1$, P is not singular and from

   $$PP^{-1} = P^{-1}P = I \quad \text{follows also}$$

   $$\det(PP^{-1}) = 1 \quad \Rightarrow \quad \det(P^{-1}) = 1$$

   and therefore $P^{-1} \in \mathbb{U}$.

   An orthogonal matrix Q with $QQ^T = 1$ is obviously unimodular, since $\det(QQ^T) = 1 \Rightarrow \det(Q) = 1$; the reverse is not true; so if e.g. the unimodular stretch matrix is given by

   $$P = \begin{pmatrix} a & 0 & 0 \\ 0 & b & 0 \\ 0 & 0 & \frac{1}{ab} \end{pmatrix} , \quad \det(P) = 1 ,$$

   this matrix is not orthogonal, except for $a = b = 1$.

7. The volume of a prism, which is spanned by three vectors a, b, c, is given by

$$V = |a \cdot (b \times c)| = |\det(a,\ b,\ c)| ,$$

where $(a,\ b,\ c)$ is a matrix composed of three column vectors a, b, c.

Through a transformation with the unimodular matrix $\boldsymbol{P}$ the new vectors $\boldsymbol{a}' = \boldsymbol{Pa}$, $\boldsymbol{b}' = \boldsymbol{Pb}$, $\boldsymbol{c}' = \boldsymbol{Pc}$ can be defined whose triple product is

$$V' = \boldsymbol{Pa} \cdot (\boldsymbol{Pb} \times \boldsymbol{Pc}) = \det(\boldsymbol{Pa},\ \boldsymbol{Pb},\ \boldsymbol{Pc})\ .$$

According to the rule of calculation of determinants this yields

$$V' = \det(\boldsymbol{Pa},\ \boldsymbol{Pb},\ \boldsymbol{Pc}) = \underbrace{\det(\boldsymbol{P})}_{=1,\ \text{unimodular}} \det(\boldsymbol{a},\ \boldsymbol{b},\ \boldsymbol{c}) = \det(\boldsymbol{a},\ \boldsymbol{b},\ \boldsymbol{c}) = V\ .$$

The unimodular transformation is volume preserving.

8. Let

$$\boldsymbol{P} = (\boldsymbol{\mathcal{Q}},\ \boldsymbol{c}) \in \mathbb{E}\ (\text{EUCLIDian transformation})$$

denote the elements of an EUCLIDian transformation , where $\boldsymbol{\mathcal{Q}}$ is an orthogonal transformation and $\boldsymbol{c}$ is a displacement vector.

a. The composition of two EUCLIDian transformations

$$\boldsymbol{P} = \boldsymbol{P}_1 \circ \boldsymbol{P}_2 = (\boldsymbol{\mathcal{Q}}_1, \boldsymbol{c}_1) \circ (\boldsymbol{\mathcal{Q}}_2, \boldsymbol{c}_2) = (\boldsymbol{\mathcal{Q}}_1\boldsymbol{\mathcal{Q}}_2,\ \boldsymbol{\mathcal{Q}}_1\boldsymbol{c}_2 + \boldsymbol{c}_1)$$

is again an EUCLIDian transformation, with the displacement vector $\boldsymbol{\mathcal{Q}}_1\boldsymbol{c}_2 + \boldsymbol{c}_1$, and the orthogonal transformation $\boldsymbol{\mathcal{Q}}_1\boldsymbol{\mathcal{Q}}_2$ is then

$$(\boldsymbol{\mathcal{Q}}_1\boldsymbol{\mathcal{Q}}_2)(\boldsymbol{\mathcal{Q}}_1\boldsymbol{\mathcal{Q}}_2)^T = \boldsymbol{\mathcal{Q}}_1 \underbrace{\boldsymbol{\mathcal{Q}}_2\boldsymbol{\mathcal{Q}}_2^T}_{\text{orthogonal}} \boldsymbol{\mathcal{Q}}_1^T = \underbrace{\boldsymbol{\mathcal{Q}}_1\boldsymbol{\mathcal{Q}}_1^T}_{\text{orthogonal}} = \boldsymbol{I}\ .$$

b. The identity of the group is $\boldsymbol{E} = (\boldsymbol{I},\ \boldsymbol{0})$, since

$$(\boldsymbol{I},\ \boldsymbol{0}) \circ \boldsymbol{P} = \boldsymbol{P} \circ (\boldsymbol{I},\ \boldsymbol{0}) = (\boldsymbol{\mathcal{Q}},\ \boldsymbol{c}) \circ (\boldsymbol{I},\ \boldsymbol{0}) = (\boldsymbol{\mathcal{Q}I},\ \boldsymbol{\mathcal{Q}0} + \boldsymbol{c}) = (\boldsymbol{\mathcal{Q}},\ \boldsymbol{c}) = \boldsymbol{P}$$

and $\boldsymbol{E} = (\boldsymbol{I},\ \boldsymbol{0}) \in \mathbb{E}$, here $\boldsymbol{I}(\boldsymbol{I})^T = \boldsymbol{I}$ holds.

c. The inverse element of an EUCLIDian transformation $\boldsymbol{P} = (\boldsymbol{\mathcal{Q}},\ \boldsymbol{c})$ is $\boldsymbol{P}^{-1} = (\boldsymbol{\mathcal{Q}}^T,\ -\boldsymbol{\mathcal{Q}}^T\boldsymbol{c})$. Therefore,

$$\begin{aligned} \boldsymbol{P} \circ \boldsymbol{P}^{-1} &= (\boldsymbol{\mathcal{Q}},\ \boldsymbol{c}) \circ (\boldsymbol{\mathcal{Q}}^T,\ -\boldsymbol{\mathcal{Q}}^T\boldsymbol{c}) \\ &= (\boldsymbol{\mathcal{Q}}\boldsymbol{\mathcal{Q}}^T, -\boldsymbol{\mathcal{Q}}\boldsymbol{\mathcal{Q}}^T\boldsymbol{c} + \boldsymbol{c}) = (\boldsymbol{I},\ \boldsymbol{0}) = \boldsymbol{E}\ , \end{aligned}$$

For the same reason $\boldsymbol{P}^{-1}$ is also an EUCLIDian transformation.

9. We choose the following notation:

$$\boldsymbol{A} := (\boldsymbol{F}, \Theta, \operatorname{Grad}\Theta)\ ,\quad \boldsymbol{H} := (\boldsymbol{P}, 1, \boldsymbol{P}) \in \mathbb{G}$$

with the product rule

$$\boldsymbol{AH} = (\boldsymbol{FP}, \Theta, \operatorname{Grad}\Theta\boldsymbol{P})\ .$$

Then the symmetry condition reads

$$\mathop{\hat{\Psi}}_{\tau=0}^{\infty}(\boldsymbol{A}(t-\tau)\boldsymbol{H}) = \mathop{\hat{\Psi}}_{\tau=0}^{\infty}(\boldsymbol{A}(t-\tau)) \quad \forall \boldsymbol{A} \quad \text{and} \quad \forall \boldsymbol{H} \in \mathbb{G}\ . \tag{5.10.4}$$

- The first is to show that

$$H_1 \in \mathbb{G}\,, \quad H_2 \in \mathbb{G} \quad \Rightarrow \quad H_1 H_2 \in \mathbb{G}\,.$$

In fact

$$\overset{\infty}{\underset{\tau=0}{\hat{\Psi}}}[(AH_1)H_2] \overset{(1)}{=} \overset{\infty}{\underset{\tau=0}{\hat{\Psi}}}(AH_1) \overset{(2)}{=} \overset{\infty}{\underset{\tau=0}{\hat{\Psi}}}(A)\,.$$

The first step (1) follows because $H_2 \in \mathbb{G}$, and the second step (2), because $H_1 \in \mathbb{G}$, where each time the property (5.10.4) is used. With this, the relation $\overset{\infty}{\underset{\tau=0}{\hat{\Psi}}}[A(H_1 H_2)] = \overset{\infty}{\underset{\tau=0}{\hat{\Psi}}}(A)$ also holds, which proves that $H_1 H_2 \in \mathbb{G}$ qed.
- Secondly, the identity $I = (I, 1, I)$ must be an element of $\mathbb{G}$, this is the case, from which it directly follows that $\overset{\infty}{\underset{\tau=0}{\hat{\Psi}}}(AI) = \overset{\infty}{\underset{\tau=0}{\hat{\Psi}}}(A)$ holds.
- Thirdly, $H \in \mathbb{G}$ implies $H^{-1} \in \mathbb{G}$, with $H^{-1} = (P^{-1}, 1, P^{-1})$ provided H^{-1} exists. In fact,

$$\overset{\infty}{\underset{\tau=0}{\hat{\Psi}}}(A) = \overset{\infty}{\underset{\tau=0}{\hat{\Psi}}}(AH^{-1}H) \overset{(1)}{=} \overset{\infty}{\underset{\tau=0}{\hat{\Psi}}}(AH^{-1})$$

holds. The first step follows because $H \in \mathbb{G}$, and with this (5.10.4) holds with A replaced by AH^{-1}. Thus

$$\overset{\infty}{\underset{\tau=0}{\hat{\Psi}}}(A) = \overset{\infty}{\underset{\tau=0}{\hat{\Psi}}}(AH^{-1})\,,$$

which implies directly $H^{-1} \in \mathbb{G}$. This proves that $\mathbb{G}$ forms a group.

10. The residual inequality of the entropy principle for a viscous, heat conducting fluid corresponding to (5.7.20) reads

$$\Gamma = \Big[-\rho^2\left(\Theta\frac{\partial s}{\partial \rho} - \frac{\partial \varepsilon}{\partial \rho}\right) I + t\Big] \cdot D - \frac{q \cdot \operatorname{grad}\Theta}{\Theta} \geq 0\,.$$

In equilibrium, Γ is a minimum relative to D and $\operatorname{grad}\Theta$, so that

$$\left(\frac{\partial \Gamma}{\partial D}\right)_{|E} = 0\,, \quad \left(\frac{\partial \Gamma}{\partial(\operatorname{grad}\Theta)}\right)_{|E} = 0$$

must hold. Introducing the derivatives yields immediately

$$\left(\frac{\partial \Gamma}{\partial D}\right)_{|E} = -\rho^2\left(\Theta\frac{\partial s}{\partial \rho} - \frac{\partial \varepsilon}{\partial \rho}\right) I + t_{|E} = 0$$

or

$$t_{|E} = -\rho^2\left(\Theta\frac{\partial s}{\partial \rho} - \frac{\partial \varepsilon}{\partial \rho}\right) I =: -pI$$

and

$$\left(\frac{\partial \Gamma}{\partial(\operatorname{grad}\Theta)}\right)_{|E} = -\frac{q_{|E}}{\Theta} = 0 \quad \Rightarrow \quad q_{|E} = 0\,.$$

11. In view of (5.4.2), the CAUCHY stress tensor for an isotropic body must fulfill the following condition of material symmetry

$$\boldsymbol{Q}\,\underset{\tau=0}{\overset{\infty}{\hat{\boldsymbol{t}}}}\Big(\boldsymbol{F}(t-\tau),\Theta(t-\tau),\operatorname{Grad}\Theta(t-\tau)\Big)\boldsymbol{Q}^T = \underset{\tau=0}{\overset{\infty}{\hat{\boldsymbol{t}}}}\Big(\boldsymbol{Q}\boldsymbol{F}(t-\tau)\boldsymbol{Q}^T,\Theta(t-\tau),\boldsymbol{Q}\operatorname{Grad}\Theta(t-\tau)\Big)\,, \tag{5.10.5}$$

$\forall$ time independent orthogonal transformations $\boldsymbol{Q}\in\mathbb{O}$.

The second PIOLA–KIRCHHOFF stress tensor satisfies, in addition, the condition (see (5.2.21))

$$\tilde{\boldsymbol{T}} = \hat{\tilde{\boldsymbol{T}}}(\boldsymbol{C}) \quad \text{with} \quad \tilde{\boldsymbol{T}} = J\boldsymbol{F}^{-1}\boldsymbol{t}\boldsymbol{F}^T\,,$$

where for an elastic body the deformation history plays no role and $\tilde{\boldsymbol{T}}$ is thus not a function of $\dot{\boldsymbol{C}}$ or higher time derivatives.

Under an orthogonal transformation $\boldsymbol{Q}^T$ of the reference system the second PIOLA–KIRCHHOFF stress tensor and the right CAUCHY–GREEN deformation tensor are transformed in accordance with

$$\begin{aligned}\tilde{\boldsymbol{T}}^* &= \det(\boldsymbol{F}^*)\boldsymbol{F}^{*-1}\boldsymbol{t}^*\boldsymbol{F}^{-T} = \det(\boldsymbol{F}\boldsymbol{Q}^T)(\boldsymbol{F}\boldsymbol{Q}^T)^{-1}\boldsymbol{t}^*(\boldsymbol{F}\boldsymbol{Q}^T)^{-T}\\ &= \det(\boldsymbol{Q}^T)\det(\boldsymbol{F})\boldsymbol{Q}\boldsymbol{F}^{-1}\boldsymbol{t}\boldsymbol{F}^{-T}\boldsymbol{Q}^T = \boldsymbol{Q}\big[\det(\boldsymbol{F})\boldsymbol{F}^{-1}\boldsymbol{t}\boldsymbol{F}^{-1}\big]\boldsymbol{Q}^T\\ &= \boldsymbol{Q}\tilde{\boldsymbol{T}}\boldsymbol{Q}^T\,,\end{aligned}$$

$$\boldsymbol{C}^* = \boldsymbol{F}^{*T}\boldsymbol{F}^* = (\boldsymbol{F}\boldsymbol{Q}^T)^T(\boldsymbol{F}\boldsymbol{Q}) = \boldsymbol{Q}\boldsymbol{F}^T\boldsymbol{F}\boldsymbol{Q} = \boldsymbol{Q}\boldsymbol{C}\boldsymbol{Q}^T\,,$$

so that the isotropy condition (5.10.5) reads

$$\boldsymbol{Q}\hat{\tilde{\boldsymbol{T}}}\boldsymbol{Q}^T = \hat{\tilde{\boldsymbol{T}}}(\boldsymbol{Q}\boldsymbol{C}\boldsymbol{Q}^T)\,.$$

The most general representation of this tensor valued isotropic function is now, with (5.4.14), written as

$$\tilde{\boldsymbol{T}} = \hat{\tilde{\boldsymbol{T}}}(\boldsymbol{C}) = \beta_1\boldsymbol{I} + \beta_2\boldsymbol{C} + \beta_3\boldsymbol{C}^2\,,$$

$$\beta_i = \hat{\beta}_i(I_{\boldsymbol{C}}, II_{\boldsymbol{C}}, III_{\boldsymbol{C}})\,,\quad i=1,2,3\,.$$

12. For the proof of objectivity of the o-derivative of an objective vector, $\boldsymbol{b}^* = \boldsymbol{O}\boldsymbol{b}$, one uses the transformation of the velocity gradient

$$\boldsymbol{L}^* = \boldsymbol{O}\boldsymbol{L}\boldsymbol{O}^T + \boldsymbol{\Omega} \quad \text{with} \quad \boldsymbol{\Omega} = \dot{\boldsymbol{O}}\boldsymbol{O}^T\,,$$

where $\boldsymbol{\Omega} = -\boldsymbol{\Omega}^T$ is skew symmetric. For the derivative one obtains

$$
\begin{aligned}
\overset{\circ}{\boldsymbol{b}^*} &= (\boldsymbol{b}^*)^{\cdot} - \tfrac{1}{2}\left(\boldsymbol{L}^* - \boldsymbol{L}^{*T}\right)\boldsymbol{b}^* \\
&= \boldsymbol{O}\dot{\boldsymbol{b}} + \dot{\boldsymbol{O}}\boldsymbol{b} - \left[\boldsymbol{O}\tfrac{1}{2}\left(\boldsymbol{L} - \boldsymbol{L}^T\right)\boldsymbol{O}^T + \boldsymbol{\Omega}\right]\boldsymbol{O}\boldsymbol{b} \\
&= \boldsymbol{O}\dot{\boldsymbol{b}} + \dot{\boldsymbol{O}}\boldsymbol{b} - \boldsymbol{O}\tfrac{1}{2}\left(\boldsymbol{L} - \boldsymbol{L}^T\right)\boldsymbol{O}^T\boldsymbol{O}\boldsymbol{b} - \dot{\boldsymbol{O}}\boldsymbol{O}^T\boldsymbol{O}\boldsymbol{b} \\
&= \boldsymbol{O}\left[\dot{\boldsymbol{b}} - \tfrac{1}{2}\left(\boldsymbol{L} - \boldsymbol{L}^T\right)\boldsymbol{b}\right] \\
&= \boldsymbol{O}\overset{\circ}{\boldsymbol{b}}\,, \quad \text{qed.}
\end{aligned}
$$

This shows that $(5.2.5)_1$ is an objective time derivative of a vector.

For the $\diamond$-derivative of an objective tensor $\boldsymbol{A}^* = \boldsymbol{O}\boldsymbol{A}\boldsymbol{O}^T$ one follows the same procedure. Hereby one further uses the identities

$$
\boldsymbol{L}^{*T} = (\boldsymbol{O}\boldsymbol{L}\boldsymbol{O}^T + \boldsymbol{\Omega})^T = \boldsymbol{O}\boldsymbol{L}^T\boldsymbol{O}^T - \boldsymbol{\Omega}
$$

and

$$
\operatorname{div}^* \boldsymbol{v}^* = \operatorname{tr}(\boldsymbol{L}^*) = \operatorname{tr}(\boldsymbol{L}) = \operatorname{div} \boldsymbol{v}\,.
$$

Then one can write

$$
\begin{aligned}
\overset{\diamond}{\boldsymbol{A}^*} &= (\boldsymbol{A}^*)^{\cdot} + \boldsymbol{A}^* \operatorname{div}^* \boldsymbol{v}^* - \boldsymbol{A}^*\boldsymbol{L}^{*T} - \boldsymbol{L}^*\boldsymbol{A}^{*T} \\
&= \left(\boldsymbol{O}\boldsymbol{A}\boldsymbol{O}^T\right)^{\cdot} + \boldsymbol{O}\boldsymbol{A}\boldsymbol{O}^T \operatorname{div} \boldsymbol{v} \\
&\quad - \boldsymbol{O}\boldsymbol{A}\boldsymbol{O}^T\left[\boldsymbol{O}\boldsymbol{L}^T\boldsymbol{O}^T - \boldsymbol{\Omega}\right] - \left[\boldsymbol{O}\boldsymbol{L}\boldsymbol{O}^T + \boldsymbol{\Omega}\right]\boldsymbol{O}\boldsymbol{A}^T\boldsymbol{O}^T \\
&= \dot{\boldsymbol{O}}\boldsymbol{A}\boldsymbol{O}^T + \boldsymbol{O}\dot{\boldsymbol{A}}\boldsymbol{O}^T + \boldsymbol{O}\boldsymbol{A}\dot{\boldsymbol{O}}^T + \boldsymbol{O}\boldsymbol{A}\boldsymbol{O}^T \operatorname{div} \boldsymbol{v} \\
&\quad - \boldsymbol{O}\boldsymbol{A}\boldsymbol{L}^T\boldsymbol{O}^T - \boldsymbol{O}\boldsymbol{A}\boldsymbol{O}^T\boldsymbol{O}\dot{\boldsymbol{O}}^T \\
&\quad - \boldsymbol{O}\boldsymbol{L}\boldsymbol{A}^T\boldsymbol{O}^T - \dot{\boldsymbol{O}}\boldsymbol{O}^T\boldsymbol{O}\boldsymbol{A}^T\boldsymbol{O}^T \\
&= \boldsymbol{O}\dot{\boldsymbol{A}}\boldsymbol{O}^T + \boldsymbol{O}\boldsymbol{A}\boldsymbol{O}^T \operatorname{div} \boldsymbol{v} - \boldsymbol{O}\boldsymbol{A}\boldsymbol{L}^T\boldsymbol{O}^T - \boldsymbol{O}\boldsymbol{L}\boldsymbol{A}^T\boldsymbol{O}^T \\
&= \boldsymbol{O}\left(\dot{\boldsymbol{A}} + \boldsymbol{A} \operatorname{div} \boldsymbol{v} - \boldsymbol{A}\boldsymbol{L}^T - \boldsymbol{L}\boldsymbol{A}^T\right)\boldsymbol{O}^T \\
&= \boldsymbol{O}\overset{\diamond}{\boldsymbol{A}}\boldsymbol{O}\,, \quad \text{qed.}
\end{aligned}
$$

13. Let the stress power $\varphi = \operatorname{tr}(\boldsymbol{L}\boldsymbol{t}^T)$ be an objective scalar, but do not assume that $\boldsymbol{t}$ is symmetric, which transforms in accordance with

$$
\varphi^* = \varphi \quad \text{or} \quad \operatorname{tr}(\boldsymbol{L}^*\boldsymbol{t}^{*T}) = \operatorname{tr}(\boldsymbol{L}\boldsymbol{t}^T)\,.
$$

With the transformation rule of the velocity gradient

$$
\boldsymbol{L}^* = \boldsymbol{O}\boldsymbol{L}\boldsymbol{O}^T + \boldsymbol{\Omega}
$$

we deduce, by assuming that the stress tensor is an objective tensor,

$$\begin{aligned}\varphi^* &= \boldsymbol{L}^* \cdot \boldsymbol{t}^{*T} \\ &= (\boldsymbol{O}\boldsymbol{L}\boldsymbol{O}^T + \boldsymbol{\Omega}) \cdot \boldsymbol{O}\boldsymbol{t}^T\boldsymbol{O}^T \\ &= \boldsymbol{L} \cdot \boldsymbol{t}^T + \boldsymbol{\Omega} \cdot (\boldsymbol{O}\boldsymbol{t}\boldsymbol{O}^T)^T .\end{aligned}$$

The invariance of the power, $\varphi^* = \varphi$, is thus only given when

$$\boldsymbol{\Omega} \cdot (\boldsymbol{O}\boldsymbol{t}\boldsymbol{O}^T)^T = 0$$

holds. Here the spin tensor is antisymmetric, $\boldsymbol{\Omega}^T = -\boldsymbol{\Omega}$, then $\boldsymbol{O}\boldsymbol{t}\boldsymbol{O}^T$ must be symmetric, i.e.,

$$\boldsymbol{O}\boldsymbol{t}\boldsymbol{O}^T = (\boldsymbol{O}\boldsymbol{t}\boldsymbol{O}^T)^T = \boldsymbol{O}\boldsymbol{t}^T\boldsymbol{O}^T \quad \Rightarrow \quad \boldsymbol{t} = \boldsymbol{t}^T , \tag{5.10.6}$$

from which the statement follows: *when the stress power is an objective scalar, then the stress tensor must necessarily be symmetric.*

14. To prove that every material law which satisfies (5.2.20) is also objective, let us first consider the internal energy. The material law thus reads

$$\hat{\varepsilon}(\boldsymbol{F}, \cdot) = \hat{\varepsilon}(\boldsymbol{U}, \cdot) , \tag{5.10.7}$$

where the other arguments $(\Theta, \operatorname{Grad}\Theta, \boldsymbol{X})$ are irrelevant for the proof, since these do not change under an EUCLIDian transformation. In addition the history dependency

$$\overset{\infty}{\underset{\tau=0}{\hat{\varepsilon}}}\Big(\boldsymbol{F}(t-\tau)\Big)$$

is not explicitly accounted for. (5.10.7) is also valid relative to a moving frame. Thus the internal energy must fulfill the relation

$$\hat{\varepsilon}(\boldsymbol{F}^*) = \hat{\varepsilon}(\boldsymbol{U}^*) = \hat{\varepsilon}(\boldsymbol{O}\boldsymbol{F}) = \hat{\varepsilon}(\boldsymbol{U}) \overset{(5.10.7)}{=} \hat{\varepsilon}(\boldsymbol{F})$$

and is objective.

The heat flux vector satisfies the material law

$$\boldsymbol{q} = \hat{\boldsymbol{q}}(\boldsymbol{F}) = \boldsymbol{R}\hat{\boldsymbol{q}}(\boldsymbol{U}) , \tag{5.10.8}$$

then, under EUCLIDian transformations with $\boldsymbol{R}^* = \boldsymbol{O}\boldsymbol{R}$,

$$\boldsymbol{q}^* = \hat{\boldsymbol{q}}(\boldsymbol{F}^*) = \boldsymbol{R}^*\hat{\boldsymbol{q}}(\boldsymbol{U}^*) = \boldsymbol{O}\boldsymbol{R}\hat{\boldsymbol{q}}(\boldsymbol{U}) \overset{(5.10.8)}{=} \boldsymbol{O}\hat{\boldsymbol{q}}(\boldsymbol{F}) = \boldsymbol{O}\boldsymbol{q} ,$$

thus one again obtains the condition (5.2.3) of material objectivity.

Analogous inferences follow for the stress tensor. If this satisfies the material law

$$\hat{\boldsymbol{t}}(\boldsymbol{F}) = \boldsymbol{R}\hat{\boldsymbol{t}}(\boldsymbol{U})\boldsymbol{R}^T \ , \tag{5.10.9}$$

then for an EUCLIDian transformation we have

$$\begin{aligned}\hat{\boldsymbol{t}}^* &= \hat{\boldsymbol{t}}(\boldsymbol{F}^*) = \boldsymbol{R}^*\hat{\boldsymbol{t}}(\boldsymbol{U}^*)\boldsymbol{R}^{*T} \\ &= \boldsymbol{O}\boldsymbol{R}\hat{\boldsymbol{t}}(\boldsymbol{U})\boldsymbol{R}^T\boldsymbol{O}^T \overset{(5.10.9)}{=} \boldsymbol{O}\hat{\boldsymbol{t}}(\boldsymbol{F})\boldsymbol{O}^T \\ &= \boldsymbol{O}\boldsymbol{t}\boldsymbol{O}^T \ .\end{aligned}$$

Therefore, the material law (5.10.9) is sufficient to satisfy the objectivity of the stress tensor.

15. We solve this exercise, as an example, for the stress tensor. The representation for the heat flux vector and the internal energy are then directly given by the calculations presented here.

The CAUCHY stress tensor of a viscoelastic body is given by

$$\boldsymbol{t} = \hat{\boldsymbol{t}}(\boldsymbol{F},\dot{\boldsymbol{F}},\Theta,\operatorname{Grad}\Theta,\boldsymbol{X}) \ . \tag{5.10.10}$$

Here, the temperature, temperature gradient and the position – and also the history $\overset{\infty}{\underset{\tau=0}{\hat{\boldsymbol{t}}}}(...)$ – do not play any role and are thus removed from the arguments. Material objectivity demands that the condition

$$\hat{\boldsymbol{t}}(\boldsymbol{O}\boldsymbol{F},(\boldsymbol{O}\boldsymbol{F})^{\boldsymbol{\cdot}}) = \boldsymbol{O}\hat{\boldsymbol{t}}(\boldsymbol{F},\dot{\boldsymbol{F}})\boldsymbol{O}^T$$

must hold for all orthogonal transformations $\boldsymbol{O}$, in particular also for the special choice $\boldsymbol{O} = \boldsymbol{R}^T$, where $\boldsymbol{R}$ is the rotational tensor of the polar decomposition $\boldsymbol{F} = \boldsymbol{R}\boldsymbol{U}$. One thus obtains

$$\hat{\boldsymbol{t}}(\boldsymbol{F},\dot{\boldsymbol{F}}) = \boldsymbol{R}\hat{\boldsymbol{t}}(\boldsymbol{U},\dot{\boldsymbol{U}})\boldsymbol{R}^T \ .$$

This is the most general form of the Cauchy stress tensor of a viscoelastic body.

The second PIOLA–KIRCHHOFF stress tensor $\tilde{\boldsymbol{T}}$ is related to the CAUCHY stress tensor by

$$\tilde{\boldsymbol{T}} = J\boldsymbol{F}^{-1}\boldsymbol{t}\boldsymbol{F}^{-T} \ .$$

With (5.10.10) we obtain, since $\boldsymbol{F} = \boldsymbol{R}\boldsymbol{U}$, $J = \det|\boldsymbol{F}| = \det\boldsymbol{U}$ and $\boldsymbol{R}$ is orthogonal,

$$\begin{aligned}\tilde{\boldsymbol{T}} &= |\det\boldsymbol{U}|\boldsymbol{U}^{-1}\boldsymbol{R}^T\boldsymbol{R}\hat{\boldsymbol{t}}(\boldsymbol{U},\dot{\boldsymbol{U}})\boldsymbol{R}^T\boldsymbol{R}\boldsymbol{U}^{-1} \\ &= |\det\boldsymbol{U}|\boldsymbol{U}^{-1}\hat{\boldsymbol{t}}(\boldsymbol{U},\dot{\boldsymbol{U}})\boldsymbol{U}^{-1} \ .\end{aligned}$$

Clearly $\tilde{\boldsymbol{T}}$ is a function of $\boldsymbol{U}$ and $\dot{\boldsymbol{U}}$,

$$\tilde{\boldsymbol{T}} = \hat{\tilde{\boldsymbol{T}}}(\boldsymbol{U},\dot{\boldsymbol{U}}) \ .$$

Because with the right CAUCHY–GREEN tensor

$$\boldsymbol{C} = \boldsymbol{U}^2 = \boldsymbol{F}^T\boldsymbol{F} ,$$

a unique correlation between $\boldsymbol{C}$ and $\boldsymbol{U}$ is established, one can also write

$$\tilde{\boldsymbol{T}} = \hat{\tilde{\boldsymbol{T}}}'(\boldsymbol{C},\dot{\boldsymbol{C}}) .$$

For the heat flux vector we have

$$\boldsymbol{Q} = J\boldsymbol{q}\boldsymbol{F}^{-T} = J\boldsymbol{F}^{-1}\boldsymbol{q} ,$$

$$\boldsymbol{q} = \hat{\boldsymbol{q}}(\boldsymbol{F},\dot{\boldsymbol{F}}) = \boldsymbol{Q}^T\hat{\boldsymbol{q}}(\boldsymbol{O}\boldsymbol{F},(\boldsymbol{O}\boldsymbol{F})^{\cdot} = \boldsymbol{R}\hat{\boldsymbol{q}}(\boldsymbol{U},\dot{\boldsymbol{U}}) .$$

Thus we likewise deduce

$$\boldsymbol{Q} = (\det\boldsymbol{U})\boldsymbol{U}^{-1}\boldsymbol{R}^T\boldsymbol{R}\hat{\boldsymbol{q}}(\boldsymbol{U},\dot{\boldsymbol{U}}) = (\det\boldsymbol{U})\boldsymbol{U}^{-1}\hat{\boldsymbol{q}}(\boldsymbol{U},\dot{\boldsymbol{U}})$$

$$= \hat{\boldsymbol{Q}}(\boldsymbol{U},\dot{\boldsymbol{U}}) = \hat{\boldsymbol{Q}}'(\boldsymbol{C},\dot{\boldsymbol{C}}) ,$$

and similarly for the internal energy,

$$\varepsilon = \hat{\varepsilon}(\boldsymbol{F},\dot{\boldsymbol{F}}) = \hat{\varepsilon}(\boldsymbol{O}\boldsymbol{F},(\boldsymbol{O}\boldsymbol{F})^{\cdot}) = \hat{\varepsilon}(\boldsymbol{U},\dot{\boldsymbol{U}}) = \hat{\varepsilon}'(\boldsymbol{C},\dot{\boldsymbol{C}}) .$$

Further, the GREEN strain tensor $\boldsymbol{G}$ is defined as $\boldsymbol{G} = \frac{1}{2}(\boldsymbol{C} - \boldsymbol{I})$ and one can write all these statements with $\boldsymbol{G}$ as independent variable,

$$\varepsilon = \hat{\varepsilon}''(\boldsymbol{G},\dot{\boldsymbol{G}}) , \quad \boldsymbol{Q} = \hat{\boldsymbol{Q}}''(\boldsymbol{G},\dot{\boldsymbol{G}}) , \quad \tilde{\boldsymbol{T}} = \hat{\tilde{\boldsymbol{T}}}''(\boldsymbol{G},\dot{\boldsymbol{G}}) , \quad \text{qed.}$$

16. The stress power in the LAGRANGE representation is written as

$$\varphi = \mathrm{Tr}\big((\mathrm{Grad}\,\boldsymbol{v})\boldsymbol{T}^T\big) = \mathrm{Tr}(\dot{\boldsymbol{F}}\boldsymbol{T}^T) = \dot{\boldsymbol{F}}\cdot\boldsymbol{T}^T \quad ;$$

it is thus equal to the scalar product of the time rate of change of the deformation gradient with the first PIOLA–KIRCHHOFF stress tensor . Expressing $\boldsymbol{T}$ with the help of the second PIOLA–KIRCHHOFF stress tensor as

$$\tilde{\boldsymbol{T}} = \boldsymbol{F}^{-1}\boldsymbol{T} \quad \text{or} \quad \boldsymbol{T} = \boldsymbol{F}\tilde{\boldsymbol{T}}$$

and using its symmetry[24] there follows

$$\varphi = \dot{\boldsymbol{F}}\cdot\boldsymbol{T}^T = \dot{\boldsymbol{F}}\cdot(\boldsymbol{F}\tilde{\boldsymbol{T}})^T = \dot{\boldsymbol{F}}\cdot\tilde{\boldsymbol{T}}\boldsymbol{F}^T$$

$$= \boldsymbol{F}^T\dot{\boldsymbol{F}}\cdot\tilde{\boldsymbol{T}} = (\boldsymbol{F}^T\dot{\boldsymbol{F}})^T\cdot\tilde{\boldsymbol{T}}^T = \dot{\boldsymbol{F}}^T\boldsymbol{F}\cdot\tilde{\boldsymbol{T}} .$$

[24] In Exercise 13, we have shown that the stress power is objective if the CAUCHY stress tensor is symmetric; from $\boldsymbol{t} = \boldsymbol{t}^T$ therefore follows

$$\tilde{\boldsymbol{T}} = J\boldsymbol{F}^{-1}\boldsymbol{t}\boldsymbol{F}^{-T} = J(\boldsymbol{F}^{-1}\boldsymbol{t}\boldsymbol{F}^{-T})^T = \tilde{\boldsymbol{T}}^T .$$

With the definition of the right CAUCHY–GREEN deformation tensor

$$C = F^T F \quad \text{with} \quad \dot{C} = F^T \dot{F} + \dot{F}^T F$$

one obtains

$$\varphi = \tfrac{1}{2}\left(F^T \dot{F} + \dot{F}^T F\right) \cdot \tilde{T} = \tfrac{1}{2}\dot{C} \cdot \tilde{T} = \tfrac{1}{2}\,\mathrm{Tr}\left(\tilde{T}\dot{C}\right) .$$

Using the GREEN strain tensor $G = \frac{1}{2}(C - I)$ in place of C, one can write the stress power also as

$$\varphi = \tfrac{1}{2}\,\mathrm{Tr}\left(\tilde{T}\dot{C}\right) = \mathrm{Tr}(\tilde{T}\dot{G}) , \quad \text{qed.}$$

17. The RIVLIN–ERICKSEN tensors are recursively defined as

$$A_{(n+1)} = \dot{A}_{(n)} + L^T A_{(n)} + A_{(n)} L , \quad A_{(0)} = I ,$$

where $A_{(n)} = A_{(n)}^T$ is symmetric. With this one obtains for $A_{(1)}$

$$A_{(1)} = \dot{A}_{(0)} + L^T A_{(0)} + A_{(0)} L = L^T + L = 2D .$$

Both $A_{(0)}$ and $A_{(1)}$ are symmetric objective tensors. By induction it can also be proved that this holds for all $A_{(n)}, n > 1$.

Let us assume that $A_{(n)}$ for a fixed chosen n is objective and symmetric,

$$A_{(n)} = A_{(n)}^T , \quad A_{(n)}^* = O A_{(n)} O^T ,$$

one can then prove the symmetry of $n + 1$ as follows:

$$\begin{aligned} A_{(n+1)}^T &= \dot{A}_{(n)}^T + (L^T A_{(n)})^T + (A_{(n)} L)^T \\ &= \dot{A}_{(n)}^T + A_{(n)}^T L + L^T A_{(n)}^T \\ &= \dot{A}_{(n)} + A_{(n)} L + L^T A_{(n)} \\ &= A_{(n+1)} . \end{aligned}$$

If one intends to demonstrate the objectivity, one has to consider the transformation of the velocity gradients

$$L^* = O L O^T + \Omega$$

along with the antisymmetric spin tensor

$$\Omega = \dot{O} O^T = -(\dot{O} O^T)^T = -O \dot{O}^T .$$

One thus obtains for the $(n + 1)$-th RIVLIN–ERICKSEN tensor

$$\begin{aligned}
\boldsymbol{A}^*_{(n+1)} &= \dot{\boldsymbol{A}}^*_{(n)} + \boldsymbol{L}^{*T}\boldsymbol{A}^*_{(n)} + \boldsymbol{A}^*_{(n)}\boldsymbol{L}^* \\
&= (\boldsymbol{O}\boldsymbol{A}_{(n)}\boldsymbol{O}^T)^{\cdot} + (\boldsymbol{O}\boldsymbol{L}\boldsymbol{O}^T + \dot{\boldsymbol{O}}\boldsymbol{O}^T)^T\boldsymbol{O}\boldsymbol{A}_{(n)}\boldsymbol{O}^T \\
&\quad + \boldsymbol{O}\boldsymbol{A}_{(n)}\boldsymbol{O}^T(\boldsymbol{O}\boldsymbol{L}\boldsymbol{O}^T + \dot{\boldsymbol{O}}\boldsymbol{O}^T) \\
&= \boldsymbol{O}\dot{\boldsymbol{A}}_{(n)}\boldsymbol{O}^T + \dot{\boldsymbol{O}}\boldsymbol{A}_{(n)}\boldsymbol{O}^T + \boldsymbol{O}\boldsymbol{A}_{(n)}\dot{\boldsymbol{O}}^T + \boldsymbol{O}\boldsymbol{L}^T\boldsymbol{O}^T\boldsymbol{O}\boldsymbol{A}_{(n)}\boldsymbol{O}^T \\
&\quad - \dot{\boldsymbol{O}}\boldsymbol{O}^T\boldsymbol{O}\boldsymbol{A}_{(n)}\boldsymbol{O}^T + \boldsymbol{O}\boldsymbol{A}_{(n)}\boldsymbol{O}^T\boldsymbol{O}\boldsymbol{L}\boldsymbol{O}^T - \boldsymbol{O}\boldsymbol{A}_{(n)}\boldsymbol{O}^T\boldsymbol{O}\dot{\boldsymbol{O}}^T \\
&= \boldsymbol{O}(\dot{\boldsymbol{A}}_{(n)} + \boldsymbol{L}^T\boldsymbol{A}_{(n)} + \boldsymbol{A}_{(n)}\boldsymbol{L})\boldsymbol{O}^T = \boldsymbol{O}\boldsymbol{A}_{(n+1)}\boldsymbol{O}^T .
\end{aligned}$$

In the derivation one uses the orthogonality of the transformation $\boldsymbol{O}\boldsymbol{O}^T = \boldsymbol{I}$, and the antisymmetry of the spin tensor. With both of these results one obtains

$$\boldsymbol{A}^*_{(n+1)} = \boldsymbol{O}\boldsymbol{A}_{(n+1)}\boldsymbol{O}^T , \quad \boldsymbol{A}^T_{(n+1)} = \boldsymbol{A}_{(n+1)} ,$$

i.e., the RIVLIN–ERICKSEN tensors are objective and symmetric tensors for all $n \geq 0$.

18. Every analytic function $\boldsymbol{f}(\boldsymbol{A})$ is expressed as a power series

$$\boldsymbol{f}(\boldsymbol{A}) = \sum_{\nu=0}^{\infty} a_\nu \boldsymbol{A}^\nu , \quad \boldsymbol{A}^0 := \boldsymbol{I} . \tag{5.10.11}$$

Let n be any entire number. Then, one can write the characteristic equation for $\boldsymbol{A}$ as

$$\boldsymbol{A}^3 - I_{\boldsymbol{A}}\boldsymbol{A}^2 + II_{\boldsymbol{A}}\boldsymbol{A} - III_{\boldsymbol{A}}\boldsymbol{I} = 0 , \tag{5.10.12}$$

multiplying by $\boldsymbol{A}^{n-3}$ and solving for $\boldsymbol{A}^n$, one obtains

$$\boldsymbol{A}^n = I_{\boldsymbol{A}}\boldsymbol{A}^{n-1} - II_{\boldsymbol{A}}\boldsymbol{A}^{n-2} + III_{\boldsymbol{A}}\boldsymbol{A}^{n-3} , \tag{5.10.13}$$

which can be reduced to a polynomial of degree 3 by recursive application of this rule:

$$\boldsymbol{A}^n = \alpha_0\boldsymbol{I} + \alpha_1\boldsymbol{A} + \alpha_2\boldsymbol{A}^2 . \tag{5.10.14}$$

Substituting this in (5.10.11), one obtains

$$\boldsymbol{f}(\boldsymbol{A}) = a_0\boldsymbol{I} + a_1\boldsymbol{A} + a_2\boldsymbol{A}^2 . \tag{5.10.15}$$

The coefficients a_0, a_1 and a_2 are expressed by a_μ ($\mu = 1, \ldots, \infty$) and $I_{\boldsymbol{A}}$, $II_{\boldsymbol{A}}$ and $III_{\boldsymbol{A}}$ are given in the form

$$a_i := a_i(I_{\boldsymbol{A}},\ II_{\boldsymbol{A}},\ III_{\boldsymbol{A}}) \quad i = 0, 1, 2 .$$

19. The objective vector $\boldsymbol{q}$ must fulfill the isotropy condition

$$\boldsymbol{q}(\mu) = \boldsymbol{O}\boldsymbol{q}(\mu)$$

for arbitrary orthogonal transformations $\boldsymbol{O}$. This equation is only fulfilled when $\lambda = 1$ is the eigenvalue of $\boldsymbol{O}$. This is, however, not possible for arbitrary orthogonal transformations, thus $\boldsymbol{q}$ must be equal to the zero vector. i.e.,

$$\boldsymbol{q} = \boldsymbol{0} \,.$$

20. In this exercise the entropy principle will be implemented for the material class of a thermoelastic solid, which is given by the material equations (see (5.1.9))

$$\Psi = \hat{\Psi}(\boldsymbol{C}, \Theta, \operatorname{Grad}\Theta) \,, \quad \Psi \in \{\Psi_\varepsilon, s, \tilde{\boldsymbol{T}}, \boldsymbol{Q}\} \,,$$

where

$$\Psi_\varepsilon = \varepsilon - \Theta s \tag{5.10.16}$$

is the HELMHOLTZ free energy.

In the LAGRANGE representation the energy and entropy balances are given by

$$\rho_R \dot{\varepsilon} = -\operatorname{Div}\boldsymbol{Q} + \operatorname{Tr}(\operatorname{Grad}\boldsymbol{v}\,\boldsymbol{T}^T) + \rho_R \mathfrak{r} \,,$$

$$\rho_R \dot{s} + \operatorname{Div}\left(\frac{\boldsymbol{Q}}{\Theta}\right) - \rho_R \frac{\mathfrak{r}}{\Theta} \geq 0 \,. \tag{5.10.17}$$

In Exercise 13, we have shown that the objective stress power can also be written in the form

$$\hat{\Psi} = \operatorname{Tr}(\operatorname{Grad}\boldsymbol{v}\,\boldsymbol{T}^T) = \tfrac{1}{2}\tilde{\boldsymbol{T}} \cdot \dot{\boldsymbol{C}} \,. \tag{5.10.18}$$

Substituting the internal energy by the HELMHOLTZ free energy in (5.10.17) and replacing the stress power by (5.10.18), one obtains

$$\rho_R(\dot{\Psi}_\varepsilon + \dot{\Theta}s + \Theta\dot{s}) = -\operatorname{Div}\boldsymbol{Q} + \tfrac{1}{2}\tilde{\boldsymbol{T}} \cdot \dot{\boldsymbol{C}} + \rho_R \mathfrak{r} \,.$$

With the help of this relation the term $\rho_R(\Theta\dot{s} - \mathfrak{r})$ can be eliminated in the entropy balance, so that

$$-\rho_R(\dot{\Psi}_\varepsilon + \dot{\Theta}s) + \tfrac{1}{2}\tilde{\boldsymbol{T}} \cdot \dot{\boldsymbol{C}} - \operatorname{Div}\boldsymbol{Q} + \Theta\operatorname{Div}\left(\frac{\boldsymbol{Q}}{\Theta}\right) \geq 0 \,.$$

Here Ψ_ε is a function of $(\boldsymbol{C}, \Theta, \operatorname{Grad}\Theta)$; then for the time derivative (considering $\boldsymbol{C} = \boldsymbol{C}^T$) we write

$$\dot{\Psi}_\varepsilon = \frac{\partial \Psi_\varepsilon}{\partial \Theta}\dot{\Theta} + \frac{\partial \Psi_\varepsilon}{\partial(\operatorname{Grad}\Theta)} \cdot (\operatorname{Grad}\Theta)^{\boldsymbol{\cdot}} + \frac{\partial \Psi_\varepsilon}{\partial \boldsymbol{C}} \cdot \dot{\boldsymbol{C}} \,.$$

Thus, the entropy balance becomes

$$-\rho_R\left(\frac{\partial\Psi_\varepsilon}{\partial\Theta}+s\right)\dot{\Theta}-\rho_R\frac{\partial\Psi_\varepsilon}{\partial(\operatorname{Grad}\Theta)}\cdot(\operatorname{Grad}\Theta)^{\boldsymbol{\cdot}}$$
$$+\left(\frac{1}{2}\tilde{\boldsymbol{T}}-\rho_R\frac{\partial\Psi_\varepsilon}{\partial\boldsymbol{C}}\right)\cdot\dot{\boldsymbol{C}}-\frac{\boldsymbol{Q}\cdot\operatorname{Grad}\Theta}{\Theta}\geq 0\,.$$

Since this inequality must be valid for all admissible thermodynamic processes, this can only be satisfied for arbitrary values of $\dot{\Theta}$, $(\operatorname{Grad}\Theta)^{\boldsymbol{\cdot}}$ and $\dot{\boldsymbol{C}}$, if the corresponding forefactors vanish identically (see the Theorem in Sect. 5.7)

$$\frac{\partial\Psi_\varepsilon}{\partial(\operatorname{Grad}\Theta)}=0\quad\Rightarrow\quad\Psi_\varepsilon=\hat{\Psi}_\varepsilon(\boldsymbol{C},\Theta)\,,$$
$$s=-\frac{\partial\Psi_\varepsilon}{\partial\Theta}\,,\quad\tilde{\boldsymbol{T}}=2\rho_R\frac{\partial\Psi_\varepsilon}{\partial\boldsymbol{C}}=\rho_R\frac{\partial\Psi_\varepsilon}{\partial\boldsymbol{G}}\,.$$

In the last term an equivalent relation for the GREEN strain tensor $\boldsymbol{G}=\frac{1}{2}(\boldsymbol{C}-\boldsymbol{I})$ has been written. The HELMHOLTZ free energy can thus not depend on the material temperature gradients, and the entropy is given as the negative change of Ψ_ε with temperature. In addition, for a thermoelastic solid, one can derive the stress tensor from the thermodynamic *potential*, Ψ_ε.

21. The extra stress tensor of a viscous material is given by

$$\boldsymbol{t}^E=\kappa(\operatorname{tr}\boldsymbol{D})\boldsymbol{I}+2\mu\boldsymbol{D}^{'}\,. \tag{5.10.19}$$

with the bulk and shear viscosities

$$\kappa=\hat{\kappa}(\rho,\Theta)\quad\text{and}\quad\mu=\hat{\mu}(\rho,\Theta)\,.$$

One wishes now to find a relation for this material in which $\boldsymbol{t}^E$ is not a function of $\boldsymbol{D}$, but given by the inverse relation

$$\boldsymbol{D}=g(\operatorname{tr}\boldsymbol{t}^E)\boldsymbol{I}+\tfrac{1}{2}f\boldsymbol{t}^{E^{'}}\,, \tag{5.10.20}$$

where the fluidities are functions of the pressure and temperature in accordance with the relations

$$g=\hat{g}(p,\Theta)\,,\quad f=\hat{f}(p,\Theta)\,.$$

Forming the trace of (5.10.19) and (5.10.20) one obtains

$$\operatorname{tr}\boldsymbol{t}^E=3\kappa\operatorname{tr}\boldsymbol{D}\,,\quad\operatorname{tr}\boldsymbol{D}=3g\operatorname{tr}\boldsymbol{t}^E\,.$$

Both these relations can simultaneously be fulfilled if

$$g=\frac{1}{9\kappa}\,.$$

Hereby in $\hat{\kappa}(\rho,\Theta)$, the thermal equation of state $p=\hat{p}(\rho,\Theta)$ and its inverse $\rho=\hat{\rho}(p,\Theta)$ is used, so that

$$g = \hat{g}(p, \Theta) = \frac{1}{9\hat{\kappa}\left(\hat{\rho}(p,\Theta),\Theta\right)}$$

is obtained.
Forming the deviators on the left and right of the material equations (5.10.19), (5.10.20), one obtains

$$\boldsymbol{t}^{E'} = \boldsymbol{t}^E - \tfrac{1}{3}(\operatorname{tr}\boldsymbol{t}^E)\boldsymbol{I} = 2\mu\boldsymbol{D}' \ , \quad \boldsymbol{D}' = \tfrac{1}{2}f\boldsymbol{t}^{E'} \ .$$

Thus, the shear fluidity follows as

$$f = \hat{f}(p, \Theta) = \frac{1}{\hat{\mu}\left(\hat{\rho}(p,\Theta),\Theta\right)} \ .$$

22.

a. The time derivative of the position vector gives

$$\boldsymbol{v}^* = \boldsymbol{O}\boldsymbol{v} + \dot{\boldsymbol{O}}\boldsymbol{x} - \dot{\boldsymbol{c}}^* \ .$$

b. The difference of two velocities at a point $\boldsymbol{x}$ is thus obtained (the same transformation is used):

$$\begin{aligned}\boldsymbol{v}_1^* - \boldsymbol{v}_2^* &= \boldsymbol{O}(\boldsymbol{v}_1 - \boldsymbol{v}_2) + \dot{\boldsymbol{O}}(\boldsymbol{x} - \boldsymbol{x}) - (\dot{\boldsymbol{c}}^* - \dot{\boldsymbol{c}}^*) \\ &= \boldsymbol{O}(\boldsymbol{v}_1 - \boldsymbol{v}_2) \ .\end{aligned}$$

Difference velocities with the same point of action are always objective vectors.

c. The deformation gradient is defined as $\boldsymbol{F} = (\partial\boldsymbol{x}/\partial\boldsymbol{X})$. Here, since the orthogonal transformation $\boldsymbol{O}$ and the translation vector $\boldsymbol{c}^*$ only depend on time, we have

$$\boldsymbol{F}^* = \frac{\partial\boldsymbol{y}^*}{\partial\boldsymbol{X}} = \boldsymbol{O}\frac{\partial\boldsymbol{x}}{\partial\boldsymbol{X}} = \boldsymbol{O}\boldsymbol{F} \ ,$$

i.e., the deformation gradient is transformed like an objective vector. (This is so because $F_{i\alpha}\boldsymbol{e}_i \otimes \boldsymbol{e}_\alpha$ is related to the basis vectors of the present configuration $\boldsymbol{e}_i$ and that of the reference configuration $\boldsymbol{e}_\alpha$ and transforms relative to the present configuration as a vector).

d. Forming the gradient of the velocity vector and using the chain rule, one obtains

$$\boldsymbol{L}^* = \frac{\partial\boldsymbol{v}^*}{\partial\boldsymbol{y}^*} = \frac{\partial}{\partial\boldsymbol{x}}\left(\boldsymbol{O}\boldsymbol{v} + \dot{\boldsymbol{O}}\boldsymbol{x} - \dot{\boldsymbol{c}}^*\right)\boldsymbol{O}^T \ ,$$

where $\dfrac{\partial\boldsymbol{x}}{\partial\boldsymbol{y}^*} = \dfrac{\partial}{\partial\boldsymbol{y}^*}\left(\boldsymbol{O}^T(\boldsymbol{y}^* + \boldsymbol{c}^*)\right) = \boldsymbol{O}^T$ is used. Here $\boldsymbol{O}$ and $\boldsymbol{c}^*$ depend only on time, implying

$$\boldsymbol{L}^* = \boldsymbol{O}\boldsymbol{L}\boldsymbol{O}^T + \dot{\boldsymbol{O}}\boldsymbol{O}^T = \boldsymbol{O}\boldsymbol{L}\boldsymbol{O}^T + \boldsymbol{\Omega}$$

with the spin tensor $\boldsymbol{\Omega} = \dot{\boldsymbol{O}}\boldsymbol{O}^T$.

e. Applying the decomposition into symmetric and antisymmetric parts

$$\boldsymbol{L} = \boldsymbol{D} + \boldsymbol{W} \quad , \qquad \boldsymbol{D} = \tfrac{1}{2}(\boldsymbol{L} + \boldsymbol{L}^T) \, , \quad \boldsymbol{W} = \tfrac{1}{2}(\boldsymbol{L} - \boldsymbol{L}^T) \, ,$$

one obtains

$$\boldsymbol{D}^* = \tfrac{1}{2}(\boldsymbol{L}^* + \boldsymbol{L}^{*T}) = \tfrac{1}{2}\boldsymbol{O}(\boldsymbol{L} + \boldsymbol{L}^T)\boldsymbol{O}^T = \boldsymbol{O}\boldsymbol{D}\boldsymbol{O}^T \, ,$$

and, since $\boldsymbol{\Omega} = -\boldsymbol{\Omega}^T$ is antisymmetric,

$$\boldsymbol{W}^* = \tfrac{1}{2}(\boldsymbol{L}^* - \boldsymbol{L}^{*T}) = \boldsymbol{O}\boldsymbol{W}\boldsymbol{O}^T + \tfrac{1}{2}(\boldsymbol{\Omega} - \boldsymbol{\Omega}^T) = \boldsymbol{O}\boldsymbol{W}\boldsymbol{O}^T + \boldsymbol{\Omega} \, .$$

f. The transformations of the left CAUCHY–GREEN deformation tensor is obtained from its definition $\boldsymbol{B} = \boldsymbol{F}\boldsymbol{F}^T$ and the transformation rule of the deformation gradients

$$\boldsymbol{B}^* = \boldsymbol{F}^*\boldsymbol{F}^{*T} = \boldsymbol{O}\boldsymbol{F}(\boldsymbol{O}\boldsymbol{F})^T = \boldsymbol{O}(\boldsymbol{F}\boldsymbol{F}^T)\boldsymbol{O}^T = \boldsymbol{O}\boldsymbol{B}\boldsymbol{O}^T \, ,$$

$\boldsymbol{B}$ is thus an objective tensor.

g. The right CAUCHY–GREEN deformation tensor transforms according to

$$\boldsymbol{C}^* = \boldsymbol{F}^{*T}\boldsymbol{F}^* = (\boldsymbol{O}\boldsymbol{F})^T\boldsymbol{O}\boldsymbol{F} = \boldsymbol{F}^T(\boldsymbol{O}^T\boldsymbol{O})\boldsymbol{F} = \boldsymbol{F}^T\boldsymbol{F} = \boldsymbol{C} \, ,$$

and is thus a scheme of nine objective scalars. Notice that $\boldsymbol{C}$ is defined in the reference configuration, and therefore a transformation of the present system can not affect it.

h. The transformation of the first PIOLA–KIRCHHOFF stress tensor results in

$$\begin{aligned} \boldsymbol{T}^* = J^* \, \boldsymbol{t}^* \, \boldsymbol{F}^{*-T} &= \det(\boldsymbol{O}\boldsymbol{F})\boldsymbol{O}\boldsymbol{t}\boldsymbol{O}^T(\boldsymbol{O}\boldsymbol{F})^{-T} \\ &= \det(\boldsymbol{F})\boldsymbol{O}\boldsymbol{t}\boldsymbol{O}^T\boldsymbol{O}\boldsymbol{F}^{-T} = \boldsymbol{O}(J\boldsymbol{t}\boldsymbol{F}^{-T}) \, , \end{aligned}$$

since the CAUCHY stress tensor is an objective tensor, i.e., $\boldsymbol{t}^* = \boldsymbol{O} \, \boldsymbol{t} \, \boldsymbol{O}^*$ holds. $\boldsymbol{T}$ is thus a scheme of three objectives vectors.

i. The second PIOLA–KIRCHHOFF stress tensor gives

$$\begin{aligned} \tilde{\boldsymbol{T}}^* = J^*\boldsymbol{F}^{*T}\boldsymbol{t}^* \, \boldsymbol{F}^{*-T} &= J(\boldsymbol{F}^T\boldsymbol{O}^T)(\boldsymbol{O}\boldsymbol{t} \, \boldsymbol{O}^T)(\boldsymbol{O}\boldsymbol{F}^{-T}) \\ &= J\boldsymbol{F}^T\boldsymbol{t} \, \boldsymbol{F}^{-T} \, , \end{aligned}$$

i.e., it acts as a scheme of nine objective scalars.

23. The stress tensor for a NEWTONian fluid is given by

$$\boldsymbol{t} = (-p + \lambda \operatorname{tr} \boldsymbol{D})\boldsymbol{I} + 2\mu\boldsymbol{D}$$

or

$$t_{ij} = -p\delta_{ij} + \lambda D_{kk}\delta_{ij} + 2\mu D_{ij} \ .$$

In the momentum balance the term $\operatorname{div}\boldsymbol{t}$ is to be calculated, which in component notation can be written as

$$t_{ij,j} = -p_{,i} + \lambda D_{ij,j} + 2\mu D_{ij,j} \ .$$

With $D_{ij} = \frac{1}{2}(v_{i,j} + v_{j,i})$ and $D_{kk} = v_{k,k}$ there follows immediately

$$t_{ij,j} = -p_{,i} + \lambda v_{k,ki} + \mu(v_{i,jj} + v_{j,ji}) \ ,$$

or, after renaming the indices,

$$t_{ij,j} = -p_{,i} + (\lambda + \mu) v_{j,ji} + \mu v_{i,jj} \ .$$

Hence,

$$\operatorname{div}\boldsymbol{t} = -\operatorname{grad} p + (\lambda + \mu)\operatorname{grad}(\operatorname{div}\boldsymbol{v}) + \mu \operatorname{div}\operatorname{grad}\boldsymbol{v} \ .$$

The NAVIER–STOKES equations for the NEWTONian fluid are thus

$$\rho\frac{\mathrm{d}\boldsymbol{v}}{\mathrm{d}t} = -\operatorname{grad} p + \rho\boldsymbol{g} + (\lambda + \mu)\operatorname{grad}(\operatorname{div}\boldsymbol{v}) + \mu \operatorname{div}\operatorname{grad}\boldsymbol{v} \ ,$$

which – if the bulk viscosity is set to zero, $\kappa = \lambda + \frac{2}{3}\mu \approx 0$ (STOKES assumption) – reads

$$\rho\frac{\mathrm{d}\boldsymbol{v}}{\mathrm{d}t} = -\operatorname{grad} p + \rho\boldsymbol{g} + \tfrac{1}{3}\mu\operatorname{grad}(\operatorname{div}\boldsymbol{v}) + \mu \operatorname{div}\operatorname{grad}\boldsymbol{v} \ .$$

This result can be simplified with the identity $\operatorname{div}\operatorname{grad}\boldsymbol{v} = \operatorname{grad}(\operatorname{div}\boldsymbol{v}) - \operatorname{curl}(\operatorname{curl}\boldsymbol{v})$ to yield

$$\rho\frac{\mathrm{d}\boldsymbol{v}}{\mathrm{d}t} = -\operatorname{grad} p + \rho\boldsymbol{g} + \tfrac{4}{3}\mu\operatorname{grad}(\operatorname{div}\boldsymbol{v}) - \mu \operatorname{curl}(\operatorname{curl}\boldsymbol{v}) \ .$$

The stress power is given by

$$\varphi = \operatorname{tr}(\boldsymbol{D}\boldsymbol{t}^T) = \boldsymbol{D}\cdot(\lambda\operatorname{tr}(\boldsymbol{D})\boldsymbol{I} + 2\mu\boldsymbol{D}) - \lambda I_{\boldsymbol{D}}^2 + 4\mu I_{\boldsymbol{D}^2} \ .$$

For a density preserving fluid $I_{\boldsymbol{D}} = \operatorname{tr}(\boldsymbol{D}) = 0$, and thus one obtains

$$\rho\frac{\mathrm{d}\boldsymbol{v}}{\mathrm{d}t} = -\operatorname{grad} p + \rho\boldsymbol{g} + \mu \operatorname{div}\operatorname{grad}\boldsymbol{v}$$

and

$$\varphi_{\text{dens.pr.}} = 4\mu I_{\boldsymbol{D}^2} \ .$$

The stress power in case of STOKES assumption is

$$\varphi_{\text{Stokes}} = 4\mu I_{\boldsymbol{D}'^2} \ ,$$

where

$$I_{\boldsymbol{D}'^2} = \boldsymbol{D}'\cdot\boldsymbol{D}' = \operatorname{tr}(\boldsymbol{D}^2) - \tfrac{1}{3}\operatorname{tr}(\boldsymbol{D})^2$$

holds.

24. Let us put the x, y coordinates in the so-called basal plane with origin in the center of the regular hexagon. The axis perpendicular to this basal plane and passing through the center is called the c-axis. As shown in Fig. 5.2, the x- and y-axes, which are the symmetry axes in the basal plane, are pairwise perpendicular to the edge and the connecting lines of the center with the edge points; by rotating this arrangement $n \times 30°$, $n = 1, 2, 3, 4, 5, 6$ about the c-axis one obtains the mirror reflections of identical configurations; since also the mirror reflections are symmetry transformations, it suffices for the determination of the ($\mathbb{C}$)-matrix to start with the orthogonally horizontal regular solid, i.e., the formula (5.5.43) and additionally to use the invariance of Φ under rotation by 30° about the c-axis. With (5.5.27) and $\phi = 30°$ we thus have

$$\boldsymbol{H} = \begin{pmatrix} \sqrt{3}/2 & 1/2 & 0 \\ -1/2 & \sqrt{3}/2 & 0 \\ 0 & 0 & 1 \end{pmatrix} \tag{5.10.21}$$

and

$$\boldsymbol{HAH}^T =$$

$$\begin{pmatrix} \frac{3}{4}A_{11} + \frac{\sqrt{3}}{2}A_{12} + \frac{1}{4}A_{22} & -\frac{\sqrt{3}}{4}A_{11} + \frac{1}{2}A_{12} + \frac{\sqrt{3}}{4}A_{22} & \frac{\sqrt{3}}{2}A_{13} + \frac{1}{2}A_{23} \\ & \frac{1}{4}A_{11} - \frac{\sqrt{3}}{2}A_{12} + \frac{3}{4}A_{22} & \frac{\sqrt{3}}{2}A_{23} + \frac{1}{2}A_{13} \\ & & A_{33} \end{pmatrix} . \tag{5.10.22}$$

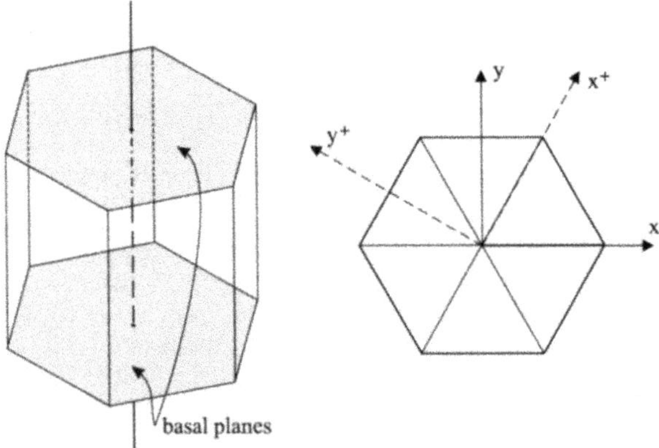

Fig. 5.2. Hexagonal crystal with definition of the c-axis and the basal plane.

Calculating from this matrix Φ^+ (where $\boldsymbol{\eta}^+ = \boldsymbol{\eta}$, because $\boldsymbol{H}$ is a symmetry transformation), one obtains, after a simple manipulation,

$$\Phi^+ = \Phi + \triangle\Phi \, , \tag{5.10.23}$$

in which $\triangle\Phi$ is given by

$$\begin{aligned}\triangle\Phi &= \mathfrak{X}(2\eta_{1212} + \eta_{1122} - \eta_{1111}) \quad , \\ \mathfrak{X} &= \tfrac{3}{16}A_{11}^2 - \tfrac{3}{4}A_{12}^2 + \tfrac{3}{16}A_{22}^2 + \tfrac{\sqrt{3}}{4}A_{12}(A_{22} - A_{11}) \\ &\quad - \tfrac{3}{8}A_{11}A_{22} \, .\end{aligned} \tag{5.10.24}$$

If $\Phi^+ = \Phi$, $\triangle\Phi$ must vanish, which is possible for all $\boldsymbol{A}$ when

$$2\eta_{1212} = \eta_{1111} - \eta_{1122} \, . \tag{5.10.25}$$

A hexagonal crystal is therefore characterized by 5 different coefficients of the matrix $(\mathbb{C})$,

$$(C_{pq}) = \begin{pmatrix} \eta_{1111} & \eta_{1122} & \eta_{1133} & 0 & 0 & 0 \\ & \eta_{1111} & \eta_{1133} & 0 & 0 & 0 \\ & & \eta_{3333} & 0 & 0 & 0 \\ & & & 4\eta_{2323} & 0 & 0 \\ & & & & 4\eta_{2323} & 0 \\ & & & & & 2(\eta_{1111} - \eta_{1122}) \end{pmatrix} .$$

6. Phase Transitions in Viscous Heat Conducting Compressible Fluids

6.1 Jump Conditions on a Phase Change Surface

As will become apparent in later chapters a large number of physical processes is so conditioned that phase change phenomena play an important role in them. As an example we might mention the water content in a glacier or the melting and freezing processes that take place at the lower boundary of an ice shelf between the shelf and the ocean. The permafrost in arctic regions or at high altitudes in the mountains often thaws in summer and forms a phase change surface between the frozen and thawed soil. If the temperature in an ice sheet is close to or at the melting point then the heat due to internal dissipation will not be used to adjust the ice temperature to the new conditions but to melt ice. For these and many other reasons it is important that the phase change processes are well understood.

Naturally, the form of the equations describing the phase change processes will likely depend upon the material behaviour of the two phases. Often one does not deal with two different substances but with two so-called phases of the same material, however, in two different forms of molecular state. In the following analysis we restrict ourselves to describing the phase change behaviour of viscous heat conducting fluids at their phase change boundaries.

Phase change surface Consider a body consisting of two distinct materials that are separated from one another by an orientable surface. Let the two subbodies be formed by the same substance in different states of aggregation: thus we shall call the two subbodies simply separate phases of the same substance. If the two phases exchange mass, then the separating surface moves with the non-material velocity $\boldsymbol{w}$. This suggests to regard the separating surface as a singular surface across which the following jump conditions of mass, momentum, energy and entropy must hold:

$$
\begin{aligned}
&[\![\rho(\boldsymbol{v}-\boldsymbol{w})]\!]\cdot\boldsymbol{n}_s = 0 \qquad \text{or} \qquad [\![\mathcal{M}]\!] = 0\,, \\
&[\![\boldsymbol{v}]\!]\mathcal{M} - [\![\boldsymbol{t}\boldsymbol{n}_s]\!] = \boldsymbol{0}\,, \\
&\left[\!\!\left[\left(\tfrac{1}{2}v^2+\varepsilon\right)\right]\!\!\right]\mathcal{M} - [\![(\boldsymbol{v}\boldsymbol{t}-\boldsymbol{q})\cdot\boldsymbol{n}_s]\!] = 0\,, \\
&[\![s]\!]\mathcal{M} + \left[\!\!\left[\frac{\boldsymbol{q}\cdot\boldsymbol{n}_s}{\Theta}\right]\!\!\right] \geq 0\,.
\end{aligned}
\tag{6.1.1}
$$

The variables arising in these relations have the usual meaning introduced in earlier chapters: ρ, $\boldsymbol{v}$, $\boldsymbol{t}$, ε, $\boldsymbol{q}$, s and Θ are the mass density, particle velocity, CAUCHY stress tensor, internal energy, heat flux vector, entropy and (absolute) temperature. Furthermore, $\boldsymbol{n}_s$ is the unit normal vector of the singular surface pointing into the positive part of the body separated by the singular surface. Equations (6.1.1) are given in the EULERian description, see also Table 3.5 on p. 109. In writing the above jump conditions use was made of the jump condition of mass in the remaining jump conditions by substituting the mass flow $\mathcal{M}$ as defined in $(6.1.1)_1$.

Definition *A phase change surface is a singular surface, not necessarily material, between two different phases of a material across which the temperature is continuous,*

$$[\![\Theta]\!] = 0 \,. \tag{6.1.2}$$

■

We would like to emphasize that in the older literature the continuity of the tangential component of the material velocity is also part of the definition of the phase change surface. Here we shall impose this assumption only at a later stage. We will then realize that this additional assumption is actually a restriction.

If (6.1.2) is used in $(6.1.1)_{2,3,4}$, then one obtains

$$\begin{aligned} &-\mathcal{M}[\![\boldsymbol{v}]\!] + [\![\boldsymbol{t}\boldsymbol{n}_s]\!] = \boldsymbol{0} \,, \\ &-\mathcal{M}\,(e + \Theta[\![s]\!]) + [\![\boldsymbol{v}]\!]_{\|} \cdot \boldsymbol{\tau} - [\![\boldsymbol{q}\cdot\boldsymbol{n}_s]\!] = 0 \,, \\ &-\mathcal{M}e + [\![\boldsymbol{v}]\!]_{\|} \cdot \boldsymbol{\tau} \geq 0 \,, \end{aligned} \tag{6.1.3}$$

in which

$$\begin{aligned} [\![\boldsymbol{v}]\!]_{\|} &:= [\![\boldsymbol{v} - (\boldsymbol{n}_s\cdot\boldsymbol{v})\,\boldsymbol{n}_s]\!] = (\boldsymbol{I} - \boldsymbol{n}_s\otimes\boldsymbol{n}_s)\,[\![\boldsymbol{v}]\!] \,, \\ \boldsymbol{\tau} &:= (\boldsymbol{I} - \boldsymbol{n}_s\otimes\boldsymbol{n}_s)\,\langle\!\langle \boldsymbol{t}\boldsymbol{n}_s\rangle\!\rangle \,, \\ e &:= [\![\varepsilon - \Theta s]\!] - \langle\!\langle \boldsymbol{n}_s\cdot\boldsymbol{t}\boldsymbol{n}_s\rangle\!\rangle [\![1/\rho]\!] \\ &= [\![\varepsilon - \Theta s]\!] - [\![(1/\rho)\,\boldsymbol{n}_s\cdot\boldsymbol{t}\boldsymbol{n}_s]\!] + \tfrac{1}{2}\left[\!\left[((\boldsymbol{w}-\boldsymbol{v})\cdot\boldsymbol{n}_s)^2\right]\!\right] . \end{aligned} \tag{6.1.4}$$

The scalar quantity e is called the *energy release rate* and the vector $\boldsymbol{\tau}$ the mean *shear stress*, whilst $[\![\boldsymbol{v}]\!]_{\|}$ is the jump of the tangential velocity or the *velocity slip*. $\boldsymbol{\tau}$ is the average of the shear stress vector on each side of the singular surface[1].

[1] Equations (6.1.4) have first been presented in this elegant form by E. FRIED [79]. Earlier derivations (see e.g. MÜLLER [165] or HUTTER [105]) were less general as they restricted considerations to cases where the tangential velocity did not suffer a jump.

To prove the above formulas (6.1.3) and (6.1.4) we first remark that $(6.1.3)_1$ follows immediately from $(6.1.1)_2$. On the other hand, $(6.1.3)_3$ follows immediately from $(6.1.3)_2$, if $[\![\boldsymbol{q}\cdot\boldsymbol{n}_s]\!]$ is eliminated with the help of $(6.1.1)_4$. Thus, there only remains the proof of $(6.1.3)_2$. To this end, note that

$$[\![\boldsymbol{v}\boldsymbol{t}\cdot\boldsymbol{n}_s]\!] = [\![\boldsymbol{t}\boldsymbol{n}_s\cdot\boldsymbol{v}]\!] = \langle\!\langle \boldsymbol{t}\boldsymbol{n}_s\rangle\!\rangle\cdot[\![\boldsymbol{v}]\!] + [\![\boldsymbol{t}\boldsymbol{n}_s]\!]\cdot\langle\!\langle\boldsymbol{v}\rangle\!\rangle\,, \tag{6.1.5}$$

from which one obtains

$$\begin{aligned}
[\![\boldsymbol{v}]\!]\cdot\langle\!\langle \boldsymbol{t}\boldsymbol{n}_s\rangle\!\rangle &= [\![\boldsymbol{t}\boldsymbol{n}_s\cdot\boldsymbol{v}]\!] - [\![\boldsymbol{t}\boldsymbol{n}_s]\!]\cdot\langle\!\langle\boldsymbol{v}\rangle\!\rangle \\
&= [\![\boldsymbol{t}\boldsymbol{n}_s\cdot\boldsymbol{v}]\!] - \mathcal{M}[\![\boldsymbol{v}]\!]\cdot\langle\!\langle\boldsymbol{v}\rangle\!\rangle \\
&= [\![\boldsymbol{t}\boldsymbol{n}_s\cdot\boldsymbol{v}]\!] - \tfrac{1}{2}\mathcal{M}[\![\boldsymbol{v}\cdot\boldsymbol{v}]\!]\,.
\end{aligned} \tag{6.1.6}$$

If, next, the energy jump condition $(6.1.1)_3$ is written in the form

$$\frac{\mathcal{M}}{2}[\![\boldsymbol{v}\cdot\boldsymbol{v}]\!] + \mathcal{M}[\![\varepsilon]\!] - [\![\boldsymbol{t}\boldsymbol{n}_s\cdot\boldsymbol{v}]\!] + [\![\boldsymbol{q}\cdot\boldsymbol{n}_s]\!] = 0 \tag{6.1.7}$$

and (6.1.6) is used, the energy jump condition takes the form

$$-\mathcal{M}[\![\varepsilon]\!] + [\![\boldsymbol{v}]\!]\cdot\langle\!\langle \boldsymbol{t}\boldsymbol{n}_s\rangle\!\rangle - [\![\boldsymbol{q}\cdot\boldsymbol{n}_s]\!] = 0\,. \tag{6.1.8}$$

Next, one uses the jump condition for the balance of mass to derive the chain of identities

$$[\![\boldsymbol{v}\cdot\boldsymbol{n}_s]\!] = [\![(\boldsymbol{v}-\boldsymbol{w})\cdot\boldsymbol{n}_s]\!] = \left[\!\!\left[\frac{1}{\rho}\rho\,(\boldsymbol{v}-\boldsymbol{w})\cdot\boldsymbol{n}_s\right]\!\!\right] = \mathcal{M}[\![1/\rho]\!]\,. \tag{6.1.9}$$

This equation, on the far left and far right, shows that the jump of the material normal speed is given by the mass flow through the singular surface multiplied with the jump of the specific volume. In (6.1.8) the velocity jump $[\![\boldsymbol{v}]\!]$ may now be decomposed into its tangential and normal components as follows

$$\begin{aligned}
[\![\boldsymbol{v}]\!] &= [\![(\boldsymbol{v}-(\boldsymbol{v}\cdot\boldsymbol{n}_s)\,\boldsymbol{n}_s) + (\boldsymbol{v}\cdot\boldsymbol{n}_s)\,\boldsymbol{n}_s]\!] \\
&= [\![(\boldsymbol{I}-\boldsymbol{n}_s\otimes\boldsymbol{n}_s)\,\boldsymbol{v}]\!] + \mathcal{M}[\![1/\rho]\!]\boldsymbol{n}_s \\
&= (\boldsymbol{I}-\boldsymbol{n}_s\otimes\boldsymbol{n}_s)\,[\![\boldsymbol{v}]\!] + \mathcal{M}[\![1/\rho]\!]\boldsymbol{n}_s \\
&= [\![\boldsymbol{v}]\!]_{||} + \mathcal{M}[\![1/\rho]\!]\boldsymbol{n}_s\,.
\end{aligned} \tag{6.1.10}$$

Substituting (6.1.10) into (6.1.8) transforms the latter equation into

$$\begin{aligned}
&-\mathcal{M}\left([\![\varepsilon]\!] - [\![1/\rho]\!]\langle\!\langle\boldsymbol{n}_s\cdot\boldsymbol{t}\boldsymbol{n}_s\rangle\!\rangle\right) + \\
&[\![\boldsymbol{v}]\!]_{||}\cdot(\boldsymbol{I}-\boldsymbol{n}_s\otimes\boldsymbol{n}_s)\,\langle\!\langle \boldsymbol{t}\boldsymbol{n}_s\rangle\!\rangle - [\![\boldsymbol{q}\cdot\boldsymbol{n}_s]\!] = 0\,,
\end{aligned} \tag{6.1.11}$$

where in the third term $\langle\!\langle \boldsymbol{t}\boldsymbol{n}_s\rangle\!\rangle$ has been replaced by $(\boldsymbol{I}-\boldsymbol{n}_s\otimes\boldsymbol{n}_s)\,\langle\!\langle \boldsymbol{t}\boldsymbol{n}_s\rangle\!\rangle$; this is permissible, because $[\![\boldsymbol{v}]\!]_{||}$ is also tangential to the singular surface. The second term in (6.1.11) can be transformed as follows:

$$
\begin{aligned}
\langle\!\langle \boldsymbol{n}_s \cdot \boldsymbol{t}\boldsymbol{n}_s \rangle\!\rangle [\![1/\rho]\!] &= [\![\boldsymbol{n}_s \cdot \boldsymbol{t}\boldsymbol{n}_s 1/\rho]\!] - [\![\boldsymbol{n}_s \cdot \boldsymbol{t}\boldsymbol{n}_s]\!] \langle\!\langle 1/\rho \rangle\!\rangle \\
[\![\boldsymbol{n}_s \cdot \boldsymbol{t}\boldsymbol{n}_s]\!] \langle\!\langle 1/\rho \rangle\!\rangle &\overset{(6.1.3)}{=} \mathcal{M} [\![\boldsymbol{v} \cdot \boldsymbol{n}_s]\!] \langle\!\langle 1/\rho \rangle\!\rangle \overset{(6.1.9)}{=} \mathcal{M}^2 [\![1/\rho]\!] \langle\!\langle 1/\rho \rangle\!\rangle = \tfrac{\mathcal{M}^2}{2} [\![1/\rho^2]\!] \\
&\overset{(6.1.9)}{=} \tfrac{1}{2} [\![\rho^2 \Big((\boldsymbol{v} - \boldsymbol{w}) \cdot \boldsymbol{n}_s \Big)^2 / \rho^2]\!] \\
&= \tfrac{1}{2} [\![\Big((\boldsymbol{v} - \boldsymbol{w}) \cdot \boldsymbol{n}_s \Big)^2]\!] \, .
\end{aligned}
\tag{6.1.12}
$$

With these results (6.1.11) can be written as

$$
\begin{aligned}
&-\mathcal{M} [\![\varepsilon - (1/\rho) \boldsymbol{n}_s \cdot \boldsymbol{t}\boldsymbol{n}_s + \tfrac{1}{2} \big((\boldsymbol{v} - \boldsymbol{w}) \cdot \boldsymbol{n}_s \big)^2]\!] \\
&+ [\![\boldsymbol{v}]\!]_{\|} \cdot (\boldsymbol{I} - \boldsymbol{n}_s \otimes \boldsymbol{n}_s) \langle\!\langle \boldsymbol{t}\boldsymbol{n}_s \rangle\!\rangle - [\![\boldsymbol{q} \cdot \boldsymbol{n}_s]\!] = 0 \, .
\end{aligned}
\tag{6.1.13}
$$

If in this relation the quantities e and $\boldsymbol{\tau}$, defined in (6.1.4) are identified, then $(6.1.3)_2$ is obtained, provided $[\![\Theta]\!] = 0$, qed.

In thermostatic equilibrium the heat flux vector vanishes, $\boldsymbol{q}_{|E} = \boldsymbol{0}$. Furthermore, no entropy must be produced. Hence, for such a situation $(6.1.1)_4$ must hold with equality sign. Equations $(6.1.3)_{2,3}$ then become

$$
\begin{aligned}
&\Big(-\mathcal{M} \left(e + \Theta [\![s]\!] \right) + [\![\boldsymbol{v}]\!]_{\|} \cdot \boldsymbol{\tau} \Big)_{|E} = 0 \, , \\
&\Big(-\mathcal{M} e \qquad\qquad\;\; + [\![\boldsymbol{v}]\!]_{\|} \cdot \boldsymbol{\tau} \Big)_{|E} = 0 \, .
\end{aligned}
\tag{6.1.14}
$$

The only physically acceptable solution of these equations is

$$
[\![\boldsymbol{v}]\!]_{\| |E} = \boldsymbol{0} \quad \text{and} \quad \mathcal{M}_{|E} = 0 \, ,
\tag{6.1.15}
$$

for any other solution would require that the entropy jump $[\![s]\!]$ vanishes, which is not reasonable. Thus, one necessarily concludes from $(6.1.15)_2$ $\boldsymbol{v} \cdot \boldsymbol{n}_s = \boldsymbol{w} \cdot \boldsymbol{n}_s$, or: *In thermodynamic equilibrium the phase change surface is material.*

A complete theoretical formulation of the thermomechanical behaviour of phase change surfaces under consideration of mass flow through the surface *and* velocity slip requires constitutive postulates for the specific energy release rate e and the shear traction $\boldsymbol{\tau}$ by sliding as functionals of $\mathcal{M}$, $[\![\boldsymbol{v}]\!]_{\|}$, Θ, and these constitutive relations must be constrained by the second law of thermodynamics $(6.1.1)_4$. Here we are less ambitious and continue the analysis with the following restrictive assumption.

Postulate: *The phase change processes at the phase boundary are frictionless and reversible.* ■

Firstly, this postulate requires that $[\![\boldsymbol{v}]\!]_{\|} = \boldsymbol{0}$; in other words, the tangential component of the material velocity is continuously changing across the phase change surface. On the other hand, because of the required reversibility $\mathcal{M}e$ must vanish, $\mathcal{M}e = 0$, which necessarily requires

$$e = 0 \quad \text{(for reversible, frictionless phase change surfaces).} \tag{6.1.16}$$

The other possibility to achieve reversibility, namely $\mathcal{M} = 0$, cannot hold, because this assumption contradicts the energy balance $(6.1.1)_3$. With the definition $(6.1.4)_3$ for the energy release rate, the requirement $e = 0$ takes the form

$$[\![s]\!] + \frac{1}{\Theta}\left[\!\!\left[\frac{\boldsymbol{n}_s \cdot \boldsymbol{t}\boldsymbol{n}_s}{\rho} - \varepsilon - \tfrac{1}{2}(\boldsymbol{w} - \boldsymbol{v}) \cdot (\boldsymbol{w} - \boldsymbol{v})\right]\!\!\right] = 0\,, \tag{6.1.17}$$

where we have set $((\boldsymbol{w} - \boldsymbol{v}) \cdot \boldsymbol{n}_s)^2 = (\boldsymbol{w} - \boldsymbol{v})^2$, since the tangential component of the velocity is continuous. Thus, instead of (6.1.17) one can also write

$$\left[\!\!\left[\tfrac{1}{2}(\boldsymbol{v} - \boldsymbol{w})^2 + \varepsilon - \Theta s - \frac{1}{\rho}(\boldsymbol{t}\boldsymbol{n}_s) \cdot \boldsymbol{n}_s\right]\!\!\right] = 0\,. \tag{6.1.18}$$

6.2 Phase Relations in Thermodynamic Equilibrium

6.2.1 Chemical Potential and Latent Heat

If the adjacent phases are each in thermodynamic equilibrium, the heat flux vector must vanish, and the stress tensor must reduce to an isotropic pressure, viz.,

$$\boldsymbol{q}_{|E} = \boldsymbol{0}\,, \quad \boldsymbol{t}_{|E} = -p_{|E}\boldsymbol{I}\,. \tag{6.2.1}$$

If one assumes that the entropies of the two phases on each side of the phase change surface are different from one another (the entropy of water is certainly larger than that of ice), then the entropy jump does not vanish, $[\![s]\!] \neq 0$. Thus the jump condition of entropy, $[\![s]\!]\rho(\boldsymbol{v} - \boldsymbol{w}) \cdot \boldsymbol{n}_s = 0$, implies that in phase equilibrium the phase change surface must be material, and so no mass exchange occurs,

$$\boldsymbol{v}_{|E} \cdot \boldsymbol{n}_s = \boldsymbol{w} \cdot \boldsymbol{n}_s \quad \Longrightarrow \quad \boldsymbol{v}_{|E} = \boldsymbol{w}\,, \tag{6.2.2}$$

where use has been made that the tangential velocity is continuous. If this condition for thermodynamic equilibrium is substituted in (6.1.18) and if one observes that the pressure in equilibrium is continuous across the singular surface, then one has

$$[\![\mu_{|E}]\!] := \left[\!\!\left[\varepsilon - \Theta s + \frac{p}{\rho}\right]\!\!\right]_{|E} = 0\,, \quad [\![p_{|E}]\!] = 0\,. \tag{6.2.3}$$

The quantity $\mu_{|E}$ is called the *chemical potential* in equilibrium or the free enthalpy or the GIBBS free energy in equilibrium. Solving this equation for the equilibrium pressure yields

$$p_{|E} = \frac{\Theta[\![s_{|E}]\!] - [\![\varepsilon_{|E}]\!]}{[\![1/\rho]\!]}\,. \tag{6.2.4}$$

The chemical potential in equilibrium and the equilibrium pressure are continuous across the phase boundary. The quantity

$$L := \Theta[\![s_{|E}]\!] = [\![\varepsilon_{|E}]\!] + p_{|E}[\![1/\rho]\!] \tag{6.2.5}$$

is called the *latent heat*. Thus the jump of the equilibrium entropy can be written as

$$[\![s_{|E}]\!] = \frac{1}{\Theta}\left([\![\varepsilon_{|E}]\!] + p_{|E}[\![1/\rho]\!]\right) = \frac{1}{\Theta}L\,. \tag{6.2.6}$$

This equation is formally exactly the same as the GIBBS equation (5.7.26).

6.2.2 CLAUSIUS–CLAPEYRON Equation

We are now in a position to derive additional relations for the thermodynamic equilibrium. Energy, entropy and pressure of a compressible viscous heat conducting fluid are merely functions of the temperature and density. If we denote the densities of the two phases by ρ_I and ρ_{II}, then the continuity requirement for the pressure and the chemical potential, which are equally also only functions of (Θ, ρ), implies

$$\begin{aligned} [\![p_{|E}]\!] &= f(\rho_I, \rho_{II}, \Theta) = 0\,, \\ [\![\mu_{|E}]\!] &= g(\rho_I, \rho_{II}, \Theta) = 0\,. \end{aligned} \tag{6.2.7}$$

The jumps of the pressure and the chemical potential across the phase change surface in equilibrium are merely functions of the densities of the two phases and the temperature and their value is zero. If one regards the function f in $(6.2.7)_1$ to be explicitly solvable for ρ_I, i.e., $\rho_I = f_I(\rho_{II}, \Theta)$, then by substituting this into $(6.2.7)_2$ one obtains an equation of the form $g(f_I(\rho_{II}), \rho_{II}, \Theta) = 0$, or when solving this equation for $\rho_{II}, \rho_{II} = \hat{\rho_{II}}(\Theta)$. In an entirely analogous way, one may also show that $\rho_I = \hat{\rho}_I(\Theta)$. Therefore the equilibrium densities of the adjacent phases are representable as

$$\rho_I = \hat{\rho}_I(\Theta)\,, \quad \rho_{II} = \hat{\rho}_{II}(\Theta)\,. \tag{6.2.8}$$

If this result is substituted in the equation for the equilibrium pressure (6.2.4) including the jump quantities for entropy and internal energy, one sees that

$$p_{|E} = \hat{p}_{|E}(\Theta)\,. \tag{6.2.9}$$

The change of the equilibrium pressure with the temperature may thus be computed from (6.2.4), and what obtains reads

$$\frac{\mathrm{d}p_{|E}}{\mathrm{d}\Theta} = \frac{[\![s_{|E}]\!]}{[\![1/\rho]\!]}\,, \tag{6.2.10}$$

see Exercise 1. If in this equation the jump of the entropy is replaced by (6.2.6), the classical form of the CLAUSIUS–CLAPEYRON *equation*, namely

$$\frac{\mathrm{d}\Theta}{\mathrm{d}p_{|E}} = \frac{\Theta[\![1/\rho]\!]}{[\![\varepsilon_{|E}]\!] + p_{|E}[\![1/\rho]\!]} = \frac{\Theta[\![1/\rho]\!]}{L} \tag{6.2.11}$$

is obtained, valid for compressible, heat conducting viscous fluids.

6.3 Phase Change Surfaces in Non-Equilibrium

In non-equilibrium the relation $e = 0$, valid for a reversible frictionless phase change surface can be written as (see (6.1.4))

$$\begin{aligned} e &= \boldsymbol{n}_s \cdot [\![\boldsymbol{M}]\!]\boldsymbol{n}_s = 0 \quad , \\ \boldsymbol{M} &:= \left((\varepsilon - \Theta s)\,\boldsymbol{I} - \frac{1}{\rho}\boldsymbol{t} \right) + \tfrac{1}{2}\,(\boldsymbol{w} - \boldsymbol{v}) \cdot (\boldsymbol{w} - \boldsymbol{v})\,\boldsymbol{I} \\ &= \boldsymbol{\Pi} + \tfrac{1}{2}\,(\boldsymbol{w} - \boldsymbol{v})\,(\boldsymbol{w} - \boldsymbol{v})\,\boldsymbol{I}\,. \end{aligned} \tag{6.3.1}$$

This relation is the generalisation of the classical condition of phase change equilibrium $[\![\mu_{|E}]\!] = 0$ to non-equilibrium thermodynamics. $\boldsymbol{M}$ is called *dynamic, tensorial chemical potential*, and it reduces in equilibrium to $\mu_{|E}\boldsymbol{I}$. Thus the quantity $\mu := \boldsymbol{n}_s \cdot \boldsymbol{M}\boldsymbol{n}_s$ could be interpreted as chemical potential in non-equilibrium. Its sign, however, depends on the orientation of the phase change surface, which is not the case for $\boldsymbol{M}$.

The tensor

$$\boldsymbol{\Pi} = (\varepsilon - \Theta s)\,\boldsymbol{I} - \frac{1}{\rho}\boldsymbol{t} = \Psi_{\varepsilon}\boldsymbol{I} - \frac{1}{\rho}\boldsymbol{t} \tag{6.3.2}$$

is called the *spatial energy–momentum tensor of* ESHELBI. A dynamical reversible phase change surface is therefore governed by the equations

$$-\mathcal{M}[\![\boldsymbol{v}]\!] + [\![\boldsymbol{t}\boldsymbol{n}_s]\!] = 0\,, \quad \boldsymbol{n}_s \cdot [\![\boldsymbol{M}\boldsymbol{n}_s]\!] = 0\,. \tag{6.3.3}$$

The exact solution of these equations is not known. However, with the restricting assumption that both the particle velocity and the velocity of the singular surface are small $((\boldsymbol{v} \sim \boldsymbol{w} \sim \boldsymbol{0}))$, one may in (6.3.3) ignore the diffusive momentum flux and the diffusive kinetic energy flux, so that $\mid \mathcal{M}[\![\boldsymbol{v}]\!] \mid \ll \mid [\![\boldsymbol{t}\boldsymbol{n}_s]\!] \mid$ and $\boldsymbol{M} \cong \boldsymbol{\Pi}$ hold. In this case (6.3.3) may be simplified to[2]

$$[\![\boldsymbol{n}_s \cdot \boldsymbol{t}\boldsymbol{n}_s]\!] \cong 0\,, \qquad [\![s]\!] + \frac{1}{\Theta}\left[\!\!\left[\frac{\boldsymbol{n}_s \cdot \boldsymbol{t}\boldsymbol{n}_s}{\rho} - \varepsilon\right]\!\!\right] \cong 0\,, \tag{6.3.4}$$

which implies, since $p_{\perp} := -\boldsymbol{n}_s \cdot \boldsymbol{t}\boldsymbol{n}_s$,

$$L := \Theta[\![s]\!] = [\![\varepsilon]\!] + p_{\perp}[\![1/\rho]\!]\,, \quad p_{\perp} = \frac{L - [\![\varepsilon]\!]}{[\![1/\rho]\!]}\,. \tag{6.3.5}$$

[2] If one ignores $\mathcal{M}[\![\boldsymbol{v}]\!]$ in comparison to $[\![\boldsymbol{t}\boldsymbol{n}_s]\!]$ then apart from (6.3.4) one also has $[\![\boldsymbol{\tau}]\!] = \boldsymbol{0}$, where $\boldsymbol{\tau} = (\boldsymbol{1} - \boldsymbol{n}_s \otimes \boldsymbol{n}_s)\,\boldsymbol{t}$ is the shear traction acting at the phase boundary.

We recognize that, contrary to the CLAUSIUS–CLAPEYRON equation in equilibrium, the pressure entering the formulas (6.3.5) is not the thermodynamic pressure but the total normal pressure, which also includes possible viscous contributions. The equations (6.3.5) are, however, formally the same as in thermostatic equilibrium. These non-equilibrium phase change expressions were derived e.g. by KAMB [122] with methods of classical thermodynamics, but were given, following essentially the above derivation, by HUTTER [105].

6.4 Density Preserving Fluids

To complete the analysis we also consider here the phase change relationships for a viscous heat conducting density preserving fluid. In this case the pressure is not described by a material equation; it is rather an independent field quantity as we have seen in Chap. 5. In thermodynamic equilibrium, the jump condition of entropy across the phase change surface is given by the relation

$$\underbrace{[\![s_{|E}]\!]}_{f(\Theta)} = \frac{1}{\Theta}\left([\![\varepsilon_{|E}]\!] + p_{|E}[\![1/\rho]\!]\right) . \tag{6.4.1}$$

This has simply been taken over from (6.2.5), because its derivation nowhere uses the fact that the material is compressible. Because in a density preserving fluid the entropy only depends on temperature but neither on pressure (which here is an independent variable) nor density, (6.4.1) can only be a valid equation, if no freely assignable term arises, i.e., if

$$[\![1/\rho]\!] = 0 \quad \Longrightarrow \quad [\![s_{|E}]\!] = \frac{1}{\Theta}[\![\varepsilon_{|E}]\!] \tag{6.4.2}$$

holds true. Thus, no CLAUSIUS–CLAPEYRON equation can exist. The supposition that water or ice be a density preserving fluid is therefore not the adequate limiting model. If one wishes to treat ice as a density preserving material and if phase change processes are significant, one must treat this model as an asymptotic limit of a compressible fluid of which the compressibility goes to zero (Exercise 2). For such a model the CLAUSIUS–CLAPEYRON equation is preserved and thus allows determination of the melting temperature as a function of pressure.

6.5 Exercises

1. In thermodynamic equilibrium for a viscous compressible fluid one has at a phase change surface

$$p_{|E} = \frac{\Theta[\![s_{|E}]\!] - [\![\varepsilon_{|E}]\!]}{[\![1/\rho_{|E}]\!]} .$$

   Derive the CLAUSIUS-CLAPEYRON equation.

2. Consider a viscous, heat conducting compressible fluid with vanishing coefficient of thermal expansion. What are the field equations and jump conditions in this limiting case? In particular show that the CLAUSIUS-CLAPEYRON equation is given by (6.2.10) and (6.2.11).

6.6 Solutions

1. In equilibrium, on a phase change surface, one has

$$[\![\Theta_{|E}]\!] = 0\,, \quad [\![p_{|E}]\!] = 0\,.$$

The change of the equilibrium pressure with the temperature may therefore be computed as follows

$$\begin{aligned}
\frac{\mathrm{d}p_{|E}}{\mathrm{d}\Theta} &= \frac{\mathrm{d}}{\mathrm{d}\Theta}\Big(\frac{\Theta[\![s]\!] - [\![\varepsilon]\!]}{[\![1/\rho]\!]}\Big) \\
&= \frac{[\![s]\!]}{[\![1/\rho]\!]} + \frac{1}{[\![1/\rho]\!]}\Big(\Theta[\![\frac{\mathrm{d}s}{\mathrm{d}\Theta}]\!] - [\![\frac{\mathrm{d}\varepsilon}{\mathrm{d}\Theta}]\!] - \frac{\Theta[\![s]\!] - [\![\varepsilon]\!]}{[\![1/\rho]\!]}[\![\frac{\mathrm{d}(1/\rho)}{\mathrm{d}\Theta}]\!]\Big) \\
&= \frac{[\![s]\!]}{[\![1/\rho]\!]} + \frac{1}{[\![1/\rho]\!]}\Big([\![\frac{\Theta\mathrm{d}s - \mathrm{d}\varepsilon}{\mathrm{d}\Theta}]\!] - p[\![\frac{\mathrm{d}(1/\rho)}{\mathrm{d}\Theta}]\!]\Big)\,,
\end{aligned}$$

where the index $_{|E}$ has been omitted for brevity. Since in equilibrium $\mathrm{d}\varepsilon = \Theta\mathrm{d}s - p\,\mathrm{d}(1/\rho)$, the term in parenthesis vanishes and one has

$$\Big(\frac{\mathrm{d}p}{\mathrm{d}\Theta}\Big)_{|E} = \frac{[\![s_{|E}]\!]}{[\![1/\rho]\!]} = \frac{L}{\Theta[\![1/\rho]\!]}\,.$$

2. The important equation is the energy balance, which has the form

$$\rho\dot{\varepsilon} = \mathrm{div}(\kappa\,\mathrm{grad}\,\Theta) + \mathrm{tr}(\boldsymbol{tD})\,.$$

With $\boldsymbol{t} = -p\boldsymbol{I} + \boldsymbol{t}^E$ and $-\,\mathrm{tr}(p\boldsymbol{ID}) = -p\,\mathrm{div}\,\boldsymbol{v} = (p/\rho)\dot{\rho}$, this equation takes the form

$$\rho\dot{\varepsilon} - \frac{p}{\rho}\dot{\rho} = \mathrm{div}(\kappa\,\mathrm{grad}\,\Theta) + \mathrm{tr}(\boldsymbol{t}^E\boldsymbol{D})\,.$$

By introducing the free enthalpy

$$\psi_h := \varepsilon + \frac{p}{\rho} - \Theta s = \psi_h(p,\Theta)\,,$$

the left-hand side can also be written as

$$\begin{aligned}
&\rho\left\{\dot{\psi}_h - \frac{\dot{p}}{\rho} + \frac{p}{\rho^2}\dot{\rho} + \dot{\Theta}s + \Theta\dot{s} - \frac{p}{\rho^2}\dot{\rho}\right\} \\
&= \rho\left\{\underbrace{\frac{\partial\psi_h}{\partial p}}_{1/\rho}\dot{p} + \underbrace{\frac{\partial\psi_h}{\partial\Theta}}_{-s}\dot{\Theta} - \frac{\dot{p}}{\rho} + \dot{\Theta}s + \Theta\dot{s}\right\} \qquad (6.6.1)\\
&= \rho\Theta\dot{s} = \rho\Theta\left\{\frac{\partial s}{\partial\Theta}\dot{\Theta} + \frac{\partial s}{\partial p}\dot{p}\right\} .
\end{aligned}$$

Now, since $\mathrm{d}\psi_h = -s\mathrm{d}\Theta + (1/\rho)\mathrm{d}p$, the MAXWELL relation

$$-\left.\frac{\partial s}{\partial p}\right|_\Theta = \left.\frac{\partial}{\partial\Theta}\left(\frac{1}{\rho}\right)\right|_p$$

holds, so that

$$\rho\Theta\dot{s} = \rho\Theta\left.\frac{\partial s}{\partial\Theta}\right|_p\dot{\Theta} - \Theta\left(\rho\left.\frac{\partial}{\partial\Theta}\left\{\frac{1}{\rho}\right\}\right|_p\right)\dot{p} = \rho c_p\dot{\Theta} - \Theta\alpha\dot{p} ,$$

in which

$$c_p = \Theta\left.\frac{\partial s}{\partial\Theta}\right|_p , \qquad \alpha = \rho\left.\frac{\partial}{\partial\Theta}\left(\frac{1}{\rho}\right)\right|_p = \left.\frac{1}{v}\frac{\partial v}{\partial\Theta}\right|_p$$

are the specific heat at constant pressure and the coefficient of thermal expansion. Therefore the energy equation takes the form

$$\rho c_p\dot{\Theta} - \Theta\alpha\dot{p} = \mathrm{div}(\kappa\,\mathrm{grad}\,\Theta) + \mathrm{tr}(\boldsymbol{t}^E\boldsymbol{D}) ,$$

in which the second term on the left-hand side is generally ignored.

7. Theory of Mixtures

7.1 General Introduction

In the preceding chapters our focus was on the study of continuum theories of simple constituent media; mixtures, i. e., materials, which are composed of several different constituents or phases, were not considered. However, the latter play a central role in continuum physics and will briefly be touched upon in this chapter.

The principal idea for the theoretical treatment of the mechanics and thermodynamics of mixtures is the supposition that the mutual interconnection of the different *components, phases* or *constituents* is conceptually idealized so as to assume that *each spatial point is simultaneously occupied by material of all constituents*[1]. Such an assumption obviously cannot be physically correct – it contradicts the atomistic structure of matter – however by posing this assumption, a *homogenisation process* is assumed which accounts for the fact that in such a theory one only attempts to describe processes of which the length scales extend over several lengths of the individual constituent elements. A further assumption that is imposed when one formulates a continuum theory of mixtures constitutes the postulate that the balance laws of mass, momentum, angular momentum, energy and entropy be valid for the individual constituents. The only difference that exists between these balance laws and those of a single constituent material is the assumption that the balance laws of mass, linear and angular momentum and energy for the individual constituents are not conserved; only their sum as a whole is supposed to formally behave just as a single constituent material; this

[1] This metaphysical principle was spelled out first by TRUESDELL [238] and forms the basis of approach in most texts dealing with the theory of mixtures. The early development of thermodynamic irreversibility in mixture theory took place in the late sixties and early seventies of the twentieth century and centered around a proper formulation of the second law of thermodynamics and its exploitation by the entropy principle. The breakthrough was accomplished by I. MÜLLER [161]. An account on the principles and controversies of the developments can be found in TRUESDELL's book on Rational Thermodynamics [241], [243] (1968, first edition, substantially extended in 1984, second edition). Further treatizes on thermodynamics, where mixtures are treated, are also by MÜLLER [163], [165]. These works mention earlier literature.

corresponds to the request that the sums of the production terms of mass, momentum, angular momentum and energy add up to zero.

As is the case for single constituent materials, the balance laws by themselves are not sufficient to uniquely determine the field quantities arising in them. Some of the field variables must be expressed in terms of others by postulating materially dependent constitutive relations, for instance the partial stress tensors can be functionally expressed in terms of the deformation fields of the constituents. Even in those cases in which the relevant equations can be explicitly written down, these relations tend to be very complicated. As a rule, simplifications are necessary, if not for any other reason simply to put the resulting system of equations into a form amenable to physical interpretation.

To achieve such simplifications mixture theories were developed, in which only a part of all possible balance laws is used. For example, if water percolates through the firn of a snow cover, it is necessary to differentiate between the temperatures of the water and the snow; this requires that (at least) two balance laws of energy are formulated separately for the constituents water and snow. On the other hand, if the water flow in temperate ice, whose temperature is exactly at the melting point, is studied, then the assumption that water and ice have the same temperature is plausible and may be used as a simplifying postulate; in this case, a single energy equation for the mixture as a whole will most likely adequately describe the energetic processes in the system. Similarly, if one wishes to describe the diffusion of a substance in a fluid (pollutant or nutrient in water) which may arise only in very small concentrations, it is permissible to consider the momentum balance of the tracer mass and the fluid together.

It is evident that this procedure of simplification leads to a *hierarchy of mixture theories*, each with its own complexity and each applicable under restricting physical conditions. The most important models possess the following structure (HUTTER [107], [108]):

- **Class I:** Here, the balance laws of mass of all constituents are used, however, only the momentum and energy balances of the mixture as a whole are employed. Often considerations of energy are left untouched, because only mechanical and no thermodynamic processes are in focus.
  These models are typical for the description of the *diffusive motion* of any particulate substance, which exists as a pollutant or tracer in another substance. The equations possess advective, diffuse and possibly reactive structure. Examples are the salinity content in the water, the distribution of a nutrient (phosphate) or a chemical element (O_2, CO_2) in the water of a lake or in the ocean and many others.
- **Class II:** These mixture theories are appropriate, if the interpenetrating constituents possess comparable concentrations of mass or specific momentum, but move with different velocities. They are also know as *multiphase theories.* In this class the balance laws of mass, momentum and angular

momentum are formulated for all constituents, but only one energy balance is used for the mixture as a whole. In other words, the individual constituents possess sufficiently distinct specific masses and momenta but the same temperature, so that to account for the thermodynamic effects the formulation of a single energy balance for the mixture as a whole suffices. These models go beyond the classical diffusion models which are only able to describe a dilution of a tracer substance, and no growth. The interaction forces that are active between the constituents are important and make it possible that the phase separation between the individual constituents can take place.
Examples of theories of this class are practically all soil mechanics theories, which describe the interaction between the granular matrix and the water. For instance, DARCY's *law* is nothing else than a mathematical ansatz for the interaction force between the granulate and the pore water. Members of this class are also all mechanical models of multiphase suspension flows. In the geophysical context these are *powder snow avalanches*, of which the constituents are air and snow. In general, a diffusion model does not suffice to describe their dynamics, because, first, snow of the snow cover is entrained into the avalanche and, second, snow is deposited in the runout zone. Both processes lead to an increase in the particle concentration. Other geophysical examples are *turbidity currents*, i. e., subaquatic slope motions similar to the *powder snow avalanches*, of which the constituents are soil and water, or *sturzstroms* and *mud flows*, mixtures of water and gravel or soil which move down a mountain slope, and finally *pyroclastic flows*, i. e., suspension flows of volcanic dust and air.
- **Class III:** The next level is occupied by the full thermodynamic mixture theories, in which the balance laws of mass, (linear and angular) momentum and energy of all constituents must be formulated. *Plasmas* belong to this class; in the geophysical context there exist, however, equally a multitude of mixture concepts, which can be assigned to this class. For instance, the creeping deformation of cold firn in the uppermost layers of an ice sheet under the influence of the percolation flow of the melt water can be described by a mixture model of this complexity: Water and snow have distinct specific masses, velocities and temperatures, and apart from these complexities there may occur phase changes between the two constituents. Problems of geothermics also belong to this class of mixtures.

The three just described classes of mixture theories may also occur in a mixed form; all the more, such mixed forms are often applied in practice. For example, the dispersion of a pollutant in the groundwater is formulated by a model which contains elements of classes I and II. The pollutant and the water form together a mixture of class I; the polluted water together with the soil a mixture of class II. All combinations are thinkable, and it lies in the talent and depth of physical understanding of the scientist who develops a model to make the choice appropriate to a given situation.

After this introduction, let us now move forward and formulate the balance laws for mixtures.

7.2 Balance Laws for the Constituents

We wish to restrict attention to the formulation of the balance laws in the present configuration. The corresponding LAGRANGE description can be derived in a much similar way and is given as one of the exercises (Exercise 1). The individual constituents are differentiated amongst each other by lower case indices; in particular a general, non-specified constituent will be characterized by the index α which takes values from $\alpha = 1$ to $\alpha = \nu$. The general balance law (2.1.16) takes here the mixture specific form

$$\frac{\mathrm{d}^\alpha}{\mathrm{d}t} \int_\omega \gamma^\alpha(\boldsymbol{x}, t)\mathrm{d}v = \int_\omega (\pi^\alpha(\boldsymbol{x}, t) + \varsigma^\alpha(\boldsymbol{x}, t))\,\mathrm{d}v - \int_{\partial\omega} \boldsymbol{\phi}^\alpha(\boldsymbol{x}, t) \cdot \boldsymbol{n}\,\mathrm{d}a\,, \quad (7.2.1)$$

with the following local counterparts

$$\frac{\partial \gamma^\alpha}{\partial t} + \mathrm{div}(\gamma^\alpha \boldsymbol{v}^\alpha) = -\,\mathrm{div}\ \boldsymbol{\phi}^\alpha + \pi^\alpha + \varsigma^\alpha\,, \quad (7.2.2)$$

$$[\![\gamma^\alpha((\boldsymbol{v}^\alpha - \boldsymbol{u}) \cdot \boldsymbol{n}_\mathfrak{s})]\!] - [\![\boldsymbol{\phi}^\alpha]\!]\boldsymbol{n}_\mathfrak{s} = \mathfrak{P}^\alpha\,. \quad (7.2.3)$$

In these equations γ^α, π^α, ς^α and $\boldsymbol{\phi}^\alpha$ denote an unspecified physical quantity of the constituent α, its production in and its supply to ω as well as the flux across the boundary $\partial\omega$; $\mathrm{d}^\alpha/\mathrm{dt}$ is the material time derivative of the constituent α, i. e., the time derivative if one follows the motion of constituent α, and $\boldsymbol{v}^\alpha$ is the velocity vector of constituent α. If $\boldsymbol{\gamma}^\alpha$ is a vectorial field, then $\boldsymbol{\gamma}^\alpha\boldsymbol{v}^\alpha$ must be interpreted as a dyadic product: $\boldsymbol{\gamma}^\alpha \otimes \boldsymbol{v}^\alpha$. In the jump condition (7.2.3) $\boldsymbol{n}_\mathfrak{s}$ denotes the unit vector perpendicular to the singular surface, $\mathfrak{P}^\alpha$ is the surface production of the constituent physical quantity γ^α, and $\boldsymbol{u}$ is the velocity with which the surface moves; it must not coincide with any of the constituents' particle velocity instantaneously sitting upon the surface. Finally, $[\![f]\!] = f^+ - f^-$ is the difference of the values of f on the positive and negative sides of the singular surface.

The various interpretations of the fields for the different physical balances are summarized in Table 7.1. In the **balance law of mass** for constituent α flux and supply vanish as they do for a single constituent material; however, there can arise a nontrivial production of mass for constituent α, $\mathfrak{c}^\alpha$; this is e. g. the case if mass of constituent α is produced by chemical reactions or phase changes. Likewise, it is so that mass of a constituent can be generated on a singular surface, if phase change processes arise there. This is described as surface production μ^α.

In the **balance law of linear momentum** the momentum density $\rho^\alpha\boldsymbol{v}^\alpha$ is balanced by the volume forces $\rho^\alpha\boldsymbol{f}^\alpha$ and the partial stress $\boldsymbol{t}^\alpha$; to these contributions, also arising in a one-component material, one must add a density of momentum production or interaction force $\mathfrak{m}^\alpha$. This is the force that is exerted on constituent α by all other constituents $1, 2, ..., \nu$ (without α). In Table 7.1 we also list a surface momentum production $\boldsymbol{\tau}^\alpha$, which is for instance necessary when a singular surface separates two mixtures with different numbers of constituents.

The **balance of angular momentum** in Table 7.1 is formulated for a so called *polar continuum*; more specifically, each entry in Table 7.1 consists of two contributions for γ^α, an intrinsic angular momentum $\rho^\alpha \boldsymbol{s}^\alpha$, called *spin* and a *moment of momentum* $\boldsymbol{x} \times \rho^\alpha \boldsymbol{v}^\alpha$. Both are obviously axial vectors[2]. In much the same way the supply of angular momentum consists of an external specific couple $\rho^\alpha \boldsymbol{l}^\alpha$ and the moment of the specific exterior volume force $\boldsymbol{x} \times \rho^\alpha \boldsymbol{f}^\alpha$; and the flux of angular momentum is obtained as the sum of the (negative) couple stresses $-\boldsymbol{m}^\alpha$ and the moment of the (negative) partial CAUCHY stresses $-\boldsymbol{x} \times \boldsymbol{t}^\alpha$. Likewise the angular momentum production is composed of two contributions, the production of the specific spin $\mathfrak{k}^\alpha$ and the moment of the momentum production $\boldsymbol{x} \times \mathfrak{m}^\alpha$. Finally, one may also compose the surface production of angular momentum in this way; however, we shall not do this here, because we see no advantages by doing so; we call $\boldsymbol{\sigma}^\alpha$ the surface production of angular momentum of constituent α.

The physical quantity in the **balance of energy**, the first law of thermodynamics, is the specific energy density, which is composed of the kinetic energy (of translation and rotation), $\frac{1}{2}\rho^\alpha(\boldsymbol{v}^\alpha \cdot \boldsymbol{v}^\alpha + \boldsymbol{s}^\alpha \cdot \boldsymbol{\omega}^\alpha)$ plus the internal energy, $\rho^\alpha \varepsilon^\alpha$. These are the kinetic and thermodynamic contributions; $\boldsymbol{v}^\alpha$ is the velocity of constituent α, and $\boldsymbol{\omega}^\alpha$ is its angular velocity; the latter is in no way related to the velocity $\boldsymbol{v}^\alpha$. Since $\boldsymbol{s}^\alpha$ and ω^α are both axial vectors, $\boldsymbol{s}^\alpha \cdot \boldsymbol{\omega}^\alpha$ is a scalar. Likewise, the energy supply and energy flux are composed of such contributions. For instance, the energy supply is given by the power of working of the specific volume forces $\boldsymbol{v}^\alpha \cdot \rho^\alpha \boldsymbol{f}^\alpha$ and specific volume couples $\rho^\alpha \boldsymbol{l}^\alpha \cdot \boldsymbol{\omega}^\alpha$ plus the specific radiation $\rho^\alpha \mathfrak{r}^\alpha$, all for constituent α. Similarly the flux of the energy of constituent α is composed of the power of working of the negative partial stresses $-\boldsymbol{v}^\alpha \boldsymbol{t}^\alpha$, and the couple stresses $-\boldsymbol{\omega}^\alpha \boldsymbol{m}^\alpha$ plus the partial heat flux $\boldsymbol{q}^\alpha$ of constituent α. The production terms of energy are given by $\mathfrak{e}^\alpha$ and ϵ^α and denote specific quantities per unit volume and surface, respectively.

The individual quantities in the **entropy balance** are not specifically identified in Table 7.1 except that the quantities carry the superscript s^α.

[2] A vector is called axial, if it changes its sign when the orientation of the basis changes.

Table 7.1. Field quantities which arise in the balance law (7.2.1) for mass, linear and angular momentum, energy and entropy for a physical quantity of constituent α. γ^α, $\boldsymbol{\phi}^\alpha$, ς^α, π^α and $\mathfrak{P}^\alpha$ denote the physical quantity, its flux, supply, volume production and surface production. Concerning the individual entries, see main text.

Balance	γ^α	$\boldsymbol{\phi}^\alpha$	ς^α	π^α	$\mathfrak{P}^\alpha$
Mass	ρ^α	0	0	$\mathfrak{c}^\alpha$	μ^α
Momentum	$\rho^\alpha \boldsymbol{v}^\alpha$	$-\boldsymbol{t}^\alpha$	$\rho^\alpha \boldsymbol{f}^\alpha$	$\mathbf{m}^\alpha$	$\boldsymbol{\tau}^\alpha$
Angular momentum	$\rho^\alpha \boldsymbol{s}^\alpha + \boldsymbol{x} \times \rho^\alpha \boldsymbol{v}^\alpha$	$-\boldsymbol{m}^\alpha - \boldsymbol{x} \times \boldsymbol{t}^\alpha$	$\rho^\alpha \boldsymbol{l}^\alpha + \boldsymbol{x} \times \rho^\alpha \boldsymbol{f}^\alpha$	$\boldsymbol{\mathfrak{k}}^\alpha + \boldsymbol{x} \times \mathbf{m}^\alpha$	$\boldsymbol{\sigma}^\alpha$
Energy	$\frac{1}{2}\rho^\alpha \boldsymbol{v}^\alpha \cdot \boldsymbol{v}^\alpha + \frac{1}{2}\rho^\alpha \boldsymbol{s}^\alpha \cdot \boldsymbol{\omega}^\alpha$ $+\rho^\alpha \varepsilon^\alpha$	$-\boldsymbol{v}^\alpha\, \boldsymbol{t}^\alpha - \boldsymbol{\omega}^\alpha \boldsymbol{m}^\alpha + \boldsymbol{q}^\alpha$	$\boldsymbol{v}^\alpha \cdot \rho^\alpha \boldsymbol{f}^\alpha + \rho^\alpha \boldsymbol{l}^\alpha \cdot \boldsymbol{\omega}^\alpha$ $+\rho^\alpha \mathfrak{r}^\alpha$	$\mathfrak{e}^\alpha$	ϵ^α
Entropy	$\rho^\alpha s^\alpha$	$\boldsymbol{\phi}^{s^\alpha}$	$\rho^\alpha \eta^{s^\alpha}$	$\rho^\alpha \pi^{s^\alpha}$	$\mathfrak{P}^{s^\alpha}$

With the identifications of Table 7.1 the local balance laws (7.2.2) and (7.2.3) take the following forms:

Mass:

$$\begin{aligned} &\frac{\partial \rho^\alpha}{\partial t} + \operatorname{div}(\rho^\alpha \cdot \boldsymbol{v}^\alpha) = \mathfrak{c}^\alpha , \\ &[\![\rho^\alpha(\boldsymbol{v}^\alpha - \boldsymbol{u}) \cdot \boldsymbol{n}_{\mathfrak{s}})]\!] = \mu^\alpha , \end{aligned} \tag{7.2.4}$$

Momentum:

$$\begin{aligned} &\frac{\partial}{\partial t}(\rho^\alpha \boldsymbol{v}^\alpha) + \operatorname{div}(\rho^\alpha \boldsymbol{v}^\alpha \otimes \boldsymbol{v}^\alpha - \boldsymbol{t}^\alpha) = \rho^\alpha \boldsymbol{f}^\alpha + \mathfrak{m}^\alpha , \\ &[\![\rho^\alpha \otimes (\boldsymbol{v}^\alpha - \boldsymbol{u}) \cdot \boldsymbol{n}_{\mathfrak{s}}]\!] + [\![\boldsymbol{t}^\alpha \boldsymbol{n}_{\mathfrak{s}}]\!] = \boldsymbol{\tau}^\alpha , \end{aligned} \tag{7.2.5}$$

Angular Momentum:

$$\begin{aligned} &\frac{\partial}{\partial t}(\rho^\alpha \boldsymbol{s}^\alpha + \boldsymbol{x}^\alpha \times \rho^\alpha \boldsymbol{v}^\alpha) + \operatorname{div}\Big((\rho^\alpha \boldsymbol{s}^\alpha + \boldsymbol{x} \times \rho^\alpha \boldsymbol{v}^\alpha) \otimes \boldsymbol{v}^\alpha \\ &\qquad - \boldsymbol{m}^\alpha - \boldsymbol{x} \times \boldsymbol{t}^\alpha\Big) = \rho^\alpha \boldsymbol{l}^\alpha + \boldsymbol{x} \times \rho^\alpha \boldsymbol{f}^\alpha + \mathfrak{k}^\alpha + \boldsymbol{x} \times \mathfrak{m}^\alpha , \\ &[\![(\rho^\alpha \boldsymbol{s}^\alpha + \boldsymbol{x}^\alpha \times \rho^\alpha \boldsymbol{v}^\alpha)(\boldsymbol{v}^\alpha - \boldsymbol{u}) \cdot \boldsymbol{n}_{\mathfrak{s}}]\!] + [\![\boldsymbol{m}^\alpha + \boldsymbol{x} \times \boldsymbol{t}^\alpha]\!] \cdot \boldsymbol{n}_{\mathfrak{s}} = \boldsymbol{\sigma}^\alpha , \end{aligned} \tag{7.2.6}$$

Energy:

$$\begin{aligned} &\frac{\partial}{\partial t}\big(\tfrac{1}{2}\rho^\alpha \boldsymbol{v}^\alpha \cdot \boldsymbol{v}^\alpha + \tfrac{1}{2}\rho^\alpha \boldsymbol{s}^\alpha \cdot \boldsymbol{\omega}^\alpha + \rho^\alpha \varepsilon^\alpha\big) + \operatorname{div}\big((\tfrac{1}{2}\rho^\alpha \boldsymbol{v}^\alpha \cdot \boldsymbol{v}^\alpha \\ &+ \tfrac{1}{2}\rho^\alpha \boldsymbol{s}^\alpha \cdot \boldsymbol{\omega}^\alpha + \rho^\alpha \varepsilon^\alpha)\boldsymbol{v}^\alpha - \boldsymbol{v}^\alpha \boldsymbol{t}^\alpha - \boldsymbol{\omega}^\alpha \boldsymbol{m}^\alpha + \boldsymbol{q}^\alpha\big) \\ &= \mathfrak{e}^\alpha + \boldsymbol{v}^\alpha \cdot \rho^\alpha \boldsymbol{f}^\alpha + \boldsymbol{\omega}^\alpha \cdot \rho^\alpha \boldsymbol{l}^\alpha + \rho^\alpha \mathfrak{r}^\alpha , \\ &[\![(\tfrac{1}{2}\rho^\alpha \boldsymbol{v}^\alpha \cdot \boldsymbol{v}^\alpha + \tfrac{1}{2}\rho^\alpha \boldsymbol{s}^\alpha \cdot \boldsymbol{\omega}^\alpha)(\boldsymbol{v}^\alpha - \boldsymbol{u}) \cdot \boldsymbol{n}_{\mathfrak{s}}]\!] \\ &-[\![\boldsymbol{v}^\alpha \boldsymbol{t}^\alpha + \boldsymbol{\omega}^\alpha \boldsymbol{m}^\alpha - \boldsymbol{q}^\alpha]\!] \cdot \boldsymbol{n}_{\mathfrak{s}} = \epsilon^\alpha , \end{aligned} \tag{7.2.7}$$

Entropy:

$$\begin{aligned} &\frac{\partial}{\partial t}(\rho^\alpha s^\alpha) + \operatorname{div}\left(\rho^\alpha s^\alpha \boldsymbol{v}^\alpha + \boldsymbol{\phi}^{s^\alpha}\right) = \rho^\alpha \eta^{s^\alpha} + \rho^\alpha \pi^{s^\alpha} , \\ &[\![\rho^\alpha s^\alpha(\boldsymbol{v}^\alpha - \boldsymbol{u}) \cdot \boldsymbol{n}_{\mathfrak{s}}]\!] - [\![\boldsymbol{\phi}^{s^\alpha}]\!] \cdot \boldsymbol{n}_{\mathfrak{s}} = \mathfrak{P}^{s^\alpha} . \end{aligned} \tag{7.2.8}$$

These laws can be further simplified, however, they then no longer appear in the classical divergence form. To this end we use the material time derivative of constituent α

$$(\cdot)^{\backslash\alpha} := \frac{\mathrm{d}^\alpha(\cdot)}{\mathrm{d}t} := \frac{\partial(\cdot)}{\partial t} + \big(\operatorname{grad}(\cdot)\big)\boldsymbol{v}^\alpha . \tag{7.2.9}$$

Introducing this in the above equations (7.2.4) - (7.2.8) and sequentially using the results of the previous equations, it can be shown (Exercise 2) that the following alternative forms hold:

Momentum:

$$\begin{aligned}
&\rho^\alpha \frac{\mathrm{d}^\alpha \boldsymbol{v}^\alpha}{\mathrm{d}t} = \operatorname{div}\boldsymbol{t}^\alpha + \rho^\alpha \boldsymbol{f}^\alpha + (\mathfrak{m}^\alpha - \mathfrak{c}^\alpha \boldsymbol{v}^\alpha) , \\
&[\![\rho^\alpha \boldsymbol{v}^\alpha \otimes (\boldsymbol{v}^\alpha - \boldsymbol{u})\cdot \boldsymbol{n}_{\mathfrak{s}}]\!] + [\![\boldsymbol{t}^\alpha \boldsymbol{n}_{\mathfrak{s}}]\!] = \boldsymbol{\tau}^\alpha ,
\end{aligned} \tag{7.2.10}$$

Angular Momentum or Spin:

$$\begin{aligned}
&\rho^\alpha \frac{\mathrm{d}^\alpha \boldsymbol{s}^\alpha}{\mathrm{d}t} = \operatorname{div}\boldsymbol{m}^\alpha + \rho^\alpha \boldsymbol{l}^\alpha - \operatorname{dual}\left(\boldsymbol{t}^\alpha - \boldsymbol{t}^{\alpha T}\right) + (\mathfrak{k}^\alpha - \mathfrak{c}^\alpha \boldsymbol{s}^\alpha) , \\
&[\![\rho^\alpha \boldsymbol{s}^\alpha \left((\boldsymbol{v}^\alpha - \boldsymbol{u})\cdot \boldsymbol{n}_{\mathfrak{s}}\right)]\!] + [\![\boldsymbol{m}^\alpha \boldsymbol{n}_{\mathfrak{s}}]\!] = \boldsymbol{\sigma}^\alpha ,
\end{aligned} \tag{7.2.11}$$

Energy:

$$\begin{aligned}
\rho^\alpha \frac{\mathrm{d}^\alpha \varepsilon^\alpha}{\mathrm{d}t} &= -\operatorname{div}\boldsymbol{q}^\alpha + \boldsymbol{\omega}^\alpha \cdot \left(\operatorname{dual}(\boldsymbol{t}^\alpha - \boldsymbol{t}^{\alpha T})\right) \\
&\quad + (\operatorname{grad}\boldsymbol{v}^\alpha \cdot \boldsymbol{t}^{\alpha T}) + (\operatorname{grad}\boldsymbol{\omega}^\alpha \cdot \boldsymbol{m}^{\alpha T}) \\
&\quad + \tfrac{1}{2}\rho^\alpha \left(\boldsymbol{\omega}^\alpha \cdot \frac{\mathrm{d}^\alpha \boldsymbol{s}^\alpha}{\mathrm{d}t} - \boldsymbol{s}^\alpha \cdot \frac{\mathrm{d}^\alpha \boldsymbol{\omega}^\alpha}{\mathrm{d}t}\right) + \rho^\alpha \mathfrak{r}^\alpha \\
&\quad + \left(\mathfrak{e}^\alpha - \mathfrak{c}^\alpha \left(\varepsilon^\alpha + \tfrac{1}{2}\boldsymbol{v}^\alpha \cdot \boldsymbol{v}^\alpha + \tfrac{1}{2}\boldsymbol{s}^\alpha \cdot \boldsymbol{\omega}^\alpha\right)\right. \\
&\qquad \left. - (\mathfrak{m}^\alpha - \mathfrak{c}^\alpha \boldsymbol{v}^\alpha)\cdot \boldsymbol{v}^\alpha - (\mathfrak{k}^\alpha - \mathfrak{c}^\alpha \boldsymbol{s}^\alpha)\cdot \boldsymbol{\omega}^\alpha\right) , \\
&[\![(\tfrac{1}{2}\rho^\alpha \boldsymbol{v}^\alpha \cdot \boldsymbol{v}^\alpha + \tfrac{1}{2}\rho^\alpha \boldsymbol{s}^\alpha \cdot \boldsymbol{\omega}^\alpha)(\boldsymbol{v}^\alpha - \boldsymbol{u})\cdot \boldsymbol{n}_{\mathfrak{s}}]\!] \\
&\qquad - [\![\boldsymbol{v}^\alpha \boldsymbol{t}^\alpha + \boldsymbol{\omega}^\alpha \boldsymbol{m}^\alpha - \boldsymbol{q}^\alpha]\!]\cdot \boldsymbol{n}_{\mathfrak{s}} = \epsilon^\alpha .
\end{aligned} \tag{7.2.12}$$

Entropy:

$$\begin{aligned}
&\rho^\alpha \frac{\mathrm{d}^\alpha s^\alpha}{\mathrm{d}t} = -\operatorname{div}\boldsymbol{\phi}^{s^\alpha} + \rho^\alpha \eta^{s^\alpha} + \rho^\alpha \pi^{s^\alpha} \quad \text{with} \quad \pi^{s^\alpha} \geq 0 , \\
&[\![\rho^\alpha s^\alpha (\boldsymbol{v}^\alpha - \boldsymbol{u})\cdot \boldsymbol{n}_{\mathfrak{s}}]\!] - [\![\boldsymbol{\phi}^{s^\alpha}]\!]\cdot \boldsymbol{n}_{\mathfrak{s}} = \mathfrak{P}^{s^\alpha} .
\end{aligned} \tag{7.2.13}$$

In the spin and internal energy balances the dual $\boldsymbol{W}$ of the skew-symmetric tensor $\boldsymbol{W}$ denotes the axial vector $\boldsymbol{w}$ defined by

$$w_i := \tfrac{1}{2}\epsilon_{ijk}W_{jk} \,. \tag{7.2.14}$$

Formally, the above local balance laws, valid in spatial points where the fields are supposed to be differentiable, can be somewhat simplified if the production terms π^α of Table 7.1 are replaced by the following new production quantities, called *intrinsic* or *non-convective production densities of momentum, angular momentum and energy,*

$$\begin{aligned}
\mathfrak{m}^\alpha_{\text{Euclid}} &:= \mathfrak{m}^\alpha - \mathfrak{c}^\alpha \boldsymbol{v}^\alpha \,, \\
\mathfrak{k}^\alpha_{\text{Euclid}} &:= \mathfrak{k}^\alpha - \mathfrak{c}^\alpha \boldsymbol{s}^\alpha \,, \\
\mathfrak{e}^\alpha{}_{\text{Euclid}} &:= \mathfrak{e}^\alpha - \mathfrak{c}^\alpha \left(\varepsilon^\alpha + \tfrac{1}{2}\boldsymbol{v}^\alpha \cdot \boldsymbol{v}^\alpha + \tfrac{1}{2}\boldsymbol{s}^\alpha \cdot \boldsymbol{\omega}^\alpha\right) \\
&\quad -(\mathfrak{m}^\alpha - \mathfrak{c}^\alpha \boldsymbol{v}^\alpha)\cdot \boldsymbol{v}^\alpha - (\mathfrak{k}^\alpha - \mathfrak{c}^\alpha \boldsymbol{s}^\alpha)\cdot \boldsymbol{\omega}^\alpha \,.
\end{aligned} \tag{7.2.15}$$

These are so defined that the combination of all production terms in (7.2.10)–(7.2.12) reduce to $\mathfrak{m}^\alpha_{\text{Euclid}}$ (momentum), $\mathfrak{k}^\alpha_{\text{Euclid}}$ (spin) and $\rho^\alpha \mathfrak{e}^\alpha{}_{\text{Euclid}}$ (energy). However, the choices (7.2.15) can also be physically motivated. Accordingly, the specific momentum production $\mathfrak{m}^\alpha$ is given by the intrinsic (i.e., non-convective) production $\mathfrak{m}^\alpha_{\text{Euclid}}$ plus the convective momentum of the specific mass production, $\mathfrak{c}^\alpha \boldsymbol{v}^\alpha$. Likewise, the spin production $\mathfrak{k}^\alpha$ is given as the sum of the intrinsic (non-convective) spin production $\mathfrak{k}^\alpha_{\text{Euclid}}$ plus the spin of the mass production. Both are quite natural definitions, and the same holds true for the energy production $\mathfrak{e}^\alpha$. This is additively composed of the intrinsic contribution, $\mathfrak{e}^\alpha{}_{\text{Euclid}}$, the production of internal energy due to the mass production, $\mathfrak{c}^\alpha \varepsilon^\alpha$, the kinetic energies of translation and rotation due to the mass production, $\frac{1}{2}\mathfrak{c}^\alpha \boldsymbol{v}^\alpha \cdot \boldsymbol{v}^\alpha$ and $\frac{1}{2}\mathfrak{c}^\alpha \boldsymbol{s}^\alpha \cdot \boldsymbol{\omega}^\alpha$, respectively, plus the powers of working of the intrinsic productions of momentum and spin, $(\mathfrak{m}^\alpha_{\text{Euclid}} \cdot \boldsymbol{v}^\alpha)$ and $(\mathfrak{k}^\alpha_{\text{Euclid}} \cdot \boldsymbol{\omega}^\alpha)$, respectively. In addition, one can show that $\mathfrak{m}^\alpha_{\text{Euclid}}$, $\mathfrak{k}^\alpha_{\text{Euclid}}$ and $\mathfrak{e}^\alpha{}_{\text{Euclid}}$ are objective quantities under EUCLIDian transformations but $\mathfrak{m}^\alpha$, $\mathfrak{k}^\alpha$ and $\mathfrak{e}^\alpha$ are not. They transform, respectively, as a polar and axial vector and as a scalar, respectively.

We derived above the physical balance laws for mixtures for *polar* constituents; these are continua of which the motion possesses independent degrees of freedom for translation and rotation; in this sense the velocities $\boldsymbol{v}^\alpha$ and angular velocities $\boldsymbol{\omega}^\alpha$ are vector fields kinematically independent of one another. Associated with these independent kinematic fields are the balances of momentum and spin. One deduces from the latter balance (7.2.11) that with

$$\boldsymbol{s}^\alpha = \boldsymbol{0}, \qquad \boldsymbol{m}^\alpha = \boldsymbol{0}, \qquad \boldsymbol{l}^\alpha = \boldsymbol{0} \,, \tag{7.2.16}$$

the CAUCHY stress of constituent α must satisfy the relation

$$\text{dual}(\boldsymbol{t}^\alpha - \boldsymbol{t}^{\alpha T}) = \mathfrak{k}^\alpha_{\text{Euclid}} \,. \tag{7.2.17}$$

A component of a mixture or a continuum for which the relations (7.2.16) are fulfilled is called a *non-polar* or *a-polar* continuum. One concludes from (7.2.17) that in non-polar mixtures the partial CAUCHY stress $\boldsymbol{t}^\alpha$ is only symmetric if the corresponding intrinsic spin production vanishes.

The local equation (7.2.11) for the spin is a balance equation that allows the following interpretations:

- Flux of spin $\equiv -\boldsymbol{m}^\alpha$ (negative couple stresses),
- Supply of spin $\equiv \rho^\alpha \boldsymbol{l}^\alpha$ (externally applied body couple),
- Production of spin $\equiv -\,\text{dual}(\boldsymbol{t}^\alpha - \boldsymbol{t}^{\alpha T}) + \mathfrak{t}^\alpha_{\text{Euclid}}$.

Likewise, the equation for the internal energy of constituent α has the form of a balance law with the interpretations

- Flux of internal energy $\equiv \boldsymbol{q}^\alpha$ (heat flux),
- Supply of internal energy $\equiv \rho^\alpha \mathfrak{r}^\alpha$ (radiation),
- Production of internal energy

$$\begin{aligned} &\equiv (\text{grad}\,\boldsymbol{v}^\alpha \cdot \boldsymbol{t}^{\alpha T}) + \boldsymbol{\omega}^\alpha \cdot \big(\text{dual}(\boldsymbol{t}^\alpha - \boldsymbol{t}^{\alpha T})\big) + (\text{grad}\,\boldsymbol{\omega}^\alpha \cdot \boldsymbol{m}^{\alpha T}) \\ &\quad + \mathfrak{e}^\alpha{}_{\text{Euclid}} + \tfrac{1}{2}\rho^\alpha \left(\boldsymbol{\omega}^\alpha \cdot \frac{\mathrm{d}^\alpha \boldsymbol{s}^\alpha}{dt} - \boldsymbol{s}^\alpha \cdot \frac{\mathrm{d}^\alpha \boldsymbol{\omega}^\alpha}{dt}\right) . \end{aligned} \tag{7.2.18}$$

Apart from the well known term for the power of working of non-polar continua $(\text{grad}\,\boldsymbol{v}^\alpha \cdot \boldsymbol{t}^{\alpha T})$, this formula contains an analogous term of the spin motion, namely $\boldsymbol{\omega}^\alpha \cdot (\text{dual}(\boldsymbol{t}^\alpha - \boldsymbol{t}^{\alpha T})) + (\text{grad}\,\boldsymbol{\omega}^\alpha \cdot \boldsymbol{m}^{\alpha T})$ plus a further term, new by its structure and given by

$$\mathfrak{S}^\alpha = \tfrac{1}{2}\rho^\alpha \left(\boldsymbol{\omega}^\alpha \cdot \frac{\mathrm{d}^\alpha \boldsymbol{s}^\alpha}{\mathrm{d}t} - \boldsymbol{s}^\alpha \cdot \frac{\mathrm{d}^\alpha \boldsymbol{\omega}^\alpha}{\mathrm{d}t}\right) . \tag{7.2.19}$$

It may be called the *micromorphic spin production of constituent* α. To fathom the significance of this term, let us be guided by rigid body dynamics of a solid body; its moment of momentum relative to its center of mass c is given by

$$\boldsymbol{D}_c = \hat{\boldsymbol{\Theta}}_c \boldsymbol{\omega} , \qquad \hat{\boldsymbol{\Theta}}_c = \hat{\boldsymbol{\Theta}}_c^T , \tag{7.2.20}$$

in which $\hat{\boldsymbol{\Theta}}_c$ denotes the *tensor of moment of inertia*, a symmetric second rank tensor. Being inspired by this, we now postulate the spin of constituent α to be given by

$$\boldsymbol{s}^\alpha = \boldsymbol{\Theta}^\alpha \boldsymbol{\omega}^\alpha , \qquad \boldsymbol{\Theta}^\alpha = \boldsymbol{\Theta}^{\alpha T} , \tag{7.2.21}$$

where $\boldsymbol{\Theta}^\alpha$ is a symmetric second rank tensor, called the *tensor of specific inertia* of constituent α. It is a measure for the distribution of mass of constituent α in a region of influence of a particle at position $\boldsymbol{x}$. In general, this distribution changes with time for any material particle, so that $\mathrm{d}^\alpha \boldsymbol{\Theta}^\alpha/\mathrm{d}t \neq 0$.

If the rotations are small for a particular deformation of the body, then the time rate of change of the distribution of mass of constituent α about the particle at position $\boldsymbol{x}$ may be ignored. One therefore distinguishes materials as follows:

- A constituent α is called *micro-polar* (hyphened here for emphasis) if the specific tensor of inertia is materially constant for that constituent,

$$\frac{\mathrm{d}^{\alpha}\boldsymbol{\Theta}^{\alpha}}{\mathrm{d}t} = \mathbf{0} \qquad \Longleftrightarrow \qquad \text{micropolar.}$$

- A constituent α is called *micro-morphic* (hyphened here for emphasis) if the tensor of inertia may vary with time,

$$\frac{\mathrm{d}^{\alpha}\boldsymbol{\Theta}^{\alpha}}{\mathrm{d}t} \neq \mathbf{0} \qquad \Longleftrightarrow \qquad \text{micromorphic.}$$

With these definitions and the representation (7.2.21) for the spin density it follows that $\mathfrak{S}^{\alpha}$ vanishes for micropolar continua, but may be different from zero for micromorphic continua.

7.3 Balance Laws for the Mixture as a Whole

If the balance laws of mass, linear and angular momentum and energy are summed over all constituents, one obtains the corresponding balance law for the mixture as a whole. If one views the body as an entity it seems plausible to assume that an observer, who looks at the body as a whole and is not aware that it may be composed of a number of constituents, will recognize this body as a one-component material and impose the postulates of a one-component continuum. This suggests that *the balance laws of mass, momentum, angular momentum and energy for the mixture as a whole must be conservation laws.* If so, the sum of the production terms over all constituents of mass, momentum, angular momentum and energy must vanish, implying that

$$\sum_{\alpha=1}^{\nu} \mathfrak{c}^{\alpha} = 0\,, \quad \sum_{\alpha=1}^{\nu} \mathbf{m}^{\alpha} = \mathbf{0}\,, \quad \sum_{\alpha=1}^{\nu} \boldsymbol{\mathfrak{k}}^{\alpha} = \mathbf{0}\,, \quad \sum_{\alpha=1}^{\nu} \mathfrak{e}^{\alpha} = 0\,. \tag{7.3.1}$$

Notice that these statements are required for the production terms of the original balance laws and not their EUCLIDian invariant partners defined in (7.2.15).

If we request that the balance laws of the mixture as a whole be obtained as the sum of the partial balances, it is only natural if we also request that *these balance laws coincide with the balance laws of a one-constituent continuum.* In order to realize this principle, which goes back to TRUESDELL [238], it is necessary that variables for the mixture as a whole are related to the variables of the constituents. Summing (7.2.2) over all constituents yields

$$\sum_{\alpha=1}^{\nu} \frac{\partial \gamma^\alpha}{\partial t} + \sum_{\alpha=1}^{\nu} \operatorname{div}(\gamma^\alpha \boldsymbol{v}^\alpha) = - \sum_{\alpha=1}^{\nu} \operatorname{div} \boldsymbol{\phi}^\alpha + \sum_{\alpha=1}^{\nu} (\pi^\alpha + \varsigma^\alpha) . \tag{7.3.2}$$

The second term on the left-hand side in this equation contains products of constituent quantities, $\gamma^\alpha \boldsymbol{v}^\alpha$. This is inconvenient; if we therefore introduce a mixture velocity $\boldsymbol{v}$ – at the present state of the knowledge it is not clear how to specify it – by simply writing $\boldsymbol{v}^\alpha = \boldsymbol{v}^\alpha - \boldsymbol{v} + \boldsymbol{v}$, then we may write

$$\sum_{\alpha=1}^{\nu} \gamma^\alpha \boldsymbol{v}^\alpha = \sum_{\alpha=1}^{\nu} \gamma^\alpha \Big(\boldsymbol{v} + \underbrace{(\boldsymbol{v}^\alpha - \boldsymbol{v})}_{\boldsymbol{u}^\alpha}\Big) = \sum_{\alpha=1}^{\nu} \gamma^\alpha \boldsymbol{u}^\alpha + \left(\sum_{\alpha=1}^{\nu} \gamma^\alpha\right) \boldsymbol{v} . \tag{7.3.3}$$

The quantity $\boldsymbol{u}^\alpha$ is called *diffusion velocity* of constituent α relative to the mixture velocity. Introducing, moreover, the definitions

$$\gamma = \sum_{\alpha=1}^{\nu} \gamma^\alpha, \quad \pi = \sum_{\alpha=1}^{\nu} \pi^\alpha, \quad \varsigma = \sum_{\alpha=1}^{\nu} \varsigma^\alpha, \quad \boldsymbol{\phi} = \sum_{\alpha=1}^{\nu} (\boldsymbol{\phi}^\alpha + \gamma^\alpha \boldsymbol{u}^\alpha) \tag{7.3.4}$$

and thus defining the physical quantity γ of the mixture, its production π, supply ς and flux $\boldsymbol{\phi}$, respectively, one obtains from (7.3.2) the new balance law

$$\frac{\partial \gamma}{\partial t} + \operatorname{div}(\gamma \boldsymbol{v}) = \operatorname{div} \boldsymbol{\phi} + \pi + \varsigma . \tag{7.3.5}$$

To guarantee that the balance law for the mixture takes the form of a balance law for a single constituent body, two things have necessarily to be satisfied. First, the physical quantity γ characterizing the mixture as a whole, its production π and its supply ς are given as the sum of the respective constituent quantities. Second, the flux $\boldsymbol{\phi}$ for the mixture is given as the sum of the constituent fluxes $\boldsymbol{\phi}^\alpha$ plus the diffusive fluxes $\gamma^\alpha \boldsymbol{u}^\alpha$.

Above, the mixture velocity $\boldsymbol{v}$ and the diffusion velocity of constituent α, $\boldsymbol{u}^\alpha = \boldsymbol{v}^\alpha - \boldsymbol{v}$ were introduced, but not defined. Their exact definition follows from the *balance law of mass.* With $\gamma^\alpha = \rho^\alpha$, $\gamma = \rho$, $\pi = 0$, $\varsigma = 0$ and $\boldsymbol{\phi}^\alpha_{Mass} = \boldsymbol{0}$ for all α, the mass flux of the mixture (7.3.4) is given by

$$\boldsymbol{\phi}_{Mass} = \sum_{\alpha=1}^{\nu} \rho^\alpha \boldsymbol{u}^\alpha,$$

which ought to vanish, if the balance law of mass for the mixture as a whole should be formally the same as the balance law of mass for a single constituent material. One therefore must request

$$\sum_{\alpha=1}^{\nu} \rho^\alpha \boldsymbol{u}^\alpha = \boldsymbol{0} \qquad \Longleftrightarrow \qquad \rho \boldsymbol{v} = \sum_{\alpha=1}^{\nu} \rho^\alpha \boldsymbol{v}^\alpha . \tag{7.3.6}$$

This condition defines the mixture velocity, and only with this choice the mass balance for the mixture takes the form

$$\frac{\partial \rho}{\partial t} + \operatorname{div}(\rho \boldsymbol{v}) = 0 \, . \tag{7.3.7}$$

The mixture velocity defined by (7.3.6) is called the *barycentric velocity* because it represents the averaged velocity, using the constituent densities as weights. Occasionally, it is also written as

$$\boldsymbol{v} = \sum_{\alpha=1}^{\nu} \xi^\alpha \boldsymbol{v}^\alpha, \qquad \xi^\alpha = \frac{\rho^\alpha}{\rho}, \tag{7.3.8}$$

where ξ^α is the mass concentration of constituent α.

For the *momentum balance* one defines

$$\begin{aligned} &\boldsymbol{\gamma} = \rho \boldsymbol{v} = \sum_{\alpha=1}^{\nu} \rho^\alpha \boldsymbol{v}^\alpha \, , \quad \boldsymbol{\varsigma} = \rho \boldsymbol{f} = \sum_{\alpha=1}^{\nu} \rho^\alpha \boldsymbol{f}^\alpha \, , \quad \boldsymbol{\pi} = \boldsymbol{0} \, , \\ &\boldsymbol{\phi} = -\boldsymbol{t} = -\sum_{\alpha=1}^{\nu} (\boldsymbol{t}^\alpha - \rho^\alpha \boldsymbol{v}^\alpha \otimes \boldsymbol{u}^\alpha) = -\sum_{\alpha=1}^{\nu} (\boldsymbol{t}^\alpha - \rho^\alpha \boldsymbol{u}^\alpha \otimes \boldsymbol{u}^\alpha) \, . \end{aligned} \tag{7.3.9}$$

Whilst the specific momentum and the specific body force of the mixture are given as the sums of the specific partial momentum and specific body forces of the constituents, the stress tensor for the mixture is composed of two contributions, the sum of the partial stress tensors over all constituents $\sum \boldsymbol{t}^\alpha$ minus the diffusive fluxes of the partial momenta $\sum(\rho^\alpha \boldsymbol{u}^\alpha) \otimes \boldsymbol{u}^\alpha$. With these identifications the balance of momentum for the mixture as a whole takes the classical form

$$\frac{\partial(\rho \boldsymbol{v})}{\partial t} + \operatorname{div}(\rho \boldsymbol{v} \otimes \boldsymbol{v}) = \operatorname{div} \boldsymbol{t} + \rho \boldsymbol{f} \, . \tag{7.3.10}$$

Often, in mixture–theory contexts the mixture stress is given as the sum of the partial stresses without accounting for the diffusive contribution as stated in $(7.3.9)_4$. It is straightforward to show that this is only justified, if the convective contributions to the acceleration are negligible, viz.,

$$\frac{\mathrm{d}\boldsymbol{v}^\alpha}{\mathrm{d}t} \approx \frac{\partial \boldsymbol{v}^\alpha}{\partial t} \quad \forall \alpha \qquad \Longleftrightarrow \qquad \boldsymbol{t} \approx \sum_{\alpha=1}^{\nu} \boldsymbol{t}^\alpha \, . \tag{7.3.11}$$

Next, let us consider the *balance law of angular momentum*, and let us define

$$
\begin{aligned}
\gamma &= \sum_{\alpha=1}^{\nu} (\rho^\alpha \boldsymbol{s}^\alpha + \boldsymbol{x} \times \rho^\alpha \boldsymbol{v}^\alpha) = \underbrace{\sum_{\alpha=1}^{\nu} \rho^\alpha \boldsymbol{s}^\alpha}_{\rho \boldsymbol{s}} + \boldsymbol{x} \times \sum_{\alpha=1}^{\nu} \rho^\alpha \boldsymbol{v}^\alpha = \rho \boldsymbol{s} + \boldsymbol{x} \times \rho \boldsymbol{v} \,, \\
\varsigma &= \sum_{\alpha=1}^{\nu} (\rho^\alpha \boldsymbol{l}^\alpha + \boldsymbol{x} \times \rho^\alpha \boldsymbol{f}^\alpha) = \underbrace{\sum_{\alpha=1}^{\nu} \rho^\alpha \boldsymbol{l}^\alpha}_{\rho \boldsymbol{l}} + \boldsymbol{x} \times \sum_{\alpha=1}^{\nu} \rho^\alpha \boldsymbol{f}^\alpha = \rho \boldsymbol{l} + \boldsymbol{x} \times \rho \boldsymbol{f} \,, \\
\pi &= \boldsymbol{0} \,, \\
\phi &= \sum_{\alpha=1}^{\nu} (-\boldsymbol{m}^\alpha - \boldsymbol{x} \times \boldsymbol{t}^\alpha + (\rho^\alpha \boldsymbol{s}^\alpha + \boldsymbol{x} \times \rho^\alpha \boldsymbol{v}^\alpha) \otimes \boldsymbol{u}^\alpha) \\
&= -\underbrace{\sum_{\alpha=1}^{\nu} (\boldsymbol{m}^\alpha - \rho^\alpha \boldsymbol{s}^\alpha \otimes \boldsymbol{u}^\alpha)}_{\boldsymbol{m}} - \boldsymbol{x} \times \underbrace{\sum_{\alpha=1}^{\nu} (\boldsymbol{t}^\alpha - \rho^\alpha \boldsymbol{u}^\alpha \otimes \boldsymbol{u}^\alpha)}_{\boldsymbol{t}} \\
&= -\boldsymbol{m} - \boldsymbol{x} \times \boldsymbol{t} \,.
\end{aligned}
\tag{7.3.12}
$$

Accordingly, the specific density of the angular momentum of the mixture is composed of the sum of the spin densities of the constituents and the moment of the linear momentum of the mixture; the supply of angular momentum is the sum of the specific body couples of the constituents minus the moment of the body forces, and the flux of angular momentum of the mixture is given as the negative sum of the couple stresses of the constituents minus the moment of the stress of the mixture. As was the case for the stress tensor of the mixture, the couple stress for the mixture is composed of the couple stresses of the constituents minus the diffusive fluxes of the specific spins of the constituents. With (7.3.12), therefore, the balance law of angular momentum for the mixture as a whole takes the form

$$
\begin{aligned}
&\frac{\partial}{\partial t}(\rho \boldsymbol{s} + \boldsymbol{x} \times \rho \boldsymbol{v}) + \operatorname{div}((\rho \boldsymbol{s} + \boldsymbol{x} \times \rho \boldsymbol{v}) \otimes \boldsymbol{v}) \\
&\qquad = \operatorname{div}(\boldsymbol{m} + \boldsymbol{x} \times \boldsymbol{t}) + \rho \boldsymbol{l} + \boldsymbol{x} \times \rho \boldsymbol{f} \,.
\end{aligned}
\tag{7.3.13}
$$

Straightforward calculation shows that this equation may alternatively also be written as

$$
\frac{\partial}{\partial t}(\rho \boldsymbol{s}) + \operatorname{div}(\rho \boldsymbol{s} \otimes \boldsymbol{v}) + \operatorname{div} \boldsymbol{m} + \operatorname{dual}(\boldsymbol{t} - \boldsymbol{t}^T) + \rho \boldsymbol{l} \,, \tag{7.3.14}
$$

which is the local spin balance. From this equation it is directly evident that for $\boldsymbol{s} = \boldsymbol{0}$, $\boldsymbol{m} = \boldsymbol{0}$ and $\boldsymbol{l} = \boldsymbol{0}$ the CAUCHY stress tensor must be symmetric.

It is worth mentioning that in the above derivation of the balance of angular momentum for the mixture the kinematic variable associated with the spin does not arise; an angular velocity for the mixture was not defined and did not need to be defined. This can nevertheless be done by connecting the constituent spin density $\boldsymbol{s}^\alpha$ with the constituent angular velocity $\boldsymbol{\omega}^\alpha$, as done in (7.2.21) via $\boldsymbol{s}^\alpha = \boldsymbol{\Theta}^\alpha \boldsymbol{\omega}^\alpha$, where $\boldsymbol{\Theta}^\alpha$ is the tensor of specific moment of inertia. If one writes now

$$\boldsymbol{\omega}^\alpha = \boldsymbol{\omega} + \underbrace{(\boldsymbol{\omega}^\alpha - \boldsymbol{\omega})}_{\Delta\boldsymbol{\omega}^\alpha} = \boldsymbol{\omega} + \Delta\boldsymbol{\omega}^\alpha \ , \tag{7.3.15}$$

where $\Delta\boldsymbol{\omega}^\alpha$ may be called *diffusive angular velocity of constituent* α, one obtains

$$\begin{aligned} \rho\boldsymbol{s} &= \sum_{\alpha=1}^{\nu} \rho^\alpha \boldsymbol{s}^\alpha = \sum_{\alpha=1}^{\nu} \rho^\alpha \boldsymbol{\Theta}^\alpha (\boldsymbol{\omega} + \Delta\boldsymbol{\omega}^\alpha) \\ &= \underbrace{\Big(\sum_{\alpha=1}^{\nu} \rho^\alpha \boldsymbol{\Theta}^\alpha\Big)}_{\rho\boldsymbol{\Theta}} \boldsymbol{\omega} + \sum_{\alpha=1}^{\nu} \underbrace{\rho^\alpha \boldsymbol{\Theta}^\alpha \Delta\boldsymbol{\omega}^\alpha}_{\Delta\boldsymbol{s}^\alpha} \ . \end{aligned} \tag{7.3.16}$$

We may now define the *angular velocity of the mixture* by requiring

$$\sum_{\alpha=1}^{\nu} \Delta\boldsymbol{s}^\alpha := \sum_{\alpha=1}^{\nu} \rho^\alpha \boldsymbol{\Theta}^\alpha \Delta\boldsymbol{\omega}^\alpha = \boldsymbol{0} \quad \Longleftrightarrow \quad \rho\boldsymbol{\Theta}\boldsymbol{\omega} = \sum_{\alpha=1}^{\nu} \rho^\alpha \boldsymbol{\Theta}^\alpha \boldsymbol{\omega}^\alpha \ , \tag{7.3.17}$$

in which

$$\rho\boldsymbol{\Theta} := \sum_{\alpha=1}^{\nu} \rho^\alpha \boldsymbol{\Theta}^\alpha \tag{7.3.18}$$

can be called the *barycentric specific moment of inertia.* With these specifications we have

$$\rho\boldsymbol{s} = \rho\boldsymbol{\Theta}\boldsymbol{\omega} \ . \tag{7.3.19}$$

Of course, different definitions of the mixture angular velocity are thinkable, however, the above one seems to be particularly natural.

For the derivation of the *balance law of energy* for the mixture we write

$$\begin{aligned} \gamma &= \sum_{\alpha=1}^{\nu} \Big(\tfrac{1}{2}\rho^\alpha \boldsymbol{v}^\alpha \cdot \boldsymbol{v}^\alpha + \tfrac{1}{2}\rho^\alpha \boldsymbol{s}^\alpha \cdot \boldsymbol{\omega}^\alpha + \rho^\alpha \varepsilon^\alpha\Big) \\ &= \sum_{\alpha=1}^{\nu} \tfrac{1}{2}\rho^\alpha \Big(\boldsymbol{v}\cdot\boldsymbol{v} + \boldsymbol{u}^\alpha\cdot\boldsymbol{v} + \boldsymbol{v}\cdot\boldsymbol{u}^\alpha + \boldsymbol{u}^\alpha\cdot\boldsymbol{u}^\alpha\Big) \\ &\quad + \sum_{\alpha=1}^{\nu} \tfrac{1}{2}\rho^\alpha \boldsymbol{\Theta}^\alpha (\boldsymbol{\omega} + \Delta\boldsymbol{\omega}^\alpha)\cdot(\boldsymbol{\omega} + \Delta\boldsymbol{\omega}^\alpha) + \sum_{\alpha=1}^{\nu} \rho^\alpha \varepsilon^\alpha \\ &= \tfrac{1}{2}\rho\boldsymbol{v}\cdot\boldsymbol{v} + \tfrac{1}{2}\rho\boldsymbol{s}\cdot\boldsymbol{\omega} \\ &\quad + \underbrace{\sum_{\alpha=1}^{\nu} \Big(\rho^\alpha \varepsilon^\alpha + \tfrac{1}{2}\rho^\alpha \boldsymbol{u}^\alpha\cdot\boldsymbol{u}^\alpha + \tfrac{1}{2}\rho^\alpha \boldsymbol{\Theta}^\alpha \Delta\boldsymbol{\omega}^\alpha \cdot \Delta\boldsymbol{\omega}^\alpha\Big)}_{\rho\varepsilon} \ , \end{aligned} \tag{7.3.20}$$

$$
\begin{aligned}
\varsigma &= \sum_{\alpha=1}^{\nu} \left(\boldsymbol{v}^\alpha \cdot \rho^\alpha \boldsymbol{f}^\alpha + \boldsymbol{\omega}^\alpha \cdot \rho^\alpha \boldsymbol{l}^\alpha + \rho^\alpha \mathfrak{r}^\alpha \right) \\
&= \sum_{\alpha=1}^{\nu} \left((\boldsymbol{v} + \boldsymbol{u}^\alpha) \cdot \rho^\alpha \boldsymbol{f}^\alpha + (\boldsymbol{\omega} + \Delta\boldsymbol{\omega}^\alpha) \cdot \rho^\alpha \boldsymbol{l}^\alpha + \rho^\alpha \mathfrak{r}^\alpha \right) \\
&= \boldsymbol{v} \cdot \rho \boldsymbol{f} + \boldsymbol{\omega} \cdot \rho \boldsymbol{l} \\
&\quad + \underbrace{\sum_{\alpha=1}^{\nu} \left(\rho^\alpha \mathfrak{r}^\alpha + \boldsymbol{u}^\alpha \cdot \rho^\alpha \boldsymbol{f}^\alpha + \Delta\boldsymbol{\omega}^\alpha \cdot \rho^\alpha \boldsymbol{l}^\alpha \right)}_{\rho\mathfrak{r}} .
\end{aligned}
\tag{7.3.21}
$$

Thus, the specific energy of the mixture, $\rho\varepsilon$ is given as the sum of the partial internal plus translational and rotational kinetic energies formed with the diffusion velocities of the constituent motions, $\boldsymbol{u}^\alpha$ and $\Delta\boldsymbol{\omega}^\alpha$, respectively. Likewise, the energy supply of the mixture, $\rho\mathfrak{r}$ is the sum of the partial supplies, $\rho^\alpha\mathfrak{r}^\alpha$ plus the power of working of the volume forces and body couples on the diffuse motion of the constituents.

The mixture energy flux is given by

$$
\begin{aligned}
\boldsymbol{\phi} = \sum_{\alpha=1}^{\nu} \Big[&-(\boldsymbol{v}^\alpha \boldsymbol{t}^\alpha + \boldsymbol{\omega}^\alpha \boldsymbol{m}^\alpha - \boldsymbol{q}^\alpha) \\
&+ (\tfrac{1}{2}\rho^\alpha \boldsymbol{v}^\alpha \cdot \boldsymbol{v}^\alpha + \tfrac{1}{2}\rho^\alpha \boldsymbol{s}^\alpha \cdot \boldsymbol{\omega}^\alpha + \rho^\alpha \varepsilon^\alpha)\boldsymbol{u}^\alpha \Big] .
\end{aligned}
\tag{7.3.22}
$$

A simple but somewhat long calculation shows that it can be written as (see Exercise 3)

$$
\boldsymbol{\phi} = -\boldsymbol{v}\boldsymbol{t} - \boldsymbol{\omega}\boldsymbol{m} + \boldsymbol{q} ,
$$

where

$$
\begin{aligned}
\boldsymbol{q} := \sum_{\alpha=1}^{\nu} \Big[&\boldsymbol{q}^\alpha - \boldsymbol{u}^\alpha \boldsymbol{t}^\alpha - \Delta\boldsymbol{\omega}^\alpha \boldsymbol{m}^\alpha \\
&+ \Big(\rho^\alpha \varepsilon^\alpha + \tfrac{1}{2}\rho^\alpha \boldsymbol{u}^\alpha \cdot \boldsymbol{u}^\alpha + \tfrac{1}{2}\rho^\alpha \boldsymbol{s}^\alpha \cdot (\Delta\boldsymbol{\omega}^\alpha - \boldsymbol{\omega}) \Big) \boldsymbol{u}^\alpha \Big] .
\end{aligned}
\tag{7.3.23}
$$

As this formula shows, the vector $\boldsymbol{q}$ cannot be interpreted as heat flux alone; energy flux is the better terminology. Evidently, the energy flux vector of the mixture is composed of the constituent heat flux vectors $\boldsymbol{q}^\alpha$, the powers of working of the stresses and couple stresses on the diffuse motion and the convective energy transport on the diffuse motion. Consequently, the balance law of energy for the mixture as a whole takes the form

$$
\begin{aligned}
&\frac{\partial}{\partial t} \left(\rho\varepsilon + \tfrac{1}{2}\rho \boldsymbol{v} \cdot \boldsymbol{v} + \tfrac{1}{2}\rho \boldsymbol{s} \cdot \boldsymbol{\omega} \right) + \operatorname{div} \left((\rho\varepsilon + \tfrac{1}{2}\rho \boldsymbol{v} \cdot \boldsymbol{v} + \tfrac{1}{2}\rho \boldsymbol{s} \cdot \boldsymbol{\omega}) \boldsymbol{v} \right) \\
&\quad = \operatorname{div}(\boldsymbol{v}\boldsymbol{t} + \boldsymbol{\omega}\boldsymbol{m} - \boldsymbol{q}) + \boldsymbol{v} \cdot \rho \boldsymbol{f} + \boldsymbol{\omega} \cdot \rho \boldsymbol{l} + \rho\mathfrak{r}
\end{aligned}
\tag{7.3.24}
$$

and reduces for a non-polar continuum with $\boldsymbol{s} = \mathbf{0}, \boldsymbol{\omega} = \mathbf{0}, \boldsymbol{l} = \mathbf{0}$ to the classical energy balance

$$\begin{aligned}&\frac{\partial}{\partial t}\left(\rho\left(\varepsilon+\tfrac{1}{2}\boldsymbol{v}\cdot\boldsymbol{v}\right)\right)+\operatorname{div}\left(\rho\left(\varepsilon+\tfrac{1}{2}\boldsymbol{v}\cdot\boldsymbol{v}\right)\boldsymbol{v}\right)\\&=\operatorname{div}(\boldsymbol{v}\boldsymbol{t}-\boldsymbol{q})+\boldsymbol{v}\cdot\rho\boldsymbol{f}+\rho\mathfrak{r}\,.\end{aligned} \tag{7.3.25}$$

Incidentally, it is interesting to see that the energy flux $\boldsymbol{q}$ contains a term proportional to $\boldsymbol{\omega}$ which cannot be eliminated; indeed, $\boldsymbol{q}$ can be written in the form

$$\begin{aligned}\boldsymbol{q}=\sum_{\alpha=1}^{\nu}&\left\{\boldsymbol{q}^{\alpha}-\boldsymbol{u}^{\alpha}\boldsymbol{t}^{\alpha}-\Delta\boldsymbol{\omega}^{\alpha}\boldsymbol{m}^{\alpha}+\left(\rho^{\alpha}\varepsilon^{\alpha}+\tfrac{1}{2}\varepsilon^{\alpha}\boldsymbol{u}^{\alpha}\cdot\boldsymbol{u}^{\alpha}\right.\right.\\&\left.\left.+\tfrac{1}{2}\Delta\boldsymbol{s}^{\alpha}\cdot\Delta\boldsymbol{\omega}^{\alpha}\right)\boldsymbol{u}^{\alpha}\right\}+\boldsymbol{s}\sum_{\alpha=1}^{\nu}\tfrac{1}{2}\rho^{\alpha}\Delta\boldsymbol{\omega}^{\alpha}\otimes\boldsymbol{u}^{\alpha}-\boldsymbol{\omega}\sum_{\alpha=1}^{\nu}\tfrac{1}{2}\rho^{\alpha}\Delta\boldsymbol{s}^{\alpha}\otimes\boldsymbol{u}^{\alpha}\,,\end{aligned} \tag{7.3.26}$$

in which the two terms that involve $\boldsymbol{s}$ and $\boldsymbol{\omega}$ do in general not cancel each other. The translational and rotational components of the motion have therefore a different effect on the mixture energy flux.

Finally, by summing the constituent *entropy balance laws* (7.2.8) one may deduce the relation

$$\frac{\partial}{\partial t}(\rho s)+\operatorname{div}(\rho s\boldsymbol{v}+\boldsymbol{\phi}^{\eta})-\rho\eta^{s}=\rho\pi^{\eta}\geq 0\,, \tag{7.3.27}$$

in which

$$\begin{aligned}&\rho s=\sum_{\alpha}\rho^{\alpha}s^{\alpha},\quad \rho\eta^{s}=\sum_{\alpha}\rho^{\alpha}\eta^{s^{\alpha}},\quad \rho\pi^{s}=\sum_{\alpha}\rho\pi^{s^{\alpha}},\\&\boldsymbol{\phi}^{s}=\sum_{\alpha}\left(\boldsymbol{\phi}^{s^{\alpha}}+s^{\alpha}\boldsymbol{u}^{\alpha}\right),\end{aligned} \tag{7.3.28}$$

as anticipated already in (7.3.4).

7.4 Summary

The local balance laws of mass, momentum, angular momentum, energy and entropy for the mixture as a whole take the forms

$$\frac{\partial \rho}{\partial t} + \operatorname{div}(\rho \boldsymbol{v}) = 0 \,,$$

$$\frac{\partial (\rho \boldsymbol{v})}{\partial t} + \operatorname{div}(\rho \boldsymbol{v} \otimes \boldsymbol{v}) = \operatorname{div} \boldsymbol{t} + \rho \boldsymbol{f} \,,$$

$$\begin{aligned} &\frac{\partial}{\partial t}(\rho \boldsymbol{s} + \boldsymbol{x} \times \rho \boldsymbol{v}) + \operatorname{div}\left((\rho \boldsymbol{s} + \boldsymbol{x} \times \rho \boldsymbol{v}) \otimes \boldsymbol{v}\right) \\ &\qquad = \operatorname{div}(\boldsymbol{m} + \boldsymbol{x} \times \boldsymbol{t}) + \rho \boldsymbol{l} + \boldsymbol{x} \times \rho \boldsymbol{f} \,, \end{aligned} \tag{7.4.1}$$

$$\begin{aligned} &\frac{\partial}{\partial t}\left(\rho \varepsilon + \tfrac{1}{2}\rho \boldsymbol{v} \cdot \boldsymbol{v} + \tfrac{1}{2}\rho \boldsymbol{s} \cdot \boldsymbol{\omega}\right) + \operatorname{div}\left(\left(\rho \varepsilon + \tfrac{1}{2}\rho \boldsymbol{v} \cdot \boldsymbol{v} + \tfrac{1}{2}\rho \boldsymbol{s} \cdot \boldsymbol{\omega}\right) \cdot \boldsymbol{v}\right) \\ &\qquad = \operatorname{div}(\boldsymbol{v}\boldsymbol{t} + \boldsymbol{\omega}\boldsymbol{m} - \boldsymbol{q}) + \boldsymbol{v} \cdot \rho \boldsymbol{f} + \boldsymbol{\omega} \cdot \rho \boldsymbol{l} + \rho \mathfrak{r} \,, \end{aligned}$$

$$\frac{\partial}{\partial t}(\rho s) + \operatorname{div}(\rho s \boldsymbol{v}) = -\operatorname{div} \boldsymbol{\phi}^s + \rho \eta^s + \rho \pi^s \,,$$

in which the field quantities of the mixture (without index α) are related to those of the constituents (with index α) according to

$$\begin{aligned} \rho &= \sum_{\alpha=1}^{\nu} \rho^\alpha \,, & \rho \boldsymbol{v} &= \sum_{\alpha=1}^{\nu} \rho^\alpha \boldsymbol{v}^\alpha \,, \\ \rho \boldsymbol{\Theta} &= \sum_{\alpha=1}^{\nu} \rho^\alpha \boldsymbol{\Theta}^\alpha \,, & \rho \boldsymbol{\Theta} \boldsymbol{\omega} &= \sum_{\alpha=1}^{\nu} \rho^\alpha \boldsymbol{\Theta}^\alpha \boldsymbol{\omega}^\alpha \,, \end{aligned} \tag{7.4.2}$$

$$\begin{aligned} \boldsymbol{u}^\alpha &= \boldsymbol{v}^\alpha - \boldsymbol{v} \,, & \Delta \boldsymbol{\omega}^\alpha &= (\boldsymbol{\omega}^\alpha - \boldsymbol{\omega}) \,, \\ \boldsymbol{t} &= \sum_{\alpha=1}^{\nu} \left(\boldsymbol{t}^\alpha - \rho^\alpha \boldsymbol{u}^\alpha \otimes \boldsymbol{u}^\alpha\right) , & \rho \boldsymbol{f} &= \sum_{\alpha=1}^{\nu} \rho^\alpha \boldsymbol{f}^\alpha \,, \end{aligned} \tag{7.4.3}$$

$$\begin{aligned} \rho \boldsymbol{s} &= \sum_{\alpha=1}^{\nu} \rho^\alpha \boldsymbol{s}^\alpha \,, & \rho \boldsymbol{l} &= \sum_{\alpha=1}^{\nu} \rho^\alpha \boldsymbol{l}^\alpha \,, \\ \boldsymbol{m} &= \sum_{\alpha=1}^{\nu} \left(\boldsymbol{m}^\alpha - \rho^\alpha \boldsymbol{s}^\alpha \otimes \boldsymbol{u}^\alpha\right) , \end{aligned} \tag{7.4.4}$$

$$\begin{aligned} \rho \eta &= \sum_{\alpha=1}^{\nu} \rho^\alpha s^\alpha, \quad \rho \varsigma = \sum_{\alpha=1}^{\nu} \rho^\alpha \eta^{s^\alpha}, \quad \rho \pi = \sum_{\alpha=1}^{\nu} \rho^\alpha \pi^{s^\alpha} \,, \\ \boldsymbol{\phi}^s &= \sum_{\alpha=1}^{\nu} \left(\phi^{s^\alpha} + s^\alpha \boldsymbol{u}^\alpha\right) , \end{aligned} \tag{7.4.5}$$

$$
\begin{aligned}
\rho\varepsilon &= \sum_{\alpha=1}^{\nu} \left(\rho^\alpha \varepsilon^\alpha + \tfrac{1}{2}\rho^\alpha \boldsymbol{u}^\alpha \cdot \boldsymbol{u}^\alpha + \tfrac{1}{2}\rho^\alpha \boldsymbol{s}^\alpha \cdot \Delta\boldsymbol{\omega}^\alpha \right) , \\
\rho\mathfrak{r} &= \sum_{\alpha=1}^{\nu} \left(\rho^\alpha \mathfrak{r}^\alpha + \boldsymbol{u}^\alpha \cdot \rho^\alpha \boldsymbol{f}^\alpha + \Delta\boldsymbol{\omega}^\alpha \cdot \rho^\alpha \boldsymbol{l}^\alpha \right) , \\
\boldsymbol{q} &= \sum_{\alpha=1}^{\nu} \Big[\boldsymbol{q}^\alpha - \boldsymbol{u}^\alpha \boldsymbol{t}^\alpha - \Delta\boldsymbol{\omega}^\alpha \boldsymbol{m}^\alpha \\
&\quad + \left(\rho^\alpha \varepsilon^\alpha + \tfrac{1}{2}\rho^\alpha \boldsymbol{u}^\alpha \cdot \boldsymbol{u}^\alpha + \tfrac{1}{2}\rho^\alpha \boldsymbol{s}^\alpha (\Delta\boldsymbol{\omega}^\alpha - \boldsymbol{\omega}) \right) \boldsymbol{u}^\alpha \Big] .
\end{aligned}
\tag{7.4.6}
$$

All these quantities were already previously defined and are repeated here for convenience. Simple computations, using product differentiation as e. g. explained in Exercise 2 for constituent α lead to the following alternative representations

$$
\begin{aligned}
&\frac{\mathrm{d}\rho}{\mathrm{d}t} + \operatorname{div}(\rho\boldsymbol{v}) = 0 , \\
&\rho\frac{\mathrm{d}\boldsymbol{v}}{\mathrm{d}t} = \operatorname{div}\boldsymbol{t} + \rho\boldsymbol{f} , \\
&\rho\frac{\mathrm{d}\boldsymbol{s}}{\mathrm{d}t} = \operatorname{div}\boldsymbol{m} - \operatorname{dual}(\boldsymbol{t} - \boldsymbol{t}^T) + \rho\boldsymbol{l} , \\
&\rho\frac{\mathrm{d}\varepsilon}{\mathrm{d}t} = -\operatorname{div}\boldsymbol{q} + \boldsymbol{\omega} \cdot \operatorname{dual}(\boldsymbol{t} - \boldsymbol{t}^T) + (\operatorname{grad}\boldsymbol{v}) \cdot \boldsymbol{t}^T + (\operatorname{grad}\boldsymbol{\omega}) \cdot \boldsymbol{m}^T + \rho\mathfrak{r} , \\
&\rho\frac{\mathrm{d}s}{\mathrm{d}t} = -\operatorname{div}\boldsymbol{\phi}^s + \rho\eta^s + \rho\pi^s .
\end{aligned}
\tag{7.4.7}
$$

In the above form, these equations hold for *polar* continua. In exactly the same fashion as done for the constituent balance laws, a mixture is defined to be *non-polar* if

$$
\boldsymbol{s} = \boldsymbol{0} , \qquad \boldsymbol{m} = \boldsymbol{0} , \qquad \boldsymbol{l} = \boldsymbol{0} , \tag{7.4.8}
$$

so that

$$
\boldsymbol{t} = \boldsymbol{t}^T . \tag{7.4.9}
$$

The CAUCHY stress tensor of a non-polar mixture is automatically symmetric. This conclusion also holds true if the partial CAUCHY stress tensors should possess skew symmetric contributions, compare equations (7.2.16) and (7.2.17).

The jump conditions, which hold across singular surfaces $\mathfrak{s}$ are

$$
\begin{aligned}
&[\![\rho(\boldsymbol{w} - \boldsymbol{v}) \cdot \boldsymbol{n}_\mathfrak{s}]\!] = 0 , \\
&[\![\rho\boldsymbol{v}(\boldsymbol{w} - \boldsymbol{v}) \cdot \boldsymbol{n}_\mathfrak{s}]\!] + [\![\boldsymbol{t}\boldsymbol{n}_\mathfrak{s}]\!] = 0 , \\
&[\![\rho\boldsymbol{s}(\boldsymbol{w} - \boldsymbol{v}) \cdot \boldsymbol{n}_\mathfrak{s}]\!] + [\![\boldsymbol{m}\boldsymbol{n}_\mathfrak{s}]\!] = 0 , \\
&[\![\rho(\varepsilon + \tfrac{1}{2}v^2 + \tfrac{1}{2}\boldsymbol{\omega} \cdot \boldsymbol{s})(\boldsymbol{w} - \boldsymbol{v}) \cdot \boldsymbol{n}_\mathfrak{s}]\!] + [\![\boldsymbol{v}\boldsymbol{t} + \boldsymbol{\omega}\boldsymbol{m} - \boldsymbol{q}]\!] \cdot \boldsymbol{n}_\mathfrak{s} = 0 , \\
&[\![\rho s(\boldsymbol{w} - \boldsymbol{v}) - \boldsymbol{\phi}^s]\!] \cdot \boldsymbol{n}_\mathfrak{s} = -\mathfrak{P}^s_\mathfrak{s} ,
\end{aligned}
\tag{7.4.10}
$$

in which $\boldsymbol{w}$ is the velocity of the singular surface.

7.5 Diffusion of Tracers in a Fluid

7.5.1 Basic Assumptions

We consider a mixture of non-polar constituents so that

$$\boldsymbol{s}^{\alpha}=\mathbf{0}\,,\quad \boldsymbol{m}^{\alpha}=\mathbf{0}\,,\quad \boldsymbol{l}^{\alpha}=\mathbf{0}\,,\quad \mathfrak{k}^{\alpha}=\mathbf{0}\,,\quad (\alpha=1,2,.....,\nu)\,. \tag{7.5.1}$$

We assume that this mixture consists of a fluid and $\nu-1$ *substances* in suspension or solution, called *tracers*, which, in comparison to the main fluid, arise only in small concentrations. Examples of such fluid mixtures are all waters on Earth (rivers, lakes, the ocean), which contain minerals, salts and pollutants, but also nutrients (for instance phosphate and the plankton) in solved or particulate form. Another example is the atmosphere – primarily the troposphere and stratosphere[3] – of which the chemical composition varies spatially and temporally, and which is loaded by *aerosols*, i.e., suspended particles (dust and industrial waste).

If the masses of the tracers are small in comparison to the mass of the main fluid, it may be assumed that the tracer particles are carried by the main fluid and therefore possess the same or nearly the same velocity as the main fluid. Likewise, one may generally assume that the tracer particles have the same temperature as the fluid particle at the same location. Consequently, the prerequisites of a *mixture theory of class I* are here given, for which the differences of the momenta and internal energies of the individual components are dynamically and thermodynamically not important; it therefore suffices to model only the momentum and internal energy of the mixture as a whole, but to follow in detail the mass evolution of each constituent[4].

It is customary to write down the balance laws of mass, momentum and energy for the mixture as a whole and to complement these by the mass balance laws for the $\nu-1$ tracers. Because of the simplifying prerequisites (7.5.1) one obtains

$$\frac{\mathrm{d}\rho}{\mathrm{d}t}+\rho\,\mathrm{div}\,\boldsymbol{v}=0\,, \tag{7.5.2}$$

$$\rho\frac{\mathrm{d}\xi^{\alpha}}{\mathrm{d}t}=\mathrm{div}\,\boldsymbol{j}^{\alpha}+\mathfrak{c},\quad (\alpha=1,2,3,....,\nu-1)\,, \tag{7.5.3}$$

$$\rho\frac{\mathrm{d}\boldsymbol{v}}{\mathrm{d}t}=\mathrm{div}\,\boldsymbol{t}+\rho\boldsymbol{f},\qquad (\boldsymbol{t}=\boldsymbol{t}^{T})\,, \tag{7.5.4}$$

$$\rho\frac{\mathrm{d}\varepsilon}{\mathrm{d}t}=-\,\mathrm{div}\,\boldsymbol{q}+(\mathrm{grad}\,\boldsymbol{v}\cdot\boldsymbol{t}^{T})+\rho\mathfrak{r}\,, \tag{7.5.5}$$

[3] The troposphere is the lowest layer of the atmosphere, approximately 8–10 km thick. The stratosphere is the atmosphere layer immediately above the troposphere and extends about to 50 km above the Earth's surface.

[4] In counting the constituents, the index $\alpha=\nu$ will be reserved to the main fluid, so that $\alpha=1,2,3,...,\nu-1$ will be used for the tracers. This convention will be retained in the sequel.

in which $\boldsymbol{v}$ denotes the barycentric velocity and

$$\xi^\alpha := \frac{\rho_\alpha}{\rho}, \quad \boldsymbol{j}^\alpha := \rho\xi^\alpha(\boldsymbol{v}^\alpha - \boldsymbol{v}), \quad (\alpha = 1, 2, 3, ..., \nu - 1) \tag{7.5.6}$$

are the *mass ratio* or *concentration* and the so-called *diffusive fluxes* of the constituents α. The first variable, ξ^α, gives the ratio of the mass of constituent α to that of the mixture as a whole, the second describes the flux of mass of constituent α relative to the barycentric motion. Equations (7.5.2), (7.5.4) and (7.5.5) are derivable from (7.4.1), and (7.5.3) follows from (7.2.4) by simple transformations, see Exercise 4. Incidentally, (7.5.2)–(7.5.5) are $[5 + \nu - 1]$ partial differential equations for the $[15 + 5(\nu - 1)]$ field variables

$$\{\rho, \xi^\alpha, \boldsymbol{v}, \boldsymbol{j}^\alpha, \mathfrak{c}^\alpha, \boldsymbol{t}, \varepsilon, \boldsymbol{q}, \theta\}, \quad (\alpha = 1, 2, ..., \nu - 1) . \tag{7.5.7}$$

The exterior body force and the radiation $\rho\mathfrak{r}$ are regarded as prescribed field quantities. Finally it is emphasized once more that $\xi^\nu, \boldsymbol{j}^\nu$ and $\mathfrak{c}^\nu$ do not constitute independent variables, since the relations

$$\sum_{\alpha=1}^{\nu} (\xi^\alpha, \boldsymbol{j}^\alpha, \mathfrak{c}^\alpha) = (1, \boldsymbol{0}, 0) \tag{7.5.8}$$

must be satisfied, so that $\xi^\nu, \boldsymbol{j}^\nu$ and $\mathfrak{c}^\nu$ are expressible as

$$(\xi^\nu, \boldsymbol{j}^\nu, \mathfrak{c}^\nu) = \left(1 - \sum_{\alpha=1}^{\nu-1} \xi^\alpha, - \sum_{\alpha=1}^{\nu-1} \boldsymbol{j}^\alpha, - \sum_{\alpha=1}^{\nu-1} \mathfrak{c}^\alpha\right) \tag{7.5.9}$$

and are thus given once these variables are prescribed for $\alpha = 1, 2, ..., \nu - 1$.

7.5.2 Material Theory for Diffusion Processes

Let us define the following variables as the $[5 + (\nu - 1)]$ independent fields:

$$\{\rho, \xi^\alpha, \boldsymbol{v}, \theta\}, \qquad (\alpha = 1, 2, ..., \nu - 1) . \tag{7.5.10}$$

This implies that, consequently,

$$\Psi := \{\boldsymbol{j}^\alpha, \mathfrak{c}^\alpha, \boldsymbol{t}, \varepsilon, \boldsymbol{q}, \}, \qquad (\alpha = 1, 2, ..., \nu - 1) \tag{7.5.11}$$

are the $[10 + 4(\nu - 1)]$ field variables for which constitutive relations must be formulated in order to close the system (7.5.2)–(7.5.5) and (7.5.11). We wish to consider here the most simple form of a mixture theory that is capable of describing the dispersion of a number of tracers in a fluid; to this end we propose a *theory of diffusion* of a *heat conducting viscous fluid* with constitutive equations for the dependent field quantities (7.5.11) of the complexity

$$\Psi = \hat{\Psi}(\rho, \xi^\alpha, \boldsymbol{v}, \theta, \operatorname{grad} \boldsymbol{v}, \operatorname{grad} \xi^\alpha, \operatorname{grad} \theta) . \tag{7.5.12}$$

Because all variables listed in (7.5.11) are objective scalars, vectors and tensors, and since it is requested that the constitutive relations (7.5.12) obey the rule of material frame indifference, it follows that $\hat{\Psi}(\cdot)$ in (7.5.12) cannot explicitly depend on $\boldsymbol{v}$; furthermore, $\hat{\Psi}(\cdot)$ can depend on grad $\boldsymbol{v}$ only via its symmetric part, sym(grad $\boldsymbol{v}$) $=:$ $\boldsymbol{D}$ (see Exercise 5). Therefore, for the considered constitutive class the material relations (7.5.12) reduce to

$$\Psi = \hat{\Psi}(\boldsymbol{\Xi}), \qquad \boldsymbol{\Xi} := (\rho, \xi^\alpha, \theta, \operatorname{grad} \xi^\alpha, \boldsymbol{D}, \operatorname{grad} \theta) . \tag{7.5.13}$$

The material relations (7.5.13) must be in conformity with the second law of thermodynamics. We wish to explore this requirement with the entropy principle of MÜLLER. Accordingly, there exists an additive quantity, the entropy, which obeys the balance law

$$\rho \frac{\mathrm{d}s}{\mathrm{d}t} + \operatorname{div} \boldsymbol{\phi}^s + \rho \eta^s = \rho \pi^s , \tag{7.5.14}$$

in which $\boldsymbol{\phi}^s, \eta^s$ and π^s are the entropy flux, the entropy supply and entropy production. According to the Second Law of Thermodynamics the entropy production must not be negative for any thermodynamic process; in other words, the inequality

$$\rho \pi^s \geq 0 \tag{7.5.15}$$

must hold for all thermodynamic processes, i. e., all solutions of (7.5.2)–(7.5.5) and all constitutive relations of the form (7.5.13) for the field variables (7.5.11). LIU has shown that instead of satisfying (7.5.15) for all fields which simultaneously also satisfy the field equations (7.5.2)–(7.5.5), (7.5.11), (7.5.13) one may also proceed as follows: One subtracts (or adds, which only changes the signs of the LAGRANGE parameters) on the left-hand side of (7.5.15) the scalar products of the field equations with the appropriate LAGRANGE parameters, viz.,

$$\begin{aligned}
&\rho \frac{\mathrm{d}\hat{s}}{\mathrm{d}t} + \operatorname{div} \hat{\boldsymbol{\phi}}^s - \rho \eta^s \\
&\quad - \Lambda^\rho \left(\frac{\mathrm{d}\rho}{\mathrm{d}t} + \rho \operatorname{div} \boldsymbol{v} \right) - \Lambda^{\xi^\alpha} \left(\rho \frac{\mathrm{d}\xi^\alpha}{\mathrm{d}t} + \operatorname{div} \hat{\boldsymbol{j}}^\alpha - \hat{\mathfrak{c}}^\alpha \right) \\
&\quad - \boldsymbol{\Lambda}^{\boldsymbol{v}} \cdot \left(\rho \frac{\mathrm{d}\boldsymbol{v}}{\mathrm{d}t} - \operatorname{div} \hat{\boldsymbol{t}} - \rho \boldsymbol{f} \right) - \Lambda^\varepsilon \left(\rho \frac{\mathrm{d}\hat{\varepsilon}}{\mathrm{d}t} + \operatorname{div} \hat{\boldsymbol{q}} - \boldsymbol{D} \cdot \hat{\boldsymbol{t}} - \rho \mathfrak{r} \right) \geq 0
\end{aligned} \tag{7.5.16}$$

and satisfies this extended inequality now for arbitrary, unconstrained fields. $\Lambda^\rho, \Lambda^{\xi^\alpha}, \boldsymbol{\Lambda}^{\boldsymbol{v}}$ and Λ^ε are LAGRANGE parameters, which must be determined along with the exploitation of the inequality (7.5.16). In this formula summation must be understood over the doubly arising index $\alpha = 1, 2, 3, ..., \nu - 1$. The dot in the expression $\boldsymbol{\Lambda}^{\boldsymbol{v}} \cdot (...)$ is to be understood as the scalar product in $\mathbb{R}^3$, and all constitutive relations are thought to be substituted, a fact which we made visible by using the hat, $\hat{(\cdot)}$, in all dependent constitutive variables.

The supplies or sources $\rho\eta^s, \rho\boldsymbol{f}$ and $\rho\mathfrak{r}$ are prescribable with any value we please.

If the constitutive relations for s, $\boldsymbol{j}^\alpha$, ξ^α, $\boldsymbol{t}$, ε and $\boldsymbol{q}$ $(\alpha = 1, 2, \ldots\ldots, \nu - 1)$ are substituted in (7.5.16) and the required differentiations are performed according to the chain rule of differentiation, one obtains after lengthy calculations an expression of the form

$$\boldsymbol{\alpha}(\boldsymbol{\Xi}) \cdot \boldsymbol{\beta} + \Gamma(\boldsymbol{\Xi}) - (\rho\eta^s - \boldsymbol{\Lambda}^v \cdot \rho\boldsymbol{f} - \Lambda^\varepsilon \rho\mathfrak{r}) \geq 0 \,. \tag{7.5.17}$$

Here, $\boldsymbol{\Xi}$ is defined in (7.5.13), and $\boldsymbol{\alpha}$ as well as $\boldsymbol{\beta}$ are vectors with the same number of components (so that the scalar product is meaningful), and $\boldsymbol{\beta}$ is given by

$$\boldsymbol{\beta} = \left(\frac{\mathrm{d}v_i}{\mathrm{d}t}; \dot{\rho}; \dot{\xi}^\alpha; \dot{\theta}; (\xi^\beta_{,k})^{\cdot}; \dot{D}_{kl}; (\theta_{,k})^{\cdot}; \xi^\beta_{,ij}; D_{kl,j}; \theta_{,kj}; \rho_{,i} \right) , \tag{7.5.18}$$

in which Cartesian index notation has been used. The inequality (7.5.17) is linear in the vectorial variable $\boldsymbol{\beta}$; i. e., none of the variables that define $\boldsymbol{\beta}$ is contained in $\boldsymbol{\Xi}$. The inequality (7.5.17) contains in its third braced term also the sources, and it is equally linear in these source terms, if Λ^ε and $\boldsymbol{\Lambda}^v$ are independent of these, a fact that we will provisionally assume but corroborate lateron.

It is plausible to request that the constitutive relations be independent of the sources, to which the material is subjected in a thermodynamic process. Since in (7.5.17) only the first two terms on the left-hand side are independent of the sources but not the third, the latter must identically vanish, so that

$$\eta^s = \Lambda^\varepsilon \mathfrak{r} + \boldsymbol{\Lambda}^v \cdot \boldsymbol{f} \,. \tag{7.5.19}$$

The entropy supply is therefore known once the LAGRANGE parameters Λ^ε and $\boldsymbol{\Lambda}^v$ are determined. It is a linear combination of the energy supply (radiation) and momentum supply (external body forces). With (7.5.19) inequality (7.5.17) reduces to

$$\boldsymbol{\alpha}(\boldsymbol{\Xi}) \cdot \boldsymbol{\beta} + \Gamma(\boldsymbol{\Xi}) \geq 0 \,, \tag{7.5.20}$$

which must be fulfilled for all values of $\boldsymbol{\Xi}$ and $\boldsymbol{\beta}$. Sufficient conditions are certainly the requirements[5]

$$\boldsymbol{\alpha}(\boldsymbol{\Xi}) = \boldsymbol{0} \qquad \text{and} \qquad \Gamma(\boldsymbol{\Xi}) \geq 0 \,. \tag{7.5.21}$$

One can show that these conditions are equally necessary (Exercise 6). Explicitly, the identities (7.5.21) read

$$\boldsymbol{\Lambda}^v \equiv \boldsymbol{0} \,, \tag{7.5.22}$$

[5] The equations $\boldsymbol{\alpha}(\boldsymbol{\Xi}) = \boldsymbol{0}$ are frequently called the LIU equations and $\Gamma(\boldsymbol{\Xi}) \geq 0$ is the residual entropy inequality.

$$\left.\begin{aligned}
&\frac{\partial \hat{s}}{\partial \rho} - \Lambda^{\varepsilon}\frac{\partial \hat{\varepsilon}}{\partial \rho} - \frac{\Lambda^{\rho}}{\rho} \equiv 0\,, \\
&\frac{\partial \hat{s}}{\partial \xi^{\alpha}} - \Lambda^{\varepsilon}\frac{\partial \hat{\varepsilon}}{\partial \xi^{\alpha}} - \Lambda^{\xi^{\alpha}} \equiv 0, \quad (\alpha = 1,2,3,...,\nu-1)\,, \\
&\frac{\partial \hat{s}}{\partial \theta} - \Lambda^{\varepsilon}\frac{\partial \hat{\varepsilon}}{\partial \theta} \equiv 0\,, \\
&\frac{\partial \hat{s}}{\partial \xi^{\beta}_{,k}} - \Lambda^{\varepsilon}\frac{\partial \hat{\varepsilon}}{\partial \xi^{\beta}_{,k}} \equiv 0, \quad (\beta = 1,2,...,\nu-1)\,, \\
&\frac{\partial \hat{s}}{\partial D_{kl}} - \Lambda^{\varepsilon}\frac{\partial \hat{\varepsilon}}{\partial D_{kl}} \equiv 0\,, \\
&\frac{\partial \hat{s}}{\partial \theta_{,k}} - \Lambda^{\varepsilon}\frac{\partial \hat{\varepsilon}}{\partial \theta_{,k}} \equiv 0\,,
\end{aligned}\right\} \qquad (7.5.23)$$

$$\left.\begin{aligned}
&\frac{\partial \phi_{(i}}{\partial \xi^{\beta}_{,j)}} - \Lambda^{\varepsilon}\frac{\partial q_{(i}}{\partial \xi^{\beta}_{,j)}} - \Lambda^{\xi^{\alpha}}\frac{\partial j^{\alpha}_{(i}}{\partial \xi^{\beta}_{,j)}} \equiv 0, \quad (\beta = 1,2,...,\nu-1)\,, \\
&\frac{\partial \phi_{i}}{\partial D_{kl}} - \Lambda^{\varepsilon}\frac{\partial q_{i}}{\partial D_{kl}} - \Lambda^{\xi^{\alpha}}\frac{\partial j^{\alpha}_{i}}{\partial D_{kl}} \equiv 0\,, \\
&\frac{\partial \phi_{(k}}{\partial \theta_{,j)}} - \Lambda^{\varepsilon}\frac{\partial q_{(k}}{\partial \theta_{,j)}} - \Lambda^{\xi^{\alpha}}\frac{\partial j^{\alpha}_{(k}}{\partial \theta_{,j)}} \equiv 0\,, \\
&\frac{\partial \phi_{i}}{\partial \rho} - \Lambda^{\varepsilon}\frac{\partial q_{i}}{\partial \rho} - \Lambda^{\xi^{\alpha}}\frac{\partial j^{\alpha}_{i}}{\partial \rho} \equiv 0\,,
\end{aligned}\right\} \qquad (7.5.24)$$

as well as

$$\begin{aligned}
\Gamma \equiv &\left\{\frac{\partial \phi_i}{\partial \xi^{\beta}} - \Lambda^{\varepsilon}\frac{\partial q_i}{\partial \xi^{\beta}} - \Lambda^{\xi^{\alpha}}\frac{\partial j^{\alpha}_i}{\partial \xi^{\beta}}\right\}\xi^{\beta}_{,i} + \left\{\frac{\partial \phi_i}{\partial \theta} - \Lambda^{\varepsilon}\frac{\partial q_i}{\partial \theta} - \Lambda^{\xi^{\alpha}}\frac{\partial j^{\alpha}_i}{\partial \theta}\right\}\theta_{,i} \\
&+\Lambda^{\xi^{\alpha}}\mathfrak{c}^{\alpha} + \Lambda^{\varepsilon}\left\{t_{ij} - \frac{\Lambda^{\rho}}{\Lambda^{\varepsilon}}\rho\delta_{ij}\right\}D_{ij} \geq 0\,, \qquad (7.5.25)
\end{aligned}$$

in which indexed round brackets denote symmetrization and the summation is from $\alpha = 1$ to $\alpha = \nu - 1$; similarly for β. Our next goal is to draw all inferences from these identities.

One may regard the identities (7.5.23)–(7.5.24) as equations for the determination of the LAGRANGE parameters $\Lambda^{\varepsilon}, \Lambda^{\rho}$, and one concludes from this, that $\Lambda^{\varepsilon}, \Lambda^{\rho}$ can only depend on the variables $\boldsymbol{\Xi}$. This is the a posteriori proof of the independence of the LAGRANGE parameters of the external source terms, that was assumed above. Indeed without such an assumption (7.5.22)–(7.5.25) would still follow and the independence of the Λ's of the sources would follow in exactly the same way. However, additional deductions are possible; with the identities (7.5.24) it is namely possible to demonstrate that Λ^{ε} is only a function of the empirical temperature,

$$\Lambda^\varepsilon = \Lambda^\varepsilon(\theta), \qquad \Theta = \frac{1}{\Lambda^\varepsilon(\theta)} \,. \tag{7.5.26}$$

This proof is long and rather involved and reserved to Exercise 7[6]. Because of the above property, Λ^ε is called *coldness function* and its inverse is called the *absolute temperature* Θ; the latter agrees with the KELVIN temperature, a fact that we shall corroborate. Alternatively, if we start from (7.5.26) as an assumption, one obtains from $(7.5.23)_{3,4}$ by cross differentiation

$$\frac{\partial^2 \hat{s}}{\partial \xi^\beta_{,k}\partial\theta} - \Lambda^\varepsilon \frac{\partial^2 \hat{\varepsilon}}{\partial \xi^\beta_{,k}\partial\theta} \equiv \frac{\partial^2 \hat{s}}{\partial\theta\partial \xi^\beta_{,k}} - \Lambda^\varepsilon \frac{\partial^2 \hat{\varepsilon}}{\partial\theta\partial \xi^\beta_{,k}} - \frac{\partial \Lambda^\varepsilon}{\partial\theta}\frac{\partial \hat{\varepsilon}}{\partial \xi^\beta_{,k}} \,, \tag{7.5.27}$$

from which one concludes, since $\partial\Lambda^\varepsilon/\partial\theta \neq 0$ that $\partial\hat{\varepsilon}/\partial\xi^\beta_{,k} = 0$ and, in view of $(7.5.23)_4$, also $\partial\hat{s}/\partial\xi^\beta_{,k} = 0$. Thus, internal energy and entropy cannot depend on $\xi^\beta_{,k}$. In much the same way one may equally show by using $(7.5.23)_{3,5,6}$ that $\hat{\varepsilon}$ and $\hat{s}$ can neither depend on $\boldsymbol{D}$ nor on $\operatorname{grad}\theta$, so that the identities (7.5.23) lead to the reduced statements

$$\begin{aligned}
\varepsilon &= \hat{\varepsilon}(\rho, \xi^\beta, \theta), \\
s &= \hat{s}(\rho, \xi^\beta, \theta), \\
\Lambda^\rho &= \rho\left(\frac{\partial \hat{s}}{\partial\rho} - \Lambda^\varepsilon \frac{\partial\hat{\epsilon}}{\partial\rho}\right) = \Lambda^\rho(\rho, \xi^\beta, \theta), \\
\Lambda^{\xi^\alpha} &= \left(\frac{\partial \hat{s}}{\partial\xi^\alpha} - \Lambda^\varepsilon \frac{\partial\hat{\epsilon}}{\partial\xi^\alpha}\right) = \Lambda^{\xi^\alpha}(\rho, \xi^\beta, \theta)\,.
\end{aligned} \tag{7.5.28}$$

The last two of these statements, together with $(7.5.23)_3$ may also be written as

$$\begin{aligned}
\frac{\partial\hat{s}}{\partial\theta} &= \Lambda^\varepsilon \frac{\partial\hat{\varepsilon}}{\partial\theta}\,, \\
\frac{\partial\hat{s}}{\partial\rho} &= \Lambda^\varepsilon \frac{\partial\hat{\varepsilon}}{\partial\rho} + \frac{1}{\rho}\Lambda^\rho = \Lambda^\varepsilon\left(\frac{\partial\hat{\varepsilon}}{\partial\rho} - \frac{\hat{p}}{\rho^2}\right), \qquad p := -\frac{\Lambda^\rho}{\Lambda^\varepsilon}\rho\,, \\
\frac{\partial\hat{s}}{\partial\xi^\alpha} &= \Lambda^\varepsilon \frac{\partial\hat{\varepsilon}}{\partial\xi^\alpha} + \Lambda^{\xi^\alpha} = \Lambda^\varepsilon\left(\frac{\partial\hat{\varepsilon}}{\partial\xi^\alpha} - \mu^{\xi^\alpha}\right), \qquad \mu^{\xi^\alpha} := -\frac{\Lambda^{\xi^\alpha}}{\Lambda^\varepsilon}\,,
\end{aligned} \tag{7.5.29}$$

from which the total differential of the entropy is seen to be expressible as

$$\begin{aligned}
\mathrm{d}s &= \Lambda^\varepsilon \left\{\frac{\partial\hat{\varepsilon}}{\partial\theta}\mathrm{d}\theta + \left(\frac{\partial\hat{\varepsilon}}{\partial\rho} + \frac{\Lambda^\rho}{\Lambda^\epsilon}\frac{1}{\rho}\right)\mathrm{d}\rho + \left(\frac{\partial\hat{\varepsilon}}{\partial\xi^\alpha} + \frac{\Lambda^{\xi^\alpha}}{\Lambda^\epsilon}\right)\mathrm{d}\xi^\alpha\right\} \\
&= \Lambda^\varepsilon\left\{\mathrm{d}\varepsilon + p\,\mathrm{d}\left(\frac{1}{\rho}\right) - \mu^{\xi^\alpha}\mathrm{d}\xi^\alpha\right\}.
\end{aligned} \tag{7.5.30}$$

[6] One essential ingredient of the proof is that $\boldsymbol{\phi}$, $\boldsymbol{q}$ and $\boldsymbol{j}^\alpha$ are isotropic functions of their arguments. In this particular case this is not a restriction because the chosen constitutive class (7.5.13) does not permit anisotropic behaviour.

This is the so-called GIBBS *relation*, which connects the total differential of the entropy with the product of the inverse of the absolute temperature $\Theta^{-1} = \Lambda^{\varepsilon}(\theta)$ and the total differential of the internal energy plus the additional contributions

$$\frac{\Lambda^{\rho}}{\rho\Lambda^{\epsilon}}\mathrm{d}\rho\,, \quad \frac{\Lambda^{\xi^\alpha}}{\Lambda^{\epsilon}}\mathrm{d}\xi^\alpha\,,$$

which with the definitions for p and μ^{ξ^α} become

$$-\frac{p}{\rho^2}\mathrm{d}\rho = p\mathrm{d}\left(\frac{1}{\rho}\right) = \frac{\Lambda^{\rho}}{\rho\Lambda^{\epsilon}}\mathrm{d}\rho\,, \quad -\mu^{\xi^\alpha}\mathrm{d}\xi^\alpha = \frac{\Lambda^{\xi^\alpha}}{\Lambda^{\epsilon}}\mathrm{d}\xi^\alpha \tag{7.5.31}$$

so that

$$\mathrm{d}s = \frac{1}{\Theta}\left\{d\varepsilon + p\mathrm{d}(1/\rho) - \mu^{\xi^\alpha}\mathrm{d}\xi^\alpha\right\}\,, \tag{7.5.32}$$

in which p is called the *thermodynamic pressure*, μ^{ξ^α} is the *chemical potential of constituent* α[7]. The latter identification must, however, still be demonstrated.

The GIBBS equation must satisfy integrability conditions, which are deduced from (7.5.29) by performing the respective cross differentiations[8]; if this is done one obtains the relations (see Exercise 8)

---

[7] If we would have written down the mass balance (7.5.3) for all constituents and if the mass fractions $\xi^\alpha(\alpha = 1, 2, ..., \nu)$ for all constituents would have been used as independent constitutive variables, then instead of (7.5.30) one would have obtained

$$\begin{aligned}\mathrm{d}s &= \frac{1}{\Theta}\left\{\mathrm{d}\varepsilon + p\mathrm{d}(1/\rho) - \sum_{\alpha=1}^{\nu}\bar{\mu}^{\xi^\alpha}\mathrm{d}\xi^\alpha\right\}\\ &= \frac{1}{\Theta}\left\{\mathrm{d}\varepsilon + p\mathrm{d}(1/\rho) - \sum_{\alpha=1}^{\nu-1}\left(\bar{\mu}^{\xi^\alpha} - \bar{\mu}^{\xi^\nu}\right)\mathrm{d}\xi^\alpha\right\}\end{aligned} \tag{7.5.33}$$

with new functions $\bar{\mu}^{\xi^\alpha}$; in the above expression $(7.5.33)_2$, the relation $\sum_{\alpha=1}^{\nu}\xi^\alpha = 1$ was used. Thus, one necessarily has

$$\mu^{\xi^\alpha} := \left(\bar{\mu}^{\xi^\alpha} - \bar{\mu}^{\xi^\nu}\right)\big|_{\xi^\nu = 1 - \sum_{\alpha=1}^{\nu-1}\xi^\alpha}\,, \tag{7.5.34}$$

where the subscript in (7.5.34) indicates that ξ^ν is replaced by $(1-\sum_{\alpha=1}^{\nu-1}\xi^\alpha)$. The variable μ^{ξ^α} is therefore the difference of the chemical potentials $\left(\bar{\mu}^{\xi^\alpha} - \bar{\mu}^{\xi^\nu}\right)$, which, however, are functions of all $\xi^\alpha(\alpha = 1, 2, ..., \nu)$.

[8] What is meant here is that all mixed second order derivatives of s are formed, e.g., $\partial^2 s/\partial\rho\partial\theta$, and for s to be uniquely determined it is required that $\partial^2 s/\partial\rho\partial\theta = \partial^2 s/\partial\theta\partial\rho$ etc. This then yields (7.5.35) and (7.5.36).

$$\frac{d(\ln \Lambda^{\varepsilon})}{d\theta} = \frac{1}{\Lambda^{\varepsilon}}\frac{d\Lambda^{\varepsilon}}{d\theta} = \frac{\partial\hat{p}/\partial\theta}{(\partial\hat{\varepsilon}/\partial\rho)\rho^2 - \hat{p}}$$
$$= \frac{\partial\hat{\mu}^{\xi^{\alpha}}/\partial\theta}{\partial\hat{\varepsilon}/\partial\xi^{\alpha} - \hat{\mu}^{\xi^{\alpha}}}, \quad (\alpha = 1, 2, ..., \nu - 1) \quad (7.5.35)$$

as well as

$$\frac{\partial\hat{\mu}^{\xi^{\alpha}}}{\partial\xi^{\beta}} = \frac{\partial\hat{\mu}^{\xi^{\beta}}}{\partial\xi^{\alpha}}, \quad \frac{\partial\hat{\mu}^{\xi^{\alpha}}}{\partial\rho} = \frac{\partial}{\partial\xi^{\alpha}}\left(\frac{\hat{p}}{\rho^2}\right), \quad (\alpha = 1, 2, ..., \nu - 1). \quad (7.5.36)$$

The identities (7.5.35) state that the pressure, chemical potentials and the internal energy must be constitutive equations of such a form that the $\nu - 1$ different fractions (7.5.35) are the same function of the empirical temperature only. Equations (7.5.36), on the other hand, indicate that the pressure, chemical potentials (and entropy) are themselves derivable from a potential; this shall be ascertained shortly. Before, let us, however, integrate $(7.5.35)_1$ to obtain

$$\ln\frac{\Lambda^{\varepsilon}}{\Lambda^{\varepsilon}_0} = -\ln\frac{\Theta}{\Theta_0} = \int_{\theta_0}^{\theta}\frac{\partial\hat{p}/\partial\theta}{((\partial\hat{\varepsilon}/\partial\rho)\rho^2 - \hat{p})}d\bar{\theta}\,,$$

where $\Theta(\theta) = 1/\Lambda^{\varepsilon}$, or

$$\frac{\Theta}{\Theta_0} = \exp\left\{-\int_{\theta_0}^{\theta}\frac{\partial\hat{p}/\partial\theta}{((\partial\hat{\varepsilon}/\partial\rho)\rho^2 - \hat{p})}d\bar{\theta}\right\}. \quad (7.5.37)$$

For an *ideal gas* with the thermal and caloric equations of state

$$\hat{p} = R\Theta(\theta)\rho, \quad \hat{\varepsilon} = \hat{\varepsilon}(\theta) \quad (7.5.38)$$

(7.5.37) leads to an identity $\Theta(\theta) \equiv \Theta(\theta)$. This can be taken as motivation to identify the empirical temperature θ with the temperature T of ideal gases (a special empirical temperature), and to request that

$$\Theta(\theta) =: T\,. \quad (7.5.39)$$

Because of the monotonicity of $\Theta(\theta)$ as a function of θ, one even may use T as a measure of temperature; from now on this will be our choice. So, the temperature is now uniquely defined, and therefore the identities (7.5.35) become constraint equations for the experimentator, which he must fulfill, if he determines p, $\mu^{\xi^{\alpha}}$ and ε as functions of their variables by experiment. Thus the chosen temperature is the ***absolute*** temperature or the KELVIN *temperature.*

The integrability conditions (7.5.35) and (7.5.36) can be satisfied identically, if the HELMHOLTZ free energy

$$\psi = \varepsilon - Ts = \hat{\psi}(\rho, \xi^{\alpha}, T) \quad (7.5.40)$$

is introduced. Computing with (7.5.40) the total differential of the internal energy

$$\mathrm{d}\varepsilon = \mathrm{d}\psi + T\mathrm{d}s + s\mathrm{d}T \tag{7.5.41}$$

and substituting this result into the GIBBS relation (7.5.32) yields

$$\left(\frac{\partial\hat{\psi}}{\partial\rho} - \frac{\hat{p}}{\rho^2}\right)\mathrm{d}\rho + \left(\frac{\partial\hat{\psi}}{\partial\xi^\alpha} - \hat{\mu}^{\xi^\alpha}\right)\mathrm{d}\xi^\alpha + \left(\frac{\partial\hat{\psi}}{\partial T} + \hat{s}\right)\mathrm{d}T \equiv 0\,, \tag{7.5.42}$$

which, as an identity, can only be satisfied, provided the following relations hold:

$$\frac{\hat{p}}{\rho^2} = \frac{\partial\hat{\psi}}{\partial\rho}, \quad \hat{\mu}^{\xi^\alpha} = \frac{\partial\hat{\psi}}{\partial\xi^\alpha}, \quad (\alpha = 1, 2, ..., \nu - 1)\,, \quad \hat{s} = -\frac{\partial\hat{\psi}}{\partial T}\,. \tag{7.5.43}$$

Accordingly, the pressure, the chemical potentials and the entropy are obtained from the HELMHOLTZ free energy by "taking the gradient" with respect to its independent variables. With the choice of $\hat{\psi}(\cdot)$ and the relations (7.5.43) the integrability conditions (7.5.35) and (7.5.36) are automatically satisfied.

Next, let us closely analyze the identities (7.5.24). They suggest the choice

$$\boldsymbol{\phi} = \Lambda^\varepsilon \boldsymbol{q} + \boldsymbol{k} = \frac{1}{T}\boldsymbol{q} + \boldsymbol{k}\,, \tag{7.5.44}$$

where (7.5.26) and (7.5.39) have been used. It is evident that the *extra entropy flux vector*, $\boldsymbol{k}$ describes that part of the entropy flux which may not be collinear to the heat flux. Incidentally, the choice (7.5.44) does not amount to a restriction. If we, furthermore, introduce

$$\mathfrak{F} = \boldsymbol{k} + \frac{\mu^{\xi^\alpha}\boldsymbol{j}^\alpha}{T} = \boldsymbol{k} - \Lambda^{\xi^\alpha}\boldsymbol{j}^\alpha\,, \tag{7.5.45}$$

then the identities (7.5.24) can equally be written as

$$\frac{\partial\mathfrak{F}_{(i}}{\partial\xi^\beta_{,j)}} = 0, \quad \frac{\partial\mathfrak{F}_i}{\partial D_{kl}} = 0, \quad \frac{\partial\mathfrak{F}_{(i}}{\partial T_{,j)}} = 0, \quad \frac{\partial\mathfrak{F}_i}{\partial\rho} = \frac{1}{T}\frac{\partial\hat{\mu}^{\xi^\alpha}}{\partial\rho}j_i^\alpha\,. \tag{7.5.46}$$

These equations can be satisfied by the ansatz

$$\mathfrak{F} = \boldsymbol{0}, \quad \psi = \psi_1(\rho, T) + \psi_2(\xi^\alpha, T)\,, \tag{7.5.47}$$

which automatically satisfies $\partial\hat{\mu}^{\xi^\alpha}/\partial\rho = 0$. (7.5.47) is not the most general solution of (7.5.46), however if one chooses it one has

$$\boldsymbol{k} = -\frac{\mu^{\xi^\alpha}\boldsymbol{j}^\alpha}{T} \quad \Rightarrow \quad \boldsymbol{\phi} = \frac{\boldsymbol{q} - \mu^{\xi^\alpha}\boldsymbol{j}^\alpha}{T}\,. \tag{7.5.48}$$

The entropy flux (multiplied with the absolute temperature) deviates from the heat flux via a vector, which is a linear combination of the diffusive fluxes; the weights of the individual diffusive fluxes are the chemical potentials of the corresponding constituents.

If the results (7.5.31), (7.5.39) and (7.5.48) are substituted into the inequality (7.5.25), one obtains

$$\begin{aligned} \pi^s = &-\frac{\boldsymbol{q} - \mu^{\xi^\alpha} \boldsymbol{j}^\alpha}{T^2} \cdot \operatorname{grad} T - \frac{\boldsymbol{j}^\alpha \cdot \operatorname{grd} \mu^{\xi^\alpha}}{T} \\ &+ \frac{(\boldsymbol{t} + p\mathbf{1}) \cdot \boldsymbol{D}}{T} - \frac{\mu^{\xi^\alpha} \mathfrak{c}^\alpha}{T} \geq 0 \,, \end{aligned} \tag{7.5.49}$$

in which $\operatorname{grd}\mu^{\xi^\alpha}$ denotes the reduced gradient of $\operatorname{grad}\mu^{\xi^\alpha}$, in which the density is regarded as constant (sometimes called frozen), i. e.,

$$\operatorname{grd}\mu^{\xi^\alpha} := \frac{\partial \hat{\mu}^{\xi^\alpha}}{\partial \xi^\beta} \operatorname{grad} \xi^\beta + \frac{\partial \hat{\mu}^{\xi^\alpha}}{\partial T} \operatorname{grad} T \,. \tag{7.5.50}$$

In evaluating (7.5.49) the fact was also used that the absolute temperature is a strictly positive function. Incidentally, the same result is also obtained, if the GIBBS equation

$$\dot{s} = \frac{1}{T} \left\{ \dot{\varepsilon} - \frac{p}{\rho^2} \dot{\rho} - \mu^{\xi^\alpha} \dot{\xi^\alpha} \right\} \tag{7.5.51}$$

is used and the rates $\dot{\rho}$ and $\dot{\xi^\alpha}$ are eliminated with the aid of the balance equations (7.5.2), (7.5.3); one then obtains

$$\rho \dot{s} = -\operatorname{div} \left(\frac{\boldsymbol{q} - \mu^{\xi^\alpha} \boldsymbol{j}^\alpha}{T} \right) + \pi^s \,, \tag{7.5.52}$$

in which π^s is given in (7.5.49). Inequality (7.5.49) suggests that entropy is produced by the inner product of the entropy flux with the temperature gradient, the diffusive fluxes with the reduced gradients of the chemical potential, the extra stresses with the rates of deformation and by the product of the mass productions with the chemical potentials.

Thermostatic equilibrium is defined as a process which does not produce any entropy. Inequality (7.5.49) shows then together with the definition of the reduced gradient of the chemical potentials, (7.5.50) that a thermostatic equilibrium process must fulfill the conditions

$$\begin{aligned} &\operatorname{grad} T = 0; \qquad \boldsymbol{D} = \mathbf{0}, \\ &\operatorname{grad} \xi^\beta = 0, \qquad \mathfrak{c}^\beta = 0, \quad (\beta = 1, 2, ..., \nu - 1) \,. \end{aligned} \tag{7.5.53}$$

Thus, the temperature and concentration fields must be homogeneous, the mass productions of all constituents must vanish – there are no chemical reactions and no phase change processes – and the barycentric velocity field is a rigid body field (or a rest field). Simultaneously, since the entropy production cannot take negative values, it assumes its minimum value, namely zero, in equilibrium. Of necessity then

$$\begin{aligned}&\left.\frac{\partial \pi^s}{\partial \mathfrak{X}}\right|_E = 0, \quad \mathfrak{X} := \left\{\operatorname{grad}\xi^\beta, \operatorname{grad}T, \boldsymbol{D}\right\},\\&\left|\frac{\partial \pi^s}{\partial \mathfrak{X}\partial \mathfrak{X}}\right|_E \quad \text{is positive–semi definite,}\end{aligned} \tag{7.5.54}$$

where the index $(\cdot)|_E$ is a reminder that the indexed quantity must be evaluated in thermostatic equilibrium.

If the differentiations indicated in $(7.5.54)_1$ are performed, one obtains

$$\boldsymbol{t}|_E = -p\boldsymbol{1}, \quad \boldsymbol{q}|_E = \boldsymbol{0}, \quad \boldsymbol{j}^\alpha|_E = \boldsymbol{0}, \quad \text{and} \quad \mathfrak{c}^\alpha|_E = 0 \tag{7.5.55}$$

for $\alpha = 1, 2, ..., \nu - 1$. Hence, the stress tensor in thermostatic equilibrium is given by the thermodynamic pressure, which itself is determined by the HELMHOLTZ free energy, see $(7.5.43)_1$. This relation now corroborates *a posteriori* the interpretation of the LAGRANGE parameter as the negative thermodynamic pressure according to $(7.5.31)_1$, $-\Lambda^\rho = \Lambda^\varepsilon p/\rho$. The heat flux and the diffusive fluxes of all tracers vanish in thermostatic equilibrium as one would expect. This property allows us also to conclude that the HELMHOLTZ free energy must at least quadratically depend on the tracer concentrations ξ^α. Finally, also all production rates of masses of the constituents must vanish.

Let us next scrutinize the second of relations (7.5.54). To exploit it, we restrict ourselves here to the simplest possible forms for the constitutive relations for $\boldsymbol{j}^\alpha, \boldsymbol{q}$ and $\boldsymbol{t}^E = \boldsymbol{t} + p\boldsymbol{1}$; furthermore, we ignore chemical reactions: $\mathfrak{c}^\beta = 0$; $\beta = 1, 2, ..., \nu - 1$. To exploit the inequality (7.5.49) we shall assume that the thermodynamic "fluxes"

$$\frac{1}{T}\left(\boldsymbol{q} - \mu^{\xi^\alpha}\boldsymbol{j}^\alpha\right) \quad \text{and} \quad \frac{\boldsymbol{j}^\alpha}{T} \tag{7.5.56}$$

are derivable from a dissipation potential Ψ_D which is expressible as a quadratic form of the corresponding thermodynamic "forces"

$$\frac{1}{T}\operatorname{grad}T, \quad \operatorname{grad}\mu^{\xi^\alpha} \tag{7.5.57}$$

according to

$$\begin{aligned}\Psi_D = {}& \tfrac{1}{2}C_{11}^{\alpha\beta}\left(\operatorname{grd}\mu^{\xi^\alpha}\right)\cdot\left(\operatorname{grd}\mu^{\xi^\beta}\right) + C_{12}^{\alpha}\left(\operatorname{grd}\mu^{\xi^\alpha}\right)\cdot\frac{\operatorname{grad}T}{T}\\&+\tfrac{1}{2}C_{22}\left(\frac{\operatorname{grad}T}{T}\right)\cdot\left(\frac{\operatorname{grad}T}{T}\right),\end{aligned} \tag{7.5.58}$$

in which $C_{11}^{\alpha\beta} = C_{11}^{\beta\alpha}$ is symmetric and where summation from 1 to $\nu - 1$ is understood over doubly repeated Greek indices. If one requests that

$$\frac{\boldsymbol{j}^\alpha}{T} = -\frac{\partial \Psi_D}{\partial \mathrm{grd}\mu^{\xi^\alpha}}, \quad \frac{1}{T}\left(\boldsymbol{q} - \mu^{\xi^\alpha}\boldsymbol{j}^\alpha\right) = -\frac{\partial \Psi_D}{\partial\,(\mathrm{grad}\,T/T)}, \tag{7.5.59}$$

(which gives Ψ_D the desired property of a potential), and if NEWTONian behaviour

$$\begin{aligned} &\boldsymbol{t} + p\mathbf{1} = \zeta(\mathrm{tr}\,\boldsymbol{D})\mathbf{1} + 2\mu\boldsymbol{D}' \\ &\Rightarrow \Phi = (\boldsymbol{t} + p\mathbf{1})\cdot\boldsymbol{D} = \zeta(\mathrm{tr}\,\boldsymbol{D})^2 + 2\mu\boldsymbol{D}'\cdot\boldsymbol{D}', \end{aligned} \tag{7.5.60}$$

is assumed, then for $\mathfrak{c}^\alpha = 0$ inequality (7.5.49) takes the form

$$\pi^s = 2\Psi_D + \frac{\Phi}{T} \geq 0\,. \tag{7.5.61}$$

In other words, Ψ_D and Φ are positive–semi–definite quadratic forms. Therefore,

$$\begin{aligned} &\begin{pmatrix} C_{11}^{\alpha\beta} & C_{12}^{\alpha} \\ C_{12}^{\alpha T} & C_{22} \end{pmatrix} \quad \text{is a positive semi–definite } \nu\times\nu \text{ matrix,} \\ &\zeta \geq 0, \quad \mu \geq 0\,. \end{aligned} \tag{7.5.62}$$

The diffusive fluxes $\boldsymbol{j}^\alpha$ and the entropy flux $(\boldsymbol{q} - \mu^{\xi^\alpha}\boldsymbol{j}^\alpha)$ are given by

$$\begin{pmatrix} \boldsymbol{j}^\alpha \\ \boldsymbol{q} - \mu^{\xi^\alpha}\boldsymbol{j}^\alpha \end{pmatrix} = -\begin{pmatrix} C_{11}^{\alpha\beta}T & C_{12}^{\alpha} \\ C_{12}^{\beta}T & C_{22} \end{pmatrix}\begin{pmatrix} \mathrm{grd}\mu^{\xi^\beta} \\ \mathrm{grad}\,T \end{pmatrix}, \tag{7.5.63}$$

which can easily be deduced from (7.5.58) and (7.5.59).

With the equation of state $\mu^{\xi^\beta} = \hat{\mu}^{\xi^\beta}(\rho, \xi^\gamma, T)$ for the chemical potentials there follows

$$\mathrm{grd}\mu^{\xi^\beta} = \frac{\partial\hat{\mu}^{\xi^\beta}}{\partial\xi^\gamma}\,\mathrm{grad}\,\xi^\gamma + \frac{\partial\hat{\mu}^{\xi^\beta}}{\partial T}\,\mathrm{grad}\,T \tag{7.5.64}$$

and, thus, owing to (7.5.63),

$$\begin{aligned} \boldsymbol{q} &= -\kappa_T\,\mathrm{grad}\,T - \kappa_{\xi^\gamma}\,\mathrm{grad}\,\xi^\gamma\,, \\ \boldsymbol{j}^\alpha &= -D_T^\alpha\,\mathrm{grad}\,T - D_{\xi^\gamma}^\alpha\,\mathrm{grad}\,\xi^\gamma\,, \end{aligned} \tag{7.5.65}$$

in which

$$
\begin{aligned}
\kappa_T &:= \mu^{\xi^\alpha}\left(C_{11}^{\alpha\beta}T\frac{\partial\mu^{\xi^\beta}}{\partial T}+C_{12}^\alpha\right)+\left(C_{12}^\beta\frac{\partial\mu^{\xi^\beta}}{\partial T}+C_{22}\right),\\
\kappa_{\xi^\gamma} &:= C_{11}^{\alpha\beta}T\mu^{\xi^\alpha}\frac{\partial\mu^{\xi^\beta}}{\partial\xi^\gamma}+C_{12}^\beta T\frac{\partial\mu^{\xi^\beta}}{\partial\xi^\gamma},\\
D_{\xi^\gamma}^\alpha &:= C_{11}^{\alpha\beta}T\frac{\partial\mu^{\xi^\beta}}{\partial\xi^\gamma},\\
D_T^\alpha &:= C_{11}^{\alpha\beta}T\frac{\partial\mu^{\xi^\beta}}{\partial T}+C_{12}^\alpha .
\end{aligned}
\qquad (7.5.66)
$$

κ_T is the coefficient of heat conduction, and $D_{\xi^\gamma}^\alpha$ represent the matrices of mass diffusivities; κ_{ξ^γ} and D_T^α are coupling coefficients and their presence in (7.5.65) is known in a special case as SORET effect if only one tracer is present.

A significant simplification of the formulas (7.5.66) is achieved by further specialisation. If, for instance, it is assumed that $C_{11}^{\alpha\beta}$ only contains non-vanishing entries in the diagonal, i. e., if $C_{11}^{\alpha\beta} = 0$ for $\alpha \neq \beta$ and if it is simultaneously assumed that also the interaction terms C_{12}^α vanish, then one obtains

$$
\begin{aligned}
\Psi_D = \sum_\alpha \tfrac{1}{2}C_{11}^{\alpha\alpha}\left(\mathrm{grd}\mu^{\xi^\alpha}\right)\cdot\left(\mathrm{grd}\mu^{\xi^\alpha}\right)\\
+\tfrac{1}{2}C_{22}\left(\frac{\mathrm{grad}\,T}{T}\right)\cdot\left(\frac{\mathrm{grad}\,T}{T}\right).
\end{aligned}
\qquad (7.5.67)
$$

Assuming furthermore the HELMHOLTZ free energy (7.5.47) in the special form

$$
\psi = \psi_1(\rho,T)+\sum_{\alpha=1}^{\nu-1}\psi_2^\alpha(\xi^\alpha,T)\,, \qquad (7.5.68)
$$

then one may easily show that $\boldsymbol{q}$ and $\boldsymbol{j}^\alpha$ take the forms

$$
\begin{aligned}
\boldsymbol{q} &= -\kappa_T\,\mathrm{grad}\,T-\sum_{\alpha=1}^{\gamma-1}\kappa_{\xi^\alpha}\,\mathrm{grad}\,\xi^\alpha\\
\boldsymbol{j}^\alpha &= -D_t^\alpha\,\mathrm{grad}\,T-D_{\xi^\alpha}^\alpha\,\mathrm{grad}\,\xi^\alpha, \quad \text{(no summation over } \alpha)
\end{aligned}
$$

with the coefficients

$$
\begin{aligned}
\kappa_T &= C_{22}+\sum_{\alpha=1}^{\nu-1}\left(\frac{\partial\psi_2^\alpha}{\partial\xi^\alpha}C_{11}^{\alpha\alpha}T\frac{\partial^2\psi_2^\alpha}{\partial\xi^\alpha\partial T}\right), \qquad & D_{\xi^\alpha}^\alpha &= C_{11}^{\alpha\alpha}T\frac{\partial^2\psi_2^\alpha}{\partial{\xi^\alpha}^2},\\
\kappa_{\xi^\alpha} &= \sum_{\alpha=1}^{\nu-1}C_{11}^{\alpha\alpha}T\frac{\partial\psi_2^\alpha}{\partial\xi^\alpha}\frac{\partial^2\psi_2^\alpha}{\partial{\xi^\alpha}^2}, & D_T^\alpha &= C_{11}^{\alpha\alpha}T\frac{\partial^2\psi_2^\alpha}{\partial\xi^\alpha\partial T}.
\end{aligned}
\qquad (7.5.69)
$$

These formulas permit significant qualitative inferences. As the temperature dependence of the chemical potentials is generally weak one may ignore the coefficients involving the factor $\partial^2\psi_2^\alpha/(\partial\xi^\alpha\partial T)$. Then (7.5.69) implies

$$\kappa_T \sim C_{22}\,, \qquad D_T^\alpha \sim 0\,,$$

whilst κ_{ξ^α} and $D_{\xi^\alpha}^{\alpha\alpha}$ are still given by (7.5.69). In this approximation the influence of the temperature gradient on the mass fluxes $\boldsymbol{j}^\alpha$ is ignored but that of the concentration gradients on the heat flux is maintained. This is the approximation usually referred to as the SORET effect.

Remark In the theory derived above the density of the mixture and the mass concentrations of the tracers $\xi^\alpha(\alpha = 1, 2, ..., \nu - 1)$ were regarded as independent fields. Furthermore, the mass balance of the main fluid was not considered as a field equation and neither was the mass concentration of this fluid regarded as an independent constitutive variable. This was done so since $\xi^\alpha \ll 1, \quad \alpha = 1, 2, ..., \nu - 1$, and $(\xi^\nu - 1) \ll 1$ were assumed; in other words the mass of the main fluid was assumed so much greater than the masses of the tracers that the dynamics of the mixture could be regarded as essentially indistinguishable from that of the main fluid. For many situations in environmental physics this is the case. However, for mixtures of gases, this need not be so. In such circumstances it may be advantageous also to incorporate ξ^ν as an independent variable in the constitutive relations; if this is done, however, one must account for the fact that $\xi^\nu = 1 - \sum_{\alpha=1}^{\nu-1}\xi^\alpha$. This identity may be taken care of by considering $\sum_{\alpha=1}^{\nu}\xi^\alpha = 1$ as a constraint equation in the exploitation of the entropy principle. In such a formulation ξ^ν is then treated as an independent field. It is clear that such a formulation of the model equations must yield a theory that is identical to that just derived; this is so but certain variables must be transformed to achieve this one-to-one correspondence.

7.6 Saturated Mixture of Non-Polar Solid and Fluid Constituents[9]

7.6.1 Motivation

There exists a large number of *heterogeneous materials*, which constitute of both solid and fluid components. In *porous media* the solid component consists of a connected coherent material or a number of such materials, and the

[9] Whereas the text of this section has independently been drafted by K. HUTTER, many of the detailed arguments have been influenced by the dissertation of G. BAUER [22]. We wish to acknowledge this source, as it has also streamlined and clarified the writings at other places of this book, even though its influence may only here directly be seen.

so-called *pore space* is filled with the fluid constituents, usually fluids and/or gases. This pore space can by itself be connected, or it consists of isolated inclusions or may possess both isolated and connected pores. In *granular media* the solid component consists of individual grains, which would fall apart, if they were not held together by an ambient pressure. Thus a granular heap can, in general, only assume special, distinct rest positions. Sand, soil and snow belong to this class. If the pore space is partly or completely connected, then the pore fluid may move under the action of gravity or pressure differences. Displacements of fluid particles may then be several orders of magnitude larger than those of the solid constituents. This is so e. g. in foams, that are used as acoustic insulators, or in the so-called "wind pumping" of the upper most snow layer in snow depositions, through which air percolates, under heavy storms sometimes with surprisingly large seepage velocities. It arises also in ground water flows. However, if the ground water current in the saturated soil is too strong, rapid earth movements may arise, which are known as *debris flows*, *sturzstroms* and *mudflows* in which the granular component also moves as a fluid. The transition from a ground water flow in practically silent sand to the catastrophic flow of a sand water mixture is almost always very abrupt and is for this reason in geotechnique and geology called an instability; the phenomenon is, however, also called "*quick sand*".

If the differential lengths of the governing processes extend over several characteristic diameters of the grains, pores and typical curvatures or twists of the composites, one may well apply a mixture concept with field variables that are thought to be homogenized over these differential lengths. In other words, within a volume of influence of a spatial point – this is called a *Representative Element Volume*, REV – all field variables are regarded as averaged quantities of true micro fields, and it is anticipated that these averaged quantities interact with each other such that balance laws of mixtures apply to them. Of course this is an assumption, which can be checked for correctness by more detailed averaging procedures; this we will, however, not do here. At last, the correctness of the procedure can be obtained by testing the complete theory against observations.

The applications in focus above are taken from the special fields of geotechnique, geology, geophysics and environmental physics; industrial applications from chemical process engineering, chemistry, civil and mechanical engineering could equally be given. Our intention is the derivation of the governing equations of a solid–fluid–interaction theory from first thermodynamic principles by using the second law of thermodynamics and exploiting it. A thermodynamic view is also necessary for the derivation of the equations describing purely mechanical processes for which the evolution of the temperature as a variable is not pursued. This is so, because the particular formulation of the equilibrium conditions, especially with regard to the equilibrium pressure and the DARCY interaction force in saturated or unsaturated soil, directly depends on the postulated constitutive equations, i. e., the material complexity and the second law of thermodynamics.

As is transparent from this discussion the physical circumstances in focus are here mixtures of (partly) *immiscible* constituents, which is quite contrary to the situation dealt with in Sect. 7.5, where the diffusion processes in a mixture of miscible constituents were studied. Furthermore, at least one of the constituents is here a solid. The prerequisites of the assumption that material of each constituent may occupy each point of the body is therefore less convincing than for miscible mixtures. Differential lengths are certainly larger and microstructural effects i.e., effects at the scale within an REV might have to be accounted for by formulating additional relationships that describe the effects of the microstructure at the macro-level. Such equations may be additional balance laws for the volume fractions occupied by the individual constituents.

The concepts of mixture theory including TRUESDELL's metaphysical principles remain valid guidelines and so TRUESDELL [238], [241], [243] and MÜLLER [165] are the pertinent references that should be mentioned. Early concepts of the dynamics of porous media are by BIOT [26], [27] and, within the context of continuum mechanics, by BOWEN [35], [36]. Additional balance laws for variables describing the effects of the porous or granular structure have e.g. been introduced by GOODMAN & COWIN [85] and PASSMAN, NUNZIATO and others (see e.g. the article "A theory of multiphase mixtures" by PASSMAN ET AL. in Appendix 5C of TRUESDELL [243], or DREW & PASSMAN [57] where a large number of references is stated).

In the above mentioned references a postulational approach towards the formulation of the mixture balance laws is taken. Derivations of the balance relations by using an averaging procedure to arrive from physical statements at the micro-scale to relations of the macro-scale have been given by HASSANIAZADEH & GRAY [95], [96], [97] and GRAY [86], an approach that is also followed by EHLERS [61], [62] and BLUHM [29]. The history of the developments from the very early stages to present is given by DE BOER [33].

We consider again a mixture of non-polar constituents[10] so that

$$\boldsymbol{s}^\alpha = \boldsymbol{0}, \quad \boldsymbol{m}^\alpha = \boldsymbol{0}, \quad \boldsymbol{l}^\alpha = \boldsymbol{0}, \quad \boldsymbol{\mathfrak{k}}^\alpha = \boldsymbol{0} \quad (\alpha = 1, 2, ..., N)\,. \tag{7.6.1}$$

Moreover, we shall assume that the individual constituents perform distinct motions, but that these motions are slow on the time scales of thermal relaxation, so that all constituents may be assumed to possess the same temperature. It follows that the prerequisites of a *mixture of class II* are fulfilled, for which the mass and momentum balance laws for all constituents must be formulated, but only the energy balance for the mixture as a whole is relevant. Consequently, the following balance laws define this mixture theory:

[10] We shall denote in this section the number of constituents by N (and not by ν) since ν_α is reserved for the volume fractions of the constituents.

$$
\begin{aligned}
&\frac{\partial \rho_\alpha}{\partial t} + \operatorname{div}(\rho_\alpha \boldsymbol{v}^\alpha) = \mathfrak{c}^\alpha, \quad (\alpha = 1, 2, ..., N)\,, \\
&\frac{\partial \rho_\alpha \boldsymbol{v}^\alpha}{\partial t} + \operatorname{div}(\rho_\alpha \boldsymbol{v}^\alpha \otimes \boldsymbol{v}^\alpha) = \operatorname{div} \boldsymbol{t}^\alpha + \rho_\alpha \boldsymbol{f}^\alpha + \mathfrak{m}^\alpha, \quad (\alpha = 1, 2, ..., N)\,, \\
&\frac{\partial \rho \varepsilon}{\partial t} + \operatorname{div}(\rho \varepsilon \boldsymbol{v}) = -\operatorname{div} \boldsymbol{q} + \operatorname{grad} \boldsymbol{v} \cdot \boldsymbol{t}^T + \rho \mathfrak{r}\,, \qquad (7.6.2) \\
&\frac{\partial \rho s}{\partial t} + \operatorname{div}(\rho s \boldsymbol{v} + \boldsymbol{\phi}^{\rho s}) - \rho \eta^{\rho s} \geq 0\,.
\end{aligned}
$$

These comprise the balance laws of mass and momentum for the constituents $(7.6.2)_{1,2}$, those of the internal energy of the mixture $(7.6.2)_3$ and of the entropy $(7.6.2)_4$. Because of the assumption (7.6.1), as follows from (7.2.17), the partial CAUCHY stress tensors $\boldsymbol{t}^\alpha$ are symmetric, $\boldsymbol{t}^\alpha = \boldsymbol{t}^{\alpha T}$; furthermore, the mixture variables are defined in (7.4.2)–(7.4.6) and could easily be written down in the simplified forms appropriate for the specification (7.6.1).

The goal in this section is not to derive a general theory of mixtures of class II, but rather to present an example of such a theory. For this reason, the following simplifying assumptions are introduced.

Assumption of Constant True Densities It shall be assumed that solid and fluid constituents are such that the density preserving property may be imposed. We shall thus assume that all constituents possess constant true densities[11]. The true independent variables are then not the mass densities ρ_α but the *constituent volume fractions* ν_α, which connect the partial densities ρ_α with the true densities $\hat{\rho}_\alpha$ according to

$$
\rho_\alpha = \nu_\alpha \hat{\rho}_\alpha, \quad \hat{\rho}_\alpha = \text{constant}, \quad (\alpha = 1, 2, 3, ..., N)\,. \qquad (7.6.3)
$$

The assumption of constant $\hat{\rho}_\alpha$ means that the materials of which the individual constituents are made, are density preserving.

If one substitutes (7.6.3) into the mass balance (7.6.2), then, after division by $\hat{\rho}_\alpha$, one obtains

$$
\frac{\partial \nu_\alpha}{\partial t} + \operatorname{div}(\nu_\alpha \boldsymbol{v}^\alpha) = \frac{1}{\hat{\rho}_\alpha} \mathfrak{c}^\alpha =: \mathfrak{n}^\alpha, \quad (\alpha = 1, 2, ..., N)\,, \qquad (7.6.4)
$$

which is an equation for the volume fractions ν_α; in other words, in this simple theory the volume fractions replace the partial densities as independent variables.

Assumption of Saturation of the Mixture We also wish to suppose that the considered mixture does not have any cavities or empty spaces; in other words, the sum of the volume fractions fills the entire space at all positions and at all times within the body, i.e.,

[11] The *true* density of constituent α is the mass of constituent α per unit volume of this constituent (and not of that the mixture). Sometimes this is also called the *apparent* density.

$$\sum_{\alpha=1}^{N} \nu_\alpha = 1 \,. \tag{7.6.5}$$

This equation implies: of the N volume fraction variables only $(N-1)$ are independent. One particular ν_α can be replaced by $(1 - \sum_{\beta\neq\alpha} \nu_\beta)$; we shall do this in the sequel for the variable ν_N.

If a function $f = f(\nu_1, ..., \nu_N)$ depends on all $\nu_1, ..., \nu_N$, then, upon elimination of $\nu_N = (1 - \sum_{\beta\neq N} \nu_\beta)$ one may write

$$\tilde{f}(\nu_1, ..., \nu_{N-1}) = f\left(\nu_1, ..., \nu_{N-1}, 1 - \sum_{\beta=1}^{N-1} \nu_\beta\right) \,. \tag{7.6.6}$$

If one must differentiate this function with respect to time or space – this will be expressed by ∂ – then one obtains

$$\partial f = \sum_{\alpha=1}^{N-1} \frac{\partial \tilde{f}}{\partial \nu_\alpha} \partial \nu_\alpha = \sum_{\alpha=1}^{N-1} \left(\frac{\partial f}{\partial \nu_\alpha} - \frac{\partial f}{\partial \nu_N} \right) \partial \nu_\alpha \,. \tag{7.6.7}$$

It follows that one must always be aware which independent variables one wants to use as field variables, either the first $N-1$ volume fractions as for $\tilde{f}$, or the N volume fractions as for f. It should also be observed, that the balance law for the dependent volume fraction still constitutes an independent equation. In fact the volume balance (7.6.4) can, for $\alpha = N$ be written as

$$-\left\{\sum_{\beta=1}^{N-1} \frac{\partial \nu_\beta}{\partial t}\right\} + \left\{\mathrm{div}\left[\left(1 - \sum_{\beta=1}^{N-1} \nu_\beta\right) \boldsymbol{v}_N\right]\right\} = -\frac{1}{\hat{\rho}_N}\left(\sum_{\gamma=1}^{N-1} \mathfrak{c}^\gamma\right) =: \mathfrak{n}_N \,, \tag{7.6.8}$$

where the saturation condition (7.6.5) and $\sum_{\alpha=1}^{N} \mathfrak{c}^\alpha = 0$ have been used. This equation is independent of the other volume balances. The *saturation condition* must be interpreted as a constraint condition, which is the cause for the fact that in the resulting equations there will be an independent equation in excess to the number of independent variables. This will lead to the introduction of an additional constraint stress, the so called *saturation pressure*.

7.6.2 Choice of the Material Class and Material Theory

The balance laws $(7.6.2)_{2,3}$ and the mass balance equations for the density preserving constituents in the form (7.6.4) constitute together $4N+1$ equations for the $N-1$ volume fractions ν_α, $3N$ velocity components of the constituents $\boldsymbol{v}^\alpha$, as well as the temperature θ, i.e., $4N$ variables. They must be complemented by phenomenological relations for the constitutive quantities

$$\Psi = \{\mathfrak{c}^\alpha, \mathfrak{m}_\alpha, \boldsymbol{t}^\alpha, \varepsilon, s, \boldsymbol{q}^\alpha, \boldsymbol{\phi}^{\rho s}\} \,, \tag{7.6.9}$$

which, according to the principle of equipresence, all are postulated to depend on the following set of constitutive variables

$$\begin{aligned}\boldsymbol{\Xi} = \{&\nu_1, ..., \nu_{N-1}, \theta, \operatorname{grad}\nu_1, ..., \operatorname{grad}\nu_{N-1},\\ &\boldsymbol{v}^1, ..., \boldsymbol{v}^N, \operatorname{grad}\boldsymbol{v}^1, ..., \operatorname{grad}\boldsymbol{v}^N, \operatorname{grad}\theta\} \ .\end{aligned} \tag{7.6.10}$$

The chosen material class is therefore a mixture of viscous, heat conducting constituents which principally possess fluid character as will become apparent in due course with the developments. Elasticity effects are left out of consideration merely for reasons of mathematical simplicity[12]. Notice, moreover, that only $(N-1)$ volume fractions are taken into account in (7.6.10), because the N-th volume fraction in a saturated mixture is determined by the other $(N-1)$ constituent volume fractions. Furthermore, the $\operatorname{grad}\nu_\alpha$-terms account for the fact that the momentum productions or interaction forces $\mathbf{m}^\alpha$ may also depend on the distribution of the constituents; such a theory is also known as a *gradient theory*, in any case it does have non-simple character[13].

We also request that the constitutive relations $\Psi = \hat{\Psi}(\boldsymbol{\Xi})$ obey the rule of material objectivity, which requires that $\hat{\Psi}(\cdot)$ satisfies the equation

$$\begin{aligned}&\hat{\Psi}\left(..., \boldsymbol{v}^1, ..., \boldsymbol{v}^N, \operatorname{grad}\boldsymbol{v}^1, ..., \operatorname{grad}\boldsymbol{v}^N, ...\right)\\ &= \hat{\Psi}\left(..., \boldsymbol{v}^1 + \boldsymbol{a}^1, ..., \boldsymbol{v}^N + \boldsymbol{a}^N, \operatorname{grad}\boldsymbol{v}^1 + \boldsymbol{\Omega}^1, .., \operatorname{grad}\boldsymbol{v}^N + \boldsymbol{\Omega}^N, ...\right)\end{aligned} \tag{7.6.11}$$

as an identity for any vector $\boldsymbol{a}^\alpha$ and any skew symmetric tensor $\boldsymbol{\Omega}^\alpha$ $(\alpha = 1, 2, ..., N)$. This requirement implies that $\hat{\Psi}$ cannot depend upon the constituent velocities and velocity gradients themselves, but only upon the differences

$$\begin{aligned}\boldsymbol{u}^\alpha &:= \boldsymbol{v}^\alpha - \boldsymbol{v}, & \boldsymbol{v} &:= \sum_{\alpha=1}^{N} \xi^\alpha \boldsymbol{v}^\alpha \ ,\\ \boldsymbol{U}^\alpha &:= \operatorname{grad}\boldsymbol{v}^\alpha - \boldsymbol{W}, & \boldsymbol{W} &:= \operatorname{skw}\operatorname{grad}\boldsymbol{v} \ ,\end{aligned} \tag{7.6.12}$$

in which $\boldsymbol{u}^\alpha$ are the diffusion velocities relative to the barycentric velocity $\boldsymbol{v}$ and $\boldsymbol{U}^\alpha$ the relative deformation rate tensors. We leave it as an exercise to the reader to corroborate that $\boldsymbol{u}^\alpha$ and $\boldsymbol{U}^\alpha$ are objective vectors and objective

---

[12] Such effects are important to be included for the solid constituent when rest positions are important for the physical processes under consideration. For such cases, one must add the deformation gradient $\boldsymbol{F}^s$ as an additional variable in (7.6.10), see e.g. SVENDSEN & HUTTER [230]. Thus, the postulate (7.6.10) is appropriate for modeling debris flows or mud flows where fluid and solid constituents are moving very much like fluids.

[13] In classical continuum mechanics a non-simple material is defined as a medium in which higher gradients of the motion $\boldsymbol{\chi}(\boldsymbol{X}, t)$ may arise as independent constitutive variables. This is the case, if $\operatorname{grad}\rho$ is an independent constitutive variable, because $\rho = \rho_0 / \det\boldsymbol{F}$ and therefore $\operatorname{grad}\rho = \rho_0 \operatorname{grad}\det\boldsymbol{F}^{-1}$, see Chap.5.

tensors[14]. With the above restrictions the constitutive postulate (7.6.10) may also be written as

$$\Psi = \hat{\Psi}\Big(\nu_1, ..., \nu_{N-1}, \theta, \operatorname{grad}\nu_1, ..., \operatorname{grad}\nu_{N-1}, \boldsymbol{u}^1, .., \boldsymbol{u}^N, \boldsymbol{U}^1, ..., \boldsymbol{U}^N, \operatorname{grad}\theta\Big). \tag{7.6.13}$$

In performing the explicit calculations it has been seen, however, that it is more convenient, to perform the calculations with the variables (7.6.10) rather than (7.6.13) and to incorporate the specialization (7.6.12) afterwards. This requires then that differentiations with respect to $\boldsymbol{u}^\alpha$ and $\boldsymbol{U}^\alpha$ must be executed with care. For instance, the reader may reproduce the following definitions and rules of differentiation (Exercise 9):

$$\left.\begin{aligned}
\operatorname{grad}\boldsymbol{v} &= \sum_{\gamma=1}^{N} [\xi^\gamma \operatorname{grad}\boldsymbol{v}^\gamma + \boldsymbol{v}^\gamma \otimes \operatorname{grad}\xi^\gamma],\\
\boldsymbol{D} &= \sum_{\gamma=1}^{N} [\xi^\gamma \boldsymbol{D}^\gamma + \operatorname{sym}(\boldsymbol{v}^\gamma \otimes \operatorname{grad}\xi^\gamma)],\\
\boldsymbol{W} &= \sum_{\gamma=1}^{N} [\xi^\gamma \boldsymbol{W}^\gamma + \operatorname{skw}(\boldsymbol{v}^\gamma \otimes \operatorname{grad}\xi^\gamma)],\\
\frac{\partial \boldsymbol{u}^\beta}{\partial \boldsymbol{v}^\alpha} &= \left(\delta^{\alpha\beta} - \xi^\alpha\right)\mathbf{1},\\
\frac{\partial \boldsymbol{U}^\beta}{\partial \boldsymbol{W}^\alpha} &= \left(\delta^{\alpha\beta} - \xi^\alpha\right)\mathbf{1}_4,\\
\frac{\partial f}{\partial \boldsymbol{v}^\alpha} &= \frac{\partial f}{\partial \boldsymbol{u}^\alpha} - \xi^\alpha \sum_{\gamma=1}^{N} \frac{\partial f}{\partial \boldsymbol{u}^\gamma},\\
\frac{\partial f}{\partial \boldsymbol{W}^\alpha} &= \frac{\partial f}{\partial \boldsymbol{U}^\alpha} - \xi^\alpha \sum_{\gamma=1}^{N} \frac{\partial f}{\partial \boldsymbol{U}^\gamma},\\
\frac{\partial f}{\partial \operatorname{grad}\boldsymbol{v}^\alpha} &= \frac{\partial f}{\partial \boldsymbol{D}^\alpha} + \frac{\partial f}{\partial \boldsymbol{U}^\alpha} - \xi^\alpha \sum_{\gamma=1}^{N} \frac{\partial f}{\partial \boldsymbol{U}^\gamma},
\end{aligned}\right\} \tag{7.6.14}$$

in which $\mathbf{1}_4$ denotes the fourth order unit tensor and $\delta^{\alpha\beta}$ the N-dimensional KRONECKER delta. Formula $(7.6.14)_6$ must e. g. be applied if the internal energy of the mixture must be differentiated with respect to $\boldsymbol{v}^\alpha$:

[14] $\boldsymbol{U}^\alpha$ contains symmetric and skewsymmetric contributions which can be separated into

$$\boldsymbol{D}^\alpha = \operatorname{sym}\boldsymbol{U}^\alpha = \operatorname{sym}(\operatorname{grad}\boldsymbol{v}^\alpha), \qquad \boldsymbol{W}^\alpha = \operatorname{skw}\boldsymbol{U}^\alpha = \operatorname{skw}(\operatorname{grad}\boldsymbol{v}^\alpha) - \boldsymbol{W}.$$

$\boldsymbol{D}^\alpha$ and $\boldsymbol{W}^\alpha$ are relative stretching and relative vorticity tensors.

$$
\begin{aligned}
\rho\varepsilon &= \underbrace{\sum_{\alpha=1}^{N} \rho_\alpha \varepsilon^\alpha}_{\rho\varepsilon_I} + \underbrace{\sum_{\alpha=1}^{N} \frac{\rho_\alpha}{2} \boldsymbol{u}^\alpha \cdot \boldsymbol{u}^\alpha}_{\rho\varepsilon_D} = \rho\varepsilon_I + \rho\varepsilon_D \,, \\
\frac{\partial(\rho\varepsilon_D)}{\partial \boldsymbol{v}^\alpha} &= \rho_\alpha \boldsymbol{u}^\alpha - \xi^\alpha \underbrace{\sum_{\gamma=1}^{N} \rho_\gamma \boldsymbol{u}^\gamma}_{0} = \rho_\alpha \boldsymbol{u}^\alpha \,, \qquad (7.6.15) \\
\frac{\partial(\rho\varepsilon)}{\partial \boldsymbol{v}^\alpha} &= \frac{\partial(\rho\varepsilon_I)}{\partial \boldsymbol{v}^\alpha} + \rho_\alpha \boldsymbol{u}^\alpha = \frac{\partial(\rho\varepsilon_I)}{\partial \boldsymbol{u}^\alpha} - \xi^\alpha \sum_{\gamma=1}^{N} \frac{\partial(\rho\varepsilon_I)}{\partial \boldsymbol{u}^\gamma} + \rho_\alpha \boldsymbol{u}^\alpha \,.
\end{aligned}
$$

The quantities ε_I and ε_D are called *inner internal energy* and *diffusive internal energy*, respectively.

A *thermodynamic process* is in this section a set of field variables ν_α ($\alpha = 1, 2, ..., N-1$), $\boldsymbol{v}_1, ..., \boldsymbol{v}_N, \theta$ which obeys the balance laws of momenta and energy $(7.6.2)_{2,3}$, the volume balances (7.6.4) for $\alpha = 1, \ldots, N-1$, and (7.6.8) for $\alpha = N$ as well as the constitutive equations $\Psi = \Psi(\Xi)$ given in (7.6.10) or (7.6.13). If we now request, as usual, that the entropy inequality $(7.6.2)_4$ be satisfied for all thermodynamic processes, then the equations just mentioned play the role of constraint conditions for the entropy inequality. According to the procedure of LIU this requirement can be fulfilled as follows: One subtracts the scalar products of these equations with so-called LAGRANGE parameters (or adds them which only changes the signs of the LAGRANGE parameters) from the entropy inequality and satisfies this extended inequality for unrestricted fields. This extended inequality reads

$$
\begin{aligned}
&\frac{\partial(\rho\hat{s})}{\partial t} + \operatorname{div}\left(\hat{\boldsymbol{\phi}}^{\rho s} + \rho s \boldsymbol{v}\right) - \eta^{\rho s} - \sum_{\alpha=1}^{N-1} \Lambda^\nu_\alpha \left(\frac{\partial \nu_\alpha}{\partial t} + \operatorname{div}(\nu_\alpha \boldsymbol{v}^\alpha) - \hat{\mathfrak{n}}_\alpha\right) \\
&-\Lambda^\nu_N \left(-\left\{\sum_{\alpha=1}^{N-1} \frac{\partial \nu_\beta}{\partial t}\right\} + \left\{1 - \sum_{\beta=1}^{N-1} \nu_\beta\right\} \cdot \operatorname{div} \boldsymbol{v}_N - \left\{\sum_{\beta=1}^{N-1} \operatorname{grad} \nu_\beta\right\} \cdot \boldsymbol{v}^N - \hat{\mathfrak{n}}_N\right) \\
&-\sum_{\alpha=1}^{N-1} \boldsymbol{\Lambda}^{\boldsymbol{v}}_\alpha \cdot \left(\frac{\partial(\rho_\alpha \boldsymbol{v}^\alpha)}{\partial t} - \operatorname{div}(\hat{\boldsymbol{t}}^\alpha - \rho_\alpha \boldsymbol{v}^\alpha \otimes \boldsymbol{v}^\alpha) - \hat{\mathfrak{m}}^\alpha - \rho_\alpha \boldsymbol{f}^\alpha\right) \\
&-\Lambda^\varepsilon \left(\frac{\partial(\rho\hat{\varepsilon})}{\partial t} + \operatorname{div}(\hat{\boldsymbol{q}} + \rho\hat{\varepsilon}\boldsymbol{v}) - \operatorname{tr}(\hat{\boldsymbol{t}}\boldsymbol{D}) - \rho\mathfrak{r}\right) \geq 0 \,, \qquad (7.6.16)
\end{aligned}
$$

in which the materially dependent variables are thought to be substituted by their constitutive relations, a fact which we have made visible by writing a hat for such variables, $\hat{(\cdot)}$. In (7.6.16) the balance laws for the volume fractions appear multiplied with the LAGRANGE multiplier Λ^ν_α; the momentum balances are scalarly multiplied with $\boldsymbol{\Lambda}^{\boldsymbol{v}}_\alpha$ and the balance of internal energy appears multiplied with the factor Λ^ε. Apart from the supply terms of entropy $\eta^{\rho s}$, of momentum $\rho_\alpha \boldsymbol{f}$ and of energy $\rho\mathfrak{r}$ inequality (7.6.16) depends explicitly (and implicitly via the constitutive quantities) upon the independent fields and or spatial and temporal derivatives of these. Hence, the LAGRANGE multipliers are in general also functionals of these quantities[15].

A first step in the exploitation of inequality (7.6.16) consists in the requirement that the material behaviour of a body, since it describes the material behaviour, does not depend upon the source terms that are applied to the body from its outside. The sum of the external supply terms in (7.6.16) must therefore add up to zero, so that

$$\eta^{\rho s} = \Lambda^\varepsilon \rho \mathfrak{r} + \sum_{\alpha=1}^{N-1} (\boldsymbol{\Lambda}^{\boldsymbol{v}}_\alpha \cdot \rho_\alpha \boldsymbol{f}^\alpha) \,; \tag{7.6.17}$$

consequently, the entropy supply is a linear combination of the energy supply (radiation) and the momentum supplies (external forces) and is fixed once the LAGRANGE parameters Λ^ε and $\boldsymbol{\Lambda}^{\boldsymbol{v}}_\alpha$ are determined.

The further procedure in exploiting the inequality (7.6.16), shortened by (7.6.17), consists now in the following steps: all differentiations are performed using the chain rule of differentiation where needed; the chain rule must in particular be used where a constitutive quantity is differentiated with respect to time or space. The inequality thus obtained is sorted and rearranged according to the independent constitutive variables (and their higher order derivatives). If this is done – a relatively lengthy procedure – one recognizes that the emerging inequality can be considerably condensed, if the following one forms are defined:

$$\mathcal{P} := \mathrm{d}(\rho s) - \Lambda^\varepsilon \mathrm{d}(\rho\varepsilon), \quad \boldsymbol{\mathcal{F}} := \mathrm{d}\phi^{\rho s} - \Lambda^\varepsilon \mathrm{d}\boldsymbol{q} + \sum_{\alpha=1}^{N} \boldsymbol{\Lambda}^{\boldsymbol{v}}_\alpha (\mathrm{d}\boldsymbol{t}^\alpha) \,, \tag{7.6.18}$$

with the coefficients

$$\mathcal{P}_{x_i} := \frac{\partial(\rho\hat{s})}{\partial x_i} - \Lambda^\varepsilon \frac{\partial(\rho\hat{\varepsilon})}{\partial x_i}, \quad \boldsymbol{\mathcal{F}}_{x_i} := \frac{\partial\hat{\phi}^{\rho s}}{\partial x_i} - \Lambda^\varepsilon \frac{\partial\hat{\boldsymbol{q}}}{\partial x_i} + \sum_{\alpha=1}^{N} \boldsymbol{\Lambda}^{\boldsymbol{v}}_\alpha \frac{\partial\hat{\boldsymbol{t}}^\alpha}{\partial x_i} \,, \tag{7.6.19}$$

so that[16]

$$\mathcal{P} = \mathcal{P}_{x_i} \mathrm{d}x_i \quad \text{and} \quad \mathcal{F} = \mathcal{F}_{x_i} \mathrm{d}x_i \,. \tag{7.6.20}$$

---

[15] It might be worth mentioning that most authors of papers dealing with thermodynamics of solid-fluid mixtures of porous or granular materials use the second law of thermodynamics in a form akin to the CLAUSIUS–DUHEM inequality with an exploitation following COLEMAN–NOLL procedure of exploitation. In these approaches the balance laws of peculiar momenta and energy are associated with arbitrary source terms.

[16] Notice that $\mathcal{P}_{x_i}$ and $\mathcal{F}_{x_i}$ do not represent derivatives of $\mathcal{P}$ and $\mathcal{F}$, since Λ^ε in (7.6.19) cannot be pulled into the differential of $\rho\varepsilon$. So, neither $\mathcal{P}$ nor $\mathcal{F}$ are complete differentials.

Furthermore, we use the abbreviation

$$\mathcal{O}_{x_i} = \boldsymbol{v} \otimes \mathcal{P}_{x_i} + \mathcal{F}_{x_i} \,. \tag{7.6.21}$$

With these notations the extended entropy inequality can be written in the following form:[17]

$$\begin{aligned}
&\mathcal{P}_\theta \frac{\partial \theta}{\partial t} + \mathcal{P}_{\operatorname{grad}\theta} \cdot \frac{\partial(\operatorname{grad}\theta)}{\partial t} \\
&+ \sum_{\alpha=1}^{N} \Big\{ (\mathcal{P}_{\nu_\alpha} - \mathcal{P}_{\nu_N}) - (\Lambda^\nu_\alpha - \Lambda^\nu_N) - (\hat{\rho}_\alpha \boldsymbol{\Lambda}^{\boldsymbol{v}}_\alpha \cdot \boldsymbol{v}^\alpha - \hat{\rho}_N \boldsymbol{\Lambda}^{\boldsymbol{v}}_N \cdot \boldsymbol{v}^N) \Big\} \frac{\partial \nu_\alpha}{\partial t} \\
&+ \sum_{\alpha=1}^{N-1} \{\mathcal{P}_{\operatorname{grad}\nu_\alpha} - \mathcal{P}_{\operatorname{grad}\nu_N}\} \cdot \frac{\partial(\operatorname{grad}\nu_\alpha)}{\partial t} \\
&+ \sum_{\alpha=1}^{N} (\mathcal{P}_{\boldsymbol{v}^\alpha} - \rho_\alpha \boldsymbol{\Lambda}^{\boldsymbol{v}}_\alpha) \cdot \frac{\partial \boldsymbol{v}^\alpha}{\partial t} + \sum_{\alpha=1}^{N} \mathcal{P}_{\operatorname{grad}\boldsymbol{v}^\alpha} \cdot \frac{\partial \operatorname{grad}\boldsymbol{v}^\alpha}{\partial t} \\
&+ \sum_{\alpha=1}^{N-1} \{\mathcal{O}_{\operatorname{grad}\nu_\alpha} - \mathcal{O}_{\operatorname{grad}\nu_N}\} \cdot \operatorname{grad}(\operatorname{grad}\nu_\alpha) \\
&+ \sum_{\alpha=1}^{N} (\mathcal{O}_{\operatorname{grad}\boldsymbol{v}^N}) \cdot \operatorname{grad}(\operatorname{grad}\boldsymbol{v}^\alpha) + (\mathcal{O}_{\operatorname{grad}\theta}) \cdot (\operatorname{grad}(\operatorname{grad}\theta)) \\[2ex]
&+ \mathcal{O}_\theta \cdot \operatorname{grad}\theta \\
&+ \sum_{\alpha=1}^{N-1} \Big\{ \mathcal{O}_{\nu_\alpha} - \mathcal{O}_{\nu_N} - (\Lambda^\nu_\alpha \boldsymbol{v}^\alpha - \Lambda^\nu_\alpha \boldsymbol{v}^N) + \Lambda^\varepsilon \frac{\boldsymbol{t}}{\rho} (\hat{\rho}_\alpha \boldsymbol{u}^\alpha - \hat{\rho}_N \boldsymbol{u}^N) \\
&- (\hat{\rho}_\alpha \boldsymbol{\Lambda}^{\boldsymbol{v}}_\alpha \cdot \boldsymbol{v}^\alpha \otimes \boldsymbol{v}^\alpha - \hat{\rho}_N \boldsymbol{\Lambda}^{\boldsymbol{v}} \cdot \boldsymbol{v}^N \otimes \boldsymbol{v}^N) \\
&+ (s - \Lambda^\varepsilon \varepsilon)(\hat{\rho}_\alpha \boldsymbol{u}^\alpha - \hat{\rho}_N \boldsymbol{u}^N) \Big\} \cdot \operatorname{grad}\nu_\alpha \\
&+ \sum_{\alpha=1}^{N} \Big\{ \mathcal{O}_{\boldsymbol{v}^\alpha} + \rho_\alpha (s - \Lambda^\varepsilon \varepsilon) \mathbf{1} + \Lambda^\varepsilon \frac{\rho_\alpha}{\rho} \boldsymbol{t} - (\boldsymbol{\Lambda}^{\boldsymbol{v}}_\alpha \nu_\alpha + \rho_\alpha \boldsymbol{\Lambda}^{\boldsymbol{v}}_\alpha \cdot \boldsymbol{v}^\alpha) \mathbf{1} \\
&- \rho_\alpha (\boldsymbol{\Lambda}^{\boldsymbol{v}} \otimes \boldsymbol{v}^\alpha) \Big\} \cdot \operatorname{grad}\boldsymbol{v}^\alpha + \sum_{\alpha=1}^{N} \{\Lambda^\nu_\alpha \mathfrak{n}_\alpha + \boldsymbol{\Lambda}^{\boldsymbol{v}}_\alpha \cdot \mathfrak{m}^\alpha\} \geq 0 \,.
\end{aligned} \tag{7.6.22}$$

This still rather formidably looking inequality is written in two blocks which are separated from one another by an empty line. The block above this empty line consists of a sum of products of two factors $\boldsymbol{\alpha} \cdot \boldsymbol{\beta}$ where

$$\boldsymbol{\beta} = \left\{ \operatorname{grad}(\operatorname{grad}\zeta), \frac{\partial \operatorname{grad}\zeta}{\partial t}, \frac{\partial \zeta}{\partial t} \right\}, \qquad \zeta = (\theta, \nu_\alpha, \boldsymbol{v}^\alpha) \tag{7.6.23}$$

[17] The derivation of this formidably looking inequality is not difficult, however it is lengthy and time consuming, and it is susceptible to error. For this text it was derived several times and has been double-checked.

and the prefactors of $\boldsymbol{\beta}$ are collected in the second vector $\boldsymbol{\alpha}$ with the same dimension; $\boldsymbol{\alpha}$ is, as one can easily see, only dependent upon the independent constitutive variables and upon $\boldsymbol{v}^\alpha$. The upper block of inequality (7.6.22) is therefore linear in $\boldsymbol{\beta}$. The lower block in (7.6.22) does not contain variables from $\boldsymbol{\beta}$ and only depends upon the independent constitutive variables and $\boldsymbol{v}^\alpha$ $(\alpha = 1, 2, ..., N)$ and $\boldsymbol{v}$; let it be collected in the scalar Γ. In summary, inequality (7.6.22) has the form

$$\boldsymbol{\alpha} \cdot \boldsymbol{\beta} + \Gamma \geq 0 \tag{7.6.24}$$

and is linear in $\boldsymbol{\beta}$.

It is relatively easy to become convinced that to any chosen $\boldsymbol{\beta}$ in a neighbourhood of a material point, there exists (at least) one admissible thermodynamic process (which satisfies the balance equations and constitutive relations). *A forteriori*, since the balance laws are accounted for in inequality (7.6.22) via the terms involving the LAGRANGE parameters, $\boldsymbol{\beta}$ may assume any value we please; of necessity then, (7.6.24) is satisfied if and only if

$$\boldsymbol{\alpha} = \mathbf{0} \quad \text{and} \quad \Gamma \geq 0 \,. \tag{7.6.25}$$

The first equations comprise the LIU equations, the latter is the residual entropy inequality. The statement $\boldsymbol{\alpha} = \mathbf{0}$ corresponds in (7.6.22) to the following conditions, which must be satisfied as identities[18]

$$\begin{aligned}
&\mathcal{P}_\theta = 0, \\
&\mathcal{P}_{\mathrm{grad}\,\theta} = \mathbf{0}\,, \\
&(\mathcal{P}_{\nu_\alpha} - \mathcal{P}_{\nu_N}) - (\Lambda^\nu_\alpha - \Lambda^\nu_N) - (\hat{\rho}_\alpha \boldsymbol{\Lambda}^{\boldsymbol{v}}_\alpha \cdot \boldsymbol{v}^\alpha - \hat{\rho}_N \boldsymbol{\Lambda}^{\boldsymbol{v}}_N \cdot \boldsymbol{v}^N) = \mathbf{0}, \\
&\qquad\qquad\qquad\qquad\qquad\qquad\qquad\qquad (\alpha = 1, 2, ..., N-1), \\
&(\mathcal{P}_{\mathrm{grad}\,\nu_\alpha} - \mathcal{P}_{\mathrm{grad}\,\nu_N}) = \mathbf{0}, \qquad (\alpha = 1, 2, ..., N-1), \\
&(\mathcal{P}_{\boldsymbol{v}^\alpha} - \rho_\alpha \boldsymbol{\Lambda}^{\boldsymbol{v}}_\alpha) = \mathbf{0}, \qquad (\alpha = 1, 2, ..., N), \\
&\mathcal{P}_{\mathrm{grad}\,\boldsymbol{v}^\alpha} = \mathbf{0}, \qquad (\alpha = 1, 2, ..., N), \\
&(\boldsymbol{\mathcal{F}}_{\mathrm{grad}\,\nu_\alpha} - \boldsymbol{\mathcal{F}}_{\mathrm{grad}\,\nu_N})_{\mathrm{sym}} = \mathbf{0}, \qquad (\alpha = 1, 2, ..., N-1), \\
&(\boldsymbol{\mathcal{F}}_{\mathrm{grad}\,\boldsymbol{v}^\alpha})_{\mathrm{sym}} = \mathbf{0}, \quad (\alpha = 1, 2, ..., N) \\
&(\boldsymbol{\mathcal{F}}_{\mathrm{grad}\,\theta})_{\mathrm{sym}} = \mathbf{0}\,.
\end{aligned} \tag{7.6.26}$$

[18] With regard to the terms $\mathcal{P}_{\nu_\alpha}, \mathcal{P}_{\mathrm{grad}\,\nu_\alpha}$ and $\mathcal{F}_{\nu_\alpha}$, the identities $(7.6.26)_{3,4,7}$ are to be understood in the sense of eq. (7.6.7). If $\mathrm{d}\nu_N$ should not be replaced by $-\sum_{\alpha=1}^{N-1} \mathrm{d}\nu_\alpha$ as in (7.6.22), then the following substitutions should be made:

$$\begin{aligned}
\mathcal{P}_{\nu_\alpha} - \mathcal{P}_{\nu_N} &\longrightarrow \mathcal{P}_{\nu_\alpha}, \\
\mathcal{P}_{\mathrm{grad}\,\nu_\alpha} - \mathcal{P}_{\nu_N} &\longrightarrow \mathcal{P}_{\mathrm{grad}\,\nu_\alpha}, \\
\mathcal{F}_{\mathrm{grad}\,\nu_\alpha} - \mathcal{F}_{\nu_N} &\longrightarrow \mathcal{F}_{\mathrm{grad}\,\nu_\alpha},
\end{aligned}$$

(later in (7.6.41)–(7.6.46) this is done so).

In the three last identities the results $(7.6.26)_{2,4,6}$ were used. The residual inequality now only consists of the last block in (7.6.22).

The identities (7.6.26) allow a number of inferences, of which the most simple ones are presented now. First, it follows from the first six identities (7.6.26) plus the definitions (7.6.18) that the *generalized* GIBBS *relation*

$$
\begin{aligned}
\mathrm{d}(\rho\hat{s}) = {} & \Lambda^{\varepsilon}\mathrm{d}(\rho\hat{\varepsilon}) \\
& + \sum_{\alpha=1}^{N-1} \{(\Lambda^{\nu}_{\alpha} - \Lambda^{\nu}_{N}) + (\hat{\rho}_{\alpha}\boldsymbol{\Lambda}^{\boldsymbol{v}}_{\alpha}\cdot\boldsymbol{v}^{\alpha} - \hat{\rho}_{N}\boldsymbol{\Lambda}^{\boldsymbol{v}}_{N}\cdot\boldsymbol{v}^{N})\}\,\mathrm{d}\nu_{\alpha} \\
& + \sum_{\alpha=1}^{N} \rho_{\alpha}\boldsymbol{\Lambda}^{\boldsymbol{v}}_{\alpha}\cdot\mathrm{d}\boldsymbol{v}^{\alpha}
\end{aligned}
\tag{7.6.27}
$$

must hold. It can equally be written as

$$
\begin{aligned}
\mathrm{d}(\rho\hat{s}) = \Lambda^{\varepsilon}\Bigg\{ & \mathrm{d}(\rho\hat{\varepsilon}) \\
& + \sum_{\alpha=1}^{N-1} \Big\{(\tilde{\Lambda}^{\nu}_{\alpha} - \tilde{\Lambda}^{\nu}_{N}) + (\hat{\rho}_{\alpha}\tilde{\boldsymbol{\Lambda}}^{\boldsymbol{v}}_{\alpha}\cdot\boldsymbol{v}^{\alpha} - \hat{\rho}_{N}\tilde{\boldsymbol{\Lambda}}^{\boldsymbol{v}}_{\alpha}\cdot\boldsymbol{v}^{N})\Big\}\,\mathrm{d}\nu_{\alpha} \\
& + \sum_{\alpha=1}^{N} \rho_{\alpha}\tilde{\boldsymbol{\Lambda}}^{\boldsymbol{v}}_{\alpha}\cdot\mathrm{d}\boldsymbol{v}^{\alpha}\Bigg\}\,,
\end{aligned}
\tag{7.6.28}
$$

in which

$$
\tilde{\Lambda}^{\nu}_{\alpha} = \frac{\Lambda^{\nu}_{\alpha}}{\Lambda^{\varepsilon}}, \quad \tilde{\boldsymbol{\Lambda}}^{\boldsymbol{v}}_{\alpha} = \frac{\boldsymbol{\Lambda}^{\boldsymbol{v}}_{\alpha}}{\Lambda^{\varepsilon}} \quad (\alpha = 1, 2, ...N)\,. \tag{7.6.29}
$$

The total differential of the entropy is thus given by the total differential of the internal energy plus additional incremental "corrections" multiplied with Λ^{ε}. These additional terms are necessary to allow the differential $\mathrm{d}(\rho\hat{s})$ to become total. If one uses also the representation

$$
\begin{aligned}
\mathrm{d}(\rho\hat{\varepsilon}) &= \mathrm{d}(\rho\hat{\varepsilon}_{I}) + \mathrm{d}\left(\sum_{\alpha=1}^{N}\frac{\rho_{\alpha}}{2}\boldsymbol{u}^{\alpha}\cdot\boldsymbol{u}^{\alpha}\right) \\
&= \mathrm{d}(\rho\hat{\varepsilon}_{I}) + \sum_{\alpha=1}^{N}(\rho_{\alpha}\boldsymbol{u}^{\alpha}\cdot\mathrm{d}\boldsymbol{u}^{\alpha} + \boldsymbol{u}^{\alpha}\cdot\boldsymbol{u}^{\alpha}\mathrm{d}\rho_{\alpha}) \\
&= \mathrm{d}(\rho\varepsilon_{I}) + \sum_{\alpha=1}^{N}\rho_{\alpha}\boldsymbol{u}^{\alpha}\cdot\mathrm{d}\boldsymbol{v}^{\alpha} + \sum_{\alpha=1}^{N}\hat{\rho}_{\alpha}\boldsymbol{u}^{\alpha}\cdot\boldsymbol{u}^{\alpha}\mathrm{d}\nu_{\alpha}\,,
\end{aligned}
\tag{7.6.30}
$$

then one may write (7.6.28) also in the form

$$
\begin{aligned}
\mathrm{d}(\rho\hat{s}) = \Lambda^{\varepsilon} \Bigg\{ \mathrm{d}(\rho\hat{\varepsilon}_I) + \sum_{\alpha=1}^{N-1} \Big\{ (\tilde{\Lambda}^{\nu}_{\alpha} - \tilde{\Lambda}^{\nu}_{N}) + \hat{\rho}_\alpha (\tilde{\boldsymbol{\Lambda}}^{\boldsymbol{v}}_{\alpha} \cdot \boldsymbol{v}^{\alpha} + \boldsymbol{u}^{\alpha} \cdot \boldsymbol{u}^{\alpha}) \\
- \hat{\rho}_N (\tilde{\Lambda}^{\boldsymbol{v}}_{N} \cdot \boldsymbol{v}^{N} + \boldsymbol{u}^{N} \cdot \boldsymbol{u}^{N}) \Big\} \mathrm{d}\nu_\alpha + \sum_{\alpha=1}^{N} \rho_\alpha (\tilde{\boldsymbol{\Lambda}}^{\boldsymbol{v}}_{\alpha} + \boldsymbol{u}^{\alpha}) \cdot \mathrm{d}\boldsymbol{u}^{\alpha} \Bigg\} .
\end{aligned}
\tag{7.6.31}
$$

This form of the generalized GIBBS relation suggests that for reasons of simplicity one should take $\tilde{\boldsymbol{\Lambda}}^{\boldsymbol{v}}_{\alpha} = -\boldsymbol{u}^{\alpha}$, a choice which we shall make lateron. This choice is also supported by the identity $(7.6.26)_5$ which after summation over all constituents leads to the equation

$$
\begin{aligned}
\sum_{\alpha=1}^{N} \rho_\alpha \boldsymbol{\Lambda}^{\boldsymbol{v}}_{\alpha} = \sum_{\alpha=1}^{N} \mathcal{P}_{\boldsymbol{v}^{\alpha}} = \sum_{\alpha=1}^{N} \left\{ \frac{\partial(\rho\hat{s})}{\partial \boldsymbol{v}^{\alpha}} - \Lambda^{\varepsilon} \frac{\partial(\rho\hat{\varepsilon})}{\partial \boldsymbol{v}^{\alpha}} \right\} \\
\overset{(7.6.14)_6}{=} \sum_{\alpha=1}^{N} \left(\frac{\partial(\rho\hat{s})}{\partial \boldsymbol{u}^{\alpha}} - \Lambda^{\varepsilon} \frac{\partial(\rho\hat{\varepsilon})}{\partial \boldsymbol{u}^{\alpha}} \right) - \underbrace{\left(\sum_{\alpha=1}^{N} \xi^{\alpha} \right)}_{=1} \sum_{\gamma=1}^{N} \left(\frac{\partial(\rho\hat{s})}{\partial \boldsymbol{u}^{\gamma}} - \Lambda^{\varepsilon} \frac{\partial(\rho\hat{\varepsilon})}{\partial \boldsymbol{u}^{\gamma}} \right) = \boldsymbol{0} ,
\end{aligned}
\tag{7.6.32}
$$

which is automatically and identically satisfied if $\boldsymbol{\Lambda}^{\boldsymbol{v}}_{\alpha} = -\Lambda^{\varepsilon} \boldsymbol{u}^{\alpha}$ is chosen.

7.6.3 Some Properties of Differential (PFAFFian) Forms

The generalized GIBBS equation (7.6.27), (7.6.28) or (7.6.31) is mathematically of the form of a PFAFF*ian differential equation*

$$
\mathrm{d}F = \sum_{i=1}^{N} X_i(x_j)\mathrm{d}x_i = \boldsymbol{X}(\boldsymbol{x}) \cdot \mathrm{d}\boldsymbol{x} , \tag{7.6.33}
$$

in which the coefficient functions X_i depend continuously and differentiably upon the arguments x_j of the so-called phase space. If one writes equation (7.6.33) in its homogeneous form $\mathrm{d}F = 0$, then it is evident that the functions X_i define a normal field that stays perpendicularly on the hypersurface[19], on which the value of F does not change. Solutions of the equation $\mathrm{d}F = 0$ are hypersurfaces on which the value of F equals a constant.

Locally, such a surface can always be determined; however, it may perhaps not possess the maximum possible dimension $(N-1)$; if it does, one says that the equation is *completely integrable*. Beginning at a particular point in phase space, it is in this case possible to construct a surface (this is the $(N-1)$ dimensional hypersurface) within which every arbitrary integration of the right-hand side of (7.6.33) furnishes the zero value. In this way one succeeds to cover the entire phase space with "shells" on which $\mathrm{d}F = 0$, and which do not mutually intersect nor touch each other. It is now possible to assign to

[19] In an n-dimensional space, manifolds with the dimension $n-1$ or smaller are called hypersurfaces.

each "shell" a value – the so-called potential value – ; furthermore, one can see to it that an integration of the differential between the different shells always delivers the difference of these potential values, independent upon, where between these shells this integration is performed. This procedure accounts for the fact that the "distance" of two neighbouring shells depends on the location within the phase space. However, once the shells are constructed and once adequate potential values are assigned to the shells, then these define uniquely a function, the *integrating factor*, with which the right-hand side of (7.6.33) must be multiplied to achieve the correct "distance" between the equipotential surfaces and to construct the desired connection between the differential $\mathrm{d}F$ and the potential, provided it is not already a priori given.

This construction indicates that there exists a certain arbitrariness in assigning potential values to the shells. However, except for this freedom, it is nevertheless possible in this way, to construct for each vector field $\boldsymbol{X}(\boldsymbol{x})$, that allows at every point in phase space to locally find an equipotential surface, an integrating function with the aid of which the vector field can be globally derived from a potential. The differential $\mathrm{d}F = \boldsymbol{X} \cdot \mathrm{d}\boldsymbol{x}$ becomes total or complete and, consequently, integrals between two fixed points in phase space assume a unique single value irrespective of the chosen path along which the integration is performed. In this sense the PFAFFian form is then called *completely integrable.* The question, how a differential expression of the form (7.6.33) must look like that one can decide whether it is completely integrable or may be put into completely integrable form by multiplying it with a factor, is the subject of the theory of differential forms and is answered by the *Theorem of* POINCARÉ and the *Condition of* FROBENIUS. In the language of analysis these two statements read (see e.g. BOWEN & WANG [38]):

Theorem of POINCARÉ *A* PFAFF*ian form* $\mathrm{d}F = \sum_{i=1}^{n} X_i(x_j)\mathrm{d}x_i$ *is total and can be derived from a potential if and only if the coefficient functions* X_i*, after a further differentiation, are crosswise equal, viz.,*

$$\frac{\partial X_i}{\partial x_j} = \frac{\partial X_j}{\partial x_i}, \quad (i, j = 1, 2, ..., n) . \tag{7.6.34}$$

■

The reader may recall from analysis of functions of several variables that for a differentiable function $F(x_i)$ the order of differentiation is irrelevant,

$$\frac{\partial^2 F}{\partial x_i \partial x_j} = \frac{\partial^2 F}{\partial x_j \partial x_i} \qquad (i, j = 1, 2, ..., n) . \tag{7.6.35}$$

Conversely, if a twice differentiable function is constructed from its first derivatives $\partial F/\partial x_i = X_i(x_i)$, it is only uniquely determined, if (7.6.35) is fulfilled i. e., if (7.6.34) holds.

If the prerequisites of the Theorem of POINCARÉ are not fulfilled for a differential form $\mathrm{d}F = X_i \mathrm{d}x_i$, one may ask whether such a form can be reached by multiplying this form by an integrating factor. The answer to this question may be given by the

Condition of FROBENIUS *Let $dF = X_i(x_j)dx_i$ be a differential form which does not satisfy the conditions (7.6.34) of the Theorem of* POINCARÉ. *Then this differential can be made a total or complete differential by multiplying it with an integrating factor, if and only if*

$$\sum_{i,j,k} \epsilon_{ijk} \left(\frac{\partial X_i}{\partial x_j} \right) X_k = 0 , \tag{7.6.36}$$

holds, in which the sum is taken over all possible combinations of the indices i, j, k and where ϵ_{ijk} denotes the completely antisymmetric tensor of rank three. ■

Notice that in (7.6.36) the order of the indices (i, j, k) is irrelevant. We urge the reader to familiarise himself with the POINCARÉ theorem and the condition of FROBENIUS by solving Exercises 10 and 11. In particular, with respect to the complexity of the results the dimensionality of the phase space is significant: In two dimensions a differential can always be made total, in higher dimensions this may or may not be possible.

7.6.4 The Differential of the Entropy

The ultimate goal in dealing with differential forms in the last section was to identify properties of the differential of the entropy as presented for instance in (7.6.27), (7.6.28) and (7.6.31). If one accounts in (7.6.27) for the fact that, because of the saturation condition, one has $\mathrm{d}\nu_N = -\sum_{\alpha=1}^{N-1} \mathrm{d}\nu_\alpha$, then (7.6.27) may equally be written as

$$\mathrm{d}(\rho s) = \Lambda^\varepsilon \mathrm{d}(\rho\varepsilon) + \sum_{\alpha=1}^{N} (\Lambda^\nu_\alpha \mathrm{d}\nu_\alpha + \boldsymbol{\Lambda}^{\boldsymbol{v}}_\alpha \cdot \mathrm{d}(\rho_\alpha \boldsymbol{v}^\alpha)) , \tag{7.6.37}$$

where now $\nu_1, ..., \nu_N$ must be regarded as the independent variables and not $\nu_1, ..., \nu_{N-1}$. The differential for the entropy, (7.6.37) is very simple; it identifies on the right-hand side in a particularly transparent way the variables upon which the entropy depends. These are the specific energy $\rho\varepsilon$, the specific volume fractions ν_α (or the specific mass fractions $(\hat{\rho}_\alpha \nu_\alpha)$ which would only change the LAGRANGE multiplier $\Lambda^\nu_\alpha \to \Lambda^\nu_\alpha / \hat{\rho}_\alpha$) and the specific partial momenta, according to the balance laws of energy, masses and momenta. A dependence on the chosen constitutive variables only arises if the free energy is assumed as a function of such variables. One could for instance easily dispense with the dependence of the internal energy upon the empirical temperature and introduce the internal energy as a measure of the coldness of the body as it is essentially customary in the kinetic theory. On the other hand, the constitutive variables (e. g. (7.6.10)) enter the differential for the internal energy only implicitly. Consequently, the entropy depends in fact only upon the variables which are contained in the quantities ν_α, $(\rho_\alpha \boldsymbol{v}^\alpha)$ and $(\rho\varepsilon)$.

With regard to the LAGRANGE multipliers Λ^ε, $\boldsymbol{\Lambda}^{\boldsymbol{v}}_\alpha$ and Λ^ν_α, we may state that they cannot be constrained any further by only using the GIBBS relation, since according to it, they are anyhow only determined to within an unspecified function of the entropy (see Exercise 12). A set of LAGRANGE multipliers, which makes the differential for the entropy complete, can always be multiplied with a suitable, but arbitrary differentiable function of the entropy without loosing the property of a complete differential. Thus, there follows the important conclusion: *With the tools of the integrability conditions it is not possible to reduce the constitutive dependence of the* LAGRANGE *multipliers without a simultaneous reduction of the constitutive dependence of the entropy and the internal energy.*

With the above considerations it was not possible to decide whether the differential (7.6.37) is complete or not. Only if we succeed in demonstrating this property, the entropy will serve as a thermodynamic potential. And even, if this should be successful, a unique definition of the entropy is mathematically not possible, since there exists an arbitrary number of factors with which (7.6.37) may be multiplied, so that the resulting differential is complete. If one requires that the right-hand side of (7.6.37) with the given Λ^ε, Λ^ν_α and $\boldsymbol{\Lambda}^{\boldsymbol{v}}_\alpha$ satisfies the FROBENIUS condition, then one guarantees thereby that the differential can be made complete at least after multiplication with another multiplier Λ^T. This last multiplication corresponds to the replacement $\Lambda^\varepsilon \leftrightarrow \Lambda^T\Lambda^\varepsilon$, $\Lambda^\nu_\alpha \leftrightarrow \Lambda^T\Lambda^v_\alpha$ and $\boldsymbol{\Lambda}^{\boldsymbol{v}}_\alpha \leftrightarrow \Lambda^T\boldsymbol{\Lambda}^{\boldsymbol{v}}_\alpha$, so that one now has

$$
\mathrm{d}(\rho s) = \Lambda^T\left\{\Lambda^\varepsilon \mathrm{d}(\rho\varepsilon) + \sum_{\alpha=1}^{N}(\Lambda^\nu_\alpha \mathrm{d}\nu_\alpha + \boldsymbol{\Lambda}^{\boldsymbol{v}}_\alpha \mathrm{d}(\rho\boldsymbol{v}^\alpha))\right\}, \tag{7.6.38}
$$

in other words, Λ^T could simply be absorbed into the other LAGRANGE multipliers. The FROBENIUS condition (7.6.36) now requires for instance that the $\Lambda's$ must satisfy the identities

$$
\begin{aligned}
&\left(\frac{\partial\Lambda^\varepsilon}{\partial\nu_\alpha} - \frac{\partial\Lambda^\nu_\alpha}{\partial(\rho\varepsilon)}\right)(\boldsymbol{\Lambda}^{\boldsymbol{v}}_\beta)_i + \left(\frac{\partial\Lambda^\nu_\alpha}{\partial(\rho v^\beta)_i} - \frac{\partial\Lambda^{\boldsymbol{v}}_{\beta_i}}{\partial\nu_\alpha}\right)\Lambda^\varepsilon \\
&+\left(\frac{\partial\Lambda^{\boldsymbol{v}}_{\beta_i}}{\partial(\rho\varepsilon)} - \frac{\partial\Lambda^\varepsilon}{\partial(\rho v^\beta)_i}\right)\Lambda^\nu_\alpha = 0\,.
\end{aligned} \tag{7.6.39}
$$

This is only one of many such relations, which can be obtained by varying $\alpha, \beta(=1,2,...,N)$ and $i(=1,2,3)$. Once all equations of the form (7.6.39) are satisfied, the POINCARÉ theorem must still be fulfilled; this yields equations of the form

$$
\frac{\partial(\Lambda^T\Lambda^\varepsilon)}{\partial\nu_\alpha} = \frac{\partial(\Lambda^T\Lambda^\varepsilon_\alpha)}{\partial(\rho\varepsilon)} \tag{7.6.40}
$$

or

$$\Lambda^T \left(\frac{\partial \Lambda^\varepsilon}{\partial \nu_\alpha} - \frac{\partial (\Lambda^\nu_\alpha)}{\partial (\rho\varepsilon)} \right) = \Lambda^\nu_\alpha \frac{\partial \Lambda^T}{\partial (\rho\varepsilon)} - \Lambda^\varepsilon \frac{\partial \Lambda^T}{\partial \nu_\alpha} \,, \quad \text{etc.}$$

All these equations can easily be satisfied, namely by choosing constant Λ's. Despite its triviality, this result is helpful, because it demonstrates the existence of *one* single potential (and therefore many others); otherwise it does not have any physical significance.

The above analysis shows that *pure mathematical considerations have not led to any constraints for the* LAGRANGE *parameters. To determine or to constrain their dependence one must rely upon assumptions and physical principles.* Such rules may be material objectivity, considerations of symmetry and other postulates as e. g. the expression of extremality of the entropy production.

In the above analysis it should have become apparent that the functional dependence of the LAGRANGE multipliers is dictated by that of the internal energy, so that functional restrictions of the Λ's can be obtained by restricting the functional dependence of the internal energy and vice versa. If one starts with the independent constitutive variables (7.6.10) and writes $\rho\varepsilon = \rho\hat{\varepsilon}(\boldsymbol{\Xi})$, then the necessary requirements are obtained by satisfying the Theorem of POINCARÉ; the identities which follow from $(7.6.26)_{1-6}$ after simple but somewhat lengthy manipulations are

$$\frac{\partial \mathcal{P}_\theta}{\partial \nu_\alpha} \equiv \frac{\partial \mathcal{P}_{\nu_\alpha}}{\partial \mathcal{P}_\theta} - \frac{\partial}{\partial \theta} \left\{ (\Lambda^\nu_\alpha - \Lambda^\nu_N) + (\hat{\rho}_\alpha \boldsymbol{\Lambda}^{\boldsymbol{v}}_\alpha \cdot \boldsymbol{v}^\alpha - \hat{\rho}_N \boldsymbol{\Lambda}^{\boldsymbol{v}}_N \cdot \boldsymbol{v}^N) \right\} \,, \tag{7.6.41}$$

$$\frac{\partial \mathcal{P}_\theta}{\partial x_i} \equiv \frac{\partial \mathcal{P}_{x_i}}{\partial \theta} \,, \qquad \frac{\partial \mathcal{P}_{x_i}}{\partial x_j} \equiv \frac{\partial \mathcal{P}_{x_j}}{\partial x_i} \,, \tag{7.6.42}$$

$$\frac{\partial \mathcal{P}_{x_i}}{\partial \boldsymbol{v}^\beta} \equiv \frac{\partial \mathcal{P}^\beta_{\boldsymbol{v}}}{\partial x_i} - \frac{\partial}{\partial x_i} (\rho_\beta \boldsymbol{\Lambda}^{\boldsymbol{v}}_\beta) \,, \tag{7.6.43}$$

$$\frac{\partial \mathcal{P}_{\nu_\alpha}}{\partial x_i} - \frac{\partial}{\partial x_i} \left\{ (\Lambda^\nu_\alpha - \Lambda^\nu_N) + (\hat{\rho}_\alpha \boldsymbol{\Lambda}^{\boldsymbol{v}}_\alpha \cdot \boldsymbol{v}^\alpha - \Lambda_N \boldsymbol{\Lambda}^{\boldsymbol{v}}_N \cdot \boldsymbol{v}^N) \right\} \equiv \frac{\partial \mathcal{P}_{x_i}}{\partial \nu_\alpha} \,, \tag{7.6.44}$$

$$\begin{aligned} &\frac{\partial \mathcal{P}_\theta}{\partial \boldsymbol{v}^\beta} \equiv \frac{\partial \mathcal{P}_{\boldsymbol{v}^\beta}}{\partial \theta} - \rho_\beta \frac{\partial \boldsymbol{\Lambda}^{\boldsymbol{v}}_\beta}{\partial \theta} \,, \\ &\frac{\partial \mathcal{P}_{\nu_\alpha}}{\partial \boldsymbol{v}^\beta} - \frac{\partial}{\partial \boldsymbol{v}^\beta} \Big((\Lambda^\nu_\alpha - \Lambda^\nu_N) + (\hat{\rho}_\alpha \boldsymbol{\Lambda}^{\boldsymbol{v}}_\alpha \cdot \boldsymbol{v}^\alpha - \hat{\rho}_N \boldsymbol{\Lambda}^{\boldsymbol{v}}_N \cdot \boldsymbol{u}^N) \Big) \\ &\qquad \equiv \frac{\partial \mathcal{P}^\beta_{\boldsymbol{v}}}{\partial \nu^\beta} - \frac{\partial}{\partial \nu_\alpha} (\rho_\beta \boldsymbol{\Lambda}^{\boldsymbol{v}}_\beta) \,, \end{aligned} \tag{7.6.45}$$

$$\frac{\partial \mathcal{P}_{\nu_\alpha}}{\partial \nu_\gamma} - \frac{\partial}{\partial \nu_\gamma}\Big((\Lambda^\nu_\alpha - \Lambda^\nu_N) + (\hat{\rho}_\alpha \boldsymbol{\Lambda}^{\boldsymbol{v}}_\alpha \cdot \boldsymbol{v}^\alpha - \hat{\rho}_N \boldsymbol{\Lambda}^{\boldsymbol{v}}_N \cdot \boldsymbol{v}^N) \Big)$$

$$\equiv \frac{\partial \mathcal{P}_{\nu_\gamma}}{\partial \nu_\alpha} - \frac{\partial}{\partial \nu_\alpha}\Big((\Lambda^\nu_\gamma - \Lambda^\nu_N) + (\hat{\rho}_\gamma \boldsymbol{\Lambda}^{\boldsymbol{v}}_\gamma \cdot \boldsymbol{v}^\gamma - \hat{\rho}_N \boldsymbol{\Lambda}^{\boldsymbol{v}}_N \cdot \boldsymbol{v}^N) \Big) , \quad (7.6.46)$$

$$\frac{\partial \mathcal{P}_{\boldsymbol{v}^\beta}}{\partial \boldsymbol{v}^\delta} - \frac{\partial}{\partial \boldsymbol{v}^\delta}\left(\rho_\beta \boldsymbol{\Lambda}^{\boldsymbol{v}}_\beta \right) \equiv \frac{\partial \mathcal{P}^\delta_{\boldsymbol{v}}}{\partial \boldsymbol{v}^\beta} - \frac{\partial}{\partial \boldsymbol{v}^\beta}\left(\rho^\delta \boldsymbol{\Lambda}^{\boldsymbol{v}}_\delta \right) ,$$

in which $(x_i, x_j) \in \{\operatorname{grad}\theta, \operatorname{grad}\nu_\alpha, \operatorname{grad}\boldsymbol{v}^\beta\}$ and $(\alpha,\gamma) = 1,2,...,N-1$ and $(\beta,\delta) = 1,2,...,N$. These identities pose a considerable number of conditions which constrain the functional dependence of the free energy and the LAGRANGE multipliersLagrange. We leave it as an exercise to the reader to corroborate the statements expressed in the following propositions.

Proposition 7.1 *Assume – in explicit violation of the rule of equipresence – that the internal energy does not depend on the variables* $\operatorname{grad}\theta, \operatorname{grad}\nu_\alpha$ *and* $\operatorname{grad}\boldsymbol{v}^\beta$, *viz.,*

$$\rho\varepsilon = \rho\hat{\varepsilon}(\theta, \nu_\alpha, \boldsymbol{v}^\beta) ; \quad (7.6.47)$$

then the LAGRANGE *multipliers take the form*

$$\begin{aligned} \Lambda^\varepsilon &= \hat{\Lambda}^\varepsilon(\theta, \nu_\alpha, \boldsymbol{v}^\beta) , \\ \lambda^\nu_\alpha &:= (\Lambda^\nu_\alpha - \Lambda^\nu_N) = \hat{\lambda}^\nu_\alpha(\theta, \nu_\alpha, \boldsymbol{v}^\beta) , \\ \boldsymbol{\Lambda}^{\boldsymbol{v}}_\beta &= \hat{\boldsymbol{\Lambda}}^{\boldsymbol{v}}_\beta(\theta, \nu_\alpha, \boldsymbol{v}^\beta) . \end{aligned} \quad (7.6.48)$$

In other words, if the internal energy does not depend on $\operatorname{grad}\theta$, $\operatorname{grad}\nu_\alpha$ *and* $\operatorname{grad}\boldsymbol{v}^\beta$ *then the* LAGRANGE *multipliers can neither be functions of these variables.* ■

Proposition 7.2 *Conversely, if* Λ^ε *does not depend on* $\operatorname{grad}\theta$, $\operatorname{grad}\nu_\alpha$ *and* $\operatorname{grad}\boldsymbol{v}^\beta$ *then* $\rho\varepsilon, \boldsymbol{\Lambda}^{\boldsymbol{v}}_\beta$ *and* λ^ν_α *can neither depend on them*[20]. ■

Proposition 7.3 *If the internal energy depends on* $\operatorname{grad}\theta$, $\operatorname{grad}\nu_\alpha$ *and* $\operatorname{grad}\boldsymbol{v}^\beta$, *then also* Λ^ε *must depend on these variables and vice versa.* ■

Proof: We shall only present a sketch of a proof, which reads as follows: With the definitions $(7.6.19)_1$ of $\mathcal{P}_{x_i}$ one may deduce from $(7.6.42)_1$, that

$$\frac{\partial \hat{\Lambda}^\varepsilon}{\partial \theta}\frac{\partial(\rho\hat{\varepsilon})}{\partial x_i} - \frac{\partial \hat{\Lambda}^\varepsilon}{\partial x_i}\frac{\partial(\rho\hat{\varepsilon})}{\partial \theta} \equiv 0 , \quad (7.6.49)$$

[20] Of the LAGRANGE multipliers Λ^ν_α only the difference $\lambda^\nu_\alpha := \Lambda^\nu_\alpha - \lambda^\nu_N$ can be determined. λ^ν_N remains undetermined and forms an independent variable of the theory.

which, provided $\hat{\Lambda}^\varepsilon$ and $\rho\hat{\varepsilon}$ are nontrivial functions[21] of θ, necessarily leads to the statement

$$\frac{\partial(\rho\hat{\varepsilon})}{\partial x_i} = 0 \qquad \Leftrightarrow \qquad \frac{\partial \hat{\Lambda}^\varepsilon}{\partial x_i} = 0\,. \tag{7.6.50}$$

Therefore, the assumption that $\rho\hat{\varepsilon}$ is not a function of x_i necessarily leads to the result that neither $\hat{\Lambda}^\varepsilon$ can depend on x_i and vice versa. With (7.6.49) the identities $(7.6.42)_2$ are now trivially satisfied, and (7.6.43) takes the form

$$\frac{\partial}{\partial x_i}\left(\rho_\beta \hat{\boldsymbol{\Lambda}}^{\boldsymbol{v}}_\beta\right) = \mathbf{0}\,, \tag{7.6.51}$$

hence, neither $\hat{\boldsymbol{\Lambda}}^{\boldsymbol{v}}_\beta$ can be functions of x_i. Substitution of the results (7.6.49) and (7.6.50) in (7.6.44) then yields

$$\frac{\partial}{\partial x_i}\left(\hat{\lambda}^\nu_\alpha\right) = 0\,, \tag{7.6.52}$$

which completes the proof of proposition 7.1.

To prove proposition 7.2 one starts from (7.6.49) and assumes that $\hat{\Lambda}^\varepsilon$ is independent of x_i and then deduces from (7.6.49) that $\rho\hat{\varepsilon}$ can neither depend on x_i, which then leads to statement (7.6.51). Finally, proposition 7.3 is clear, since no simplification emerges from (7.6.49) if either a dependence on x_i is assumed for the internal energy or the LAGRANGE multiplier Λ^ε.

From computations in Chap. 5 and from Sect. 7.5.1 of this Chapter it is clear that one would wish $\hat{\Lambda}^\varepsilon$ to be a function of the empirical temperature alone whose inverse may be identifiable with the absolute temperature. This interpretation is supported if the HELMHOLTZ free energy

$$\psi := \varepsilon - \frac{s}{\Lambda^\varepsilon} \tag{7.6.53}$$

is introduced, and this energy is split according to

$$\rho\psi = \rho\psi_I + \tfrac{1}{2}\sum_{\alpha=1}^{N} \rho_\alpha \boldsymbol{u}^\alpha \cdot \boldsymbol{u}^\alpha \tag{7.6.54}$$

into an inner contribution ψ_I, the *inner* HELMHOLTZ *free energy*, and a diffusive part. With (7.6.53) and (7.6.54) the identities $(7.6.26)_{3,5}$ become

$$\begin{aligned}
&-\frac{\partial\rho\psi_I}{\partial \boldsymbol{v}^\alpha} - \frac{1}{(\Lambda^\varepsilon)^2}\frac{\partial\Lambda^\varepsilon}{\partial \boldsymbol{v}^\alpha}\rho s = -\rho_\alpha\left(\frac{\boldsymbol{\Lambda}^{\boldsymbol{v}}_\alpha}{\Lambda^\varepsilon} + \boldsymbol{u}^\alpha\right)\,,\\
&-\frac{\partial\rho\psi_I}{\partial\nu_\alpha} + \frac{\partial\rho\psi_I}{\partial\nu_N} + \frac{1}{(\Lambda^\varepsilon)^2}\left(\frac{\partial\Lambda^\varepsilon}{\partial\nu_\alpha} - \frac{\partial\Lambda^\varepsilon}{\partial\nu_N}\right)\rho s\\
&\qquad = \frac{\lambda^\nu_\alpha}{\Lambda^\varepsilon} + \left(\hat{\rho}_\alpha \boldsymbol{\Lambda}^{\boldsymbol{v}}_\alpha\cdot\boldsymbol{v}^\alpha - \hat{\rho}_N\boldsymbol{\Lambda}^{\boldsymbol{v}}_N\cdot\boldsymbol{v}^N\right)\,,
\end{aligned} \tag{7.6.55}$$

[21] This is a very natural assumption, because earlier experience has shown us that $1/\Lambda^\varepsilon$ has the meaning of absolute temperature.

in the derivation of which also $(7.6.14)_6$ was used. These formulas are massively simplified, if one requests that Λ^ε is merely a function of the temperature,

$$\Lambda^\varepsilon = \Lambda^\varepsilon(\theta) \, . \tag{7.6.56}$$

If, furthermore, it is also requested that

$$\boldsymbol{\Lambda}^{\boldsymbol{v}}_\alpha = -\Lambda^\varepsilon(\theta)\boldsymbol{u}^\alpha \, , \tag{7.6.57}$$

one recognizes that the inner part ψ_I of the HELMHOLTZ free energy cannot be a function of $\boldsymbol{v}^\alpha$ $(\alpha = 1, 2, ..., N)$. The same conclusion could also have been drawn from (7.6.31). Finally, it follows with (7.6.56) and (7.6.57) from $(7.6.55)_2$, that the LAGRANGE multipliers λ^ν_α are given by

$$\begin{aligned} \lambda^\nu_\alpha := \lambda^\nu_\alpha - \lambda^\nu_N = -\Lambda^\varepsilon(\theta)\Big\{ \Big(\frac{\partial \rho\psi_I}{\partial \nu_\alpha} - \frac{\partial \rho\psi_I}{\partial \nu_N} \Big) \\ +(\hat{\rho}_\alpha \boldsymbol{u}^\alpha \cdot \boldsymbol{v}^\alpha - \hat{\rho}_N \boldsymbol{u}^N \cdot \boldsymbol{v}^N) \Big\} \quad (\alpha = 1, 2, ..., N-1) \, . \end{aligned} \tag{7.6.58}$$

With this result and with $\Lambda^\varepsilon = \hat{\Lambda}^\varepsilon(\theta)$ all LAGRANGE multipliers are determined except λ^ν_N. Conversely, with the postulated or established results (7.6.56), (7.6.57) and (7.6.58), it is straightforward to demonstrate that the remaining integrability conditions (7.6.41),(7.6.45) and (7.6.46) are identically satisfied.

As the last assumption we will now relate $\Lambda^\varepsilon(\theta)$ with the absolute temperature and identify it with the KELVIN temperature

$$\Lambda^\varepsilon(\theta) = \frac{1}{T} \, . \tag{7.6.59}$$

With this choice one obtains from $(7.6.26)_1$

$$s = -\frac{\partial \psi_I}{\partial T} \, , \quad \rightarrow \quad s = \hat{s}(T, \nu_\alpha) \, , \tag{7.6.60}$$

whence the classical relationship, one would have expected.

The above analysis shows that all restrictions on the constitutive relations of ε, s and ψ are at last tied to two basic assumptions, *first the hypothesis that the* LAGRANGE *multiplier for the internal energy,* Λ^ε, *is only a function of the empirical temperature and second, that the inner part of the* HELMHOLTZ *free energy* ψ_I *is not a function of the velocities.* With these two assumptions all LAGRANGEian multipliers could be determined with the exception of Λ^ν_N, which will play the role of an independent variable and which represents the constraint variable for the saturation condition. Since, moreover, we succeeded with the help of (7.6.56), (7.6.57) and (7.6.58) to satisfy all conditions of the POINCARÉ theorem, the differential of the entropy, as given by the GIBBS relation is complete and the entropy therefore a thermodynamic potential. Alternatively, one also knows (Exercise 12) that one may multiply the right-hand side of the GIBBS relation with an arbitrary differentiable function $\Lambda^T(\rho s)$ and that this new differential may also represent a possible entropy function. *We now* wish to *declare that only those functions be used as thermodynamic entropy function which depend merely upon the empirical temperature.* This corresponds to the selection of the LAGRANGE multiplier $\Lambda^\varepsilon(\theta)$, its identification with the coldness function and its inverse with the absolute temperature according to (7.6.59).

Of the identities (7.6.26) there still remain the identities $(7.6.26)_{7-9}$, which concern the entropy flux. If one introduces the extra entropy flux vector

$$\boldsymbol{k} := \Lambda^\varepsilon \boldsymbol{q} - \sum_{\alpha=1}^{N} \boldsymbol{\Lambda}^{\boldsymbol{v}}_{\alpha} \boldsymbol{t}^\alpha - \boldsymbol{\phi}^{\rho s} \,, \tag{7.6.61}$$

then the identities $(7.6.26)_{7-9}$ take the forms

$$\begin{aligned} &\left(\frac{\partial \boldsymbol{k}}{\partial \operatorname{grad} T}\right)_{\mathrm{sym}} = \boldsymbol{0} \,, \\ &\left(\frac{\partial \boldsymbol{k}}{\partial \operatorname{grad} \nu_\alpha} - \frac{\partial \boldsymbol{k}}{\partial \operatorname{grad} \nu_N}\right)_{\mathrm{sym}} = \boldsymbol{0} \,, \\ &\left(\frac{\partial \boldsymbol{k}}{\partial \operatorname{grad} \boldsymbol{v}^\beta}\right)_{\mathrm{sym}(1,3)} = \boldsymbol{0} \,, \end{aligned} \tag{7.6.62}$$

in which use has been made of (7.6.56). In simpler theories one can start from isotropic representations for $\boldsymbol{q}$ and $\boldsymbol{\phi}^{\rho s}$ and *prove* with such representations from $(7.6.26)_{7-9}$ that Λ^ε can only be a function of θ. The identities $(7.6.26)_{7-9}$ therefore provide for such materials restrictions to the LAGRANGE multiplier Λ^ε, which were introduced above as an assumption. For the present constitutive class the corresponding analysis is so complex that we were unable to deduce the result $\Lambda^\varepsilon = \hat{\Lambda}^\varepsilon(\theta)$. If we introduce this now as an assumption there still remain the identities (7.6.62) to be fulfilled, which could identically be satisfied by $\boldsymbol{k} = \boldsymbol{0}$ which, however, we do not wish to select for reasons that will become apparent below.

Of inequality (7.6.22) the upper block is now zero, at least when all conditions derived above are fulfilled. There still remains the lower block, which comprises what commonly is called the *residual entropy inequality.* If the definitions (7.6.19), (7.6.21), (7.6.61) and the results (7.6.26) are substituted and if $\boldsymbol{v}^\alpha$ is replaced by $\boldsymbol{v} + \boldsymbol{u}^\alpha$ wherever possible and Λ^ε is replaced by $1/\theta$ and θ by T, then this residual entropy inequality takes the following form:

$$
\begin{aligned}
\pi^s = & -\left\{\frac{\partial \boldsymbol{k}}{\partial T}+\frac{1}{T^2}\left(\boldsymbol{q}+\sum_{\alpha=1}^{N} \boldsymbol{u}^\alpha \cdot \boldsymbol{t}^\alpha\right)\right\} \cdot \operatorname{grad} T \\
& +\sum_{\alpha=1}^{N-1}\left\{\left(\Lambda_N^\nu \boldsymbol{u}^N-\Lambda_\alpha^\nu \boldsymbol{u}^\alpha\right)+\frac{1}{T}\left(\hat{\rho}_\alpha\left(\boldsymbol{u}^\alpha \cdot \boldsymbol{v}^\alpha\right) \boldsymbol{u}^\alpha-\hat{\rho}_N\left(\boldsymbol{u}^N \cdot \boldsymbol{v}^N\right) \boldsymbol{u}^N\right)\right. \\
& -\left(\frac{\partial \boldsymbol{k}}{\partial \nu_\alpha}-\frac{\partial \boldsymbol{k}}{\partial \nu_N}\right) \\
& \left.+\frac{1}{\rho T}\left(\hat{\rho}_N \boldsymbol{u}^N-\hat{\rho}_\alpha \boldsymbol{u}^\alpha\right)\left(\sum_{\beta=1}^{N} \rho_\beta\left(\boldsymbol{u}^\beta \otimes \boldsymbol{u}^\beta\right)+\rho \psi \mathbf{1}\right)\right\} \cdot \operatorname{grad} \nu_\alpha \\
& +\sum_{\alpha=1}^{N}\left\{-\frac{\partial \boldsymbol{k}}{\partial \boldsymbol{v}^\alpha}+\frac{1}{T} \boldsymbol{t}^\alpha+\frac{\rho_\alpha}{T}\left(\boldsymbol{u}^\alpha \otimes \boldsymbol{v}^\alpha-\boldsymbol{v} \otimes \boldsymbol{u}^\alpha\right)\right. \\
& \left.-\left(\frac{\rho_\alpha}{T} \psi+\Lambda_\alpha^\nu \nu_\alpha-\frac{\rho_\alpha}{T}\left(\boldsymbol{u}^\alpha \cdot \boldsymbol{v}^\alpha\right)\right) \mathbf{1}-\frac{\rho_\alpha}{T} \sum_{\beta=1}^{N} \rho_\beta\left(\boldsymbol{u}^\beta \otimes \boldsymbol{u}^\beta\right)\right\} \cdot \operatorname{grad} \boldsymbol{v}^\alpha \\
& +\sum_{\alpha=1}^{N}\left\{\Lambda_\alpha^\nu \mathfrak{n}^\alpha-\frac{1}{T} \boldsymbol{u}^\alpha \cdot \mathfrak{m}^\alpha\right\} \geq 0 . \qquad (7.6.63)
\end{aligned}
$$

This formidably looking inequality must be identically satisfied for all constitutive equations for $\boldsymbol{k}$, $\boldsymbol{q}, \mathfrak{m}^\alpha$, $\boldsymbol{t}^\alpha$ and ψ, and thus constrains them, but because of its complexity this is done in general for a restricted class of processes only, those describing, *thermodynamic equilibrium* and processes in its neighbourhood. This will be our next task.

7.6.5 Thermodynamic Equilibrium

Thermodynamic equilibrium is a process for which the entropy production is a minimum, namely zero. According to (7.6.63) this is the case, if the constituent velocities $\boldsymbol{v}^\alpha$, their gradients, $\operatorname{grad} \boldsymbol{v}^\alpha$, the temperature gradient $\operatorname{grad} T$ and the volume exchange terms $\mathfrak{n}^\alpha$ vanish identically for all $\alpha = 1, ..., N$. Under such conditions, of all the terms in the inequality (7.6.63) only the derivatives of $\boldsymbol{k}$ with respect to ν_α remain, $(\partial \boldsymbol{k}/\partial \nu_\alpha - \partial \boldsymbol{k}/\partial \nu_N)$. To assure that these terms also vanish in equilibrium, the functional form of $\boldsymbol{k}$ must be determined. It can be shown that close to equilibrium,

$$
\boldsymbol{k} = \sum_{\alpha=1}^{N}\left(\frac{\partial \boldsymbol{k}}{\partial \boldsymbol{v}^\alpha}\right)^{\text{equil}} \boldsymbol{v}^\alpha = \sum_{\alpha=1}^{N}\left(\frac{\partial \boldsymbol{k}}{\partial \boldsymbol{u}^\alpha}\right)^{\text{equil}} \boldsymbol{u}^\alpha , \qquad (7.6.64)
$$

which indeed vanishes in equilibrium. The proof of (7.6.64) is nontrivial and relatively lengthy and makes use of the identities (7.6.62). It may also be mentioned that the first of (7.6.64) is the result of this proof when the rule of material objectivity is not enforced; the second is its objective counterpart.

The proof is given in Exercise 13. It guaranties that $\partial \boldsymbol{k}/\partial \nu_\alpha|_{\text{equil}} = \mathbf{0}$ for all $\alpha = 1, ..., N$ [22].

The left hand side of (7.6.63) represents the entropy production π^s, and we just proved above that π^s as a function of the nonequilibrium variables $\operatorname{grad} T$, $\boldsymbol{v}^\alpha$ and $\operatorname{grad} \boldsymbol{v}^\alpha$ assumes its minimum, $(\pi^s)^{\text{equil}} = 0$. Since π^s is a continuously differentiable function of these variables, of necessity then

$$\begin{aligned} &\left(\frac{\partial \pi^s}{\partial \boldsymbol{X}}\right)^{\text{equil}} = \mathbf{0}, \quad \boldsymbol{X} = \{\operatorname{grad} T, \boldsymbol{v}^\alpha, \operatorname{grad} \boldsymbol{v}^\alpha\}, \\ &\left(\frac{\partial^2 \pi^s}{\partial \boldsymbol{X}\, \partial \boldsymbol{X}}\right)^{\text{equil}}, \quad \text{is positive semidefinite.} \end{aligned} \tag{7.6.65}$$

We shall first draw the inferences implied by the first of (7.6.65).

If one evaluates $\partial \pi^s / \partial \operatorname{grad} T$ in thermodynamic equilibrium and sets the result equal to zero one obtains

$$\boldsymbol{q}^{\text{equil}} = T^2 \Lambda^\nu_\alpha \left(\frac{\partial \mathfrak{n}^\alpha}{\partial \operatorname{grad} T}\right)^{\text{equil}} = \mathbf{0}\,. \tag{7.6.66}$$

In the derivation of this result use was made of the result $(\partial \boldsymbol{k}/\partial T)^{\text{equil}} = \mathbf{0}$, which also follows from (7.6.64); in addition, in view of (7.6.48) Λ^ν_α is not a function of $\operatorname{grad} T$, so that Λ^ν_α could be pulled out of the differentiation in (7.6.66). Furthermore, as indicated above, the remaining terms on the right-hand side of (7.6.66) also vanish. To prove this, let us consider the volume exchange terms more closely. It will be assumed that the thermodynamic equilibrium under consideration is distant from any phase (change) equilibrium of the constituents. A small deviation of the temperature from this equilibrium will then not cause melting or evaporation processes of any constituent[23]. If such a situation prevails all derivatives of $\mathfrak{n}^\alpha$ as well as $\mathfrak{n}^\alpha$ itself must vanish throughout the entire domain where the thermodynamic equilibrium exists. This is obvious, since $\mathfrak{n}^\alpha$ must vanish according to (7.6.4) everywhere in the domain; so it is the zero function over a region with nonvanishing measure, implying that also its derivatives are zero. Thus we have in particular

$$\left(\frac{\partial \mathfrak{n}^\alpha}{\partial \operatorname{grad} T}\right)^{\text{equil}} = 0 \qquad \text{for } \alpha = 1, ..., N. \tag{7.6.67}$$

The mixture heat flux thus vanishes in thermodynamic equilibrium. BAUER [22] proves this to hold true also in the case when phase changes do occur.

[22] The reader may recall (7.6.7), which explains the peculiarities when differentiating functions (here $\boldsymbol{k}$) of $\nu_1, ..., \nu_{N-1}$ and $\nu_1, ..., \nu_N$, respectively.

[23] This, for instance, excludes the situation of simultaneous existence of ice and water and mass exchanges between these two constituents as it occurs e.g. in temperate ice. Solidification of rock or the mushy behaviour of the interior core of the Earth would also be examples where such phase changes occur. This more general case is treated by BAUER [22].

Next we evaluate the inferences that are implied by $\partial\pi^\alpha/\partial\boldsymbol{v}^\alpha = \mathbf{0}$. The reader may deduce the validity of the following $N-1$ relations for the equilibrium interaction forces:

$$
\begin{aligned}
(\mathbf{m}^\alpha)^{\text{equil}} = \sum_{\beta=1}^{N-1} \Bigg\{ & T\Lambda_N^\nu \left(\delta^{N\alpha} - \xi^\alpha\right) - T\Lambda_\beta^\nu \left(\delta^{\alpha\beta} - \xi^\alpha\right) \\
& - T\left(\frac{\partial}{\partial\nu_\beta} - \frac{\partial}{\partial\nu_N}\right)\left(\frac{\partial \boldsymbol{k}}{\partial \boldsymbol{v}^\alpha}\right)^{\text{equil}} \\
& - \psi\mathbf{1}\left[\hat{\rho}_\beta\left(\delta^{\beta\alpha} - \xi^\alpha\right) - \hat{\rho}_N\left(\delta^{N\alpha} - \xi^\alpha\right)\right] \Bigg\} \operatorname{grad}\nu_\beta \\
& + T\sum_{\beta=1}^{N} \Lambda_\beta^\nu \underbrace{\left(\frac{\partial \mathfrak{n}^\beta}{\partial \boldsymbol{v}^\alpha}\right)^{\text{equil}}}_{\mathbf{0}} ,
\end{aligned}
\tag{7.6.68}
$$

in the derivation of which excessive use of $(7.6.14)_4$ was made, and where $(\partial\mathfrak{n}^\beta/\partial\boldsymbol{v}^\alpha)^{\text{equil}}$ vanishes for the same reason as (7.6.67) was found to be true. The relations (7.6.68) hold for the $N-1$ first interaction forces $\mathbf{m}^\alpha$; $\mathbf{m}^N$ is obtained via the condition $\sum_{\alpha=1}^{N} \mathbf{m}^\alpha = \mathbf{0}$.

It is worth pointing out that the above expressions involve a term that depends on the extra entropy flux $\boldsymbol{k}$. It will be interesting to see how this term combines with the divergence of the stress term $\operatorname{div}\boldsymbol{t}^\alpha$.

If the term $(\partial\pi^s/\partial \operatorname{grad}\boldsymbol{v}^\alpha)^{\text{equil}} = \mathbf{0}$ is evaluated the partial equilibrium stress is obtained,

$$
\begin{aligned}
(\boldsymbol{t}^\alpha)^{\text{equil}} = & \left(\rho_\alpha\psi + T\Lambda_\alpha^\nu\nu_\alpha\right)\mathbf{1} \\
& + \sum_{\beta=1}^{N-1}\left\{ T\left(\frac{\partial}{\partial\nu_\beta} - \frac{\partial}{\partial\nu_N}\right)\left(\frac{\partial\boldsymbol{k}}{\partial \operatorname{grad}\boldsymbol{v}^\alpha}\right)^{\text{equil}}\right\}\operatorname{grad}\nu_\beta \\
& + T\left(\frac{\partial\boldsymbol{k}}{\partial\boldsymbol{v}^\alpha}\right)^{\text{equil}} - T\sum_{\beta=1}^{N}\Lambda_\beta^\nu \underbrace{\left(\frac{\partial\mathfrak{n}^\beta}{\partial \operatorname{grad}\boldsymbol{v}^\alpha}\right)^{\text{equil}}}_{\mathbf{0}} ,
\end{aligned}
\tag{7.6.69}
$$

where we have not yet accounted for the fact that the partial stress $\boldsymbol{t}^\alpha$ is symmetric. Neither are the requirements of objectivity yet inserted. Furthermore, the last term vanishes for the same reason as (7.6.67) holds true; and the term that is multiplied with $\operatorname{grad}\nu_\beta$ vanishes since close to equilibrium

$$
\frac{\partial}{\partial\nu_\beta}\left(\frac{\partial\boldsymbol{k}}{\partial \operatorname{grad}\boldsymbol{v}^\alpha}\right) = \frac{\partial^2}{\partial\nu_\beta\,\partial \operatorname{grad}\boldsymbol{v}^\alpha}\left(\sum_{\gamma=1}^{N}\left(\frac{\partial\boldsymbol{k}}{\partial\boldsymbol{v}^\gamma}\right)^{\text{equil}}\boldsymbol{v}^\gamma\right) = \mathbf{0} ,
$$

in which use was also made of (7.6.7). Therefore,

$$
(\boldsymbol{t}^\alpha)^{\text{equil}} = \left(\rho_\alpha\psi + T\Lambda_\alpha^N\nu_\alpha\right)\mathbf{1} + T\left(\frac{\partial\boldsymbol{k}}{\partial\boldsymbol{v}^\alpha}\right)^{\text{equil}} . \tag{7.6.70}
$$

The constituent stress tensor $\boldsymbol{t}^\alpha$ in thermodynamic equilibrium contains a term that depends on the extra entropy flux $\boldsymbol{k}$. In the local balance law of momentum for the constituent α the stress and the interaction force contribute with the combination $\operatorname{div} \boldsymbol{t}^\alpha + \mathbf{m}^\alpha$. It is easy to show that by means of (7.6.68), (7.6.70) that

$$k\text{–dependent term } \{\operatorname{div} \boldsymbol{t}^\alpha + \mathbf{m}^\alpha\} = \mathbf{0} \, ;$$

so, *the entropy-flux-dependent terms in the local balance of the linear momentum of the constituents in thermodynamic equilibrium cancel out. They can enter the momentum balance at most through the stress boundary conditions.*

The above three equations (7.6.66), (7.6.68) and (7.6.70) exhaust the thermodynamic equilibrium conditions as far as the first derivatives $(7.6.65)_1$ are concerned. Since $\pi^s = \pi^s(\boldsymbol{X}_E, \boldsymbol{X}_{NE})$, where $\boldsymbol{X}_E = \{T, \nu_\alpha, \operatorname{grad} \nu_\alpha\}$, $\boldsymbol{X}_{NE} = \{\operatorname{grad} T, \boldsymbol{v}_\alpha, \operatorname{grad} \boldsymbol{v}_\alpha\}$ and

$$\pi^s(\boldsymbol{X}_E, \boldsymbol{X}_{NE} = \mathbf{0}) = 0, \quad \text{for all } \boldsymbol{X}_E \, ,$$

one necessarily also has

$$\left(\frac{\partial \pi^s}{\partial \boldsymbol{X}_E}\right)^{\text{equil}} = \mathbf{0}, \; \left(\frac{\partial^2 \pi^s}{\partial \boldsymbol{X}_E \, \partial \boldsymbol{X}_E}\right)^{\text{equil}} = \mathbf{0}, \quad \text{etc.} \qquad (7.6.71)$$

or according to (7.6.63)

$$\begin{aligned}
&\left\{\sum_{\alpha=1}^N \Lambda^\nu_\alpha \frac{\partial \mathfrak{n}^\alpha}{\partial T}\right\}^{\text{equil}} = 0, \\
&\left\{\sum_{\beta=1}^N \Lambda^\nu_\beta \frac{\partial \mathfrak{n}^\beta}{\partial \nu_\alpha}\right\}^{\text{equil}} = 0, \qquad \alpha = 1, ..., N-1, \qquad (7.6.72) \\
&\left\{\sum_{\beta=1}^N \Lambda^\nu_\beta \frac{\partial \mathfrak{n}^\beta}{\partial \operatorname{grad} \nu_\alpha}\right\}^{\text{equil}} = \mathbf{0}, \quad \alpha = 1, ..., N-1.
\end{aligned}$$

Similar expressions also hold for all higher derivative expressions. They do not express any particularly interesting fact other than if a function is identically zero, so are all its derivatives.

In a theory as complicated as this one, it is generally very difficult to exploit the conditions $(7.6.65)_2$, saying that the matrix of the second derivatives is positive semidefinite. A quadratic form for the symmetric matrix $\boldsymbol{A} = \boldsymbol{A}^t$ of dimension $m \times m$ is positive semidefinite, $\boldsymbol{x} \cdot \boldsymbol{A}\boldsymbol{x} \geq 0$ for all $\boldsymbol{x}$, if and only if all its principal minors are non-negative. These principal minors are all determinants of submatrices, two corners of which are positioned on the principal diagonal as shown here:

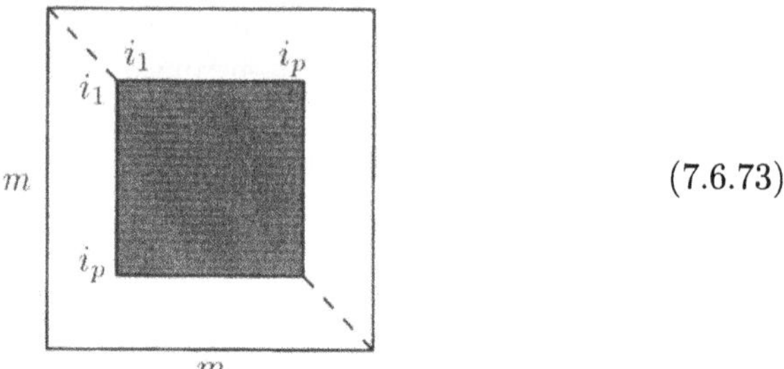

(7.6.73)

The diagonal elements are also principal minors, and they correspond in our case to the non-mixed second derivatives of π^s. We shall not derive these relations here, but only quote the results obtained by BAUER [22]:

$$\frac{\partial^2 \pi^s}{\partial(\operatorname{grad} T)^2}\Big|_{\text{equil}} \geq 0 \quad \Longrightarrow \quad -\frac{2}{T^2}\left(\frac{\partial \boldsymbol{q}}{\partial \operatorname{grad} T}\right)^{\text{equil}} \geq 0\,, \tag{7.6.74}$$

$$\frac{\partial^2 \pi^s}{\partial(\operatorname{grad} \boldsymbol{v}^\alpha)^2}\Big|_{\text{equil}} \geq 0 \quad \Longrightarrow$$
$$\left\{\frac{\partial}{\partial \operatorname{grad} \boldsymbol{v}^\alpha}\left(-T\frac{\partial \boldsymbol{k}}{\partial \boldsymbol{v}^\alpha} + \boldsymbol{t}^\alpha - (\rho\psi + T\Lambda^\nu_\alpha \nu_\alpha)\mathbf{1}\right)\right\}_{\text{sym}(12)(34)} \geq 0\,, \tag{7.6.75}$$

$$\frac{\partial^2 \pi^s}{\partial \boldsymbol{v}^{\alpha 2}}\Big|_{\text{equil}} \geq 0 \quad \Longrightarrow$$
$$\sum_{\beta=1}^{N-1}\left\{\left[2\left[\frac{\partial \Lambda^\nu_N}{\partial \boldsymbol{v}^\alpha}\left(\delta^{N\alpha} - \xi^\alpha\right) - \frac{\partial \Lambda^\nu_\beta}{\partial \boldsymbol{v}^\alpha}\left(\delta^{\beta\alpha} - \xi^\alpha\right)\right]\right.\right.$$
$$\left.\left. + \frac{2}{\rho T}\left[\hat{\rho}_N\left(\delta^{N\alpha} - \xi^\alpha\right) - \hat{\rho}_\beta\left(\delta^{\beta\alpha} - \xi^\alpha\right)\right]\frac{\partial(\rho\psi)}{\partial \boldsymbol{v}^\alpha}\right] \otimes \operatorname{grad} \nu_\beta\right\}_{\text{sym}}$$
$$-\sum_{\beta=1}^{N-1}\frac{2}{T}\left\{\left(\delta^{\beta\alpha} - \xi^\alpha\right)\left(\frac{\partial \mathbf{m}^\beta}{\partial \boldsymbol{v}^\alpha}\right)^{\text{equil}}_{\text{sym}}\right\} \geq 0\,. \tag{7.6.76}$$

Of these relations only (7.6.74) allows a direct and easy interpretation, namely,

$$\left(\frac{\partial \boldsymbol{q}}{\partial \operatorname{grad} T}\right)^{\text{equil}} \leq 0\,. \tag{7.6.77}$$

If one assumes close to thermodynamic equilibrium a linear relation between the heat flux of the mixture, $\boldsymbol{q}$, and the temperature gradient, $\operatorname{grad} T$, then $\boldsymbol{q} = -\boldsymbol{\kappa} \operatorname{grad} T$, and $\boldsymbol{\kappa}$ must be a positive semidefinite matrix. The heat flux is in this case directed towards falling temperature. For the other two relations (7.6.75) and (7.6.76) directly interpretable results have not been found.

It is advisable in most cases to await satisfaction of the positive semidefiniteness of the principal minors of $(7.6.65)_2$ until explicit constitutive relations are formulated for a concrete case. These are usually much simpler so that thermodynamic compatibility with $(7.6.65)_2$ becomes equally somewhat easier.

7.6.6 Extension to Non-Equilibrium States

As the ultimate goal is the determination of non-equilibrium states, we shall subsequently propose a first possibility to extend the constitutive relations valid in thermodynamic equilibrium to non-equilibrium. Such an extension cannot be justified in all details but must be regarded as an approximation. Furthermore, explicit expressions of non-equilibrium constitutive relations are always proposed with a certain application of the emerging theory in mind. Of course, this application is already anticipated in (7.6.10) which shows that the present model gives rise to heat conduction and viscous effects but no elasticity and may therefore be approximate to describe the viscous motion under isothermal or nonisothermal conditions. Such conditions prevail for instance in debris and mud flows of a gravel water mixture. In what follows, only the most simple extensions of the equilibrium expressions to non-equilibrium are presented.

For the *heat flux* $\boldsymbol{q}^\alpha$ of constituent α a FOURIER *type* relation

$$\boldsymbol{q}^\alpha = -c^q_\alpha \operatorname{grad} T \tag{7.6.78}$$

is suggested for which (7.3.23) yields

$$\boldsymbol{q} - -\sum_{\alpha=1}^{N} \left\{ c^q_\alpha \operatorname{grad} T + \boldsymbol{u}^\alpha \boldsymbol{t}^\alpha - \rho_\alpha \left(\varepsilon + \frac{1}{2} \boldsymbol{u}^\alpha \cdot \boldsymbol{u}^\alpha \right) \boldsymbol{u}^\alpha \right\} . \tag{7.6.79}$$

The *interaction forces* $\mathfrak{m}^\alpha$ are extended by terms which depend upon the difference velocities to the other constituents of the mixture. In so doing it must be remembered that $\mathfrak{m}^\alpha$ in non-equilibrium is not an objective vector valued variable, but $\mathfrak{m}^\alpha_{\text{Euclid}}$ defined by $(7.2.15)_1$.Thus, we may set

$$\begin{aligned} \mathfrak{m}^\alpha_{\text{Euclid}} &= \mathfrak{m}^\alpha - \rho_\alpha \mathfrak{n}^\alpha \boldsymbol{v}^\alpha = (\mathfrak{m}^\alpha)^{\text{equil}} + \sum_{\beta \neq \alpha}^{N} c^{\boldsymbol{v}}_{\alpha\beta} (\boldsymbol{v}^\beta - \boldsymbol{v}^\alpha) , \\ c^{\boldsymbol{v}}_{\alpha\beta} &= c^{\boldsymbol{v}}_{\beta\alpha} . \end{aligned} \tag{7.6.80}$$

The last term in this expression is linear in the difference velocities, if $c^{\boldsymbol{v}}_{\alpha\beta}$ are independent of $|\boldsymbol{v}^\beta \cdot \boldsymbol{v}^\alpha|$. In that form this summation term is then reminiscent of DARCY's *law* and $c^{\boldsymbol{v}}_{\alpha\beta}$ are permeabilities, but obviously, this analogy is complete only for a binary mixture. It is also natural to assume that $c^{\boldsymbol{v}}_{\alpha\beta} = c^{\boldsymbol{v}}_{\beta\alpha}$, for in that case the "DARCY-term" satisfies the condition

$$\sum_{\alpha=1}^{N} \sum_{\beta \neq \alpha}^{N} c^{\boldsymbol{v}}_{\alpha\beta}(\boldsymbol{v}^{\beta} - \boldsymbol{v}^{\alpha}) = \mathbf{0} \, ,$$

which must necessarily hold, if $\sum_{\alpha=1}^{N} \mathfrak{m}^{\alpha} = \mathbf{0}$. Furthermore the "DARCY-term" is not restricted to a linear dependence in the difference velocities. The coefficients $c^{\boldsymbol{v}}_{\alpha\beta}$ may well depend upon scalar variables such as $|\boldsymbol{v}^{\beta} - \boldsymbol{v}^{\alpha}|$. In ground water flow a quadratic dependence of the "DARCY term" is known as the FORCHHEIMER *law*; it is known to better approximate the viscous effects of the flow of water through the pore space of soil if this flow is turbulent, and it is believed that this quadratic dependence also models somewhat the tortuosity effects of the pore space. This is plausible as the *tortuosity* enhances the onset of turbulence in the flow of the interstitial fluid.

For a mixture of viscous constituents objective tensor variables with the notion of viscous behaviour and deduceable from (7.6.10) are

$$\boldsymbol{D}^{\alpha} = \mathrm{sym}\,\mathrm{grad}\,\boldsymbol{v}^{\alpha}, \quad \boldsymbol{W}^{\alpha} = \mathrm{skw}\ \mathrm{grad}\,\boldsymbol{v}^{\alpha} - \boldsymbol{W}, \quad \boldsymbol{V}^{\alpha\beta} = \boldsymbol{u}^{\alpha} \otimes \boldsymbol{u}^{\beta} \, , \tag{7.6.81}$$

where $\boldsymbol{W} = \mathrm{skw}\ \mathrm{grad}\,\boldsymbol{v}$. So, the non-equilibrium stress $\boldsymbol{t}^{\alpha}$ may be assumed in the form

$$\boldsymbol{t}^{\alpha} - (\boldsymbol{t}^{\alpha})^{\mathrm{equil}} = \boldsymbol{t}^{\alpha}_{NE}(\cdot, \boldsymbol{D}^{\alpha}, \boldsymbol{W}^{\alpha}, \boldsymbol{V}^{\alpha\beta}) \, . \tag{7.6.82}$$

The dot indicates additional dependences of (7.6.10) not explicitly stated in (7.6.82). It is thought that the dominant dependence of $\boldsymbol{t}^{\alpha}$ on the variables (7.6.81) is through $\boldsymbol{D}^{\alpha}$, and so a first "guess" may have the form

$$\boldsymbol{t}^{\alpha}_{NE} = a\mathbf{1} + b\boldsymbol{D}^{\alpha} + c(\boldsymbol{D}^{\alpha})^{2} \, , \tag{7.6.83}$$

in which a, b, c may depend on the invariants of $\boldsymbol{D}^{\alpha}$ (and other scalar variables if needed). The ansatz (7.6.83) is usually thought to be too complicated. A popular relation is a reduced version of the form

$$\boldsymbol{t}^{\alpha}_{NE} = c^{\boldsymbol{t}}_{\alpha}\left(II_{\boldsymbol{D}^{\alpha}}\right)\boldsymbol{D}^{\alpha}, \quad II_{\boldsymbol{D}^{\alpha}} = \tfrac{1}{2}\,\mathrm{tr}\left(\boldsymbol{D}^{\alpha 2}\right) \, . \tag{7.6.84}$$

If

$$c^{\boldsymbol{t}}_{\alpha} = \left(II_{\boldsymbol{D}^{\alpha}}\right)^{m} , \quad 0 < m < 1 \, , \tag{7.6.85}$$

then (7.6.84) is called a power law and the constituent body a power law material. Other denotations are GLEN's flow law (glaciology), NORTON's law (metallurgy) or OSWALT DE WAELE flow law (rheology). In a binary mixture of soil and water $c^{\boldsymbol{t}}_{s}$ for the solid is non-zero whilst $c^{\boldsymbol{t}}_{w}$ for the water is often set to zero. A somewhat more general form of (7.6.84) based on (7.6.83) is given in Exercise 14.

Apart from the above representations, a complete postulation of the constitutive relations also requires the postulation of an expression of the internal energy ε^α (it is needed in the parameterization of the heat flux vector). The most common assumption is

$$\varepsilon^\alpha = \varepsilon_0^\alpha + \int_{T_0}^{T} c_\alpha^\varepsilon(\theta)\mathrm{d}\theta = \varepsilon_0^\alpha + c_\alpha^\varepsilon(T - T_0) , \tag{7.6.86}$$

where the second formula applies if c_α^ε may be considered to be constant. Finally an expression for the inner free energy ψ_I is needed. Since $\psi_I = \hat{\psi}_I(T, \nu_\alpha)$ – a dependence on $\boldsymbol{v}^\alpha$ has been excluded – we may choose as the simplest possibility for the internal HELMHOLTZ free energy

$$\psi_I = \psi_0(\nu_\alpha) + c_\psi^1(T - T_0) + \tfrac{1}{2}c_\psi^2(T - T_0)^2 , \tag{7.6.87}$$

implying that

$$\begin{aligned} s &= -\frac{\partial \psi_I}{\partial T} = c_\psi^1 + c_\psi^2(T - T_0), \\ \rho\varepsilon_I &= \sum_{\alpha=1}^{N} \rho^\alpha \varepsilon_0^\alpha + \sum_{\alpha=1}^{N} \rho^\alpha c_\alpha^\varepsilon(T - T_0) = \rho(\psi_I - sT) , \end{aligned} \tag{7.6.88}$$

the second of which imposes restrictions on the coefficients c_α^ε, $c_\psi^{1,2}$.

This completes the formulation of the model equations in one case, namely when the constitutive relations are chosen as suggested above. The next step would now be the numerical determination of all the parameters. This step is called *parameter identification* and involves performing experiments for typical deformation fields and their optimal imitation with the corresponding boundary value problem of this theory. This is generally a similarly difficult problem as the derivation of the thermodynamic model equations itself.

One particular problem in this process is the selection of the extra entropy flux vector $\boldsymbol{k}$, which we know vanishes in equilibrium with many of its derivatives, see Exercise 13. In fact its form close to thermodynamic equilibrium is given by (7.6.64). In the local balance laws of the constituent momenta in equilibrium it has been shown that its contribution equally vanishes, and so does it in non-equilibrium with the choices of the non-equilibrium stresses $\boldsymbol{t}_{NE}^\alpha$ and the interaction forces $\boldsymbol{m}^\alpha$ as selected in (7.6.50), (7.6.83) or (7.6.84). Thus, as far as purely mechanical equations are concerned, a non-trivial $\boldsymbol{k}$ manifests itself only in the boundary conditions, and if so, in dynamical processes. All these reasons may serve as motivation to select

$$\boldsymbol{k} = \boldsymbol{0} . \tag{7.6.89}$$

This then yields, on using (7.6.61),

$$\phi^{\rho s} = \frac{1}{T}\left(\boldsymbol{q} + \sum_{\alpha=1}^{N} \boldsymbol{u}^\alpha \boldsymbol{t}^\alpha\right) , \tag{7.6.90}$$

the classical expression in mixture theory.

7.7 Exercises

1. Construct a continuum theory for the constituents of a mixture in LAGRANGEian description. Derive the general balance laws in global and local form and specialize these for the balances of mass, momentum, angular momentum, energy and entropy[24].
2. Prove the formulas (7.2.10)–(7.2.12).
3. Check if the formulas (7.3.23) are correct.
4. Starting with the mass balance (7.2.4) derive the tracer–mass–balance equation

$$\rho\frac{\mathrm{d}\xi^\alpha}{\mathrm{d}t} = -\operatorname{div}\boldsymbol{j}^\alpha + \mathfrak{c}^\alpha, \tag{7.7.1}$$

$$\xi^\alpha := \frac{\rho_\alpha}{\rho}, \quad \boldsymbol{j}^\alpha := \rho_\alpha(\boldsymbol{v}^\alpha - \boldsymbol{v}), \tag{7.7.2}$$

   in which ξ^α are the (mass) concentrations and $\boldsymbol{j}^\alpha$ the diffusive fluxes of the constituents.
5. Show that the rule of material frame indifference (material objectivity) transforms the constitutive relations for the fields (7.5.11) in the form (7.5.12) into the form given in (7.5.13).
6. Prove that under the conditions stated in the main text, inequality (7.5.20) is satisfied by (7.5.21) both as a necessary and sufficient condition.
7. Show with the use of (7.5.23), that Λ^ε is only a function of the temperature.
8. Prove the relations (7.5.35) and (7.5.36).
9. Corroborate the formulas, respectively, rules of differentiation (7.6.14).
10. Corroborate the correctness of the following statements: (i) In one dimension ($n = 1$) every differential is total or complete. (ii) in two dimensions ($n = 2$) every differential can be transformed into a total differential by multiplying it with an integrating factor. (iii) In three-dimensional space ($n = 3$) the condition that an arbitrary differential $\mathrm{d}F = X_i\,\mathrm{d}x_i$ can be made complete is the requirement that the vector field $\boldsymbol{X}$ stays perpendicular to its vorticity, curl $\boldsymbol{X}$.
11. Prove the following statement: If a total differential of the form

$$\mathrm{d}F = \sum_{i=1}^{n} X_i\,\mathrm{d}x_i \quad \text{with} \quad X_k = 0 \quad \forall\, k \in (k_1, \dots, k_m), \quad m < n$$

   possesses coefficients which vanish identically, then the non-vanishing coefficients equally depend only on the variables associated to them, i.e., $X_j \neq 0$, $j > m$, implies $X_j = X_i(x_l) \quad l > m$.

[24] This LAGRANGEian description was introduced by WILMANSKI [256].

12. If the differential $\mathrm{d}f$ is an incomplete differential of the variables x_i, and if g is an integrating denominator which makes

$$\mathrm{d}F = \frac{\mathrm{d}f}{g} \tag{7.7.3}$$

a total differential, then prove that every differentiable function $G(F)$ makes the differential

$$\mathrm{d}H := G(F)\frac{\mathrm{d}f}{g} \tag{7.7.4}$$

equally complete.
13. Prove that close to thermodynamic equilibrium, $\boldsymbol{k}$, defined in (7.6.61), can be represented as shown in (7.6.64).
14. In soil mechanics, the total stress $\boldsymbol{t}$ is sometimes divided into an elastic, $\boldsymbol{t}_E$, and viscous, $\boldsymbol{t}_v$, part, where for isotropy the viscous part is described by

$$\boldsymbol{D} = a\mathbf{1} + b\boldsymbol{t}_v + c\boldsymbol{t}_v^2, \tag{7.7.5}$$

in which $\boldsymbol{D}$ is the stretching tensor and where a, b, c are coefficients which may depend on the invariants of $\boldsymbol{t}_v$. If $\boldsymbol{t}_v$ is divided into a pressure, p_v, and deviator $\boldsymbol{\sigma}'$, viz., $\boldsymbol{t}_v = -p_v\mathbf{1} + \boldsymbol{\sigma}'$, $\operatorname{tr}\boldsymbol{\sigma}' = 0$, show that

$$\begin{aligned} \boldsymbol{E} &:= \boldsymbol{D} - \tfrac{1}{3}(\operatorname{tr}\boldsymbol{D})\mathbf{1} = b\boldsymbol{\sigma}' + c\left[\boldsymbol{\sigma}'^2 - 2p_v\boldsymbol{\sigma}' - \tfrac{2}{3}II_{\boldsymbol{\sigma}'}\mathbf{1}\right], \\ \operatorname{tr}\boldsymbol{D} &= 3(a - bp_v + cp_v^2) + 2cII_{\boldsymbol{\sigma}'}, \end{aligned} \tag{7.7.6}$$

in which $II_{\boldsymbol{\sigma}'} = \frac{1}{2}\operatorname{tr}\left(\boldsymbol{\sigma}'^2\right)$ and

$$(a, b, c) = fcts(\cdot, p_v, II_{\boldsymbol{\sigma}'}, III_{\boldsymbol{\sigma}'}). \tag{7.7.7}$$

For $c = 0$, show, moreover, that simple shear and uniaxial compression tests do not suffice in determining the scalar functions a and b, provided a and b do not depend on $III_{\boldsymbol{\sigma}'}$.

7.8 Solutions

1. The difficulty – or better the unusual situation – consists here in the fact that every constituent performs its own motion. We isolate the motion of one constituent – perhaps a fictitious one – and use the index s. In applications one usually identifies it with a solid constituent (s for solid). Its motion is described by the diffeomorphic map

$$\boldsymbol{x} = \boldsymbol{\chi}^s(\boldsymbol{X}, t), \tag{7.8.1}$$

where $\chi^s(\cdot,t) : \boldsymbol{X} \longmapsto \boldsymbol{x}$ maps the particle position of the solid constituent particle $\boldsymbol{X}$ in the reference configuration onto its position $\boldsymbol{x}$ in the present configuration. Notice that $\boldsymbol{x}$ is also occupied by all other constituents, which in the reference configuration may have other positions than constituent s, and in particular they may move in the reference configuration relative to $K(s)$, whilst $K(s)$ is at rest[25].
The velocity and deformation gradient of constituent s are given by

$$\boldsymbol{x}^{\backslash s} := \frac{\partial \chi^s(\boldsymbol{X},t)}{\partial t}, \qquad \boldsymbol{F}^s := \mathrm{Grad}\,\chi^s(\boldsymbol{X},t)\,. \tag{7.8.2}$$

Here, $(\cdot)^{\backslash\alpha} = d^\alpha(\cdot)/dt$. Any constituent $\alpha \neq s$, to which we preferably will assign fluid properties moves in the present configuration with the velocity[26]

$$\boldsymbol{v}^\alpha := \tilde{\boldsymbol{v}}^\alpha(\boldsymbol{x},t) \qquad (\alpha \dot{\underline{=}} 1,2,\dots,\nu) \tag{7.8.3}$$

and may, with the aid of (7.8.1) be referred to the reference configuration

$$\boldsymbol{x}^{\backslash\alpha} := \tilde{\boldsymbol{v}}^\alpha(\chi^s(\boldsymbol{X},t),t) = \hat{\boldsymbol{x}}^{\backslash\alpha}(\boldsymbol{X},t) \tag{7.8.4}$$

to define a field on the reference configuration. Apart from these "natural" velocity fields we also need the kinematics of the material domains which are occupied by the fluid constituents in the reference configuration. In any spatial point $\boldsymbol{x}$ of the present configuration a fluid particle in $K(\alpha)$ moves relative to $K(s)$ with the (relative) velocity $(\boldsymbol{x}^{\backslash\alpha} - \boldsymbol{x}^{\backslash s})$. Its pre-image in the reference configuration is given by

$$\boldsymbol{X}^{\backslash\alpha} := (\boldsymbol{F}^s)^{-1}(\boldsymbol{x}^{\backslash\alpha} - \boldsymbol{x}^{\backslash s}) = \hat{\boldsymbol{X}}^{\backslash\alpha}(\boldsymbol{X},t)\,. \tag{7.8.5}$$

For $\alpha = s$ we have $\boldsymbol{X}^{\backslash s} = \boldsymbol{0}$, which one would expect; every fluid component $K(\alpha)$, $\alpha \dot{\underline{=}} 1,2,\dots,\nu$ moves in the reference configuration with the velocity (7.8.5).

With the motion (7.8.1) the volume and surface elements in the present and reference configuration, respectively, are connected by

$$\mathrm{d}v = \det\,\boldsymbol{F}^s\mathrm{d}V = J^s\mathrm{d}V, \quad \mathrm{d}\boldsymbol{a} = J^s((\boldsymbol{F}^s)^{-1})^T\mathrm{d}\boldsymbol{A}\,. \tag{7.8.6}$$

The field densities in the balance laws in the EULERian

$$\frac{\mathrm{d}^\alpha}{\mathrm{d}t}\int_\omega \gamma^\alpha \mathrm{d}v = -\int_{\partial\omega} \boldsymbol{\phi}^\alpha \cdot \mathrm{d}\boldsymbol{a} + \int_\omega (\zeta^\alpha + \pi^\alpha)\mathrm{d}v \tag{7.8.7}$$

and in the LAGRANGEian description

[25] In what follows we shall frequently use $K(s)$ to denote the reference configuration of the solid constituent.

[26] The index α runs here over all natural numbers $1,2,\dots,\nu$ except $\alpha = s$, which we express by the symbol "$\dot{\underline{=}}$".

$$\frac{\mathrm{d}^\alpha}{\mathrm{d}t}\int\limits_{\Omega}\Gamma^\alpha \mathrm{d}V = -\int\limits_{\partial\Omega}\boldsymbol{\Phi}^\alpha\cdot\mathrm{d}\boldsymbol{A} + \int\limits_{\Omega}(\Xi^\alpha + \Pi^\alpha)\mathrm{d}V \tag{7.8.8}$$

are therefore connected by

$$\begin{aligned}(\Gamma^\alpha,\, \Xi^\alpha,\, \Pi^\alpha) &= J^s(\gamma^\alpha,\, \zeta^\alpha,\, \pi^\alpha)\,,\\ \boldsymbol{\Phi}^\alpha &= J^s\boldsymbol{F}^{s-1}\boldsymbol{\phi}^\alpha\,.\end{aligned} \tag{7.8.9}$$

These relations are obtained by substituting (7.8.6) into (7.8.7) and comparing the result with (7.8.8). In these relations it was tacitly assumed that $J^s > 0$, and $\mathrm{d}^\alpha/\mathrm{d}t$ denotes the time derivative following the motion of $K(\alpha)$, viz.

$$\frac{\mathrm{d}^\alpha(\cdot)}{\mathrm{d}t} = \begin{cases}\dfrac{\partial(\cdot)}{\partial t}\,, & \alpha = s\,,\\[2ex] \dfrac{\partial(\cdot)}{\partial t} + (\mathrm{grad}\,(\cdot))\boldsymbol{X}'^\alpha, & \alpha \,\check{=}\, 1,2,...,\nu\,.\end{cases} \tag{7.8.10}$$

For the derivation of the local balance laws in the LAGRANGE description, we select a material volume in the reference configuration which may be separated by a surface on which the physical fields may suffer jump discontinuities into a positive, Ω^+, and negative, Ω^-, part such that $\Omega^+\cap\Omega^- = \emptyset$. The singular surface $\mathfrak{S}$ is supposed to move with the speed of propagation

$$U = \boldsymbol{W}\cdot\boldsymbol{N}_{\mathfrak{S}}\,. \tag{7.8.11}$$

In such circumstances one obtains then for the divergence theorem

$$\int\limits_{\partial\Omega}\boldsymbol{\Phi}^\alpha\cdot\boldsymbol{N}\mathrm{d}A = \int\limits_{\Omega}\mathrm{Div}\,\boldsymbol{\Phi}^\alpha\mathrm{d}\boldsymbol{V} + \int\limits_{\mathfrak{S}}[\![\boldsymbol{\Phi}^\alpha\cdot\boldsymbol{N}_{\mathfrak{S}}]\!]\mathrm{d}A\,, \tag{7.8.12}$$

and for the REYNOLDS transport theorem

$$\begin{aligned}\frac{\mathrm{d}^\alpha}{\mathrm{d}t}\int\limits_{\Omega}\Gamma^\alpha\mathrm{d}V &= \int\limits_{\Omega}\frac{\partial\Gamma^\alpha}{\partial t}\mathrm{d}V + \int\limits_{\partial\Omega}\Gamma^\alpha\boldsymbol{X}'^\alpha\cdot\boldsymbol{N}\mathrm{d}A - \int\limits_{\mathfrak{S}}[\![\Gamma^\alpha\overbrace{\boldsymbol{W}\cdot\boldsymbol{N}_{\mathfrak{S}}}^{U}]\!]\mathrm{d}A\\ &= \int\limits_{\Omega}\left(\frac{\partial\Gamma^\alpha}{\partial t} + \mathrm{Div}(\Gamma^\alpha\boldsymbol{X}'^\alpha)\right)\mathrm{d}V + \int\limits_{\mathfrak{S}}[\![\Gamma^\alpha\underbrace{(\boldsymbol{X}'^\alpha - \boldsymbol{W})\cdot\boldsymbol{N}_{\mathfrak{S}}}_{(\boldsymbol{X}'^\alpha\cdot\boldsymbol{N}_{\mathfrak{S}} - U)}]\!]\mathrm{d}A\,.\end{aligned} \tag{7.8.13}$$

If these representations are substituted into (7.8.8), the balance law (7.8.8) in the LAGRANGE description takes the form

$$\begin{aligned}&\int\limits_{\Omega}\left(\frac{\partial\Gamma^\alpha}{\partial t} + \mathrm{Div}(\Gamma^\alpha\boldsymbol{X}'^\alpha + \boldsymbol{\Phi}^\alpha) - \Xi^\alpha - \Pi^\alpha\right)\mathrm{d}V\\ &\quad + \int\limits_{\mathfrak{S}}\Big([\![\Gamma^\alpha\underbrace{(\boldsymbol{X}'^\alpha - \boldsymbol{W})\cdot\boldsymbol{N}_{\mathfrak{S}}}_{(\boldsymbol{X}'^\alpha\cdot\boldsymbol{N}_{\mathfrak{S}} - U)}]\!] + [\![\boldsymbol{\Phi}^\alpha\cdot\boldsymbol{N}_{\mathfrak{S}}]\!]\Big)\mathrm{d}A = 0\,,\end{aligned} \tag{7.8.14}$$

in which $\int_\Omega$ must be interpreted as $\int_{\Omega+} + \int_{\Omega-}$. Since (7.8.14) is valid for all material volumes (with or without singular surfaces) there must necessarily follow as local consequences of the global balance law the statements

$$\begin{aligned} &\frac{\partial \Gamma^\alpha}{\partial t} + \operatorname{Div}(\Gamma^\alpha \boldsymbol{X}^{\backslash\alpha} + \boldsymbol{\Phi}^\alpha) - \Xi^\alpha - \Pi^\alpha = 0 \, , \\ &[\![\Gamma^\alpha \underbrace{(\boldsymbol{X}^{\backslash\alpha} - \boldsymbol{W}) \cdot \boldsymbol{N}_{\mathfrak{S}}}_{(\boldsymbol{X}^{\backslash\alpha}\cdot \boldsymbol{N}_{\mathfrak{S}} - U)}]\!] + [\![\boldsymbol{\Phi}^\alpha \cdot \boldsymbol{N}_{\mathfrak{S}}]\!] = 0 \, , \end{aligned} \tag{7.8.15}$$

valid in spatial points where the fields are differentiable or suffer a jump, respectively. The first of these is referred to as the *local balance law*, the second as the *compatibility condition* on $\mathfrak{S}$. These relations hold for $\alpha = 1, 2, ..., \nu$, inclusively $\alpha = s$; in this latter case $\boldsymbol{X}^{\backslash s} = \boldsymbol{0}$.
The individual terms that must be applied in (7.8.15) for the physical balance laws are listed in the entries of Table 7.2, which corresponds to Table 7.1 given earlier. In the **balance law of mass** the physical quantity of constituent α is the density ϱ^α_R in the reference configuration; flux and supply vanish, but there may be a non-vanishing density of mass production of constituent $\mathfrak{c}^\alpha_R$. Since $\boldsymbol{X}^{\backslash s} = \boldsymbol{0}$, the local balance laws for mass of constituent $\alpha = s$ differ from those for $\alpha \dot{=} 1, 2, ..., \nu$. In summary these laws have the form

$$\begin{aligned} &\frac{\partial \varrho^s_R}{\partial t} = \mathfrak{c}^s_R \, , && [\![\varrho^s_R \underbrace{\boldsymbol{W} \cdot \boldsymbol{N}_{\mathfrak{S}}}_{U}]\!] = [\![\varrho^s_R U]\!] = 0 \, , \\ &\frac{\partial \varrho^\alpha_R}{\partial t} + \operatorname{Div}(\varrho^\alpha_R \boldsymbol{X}^{\backslash\alpha}) = \mathfrak{c}^\alpha_R \, , && [\![\varrho^\alpha_R (\boldsymbol{X}^{\backslash\alpha} \cdot \boldsymbol{N}_{\mathfrak{S}} - U)]\!] = 0 \, , \\ & && (\alpha \dot{=} 1, 2, ..., \nu) \, . \end{aligned} \tag{7.8.16}$$

With the transformation rules (7.8.9) it is also easy to corroborate the connections

$$\rho^\alpha_R = J^\alpha \rho^\alpha \, , \qquad \mathfrak{c}^\alpha_R = J^\alpha \mathfrak{c}^\alpha \tag{7.8.17}$$

between the densities and specific mass productions in the reference and present configurations, respectively. With them the balance law

$$\frac{\partial \rho^\alpha}{\partial t} + \operatorname{div}(\rho^\alpha \boldsymbol{v}^\alpha) = \mathfrak{c}^\alpha \tag{7.8.18}$$

in the EULERian description may easily be recovered.

In the **balance law of linear momentum** we see that of the various terms only $\boldsymbol{P}^\alpha$ is new; this is the partial first PIOLA–KIRCHHOFF stress tensor which is expressible in terms of the partial CAUCHY stress tensor $\boldsymbol{t}^\alpha$ according to

$$\boldsymbol{P}^\alpha = J^\alpha \boldsymbol{F}^{s^{-1}} \boldsymbol{t}^\alpha \, . \tag{7.8.19}$$

Substituting the respective entries of Table 7.2 into (7.8.15) yields

$$
\begin{aligned}
&\frac{\partial(\overset{s}{\varrho}_R \boldsymbol{x}'^{s})}{\partial t} = \operatorname{Div} \boldsymbol{P}^s + \overset{s}{\varrho}_R \boldsymbol{f}^s + \mathfrak{m}^{\alpha}_R\,,\\
&[\![\overset{s}{\varrho}_R \boldsymbol{x}'^{s} U]\!] + [\![\boldsymbol{P}^s \boldsymbol{N}_{\mathfrak{S}}]\!] = \boldsymbol{0}\,,\\
&\frac{\partial(\overset{\alpha}{\varrho}_R \boldsymbol{x}'^{\alpha})}{\partial t} + \operatorname{Div}(\overset{\alpha}{\varrho}_R \boldsymbol{x}'^{\alpha} \otimes \boldsymbol{X}'^{\alpha} - \boldsymbol{P}^{\alpha}) = \overset{\alpha}{\varrho}_R \boldsymbol{f}^{\alpha} + \mathfrak{m}^{\alpha}_R\,,\\
&[\![\overset{\alpha}{\varrho}_R \boldsymbol{x}'^{\alpha}(\boldsymbol{X}'^{\alpha}\cdot \boldsymbol{N}_{\mathfrak{S}} - U)]\!] - [\![\boldsymbol{P}^{\alpha} \boldsymbol{N}_{\mathfrak{S}}]\!] = \boldsymbol{0}\,, \quad (\alpha \doteq 1,2,\dots,\nu)\,,
\end{aligned}
\tag{7.8.20}
$$

or, if the balance law of mass (7.8.16) is used

$$
\begin{aligned}
&\overset{s}{\varrho}_R \frac{\partial \boldsymbol{x}'^{s}}{\partial t} = \operatorname{Div} \boldsymbol{P}^s + \overset{s}{\varrho}_R \boldsymbol{f}^s + (\mathfrak{m}^{s}_R - \mathfrak{c}^{s}_R \boldsymbol{x}'^{s})\,,\\
&\overset{s}{\varrho}_R U [\![\boldsymbol{x}'^{s}]\!] + [\![\boldsymbol{P}^s \boldsymbol{N}_{\mathfrak{S}}]\!] = \boldsymbol{0}\,,\\
&\overset{\alpha}{\varrho}_R \frac{\mathrm{d}^{\alpha} \boldsymbol{x}'^{\alpha}}{\mathrm{d}t} = \operatorname{Div} \boldsymbol{P}^{\alpha} + \overset{\alpha}{\varrho}_R \boldsymbol{f}^{\alpha} + (\mathfrak{m}^{\alpha}_R - \mathfrak{c}^{\alpha}_R \boldsymbol{x}'^{\alpha})\,,\\
&\overset{\alpha}{\varrho}_R (\boldsymbol{X}'^{\alpha}\cdot \boldsymbol{N}_{\mathfrak{S}} - U)[\![\boldsymbol{x}'^{\alpha}]\!] - [\![\boldsymbol{P}^{\alpha} \boldsymbol{N}_{\mathfrak{S}}]\!] = 0\,, \quad (\alpha \doteq 1,2,\dots,\nu)\,.
\end{aligned}
\tag{7.8.21}
$$

In the **balance law of angular momentum** Table 7.2 shows that of all the physical quantities only the flux quantity $\boldsymbol{M}$ appears to be new when comparing with Table 7.1; this quantity will be called the first PIOLA–KIRCHHOFF couple stress tensor, which is related to $\boldsymbol{m}^{\alpha}$ by

$$
\boldsymbol{M}^{\alpha} = J^{\alpha} \boldsymbol{F}^{s\,-1} \boldsymbol{m}^{\alpha}\,. \tag{7.8.22}
$$

Else, the quantities in the entries of Table 7.2 are structured as in Table 7.1. Substitution into (7.8.15) and use of the mass balance (7.8.16) as well as the momentum balance (7.8.20) leads, after a somewhat lengthy calculations, to the local spin balances[27]

$$
\begin{aligned}
&\overset{s}{\varrho}_R \frac{\partial \boldsymbol{s}^{s}}{\partial t} = \operatorname{Div} \boldsymbol{M}^s + \overset{s}{\varrho}_R \boldsymbol{l}^s - \operatorname{dual}(\boldsymbol{F}^s \boldsymbol{P}^{sT} - \boldsymbol{P}^s \boldsymbol{F}^{sT}) + (\mathfrak{k}^{s}_R - \mathfrak{c}^{s}_R \boldsymbol{s}^{s})\,,\\
&\overset{s}{\varrho}_R U [\![\boldsymbol{s}^{s}]\!] + [\![\boldsymbol{M}^s \boldsymbol{N}_{\mathfrak{S}}]\!] = \boldsymbol{0}\,,\\
&\overset{\alpha}{\varrho}_R \frac{\mathrm{d}^{\alpha} \boldsymbol{s}^{\alpha}}{\mathrm{d}t} = \operatorname{Div} \boldsymbol{M}^{\alpha} + \overset{\alpha}{\varrho}_R \boldsymbol{l}^{\alpha} - \operatorname{dual}(\boldsymbol{F}^{\alpha} \boldsymbol{P}^{\alpha T} - \boldsymbol{P}^{\alpha} \boldsymbol{F}^{\alpha T}) + (\mathfrak{k}^{\alpha}_R - \mathfrak{c}^{\alpha}_R \boldsymbol{s}^{\alpha})\,,\\
&\overset{\alpha}{\varrho}_R (\boldsymbol{X}'^{\alpha}\cdot \boldsymbol{N}_{\mathfrak{S}} - U)[\![\boldsymbol{s}^{\alpha}]\!] - [\![\boldsymbol{M}^{\alpha} \boldsymbol{N}_{\mathfrak{S}}]\!] = \boldsymbol{0}\,, \quad (\alpha \doteq 1,2,\dots,\nu)\,.
\end{aligned}
\tag{7.8.23}
$$

[27] The term written as $\operatorname{dual}(\cdot)$ reads in Cartesian components $-\epsilon_{ijk}\,(\partial x_j/\partial X_A)\,P^{\alpha}_{kA}$. Note further, that x_j is given by the motion map $x_j = \overset{s}{\chi}_j(\boldsymbol{X}, t)$, so that $\partial x_j/\partial X_A$ denotes the (jA)-component of $\boldsymbol{F}^s$. A term of the form $(\partial x/\partial t) \times \overset{\alpha}{\rho}_R \boldsymbol{x}'^{s}$, arising in the derivation of (7.8.23) must therefore vanish, since $\partial x/\partial t = \boldsymbol{x}'^{s}$. Analogously, we also have $(\mathrm{d}^{\alpha}\boldsymbol{x}^{\alpha}/\mathrm{d}\,t)\times \boldsymbol{x}'^{\alpha} = 0$.

For the *energy balance law*, finally, Table 7.2 lists the relevant terms in the LAGRANGE description, in which the LAGRANGEian heat flux $\boldsymbol{Q}^\alpha$ of constituent α is new and related to its EULERian counterpart via

$$\boldsymbol{Q}^\alpha = J^\alpha \boldsymbol{F}^{s-1} \boldsymbol{q}^\alpha \,. \tag{7.8.24}$$

Substituting the relevant expressions of Table 7.2 into (7.8.15) then leads to the local **balances of the internal energy** for constituent α[28]

$$\begin{aligned}
\varrho^s_R \frac{\partial \varepsilon^s}{\partial t} &= -\operatorname{Div} \boldsymbol{Q}^s + \boldsymbol{\omega}^s \cdot \operatorname{dual}(\boldsymbol{F}^s \boldsymbol{P}^{sT} - \boldsymbol{P}^s \boldsymbol{F}^{sT}) + \boldsymbol{F}^{\backslash s} \cdot \boldsymbol{P}^{sT} \\
&\quad + (\operatorname{grad} \boldsymbol{\omega}^s) \cdot \boldsymbol{M}^{sT} + \mathfrak{e}^s_R - \mathfrak{c}^s_R \varepsilon^s - (\mathfrak{m}^s_R - \tfrac{1}{2} \mathfrak{c}^s_R \boldsymbol{x}^{\backslash s}) \cdot \boldsymbol{x}^{\backslash s} \\
&\quad - (\mathfrak{k}^s_R - \tfrac{1}{2} \mathfrak{c}^s_R \boldsymbol{s}^s) \cdot \boldsymbol{\omega}^s + \tfrac{1}{2} \varrho^s_R (\boldsymbol{\omega}^s \cdot \frac{\partial \boldsymbol{s}^s}{\partial t} - \boldsymbol{s}^s \cdot \frac{\partial \boldsymbol{\omega}^s}{\partial t}) + \varrho^\alpha_R \mathfrak{r}^s_R \,, \\
\varrho^s_R U [\![\tfrac{1}{2} \boldsymbol{x}^{\backslash s} \cdot \boldsymbol{x}^{\backslash s} &+ \tfrac{1}{2} \boldsymbol{s}^s \cdot \boldsymbol{\omega}^s + \varepsilon^s]\!] + [\![\boldsymbol{x}^{\backslash s} \boldsymbol{P}^s + \boldsymbol{\omega}^s \boldsymbol{M}^s - \boldsymbol{Q}^s]\!] \cdot \boldsymbol{N}_\mathfrak{S} = 0 \,,
\end{aligned}$$

$$\begin{aligned}
\varrho^\alpha_R \frac{\mathrm{d}^\alpha \varepsilon^\alpha}{\mathrm{d}t} &= -\operatorname{Div} \boldsymbol{Q}^\alpha + \boldsymbol{\omega}^\alpha \cdot \operatorname{dual}(\boldsymbol{F}^\alpha \boldsymbol{P}^{\alpha T} - \boldsymbol{P}^\alpha \boldsymbol{F}^{\alpha T}) + \boldsymbol{F}^{\backslash \alpha} \cdot \boldsymbol{P}^{\alpha T} \\
&\quad + (\operatorname{grad} \boldsymbol{\omega}^\alpha) \cdot \boldsymbol{M}^{\alpha T} + \mathfrak{e}^\alpha_R - \mathfrak{c}^s_R \varepsilon^s - (\mathfrak{m}^\alpha_R - \tfrac{1}{2} \mathfrak{c}^\alpha_R \boldsymbol{x}^{\backslash \alpha}) \boldsymbol{x}^{\backslash \alpha} \\
&\quad - (\mathfrak{k}^\alpha_R - \tfrac{1}{2} \mathfrak{c}^\alpha_R \boldsymbol{s}^\alpha) \cdot \boldsymbol{\omega}^\alpha + \tfrac{1}{2} \varrho^\alpha_R (\boldsymbol{\omega}^\alpha \cdot \frac{\mathrm{d}^\alpha \boldsymbol{s}^\alpha}{\mathrm{d}t} - \boldsymbol{s}^\alpha \cdot \frac{\mathrm{d}^\alpha \boldsymbol{\omega}^\alpha}{\mathrm{d}t}) + \varrho^\alpha_R \mathfrak{r}^\alpha_R , \\
\varrho^\alpha_R (\boldsymbol{X}^{\backslash \alpha} &\cdot \boldsymbol{N}_\mathfrak{S} - U) [\![\tfrac{1}{2} \boldsymbol{x}^{\backslash \alpha} \cdot \boldsymbol{x}^{\backslash \alpha} + \tfrac{1}{2} \boldsymbol{s}^\alpha \cdot \boldsymbol{\omega}^\alpha + \varepsilon^\alpha]\!] \\
&- [\![\boldsymbol{x}^{\backslash \alpha} \boldsymbol{P}^\alpha + \boldsymbol{\omega}^\alpha \boldsymbol{M}^\alpha - \boldsymbol{Q}^\alpha]\!] \cdot \boldsymbol{N}_\mathfrak{S} = 0 \,, \quad (\alpha \,\check{=}\, 1, 2, ..., \nu) \,.
\end{aligned} \tag{7.8.25}$$

[28] In these equations the following interpretations must be observed: $\boldsymbol{F}^{\backslash \alpha} \cdot \boldsymbol{P}^{\alpha T} = (\partial^\alpha x^\alpha_i / \partial X_A)\, P_{iA}$; $(\operatorname{grad} \boldsymbol{\omega}^\alpha) \cdot \boldsymbol{M}^{\alpha T} = (\partial^\alpha \omega^\alpha_i / \partial X_A)\, M^\alpha_{iA}$.

Table 7.2. Physical quantity Γ^α, its flux Φ^α, supply $\mathcal{Z}^\alpha$ and production Π^α as given in the balance laws (7.8.14) and (7.8.15), respectively, and their identifications for the balances of mass, momentum, angular momentum energy and entropy. For the respective definitions, see also the main text.

Balance	Γ^α	$\tilde{\Phi}^\alpha$	$\mathcal{Z}^\alpha$	Π^α
Mass	ρ_R^α	0	0	$\mathfrak{c}_R^\alpha$
Momentum	$\rho_R^\alpha \boldsymbol{x}'^\alpha$	$-\boldsymbol{P}^\alpha$	$\rho_R^\alpha \boldsymbol{f}^\alpha$	$\mathfrak{m}_R^\alpha$
Angular momentum	$\boldsymbol{x} \times \rho_R^\alpha \boldsymbol{x}'^\alpha + \rho_R^\alpha \boldsymbol{s}^\alpha$	$-\boldsymbol{x} \times \boldsymbol{P}^\alpha - \boldsymbol{M}^\alpha$	$\boldsymbol{x} \times \rho_R^\alpha \boldsymbol{f}^\alpha + \rho_R^\alpha \boldsymbol{l}^\alpha$	$\boldsymbol{x} \times \mathfrak{m}_R + \mathfrak{k}_R^\alpha$
Energy	$\frac{1}{2}\rho_R^\alpha \boldsymbol{x}'^\alpha \cdot \boldsymbol{x}'^\alpha + \frac{1}{2}\rho_R^\alpha \boldsymbol{s}^\alpha \cdot \boldsymbol{\omega}^\alpha$ $+\rho_R^\alpha \varepsilon^\alpha$	$-\boldsymbol{x}'^\alpha\ \boldsymbol{P}^\alpha - \boldsymbol{\omega}^\alpha \boldsymbol{M}^\alpha + \boldsymbol{Q}^\alpha$	$\boldsymbol{x}'^\alpha \cdot \rho_R^\alpha \boldsymbol{f}^\alpha + \boldsymbol{\omega}^\alpha \cdot \rho_R^\alpha \boldsymbol{l}^\alpha$ $+\rho_R^\alpha \mathfrak{r}^\alpha$	$\mathfrak{e}_R^\alpha$
Entropy	$\rho_R^\alpha \eta$	$\boldsymbol{\Phi}^{\eta\alpha}$	$\rho_R^\alpha \varsigma^{\eta\alpha}$	$\rho_R^\alpha \pi^{\eta\alpha}$

These equations are obtained after rather lengthy manipulations using the balances of mass, linear and angular momenta. In equations (7.8.21)–(7.8.25) the following new quantities can be identified

$$
\begin{aligned}
\mathfrak{m}^{\alpha}_{R\text{Euclid}} &= \left(\mathfrak{m}^{\alpha}_{R} - \mathfrak{c}^{\alpha}_{R}\boldsymbol{x}^{\backslash\alpha}\right) , \\
\mathfrak{k}^{\alpha}_{R\text{Euclid}} &= \left(\mathfrak{k}^{\alpha}_{R} - \mathfrak{c}^{\alpha}_{R}\boldsymbol{s}^{\alpha}\right) , \\
\mathfrak{e}^{\alpha}_{R\text{Euclid}} &= \mathfrak{e}^{\alpha}_{R} - \varepsilon^{\alpha}\mathfrak{c}^{\alpha}_{R} - \left(\mathfrak{m}^{\alpha}_{R} - \tfrac{1}{2}\mathfrak{c}^{\alpha}_{R}\boldsymbol{x}^{\backslash\alpha}\right)\cdot\boldsymbol{x}^{\backslash\alpha} \\
&\quad -\left(\mathfrak{k}^{\alpha}_{R} - \tfrac{1}{2}\mathfrak{c}^{\alpha}_{R}\boldsymbol{s}^{\alpha}\right)\cdot\boldsymbol{\omega}^{\alpha} \\
&= \mathfrak{e}^{\alpha}_{R} - \mathfrak{c}^{\alpha}_{R}\left(\varepsilon^{\alpha} + \tfrac{1}{2}\boldsymbol{x}^{\backslash\alpha}\cdot\boldsymbol{x}^{\backslash\alpha} + \tfrac{1}{2}\boldsymbol{s}^{\alpha}\cdot\boldsymbol{\omega}^{\alpha}\right) , \\
&\quad -\left(\mathfrak{m}^{\alpha}_{R} - \mathfrak{c}^{\alpha}_{R}\boldsymbol{x}^{\backslash\alpha}\right)\cdot\boldsymbol{x}^{\backslash\alpha} - \left(\mathfrak{k}^{\alpha}_{R} - \mathfrak{c}^{\alpha}_{R}\boldsymbol{s}^{\alpha}\right)\cdot\boldsymbol{\omega}^{\alpha} ,
\end{aligned}
\tag{7.8.26}
$$

which, under EUCLIDian transformations transform as objective tensor quantities; they correspond in the EULERian description to the quantities introduced in (7.2.15).

This completes the mixture formulation in the LAGRANGEian description. When

$$
\boldsymbol{s}^{\alpha} = \boldsymbol{0} , \quad \boldsymbol{M}^{\alpha} = \boldsymbol{0} , \quad \boldsymbol{l}^{\alpha} = \boldsymbol{0} , \quad \boldsymbol{s}^{\alpha} = \boldsymbol{0} , \quad \mathfrak{k}^{\alpha}_{R} = \boldsymbol{0} , \tag{7.8.27}
$$

the above equations reduce to the physical balance laws for non-polar mixtures in the LAGRANGEian description.

2. *Momentum balance*: We use in the first two terms of $(7.2.5)_1$ product differentiation and obtain

$$
\underbrace{\boldsymbol{v}^{\alpha}\left(\frac{\partial\rho_{\alpha}}{\partial t} + \operatorname{div}(\rho_{\alpha}\boldsymbol{v}^{\alpha})\right)}_{\mathfrak{c}^{\alpha}} + \underbrace{\rho_{\alpha}\left(\frac{\partial\boldsymbol{v}^{\alpha}}{\partial t} + (\operatorname{grad}\boldsymbol{v}^{\alpha})\boldsymbol{v}^{\alpha}\right)}_{\mathrm{d}^{\alpha}\boldsymbol{v}^{\alpha}/\mathrm{d}t}
$$

$$
- \operatorname{div}\boldsymbol{t}^{\alpha} = \rho_{\alpha}\boldsymbol{f}^{\alpha} + \mathfrak{m}^{\alpha} . \tag{7.8.28}
$$

The first two terms are obtained as indicated with the parentheses because of (7.2.4) and (7.2.9); therefore we have

$$
\rho_{\alpha}\frac{\mathrm{d}\boldsymbol{v}^{\alpha}}{\mathrm{d}t} = \operatorname{div}\boldsymbol{t}^{\alpha} + \rho_{\alpha}\boldsymbol{f}^{\alpha} + \left(\mathfrak{m}^{\alpha} - \mathfrak{c}^{\alpha}\boldsymbol{v}^{\alpha}\right) , \tag{7.8.29}
$$

which is as listed in (7.2.10).

Spin balance: Here, we start analogously from $(7.2.6)_1$ and obtain, on using product differentiation,

$$
\underbrace{s^\alpha \left(\frac{\partial \rho_\alpha}{\partial t} + \operatorname{div}(\rho_\alpha \boldsymbol{v}^\alpha)\right)}_{\mathfrak{c}^\alpha} + \rho_\alpha \underbrace{\left(\frac{\partial \boldsymbol{s}^\alpha}{\partial t} + (\operatorname{grad} \boldsymbol{s}^\alpha)\boldsymbol{v}^\alpha\right)}_{\mathrm{d}^\alpha \boldsymbol{s}^\alpha/\mathrm{d}t}
$$

$$
+\boldsymbol{x}^\alpha \times \Bigg(\underbrace{\frac{\partial \rho_\alpha \boldsymbol{v}^\alpha}{\partial t} + \operatorname{div}\left(\rho_\alpha \boldsymbol{v}^\alpha \otimes \boldsymbol{v}^\alpha - \boldsymbol{t}^\alpha - \rho_\alpha \boldsymbol{f}^\alpha - \mathfrak{m}^\alpha\right)}_{=0 \text{ (momentum balance } (7.2.5)_1)}\Bigg)
$$

$$
+\operatorname{dual}\left(\boldsymbol{t}^\alpha - \boldsymbol{t}^{\alpha T}\right) - \operatorname{div} \boldsymbol{m}^\alpha = \rho_\alpha \boldsymbol{l}^\alpha + \mathfrak{k}^\alpha , \tag{7.8.30}
$$

from which we deduce

$$
\rho_\alpha \frac{\mathrm{d}\boldsymbol{s}^\alpha}{\mathrm{d}t} = \operatorname{div} \boldsymbol{m}^\alpha - \operatorname{dual}\left(\boldsymbol{t}^\alpha - \boldsymbol{t}^{\alpha T}\right) + \rho_\alpha \boldsymbol{l}^\alpha + (\mathfrak{k}^\alpha - \mathfrak{c}^\alpha \boldsymbol{s}^\alpha), \tag{7.8.31}
$$

which agrees with $(7.2.11)_1$.

Energy balance: Here we start from $(7.2.7)_1$ and write, again by using product differentiation

$$
\tfrac{1}{2}\boldsymbol{v}^\alpha \cdot \boldsymbol{v}^\alpha \underbrace{\left(\frac{\partial \rho_\alpha}{\partial t} + \operatorname{div}(\rho_\alpha \boldsymbol{v}^\alpha)\right)}_{\mathfrak{c}^\alpha}
$$

$$
+\boldsymbol{v}^\alpha \cdot \underbrace{\left[\rho_\alpha \left(\frac{\partial \boldsymbol{v}^\alpha}{\partial t} + (\operatorname{grad} \boldsymbol{v}^\alpha)\boldsymbol{v}^\alpha\right) - \operatorname{div} \boldsymbol{t}^\alpha - \rho_\alpha \boldsymbol{f}^\alpha\right]}_{\mathfrak{m}^\alpha - \mathfrak{c}^\alpha \boldsymbol{v}^\alpha}
$$

$$
- \operatorname{grad} \boldsymbol{v}^\alpha \cdot \boldsymbol{t}^{\alpha T} + \tfrac{1}{2}\boldsymbol{s}^\alpha \cdot \boldsymbol{\omega}^\alpha \underbrace{\left(\frac{\partial \rho_\alpha}{\partial t} + \operatorname{div}(\rho_\alpha \boldsymbol{v}^\alpha)\right)}_{\mathfrak{c}^\alpha} - \operatorname{grad} \boldsymbol{\omega}^\alpha \cdot \boldsymbol{m}^{\alpha T}
$$

$$
+\tfrac{1}{2}\boldsymbol{\omega}^\alpha \cdot \Bigg[\rho_\alpha \underbrace{\left(\frac{\partial \boldsymbol{s}^\alpha}{\partial t} + (\operatorname{grad} \boldsymbol{s}^\alpha)\boldsymbol{v}^\alpha\right)}_{\mathrm{d}^\alpha \boldsymbol{s}^\alpha/\mathrm{d}t} - 2 \operatorname{div} \boldsymbol{m}^\alpha - 2\rho_\alpha \boldsymbol{l}^\alpha\Bigg]
$$

$$
+\tfrac{1}{2}\rho_\alpha \boldsymbol{s}^\alpha \cdot \underbrace{\left(\frac{\partial \boldsymbol{\omega}^\alpha}{\partial t} + (\operatorname{grad} \boldsymbol{\omega}^\alpha)\boldsymbol{v}^\alpha\right)}_{\mathrm{d}^\alpha \boldsymbol{\omega}^\alpha/\mathrm{d}t} + \varepsilon^\alpha \underbrace{\left(\frac{\partial \rho_\alpha}{\partial t} + \operatorname{div}(\rho_\alpha \boldsymbol{v}^\alpha)\right)}_{\mathfrak{c}^\alpha}
$$

$$
+\rho_\alpha \underbrace{\left(\frac{\partial \varepsilon^\alpha}{\partial t} + (\operatorname{grad} \varepsilon^\alpha)\boldsymbol{v}^\alpha\right)}_{\mathrm{d}^\alpha \varepsilon^\alpha/\mathrm{d}t} + \operatorname{div} \boldsymbol{q}^\alpha = \mathfrak{e}^\alpha + \rho_\alpha \mathfrak{r}^\alpha . \tag{7.8.32}
$$

This formula can still be simplified, if we add and subtract on the left-hand side the term

$$
\tfrac{1}{2}\boldsymbol{\omega}^\alpha \cdot \rho_\alpha \frac{\mathrm{d}\boldsymbol{s}^\alpha}{\mathrm{d}t}
$$

and use the fact that according to (7.8.31)

$$\tfrac{1}{2}\boldsymbol{\omega}^\alpha \cdot \left[2\rho_\alpha \frac{\mathrm{d}\boldsymbol{s}^\alpha}{\mathrm{d}t} - 2\,\mathrm{div}\,\boldsymbol{m}^\alpha - 2\rho_\alpha \boldsymbol{l}^\alpha\right]$$
$$= \boldsymbol{\omega}^\alpha \cdot \left((\mathfrak{k}^\alpha - \mathfrak{c}^\alpha \boldsymbol{s}^\alpha) - \mathrm{dual}\left(\boldsymbol{t}^\alpha - \boldsymbol{t}^{\alpha T}\right)\right) .$$

(7.8.32) then yields after a short calculation

$$\begin{aligned}\rho_\alpha \frac{\mathrm{d}^\alpha \varepsilon^\alpha}{\mathrm{d}t} = &- \mathrm{div}\,\boldsymbol{q}^\alpha + \boldsymbol{\omega}^\alpha \cdot (\mathrm{dual}(\boldsymbol{t}^\alpha - \boldsymbol{t}^{\alpha T})) + \mathrm{grad}\,\boldsymbol{v}^\alpha \cdot \boldsymbol{t}^{\alpha T}\\ &+ \mathrm{grad}\,\boldsymbol{\omega}^\alpha \cdot \boldsymbol{m}^{\alpha T} + \tfrac{1}{2}\rho_\alpha \left(\boldsymbol{\omega}^\alpha \cdot \frac{\mathrm{d}^\alpha \boldsymbol{s}^\alpha}{\mathrm{d}t} - \boldsymbol{s}^\alpha \cdot \frac{\mathrm{d}^\alpha \boldsymbol{\omega}^\alpha}{\mathrm{d}t}\right) + \rho_\alpha \mathfrak{r}^\alpha\\ &+ \Big(\mathfrak{e}^\alpha - (\boldsymbol{m}^\alpha - \mathfrak{c}^\alpha \boldsymbol{v}^\alpha) \cdot \boldsymbol{v}^\alpha - (\mathfrak{k}^\alpha - \mathfrak{c}^\alpha \boldsymbol{s}^\alpha) \cdot \boldsymbol{\omega}^\alpha\\ &- \mathfrak{c}^\alpha \left(\varepsilon^\alpha + \tfrac{1}{2}\boldsymbol{v}^\alpha \cdot \boldsymbol{v}^\alpha + \tfrac{1}{2}\boldsymbol{s}^\alpha \cdot \boldsymbol{\omega}^\alpha\right)\Big) ,\end{aligned} \tag{7.8.33}$$

which corresponds to $(7.2.12)_1$, qed.

3. To prove formula (7.3.23), it is advantageous to divide the entropy flux (7.3.22) into three parts

$$\begin{aligned}\boldsymbol{\phi}_1 &= \sum_{\alpha=1}^{\nu} \left(-\boldsymbol{v}^\alpha \boldsymbol{t}^\alpha + \tfrac{1}{2}(\rho_\alpha \boldsymbol{v}^\alpha \cdot \boldsymbol{v}^\alpha)\boldsymbol{u}^\alpha\right) = -\boldsymbol{v} \sum \boldsymbol{t}^\alpha\\ &- \sum \boldsymbol{u}^\alpha \boldsymbol{t}^\alpha + \tfrac{1}{2} \underbrace{\sum \rho_\alpha (\boldsymbol{v}\cdot\boldsymbol{v})\boldsymbol{u}^\alpha}_{0} + \tfrac{1}{2}\sum (\rho_\alpha(\boldsymbol{v}\cdot\boldsymbol{u}^\alpha + \boldsymbol{u}^\alpha \cdot \boldsymbol{v})\boldsymbol{u}^\alpha)\\ &+ \tfrac{1}{2}\sum (\rho_\alpha \boldsymbol{u}^\alpha \cdot \boldsymbol{u}^\alpha)\boldsymbol{u}^\alpha = -\boldsymbol{v}\boldsymbol{t} - \sum \boldsymbol{u}^\alpha \boldsymbol{t}^\alpha - \underbrace{\boldsymbol{v}\sum \rho_\alpha \boldsymbol{u}^\alpha \otimes \boldsymbol{u}^\alpha}_{\sum \rho_\alpha (\boldsymbol{v}\cdot\boldsymbol{u}^\alpha)\boldsymbol{u}^\alpha}\\ &+ \sum \rho_\alpha (\boldsymbol{v}\cdot\boldsymbol{u}^\alpha)\boldsymbol{u}^\alpha + \tfrac{1}{2}\sum (\rho_\alpha \boldsymbol{u}^\alpha \cdot \boldsymbol{u}^\alpha)\boldsymbol{u}^\alpha\\ &= -\boldsymbol{v}\boldsymbol{t} - \sum \boldsymbol{u}^\alpha \boldsymbol{t}^\alpha + \tfrac{1}{2}\sum \rho_\alpha (\boldsymbol{u}^\alpha \cdot \boldsymbol{u}^\alpha)\boldsymbol{u}^\alpha ,\end{aligned} \tag{7.8.34}$$

$$\phi_2 = \sum_{\alpha=1}^{\nu} \left(-\omega^\alpha m^\alpha + \tfrac{1}{2}\rho_\alpha s^\alpha \cdot \omega^\alpha\right) = -\omega \sum m^\alpha - \sum \Delta\omega^\alpha m^\alpha$$

$$+\tfrac{1}{2}\sum \rho_\alpha (s^\alpha \cdot \omega^\alpha) u^\alpha = -\omega m - \sum \Delta\omega^\alpha m^\alpha$$

$$\underbrace{-\omega \sum \rho_\alpha s^\alpha \otimes u^\alpha}_{\sum \rho_\alpha (\omega \cdot s^\alpha) u^\alpha} + \tfrac{1}{2}\sum \rho_\alpha (s^\alpha \cdot \omega^\alpha) u^\alpha$$

$$= \omega m - \sum \Delta\omega^\alpha m^\alpha + \tfrac{1}{2}\sum \rho_\alpha s^\alpha \cdot (\Delta\omega^\alpha - \omega) u^\alpha , \qquad (7.8.35)$$

$$\phi_3 = \sum_{\alpha=1}^{\nu} (-q^\alpha + \rho_\alpha \varepsilon^\alpha u^\alpha) . \qquad (7.8.36)$$

If these three contributions are added, one obtains

$$\phi = \phi_1 + \phi_2 + \phi_3 = -vt - \omega m + q , \qquad (7.8.37)$$

where q is defined in (7.3.23).

4. We start with the balance law of mass

$$\frac{\partial \rho_\alpha}{\partial t} + \operatorname{div}(\rho_\alpha v^\alpha) = \mathfrak{c}^\alpha \qquad (7.8.38)$$

and use $\rho_\alpha = \xi^\alpha \rho$. Product differentiation then yields

$$\frac{\partial \rho}{\partial t}\xi^\alpha + \rho \frac{\partial \xi^\alpha}{\partial t} + \xi^\alpha \operatorname{div}(\rho v^\alpha) + (\operatorname{grad} \xi^\alpha) \cdot \rho v^\alpha = \mathfrak{c}^\alpha . \qquad (7.8.39)$$

Adding and subtracting on the left-hand side the terms $\xi^\alpha \operatorname{div}(\rho v)$ and $(\operatorname{grad} \xi^\alpha) \cdot \rho v$, one obtains, upon combining the respective terms

$$\underbrace{\left(\frac{\partial \rho}{\partial t} + \operatorname{div}(\rho v)\right)}_{0} \xi^\alpha + \rho \underbrace{\left(\frac{\partial \xi^\alpha}{\partial t} + (\operatorname{grad} \xi^\alpha) v\right)}_{\mathrm{d}\xi^\alpha/\mathrm{d}t}$$

$$+ \operatorname{div} \underbrace{(\rho \xi^\alpha (v^\alpha - v))}_{j^\alpha} = \mathfrak{c}^\alpha , \qquad (7.8.40)$$

where $v = \sum_\alpha \xi^\alpha v^\alpha$ is the barycentric velocity. So (7.8.40) implies

$$\rho \frac{\mathrm{d}\xi^\alpha}{\mathrm{d}t} = -\operatorname{div} j^\alpha + \mathfrak{c}^\alpha , \qquad (\alpha = 1, 2, \ldots.\nu) , \qquad (7.8.41)$$

which is valid for all constituents.

5. With the EUCLIDian transformation

$$\boldsymbol{x}^\star = \boldsymbol{O}^\star(t)\boldsymbol{x} + \boldsymbol{c}^\star(t), \quad \boldsymbol{O}^\star \boldsymbol{O}^{\star^T} = \boldsymbol{O}^{\star^T} \boldsymbol{O}^\star = \mathbf{1} \,,$$

one obtains

$$\begin{aligned} \boldsymbol{v}^\star \quad &= \boldsymbol{O}^\star \boldsymbol{v} + \dot{\boldsymbol{O}}^\star \boldsymbol{x} + \dot{\boldsymbol{c}}^\star, \\ \operatorname{grad}^\star \boldsymbol{v}^\star &= \boldsymbol{O}^\star \operatorname{grad} \boldsymbol{v} \boldsymbol{O}^{\star T} + \dot{\boldsymbol{O}}^\star \boldsymbol{O}^{\star T} \,. \end{aligned}$$

So neither $\boldsymbol{v}$ nor $\operatorname{grad} \boldsymbol{v}$ are objective quantities under EUCLIDian transformations. However, one may easily prove that

$$\begin{aligned} \operatorname{sym} \operatorname{grad} \boldsymbol{v}^\star &= \boldsymbol{D}^\star = \boldsymbol{O}^\star \boldsymbol{D} \boldsymbol{O}^{\star T}, \\ \operatorname{skw} \operatorname{grad} \boldsymbol{v}^\star &= \boldsymbol{W}^\star = \boldsymbol{O}^\star \boldsymbol{W} \boldsymbol{O}^{\star T} + \dot{\boldsymbol{O}}^\star \boldsymbol{O}^{\star T} \,. \end{aligned}$$

Therefore, $\boldsymbol{D}$ is objective, but $\boldsymbol{W}$ is not. It follows that $\boldsymbol{v}$ and $\boldsymbol{W}$ cannot arise as independent constitutive variables, but $\boldsymbol{D}$ can do so. It is also easy to see that ρ, ξ^α, θ are objective scalars, whilst $\operatorname{grad} \xi^\alpha$, $\operatorname{grad} \theta$ are objective vectors. Therefore the constitutive relations are of the form $\psi = \hat{\psi}(\boldsymbol{\Xi})$, with $\boldsymbol{\Xi} = (\rho, \xi^\alpha, \theta, \operatorname{grad} \xi^\alpha, \boldsymbol{D}, \operatorname{grad} \theta)$, qed.

6. According to the prerequisites, $\boldsymbol{\alpha}$ and Γ are independent of $\boldsymbol{\beta}$, and $\boldsymbol{\beta}$ can assume arbitrary values. We wish to demonstrate that necessarily $\boldsymbol{\alpha} = \mathbf{0}$. To this end, assume that $\boldsymbol{\alpha} \neq \mathbf{0}$ with $\alpha_s \neq 0$ (being the s-th component of $\boldsymbol{\alpha}$) and $\alpha_j = 0$ for $j \neq s$. Then one may write $\boldsymbol{\alpha} = (\check{\boldsymbol{\alpha}}, \alpha_s)$, $\quad \boldsymbol{\beta} = (\check{\boldsymbol{\beta}}, \beta_s)$ where $\check{\boldsymbol{\alpha}}$ and $\check{\boldsymbol{\beta}}$ are "shortened" by the components α_s, β_s, respectively. With this notation, inequality (7.5.20) becomes

$$\check{\boldsymbol{\alpha}} \cdot \check{\boldsymbol{\beta}} + \alpha_s \beta_s + \Gamma \geq 0 \,. \tag{7.8.42}$$

Since the value of β_s can be freely chosen, we select

$$\beta_s = -[(\Gamma + \check{\boldsymbol{\alpha}} \cdot \check{\boldsymbol{\beta}}) + \varepsilon]/\alpha_s, \quad \varepsilon > 0 \,. \tag{7.8.43}$$

This choice is admissible, since $\alpha_s \neq 0$. Substituting (7.8.43) into (7.8.42) yields

$$-\varepsilon \geq 0,$$

a contradiction to the assumption that ε is positive, thus, α_s must vanish. This argument can be repeated for each component of $\boldsymbol{\alpha}$, so that

$$\boldsymbol{\alpha} = \mathbf{0}, \quad \text{and therefore} \quad \Gamma \geq 0 \,. \quad \text{qed.}$$

7. This proof is long and cumbersome. So read Chap. 6 of I. MÜLLER's book [165].

8. From $(7.5.29)_{1,2}$ on differentiating with respect to ρ and θ, one obtains

$$\frac{\partial^2 \hat{s}}{\partial\rho\,\partial\theta} = \Lambda^\varepsilon \frac{\partial^2 \hat{\varepsilon}}{\partial\rho\,\partial\theta} \tag{7.8.44}$$

and

$$\frac{\partial^2 \hat{s}}{\partial\theta\partial\rho} = \Lambda^\varepsilon \frac{\partial^2 \hat{\varepsilon}}{\partial\theta\partial\rho} + \frac{\partial\Lambda^\varepsilon}{\partial\theta}\left(\frac{\partial\hat{\varepsilon}}{\partial\rho} - \frac{p}{\rho^2}\right) - \Lambda^\varepsilon \frac{\partial\hat{p}}{\partial\theta}\frac{1}{\rho^2}\,. \tag{7.8.45}$$

Because the mixed derivatives $\partial^2/\partial\rho\,\partial\theta$ and $\partial^2/\partial\theta\,\partial\rho$ must be equal, if $\hat{s}$ and $\hat{\varepsilon}$ are unique functions of their arguments, there follows, by equating these expressions,

$$\frac{1}{\Lambda^\varepsilon}\frac{\partial\Lambda^\varepsilon}{\partial\theta} = \frac{\partial\hat{p}/\partial\theta}{(\partial\hat{\varepsilon}/\partial\rho)\rho^2 - \hat{p}}\,, \tag{7.8.46}$$

which agrees with $(7.5.35)_1$. Analogously, one can corroborate the identities $(7.5.35)_2$; to this end cross differentiations of $(7.5.29)_2$ and $(7.5.29)_3$ with respect to θ and ξ^α are needed, viz.,

$$\frac{\partial^2 \hat{s}}{\partial\xi^\alpha\,\partial\rho} = \Lambda^\varepsilon \frac{\partial^2 \hat{\varepsilon}}{\partial\xi^\alpha\,\partial\rho} - \frac{\partial}{\partial\xi^\alpha}\left(\frac{\Lambda^\varepsilon p}{\rho^2}\right), \quad \frac{\partial^2 \hat{s}}{\partial\rho\,\partial\xi^\alpha} = \Lambda^\varepsilon \frac{\partial^2 \hat{\varepsilon}}{\partial\rho\,\partial\xi^\alpha} - \Lambda^\varepsilon \frac{\partial\mu^{\xi^\alpha}}{\partial\rho}\,; \tag{7.8.47}$$

they yield

$$\frac{\partial\mu^{\xi^\alpha}}{\partial\rho} = \frac{\partial}{\partial\xi^\alpha}\left(\frac{p}{\rho^2}\right), \quad \forall\alpha = 1, ..., \nu - 1\,. \tag{7.8.48}$$

Similarly, from $(7.5.29)_3$ one may deduce

$$\frac{\partial\mu^{\xi^\alpha}}{\partial\xi^\beta} = \frac{\partial\mu^{\xi^\beta}}{\partial\xi^\alpha}, \quad \forall(\alpha,\beta) = 1, ..., \nu - 1\,. \tag{7.8.49}$$

Relations (7.8.47) and (7.8.48) corroborate $(7.5.36)_{1,2}$. Finally, from $(7.5.29)_{1,3}$ we deduce

$$\begin{aligned}\frac{\partial^2 \hat{s}}{\partial\xi^\alpha\,\partial\theta} &= \Lambda^\varepsilon \frac{\partial^2 \hat{\varepsilon}}{\partial\xi^\alpha\,\partial\theta} = \frac{\partial^2 \hat{s}}{\partial\theta\partial\xi^\alpha} \\ &= \Lambda^\varepsilon\left(\frac{\partial^2\hat{\varepsilon}}{\partial\theta\partial\xi^\alpha} - \frac{\partial\hat{\mu}^{\xi^\alpha}}{\partial\theta}\right) + \frac{d\Lambda^\varepsilon}{d\theta}\left(\frac{\partial\hat{\varepsilon}}{\partial\xi^\alpha} - \hat{\mu}^{\xi^\alpha}\right)\end{aligned}$$

from which follows

$$\frac{1}{\Lambda^2}\frac{d\Lambda^\varepsilon}{d\theta} = \frac{\partial\hat{\mu}^{\xi^\alpha}/\partial\theta}{\partial\hat{\varepsilon}/\partial\xi^\alpha - \hat{\mu}^{\xi^\alpha}}\,, \tag{7.8.50}$$

corroborating $(7.5.36)_2$.

9. (i) The formulas $(7.6.14)_{1,2,3}$ follow from the definition of the barycentric velocity, $\boldsymbol{v} = \sum_{\gamma=1}^{N} \xi^\gamma \boldsymbol{v}^\gamma$ by applying the product rule of differentiation; for instance

$$\operatorname{grad} \boldsymbol{v} = \sum_{\gamma=1}^{N} \left[\xi^\gamma \operatorname{grad} \boldsymbol{v}^\gamma + \boldsymbol{v}^\gamma \otimes \operatorname{grad} \xi^\gamma\right] , \tag{7.8.51}$$

from which expressions for $\boldsymbol{D}$ and $\boldsymbol{W}$ may be derived by applying the operators sym and skw.

(ii) From the definition of the diffusion velocity there follows

$$\frac{\partial \boldsymbol{u}^\beta}{\partial \boldsymbol{v}^\alpha} = \frac{\partial}{\partial \boldsymbol{v}^\alpha}\left(\boldsymbol{v}^\beta - \sum_{\gamma=1}^{N} \xi^\gamma \boldsymbol{v}^\gamma\right) = \delta^{\alpha\beta}\mathbf{1} - \xi^\alpha \mathbf{1} = \left(\delta^{\alpha\beta} - \xi^\alpha\right)\mathbf{1} \,. \tag{7.8.52}$$

In much the same way, using the definitions (7.6.12) and (7.6.14) we may derive

$$\begin{aligned}\frac{\partial \boldsymbol{U}^\beta}{\partial \boldsymbol{W}^\alpha} &= \frac{\partial}{\partial \boldsymbol{W}^\alpha}\left[\operatorname{grad} \boldsymbol{v}^\beta - \sum_{\gamma=1}^{N}\left(\xi^\gamma \boldsymbol{W}^\gamma + \operatorname{skw}(\boldsymbol{v}^\gamma \otimes \operatorname{grad} \xi^\gamma)\right)\right] \\ &= \delta^{\alpha\beta}\mathbf{1}_4 - \xi^\alpha \mathbf{1}_4 = (\delta^{\alpha\beta} - \xi^\alpha)\mathbf{1}_4 \,,\end{aligned} \tag{7.8.53}$$

in which $\mathbf{1}_4$ is the fourth order unit tensor, $(\mathbf{1}_4)_{ijkl} = \delta_{ik}\delta_{jl}$.

(iii) With the results derived in (7.8.52) and (7.8.53) we may now deduce

$$\begin{aligned}\frac{\partial f}{\partial \boldsymbol{v}^\alpha} &= \sum_\beta \frac{\partial f}{\partial \boldsymbol{u}^\beta}\frac{\partial \boldsymbol{u}^\beta}{\partial \boldsymbol{v}^\alpha} = \sum_\beta \frac{\partial f}{\partial \boldsymbol{u}^\beta}(\delta^{\alpha\beta} - \xi^\alpha)\mathbf{1} \\ &= \frac{\partial f}{\partial \boldsymbol{u}^\beta} - \xi^\alpha \sum_\beta \frac{\partial f}{\partial \boldsymbol{u}^\beta}\end{aligned} \tag{7.8.54}$$

and

$$\begin{aligned}\frac{\partial f}{\partial \boldsymbol{W}^\alpha} &= \sum_\beta \frac{\partial f}{\partial \boldsymbol{U}^\beta}\frac{\partial \boldsymbol{U}^\beta}{\partial \boldsymbol{W}^\alpha} = \sum_\beta \frac{\partial f}{\partial \boldsymbol{U}^\beta}(\delta^{\alpha\beta} - \xi^\alpha)\mathbf{1}_4 \\ &= \frac{\partial f}{\partial \boldsymbol{U}^\alpha} - \xi^\alpha \sum_\beta \frac{\partial f}{\partial \boldsymbol{U}^\beta} \,.\end{aligned} \tag{7.8.55}$$

This latter result can now be used in the evolution of $\partial f / \partial \operatorname{grad} \boldsymbol{v}^\alpha$, viz.,

$$\begin{aligned}\frac{\partial f}{\partial \operatorname{grad} \boldsymbol{v}^\alpha} &= \frac{\partial f}{\partial \boldsymbol{D}^\alpha} + \frac{\partial f}{\partial \boldsymbol{W}^\alpha} \\ &\overset{(7.8.55)}{=} \frac{\partial f}{\partial \boldsymbol{D}^\alpha} + \frac{\partial f}{\partial \boldsymbol{U}^\alpha} - \xi^\alpha \sum_\beta \frac{\partial f}{\partial \boldsymbol{U}^\beta} \,,\end{aligned} \tag{7.8.56}$$

where $\operatorname{grad} \boldsymbol{v}^\alpha = \boldsymbol{D}^\alpha + \boldsymbol{W}^\alpha$ has been used, qed.

10. (i) In one dimension there is no possibility to select between different paths of integration between two points to reach point b from point a . Every RIEMANN integral from a to a certainly vanishes.
(ii) In two dimensions we have $\mathrm{d}F = X_1\,\mathrm{d}x_1 + X_2\,\mathrm{d}x_2$. If POINCARÉ's *theorem* holds, then

$$\frac{\partial X_1}{\partial x_2} = \frac{\partial X_2}{\partial x_1} \tag{7.8.57}$$

and F is a potential such that

$$X_1 = \frac{\partial F}{\partial x_1}\,, \quad X_2 = \frac{\partial F}{\partial x_2} \overset{(7.8.57)}{\Longrightarrow} \frac{\partial^2 F}{\partial x_2 \partial x_1} = \frac{\partial^2 F}{\partial x_1 \partial x_2}\,. \tag{7.8.58}$$

It can be determined by integrating from $\boldsymbol{x}_0$ to $\boldsymbol{x}$ along any continuous curve according to

$$\boldsymbol{F}(\boldsymbol{x}) = \int_{\boldsymbol{x}_0}^{\boldsymbol{x}} (X_1(x_1,x_2)\mathrm{d}x_1 + X_2(x_1,x_2)\mathrm{d}x_2)\,. \tag{7.8.59}$$

It follows that the integral along any curve from $\boldsymbol{a}$ to $\boldsymbol{b}$ yields

$$\int_{\boldsymbol{a}}^{\boldsymbol{b}} X_1(x_1,x_2)\mathrm{d}x_1 + X_2(x_1,x_2)\mathrm{d}x_2 = \int_{\boldsymbol{a}}^{\boldsymbol{b}} \mathrm{d}F = F(\boldsymbol{b}) - F(\boldsymbol{a})\,. \tag{7.8.60}$$

If the POINCARÉ theorem is not fulfilled one may try to reach this by an integrating factor. In that case one writes

$$\mathrm{d}F_\lambda = \lambda X_1\,\mathrm{d}x_1 + \lambda X_2\,\mathrm{d}x_2 \tag{7.8.61}$$

and obtains the following condition to satisfy the POINCARÉ theorem

$$\frac{\partial \lambda X_1}{\partial x_2} - \frac{\partial \lambda X_2}{\partial x_1} = 0$$

or

$$\lambda \underbrace{\left\{\frac{\partial X_1}{\partial x_2} - \frac{\partial X_2}{\partial x_1}\right\}}_{\neq 0} = \frac{\partial \lambda}{\partial x_1} X_2 - \frac{\partial \lambda}{\partial x_2} X_1\,, \tag{7.8.62}$$

in which the expression in curly bracket differs from zero, since the starting equation does not satisfy the POINCARÉ condition. It is evident from (7.8.62) that $\lambda = \text{const}$ is not a possible integrating factor. However, the determination of the integrating factor is not unique. If $X_2 \neq 0$, one may for instance select $\lambda = \lambda(x_1)$; then it follows from (7.8.62) that

$$\frac{\mathrm{d}\lambda}{\mathrm{d}x_1} - \frac{1}{X_2}\left\{\frac{\partial X_1}{\partial x_2} - \frac{\partial X_2}{\partial x_1}\right\}\lambda = 0$$

with the solution

$$\lambda_1 = \exp\left\{\int_0^{x_1} \frac{1}{X_2}\left\{\frac{\partial X_1}{\partial x_2} - \frac{\partial X_2}{\partial x_1}\right\}\mathrm{d}\bar{x}_1\right\} . \tag{7.8.63}$$

This formula shows that λ_1 is a function of x_1 alone, if the integrand function is independent of x_2.
An analogous solution is obtained, if $X_1 \neq 0$ and $\lambda = \lambda(x_2)$ is assumed:

$$\lambda_2 = \exp\left\{\int_0^{x_2} -\frac{1}{X_2}\left\{\frac{\partial X_1}{\partial x_2} - \frac{\partial X_2}{\partial x_1}\right\}\mathrm{d}\bar{x}_2\right\} , \tag{7.8.64}$$

which shows that $\lambda = \lambda(x_2)$ if the integrand function is independent of x_1. In general λ must be determined by solving the partial differential equation (7.8.62).
(iii) In three dimensions it is known that a vector field $\boldsymbol{X}$ can be written as the gradient of a scalar potential, if and only if $\operatorname{curl}\boldsymbol{X} = \boldsymbol{0}$. If this is written in component form, then the POINCARÉ condition

$$\frac{\partial X_i}{\partial x_j} - \frac{\partial X_j}{\partial x_i} = 0 , \qquad (i,j = 1,2,3) .$$

is obtained. If $\operatorname{curl}\boldsymbol{X} \neq \boldsymbol{0}$, one may try to construct a total differential

$$\mathrm{d}F_\lambda = \lambda X_1 \mathrm{d}x_1 + \lambda X_2 \mathrm{d}x_2 + \lambda X_3 \mathrm{d}x_3 \tag{7.8.65}$$

by multiplying $\mathrm{d}F$ by an integrating factor. The condition for this to be successful is

$$\boldsymbol{0} \stackrel{!}{=} \operatorname{curl}(\lambda\boldsymbol{X}) = \lambda(\operatorname{curl}\boldsymbol{X}) + (\operatorname{grad}\lambda) \times \boldsymbol{X} . \tag{7.8.66}$$

By multiplying (7.8.66) scalarly with the vector $\boldsymbol{X}$, one obtains

$$(\operatorname{curl}\boldsymbol{X}) \cdot \boldsymbol{X} = 0 . \tag{7.8.67}$$

Therefore, $\boldsymbol{X}$ and $\operatorname{curl}\boldsymbol{X}$ must necessarily be orthogonal, qed. In Cartesian components this reads

$$\epsilon_{ijk} X_{k,j} X_i = 0 , \tag{7.8.68}$$

which is nothing else than the FROBENIUS condition. For a plane vector field, (7.8.67) is always satisfied, a fact that corroborates the statement that in two dimensions an integrating factor always exists.

11. This proposition essentially says that the problem lies in the space with dimension $n - m$.
Proof: We perform it by disproving the opposite assumption. So, assume that X_l, $l \neq k$, depends on x_k, and that $X_k = 0$. Then we can find a

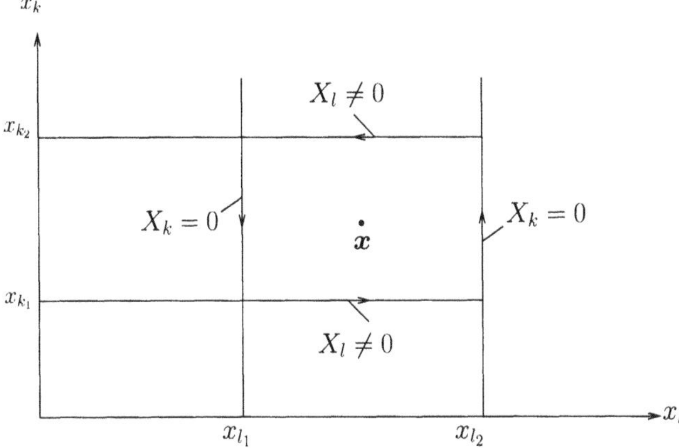

Fig. 7.1. Illustration explaining the proof of Problem 11.

closed path in phase space for which the integral of $\mathrm{d}F$ does not vanish: To this end, consider the plane (x_k, x_l) and construct a rectangle around point $\boldsymbol{x}$ as shown in Fig 7.1. Then, on using TAYLOR series expansion one has

$$\begin{aligned}\mathrm{d}F &= X_l(....x_{k_1}....)\mathrm{d}x_l + X_k(....x_{l_2}....)\mathrm{d}x_k \\ &\quad -X_l(....x_{k_2}....)\mathrm{d}x_l - X_k(....x_{l_1}....)\mathrm{d}x_k \\ &= -\frac{\partial X_l}{\partial x_k}\mathrm{d}x_k\mathrm{d}x_l + \text{higher order terms} \neq 0\,.\end{aligned}$$

This result contradicts the assumption that $\mathrm{d}F$ is a complete differential unless x_l is not a function of x_k, qed.

12. For the differential (7.7.4)

$$\mathrm{d}H = \frac{G(F)}{g}\left\{.... + \frac{\partial f}{\partial x_i}\mathrm{d}x_i + ... + \frac{\partial f}{\partial x_j}\mathrm{d}x_j + ...\right\} \qquad (7.8.69)$$

one may check with the POINCARÉ theorem that (7.8.69) is again a complete differential. To this end one forms the mixed derivatives of $\mathrm{d}H$ and checks whether the result is independent of the order of differentiation. From (7.8.69) one gets

$$\frac{\partial H}{\partial x_i} = \frac{G(F)}{g}\frac{\partial f}{\partial x_i}, \qquad \frac{\partial H}{\partial x_j} = \frac{G(F)}{g}\frac{\partial f}{\partial x_j}, \qquad (7.8.70)$$

from which the mixed derivatives take the form

$$\frac{\partial^2 H}{\partial x_i \partial x_j} = \left\{ \frac{\partial G}{\partial F} \frac{\partial F}{\partial x_i} \frac{1}{g} \frac{\partial f}{\partial x_i} + G(F) \frac{\partial}{\partial x_i} \left(\frac{1}{g} \frac{\partial f}{\partial x_j} \right) \right\} ,$$
$$\frac{\partial^2 H}{\partial x_j \partial x_i} = \left\{ \frac{\partial G}{\partial F} \frac{\partial F}{\partial x_j} \frac{1}{g} \frac{\partial f}{\partial x_i} + G(F) \frac{\partial}{\partial x_j} \left(\frac{1}{g} \frac{\partial f}{\partial x_i} \right) \right\} . \tag{7.8.71}$$

The last terms in the curly brackets of these expressions are equal. Since $\mathrm{d}F = \mathrm{d}f/g$ is a complete differential by prerequisite, this is also true for the first terms, since

$$\frac{\partial G}{\partial F} \frac{\partial F}{\partial x_i} \left(\frac{1}{g} \frac{\partial f}{\partial x_j} \right) = \frac{\partial G}{\partial F} \frac{\partial F}{\partial x_i} \frac{\partial F}{\partial x_j} ,$$
$$\frac{\partial G}{\partial F} \frac{\partial F}{\partial x_j} \left(\frac{1}{g} \frac{\partial f}{\partial x_i} \right) = \frac{\partial G}{\partial F} \frac{\partial F}{\partial x_j} \frac{\partial F}{\partial x_i} . \tag{7.8.72}$$

Therefore we have proved that the two mixed derivatives in (7.8.71) have the same value, so that $\mathrm{d}H$ is necessarily a complete differential if $\mathrm{d}F$ is one, qed.

13. So far we know that $\boldsymbol{k}$ is an objective vector of the variables (7.6.19) which is subject to the conditions (7.6.62). To exploit these restrictions we follow an argument of GURTIN [89] as presented by BAUER [22]. The diagonal elements of the derivatives of $\boldsymbol{k}$ with respect to $\operatorname{grad} T$, $\operatorname{grad} \nu_\alpha$ and $\operatorname{grad} \boldsymbol{v}^\alpha$ vanish in view of (7.6.62). Henceforth, instead of the derivatives $\partial \boldsymbol{k}/\partial \operatorname{grad} \boldsymbol{v}^\alpha$, we shall consider three derivatives with respect to the components of $\boldsymbol{v}^\alpha$, viz.,

$$\frac{\partial \boldsymbol{k}}{\partial (\operatorname{grad} v_1^\alpha)} , \qquad \frac{\partial \boldsymbol{k}}{\partial (\operatorname{grad} v_2^\alpha)} , \qquad \frac{\partial \boldsymbol{k}}{\partial (\operatorname{grad} v_3^\alpha)} , \tag{7.8.73}$$

each of which is a second order tensor. Next, if one numerates the direct vector variables as

$$\operatorname{grad} T = w^1, \quad \operatorname{grad} \nu_\alpha = w^{1+\alpha}, \quad \operatorname{grad} v_i^\alpha = w^{(N+3(\alpha-1)+i)}$$

and writes them as components of a vector $\boldsymbol{w}$, of which the components are indexed by capital latin letters, then one has

$$\frac{\partial^3 k_i}{\partial w_j^A \partial w_k^B \partial w_l^C} = 0 , \tag{7.8.74}$$

identically for $i, j, k, l \in (1,2,3)$, $A, B, C \in (1, 2, \ldots\ldots, N)$. Equation (7.8.74) holds, since either one of the indices, j, k, l, agrees with i and $\partial k_i / \partial w_i^A = 0$, according to (7.6.62), or else one of the indices j, k, l arises twice, in which case one of them may be "exchanged from the denominator to the numerator" (since $\partial k_i / \partial w_j^A = \partial k_j / \partial w_i^A$, again according to (7.6.62)), yielding again the first case, qed. Thus, writing (7.8.74) as

$$\underbrace{\frac{\partial^2}{\partial w_j^A \, \partial w_k^B} \left(\frac{\partial k_i}{\partial w_l^C} \right)}_{\mathcal{K}_{il}^C} = 0 \tag{7.8.75}$$

shows that the tensors $\mathcal{K}_{il}^C$ are skewsymmetric (because of (7.6.62)) and linear in $\boldsymbol{w}$. Now, according to (7.6.10) $\mathcal{K}_{il}^C$ is an isotropic function of the vectors $\boldsymbol{w}^A$, $\boldsymbol{v}^\alpha$ (and scalars T, ν_α) which means it is a linear combination of the unit tensors $\mathbf{1}$ and the quadratic products $\boldsymbol{w}^A \otimes \boldsymbol{w}^B$, $\boldsymbol{v}^\alpha \otimes \boldsymbol{w}^B$, $\boldsymbol{v}^\alpha \otimes \boldsymbol{v}^\beta$ with scalar coefficient functions. A dependence on $\mathbf{1}$ is not possible because it is symmetric, and the above tensor products cannot arise because they are either quadratic in $\boldsymbol{w}$ or involve $\boldsymbol{v}^\alpha$ which must vanish in thermodynamic equilibrium. Thus $\left(\mathcal{K}_{il}^C\right)^{\text{equil}} = 0$, which in the notation of (7.6.62) takes the form

$$\begin{aligned} \left(\frac{\partial \boldsymbol{k}}{\partial \operatorname{grad} T} \right)^{\text{equil}} &= \mathbf{0}\,, \\ \left(\frac{\partial \boldsymbol{k}}{\partial \operatorname{grad} \nu_\alpha} - \frac{\partial \boldsymbol{k}}{\partial \operatorname{grad} \nu_N} \right)^{\text{equil}} &= \mathbf{0}\,, \\ \left(\frac{\partial \boldsymbol{k}}{\partial \operatorname{grad} \boldsymbol{v}^\alpha} \right)^{\text{equil}} &= \mathbf{0}\,. \end{aligned} \tag{7.8.76}$$

These are stronger constraints on $\boldsymbol{k}$ than (7.6.62), but they are equally also restricted to thermodynamic equilibrium.
Since $\boldsymbol{k}$ is an isotropic vector function of its arguments it must necessarily be an explicit function of the form

$$\boldsymbol{k} = a_\alpha \boldsymbol{v}^\alpha + b \operatorname{grad} T + c_\alpha \operatorname{grad} \nu_\alpha\,, \tag{7.8.77}$$

in which a_α, b, c_α depend on scalar combinations of the variables listed in (7.6.10). In thermodynamic equilibrium $\boldsymbol{v}^\alpha = \mathbf{0}$, $\operatorname{grad} T = 0$ and $c_\alpha|_{\text{equil}} = 0$ because of $(7.8.76)_2$. Since also $\partial \boldsymbol{k}/\partial \nu_\alpha$ and $\partial \boldsymbol{k}/\partial T$ have representations of the form (7.8.77), we may summarise

$$\begin{gathered} \left(\frac{\partial \boldsymbol{k}}{\partial \nu_\alpha} \right)^{\text{equil}} = \mathbf{0}, \quad \left(\frac{\partial \boldsymbol{k}}{\partial T} \right)^{\text{equil}} = \mathbf{0}, \\ \boldsymbol{k}^{\text{equil}} = \boldsymbol{k}(T, \nu_\alpha, \boldsymbol{v}^\alpha = \mathbf{0}, \operatorname{grad} T = \mathbf{0}, \operatorname{grad} \nu_\alpha, \operatorname{grad} \boldsymbol{v}^\alpha = \mathbf{0}) = \mathbf{0}\,. \end{gathered} \tag{7.8.78}$$

It follows that close to thermodynamic equilibrium the extra entropy flux vector may be represented as

$$\boldsymbol{k} \simeq \sum_{\alpha=1}^{N} \left(\frac{\partial \boldsymbol{k}}{\partial \boldsymbol{v}^\alpha} \right)^{\text{equil}} \boldsymbol{v}^\alpha\,, \tag{7.8.79}$$

in which the coefficient matrices are functions of T and ν_α. With the aid of (7.6.14), this may alternatively be written as

$$\begin{aligned}
\boldsymbol{k} &= \sum_{\alpha=1}^{N} \left\{ \frac{\partial \boldsymbol{k}}{\partial \boldsymbol{u}^\alpha} - \xi^\alpha \left(\sum_{\gamma=1}^{N} \frac{\partial \boldsymbol{k}}{\partial \boldsymbol{u}^\gamma} \right) \right\}^{\text{equil}} (\boldsymbol{u}^\alpha + \boldsymbol{v}) \\
&= \sum_{\alpha=1}^{N} \left(\frac{\partial \boldsymbol{k}}{\partial \boldsymbol{u}^\alpha} \right)^{\text{equil}} \boldsymbol{u}^\alpha - \underbrace{\left(\sum_{\alpha=1}^{N} \xi^\alpha \boldsymbol{u}^\alpha \right)}_{0} \left(\sum_{\gamma=1}^{N} \frac{\partial \boldsymbol{k}}{\partial \boldsymbol{u}^\gamma} \right)^{\text{equil}} \\
&\quad + \left(\sum_{\alpha=1}^{N} \frac{\partial \boldsymbol{k}}{\partial \boldsymbol{u}^\alpha} \right)^{\text{equil}} \boldsymbol{v} - \underbrace{\left(\sum_{\alpha=1}^{N} \xi^\alpha \right)}_{1} \left(\sum_{\gamma=1}^{N} \frac{\partial \boldsymbol{k}}{\partial \boldsymbol{u}^\gamma} \right)^{\text{equil}} \boldsymbol{v} \\
&= \sum_{\alpha=1}^{N} \left(\frac{\partial \boldsymbol{k}}{\partial \boldsymbol{u}^\alpha} \right)^{\text{equil}} \boldsymbol{u}^\alpha .
\end{aligned} \tag{7.8.80}$$

This completes the proof of (7.6.64).

14. It is a simple matter to prove relation (7.7.6) if $\boldsymbol{t}_v = -p_v \mathbf{1} + \boldsymbol{\sigma}'$ is substituted in (7.7.5). Setting $c = 0$ in (7.7.6) yields

$$\boldsymbol{E} = b\boldsymbol{\sigma}', \ \operatorname{tr} \boldsymbol{D} = 3(a - bp_v) . \tag{7.8.81}$$

These formulas show that two independent experiments suffice, in principle, to determine the functions a and b.

Consider *simple shearing*, for which

$$\boldsymbol{E} = \boldsymbol{D} = \begin{pmatrix} 0 & \dot{\gamma}/2 & 0 \\ \dot{\gamma}/2 & 0 & 0 \\ 0 & 0 & 0 \end{pmatrix}, \ \operatorname{tr} \boldsymbol{D} = 0 . \tag{7.8.82}$$

It is feasible that for such a special motion a viscous pressure cannot develop, implying that $p_v = 0$. It follows that under those circumstances $a = 0$. It may therefore be advantageous to write $a = \tilde{a} p_v$ and thus

$$\boldsymbol{E} = b\boldsymbol{\sigma}', \ \operatorname{tr} \boldsymbol{D} = 3(\tilde{a} - b)p_v . \tag{7.8.83}$$

With this choice we have for simple shearing

$$II_E = \frac{\hat{\gamma}^2}{4}, \ \ III_E = 0, \ \ II_{\boldsymbol{\sigma}'} = \tau^2, \ \ III_{\boldsymbol{\sigma}'} = 0 \tag{7.8.84}$$

and therefore

$$b(\cdot, 0, \tau^2) = \frac{\dot{\gamma}}{2\tau} . \tag{7.8.85}$$

Thus measuring $\dot{\gamma}$ and τ determines the function b only when $p_v = 0$.

For *uniaxial compression* we have

$$D = \begin{pmatrix} \dot{\varepsilon} & 0 & 0 \\ 0 & 0 & 0 \\ 0 & 0 & 0 \end{pmatrix}, \quad E = \begin{pmatrix} \frac{2}{3}\dot{\varepsilon} & 0 & 0 \\ 0 & -\frac{1}{3}\dot{\varepsilon} & 0 \\ 0 & 0 & -\frac{1}{3}\dot{\varepsilon} \end{pmatrix}, \quad E^2 = \begin{pmatrix} \frac{4}{9}\dot{\varepsilon}^2 & 0 & 0 \\ 0 & \frac{1}{9}\dot{\varepsilon}^2 & 0 \\ 0 & 0 & \frac{1}{9}\dot{\varepsilon}^2 \end{pmatrix}, \tag{7.8.86}$$

(7.8.87)

so that

$$\operatorname{tr} D = \dot{\varepsilon}, \qquad II = \tfrac{1}{3}\dot{\varepsilon}^2 \,. \tag{7.8.88}$$

Alternatively,

$$t_v = \begin{pmatrix} -\sigma & 0 & 0 \\ 0 & 0 & 0 \\ 0 & 0 & 0 \end{pmatrix} = \underbrace{-\frac{1}{3}\sigma \begin{pmatrix} 1 & 0 & 0 \\ 0 & 1 & 0 \\ 0 & 0 & 1 \end{pmatrix}}_{-p_v \mathbf{1}} + \underbrace{\begin{pmatrix} -\frac{2}{3}\sigma & 0 & 0 \\ 0 & \frac{1}{3}\sigma & 0 \\ 0 & 0 & \frac{1}{3}\sigma \end{pmatrix}}_{\boldsymbol{\sigma}'},$$

$$p_v = \tfrac{1}{3}\sigma, \quad II_{\boldsymbol{\sigma}'} = \tfrac{1}{3}\sigma^2 \,. \tag{7.8.89}$$

Thus, from (7.8.83)

$$b\left(\cdot, \tfrac{1}{3}\sigma, \tfrac{1}{3}\sigma^2\right) = \frac{\dot{\varepsilon}}{\sigma}, \quad \tilde{a}\left(\cdot, \tfrac{1}{3}\sigma, \tfrac{1}{3}\sigma^2\right) = 2\frac{\dot{\varepsilon}}{\sigma} \,, \tag{7.8.90}$$

which shows that for uniaxial compression the functions $\tilde{a}$ and b are not independent if σ and $\dot{\varepsilon}$ are measured. An independent determination of $\tilde{a}$ and b requires bi- and triaxial experiments.

Part II

Dimensional Analysis

8. Theoretical Foundation of Dimensional Analysis[1]

Physical problems are described by relationships, which are dominated by quantities having a certain dimension, such as length, time, mass, force, temperature etc. These relations must be so structured that dependent and independent quantities are combined so as to yield dimensionally correct formulas. For instance, a physically correctly written formula must possess on each of its sides, left and right, the same physical dimension. Similarly, in an equation which describes a physical fact, quantities with different dimensions cannot be added. Such properties are connected with what is called *dimensional homogeneity.* It holds for all mathematical expressions describing physical facts. In other applied sciences, for instance mathematical economy, dimensional homogeneity is not requested to hold, a fact that allows equations with more general structure.

Dimensional analysis is a method with the aid of which one may for instance test a formula for its dimensional correctness[2]. It leads to a first understanding of the solution of a physical problem and yields a precise information about the number of variables that are necessary to describe it, a fact that is particularly important when experiments are being performed. Very often dimensional analysis reduces the number of variables upon which a physical problem was initially surmised to depend. If for instance the quantity y depends upon x_1, x_2, ..., x_n, where all quantities have a certain physical dimension, then dimensional analysis shows that y can only depend upon certain products of powers of x_1, ..., x_n, a fact that corresponds regularly to a considerable reduction of the number of variables. Naturally then, experiments may more simply or more economically be performed than without

[1] This and the following chapters are thoroughly revised and extended versions of a chapter on dimensional analysis and model theory of HUTTER: "Fluid- und Thermodynamik - eine Einführung" which appeared in the German Language by Springer Verlag, Berlin, etc. [109].

[2] The topic, presented here in a relatively brief chapter, is a popular theme in fluid mechanics and is the subject of several books, e. g., BARENBLATT [19], GÖRTLER [84], LANGHAAR [125], SPURK [227] and others. A mathematical theory based on a system of axioms but equally judiciously constructed with a constant view towards the best practices of dimensional analysis is given by CARLSON [43], [44], who gives a extensive list of related references.

knowledge of this fact. It shall be seen in the sequel, how this statement must be interpreted.

8.1 Notation

Variables for physical quantities possess normally physical dimensions. If we consider the dimension of a variable f, then we shall in this book use the symbol $[f]$. Thus the velocity V has dimension $[V] = [L][T^{-1}]$,where L is a length and T a time. Dimensionless quantities, which are computed from quantities with dimension, will be written as hallow letters, $\mathbb{R}$, $\mathbb{W}$, etc.

List of the Symbols Used

[A]	......	AMPERÈ, unit for the electric field strength
A	....	Area
$[A]_j$	.....	Derived unit
a	....	Acceleration
a_{ij}	......	Element of a matrix
$\mathbb{B}$	....	BOUSSINESQ number
$\mathbb{B}_r$	....	BRINKMANN number
[cd]	......	Candela, unit for light intensity
C_D	......	Drag coefficient
D	....	Typical diameter/diffusion coefficient
dyne	....	Force unit in the CGS–system, it corresponds to a force, which accelerates one gram of mass by 1 $\mathrm{cm\,s^{-2}}$
$\mathfrak{d}$	....	Diameter of a molecule
$E^{(\mathrm{j})}$	.....	$(j = 3, 2, 1)$ energy of a point, line and surface sources due to explosion
$\mathbb{E}_c$	....	ECKERT number
$\mathbb{E}_d$	....	Dissipation number
$\mathbb{E}_k^v$	....	EKMAN number
$\mathbb{E}_u$	......	EULER number
e	....	Excentricity
f	....	Symbol for a general function
$\mathbb{F}, \mathbb{F}_r$	....	FROUDE number
$1/\mathbb{F}_r$	.....	RICHARDSON number
$\mathfrak{f}$	....	Intermolecular interaction force
$[G]_j$	.....	Fundamental unit, or basic unit, e. g., length, time, mass, temperature
g	....	Gravity constant $\sim 9,81\ \mathrm{m\,s^{-2}}$
$\mathfrak{G}$	....	Constant arising in the law describing the intermolecular forces
h	....	Height, piezometric height
J	....	JOULE = NEWTON*meter* = N m, unit for work

[K]	KELVIN, unit for absolute temperature
K	Force
K_j	Transformation factor for dimensional quantities
k_j	Exponent of dimensional quantities
$\mathfrak{k}$	Exponent in the intermolecular force law
kg*	1 kg* = 1 kp, force of a kg-mass under the action of gravity
[kg]	Kilogram mass
L, l	Typical lengths
M	Moment of a bearing
$\mathcal{M}$	Non-dimensionalized moment of a bearing
$\mathbb{M}$	MACH number
[m]	Meter, unit of length in the International System
[mol]	Unit of amount of substance
$\mathfrak{m}$	Mass of a molecule
N	NEWTON, unit of force, corresponds to a force, which accelerates 1 kg mass by 1 $\mathrm{m\,s^{-2}}$
$\mathbb{N}$	NUSSELT number
P, p	Pressure
$\mathbb{P}$	Pressure coefficient
$\mathbb{P}_e$	PECLET number
$\mathbb{P}_r$	PRANDTL number
Q	Discharge rate
Q_j	General, dimensional variable
$\mathbb{R}$	REYNOLDS number
$\mathbb{R}_l$, $\mathbb{R}_*$ ...	Local REYNOLDS numbers
$\mathbb{R}_i$	RICHARDSON number
$\mathbb{R}_a$	RAYLEIGH number, radiation number
$\mathbb{R}_o$	ROSSBY number
$R_f^{(j)}$	$(j = 3, 2, 1)$ distance of a shock front from its explosion centre in $3, 2, 1$ dimensions
r	Radius, distance
$\mathbb{S}_0$	SOMMERFELD number
$\mathbb{S}^{\alpha\beta}$	SCHMIDT numbers
$\mathbb{S}_t$	STROUHAL number
[s]	Second, unit for time
s	Diameter of a cloud
$\langle s \rangle$	Standard deviation of s
$[T]$	Unit of time
T	Period
t	Time
U, u, $\bar{u}$ ..	Velocity, mean velocity
u^*	$= \sqrt{\tau/\rho}$ shear velocity
u_w	Circumferential velocity

V, v ..... Velocity, approaching velocity
W .... WATT, $1\ \mathrm{W} = 1\,\mathrm{N\,m\,s^{-1}}$ unit for power of working in the International System of units (MKS)
$\mathbb{W}$ .... WEBER number
x_j .... Value of a derived quantity $[A]_j$
y .... Derived quantity,
x, y, z ... Cartesian coordinates
α .... Typical angle, coefficient of thermal expansion
α_k .... Ratio of old basic fundamental unit to new basic unit
Γ .... $= 6.7 \times 10^{-11}\ \left[\mathrm{m^3\,kg^{-1}\,s^{-2}}\right]$ universal gravity constant
$\triangle$ .... Symbol for a small quantity, determinant
δ .... Boundary layer thickness
ϵ .... Internal energy
ε .... Turbulent dissipation rate
η .... Dynamic viscosity
ν .... Kinematic viscosity
θ .... Temperature
κ .... Coefficient of thermal diffusion
λ .... Drag coefficient in pipes
μ .... Dynamic viscosity
Π, Π_j ... Dimensionless power product of dimensional quantities
ρ .... Mass density
σ .... Surface tension
$[\sigma]$ .... Density anomaly
τ .... Shear stress

8.2 Systems of Physical Units and Dimensions

Physical quantities have dimensions such as length, time, mass, force, temperature, velocity, etc. Some of these must be introduced as *basic or fundamental units*, others are then obtained as derived quantities. Depending upon which fundamental units are chosen, a different system characterising the physical quantities is chosen. As one of the most important laws we mention NEWTON's second law which may in this context be stated as

$$\text{Force} = \text{Mass} \times \text{Acceleration}$$

and is not exactly correct in this form but sufficient to illustrate the point. With this law the most important transformations can be performed. If the mass is set equal to unity, 1, and the acceleration equal to the acceleration due to gravity, g, then the force equals g. It follows: *The weight of a unit mass equals exactly g units of the forces.*

Until the 60s of the last century and also later, many different systems of units were in use. Today the *International System of Units* is used almost exclusively. However, still other systems are customary especially for daily use which is why we briefly mention here a few.

- **The CGS–system** Here the fundamental units are the centimeter, the gram (of mass) and the second. According to its definition 1 g mass is the mass of 1 cm^3 water at 4° C at normal pressure of 1 atmosphere. This definition makes the mass a unit that is independent of position. The mass of different bodies can be compared with cantilever balances. According to NEWTON's law the unit of force is that force which assigns to 1 gram mass an acceleration 1 $\mathrm{cm\,s}^{-2}$. The force unit is called 1 "*dyne*". The weight of a unit mass is therefore $\sim$ 981 dyne, and it is obviously position dependent, since the acceleration due to gravity is position dependent. The unit of work is "*dyne cm*" also called "*erg*".
- **The MKS-system** Contrary to the CGS-system, the units are here the meter, kilogram mass and second. The unit of force is here that force, which assigns to 1 kg mass an acceleration of 1 $\mathrm{m\,s}^{-2}$ and it is called "*Newton*" (N). The unit of work is 1 "*Newtonmeter*" (N m) or 1 "*Joule*" (J), that of the power 1 "*Watt*" ($\mathrm{W} = 1\,\mathrm{N\,m\,s}^{-1}$).
- **The MKS–Force–system** As fundamental unit the force unit $\mathrm{kg}^* = \mathrm{kp}$ is chosen instead of the unit of mass. Hereby 1 kg^* is defined as the weight of 1 kg mass subject to the acceleration of gravity. This definition makes the MKS–Force system position dependent, so it is actually an inconvenient system. According to NEWTON's law the unit of mass takes the dimension $\mathrm{kg}^*\,\mathrm{m}^{-1}\,\mathrm{s}^{-2}$.
  Further systems of units are still used in the United States of America and for daily use in some English speaking countries, however they enjoy less and less popularity and will eventually disappear. These systems are based on the "pound" as the unit of mass or force and the "inch" as unit of length.
- **The International System of Units** The SI–System is the extended MKS–system with the following fundamental units:

Meter	[m]	as unit of lengths,
Second	[s]	as unit of time,
Kilogram	[kg]	as unit of mass,
Kelvin	[K]	as unit for the absolute temperature,
Ampère	[A]	as unit of electric current,
Mol	[mol]	as unit of substance,
Candela	[cd]	as unit of light intensity.

When expressing physical quantities in SI–units often large or small numbers arise. To avoid this, one employs in such cases multiples or parts of the SI–units in powers of ten and denotes them by a prefix; Table 8.1 collects these standard *prefix-notations*. For every dimension of a physical quantity, such as length, velocity, acceleration, temperature, specific heat, etc, a unit must be chosen (e. g., meter, inch, or kilogram or pound), which to a certain

Table 8.1. Prefixes to characterize powers of 10 of units in the SI system

Prefix	Symbol	Power	Prefix	Symbol	Power
Exa	E	10^{18}	Dezi	d	10^{-1}
Peta	P	10^{15}	Zenti	c	10^{-2}
Tera	T	10^{12}	Milli	m	10^{-3}
Giga	G	10^{9}	Mikro	μ	10^{-6}
Mega	M	10^{6}	Nano	n	10^{-9}
Kilo	K	10^{3}	Piko	p	10^{-12}
Hekto	h	10^{2}	Femto (Fermi)	f	10^{-15}
Deka	da	10	Atto	a	10^{-18}

extent may freely be chosen. However, we tend to think of mass, length, time, temperature and electric field strength to be freely assignable, so that their units can be arbitrarily assigned, whilst all units of the other dimensions depend on these. There is no apparent reason to fix the fundamental units, as it is done in this text; with similarly convincing justification, other dimensions could be chosen as fundamental, for instance force instead of mass, as these are related to one another by NEWTON's law. Likewise, the fact that seven dimensions are chosen as independent fundamental units is a consequence of more or less arbitrary definition. For instance, the kinetic theory of gases teaches us that the temperature of a gas is proportional to the kinetic energy of the fluctuating motion of the molecules. Thus temperature can be reduced to the dimensions of length and time. This fact is clearly recognisable in the history of the theory of heat, in which the equivalence of heat and energy first had to be recognised; it led to the first law of thermodynamics – and even more conspicuously – to the determination of the equivalence of heat and energy by ROBERT MAYER and PRESCOTTE JOULE[3]. A further example is the astronomical system of units. According to NEWTON's law of gravitation between two bodies with masses m and m', which are a distance r apart from each other, the force by which these bodies attract each other is given by

$$K = \Gamma \frac{m\, m'}{r^2} , \quad \Gamma = 6.7 \times 10^{-11} \; \left[\mathrm{m}^3\, \mathrm{kg}^{-1}\, \mathrm{s}^{-2}\right] , \tag{8.2.1}$$

where Γ is the gravitational constant and K the modulus of the force. The astronomical system of units is obtained if $\Gamma = 1$ is chosen. From NEWTON's law in this system one obtains

[3] JULIUS ROBERT MAYER (1814–1878) was physician but published between 1842–1848 on the balance of energy, culminating into the first law of thermodynamics. JAMES PRESCOTTE JOULE (1818–1889) was an independent private scientist and performed experiments on the thermal equivalent, see e.g. MAYER [152] and JOULE [117].

$$[m\,a] = \left[m\,m'\,r^{-2}\right] \qquad \Rightarrow \qquad \left[MLT^{-2}\right] = \left[M^2L^{-2}\right] \tag{8.2.2}$$

or

$$[M] = \left[L^3T^{-2}\right] \tag{8.2.3}$$

and therefore

$$[K] = \left[L^4T^{-4}\right] . \tag{8.2.4}$$

Mass and force are therefore already given if the dimensions length and time are assigned. This is a result one would not have expected.

The above examples demonstrate that the number of independent fundamental units in different systems must by no means be the same. Depending on one's choice one may expect a different number of independent units. In the most common system the fundamental units are restricted to seven basic dimensions. We shall restrict ourselves henceforth to this case.

8.3 Theory of Dimensional Equations

8.3.1 Dimensional Homogeneity

In this section a preliminary definition of *dimensional homogeneity* will be proposed. A more precise definition will be given later.

Definition 8.1 *An equation is called homogeneous in its dimensions or dimensionally homogeneous, if the form of this equation does not depend upon the choice of the fundamental units.* ■

As an example, consider the period of a mathematical pendulum; it is given by

$$T = 2\pi\sqrt{\frac{L}{g}} . \tag{8.3.1}$$

This equation holds in this form irrespective in which units the length L of the pendulum and the Earth's acceleration, g, is measured, in km, cm, miles and second, hours etc; its value is always correctly obtained in units of the dimension "time" that was chosen. If, instead, we choose $g = 9.81\,\mathrm{m\,s^{-2}}$ and substitute this above, then

$$T = \frac{2\pi}{\sqrt{9.81}}\sqrt{L} . \tag{8.3.2}$$

This form of the equation is still correct, however, it is restricted in its application to the Earth's surface at positions where g $= 9.81$ $\mathrm{ms^{-2}}$. Furthermore, it restricts itself to a particular system of units. Only if the value for L is substituted in [m] and the emerging value for T is interpreted in seconds, a correct answer is obtained from this second formula. This is obviously so because the factor $2\pi/\sqrt{9.81}$ is fraught with a physical dimension expressed in SI–units.

From a mathematical point of view the dimensionally homogeneous functions are a special class of functions. However, the applications of dimensional analysis to practical problems are based on the hypothesis that the solution of a problem leads to dimensionally homogeneous functions if only the independent variables are correctly selected. This hypothesis is simply justified by the fact that the *fundamental equations of physics are dimensionally homogeneous* and that deductions performed with such equations again lead to dimensionally homogeneous equations. However, there is no a priori reason to assume that unknown equations are equally dimensionally homogeneous, except if all parameters which describe the physical phenomenon are taken into account. For instance the drag force K of a body submerged in a moving fluid and held fixed depends upon the density of the fluid, its viscosity, the upstream velocity V and a characteristic length of the body D. For a given fluid, the density and viscosity of the fluid are fixed and need not be listed among the variables defining the functional dependence of the force, so that

$$K = f(V, D) .$$

This equation is not dimensionally homogeneous, because by combining V and D a quantity with the dimension of a force can never be formed. Despite this fact it is meaningful for a given fluid.

The first step in a dimensional analysis consists in the listing of the parameters, which influence a physical problem. This step is very decisive. If too many variables are listed that may describe a physical problem, then the final equations will contain superfluous variables, if too few variables are introduced, incomplete equations may emerge, which results in incomplete equations or "more often" false inferences or the result can not be expressed in terms of dimensionally homogeneous functions.

Let us explain these thoughts with the aid of a simple example; we ask, what kind of functional relation defines the drag force on a body, submerged in a steadily moving fluid that is held fixed. Typical variables are in this case the force K, a typical length L, a typical velocity V, the density ρ and the dynamic viscosity η of the fluid, the speed of sound c of the fluid, the acceleration of gravity g, the surface tension σ and the angle α between the directions of the approaching fluid and a chosen distinct orientation of the body. Since this latter variable is already dimensionless, we shall omit it in the subsequent analysis. Consequently, one may postulate the following functional relation

$$K = f(L, V, \rho, g, \eta, c, \sigma) . \tag{8.3.3}$$

If the number of independent variables on the RHS of this equation is complete, then the dimension of f must be that of a force. With the eight independent variables the following *dimensionless quantities* can be formed

$$
\begin{array}{llll}
\text{Reynolds number} & \mathbb{R} & = \dfrac{VL\rho}{\eta} & = \dfrac{VL}{\nu} \,, \\
\text{Froude number} & \mathbb{F} & & = \dfrac{V^2}{Lg} \,, \\
\text{Mach number} & \mathbb{M} & & = \dfrac{V}{c} \,, \\
\text{Weber number} & \mathbb{W} & & = \dfrac{\rho V^2 L}{\sigma} \,, \\
\text{Pressure ratio} & \mathbb{P} & & = \dfrac{K}{\rho V^2 L^2} \,.
\end{array}
$$

Hence, one may rewrite the above equation as

$$K = \rho V^2 L^2 \tilde{f}(\mathbb{R}, \mathbb{F}, \mathbb{M}, \mathbb{W}) \,. \tag{8.3.4}$$

Accordingly, the drag force K of a still body submerged in a steady moving fluid is known once $\tilde{f}$ is known. Contrary to (8.3.3) $\tilde{f}$ is a function of only four variables instead of the previously seven. The number of variables can still be further reduced under somewhat simplified situations. For small upstream velocities V the MACH number is small, $\mathbb{M} \ll 1$; thus the dependence on it may be dropped without essential restriction. The same inferences hold true for $\mathbb{F}$ and und $\mathbb{W}$ if the surface effects are not relevant and the body is completely submerged. Approximately one therefore obtains

$$K = \rho V^2 L^2 f(\mathbb{R}) \,. \tag{8.3.5}$$

The function $f(\mathbb{R})$ only depends on a single variable and satisfies, since the direction of flow changes the direction of the force, the condition $f(-\mathbb{R}) = -f(\mathbb{R})$.

One can demonstrate, and this will be the subject of this chapter, that the introduced dimensionless quantities are complete; this means that any other combination of dimensionless products of powers of the quantities L, V, ρ, η, g, c, σ, K is reducible to the dimensionless products $\mathbb{R}$, $\mathbb{F}$, $\mathbb{M}$, $\mathbb{W}$ and $\mathbb{P}$. In fact

$$\frac{V^3\rho}{\eta g} = \mathbb{R}\mathbb{F} \,, \quad \frac{\rho K}{\eta^2} = \mathbb{R}^2\mathbb{P} \,, \tag{8.3.6}$$

and so forth. It will later be proved, that every dimensionless product of the above set of dimensional variables can be represented as

$$\mathbb{R}^{a_1}\mathbb{P}^{a_2}\mathbb{F}^{a_3}\mathbb{M}^{a_4}\mathbb{W}^{a_5} \,, \tag{8.3.7}$$

where a_1, ..., a_5 are real numbers. On the other hand, one can show that $\mathbb{R}$, $\mathbb{F}$, $\mathbb{M}$, $\mathbb{W}$ and $\mathbb{P}$ are independent of each other. One can see this simply, because η is contained alone in the definition of $\mathbb{R}$, g arises only in $\mathbb{F}$, c only in $\mathbb{M}$, σ only in $\mathbb{W}$ and K only in $\mathbb{P}$. This discussion therefore suggests the following definition:

Definition 8.2 *A set of dimensionless products of given variables is complete, if each product in this set is independent of any other and if any such product of dimensional variables that does not belong to the set, can be expressed as a product of powers of the dimensionless products of the set.*

8.3.2 Theorem of BUCKINGHAM

In the last subsection a preliminary definition of the term dimensional homogeneity was given. If an equation is formed by terms all of which are dimensionally homogeneous, then this equation is trivially *dimensionally homogeneous*, because it does not depend upon the choice of the fundamental units by prerequisite. Therefore the following statement holds: *Sufficient condition for an equation to be dimensionally homogeneous is, that this equation can be reduced to an equation of dimensionless products.* An exact proof of this will be given later. BUCKINGHAM[4] has now shown that the above statement is also necessary. We therefore have the following theorem:

Theorem 8.3.1 (BUCKINGHAM). *If an equation is dimensionally homogeneous, it can be reduced to a relation of dimensionless products.* ■

This theorem gave dimensional analysis the tremendous impetus which it still enjoys. However already in the 19th century RAYLEIGH[5] conducted studies involving dimensional analysis. They differ from those of BUCKINGHAM and are mathematically more restrictive but yield essentially the same result.

Example 8.1 Drag Force of a Body In order to compare the two methods, let us look once more at the problem of *the drag force of a still body submerged in a steadily moving fluid.* The variables are the same as before. Choose the following ansatz

$$K = V^a D^b \rho^c \nu^d \, , \tag{8.3.8}$$

in which V, D, ρ, ν and K are a typical velocity, length, the fluid density, kinematic viscosity and the drag force, and a through d are exponents to be determined. To construct a dimensionally homogeneous formula, we write on the left and right the dimensions of the variables, i.e.,

$$[M\,L\,T^{-2}] \;=\; [L\,T^{-1}]^a\,[L]^b\,[M\,L^{-3}]^c\,[L^2\,T^{-1}]^d \tag{8.3.9}$$

and obtain by comparing the exponents of the quantities with the same physical dimension

[4] BUCKINGHAM, E. [39]. According to LANGHAAR [125], "BUCKINGHAM himself did not rigorously prove the theorem, although he presented evidence to make its truth seem plausible". Today there are several proofs available. We follow in spirit LANGHAAR [125]. Other proofs may e.g. be found in GÖRTLER [84] and SPURK [227].

[5] LORD RAYLEIGH (STRUTT, J.W.S.) [191] [192].

$$\begin{aligned} M &: \quad 1 = c\,, \\ L &: \quad 1 = a + b - 3c + 2d\,, \\ T &: -2 = -a - d\,, \end{aligned} \tag{8.3.10}$$

where M, L, T stand for mass, length and time, respectively. These equations possess the solution $c = 1, b = a, d = 2 - a$, so that

$$K = \rho\, V^a\, D^a\, \nu^{2-a} \tag{8.3.11}$$

or

$$K = \rho\, V^2\, D^2 \left(\frac{V\,D}{\nu}\right)^{a-2} = \rho\, V^2\, D^2\, \mathbb{R}^n\,, \quad n = a - 2\,.$$

n is an unspecified numerical coefficient, which we choose as an integer. As this number can still be arbitrarily assigned, we may take a linear combination of all positive powers of n, so that the drag force may, more generally, be written as

$$K = \rho V^2 D^2 \sum_{n=1}^{\infty} A_n \mathbb{R}^n = \rho V^2 D^2 f(\mathbb{R})\,. \tag{8.3.12}$$

$f(\mathbb{R})$ is an unspecified analytic function of the REYNOLDS number and can be determined by experiment. Prerequisite of the correctness of this argumentation is, however, the assumption, that $f(\mathbb{R})$ is representable as a TAYLOR series expansion of $\mathbb{R}$. This restricts $f(\mathbb{R})$ mathematically to analytic functions yet, physically this does not comprise a true limitation of the method.

According to the method of BUCKINGHAM, the dimensionless products of the variables K, V, D, ρ and ν are determined. These are given by

$$\mathbb{P} = \frac{K}{\rho V^2 D^2}\,, \quad \mathbb{R} = \frac{VD}{\nu}\,, \tag{8.3.13}$$

so that a relation of the form

$$f(\mathbb{P}, \mathbb{R}) = 0 \tag{8.3.14}$$

must exist. Explicitly written this may alternatively be expressed as

$$\mathbb{P} = f(\mathbb{R}) \text{ or } K = \rho V^2 D^2 f(\mathbb{R})\,, \tag{8.3.15}$$

which is the same result as before, that was deduced with the RAYLEIGH method. The two approaches are different in their mathematical execution but not in spirit. Nevertheless, BUCKINGHAM's theorem frees us from the construction of an infinite series and the supposition that this series is convergent.

Example 8.2 Drag Force of a Ship As a further example we wish to determine the drag force exerted on the hull of a ship by the water if the ship is steadily moving through still water. This force, K, depends upon the velocity V of the ship, a characteristic length L, the density ρ and dynamic viscosity of the water η and the acceleration of the Earth g. Effects of surface tension are neglected in a first step. Thus an equation of the form

$$f(K, V, L, \rho, \eta, g) = 0 \tag{8.3.16}$$

must hold. A complete set of dimensionless products which follows from the above six variables is

$$\mathbb{P} = \frac{K}{\rho V^2 L^2}\,, \quad \mathbb{R} = \frac{VL}{\eta/\rho}\,, \quad \mathbb{F} = \frac{V^2}{Lg}\,. \tag{8.3.17}$$

Therefore, one may equally write

$$K = \rho V^2 L^2 f(\mathbb{R}, \mathbb{F})\,. \tag{8.3.18}$$

If one replaces L^2 as the square of a characteristic length by the maximum cross sectional area A, drawn by the ship perpendicular to the ship's axis, then one may also write

$$K = \tfrac{1}{2} C_D \rho V^2 A\,, \tag{8.3.19}$$

in which

$$C_D = C_D(\mathbb{F}, \mathbb{R})\,. \tag{8.3.20}$$

It is seen that the same formula holds as for the completely submerged body, however with the difference that now also the FROUDE number enters the formula. If one writes

$$C_D = C_D' + C_D''\,, \tag{8.3.21}$$

where C_D' models the influence of the frictional resistance of the walls and C_D'' those of the gravity, then one may assume C_D' to depend only on the REYNOLDS number $\mathbb{R}$ and C_D'' only on the FROUDE number $\mathbb{F}$. Even though this is not exact, the approximation is reasonable. Both coefficients can be separately determined by experiment.

Example 8.3 Pressure Drop in Pipes We consider the pressure drop Δp that one measures between two cross sections in a straight circular pipe that transports an incompressible fluid under steady conditions. We denote by L the distance between the two cross sections. D is the internal diameter of the pipe, k the mean roughness length of the interior wall, $\overline{u}$ the mean velocity of the fluid within the cross section, ρ the fluid density and η its dynamic viscosity. With these one obtains the relation

$$f(\Delta p, L, D, k, \overline{u}, \rho, \eta) = 0\,. \tag{8.3.22}$$

A complete set of dimensionless products, which can be formed with the above seven dimensional quantities, is

$$\mathbb{P} = \frac{\Delta p}{\frac{1}{2}\rho V^2}\,, \quad \mathbb{R} = \frac{VD}{\eta/\rho}\,, \quad \frac{L}{D}\,, \quad \frac{k}{D}\,, \tag{8.3.23}$$

so that the last relation may also be written as

$$F\left(\mathbb{P}, \mathbb{R}, \frac{L}{D}, \frac{k}{D}\right) = 0 \tag{8.3.24}$$

or more explicitly

$$\mathbb{P} = \hat{F}\left(\mathbb{R}, \frac{L}{D}, \frac{k}{D}\right) . \tag{8.3.25}$$

The dimensionless pressure drop depends on three dimensionless quantities, the REYNOLDS number, a length to diameter ratio and a dimensionless roughness k/D. Now, the pressure drop along a pipe is caused by the wall shear traction which in steady state is constant along the pipe. As a result $\hat{F}(\cdot)$ must be proportional to L and therefore

$$\Delta p = \frac{1}{2}\rho V^2 \frac{L}{D} \lambda\left(\mathbb{R}, \frac{k}{D}\right) . \tag{8.3.26}$$

The function $\lambda(\mathbb{R}, k/D)$ has been experimentally determined. A standardisation of this, based on formulas derived by PRANDTL[6] and NIKURADSE[7] is given in Fig. 8.1.

8.3.3 Systematic Computation of Dimensionless Products

In the preceding sections we stated, starting from a number of dimensional quantities, the emerging possible number of independent dimensionless products. A rational procedure was, however, missing that would allow the determination of the complete number of dimensionless products on the one hand, and of the dimensionless products themselves on the other hand. This rule will now be given; to derive it, some facts of linear algebra will be needed which we shall assume to be known to the reader.

Theorem 8.1 *The number of dimensionless products in a complete set of dimensional variables equals the total number of variables minus the maximum of that number of variables with which no dimensionless product can be formed.* ■

This theorem will subsequently be proved in a number of steps. It is heavy going in its application and impractical, but will be replaced by a theorem that is practically more easily accessible. To simplify the algebraic computations, we arrange the variables which arise in a particular product in matrix form such that each column lists the fundamental dimensions of the variable it represents. As fundamental dimensions we choose mass, length, time and temperature (as well as electric field strength). For a relation of the form

$$f(V, L, K, \rho, \eta, g) = 0 \tag{8.3.27}$$

[6] PRANDTL [187].

[7] NIKURADSE [172], [173], [174].

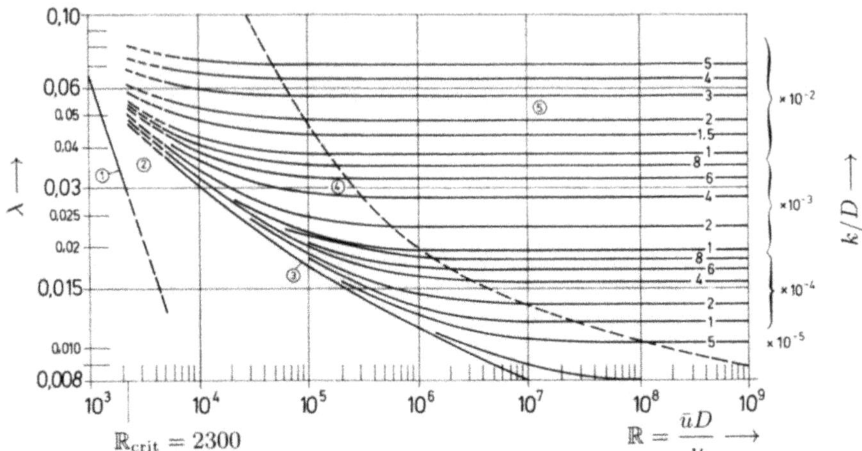

① laminar $\lambda = 64/R$
② laminar–turbulent transition zone
③ turbulent (hydraulic smooth pipe)
④ smooth–rough transition zone
⑤ fully–developed roughness region

Fig. 8.1. *Moody Diagramm*, MOODY [158]. Drag coefficient λ as a function of the REYNOLDS number $\mathbb{R}$ and the relative roughness k/D according to COLEBROOK [49]:

$$\lambda = 1/(2\log_{10}(D/k) + 1.74)^2 \, .$$

The dashed curve gives the boundary beyond which λ may be assumed constant.

the matrix of its dimensions or, as one usually says its **dimensional matrix** possesses the form

	V	L	K	ρ	η	g
M	0	0	1	1	1	0
L	1	1	1	-3	-1	1
T	-1	0	-2	0	-1	-2.

(8.3.28)

Such dimensional matrices are generally rectangular but not quadratic. Recall that the rank of the matrix is therefore at most equal to the smaller of the number of rows or columns. For the rank of a matrix is defined as the number of rows or columns of that largest quadratic submatrix of which the determinant does not vanish. With this information Theorem 8.1 can be restated as follows:

Theorem 8.2 *The number of dimensionless products in a complete set of variables equals the total number of variables minus the rank of the dimensional matrix.* ■

In the last example the rank of the matrix is $h = 3$ and the number of dimensionless products therefore three.

a) An Introductory Example Before we give the proofs for these theorems, we present a computation which shows how one finds these dimensionless products. A dimensionless product Π of the introduced variables must have the form

$$\Pi = V^{k_1} L^{k_2} K^{k_3} \rho^{k_4} \eta^{k_5} g^{k_6} \, . \tag{8.3.29}$$

The equation of the dimensions of this equation is given by

$$[\Pi] \;=\; [LT^{-1}]^{k_1} [L]^{k_2} [MLT^{-2}]^{k_3} [ML^{-3}]^{k_4} [ML^{-1}T^{-1}]^{k_5} [LT^{-2}]^{k_6} \, . \tag{8.3.30}$$

The dimension of Π is 1; therefore, the exponents of M, L and T on the right-hand side must vanish; this yields the equations

$$\begin{aligned} k_3 + k_4 + k_5 &= 0, \\ k_1 + k_2 + k_3 - 3k_4 - k_5 + k_6 &= 0, \\ - k_1 - 2k_3 - k_5 - 2k_6 &= 0 \end{aligned}$$

for the six unknowns $k_1, ..., k_6$. Notice that the matrix of this system of equations is the same as the dimensional matrix in (8.3.28). One could, consequently, have obtained the equations for the k's directly from the dimensional matrix. Naturally, the solutions for the system are not unique; for three of the six k's must be chosen to uniquely determine the other three k's. Since the rank of the system of equations is three, there are three linearly distinct choices for k_1, k_2 and k_3. These are for instance

$$\begin{aligned} &1)\; k_1 = 1 \, , \; k_2 = k_3 = 0 \, , \\ &2)\; k_2 = 1 \, , \; k_1 = k_3 = 0 \, , \\ &3)\; k_3 = 1 \, , \; k_1 = k_2 = 0 \, . \end{aligned}$$

If one consecutively substitutes these, one obtains three different solutions for k_4, k_5 and k_6. They are

$$\begin{aligned} &1) \quad k_1 = 1 \, , \; k_2 = k_3 = 0 \, , \; k_4 = -k_5 - -k_6 = \tfrac{1}{3} \, , \\ &2) \quad k_2 = 1 \, , \; k_1 = k_3 = 0 \, , \; k_4 = -k_5 = \tfrac{2}{3} \, , k_6 = \tfrac{1}{3} \, , \\ &3) \quad k_3 = 1 \, , \; k_1 = k_2 = 0 \, , \; k_4 = 1 \, , k_5 = -2 \, , k_6 = 0 \, . \end{aligned} \tag{8.3.31}$$

Each of these solutions is called a fundamental solution. If one collects these solutions in a matrix, one obtains

	V	L	K	ρ	η	g
	k_1	k_2	k_3	k_4	k_5	k_6
Π_1	1	0	0	$\frac{1}{3}$	$-\frac{1}{3}$	$-\frac{1}{3}$
Π_2	0	1	0	$\frac{2}{3}$	$-\frac{2}{3}$	$\frac{1}{3}$
Π_3	0	0	1	1	-2	0

.

The individual rows of this matrix yield the exponents of the variables indicated on the left of the above scheme, namely

$$\Pi_1 = V\rho^{\frac{1}{3}}\eta^{-\frac{1}{3}}g^{-\frac{1}{3}}\ , \quad \Pi_2 = L\rho^{\frac{2}{3}}\eta^{-\frac{2}{3}}g^{\frac{1}{3}}\ , \quad \Pi_3 = K\rho\eta^{-2}\ . \tag{8.3.32}$$

With this, three dimensionless products are found which are independent of each other.

b) The General Procedure Now that this introductory example has shown us how one can proceed we wish to present the general theory of the computation of the dimensionless products. To this end a theorem of linear algebra is needed. *A homogeneous, linear system of equations of n variables possesses $(n - r)$ linearly independent non-trivial solutions, where r is the rank of the coefficient matrix.*

Let $Q_1, Q_2, ..., Q_n$ be the variables characterising a certain physical problem. Let M, L and T be fundamental dimensions arising in the problem. The dimensional matrix of this problem possesses then the form

$$\begin{array}{c|ccccc} & Q_1 & Q_2 & Q_3 & \dots & Q_n \\ \hline M & a_{11} & a_{12} & a_{13} & \dots & a_{1n} \\ L & a_{21} & a_{22} & a_{23} & \dots & a_{2n} \\ T & a_{31} & a_{32} & a_{33} & \dots & a_{3n} \end{array} .$$

A dimensionless product of the variables $Q_1, Q_2, ..., Q_n$ must be a product of powers of these variables, so that one has

$$\Pi = Q_1^{k_1} Q_2^{k_2} Q_3^{k_3} ... Q_n^{k_n}\ , \tag{8.3.33}$$

or if this is written as an equation of its dimensions

$$[\Pi] = [M^{a_{11}}L^{a_{21}}T^{a_{31}}]^{k_1} \times [M^{a_{12}}L^{a_{22}}T^{a_{32}}]^{k_2} \times \ldots \times [M^{a_{1n}}L^{a_{2n}}T^{a_{3n}}]^{k_n}\ ,$$

which implies, via a comparison of exponents,

$$\begin{aligned} k_1a_{11} + k_2a_{12} + \ldots + k_na_{1n} &= 0\ , \\ k_1a_{21} + k_2a_{22} + \ldots + k_na_{2n} &= 0\ , \\ k_1a_{31} + k_2a_{32} + \ldots + k_na_{3n} &= 0\ . \end{aligned} \tag{8.3.34}$$

Here too, as before, the equations are immediately obtained from the dimensional matrix (8.3.34). Thus, we have:

Theorem 8.3 *The homogeneous, linear system of equations for the exponents k_j of a dimensional product*

$$\Pi = \prod_{j=1}^{n} Q_j^{k_j}\ , \tag{8.3.35}$$

is given by the dimensional matrix of the problem. ∎

Let the system of equations (8.3.34) for the n unknowns k_j have the rank r (≤ 3). Consequently, the equations possess $(n-r)$ linearly independent solutions. If one substitutes for k_1 to k_{n-r} the linearly independent arbitrary choices

$$\begin{aligned} k_1 &= 1\,, \quad k_2 = k_3 = \ldots = k_{n-r} = 0 \quad , \\ k_2 &= 1\,, \quad k_1 = k_3 = \ldots = k_{n-r} = 0 \quad , \\ &\vdots \\ k_{n-r} &= 1\,, \quad k_1 = k_2 = \ldots = k_{n-r-1} = 0 \quad , \end{aligned} \tag{8.3.36}$$

the remaining k_j $(j = n-r+1, \ldots, n)$ can be computed. The solution for the dimensionless products can then be represented by the following array

	k_1	k_2	k_3	$\ldots$	k_{n-r}	k_{n-r+1}	$\ldots$	k_n
Π_1	1	0	0	$\ldots$	0	$\alpha_{1,n-r+1}$	$\ldots$	$\alpha_{1,n}$
Π_2	0	1	0	$\ldots$	0	$\alpha_{2,n-r+1}$	$\ldots$	$\alpha_{2,n}$
Π_3	0	0	1	$\ldots$	0	$\alpha_{3,n-r+1}$	$\ldots$	$\alpha_{3,n}$
$\vdots$	$\vdots$	$\vdots$	$\vdots$		$\vdots$	$\vdots$		$\vdots$
Π_{n-r}	0	0	0	$\ldots$	1	$\alpha_{n-r,n-r+1}$	$\ldots$	$\alpha_{n-r,n}$

This is the fundamental solution of the system (8.3.36). Thus, the general computational scheme for the determination of the dimensionless products is complete.

c) Practical Suggestions The preceding computations allow determination of a complete set of dimensionless products. However, there are an infinite number of such sets. Therefore, one may feel the desire to determine a set that is optimal in some sense. There does not exist a strict rule according to which the best choice of dimensionless products could be determined; however, recommendations can be given. In this regard it is advantageous to focus e. g. at those variables which can be varied in an experiment. BUCKINGHAM recommends

Rule 1 *One reaches the maximum of experimental controllability, if each of the dimensionless products contains only one variable that can be controlled in an experiment.* ∎

For instance, if in an experiment the variable V can easily be varied, this variable should preferably arise only in one dimensionless product. The realisation of this rule in practice is not difficult. Because the first $(n-r)$ variables of the dimensional matrix arise only in one dimensionless product, the above rule can be formulated as follows:

Rule 2 *In the dimensional matrix one preferably chooses as the first variable the dependent variable. The second, third, etc. variables are to be arranged such that they can, in consecutive order, experimentally best, second to best, etc. be controlled.* ∎

With this rule practice has shown that the best arrangement of the variables in the dimensional matrix can be achieved.

Of interest are also transformations of some dimensionless products into others. Occasionally such transformations become necessary. Let it be found experimentally or by other means that a particular dimensionless product does not exercise an influence on the physical process studied. Let V be a variable that is practically without influence to the analysed results. If this variable arises only in one dimensionless product, then this product can simply be omitted from the list of influencing Π-products. However, should it arise in several dimensionless products, then these products must be so transformed that V will only arise in a single Π-product. We illustrate the computational procedure with an example. Let

$$f(\Pi_1, \Pi_2, \Pi_3) = 0 \tag{8.3.37}$$

with

$$\Pi_1 = \frac{\rho K}{\eta^2}, \quad \Pi_2 = V\left(\frac{\rho}{\eta g}\right)^{\frac{1}{3}}, \quad \Pi_3 = L\left(\frac{\rho^2 g}{\eta^2}\right)^{\frac{1}{3}}.$$

Assume, moreover, that η, the dynamic viscosity, does not exercise an influence on the studied process. Since η occurs in each product, other dimensionless products must be so defined that η arises only in one product. Such products are

$$\mathbb{P} = \frac{\Pi_1}{{\Pi_2}^2{\Pi_3}^2} = \frac{K}{\rho V^2 L^2}, \quad \mathbb{R} = \Pi_2\Pi_3 = \frac{VL}{\eta/\rho}, \quad \mathbb{F} = \frac{\Pi_2^2}{\Pi_3} = \frac{V^2}{Lg}.$$

Evidently, the viscosity only arises in the REYNOLDS number. Consequently, the relation (8.3.37) can be reduced to the simple form

$$\mathbb{P} = f(\mathbb{F})$$

not involving the viscosity of the problem.

Example 8.4 Discharge from a Basin (Fig. 8.2) Consider a basin filled with a density preserving fluid, which is emptied through a pipe of length l. The cross section of the basin is denoted by A_1, and it is assumed to be very large in comparison to the cross section of the pipe, A_2. We seek a formula for the mean velocity V over the cross section at point 2 that is established shortly after the opening of the valve as a function of the fill height h, the length l of the pipe and the two cross sections A_1 and A_2 under the action of the gravity of the fluid and the time t.

With reference to Fig. 8.2 the dimensional matrix takes the form

$$\begin{array}{c|ccccccc} & g & V & h & l & A_1 & A_2 & t \\ \hline L & 1 & 1 & 1 & 1 & 2 & 2 & 0 \\ T & -2 & -1 & 0 & 0 & 0 & 0 & 1 \end{array}\,. \tag{8.3.38}$$

a)

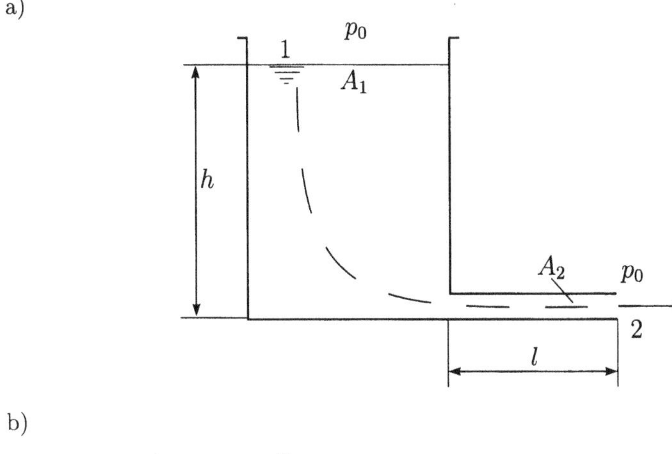

b)

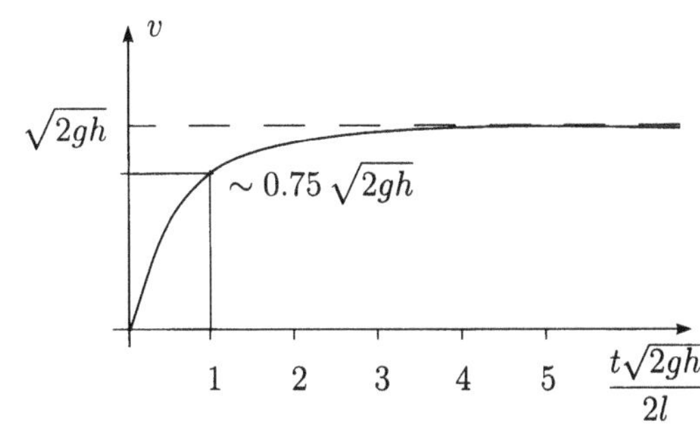

Fig. 8.2. *Outflow from a container.* a) Principal sketch with stream line, b) The evolution of the velocity with time at the end of the pipe is given by h and l.

It possesses rank 2, so that five dimensionless products must exist. They can be written as

$$\begin{aligned} \Pi_1 &= \frac{V}{\sqrt{2gh}} , \quad \Pi_2 = \frac{\sqrt{2gh}}{2l} t , \\ \Pi_3 &= \frac{h}{l} , \quad \Pi_4 = \frac{A_1}{A_2} , \quad \Pi_5 = \frac{hl}{A_1} . \end{aligned} \tag{8.3.39}$$

Since A_1 is assumed to be very much larger than hl and A_2, one may suppose that $\Pi_4 \to \infty$ and $\Pi_5 \to 0$. Of the general dependence

$$\Pi_1 = f(\Pi_2, \Pi_3, \Pi_4, \Pi_5) \tag{8.3.40}$$

there then remains the simplified dependence

$$\Pi_1 = f(\Pi_2, \Pi_3) . \tag{8.3.41}$$

A further reduction and specification, respectively, of the function (8.3.41) can not be reached with methods of dimensional analysis alone. At this point the theoretical analysis or the experimental study commences in order to obtain further restrictions. For instance, the laboratory engineer may perform experiments in which $\Pi_3 = h/l$ is systematically varied. If pipes with very smooth walls are used, such experiments show that the function $f(\cdot)$ in (8.3.41) is insensitive against changes of Π_3; in other words, $f(\cdot)$ is independent of Π_3 and one obtains

$$\Pi_1 = f(\Pi_2) \quad \Rightarrow \quad \frac{V}{\sqrt{2gh}} = f\left(\frac{\sqrt{2gh}}{2l}t\right) . \tag{8.3.42}$$

The same experiments will also show that the value of the function $f(\cdot)$ is zero for $t = 0$ and 1 for very large times. Apart from this one may suppose a monotonic behaviour from 0 at $t = 0$ to the asymptotic value 1 at $t \to \infty$. The limit value

$$V_\infty = \sqrt{2gh} \tag{8.3.43}$$

corresponds to the TORICELLI[8] formula, the exact formula $f(x) = \tanh(x)$ can be obtained by applying the BERNOULLI equation. The latter yields the differential equation

$$\frac{\mathrm{d}V}{\mathrm{d}t} + \frac{V^2}{2l} - \frac{gh}{l} = 0 \tag{8.3.44}$$

with the solution

$$\frac{V}{\sqrt{2gh}} = \tanh\left(\frac{\sqrt{2gh}}{2l}t\right) . \tag{8.3.45}$$

8.4 Algebraic Theory of Dimensional Analysis

8.4.1 Transformation of Basic Units

In this section a clear and axiomatic theory of dimensional analysis will be given. The theory culminates in the *Theorem of* BUCKINGHAM.

Let m be the number of independent dimensions, i.e., fundamental dimensions such as length $[L]$, mass $[M]$, time $[T]$, temperature $[\theta]$ and electrical field strength $[A]$. With the fundamental dimensions one may deduce derived dimensions which are built by products of the fundamental dimensions. In the following we shall denote the dimension of a **fundamental unit** by the symbol

$$[G]_j \, , \quad (j = 1, 2, \ldots, m) \, , \tag{8.4.1}$$

which may, for example, mean length, mass, time, etc. The dimensions of the **derived units** A_j may then be written in the form[9]

$$[A]_j = [G]_1^{a_{1j}} [G]_2^{a_{2j}} \cdots [G]_m^{a_{mj}} \, , \quad (j = 1, 2, \ldots, n) \tag{8.4.2}$$

[8] EVANGELISTA TORICELLI (1608–1647) was a pupil of GALILEO GALILEI (1570–1642).

[9] A_j is not the electrical field strength, introduced earlier.

or more briefly

$$[A]_j = \prod_{i=1}^{m} [G]_i^{a_{ij}} , \quad (j = 1, 2, \ldots, n) . \tag{8.4.3}$$

The product extends from $i = 1$ to $i = m$. Since all fundamental units are scalar quantities, the derived units are equally scalars, since they are given as products of powers of the former. This makes it possible to represent all dimensions, the fundamental and the derived quantities on positive real lines and to assign to each point on these axes a value of the respective dimension in its units. It is meaningless in this process to extend the domain also to the negative real numbers, since negative basic units are never introduced; so negative values can never arise. Values of dimensions in some units are by definition positive.

Let the value of the derived quantity $[A]_j$ in (8.4.3) in particularly chosen units be given by the positive real number x_j. Let the value of the fundamental basic unit $[G]_k$ in one set of units be $[G]_k^n$ and in another set be G_k^o (o and n stand for "old" and "new"). Then there must hold a relation

$$1 G_k^o = \alpha_k G_k^n \tag{8.4.4}$$

between the old and new fundamental units. Let the value of $[A_j]$ in the old and new system be x_j and $\overline{x}_j$, respectively. The old and new values of the derived quantity are then related by

$$\begin{aligned} & x_j [(G_1^o)^{a_{1j}} (G_2^o)^{a_{2j}} \ldots (G_m^o)^{a_{1m}}] \\ &= x_j [\alpha_1^{1j} (G_1^n)^{a_{1j}} \alpha_2^{2j} (G_2^n)^{a_{2j}} \ldots \alpha_m^{mj} (G_m^n)^{a_{1m}}] \\ &= \overline{x}_j [(G_1^n)^{a_{1j}} (G_2^n)^{a_{2j}} \ldots (G_m^n)^{a_{1m}}] , \end{aligned}$$

so that

$$\overline{x}_j = x_j \prod_{k=1}^{m} \alpha_k^{a_{kj}} . \tag{8.4.5}$$

This represents a formula which allows to compute the value of a derived quantity in its dimensional units, if units of the fundamental dimensions have been changed. The result can be formulated as the following.

Theorem 8.4 *If x_j is the value of a derived dimension $[A]_j = \prod_{k=1}^{m} [G]_k^{a_{kj}}$ $(j = 1, 2, \ldots, m)$ and if an old unit of the fundamental dimension $[G]_k$ is α_k times the value of this dimension in new units, then the new value $\overline{x}_j$ of the derived unit $[A]_j$ is given by*

$$\overline{x}_j = x_j \prod_{k=1}^{m} \alpha_k^{a_{kj}} , \quad (j = 1, 2, \ldots, n) . \tag{8.4.6}$$

■

8.4.2 Exact Definition of Dimensional Homogeneity

Let y be a function of n variables i.e., $y = f(x_1, \ldots, x_n)$. If the units of the basic dimensions are changed, then y and x_j become $\overline{y}$ and $\overline{x}_j$. An equation is now called dimensionally homogeneous, if $y = f(x_1, \ldots, x_n)$ can be transformed to

$$\overline{y} = f(\overline{x}_1, \overline{x}_2, \ldots, \overline{x}_n) , \tag{8.4.7}$$

in which f is the same function as before. Mathematically this means that the equation $y = f(x_1, \ldots, x_n)$ is *invariant under the group of transformations which is generated by all possible changes of units of the fundamental dimensions.* This group of transformations[10] is described by (8.4.5), in which α_k may be arbitrary positive constants. For the dependent and independent variables relation (8.4.5) can symbolically be summarised by the dimensional matrix

$$\begin{array}{c|c|ccccc} & y & x_1 & x_2 & x_3 & \ldots & x_n \\ \hline [G]_1 & a_{10} & a_{11} & a_{12} & a_{13} & \ldots & a_{1n} \\ [G]_2 & a_{20} & a_{21} & a_{22} & a_{23} & \ldots & a_{2n} \\ \vdots & \vdots & \vdots & \vdots & \vdots & & \vdots \\ [G]_m & a_{m0} & a_{m1} & a_{m2} & a_{m3} & \ldots & a_{mn} \end{array} .$$

If one writes

$$\overline{y} = K_o y , \qquad \overline{x}_j = K_j x_j ,$$

one obtains, in view of (8.4.5),

$$K_o = \prod_{k=1}^{m} \alpha_k^{a_{ko}} , \quad K_j = \prod_{k=1}^{m} \alpha_k^{a_{kj}} .$$

Therefore, the following identity must hold

$$\overline{y} = K_o y = K_o f(x_1, x_2, \ldots, x_n) = f(K_1 x_1, K_2 x_2, \ldots, K_n x_n) . \tag{8.4.8}$$

Thus we have

Proposition 8.1 *The function $f(x_1, \ldots, x_n)$ is dimensionally homogeneous if and only if the equation*

$$K_o f(x_1, x_2, \ldots, x_n) = f(K_1 x_1, K_2 x_2, \ldots, K_n x_n) \tag{8.4.9}$$

with

$$K_o = \prod_{k=1}^{m} \alpha_k^{a_{ko}} , \quad K_j = \prod_{k=1}^{m} \alpha_k^{a_{kj}} \tag{8.4.10}$$

is identically fulfilled in the variables $x_1, x_2, \ldots, x_n, \alpha_1, \alpha_2, \ldots, \alpha_m$. ■

Notice that all $K's$ are fixed if the $\alpha_k's$ and the dimensional matrix are known.

---

[10] We leave it to the reader to prove that the group properties (as mathematical statements) are satisfied.

Example 8.5 Drag Force of a Body Submerged in a Moving Fluid. With $F = f(V, D, \rho, \eta)$, where V is a typical velocity, D a characteristic cross sectional length, ρ the density of the fluid and η its dynamic viscosity, (8.4.9) takes the form

$$K_o F = f(K_1 V, K_2 D, K_3 \rho, K_4 \eta) . \tag{8.4.11}$$

and the dimensional matrix is

	F	V	D	ρ	η
M	1	0	0	1	1
L	1	1	1	-3	-1
T	-2	-1	0	0	-1

From this we deduce

$$K_o = \alpha_1^1 \alpha_2^1 \alpha_3^{-2} , \quad K_1 = \alpha_1^0 \alpha_2^1 \alpha_3^{-1} , \quad K_2 = \alpha_1^0 \alpha_2^1 \alpha_3^0 ,$$

$$K_3 = \alpha_1^1 \alpha_2^{-3} \alpha_3^0 , \quad K_4 = \alpha_1^1 \alpha_2^{-1} \alpha_3^{-1} .$$

Equation (8.4.11) is therefore given by

$$\alpha_1 \alpha_2 \alpha_3^{-2} F = f(\alpha_2 \alpha_3^{-1} V, \, \alpha_2 D, \, \alpha_1 \alpha_2^{-3} \rho, \, \alpha_1 \alpha_2^{-1} \alpha_3^{-1} \eta) ,$$

and it is easy to see that this relation is fulfilled for all $\alpha_k > 0$, if

$$F = \rho V^2 D^2 f\left(\frac{VD}{\eta/\rho}\right) . \tag{8.4.12}$$

Let us give *two important applications* of Proposition 8.1; consider first

$$(i) \quad y = f(x_1, x_2, \ldots, x_n) = x_1 + x_2 + \ldots + x_n .$$

The function $f(x_1, \ldots, x_n)$ is here the sum of its independent variables; in this case (8.4.9) reads

$$K_0(x_1 + x_2 + \ldots + x_n) = K_1 x_1 + K_2 x_2 + \ldots + K_n x_n ,$$

and since this equation must hold identically for all x_j, one must have

$$K_0 = K_1 = K_2 = \cdots = K_n \tag{8.4.13}$$

or because of (8.4.10)

$$a_{k_0} = a_{k_1} = a_{k_2} = \cdots = a_{k_n} , \quad (k = 1, 2, \ldots, m) .$$

Thus we have the

Lemma 8.1 *A sum*

$$y = x_1 + x_2 + \ldots + x_n$$

is dimensionally homogeneous if and only if all of its members have the same dimension. ■

Consider as a second example the composition

$$(ii) \quad y = x_1^{k_1} \times x_2^{k_2} \times \cdots \times x_n^{k_n} = \prod_{j=1}^{n} x_j^{k_j}$$

with arbitrary k_j, $(j = 1, \ldots, n)$. Such compositions are products of powers of the variables $x_1, \ldots, x_n$. We call them for simplicity products. Then we have

Lemma 8.2 *The product*

$$y = \prod_{j=1}^{n} x_j^{k_j} \tag{8.4.14}$$

is dimensionally homogeneous if and only if the exponents are solutions of the linear equations

$$\sum_{j=1}^{n} a_{ij} k_j = a_{i0} \,, \quad (i = 1, 2, \ldots, m) \,. \tag{8.4.15}$$

■

Proof The matrix a_{ij} is the dimensional matrix of the independent variables; the vector a_{i0} is the column of the dependent variables. To prove the Lemma we assume that the product of powers (8.4.14) is dimensionally homogeneous; in a change of the fundamental units it will thus obey (8.4.9), so that

$$K_0 \left(\prod_{j=1}^{n} x_j^{k_j} \right) = \prod_{j=1}^{n} (K_j x_j)^{k_j} = \left(\prod_{j=1}^{n} K_j^{k_j} \right) \times \left(\prod_{j=1}^{n} x_j^{k_j} \right)$$

is fulfilled, from which

$$K_0 = \prod_{j=1}^{n} K_j^{k_j}$$

is obtained. If one also takes (8.4.10) into account, one obtains

$$\prod_{k=1}^{m} \alpha_k^{a_{k0}} = \left(\prod_{k=1}^{m} \alpha_k^{a_{k1}} \right)^{k_1} \times \cdots \times \left(\prod_{k=1}^{m} \alpha_k^{a_{kn}} \right)^{k_n} = \prod_{k=1}^{m} \alpha_k^{\sum_{j=1}^{n} a_{kj} k_j} \,,$$

and by comparison of the exponents

$$a_{k0} = \sum_{j=1}^{n} a_{kj} k_j \,, \quad (k = 1, 2, \ldots, m) \,.$$

The exponents of a dimensionally homogeneous power product (8.4.14) thus necessarily satisfy (8.4.15). To show the converse, namely that the power product (8.4.14) is dimensionally homogeneous, if the exponents obey (8.4.15), we state (8.4.14), written in terms of the old and new fundamental units, as

$$y = \prod_{j=1}^{n} x_j^{k_j} , \quad \overline{y} = \prod_{j=1}^{n} \overline{x}_j^{k_j} .$$

With the use of the transformation formulas for $\overline{y}$ and $\overline{x}_j$ the last formula takes the form

$$\left(\prod_{i=1}^{m} \alpha_i^{a_{i0}}\right) y = \prod_{j=1}^{n} \left(\prod_{i=1}^{m} \alpha_i^{a_{ij}} x_j\right)^{k_j} = \prod_{j=1}^{n} \left(\prod_{i=1}^{m} \alpha_i^{a_{ij} k_j}\right) x_j^{k_j}$$
$$= \left(\prod_{i=1}^{m} \alpha_i^{\sum_{j=1}^{n} a_{ij} k_j}\right) \underbrace{\prod_{j=1}^{n} x_j^{k_j}}_{y}$$

or

$$\frac{\prod_{i=1}^{m} \alpha_i^{\sum_{j=1}^{n} a_{ij} k_j}}{\prod_{i=1}^{m} \alpha_i^{a_{i0}}} = \prod_{i=1}^{m} \alpha_i^{\sum_{j=1}^{n} a_{ij} k_j - a_{i0}} = 1 ,$$

which is satisfied for arbitrary values of α_i if (8.4.15) are fulfilled, qed.

8.4.3 Calculus of Dimensionless Products

Now that the basic theorems and propositions are known we shall proceed to present the "algebra" of the dimensionless products. Let the following power products

$$\begin{aligned} \Pi_{(1)} &= x_1^{k_1^{(1)}} \times x_2^{k_2^{(1)}} \times \cdots \times x_n^{k_n^{(1)}} , \\ \Pi_{(2)} &= x_1^{k_1^{(2)}} \times x_2^{k_2^{(2)}} \times \cdots \times x_n^{k_n^{(2)}} , \\ &\vdots \\ \Pi_{(p)} &= x_1^{k_1^{(p)}} \times x_2^{k_2^{(p)}} \times \cdots \times x_n^{k_n^{(p)}} \end{aligned} \tag{8.4.16}$$

be dimensionless. In the ensuing analysis we shall denote dimensionless products by the letter Π. These products can be arranged in a matrix as follows:

$$\begin{array}{c|cccc} & x_1 & x_2 & \cdots & x_n \\ \hline \Pi_{(1)} & k_1^{(1)} & k_2^{(1)} & \cdots & k_n^{(1)} \\ \Pi_{(2)} & k_1^{(2)} & k_2^{(2)} & \cdots & k_n^{(2)} \\ \vdots & \vdots & \vdots & & \vdots \\ \Pi_{(p)} & k_1^{(p)} & k_2^{(p)} & \cdots & k_n^{(p)} \end{array} . \tag{8.4.17}$$

If $\Pi_{(1)}$ and $\Pi_{(2)}$ are lineary dependent of each other, then a certain power of $\Pi_{(1)}$ will be equal to $\Pi_{(2)}$, or more generally $\Pi_{(1)}^{h_1} \Pi_{(2)}^{h_2} = 1$ for some non-vanishing h_1 and h_2. Extended to the dimensionless products $\Pi_{(1)}, \ldots, \Pi_{(p)}$

this implies: the dimensionless products $\Pi_{(1)}, \ldots, \Pi_{(p)}$ are dependent of each other, if there exist constants $h_1, h_2, \ldots, h_p$, not all of which vanish such that

$$\Pi_{(1)}^{h_1} \times \Pi_{(2)}^{h_2} \times \cdots \times \Pi_{(p)}^{h_p} = 1 .$$

This statement can be reformulated in the following

Proposition 8.2 *Necessary and sufficient condition that the products $\Pi_{(1)}, \ldots, \Pi_{(p)}$ are independent of each other is the fact that the rows of the matrix (8.4.17) or*

$$\begin{pmatrix} k_1^{(1)} & k_2^{(1)} & \ldots & k_n^{(1)} \\ k_1^{(2)} & k_2^{(2)} & \ldots & k_n^{(2)} \\ \vdots & \vdots & & \vdots \\ k_1^{(p)} & k_2^{(p)} & \ldots & k_n^{(p)} \end{pmatrix} \tag{8.4.18}$$

are linearly independent. ■

Proof To demonstrate necessity, let us assume that the rows of the matrix (8.4.18) are linearly dependent but the products (8.4.16) are independent. Then there must exist constants $h_1, \ldots, h_p$ not all of which are identically zero, which satisfy the relation

$$h_1 k_i^{(1)} + h_2 k_i^{(2)} + \cdots + h_p k_i^{(p)} = 0 , \quad (i = 1, 2, \ldots, n) . \tag{8.4.19}$$

This, however, implies, in view of (8.4.16),

$$\Pi_{(1)}^{h_1} \times \Pi_{(2)}^{h_2} \times \cdots \times \Pi_{(p)}^{h_p} = x_1^{\sum_{j=1}^{p} h_j k_1^{(j)}} \times x_2^{\sum_{j=1}^{p} h_j k_2^{(j)}} \times \cdots \times x_n^{\sum_{j=1}^{p} h_j k_n^{(j)}} .$$

If one substitutes here (8.4.19), there follows

$$\Pi_{(1)}^{h_1} \times \Pi_{(2)}^{h_2} \times \cdots \times \Pi_{(p)}^{h_p} = x_1^0 x_2^0 \cdots x_n^0 = 1 , \tag{8.4.20}$$

which is in conflict with the assumption that the products are independent. Thus the rows of the matrix (8.4.18) must be linearly independent.

Sufficiency can be proved as follows: if the rows of the matrix (8.4.18) are linearly independent and the dimensionless products are dependent, an equation of the form

$$\Pi_{(1)}^{h_1} \times \Pi_{(2)}^{h_2} \times \cdots \times \Pi_{(p)}^{h_p} = 1$$

must exist for the exponents h_j, not all of which vanish. Then, however, (8.4.16) implies

$$x_1^{\sum_{j=1}^{p} h_j k_1^{(j)}} \times x_2^{\sum_{j=1}^{p} h_j k_2^{(j)}} \times \cdots \times x_n^{\sum_{j=1}^{p} h_j k_n^{(j)}} = 1 , \tag{8.4.21}$$

which can only be correct, if all exponents vanish, which is a contradiction to the assumed linear independence of the matrix (8.4.18), qed.

Conversely, we may conclude that a power product is dimensionless if and only if the equations

$$\sum_{j=1}^{n} a_{ij}k_j = 0\,, \quad (i = 1, 2, \ldots, m) \tag{8.4.22}$$

hold. This follows immediately from Lemma 8.2 with $a_{i0} = 0$. The linear, homogeneous equations (8.4.22) possess $(n-r)$ linearly independent solutions, which we shall denote by

$$k_i^{(1)}, k_i^{(2)}, \ldots, k_i^{(n-r)}\,, \quad (i = 1, 2, \ldots, n)\,. \tag{8.4.23}$$

Here, r denotes the rank of the matrix (a_{ij}). According to proposition 8.2 the solution vectors furnish the exponents for all dimensionless products. There are no additional ones, so that one can formulate

Proposition 8.3 *Every fundamental system of solutions of the equations*

$$\sum_{j=1}^{n} a_{ij}k_j = 0\,, \qquad (i = 1, 2, \ldots, m) \tag{8.4.24}$$

determines the $n - r$ exponents of a complete set of dimensionless products of the variables $x_1, \ldots, x_n$. Conversely, the exponents of a complete set of dimensionless products of the variables $x_1, \ldots, x_n$ form a fundamental system of solutions of the above equations. ■

This proposition immediately also implies

Proposition 8.4 *The number of independent products in a complete set of dimensionless products of the variables $x_1, \ldots, x_n$ is $(n - r)$, where r denotes the rank of the dimensional matrix.* ■

We now consider dimensional products and assume that in the dimensional matrix not all a_{i0} are zero. Then the system of equations for the exponents k_j reads

$$\sum_{j=1}^{n} a_{ij}k_j = a_{i0}\,, \quad (i = 1, 2, \ldots, m)\,, \tag{8.4.25}$$

and we have

Proposition 8.5 *If y is not dimensionless, then there exists a product of the form*

$$y = x_1^{k_1} \cdot x_2^{k_2} \cdots x_n^{k_n} = \prod_{j=1}^{n} x_j^{k_j}\,, \tag{8.4.26}$$

if and only if the dimensional matrix of the variables $x_1, \ldots, x_n$ possesses the same rank as the dimensional matrix of the variables $y, x_1, \ldots, x_n$. ■

Proof This follows if the system

$$\sum_{j=1}^{n} a_{ij} k_j = a_{i0} \,, \quad (i = 1, 2, \ldots, m) \tag{8.4.27}$$

is considered. In books on linear algebra it is shown that solutions to the above equations exist, if the rank of the matrix (a_{ij}) is the same as the rank of the matrix (a_{ij}, a_{io}), augmented by the column of the right-hand side.

With the aid of proposition 8.5 we now may prove

Proposition 8.6 *If $y = f(x_1, x_2, \ldots, x_n)$ is a dimensionally homogeneous equation and if y is dimensional, then there exists a product of powers of x_j, which has the same dimension as y.* ■

Proof We assume that $y = f(x_1, \ldots, x_n)$ is dimensionally homogeneous, but there does not exist a product of powers of x_j with the dimension of y. Assume, moreover, that the dimensional matrix (a_{i0}, a_{ij}) possesses the rank R. The assumption that no x_j-product exists with the same dimension as y implies, because of proposition 8.5, that the rank of (a_{ij}) must be smaller than R. Without limitation of generality we may assume that a non-vanishing determinant of (a_{i0}, a_{ij}) lies in the upper left corner of this matrix; its rank is R. If $R = m$ (m is the number of independent fundamental dimensions), then this determinant is given by

$$\triangle = \begin{vmatrix} a_{10} & a_{11} & a_{12} & a_{13} & \cdots & a_{1(m-1)} \\ a_{20} & a_{21} & a_{22} & a_{23} & \cdots & a_{2(m-1)} \\ \vdots & & & & & \vdots \\ a_{m0} & a_{m1} & a_{m2} & a_{m3} & \cdots & a_{m(m-1)} \end{vmatrix} \neq 0 \,. \tag{8.4.28}$$

If A_{i0} are the algebraic complements or co-factors of a_{i0} of this matrix, then one may also write

$$\triangle = A_{10} a_{10} + \cdots + A_{m0} a_{m0} = \sum_{i=1}^{m} A_{i0} a_{i0} \,. \tag{8.4.29}$$

Alternatively, in the theory of determinants one proves that

$$\sum_{i=1}^{m} A_{i0} a_{ik} = 0 \,, \quad (\forall k = 1, 2, \ldots, n) \,. \tag{8.4.30}$$

(This result, incidentally, follows for $k = 1, 2, \ldots, m-1$ from the fact that the value of the determinant is zero if only two columns of a matrix are the same.) Thus for $k = 1, 2, \ldots, m-1$ (8.4.30) is correct. For $k \geq m$ it holds because the rank of the dimensional matrix is equal to $(a_{ij}) = R = m$.

Since y has been assumed as dimensionally homogeneous, (8.4.9), (8.4.10) hold as identities in the variables α_j $(j = 1, 2, \ldots, m)$. We therefore choose now new fundamental units, such that

$$\alpha_i = G^{A_{i0}} , \quad (i = 1, 2, \ldots, m) \tag{8.4.31}$$

holds with arbitrary positive real G. The factors of transformation K_j are then computed according to (8.4.10) and yield

$$\begin{aligned} K_j &= \prod\nolimits_{i=1}^{m} \alpha_i^{a_{ij}} = \prod\nolimits_{i=1}^{m} (G^{A_{i0}})^{a_{ij}} = \prod\nolimits_{i=1}^{m} G^{A_{i0}a_{ij}} \\ &= G^{\sum_{i=1}^{m} A_{i0}a_{ij}} = 1 , \quad (j = 1, 2, \ldots, n) . \end{aligned} \tag{8.4.32}$$

Thus, by specially choosing α_i all $K_j \quad (j = 1, 2, \ldots, n)$ have been made equal to unity. For K_0, one obtains

$$K_0 = \prod_{i=1}^{m} \alpha_i^{a_{i0}} = G^{\sum_{i=1}^{m} A_{i0}a_{i0}} \neq 1 . \tag{8.4.33}$$

Consequently, (8.4.9) takes the form

$$K_0\, y = f(x_1, x_2, \ldots, x_n) , \quad K_0 = G^{\sum_{i=1}^{m} A_{i0}a_{i0}} , \tag{8.4.34}$$

in which K_0 can be arbitrarily assigned since $G > 0$ was already freely chosen (because it is an arbitrary transformation of the units of the fundamental dimensions). With the possibility to arbitrarily choose K_0, it is now also shown that

$$K_0\, y = f(x_1, x_2, \ldots, x_n) \tag{8.4.35}$$

cannot be a function. This is in contradiction with the assumption that $y = f(x_1, \ldots, x_n)$ is a dimensionally homogeneous function.[11] The initial assumption that no power product of x_j with the dimension $[y]$ can exist, was therefore wrong. The proposition is therefore proved for $R = m$.

If $R < m$, e.g. $R = r$, then $\triangle$ is of size r, there is now an $r \times r$ matrix with non-vanishing determinant

$$\triangle = \begin{vmatrix} a_{10} & a_{11} & a_{12} & a_{13} & \ldots & a_{1(r-1)} \\ a_{20} & a_{21} & a_{22} & a_{23} & \ldots & a_{2(r-1)} \\ \vdots & \vdots & \vdots & \vdots & & \vdots \\ a_{r0} & a_{r1} & a_{r2} & a_{r3} & \ldots & a_{r(r-1)} \end{vmatrix} , \quad r < m . \tag{8.4.36}$$

In a way analogous to that before one now concludes that

$$\begin{aligned} &\triangle = \sum\nolimits_{i=1}^{r} A_{i0}a_{i0} , \\ &\sum\nolimits_{i=1}^{r} A_{i0}a_{ik} = 0 , \quad (k = 1, 2, \ldots, n) , \end{aligned} \tag{8.4.37}$$

where A_{i0} are again the algebraic complements of a_{i0}. With $\alpha_i = G^{A_{i0}}$ $(i = 1, \ldots, r)$, $\alpha_j = 1$ $(r < j < m)$ one now obtains

[11] Dimensional homogeneity is not important, since f does not even satisfy the prerequisites of a function.

$$K_j = G^{\sum_{i=1}^r A_{i0}a_{ij}} = G^0 = 1 \,, \quad K_0 = G^{\sum_{i=1}^r A_{i0}a_{i0}} \neq 1 \,, \tag{8.4.38}$$

so that one arrives at the same conclusion as before. Setting $a_j = 1$ for all $j > r$ corresponds again to a permissible change of the fundamental units. Proposition 8.6 therefore says nothing else than that a dimensionally homogeneous equation of the form $y = f(x_1, \dots, x_n)$ can always be brought to the form

$$\Pi = F(x_1, \dots, x_n) \,, \tag{8.4.39}$$

in which Π is dimensionless and F is a new function, qed.

8.5 Buckingham's Theorem

8.5.1 Proof of Buckingham's Theorem

In closing we wish in this section to prove Buckingham's *theorem* or Buckingham's *Π–theorem* [39] according to which a dimensionally homogeneous equation of several variables can be reduced to a relation only involving dimensionless quantities. The number of these new variables is usually smaller than the original number of dimensional variables.

Notice in particular that the independent variables in a problem of dimensional analysis are always real and positive quantities. If this were not so, then dimensionless products with fractional exponents would become complex valued. We shall also see that Buckingham's theorem can only hold, if the independent variables are restricted to positive quantities.

Let $x_1, \dots, x_n$ be the independent variables of a physical problem. These variables represent entities such as velocity, force, moment, temperature, heat flux etc. They may be regarded as the Cartesian coordinates of an Euclidian space $\mathcal{E}$. Let, moreover, α_i $(i = 1, 2, \dots, m)$ be positive constants and K_j $(j = 1, 2, \dots, n)$ variables, defined by

$$K_j = \prod_{i=1}^m \alpha_i^{a_{ij}} \,, \quad (j = 1, 2, \dots, n) \,, \tag{8.5.1}$$

in which a_{ij} is the dimensional matrix corresponding to the $x_j^{'}$s. The equation

$$x_j = K_j x_j^{'} \,, \quad (j = 1, 2, \dots, n) \tag{8.5.2}$$

then corresponds in the space $\mathcal{E}$ to a coordinate or point transformation; it assigns to each point $x_j^{'}$ a point x_j and vice versa. This point transformation shall subsequently be called K-transformation, and it is easy to show that all K-transformations in $\mathcal{E}$ build a group. Indeed,

(i) With $x_j = K_j^* x_j^{'} \,, \quad x_j^{'} = K_j^{**} x_j^{''}$ there follows

$$
\begin{aligned}
x_j &= (K_j^* K_j^{**}) x_j^{''} = K_j x_j^{''} \, , \\
K_j &= \left(\prod_{i=1}^{m} \alpha_i^{*a_{ij}}\right) \left(\prod_{i=1}^{m} \alpha_i^{**a_{ij}}\right) \\
&= \prod_{i=1}^{m} (\alpha_i^* \alpha_i^{**})^{a_{ij}} \, ,
\end{aligned}
\tag{8.5.3}
$$

i.e., the composition of two K-transformations is again a K-transformation.

(ii) There exists a unit–element, namely the identity transformation $x_j = x_j$

(iii) Since K_j differs from zero, there exists to every K_j an inverse transformation K_j^{-1}. With $x_j = K_j x_j^{'}$ and $x_j^{'} = K_j^{-1} x_j$ one may deduce

$$
\begin{aligned}
& x_j = K_j \left(K_j^{-1} x_j\right) = K_j K_j^{-1} x_j \Longrightarrow K_j K_j^{-1} = 1 \\
& \Longrightarrow K_j^{-1} = \frac{1}{K_j} = \frac{1}{\prod_{i=1}^{m} \alpha_i^{a_{ij}}} = \prod_{i=1}^{m} \left(\frac{1}{\alpha_i}\right)^{a_{ij}} \, .
\end{aligned}
\tag{8.5.4}
$$

This defines to each K-transformation its inverse.

One may interpret the entities $K_1, \ldots, K_n$, which are generated by $x_1, \ldots, x_n$ by all possible K-transformations, as a point in an n dimensional space; this space (or its extension by the K_0-axis) shall simply be called the K-space.

As a preparation to the BUCKINGHAM theorem we now prove the following lemmas:

Lemma 8.1 *A dimensionally homogeneous dimensionless function*

$$
\Pi = f(x_1, \ldots, x_n)
$$

is constant in each K-space. ∎

Proof Since Π is dimensionless the exponents of the fundamental dimensions on the left-hand side, a_{io}, must all vanish. Therefore, because of (8.4.10) $K_0 = 1$, and thus (8.4.9) reads

$$
\Pi = f(K_1 x_1, \ldots, K_n x_n) \, . \tag{8.5.5}
$$

In the K-space that is generated by $x_1, \ldots, x_n$, the value of Π must therefore be constant irrespective of the value of the element $(K_1, \ldots, K_n)$, qed.

Lemma 8.1 implies that every dimensionless product of the variables $x_1, \ldots, x_n$ is constant in each K-space. If, therefore, $(\Pi_1, \Pi_2, \ldots, \Pi_p)$ is a complete set of dimensionless products, then for fixed $x_1, \ldots, x_n$ this set does not change its value for all K_j-values ($j = 1, \ldots, n$) in K-space.

Lemma 8.2 *If $\{\Pi_1, \Pi_2, \ldots, \Pi_p\}$ is a complete set of dimensionless products of the variables $(x_1, \ldots, x_n)$, then to each set of values that is assigned to the dimensionless products $\{\Pi_1, \Pi_2, \ldots, \Pi_p\}$, there belongs one and only one K-space. In other words, two elements $\{x_j^{'}\}$ and $\{x_j^{''}\}$ of the K-space can only differ by a K-transformation.* ∎

Proof Let $\{\Pi_1', \Pi_2', \dots, \Pi_p'\}$ be a set of values of dimensionless products $\{\Pi_1, \Pi_2, \dots, \Pi_p\}$, and let $\{x_j'\}$ and $\{x_j''\}$ be two points in the $\mathcal{E}$-space, which belong to the values $\{\Pi_1', \Pi_2', \dots, \Pi_p'\}$. Then we have

$$\Pi_\nu' = (x_1')^{k_1^{(\nu)}} \times \dots \times (x_n')^{k_n^{(\nu)}} = (x_1'')^{k_1^{(\nu)}} \times \dots \times (x_n'')^{k_n^{(\nu)}} .$$

Since all x_j' on the right-hand side of this equation are positive, one may take the logarithm and obtains, after obvious rearrangements,

$$r_1 k_1^{(\nu)} + r_2 k_2^{(\nu)} + \dots + r_n k_n^{(\nu)} = 0 \,, \quad (\nu = 1, 2, \dots, p) \tag{8.5.6}$$

with $r_j = \ln(x_j'/x_j'')$. Notice that it is here that we assume the x_j to be positive. Thus, BUCKINGHAM's Π–Theorem is only provable for positive $x_j > 0$. Since $\{\Pi_1, \Pi_2, \dots, \Pi_p\}$ is complete, the exponents $k_j^{(1)}, \dots, k_j^{(r)}$ are solutions of the system

$$\sum_{j=1}^{n} a_{ij} k_j^{(\nu)} = 0 \,, \quad (\nu = 1, 2, \dots, p) \,, \quad (i = 1, 2, \dots, m) \,. \tag{8.5.7}$$

This is a consequence of Proposition 8.3. Since, however, the solutions of (8.5.7) are also solutions of (8.5.6), the coefficients in (8.5.6) must linearly depend upon those of (8.5.7). Therefore, there must exist non-zero numbers a_j^* $(j = 1, \dots, m)$, which satisfy the equation

$$\sum_{j=1}^{m} \alpha_j^* a_{ji} = r_i = \ln\left(\frac{x_i'}{x_i''}\right) \,, \quad (i = 1, 2, \dots, n) \,. \tag{8.5.8}$$

The last equation implies

$$x_i' = x_i'' \exp\left(\sum_{j=1}^{m} \alpha_j^* a_{ji}\right) = x_i'' \prod_{j=1}^{m} \left(e^{\alpha_j^* a_{ji}}\right) . \tag{8.5.9}$$

If for simplicity we write $\alpha_j = e^{\alpha_j^*}, (j = 1, 2, \dots, m)$, then there follows

$$x_i' = \left(\prod_{j=1}^{m} \alpha_j^{a_{ji}}\right) x_i'' = K_i x_i'' \;\; (i = 1, 2, \dots, n) \,. \tag{8.5.10}$$

This result shows that x_i' and x_i'' belong to the same K-space, *qed*. With all these results we may now prove the following proposition.

Proposition 8.7 (BUCKINGHAM **Theorem)** *Every dimensionally homogeneous equation can be transformed into an equation involving only dimensionless products.* ■

Proof According to Proposition 8.6 every dimensionally homogeneous equation $y = f(x_1 \dots x_n)$ can be brought into the form $\Pi = F(x_1, \dots, x_n)$ in which Π is dimensionless. Let $\{\Pi_1, \Pi_2, \dots, \Pi_p\}$ be a complete set of dimensionless products belonging to $(x_1, \dots, x_n)$. Then, according to Lemma 8.1, to every set of values of $\{\Pi_1, \Pi_2, \dots, \Pi_p\}$ there is only one single K-space. According to Lemma 8.1 to every K-space there is only one single value of Π. Therefore, to every set of values of $\{\Pi_1, \Pi_2, \dots, \Pi_p\}$ there is only a single value of Π i.e., Π is a unique function of $\{\Pi_1, \Pi_2, \dots, \Pi_p\}$. It follows that an arbitrary dimensionally homogeneous function $y = f(x_1, \dots, x_n)$ can be reduced to the form $\Pi = F(\Pi_1, \dots, \Pi_p)$. According to proposition 8.4, $p = (n - r)$, where r is the rank of the dimensional matrix. The converse of the theorem is equally true, i.e., an equation of dimensionless products is dimensionally homogeneous. However, this statement is trivial, qed.

8.5.2 Applications of the Theory and Π-theorem

Example 8.6 Consider the equation

$$y = x_1 \cdot x_2 \cdot x_3 \,. \tag{8.5.11}$$

What are the conditions that this equation is dimensionally homogeneous? If we apply Proposition 8.1, then there follows

$$K_0 y = K_1 x_1 \cdot K_2 x_2 \cdot K_3 x_3 \quad \text{with}$$
$$K_0 = \prod_{j=1}^{m} \alpha_j^{a_{j0}} \,, \quad K_i = \prod_{j=1}^{m} \alpha_j^{a_{ij}}$$

and therefore by substitution

$$\begin{aligned} &\left(\prod_{j=1}^{m} \alpha_j^{a_{j0}}\right) x_1 \cdot x_2 \cdot x_3 \\ &= \left(\prod_{j=1}^{m} \alpha_j^{a_{j1}}\right) x_1 \cdot \left(\prod_{j=1}^{m} \alpha_j^{a_{j2}}\right) x_2 \cdot \left(\prod_{j=1}^{m} \alpha_j^{a_{j3}}\right) x_3 \\ &= \left(\prod_{j=1}^{m} \alpha_j^{a_{j1}+a_{j2}+a_{j3}}\right) x_1 \cdot x_2 \cdot x_3 \,, \end{aligned}$$

or

$$\alpha_{j0} = \alpha_{j1} + \alpha_{j2} + \alpha_{j3} \,. \tag{8.5.12}$$

This is the condition of dimensional homogeneity.

Example 8.7 Let three dimensionless products be given by

$$\begin{aligned} \Pi_1 &= x_1 \times x_2^{1/2} \times x_4^{-1/2} \times x_5^2 \,, \\ \Pi_2 &= x_1^{-3/2} \times x_2^2 \times x_3^{5/2} \times x_4^{-4} \times x_5^6 \,, \\ \Pi_3 &= x_1^{-4} \times x_2^{7/2} \times x_3 \times x_4^{-15/2} \times x_5^{10} \,. \end{aligned} \tag{8.5.13}$$

Are these products independent? To clear this, we form the matrix of the exponents

$$\begin{array}{c|ccccc} & x_1 & x_2 & x_3 & x_4 & x_5 \\ \hline \Pi_1 & 1 & \frac{1}{2} & 0 & -\frac{1}{2} & 2 \\ \Pi_2 & -\frac{3}{2} & 2 & \frac{5}{2} & -4 & 6 \\ \Pi_3 & -4 & \frac{7}{2} & 5 & -\frac{15}{2} & 10 \end{array} . \tag{8.5.14}$$

According to Proposition 8.2 its rows must be linearly independent, if Π_1, Π_2 and Π_3 are to be independent. However, this is not the case, since the determinant of every 3×3 matrix of (8.5.14) vanishes, as one can easily corroborate. This can also be verified, if one recognised that the third row z_3 can be obtained from the first and second, z_1 and z_2, by forming $z_3 = -z_1 + 2z_2$. This is equivalent to the dependence $\Pi_3 = \Pi_2^2/\Pi_1$.

Example 8.8 Assume the dimensional matrix of five variables takes the form

$$\begin{array}{c|ccccc} & y & x_1 & x_2 & x_3 & x_4 \\ \hline M & 1 & 1 & -1 & 2 & 0 \\ L & 3 & -2 & 4 & 1 & -1 \\ T & 2 & -1 & 3 & 3 & -1 \end{array} . \tag{8.5.15}$$

We ask whether there exists a dimensionally homogeneous product of the form

$$y = x_1^{k_1} \cdot x_2^{k_2} \cdot x_3^{k_3} \cdot x_4^{k_4} .$$

This question may be answered by computing the 3×3 determinants of (8.5.15); that one formed by the columns $\{y, x_1, x_2\}$ differs from zero, but every determinant of a (3×3) submatrix with the columns $\{x_i, x_k, x_l\}$ vanishes. Therefore, the rank of the matrix that is enlarged by the column y is 3, that of the 3×3 matrices formed with the x-columns, however, smaller, namely 2; that is, there is no such dimensionally homogeneous product.

Example 8.9 Let $y = f(x_1, x_2, x_3)$ be a dimensionally homogeneous function with dimensional matrix

$$\begin{array}{c|cccc} & y & x_1 & x_2 & x_3 \\ \hline M & 1 & 1 & 2 & -1 \\ L & 3 & -1 & 0 & 2 \\ T & -2 & -3 & -2 & 2 \end{array} . \tag{8.5.16}$$

Assume, moreover, that in a physical model the quantities x_1, x_2, x_3 are to be reduced in size, x_1 to a fifth, x_2 to a tenth and x_3 to a fourth of their values in nature. What is the change in scale for the variable y in the case? According to Proposition 8.1 one has

$$K_0 y = f(K_1 x_1, K_2 x_2, K_3 x_3)$$

with

$$K_0 = \prod_{j=1}^{m} \alpha_j^{a_{j0}} , \quad K_i = \prod_{j=1}^{m} \alpha_j^{a_{ji}} . \tag{8.5.17}$$

This yields for α_j $(j = 1, 2, 3)$ the nonlinear equations

$$K_1 = \alpha_M \, \alpha_L^{-1} \, \alpha_T^{-3} = \tfrac{1}{5} \,, \qquad (8.5.18)$$

$$K_2 = \alpha_M^2 \, \alpha_L^0 \, \alpha_T^{-2} = \tfrac{1}{10} \,, \qquad (8.5.19)$$

$$K_3 = \alpha_M^{-1} \, \alpha_L^2 \, \alpha_T^2 = \tfrac{1}{4} \,. \qquad (8.5.20)$$

Multiplying the first and the third together, implies

$$\alpha_L \, \alpha_T^{-1} = \tfrac{1}{20} \,, \qquad (8.5.21)$$

and the second and third yield by elimination of α_M

$$\alpha_L^4 \, \alpha_T^2 = \tfrac{1}{160} \,. \qquad (8.5.22)$$

If one eliminates from these last two equations α_T, one obtains

$$\alpha_L^6 = \frac{1}{160 \times 400} = \frac{1}{64} \times 10^{-3} \quad \Rightarrow \quad \alpha_L = \tfrac{1}{2} \times 10^{-1/2} \qquad (8.5.23)$$

and after back–substitution

$$\alpha_T = 10^{1/2} \,, \quad \alpha_M = 1 \,. \qquad (8.5.24)$$

Thus, the change of scale for y is given by

$$\begin{aligned} K_0 &= \alpha_M \alpha_L^3 \alpha_T^{-2} = 1 \times \tfrac{1}{8} \times 10^{-3/2} \times 10^{-1} = \tfrac{1}{8} \times 10^{-5/2} \\ &= 0{,}000395 \,. \end{aligned} \qquad (8.5.25)$$

In the model the values of y are reduced in comparison to those in nature by a factor of approximately 4×10^{-4}.

Example 8.10 KELVIN–HELMHOLTZ **Instability** Consider a vertically stratified fluid with density profile $\rho_0(z)$. Let this fluid be subject to a horizontal shear flow $U(z)$. It is well known that such shear flows may become unstable depending upon how large the shearing dU/dz and the vertical density gradient $d\rho_0/dz$ are. We shall use methods of dimensional analysis to clarify which parameter governs this KELVIN–HELMHOLTZ instability[12].

A set of parameters upon which the stability criterion may depend, is

$$\frac{dU}{dz} \,, \; \frac{d\rho_0}{dz} \,, \; \rho_0 \,, \; g \,. \qquad (8.5.26)$$

dU/dz measures the shearing, $d\rho_0/dz$ the stratification, ρ_0 is a reference density and g is the acceleration due to gravity. (8.5.26) yield the dimensional matrix

[12] See e.g. CHANDRASEKHAR [45], Chap. 11; ACHESON [1], Chap. 9; LEBLOND & MYSAK [129], Sect. 42, ZEYTOUNIAN [263], Sect. 38.

$$\begin{array}{c|cccc} & dU/dz & d\rho_0/dz & \rho_0 & g \\ \hline M & 0 & 1 & 1 & 0 \\ L & 0 & -4 & -3 & 1 \\ T & -1 & 0 & 0 & -2 \end{array} \tag{8.5.27}$$

which possesses the rank $r = 3$. There is therefore exactly one dimensionless product of the above four variables. If one would have omitted the Earth's acceleration g as a possible variable, the rank of the remaining 3×3 matrix would still be $r = 3$, and no dimensionless product could be formed with the first three variables. Hence, the addition of the fourth variable – gravity – is important. One could, in place of g, have also added another variable, e.g. $U(z)$ with the same result that would allow the formation of a dimensionless product. In the two cases the following dimensionless products can be formed:

$$\mathbb{R}_i = \Pi_1 = \frac{-g\dfrac{1}{\rho_0}\dfrac{d\rho_0}{dz}}{\left(\dfrac{dU}{dz}\right)^2} \quad \text{or} \quad \Pi_1' = \frac{\dfrac{1}{\rho_0}\dfrac{d\rho_0}{dz}}{\dfrac{1}{U}\dfrac{dU}{dz}} \,. \tag{8.5.28}$$

The first of these Π-products is relevant (which is supported both by observations and computations); $\Pi_1 = \mathbb{R}_i$ is called the RICHARDSON number, and one can prove that a pure shear flow in a linearly vertically stratified fluid is stable if $\mathbb{R}_i > 1/4$ and possibly unstable if $\mathbb{R}_i < 1/4$.[13]

Such instabilities arise in the atmosphere, in the ocean and in lakes in regions of small vertical density gradients (and large shearing). It is furthermore so that the diffusive transport of tracer substances in the ocean or in lakes is strongly dependent upon the stratification. One knows e.g. that the transport of oxygen down to larger depths depends on the stratification. If $D(z)$ is the turbulent diffusion coefficient with $[D(z)] = [L^2T^{-1}]$ at depth z and D_0 a reference value, say at the surface, then it is reasonable to assume that the value of $D(z)$ will depend upon how stable the shear flow is in the region of study. Thus, one might well start with the variables listed in the following dimensional matrix

$$\begin{array}{c|cccccc} & D & D_0 & dU/dz & d\rho_0/dz & \rho_0 & g \\ \hline M & 0 & 0 & 0 & 1 & 1 & 0 \\ L & 2 & 2 & 0 & -4 & -3 & 1 \\ T & -1 & -1 & -1 & 0 & 0 & -2 \end{array} \tag{8.5.29}$$

of which the rank is $r = 3$. For this set of variables there are therefore $6-3 = 3$ dimensionless products, namely

[13] See MILES [154] and HOWARD [102]. Actually $\mathbb{R}_i > 1/4$ is a sufficient condition for stability. It is also plausible that Π_1' cannot possibly be a relevant Π-product, because $U(z)$ hardly be relevant, because for each z it can be removed by a GALILEI transformation.

$$\Pi_1 = \mathbb{R}_i = \frac{-g\dfrac{1}{\rho_0}\dfrac{d\rho_0}{dz}}{\left(\dfrac{dU}{dz}\right)^2}\,, \quad \Pi_2 = \frac{D(z)}{D_0}\,, \quad \Pi_3 = \frac{D_0\dfrac{dU}{dz}}{g\rho_0}\frac{d\rho_0}{dz}\,.$$

There is no necessity to construct these products formally, often it is advantageous to guess them; one then must merely ascertain that they are independent, which is the case here, since Π_1 does not depend on (D, D_0), Π_2 does not depend on $(g, d\rho_0/d\rho)$ and Π_3 is independent of $(D, g/\rho_0)$. Thus, we may now write

$$D(z) = D_0 f(\mathbb{R}_i, \Pi_3)\,. \tag{8.5.30}$$

A dependence upon Π_3 is not likely since Do is a reference diffusivity, which can be chosen arbitrary. Thus an ansatz

$$D(z) = D_0 f(\mathbb{R}_i) \tag{8.5.31}$$

is sufficient. Such dependences have indeed been experimentally verified.

Example 8.11 BÉNARD **Convection** A similar problem is the stability problem of the RAYLEIGH–BÉNARD convection[14]. Consider a viscous BOUSSINESQ fluid, which is kept between two infinitely long rigid plates at a distance H and is at rest. The two plates are heated and cooled, respectively, and the temperature at the upper plate T_u is smaller than the temperature of the lower plate, T_l. The coefficient of thermal expansion of the incompressible fluid and the kinematic viscosity are given by α $[K^{-1}]$ and ν $[L^2T^{-1}]$, respectively. Observations show that, depending on the kind of fluid and the gap width H, the fluid is at rest, provided the temperature difference $\triangle T = T_l - T_u$ is sufficiently small. In this state the transport of heat from the lower to the upper plate is by conduction, whence molecular heat conduction. When conditions are right, i.e., if the temperature difference $\triangle T$ is sufficiently large, a cellular convection flow is formed. If side boundaries are absent, convective rolls with horizontal axes are formed, if side boundaries are present, hexagonal cells are established, in which the fluid is vertically circulating with an eddy-type structure, see Fig. 8.3. As variables, which might describe this phenomenon, we have

	H	g	ν	α	$\triangle T$	κ
M	0	0	0	0	0	0
L	1	1	2	0	0	2
T	0	−2	−1	0	0	−1
Θ	0	0	0	−1	1	0 .

(8.5.32)

The gap width H is the only vertical length of the problem. g is the Earth's acceleration and necessarily influences the problem, since the convective flow

[14] The RAYLEIGH–BÉNARD problem is treated at length by CHANDRASEKHAR [45] for the linear theory and by STRAUGHAN [229] with non-linear energy stability techniques.

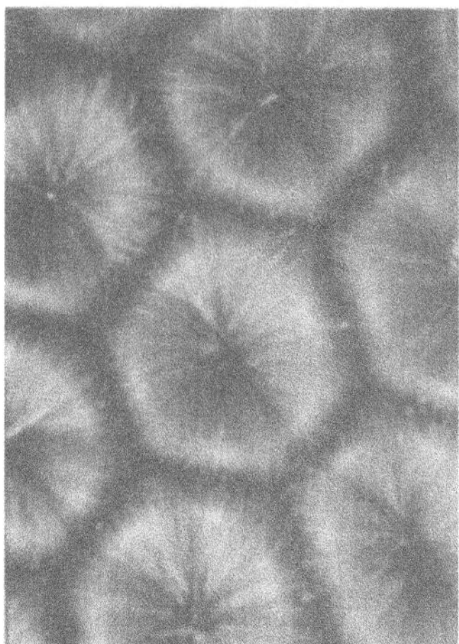
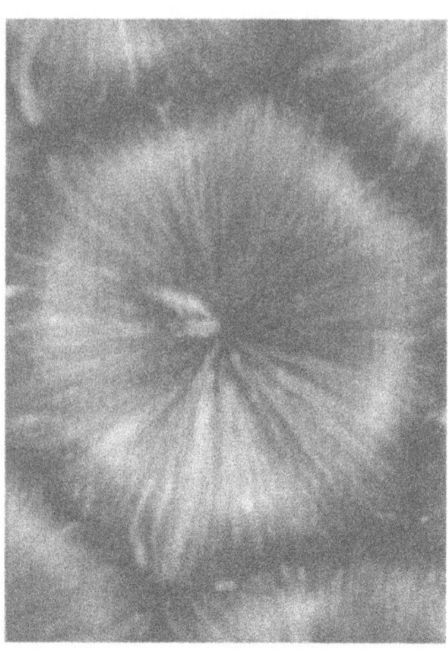

Fig. 8.3. RAYLEIGH–BÉNARD cells can most simply be made visible in shallow shells using silicon oil and aluminium powder. They are, however also briefly visible when eggs are fried (shortly before the white of the egg is coagulating). [Courtesy of M.G. VELARDE and C. NORMAND [248].]

works against gravity. The coefficient of thermal expansion α, the coefficient of thermal diffusion κ and the kinematic viscosity ν define the physical parameters of the fluid. The coefficient of thermal expansion is important since for $\alpha \equiv 0$ a convective flow due to the temperature difference $\triangle T$ can not even arise. Heat conduction is in this case the only energy transport from the lower to the upper plate and flow can not set in.

The dimensional matrix (8.5.32) has the rank $r = 3$; therefore, there are three independent power products,

$$\Pi_1 = \mathbb{R}_a = \frac{g\alpha\triangle T}{\kappa\nu} H^3 , \quad \Pi_2 = \alpha\triangle T , \quad \Pi_3 = \mathbb{P}_r = \frac{\nu}{\kappa} . \tag{8.5.33}$$

The first is the so-called RAYLEIGH number, the second has no name, the third is the PRANDTL number and is a pure material coefficient. It is customary to use in place of (8.5.33) the alternative dimensionless products

$$\mathbb{R}_a = \frac{g\alpha\triangle T}{\kappa\nu} H^3 , \quad \mathbb{R} = \sqrt{\frac{\Pi_1}{\Pi_2\Pi_3}} = \frac{\sqrt{gH}H}{\nu} , \quad \mathbb{P}_r = \frac{\nu}{\kappa} . \tag{8.5.34}$$

It follows that the transition from the pure heat conduction of the fluid at rest to the BÉNARD convection is described by a relation of the form

$$\mathbb{R}_a = f(\mathbb{R}, \mathbb{P}_r) . \tag{8.5.35}$$

Traces of BÉNARD-type convection can be seen on Earth in breathtaking beauty. Figure 8.4 shows an air photograph of a stretch of land of only a few hundred meters width between the sea in North-East Siberia. It shows a tundra landscape with more developped bushes arranged at cell boundaries and meagre growth of moss and herps in the interior of the cells. The cell geometry is not exactly hexagonal, but many hexagons can still be identified with little effort. The more efficient growth of the bushes at the cell boundaries must be due to a better transport of nutrients (nitrate, phosphate) to the roots of these plants and obviously a lack of them inside the cells. If we conjecture a BÉNARD-type convection in the groundwater of the saturated soil below, then this flow is upwards at the cell boundaries and downwards at the cell centres (see Fig. 8.3). This groundwater carries the nutrients as traces, the bushes are nourished with them, and what is left as nutrients in the centres of the cells allows only meagre growth of tundra.

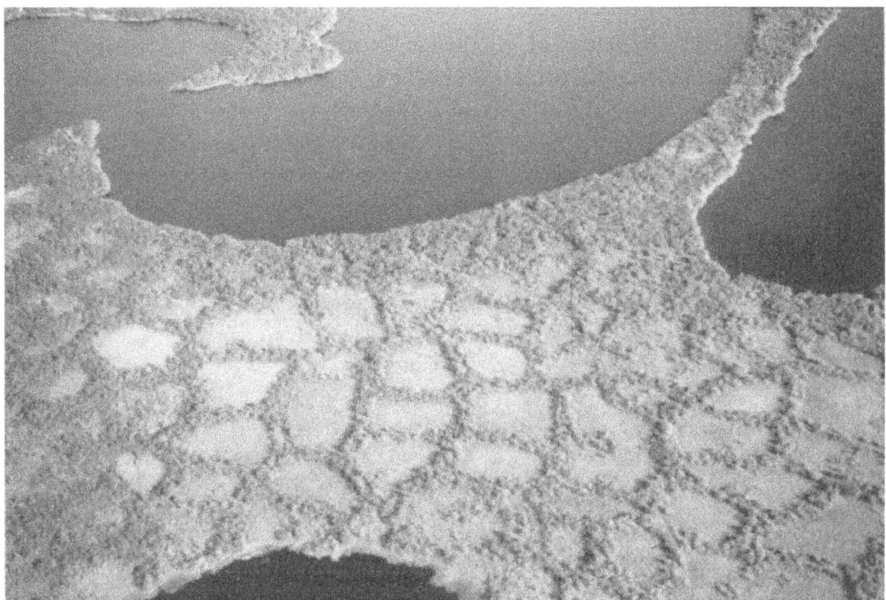

Fig. 8.4. Air photograph of a small land-ocean region in North-East Siberia, showing a tundra landscape with nearly hexagonal cell structure, with bushes at the cell boundaries and moss and herps in the cell centres due to differences in available nutrients of the convective growndwater flow. Photo: WERNER H. SCHOCH, Labor für quartäre Hölzer, 8135 Langnau, Switzerland.

Now, how does such a convection flow arise in the saturated soil and why is the permafrost thawed in this region? Here one can only guess, since no measurements were conducted that would support this conjecture. In the polar regions the subsea permafrost is often thawed in a layer below the ocean and a lower phase change boundary. This phase–change surface reaches the free surface on land, since even today the land area is exposed to the much colder climate (of the atmosphere) than the ocean ground. In this particular situation, the ocean is surrounding all sides of the land stretch. Thus, it is likely that the permafrost, covering during the last Ice Age also the ocean region, has thawed to a certain depth also within the land. This thawing process of the permafrost is still going on. It provides a fresh water source at the lower boundary of the otherwise salty groundwater and generates the prerequisites of the BÉNARD convection (now due to salt and less temperature variation) of the groundwater. The nutrients for the bushes are probably provided by salt water exchange between the ocean and the groundwater.

This interpretation has not been verified in this case. It at least provides a possible explanation of the phenomenon. Its beauty beyond that of the landscape texture lies in the botanic representation of a purely physical phenomenon.

Example 8.12[15] Motion of a Shock Front After an (Atomic) Explosion Close to the Ground Consider a half space (e.g. the atmosphere bounded by the plane ground). Let at a point on the ground, at time $t = 0$ a large quantity of energy $E^{(3)}$ be released. In an actual explosion (of an atomic bomb, say) this release is accomplished in a small volume during a very small time. As an idealisation, we ignore the small volume of the bomb and the duration of the explosion. As a consequence of the explosion a half spherical shock wave will form. Its front will at time $t > 0$ be a distance $R_f^{(3)}(t)$ away from the centre. We wish to find a relationship for this front position in

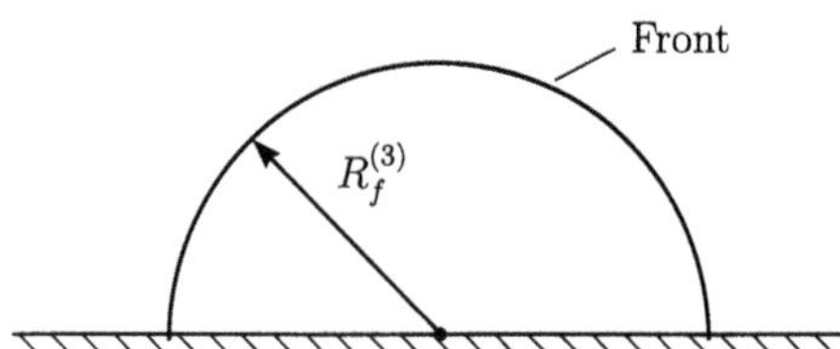

Fig. 8.5. Semi-spherical fire ball with radius $R_f^{(3)}$ due to a point explosion at the centre.

terms of the quantities which influence this process; these are likely $E^{(3)}$, t, the density of air, ρ_0 and $R_f^{(3)}(t)$. Thus, the dimensional matrix is

	t	$R_f^{(3)}$	ρ_0	$E^{(3)}$
M	0	0	1	1
L	0	1	-3	2
T	1	0	0	-2

.

[15] This example is reported in BARENBLATT [19].

It has rank $r = 3$, and so there is one dimensionless product

$$\Pi = \frac{R_f^{(3)}}{(E^{(3)}/\rho_0)^{1/5} t^{2/5}} = \text{Const}\ , \tag{8.5.36}$$

which must be constant. If this constant is known, we have

$$R_f^{(3)}(t) = \text{Const} \times \left(\frac{E^{(3)}}{\rho_0}\right)^{1/5} t^{2/5}\ . \tag{8.5.37}$$

The expansion of the semi spherical front grows with $t^{2/5}$. The result (8.5.37) shows that the "Const" can be determined, if one measures the radius of the front $R_f^{(3)}$ at different times. This is best done in doubly logarithmic representation, i.e.,

$$\ln R_f^{(3)} = \ln \text{Const} + \frac{1}{5} \ln\left(\frac{E^{(3)}}{\rho_0}\right) + \frac{2}{5} \ln t\ . \tag{8.5.38}$$

In a plot with $x \equiv \ln t$ and $y = \ln R_f^{(3)}$ (8.5.38) represents a straight line with inclination $\frac{2}{5}$; it crosses the y-axis at $\ln \text{Const} + \frac{1}{5}\ln(E^{(3)}/\rho_0)$. Incidentally, the solution of the gas dynamical problem is known and shows that Const $\cong 1$. If one knows this, then experimental determination of $R_f^{(3)}(t)$ allows estimation of the strength of the explosion. Exactly this was done by G. I. TAYLOR by using a movie film of the nuclear test in the desert of New Mexico, when the Americans were testing their atomic bombs in their Manhattan Project during the Second World War. For the nuclear agency of the USA this caused much embarrassment as TAYLOR said, since the strength of the bomb was kept secret, whilst the movie was not classified.[16]

It is interesting that in the above result the dimension of the semi-space in which the explosion takes place does not seem to explicitly arise. However, this is not so; indeed the reader may show by himself that in the two- and one-dimensional case the formulas are

$$\begin{aligned} R_f^{(2)}(t) &= \text{Const} \times \left(\frac{E^{(2)}}{\rho_0}\right)^{1/4} t^{2/4}\ , \\ R_f^{(1)}(t) &= \text{Const} \times \left(\frac{E^{(1)}}{\rho_0}\right)^{1/3} t^{2/3}\ . \end{aligned} \tag{8.5.39}$$

The speed of expansion therefore changes with the dimension of space in which the wave expands.

[16] G. I. TAYLOR. The formation of a blast wave by a very intense explosion, Part I. Theoretical discussion. Part II: The atomic explosion of 1945. Proc. R. Soc. London **A 201**, 159–186 [234]. The original publication is G. I. TAYLOR: The formation of a blast wave by a very intensive explosion. Civil Reference Research Committee, Report RC 210, 27, Juni 1941.

Example 8.13 Velocity Distribution in the Wall Near Boundary Layer of a Turbulent Flow We consider a turbulent flow of which the mean velocity does not change, neither with time nor with the coordinate in the direction of the flow, but that such a change occurs perpendicular to it. This is called a shear flow which is bounded by a wall. If the origin of a coordinate z perpendicular to the flow is positioned in the wall, then we may ask how the wall–parallel velocity changes in the direction perpendicular to the wall. It describes the near–wall turbulent velocity profile in a straight channel or in a circular pipe.

In the neighbourhood of the wall bounding the flow one may assume that the shear stress parallel to the wall is constant, thus independent of the z-coordinate. One may then suppose that the transverse velocity gradient $\partial u/\partial z = u'$ at a distance z depends upon the wall–shear stress τ, the density ρ of the fluid, its kinematic viscosity ν and the boundary layer thickness δ of the near wall layer. The dimensional matrix of these variables is given by

$$\begin{array}{c|cccccc} & u' & z & \delta & \rho & \nu & \tau \\ \hline M & 0 & 0 & 0 & 1 & 0 & 1 \\ L & 1 & 1 & 1 & -3 & 2 & -1 \\ T & -1 & 0 & 0 & 0 & -1 & -2 \end{array} , \qquad (8.5.40)$$

has rank 3, thus yielding three independent Π-products

$$\Pi_1 = \frac{zu'}{u^*} , \quad \Pi_2 = \frac{u^* z}{\nu} = \mathbb{R}_l , \quad \Pi_1 = \frac{u^* \delta}{\nu} = \mathbb{R}_* , \qquad (8.5.41)$$

in which the *shear stress velocity* u^* is defined by

$$u^* = \sqrt{\tau/\rho} , \qquad (8.5.42)$$

where $\mathbb{R}_l$ is a local and $\mathbb{R}_*$ a global REYNOLDS number. With (8.5.41) the BUCKINGHAM theorem yields the general equation

$$\frac{zu'}{u^*} = f(\mathbb{R}_l, \mathbb{R}_*) . \qquad (8.5.43)$$

If one assumes that the two REYNOLDS numbers are large (in pipe and channel flows one can assume that $\mathbb{R}_l \geq 100$ and $\mathbb{R}_* \geq 10^4$), $f = \text{const}$ may be an acceptable hypothesis, because it supposes that for large REYNOLDS numbers, Π_1 is independent of the REYNOLDS numbers. One may then integrate (8.5.43) and obtains

$$u(z) = \frac{u^*}{\kappa} \ln(z) + \text{Const} , \qquad (8.5.44)$$

in which $\kappa \approx 0.41$ is the so-called VON KÁRMÁN constant, the value of $1/f$ for large REYNOLDS numbers. Thus, with dimensional analysis and an assumption about the value of $f(\infty, \infty)$ we succeeded in determining the *universal logarithmic velocity* profile, which is experimentally well verified. Nevertheless, one may justly ask the question whether the VON KÁRMÁN constant is in fact a constant, so that systematic deviations from the logarithmic law would exist. Measurements e.g. by NIKURADSE suggest equally convincingly a power law dependence z^α of the mean velocity in the near wall region, see SCHLICHTING[17]. If one chooses

$$f(\mathbb{R}_l, \mathbb{R}_*) = \mathbb{R}_l^\lambda \tilde{f}(\mathbb{R}_*) = \frac{\mathbb{R}_l^\lambda}{\kappa(\mathbb{R}_*)} = \left(\frac{u^* z}{\nu}\right)^\lambda \frac{1}{\kappa(\mathbb{R}_*)}, \tag{8.5.45}$$

this power dependence is explicitly put in evidence, and one obtains from (8.5.43) the equation

$$u' = \frac{u_*^{\lambda+1}}{\nu^\lambda \kappa(\mathbb{R}_*)} z^{\lambda-1}, \tag{8.5.46}$$

or after an integration

$$u(z) = \frac{u_*^{\lambda+1}}{\lambda \nu^\lambda \kappa(\mathbb{R}_*)} z^\lambda + \underbrace{\text{Const}}_{0}, \tag{8.5.47}$$

in which the constant of integration is set equal to zero to fulfill the boundary condition $u(0) = 0$.

For every comparison between measured and thus parameterized velocity profiles the power law fit and the logarithmic fit are equally convincing. In practice, however, the logarithmic law is used.[18]

Example 8.14 Viscosity in a Kinetic Gas[19] In the kinetic theory of gases the balance laws are deduced by building moments of the BOLTZMANN equation; likewise it is possible to obtain the functional dependence of the shear viscosity from the BOLTZMANN equation. The decisive element in this derivation is the collision operator, the form of which depends upon the law of interaction in binary collisions. Three mechanical parameters describe this interaction,

$\mathfrak{m}$: mass of the molecule,

$\mathfrak{d}$: typical dimension of the molecule,

$\mathfrak{f}$: the intermolecular "force" (interaction force) at a unit distance of the molecules,

and these may influence the interaction only in certain combinations.

---

[17] H. SCHLICHTING [209].

[18] Recently, BARENBLATT [20] has elaborated on this and determined the dependency $\kappa(\mathbb{R}_*)$.

[19] See TRUESDELL & MUNCASTER [242], where this example is presented with less detail.

It is known that the viscosity of a gas depends upon the density ρ and the temperature Θ. Since the latter is a measure for the kinetic energy of the fluctuating motion of the molecules and since the latter is denoted as internal energy ε which is proportional to the temperature, one may assume the viscosity in the form

$$\mu = \mu(\rho, \varepsilon; \mathfrak{m}, \mathfrak{d}, \mathfrak{f}) , \tag{8.5.48}$$

in which the last three variables characterise the dependence of the viscosity upon the properties of the interaction of the molecules during collisions.

The dimensional matrix for (8.5.48) is given by

	μ	ρ	ε	$\mathfrak{m}$	$\mathfrak{d}$	$\mathfrak{f}$
M	1	1	0	1	0	1
L	−1	−3	2	0	1	1
T	−1	0	−2	0	0	−2

and possesses rank 3, so that according to the Π-theorem three independent dimensionless Π-products must exist, namely for instance

$$\frac{\mu\mathfrak{d}^2}{\mathfrak{m}\sqrt{\varepsilon}} , \quad \frac{\rho\mathfrak{d}^3}{\mathfrak{m}} , \quad \frac{\mathfrak{f}\mathfrak{d}}{\mathfrak{m}\varepsilon} . \tag{8.5.49}$$

With these the viscosity can be written as

$$\mu = \frac{\mathfrak{m}\sqrt{\varepsilon}}{\mathfrak{d}^2} f\left(\frac{\rho\mathfrak{d}^3}{\mathfrak{m}}, \frac{\mathfrak{f}\mathfrak{d}}{\mathfrak{m}\varepsilon}\right) , \tag{8.5.50}$$

in which f is an arbitrary function.

This formula allows us to study qualitatively the behaviour of the viscosity when the three dimensional parameters of the molecule are varied. For instance, because $\rho = \mathfrak{m}n$, where n denotes the number density of molecules, the first argument in f is given by $n\mathfrak{d}^3$ and hence independent of the mass of the molecules and equal to the volume density of the molecules. Furthermore, provided the molecules do not exercise any forces upon each other, if their distance is finite, the last column of the above dimensional matrix may be dropped. The thus reduced dimensional matrix still possesses rank 3 and μ can be expressed as

$$\mu = \frac{\mathfrak{m}\sqrt{\varepsilon}}{\mathfrak{d}^2} \hat{f}\left(\frac{\rho\mathfrak{d}^3}{\mathfrak{m}}\right) . \tag{8.5.51}$$

In other words, in a gas in which the molecules do not execute forces upon each other at finite distances, the viscosity grows with the square root of the temperature (or internal energy). This is e.g. the case for molecules, which may be treated as frictionless perfect elastic (hard) spheres. If the volume of the molecule $n\mathfrak{d}^3$ is vanishingly small, then $\hat{f}$ in (8.5.51) is a constant and the viscosity becomes

$$\mu \propto \frac{\mathfrak{m}\sqrt{\varepsilon}}{\mathfrak{d}^2} \quad \text{(ideal spheres)} . \tag{8.5.52}$$

Another interesting case prevails, if the molecules repel each other by a force, of which the modulus is proportional to the $\mathfrak{k}$-th power of the inverse distance of the molecules; NEWTON's law then reads

$$\mathfrak{m} f = \frac{\mathfrak{G}}{r^{\mathfrak{k}}} , \quad \mathfrak{k} > 0 , \quad \mathfrak{G} > 0 . \tag{8.5.53}$$

These are the inverse *power molecules*. The dimension of $\mathfrak{k}$ is 1 and $[\mathfrak{G}] = ML^{\mathfrak{k}+1}T^{-2}$. We leave it as an exercise to the reader to show that for such molecules the viscosity satisfies the general equation

$$\mu = \mathfrak{m}\sqrt{\varepsilon}\left(\frac{\mathfrak{m}\varepsilon}{\mathfrak{G}}\right)^{\frac{2}{\mathfrak{k}-1}} \hat{f}\left(\frac{\mathfrak{m}\varepsilon}{\mathfrak{G}}\left(\frac{\rho}{\mathfrak{m}}\right)^{(1-\mathfrak{k})/3}\right) . \tag{8.5.54}$$

This equation shows that only inverse power molecules with $\mathfrak{k} \neq 1$ are meaningful. Moreover, if one can show e.g. by experiment that the viscosity is independent of the density, then $\hat{f}(\cdot)$ in (8.5.54) must be a constant. For such a case

$$\mu \propto \mathfrak{m}\sqrt{\varepsilon}\left(\frac{\mathfrak{m}\varepsilon}{\mathfrak{G}}\right)^{\frac{2}{\mathfrak{k}-1}} . \tag{8.5.55}$$

The kinetic theory of gases (which are ideal and dilute) presupposes such conditions. Equation (8.5.54) is, however, more general. Finally, one obtains from (8.5.55) in the limit $\mathfrak{k} \to \infty$, $\mu \propto \mathfrak{m}\sqrt{\varepsilon}$, which agrees with the result (8.5.52) of ideal hard spheres.

Example 8.15 Rising Gas Bubbles Consider a gas bubble which is rising in a quiescent viscous fluid. Such bubbles reach soon after they have been formed a constant velocity, see Fig. 8.6. We wish to use dimensional analysis to derive a formula for this rising velocity. The physical quantities upon which the rising of bubbles may depend are the rising velocity U, the Earth's acceleration g, a typical diameter D of the bubbles, the density of the fluid ρ, the kinematic viscosity ν of the fluid, and for small bubbles the surface tension, σ. We neglect the kinematic viscosity of the gas in the bubbles as well as the density difference $|(\rho_{Gas} - \rho_{Fluid})/\rho_{Fluid}| \simeq 1$. Finally we also ignore thermal effects and the exchange of matter between the bubble and the fluid. Thus, the dimensional matrix is given by

	U	g	D	ν	σ	ρ
M	0	0	0	0	1	1
L	1	1	1	2	−1	−3
T	−1	−2	0	−1	−2	0

(8.5.56)

It possesses rank $r = 3$ and gives rise to three dimensionless products, namely

$$\mathbb{F}_r = \frac{U}{\sqrt{gD}} , \quad \mathbb{R} = \frac{UD}{\nu} , \quad \mathbb{W} = \frac{\rho U^2 D}{\sigma} , \tag{8.5.57}$$

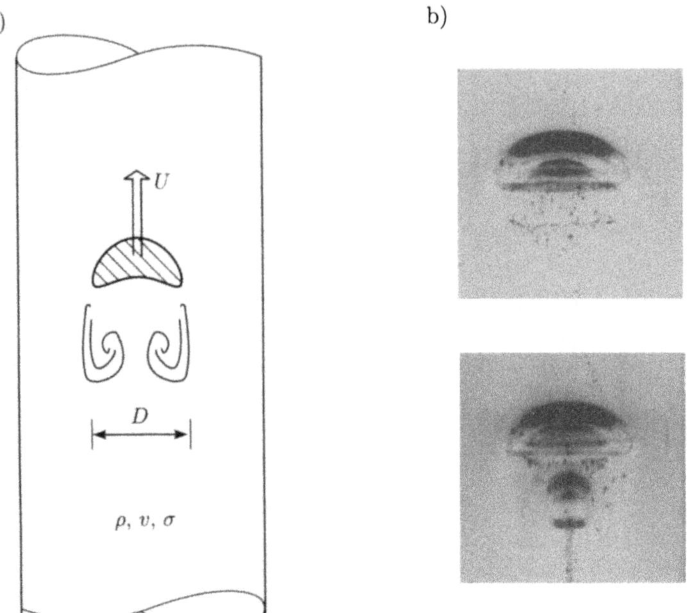

Fig. 8.6. a) Sketch of a bubble with trailing vortex ring. The bubble diameter is D and it is assumed to be small in comparison to the pipe diameter. b) Photos of rising bubbles in silicon oil (Model of demonstration, Institute of Mechanics, Darmstadt University of Technology, Photo and model courtesy of Prof. K. G. ROESNER).

the FROUDE, REYNOLDS and WEBER numbers. Therefore an equation of the form

$$\mathbb{F}_r = f(\mathbb{R}, \mathbb{W}) \tag{8.5.58}$$

must hold; for not too small bubbles a dependence on theWEBER number may be dropped; in which case one has the simplified formula

$$\mathbb{F}_r = f(\mathbb{R}) \, . \tag{8.5.59}$$

If one postulates a power law,

$$\mathbb{F}_r \equiv C\mathbb{R}^\alpha \tag{8.5.60}$$

with unknown parameters C and α, then one may, after using the definitions for $\mathbb{F}_r$ and $\mathbb{R}$, deduce the equation

$$U = C^{1/(1-\alpha)} \; g^{1/(2(1-\alpha))} \; \nu^{\alpha/(\alpha-1)} \; D^{(2\alpha+1)/(2(1-\alpha))} \, . \tag{8.5.61}$$

Observations of rising bubbles with various diameters indicate that U varies linearly with D, so that

$$\frac{2\alpha+1}{2(1-\alpha)} = 1 \quad \Longrightarrow \quad \alpha = \tfrac{1}{4}$$

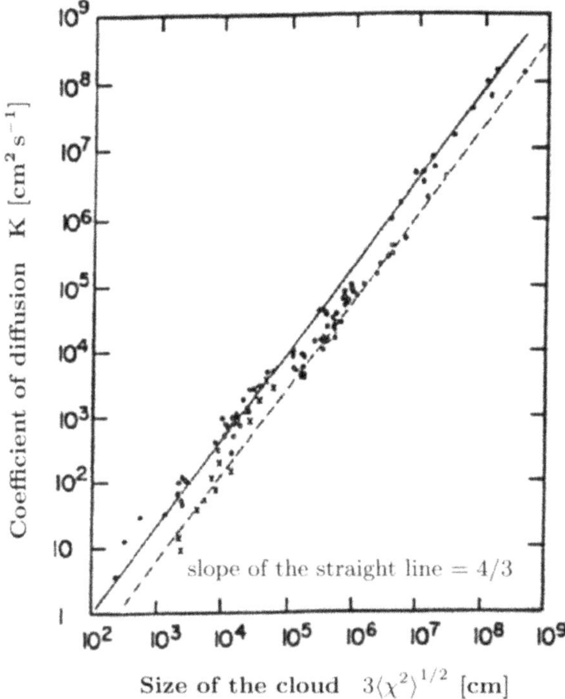

Fig. 8.7. Turbulent diffusion coefficient as a function of the diameter of the cloud. The experimental points correspond to observations taken as follows: • dispersion at the surface, ∘ in the upper layer, × in the thermocline region close to the coast, in 300m depth. The quantity $3\langle\chi^2\rangle^{1/2}$ corresponds to the mean diameter s; the solid line corresponds to K=0,01 $\langle\chi^2\rangle^{2/3}$ and the dashed line corresponds to $K = 0.002\langle\chi^2\rangle^{2/3}$. [Redrawn from OKUBO, A. [181]].

or

$$U = C^{4/3} g^{2/3} \nu^{-1/3} D \, . \tag{8.5.62}$$

The only 'free' constant is now C; it can be determined by experiment and has the value $C = 2/(3 \times 10^{1/4})$.

Example 8.16 Cloud of Pollutant in a Lake or in the Ocean Consider a cloud of a *pollutant* on the surface of a lake or on the ocean. The spreading of such a cloud depends on the turbulent intensity that exists at the location of the cloud. This turbulent intensity depends in turn on the turbulent dissipation rate ε $[L^2T^{-3}]$, i.e., the specific energy which is dissipated per unit time, and the kinematic molecular viscosity ν $[L^2T^{-1}]$ (for more details on this, see Chaps. 10 and 11). We assume the numerical values for these quantities are known. Now, at time $t = 0$, the cloud has a typical diameter s_0. As time proceeds the cloud will at time t have reached a diameter s. Several realizations of the cloud with mean initial diameter s_0 lead at time

t to clouds with mean diameters s and a standard deviation which we shall denote by $\langle s^2 \rangle$. We wish to derive a formula for the time derivative of the standard deviation $d\langle s^2 \rangle/dt$, of the mean diameters s of the cloud at time t. The following variables arise: s_0, t, $d\langle s^2 \rangle/dt$, ε and ν with the dimensional matrix

$$\begin{array}{c|ccccc} & s_0 & t & \frac{d\langle s^2 \rangle}{dt} & \varepsilon & \nu \\ \hline L & 1 & 0 & 2 & 2 & 2 \\ T & 0 & 1 & -1 & -3 & -1 \end{array} . \tag{8.5.63}$$

Since it possesses rank $r = 2$, there are three independent dimensional products, for example

$$\Pi_1 = \frac{d\langle s^2 \rangle/dt}{\varepsilon t^2}, \quad \Pi_2 = \frac{s_0}{\varepsilon^{1/2} t^{3/2}}, \quad \Pi_3 = \frac{t\varepsilon^{1/2}}{\nu^{1/2}},$$

so that

$$\frac{d\langle s^2 \rangle}{dt} = \varepsilon\, t^2\, f(\Pi_2, \Pi_3)\, . \tag{8.5.64}$$

If the cloud diameter is larger than the smallest turbulent eddies, then the molecular viscosity cannot play a role. For large times the initial configuration is equally likely forgotten, i.e., s_0 should neither arise in (8.5.64). Thus, f is a constant, and we have

$$\frac{d\langle s^2 \rangle}{dt} \sim \varepsilon\, t^2\, , \tag{8.5.65}$$

from which after integration we obtain

$$\langle s^2 \rangle \sim \varepsilon\, t^3 \quad \Longrightarrow \quad t \sim \varepsilon^{-1/3} \langle s^2 \rangle^{1/3} \tag{8.5.66}$$

or upon substitution of the latter into the former

$$\frac{d\langle s^2 \rangle}{dt} \sim \varepsilon^{1/3} \langle s^2 \rangle^{2/3} \quad \Longrightarrow \quad \frac{d\langle s^2 \rangle}{dt} \sim s^{4/3}\, . \tag{8.5.67}$$

It is shown in turbulence theory that $d\langle s^2 \rangle/dt$ is proportional to the diffusion coefficient. The result (8.5.67) therefore expresses the fact, that the coefficient of turbulent viscosity grows with the 4/3 power of the cloud diameter. This result has been experimentally verified in the ocean in a diameter range between 10 m and 1000 km with an exceptional accuracy, see Fig. 8.7.

8.6 Exercises

1. Show that all transformations which describe the change of the units of the fundamental dimensions form a group.
2. Corroborate the expansion laws (8.5.39) for the two- and one-dimensional cases.

3. Consider a molecule obeying an inverse power law (8.5.53) for the interaction force

$$\mathfrak{m}f = \frac{\mathfrak{G}}{r^{\mathfrak{k}}}, \quad \mathfrak{k} > 0, \quad \mathfrak{G} > 0,$$

where

$$[\mathfrak{k}] = 1 \quad \text{and} \quad [\mathfrak{G}] = ML^{\mathfrak{k}+1}T^{-2}$$

and assume that the viscosity of the gas shows the functional dependence

$$\mu = \mu(\rho, \epsilon; \mathfrak{m}, \mathfrak{G}), \tag{8.6.1}$$

where ρ is the density and ϵ the specific internal energy. Prove for this formula (8.5.54).

4. **Lubrication in a Cylindrical Bearing** Consider a shaft of radius r, which rotates with steady circumferential velocity u_w ; let the shaft be embedded in a cylindrical bearing with radius $r' = r + a$. The gap between the shaft and the bearing is filled with a lubricating liquid of viscosity η, see Fig. 8.8. If a (transverse) force K is applied perpendicular to the shaft then the shaft will be excentrically positioned relative to the bearing with excentricity e. Whereas the wall shear stresses do not give rise to a moment if the arrangement is centric, there will act a torque M on the shaft, if the shaft rotates eccentrically. Derive a relation between K, M, and e.

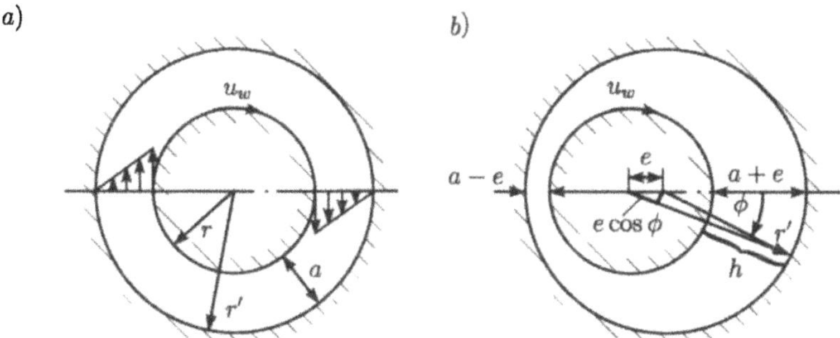

Fig. 8.8. Cylindrical bearing
a) Rotating shaft with radius r and circumferential velocity u_w concentrically positioned relative to the bearing. The gap width a is constant. b) The same with excentric position with excentricity e.

5. Give a formula for the steady discharge Q through a THOMPSON and POINCELET overfall weir, see Fig. 8.9.

a)

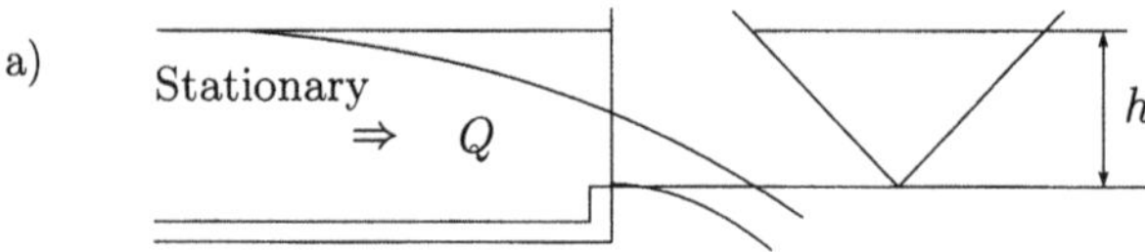

b)

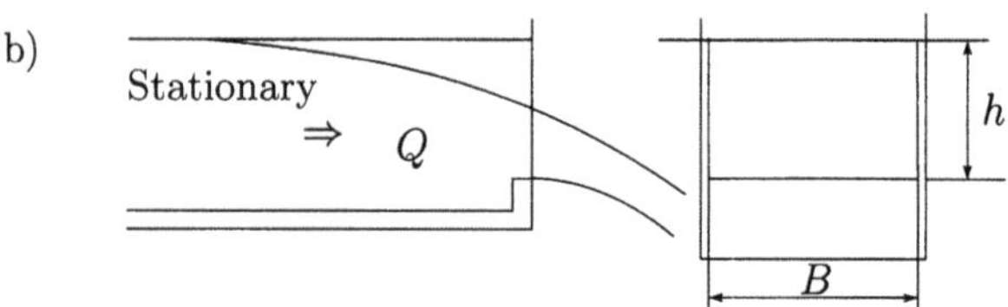

Fig. 8.9. Overfall weirs under stationary conditions. a) THOMPSON, b) POINCELET.

8.7 Solutions

1. The transformations are given by

$$\bar{y} = K_0 y \ , \quad \bar{x}_j = K_j x_j \ ,$$
$$K_0 = \prod_{k=1}^{m} \alpha_k^{a_{k0}} \ , \quad K_j = \prod_{k=1}^{m} \alpha_k^{a_{kj}} \ . \tag{8.7.1}$$

(a_{k_0}, a_{k_j}) $(j = 1, 2, \ldots, n; \ k = 1, 2, \ldots, m)$ is the dimensional matrix and $\alpha_k > 0$ are factors, which describe the ratio of the fundamental dimension k in the old to the new units. The following properties must be verified:

(i) The identity is a permissible transformation. Indeed, if $\alpha_k = 1, \ \forall \ k \in [1, m]$ no unit of the fundamental dimensions changes, so one has $K_0 = 1, \ K_j = 1, \ \forall \ j$ and therefore $\bar{y} = y, \ \bar{x}_j = x_j$.

(ii) The inverse of a K-transformation is also a K-transformation. Indeed, if we define

$$(K_0)^{-1} := \frac{1}{K_0} \ , \quad (K_j)^{-1} := \frac{1}{K_j} \ , \tag{8.7.2}$$

then we deduce form (8.7.1)

$$y = (K_0)^{-1}\bar{y} = (K_0)^{-1} K_0 y \overset{(8.7.2)}{=} y \ ,$$
$$x_j = (K_j)^{-1}\bar{x}_j = (K_j)^{-1} K_j x_j \overset{(8.7.2)}{=} x_j \ .$$

Furthermore, because of (8.7.2) and (8.7.1)

$$
\begin{aligned}
(K_0)^{-1} &= \frac{1}{\prod\limits_{k=1}^{m} (\alpha_k)^{a_{k0}}} = \prod_{k=1}^{m} \left(\frac{1}{\alpha_k}\right)^{a_{k0}} , \\
(K_j)^{-1} &= \frac{1}{\prod\limits_{k=1}^{m} (\alpha_k)^{a_{kj}}} = \prod_{k=1}^{m} \left(\frac{1}{\alpha_k}\right)^{a_{kj}} .
\end{aligned}
\tag{8.7.3}
$$

Since $\alpha_k > 0, \ \forall \ k$ the formulas (8.7.3) define the inverse K-transformation to K.

(iii) The composition of two transformations (8.7.1) is again a K-transformation. This can be proved as follows: We use (8.7.1) and

$$
\begin{aligned}
&\bar{\bar{y}} = \bar{K}_0 \bar{y} , \quad \bar{\bar{x}}_j = \bar{K}_j \bar{x}_j , \\
&\bar{K}_0 = \prod_{k=1}^{m} \bar{\alpha}_k^{a_{k0}} , \quad \bar{K}_j = \prod_{k=1}^{m} \bar{\alpha}_k^{a_{kj}}
\end{aligned}
\tag{8.7.4}
$$

and then obtain

$$
\bar{\bar{y}} = \bar{K}_0 K_0 y , \quad \bar{\bar{x}}_j = \bar{K}_j K_j x_j , \tag{8.7.5}
$$

in which

$$
\begin{aligned}
\bar{K}_0 K_0 &= \left(\prod_{k=1}^{m} (\alpha_k)^{a_{k0}}\right) \left(\prod_{l=1}^{m} (\bar{\alpha}_l)^{a_{l0}}\right) = \prod_{k=1}^{m} (\alpha_k \bar{\alpha}_k)^{a_{k0}} , \\
\bar{K}_j K_j &= \left(\prod_{k=1}^{m} (\alpha_k)^{a_{kj}}\right) \left(\prod_{l=1}^{m} (\bar{\alpha}_l)^{a_{lj}}\right) = \prod_{k=1}^{m} (\alpha_k \bar{\alpha}_k)^{a_{kj}} .
\end{aligned}
\tag{8.7.6}
$$

These formulas have the same structure as (8.7.1) qed.

2. The two-dimensional case can be seen as the expansion of a cylindrical shock front due to a line source of explosives in a three-dimensional half space. The one-dimensional case corresponds to an explosion in a tunnel. If the dimension of the propagation space is left free (n = 3, 2, 1), then the dimensional matrix is given by

	t	$R_f^{(n)}$	ρ_0	E^n
M	0	0	1	1
L	0	1	-3	$n-1$
T	1	0	1	-2

It has rank 3; so the only dimensionless product is given by

$$
\Pi = \frac{R_f^{(n)}}{\left(\dfrac{E^{(n)}}{\rho_0}\right)^{1/(n+2)} t^{2/(n+2)}} = \text{Const} .
$$

Solving this for $R_f^{(n)}$ yields

$$R_f^{(n)}(t) = \text{Const} \times \left(\frac{E^{(n)}}{\rho_0}\right)^{\frac{1}{(n+2)}} t^{\frac{2}{(n+2)}} ,$$

which gives the propagation law for any space dimension.

3. The dimensional matrix

$$\begin{array}{c|ccccc} & \mu & \rho & \epsilon & \mathfrak{m} & \mathfrak{G} \\ \hline M & 1 & 1 & 0 & 1 & 1 \\ L & -1 & -3 & 2 & 0 & \mathfrak{k}+1 \\ T & -1 & 0 & -2 & 0 & -2 \end{array} \tag{8.7.7}$$

has rank 3, implying that two dimensionless products exist, which can easily be deduced to be

$$\frac{\mathfrak{m}\sqrt{\epsilon}}{\mu}\left(\frac{\mathfrak{m}\epsilon}{\mathfrak{G}}\right)^{2/(\mathfrak{k}-1)} , \quad \frac{\mathfrak{m}\epsilon}{\mathfrak{G}}\left(\frac{\rho}{\mathfrak{m}}\right)^{(1-\mathfrak{k})/3} .$$

Therefore,

$$\mu = \mathfrak{m}\sqrt{\epsilon}\left(\frac{\mathfrak{m}\epsilon}{\mathfrak{G}}\right)^{2/(\mathfrak{k}-1)} \hat{f}\left(\frac{\mathfrak{m}\epsilon}{\mathfrak{G}}\left(\frac{\rho}{\mathfrak{m}}\right)^{(1-\mathfrak{k})/3}\right) .$$

4. The physical quantities arising in the problem formulation can be arranged in the dimensional matrix

$$\begin{array}{c|ccccccc} & K & M & u_w & \eta & a & e & r \\ \hline M & 1 & 1 & 0 & 1 & 0 & 0 & 0 \\ L & 1 & 2 & 1 & -1 & 1 & 1 & 1 \\ T & -2 & -2 & -1 & -1 & 0 & 0 & 0 \end{array} .$$

It possesses rank $r = 3$, so four independent Π-products exist, namely e.g.

$$\begin{aligned} \mathbb{S}_0 &:= \frac{K}{\eta u_w}\frac{a^2}{r^2} , & \mathcal{M} &:= \frac{M}{Ka} , \\ \varepsilon &:= \frac{e}{a} , & d &:= \frac{a}{r} . \end{aligned} \tag{8.7.8}$$

$\mathbb{S}_0$ is called SOMMERFELD number.

Purely mechanical arguments and considerations of symmetry show that the force K is directed perpendicularly to the direction of excentricity; this follows from a force balance perpendicular to the axis of symmetry of the excentric arrangement of the shaft and bearing. This condition does not involve the moment M exerted by the shaft. The condition of moments relative to the axis of the shaft instead involves the force K. Therefore, there must exist two functions $f(\cdot)$ and $g(\cdot)$ such that

$$\mathcal{M} = f(\mathbb{S}_0,\, \varepsilon,\, d)\ ,$$
$$\mathbb{S}_0 = g(\varepsilon,\, d)\ , \quad \text{or} \quad \varepsilon = g^{-1}(\mathbb{S}_0,\, d) \tag{8.7.9}$$

holds. For a given load $\mathbb{S}_0$ and geometry d, and to every excentricity one may compute the moment $\mathcal{M}$.

The RAYLEIGH–SOMMERFELD theory of lubrication in cylindrical bearings shows that neither f nor g depend on d . In fact one can show that[20]

$$f(\varepsilon) = \frac{6\pi\varepsilon}{\sqrt{1-\varepsilon^2}\,(1+\varepsilon^2/2)}\ , \quad g(\varepsilon) = \frac{1}{3\varepsilon} + \frac{2}{3}\varepsilon \tag{8.7.10}$$

So, given $\mathbb{S}_0$ one first computes ε and with it the moment.

5. We perform the computation for an overfall weir of THOMPSON type. Variables which affect the discharge Q are the overfall height h, the density ρ, the Earth's acceleration g and the dynamic viscosity η. With these quantities the dimensional matrix

	Q	h	ρ	g	η
M	0	0	1	0	1
L	3	1	−3	1	−1
T	−1	0	0	−2	−1

(8.7.11)

can be formed which has rank 3. Thus $5-3=2$ independent dimensionless products exist such as e.g.

$$\Pi_1 = \frac{Q^2}{2gh^5}\ , \quad \Pi_2 = \frac{gh^3}{\nu^2}\ . \tag{8.7.12}$$

The steady discharge of a THOMPSON overfall weir is therefore given by

$$Q = \sqrt{2g}\, h^{5/2}\, f\left(\frac{gh^3}{\nu^2}\right)\ . \tag{8.7.13}$$

If one assumes that the viscosity of the fluid does not play a role in the determination of the discharge, one has

$$Q \sim \sqrt{2g}\, h^{5/2},\ (\text{THOMPSON})\ . \tag{8.7.14}$$

For an overfall weir by POINCELET a further variable is added, the weir width B. Thus, one obtains a dimensional matrix which is enlarged by one column in comparison to (8.7.11). The analysis of this matrix leads to the Π-products (8.7.12) plus $\Pi_3 = B/h$, so that on has instead of (8.7.13)

$$Q = \sqrt{2g}\, h^{5/2}\, f\left(\frac{gh^3}{\nu^2},\, \frac{B}{h}\right)\ . \tag{8.7.15}$$

[20] See e.g. K. HUTTER [109] pp. 166–172.

It is plausible to request a linear dependence of Q upon B. With this, instead of (8.7.14) one may deduce the formula

$$Q \sim \sqrt{2g}\, B\, h^{3/2} \quad (\text{Poincelet}). \tag{8.7.16}$$

Both formulas have been thoroughly tested, and overfall weirs are used in hydraulic laboratories to measure discharge.

9. Similitude and Model Experiments

9.1 Motivation

Somewhat unprecise and stretching the correct definition, a physical *model* is a mapping of Nature or at least of a subprocess that occurs in Nature or in our world of experience to smaller scale. "Projection" instead of "mapping" would be the better denotation, since some information is lost in the mapping. It is not necessary that the processes describe a physical incidence, however in this chapter we shall limit ourselves to physical systems. Naturally, every theory that describes processes of any kind, already represents a model; the theory would better be called a model, and it is imputed that the theory describes the processes in a certain sense and is able to also predict them adequately. The restriction that is imposed by such a theory is the fact, that it only adequately describes partial facts of the reality, those which are important, and that it ignores unimportant facets or even may describe these erroneously. In this sense, every theory or model is a projection of Nature, which enjoys only some *similarity* with the latter. Its intention is the determination and possibly the forecast of those physical quantities which are thought to be important as derivatives on the basis of the presentation of external sources and of initial conditions. The problems are generally formulated as initial boundary value problems.

Often the scientist or engineer presumes as if the model equations would describe Nature *per se*. This interpretation is for instance often given to the NAVIER–STOKES equations; in turbulence modeling it finds a very enthusiastic defender, specially in **D***irect* **N***umerical* **S***imulations*. Accordingly, turbulent modelers are of the opinion that every turbulent flow of a fluid can be modeled by the NAVIER–STOKES equations, if one simply succeeds in constructing solutions on all scales, even the smallest possible ones. Naturally, also the NAVIER–STOKES equations find their limitations as a model of Nature, namely, certainly at the lengthscales of the molecules themselves, when classical continuum physics looses its validity. The apprehension that a set of equations for the description of certain processes is correct, suggests, on the other hand, the following question: Is it possible to identify certain structures in the equations, which suggest certain invariance properties? For instance, is it possible by a scale analysis of the physical variables to non-dimensionalize the field equations and boundary conditions? If yes, dimensionless quantities will enter the field equations and boundary conditions as parameters; these quantities are nothing else than the Π-products that are realized by the model equations. For every set of values for these Π-products a *class of solutions* of the governing equations is defined. This suggests a *principle of similitude*: Physical processes which are characterized by different length–, time–, velocity–, acceleration scales, etc., for which, however, the Π-products underlying the governing equations, have the same values, are similar to each other, at least insofar as it concerns the processes that are described by the equations.

There is a further form of the notion of similitude, which is connected to the field equations and boundary conditions of a model. Often the physical processes are described by partial differential equations, and an initial boundary value problem describes the variables of the problem as functions of space and time, e.g., $y = f(x,t)$. For each fixed time τ_j the equation $y = f(x, \tau_j)$ describes a curve in the (x, y)-plane. Provided one succeeds by combination of all variables x, y, t to write the solution in the form $\Pi_y = \bar{f}(\Pi_{x,t})$, in which Π_y and $\Pi_{x,t}$ are variables which involve only y and x, t respectively, then all curves $y = f(x,t)$ in (x, y) space will be mapped into one single curve $\Pi_y = \bar{f}(\Pi_{x,t})$ in the space $(\Pi_y, \Pi_{x,t})$. If the solutions of initial value problems possess this property, they are called *self similar*. If the field equations of the physical problem are partial differential equations, the *self similar* solutions are often constructable, if one succeeds to transform the partial differential equations to ordinary differential equations. Such solutions are physically important, because they contain the minimum information that one must know, to gain insight into the phenomenon.

Whereas in the above the notion *model* was interpreted as a mapping of Nature to a scientific construct that obeys physical and mathematical laws, the every–day understanding and often also the understanding in the Engineering Sciences of the notion *model* is that of a *physical model*, i.e., the "repetition" of a process arising in Nature at reduced scale; this reduced scale process is, in most cases, performed as a laboratory experiment. In such attempts an important question that arises is, how the experimental arrangement must be chosen, if the experiment is performed in a scale different from 1:1. Such physical models are also mappings, of which the rules of down–scaling must be known, if one ought to be able to deduce inferences from the results obtained with the model to those of the prototype. Model experiments are performed in practically every branch of the Engineering Sciences, generally whenever it appears to be hopeless that, because of the complexity of the geometry and/or processes, one may gain sufficient insight into the behaviour of the system to be analyzed with numerical solutions of the governing equations. Often one also employs a combination of experiments and computations.

9.2 Theory of Physical Models

In this section we will be concerned with the rules of building physical models, the execution of experiments and the transposition of answers of models to the prototype. In model theory one differentiates between *geometrically similar* and *geometrically distorted models.* Commonly, models are constructed in geometrically similar reduction. Such models are formed if all three space directions are reduced by the same scale. Distorted models are obtained if this is not the case; for instance, all superelevated models are distorted. Such superelevations are necessary e.g. in hydraulics, if the water depths in the model become so small, that effects of surface tension can be recognized, whilst they are without any effect in the prototype.

Generally, there exists a point to point correlation between the *model* and the *prototype.* In geometrical language corresponding points between model and Nature (prototype) are called *homologous points.* Several homologous points form an agglomeration of homologous points, and a set of homologous points will eventually lead to homologous regions and domains. If time dependent processes are analyzed, one must introduce the notion of homologous times. To this end NEWTON second law is used; accordingly, differences of times are declared as homologous if a material point on homologous trajectories passes two homologous points. Analogously, one can speak of homologous distributions of mass, velocities, moments of inertia, etc. When constructing models it is, however, not necessary that all the homologies are preserved. One may restrict oneself to those, which have an influence on the physical quantity under study. This, for instance, is so for an airplane wing of which the drag force is sought; it is not needed for its determination to reproduce a homologous distribution of mass, however, this is necessary if one wishes to determine the yield stress of the wing or its behaviour in flow induced vibrations.

Before a problem is subjected to a model study, it is advantageous to first contemplate about which variables might have an influence upon the processes to be studied. Via a dimensional analysis one thus determines the number of dimensionless products, which correspond to the chosen variables. If Π denotes the dependent dimensionless variable and $\Pi_1, \ldots, \Pi_p$ the independent dimensionless products then

$$\Pi = f(\Pi_1, \ldots, \Pi_p) \tag{9.2.1}$$

gives the physical variable to be analyzed as a function of all dimensionless variables $\Pi_1, \ldots, \Pi_p$. Rules of selection of the suitable Π-products in a given problem are given in Sect. 8.3.3.

If a model has to reproduce the processes arising in the prototype correctly, then the values of $\Pi_1, \ldots, \Pi_p$ are not allowed to change when going from the prototype conditions to those of the model; for only then does the function $f(\Pi_1, \ldots, \Pi_p)$ for Π deliver the same result for the dependent variable. If such conditions are satisfied, the model is called *completely similar.* Therefore,

Theorem 9.1 *A model is capable to reproduce a process in Nature with complete similarity, if all dimensionless products which describe the process, have the same values in the model as well as in Nature.* ■

Even though this theorem embraces the entire content of *model similarity*, it contains a club–foot; it namely supposes, that the engineer or physicist recognises all variables which describe a physical process. This need not be the case, if the insight into the physical problem is incomplete. Conversely, it may occasionally not be possible to map a process in Nature completely similarly to the model, because one does not always succeed to reproduce a process in Nature at small scale and thereby preserve all values of the Π-products in the last theorem. With utmost rigour this is in fact never possible, as we will shortly see, unless, of course, the physical model is not down–scaled at all and the scale is 1:1; but this is rather senseless since it would mean performance of the process in Nature itself. Physical processes in Nature and in the model can completely similarly be mapped into one another at least mathematically if all Π-products that describe the processes remain invariant in the mapping from Nature to the model. This is successful only in a rear number of cases; one is regularly forced to hold only a reduced number of Π-products constant when performing the mapping from the prototype to the model and to let the remaining Π-products vary as dictated by the laws of the mapping. In these circumstances it is hoped, and often this can also be proven, that the Π-products which do not remain invariant in the mapping will not, or at least not much, influence the physical processes that are studied. If they nevertheless should do this, then one speaks in such cases of *scale effects.* If a process depends only on Π-products which all remain invariant in a model mapping, then this process is called *scale invariant.*

The drag force K of a ship that moves with constant velocity in quiescent water depends on ρ, η, g, L, V, i.e., on the density of water, its dynamic viscosity, the Earth's acceleration, a typical length of the cross section below the water line and the velocity. A possible dependence on the surface tension is ignored here. In Chap. 8, Example 8.2, it was show that the drag coefficient C_D of this resistive force is given by

$$C_D = C_D(\mathbb{F}, \mathbb{R}) \,, \tag{9.2.2}$$

where

$$\mathbb{F} := \frac{V^2}{gL} \,, \quad \mathbb{R} := \frac{VL}{\eta/\rho} \tag{9.2.3}$$

are the FROUDE– and REYNOLDS numbers.

Let overbarred quantities denote physical quantities[1] in the model, then invariance of the FROUDE number requires

$$\frac{V^2}{gL} = \frac{\overline{V}^2}{g\overline{L}} \quad \longrightarrow \quad \frac{\overline{V}}{V} = \sqrt{\frac{\overline{L}}{L}} =: \sqrt{\lambda}\,. \tag{9.2.4}$$

If one denotes by $\lambda = \overline{L}/L$ the scale of length of a geometrically similar model, then the scale of velocity is given by the square root of the scale of length. With V and L a typical time may be defined by $T = L/V$; its rule of scale reduction is obtained as follows:

$$\frac{\overline{T}}{T} = \frac{\overline{L}\,V}{\overline{V}\,L} = \lambda\frac{1}{\sqrt{\lambda}} = \sqrt{\lambda} \tag{9.2.5}$$

and shows that homologous times transform according to the square root of the scale of length. Accelerations can also be transformed in a FROUDE model; indeed, with $A = V/T$ a typical acceleration is described, so that

$$\frac{\bar{A}}{A} = \frac{\overline{V}\,T}{\overline{T}\,V} = \frac{\overline{V}\,T}{V\,\overline{T}} = \sqrt{\lambda}\frac{1}{\sqrt{\lambda}} = 1\,. \tag{9.2.6}$$

In other words the accelerations remain unchanged. In model experiments, in which the FROUDE number remains invariant, one speaks of FROUDE *similitude.*

For complete similarity of the law (9.2.2) the REYNOLDS number must also be kept invariant; this requirement leads to

$$\frac{\overline{V}\,\overline{L}}{\overline{\nu}} = \frac{VL}{\nu} \quad \longrightarrow \quad \frac{\overline{V}}{V} = \frac{L\,\overline{\nu}}{\overline{L}\,\nu} = \frac{1}{\lambda}K_\nu\,, \tag{9.2.7}$$

where K_ν is the scale of the kinematic viscosity. Similarly, one has for the transformations of the time and acceleration

$$\frac{\overline{T}}{T} = \frac{\overline{L}\,V}{L\,\overline{V}} = \frac{\lambda^2}{K_\nu}\,, \quad \frac{\bar{A}}{A} = \frac{K_\nu^2}{\lambda^3}\,. \tag{9.2.8}$$

Models, for which the REYNOLDS number remains invariant in a mapping from the prototype to the model, are called REYNOLDS *models*, and the associated rule of similitude is called REYNOLDS *similitude.*

For complete similarity of (9.2.2) FROUDE and REYNOLDS similitude must simultaneously hold. Because of (9.2.4) and (9.2.7) one therefore has

$$K_\nu = \lambda^{3/2}\,. \tag{9.2.9}$$

[1] Notice that we have set $\overline{g} = g$, this because it is assumed that the model experiment is also performed on Earth at appr. the same position and not in a satellite or on another planet.

Table 9.1. Rules of transformation for FROUDE and REYNOLDS similitude if the length scale is λ. It is assumed that in the prototype and the model the same fluid medium is used.

	FROUDE Similitude	REYNOLDS Similitude
Length	λ	λ
Velocity	$\sqrt{\lambda}$	λ^{-1}
Time	$\sqrt{\lambda}$	λ^2
Acceleration	1	λ^{-3}

The viscosity scale is therefore necessarily given by the length scale. Practically, this can never be achieved; so, FROUDE and REYNOLDS similitude cannot be reached together. Ordinarily, one uses the same fluid medium for the model experiments as for the prototype. For this case, the transformation rules are summarized in Table 9.1. This table indicates that for FROUDE similitude homologous velocities and homologous time differences are scaled by the square root of the length scale, whilst the accelerations remain the same. The latter is plausible, since the gravity forces dominate the FROUDE similitude and since these forces remain the same. By contrast, when REYNOLDS similitude is implemented homologous velocities in the model are larger than in the prototype by a factor that equals the inverse of the length scale, homologous time differences are shortened by the square of the length scale and accelerations are enlarged by the third power of the inverse length scale. In order to develop the general theory, let us embed Nature and model in EUCLIDian spaces with Cartesian Coordinates (x, y, z) and $(\overline{x}, \overline{y}, \overline{z})$, respectively. Homologous points and homologous times are then given by

$$\overline{x} = K_x x\,, \quad \overline{y} = K_y y\,, \quad \overline{z} = K_z z\,, \quad \overline{t} = K_t t\,. \tag{9.2.10}$$

K_x, K_y, K_z are the scale factors in the spatial directions x, y, z and K_t is that for time t. For $K_x = K_y = K_z$ the model is geometrically similar; otherwise the model is distorted. K_t can be chosen as the ratio of the times that elapse when a material point tracts the distance between two homologous points in the model and prototype, respectively. If $f(x, y, z, t)$ and $f(\overline{x}, \overline{y}, \overline{z}, \overline{t})$ describe a physical process in Nature and in the model, respectively, then the principal expression of similarity is:

Definition 9.1 *The function f is similar to the function $\overline{f}$, if the ratio $\overline{f}/f$ is constant, provided for the arguments (x, y, z, t) and $(\overline{x}, \overline{y}, \overline{z}, \overline{t})$ homologous points and times are chosen. The ratio $\overline{f}/f = K_f$ is called the scale of f.*

In the following we shall discuss the various rules of similitude. Important are the notions of *kinematic* and *dynamic* similitude.

Definition 9.2 *Two systems are called* **kinematically similar** *if their motions are similar, i.e., if homologous particles are to be found at homologous times in homologous points.*

If kinematic similitude prevails, then corresponding velocities and accelerations are similar. The scale factors are easily computable from

$$\overline{u} = \frac{d\overline{x}}{d\overline{t}}, \quad \overline{v} = \frac{d\overline{y}}{d\overline{t}}, \quad \overline{w} = \frac{d\overline{z}}{d\overline{t}}. \qquad (9.2.11)$$

Since $d\overline{x} = K_x dx, \ldots, d\overline{t} = K_t dt$ one obtains

$$\overline{u} = \frac{K_x}{K_t}\frac{dx}{dt} = \frac{K_x}{K_t}u, \quad \overline{v} = \frac{K_y}{K_t}v, \quad \overline{w} = \frac{K_z}{K_t}w. \qquad (9.2.12)$$

The scale factors for the velocity are therefore

$$K_u = \frac{K_x}{K_t}, \quad K_v = \frac{K_y}{K_t}, \quad K_w = \frac{K_z}{K_t}, \qquad (9.2.13)$$

and for the *accelerations* one obtains in an analogous manner

$$K_{a_x} = \frac{K_x}{K_t^2}, \quad K_{a_y} = \frac{K_y}{K_t^2}, \quad K_{a_z} = \frac{K_z}{K_t^2}. \qquad (9.2.14)$$

This should show how homologous points are computed.

Definition 9.3 *Two systems are called* **dynamically similar**, *if homologous parts of the system are subject to similar forces, i.e., if the force scale is invariant.*

With similar distribution of masses according to $\overline{m} = K_m m$ there follows from NEWTON's second law

$$\overline{F}_x = \overline{m}\overline{a}_x, \quad \overline{F}_y = \overline{m}\overline{a}_y, \quad \overline{F}_z = \overline{m}\overline{a}_z, \qquad (9.2.15)$$

or

$$\frac{\overline{F}_x}{F_x} = \frac{\overline{m}\overline{a}_x}{ma_x} = K_m\frac{K_x}{K_t^2}, \ldots\ldots. \qquad (9.2.16)$$

The scale factors for the forces are in NEWTONian mechanics given by

$$K_{F_x} = K_m\frac{K_x}{K_t^2}, \quad K_{F_y} = K_m\frac{K_y}{K_t^2}, \quad K_{F_z} = K_m\frac{K_z}{K_t^2}. \qquad (9.2.17)$$

It is evident from formulas (9.2.12) and (9.2.14) that the scales for the velocities and accelerations are not freely assignable, but must be computed from the scale factors of geometry and time. Analogously, for dynamic similitude the scale factor for the forces (9.2.17) is obtained automatically from the scale factors for length, time and mass. These dependences were already used when the FROUDE and REYNOLDS similarity rules were discussed.

It has already been said that in a model experiment of a fluid mechanical problem the REYNOLDS and the FROUDE number cannot both simultaneously be held invariant. We thus ask for a rule, which will allow us to select which of the two numbers should be kept invariant in a particular situation to reach at least approximate similarity. The decision is easy, if one asks the question whether the gravity force has a decisive influence on the flow. The acceleration of the Earth namely arises only in the definition of the FROUDE number. If in a hydrodynamic problem the bounding walls are rigid and prescribed, as is e.g. the case for pipe flow that is driven by pumps, then the piezometric pressure $P = p + \rho g z$ as a whole is the unknown quantity (and not p and $\rho g z$ individually). Gravity does not arise as an independent variable; it follows that pipe flow is governed by REYNOLDS similitude. If the fluid is bounded by a free surface, then the variable z in the piezometric pressure is an unknown and gravity will affect the flow field. If, in addition surface tension is active to the extent that it should not be ignored, then besides the FROUDE and REYNOLDS numbers also the WEBER number will affect the similarity. We thus have the following

Rule *For dynamical "similitude" of flows of density preserving fluids it is sufficient in a model reproduction that*

- *in regions with* **fixed boundaries** *and geometrically similar boundary values the* REYNOLDS *number is kept invariant whilst*
- *in regions with* **free boundaries** *and geometrically similar boundary values the* REYNOLDS, FROUDE *and* WEBER *numbers must be the same.* ■

Now, we have already seen, when discussing the drag force of a ship, that simultaneous satisfaction of the invariance of the FROUDE as well as the REYNOLDS number is not possible. If therefore one imputes FROUDE similitude, then the REYNOLDS number will assume a different value than in Nature. However, if the quantities to be measured should not depend upon the REYNOLDS number (or at least not in observable magnitude), then FROUDE similitude is applicable without special precaution. Else, scale effects will arise with which one may cope as follows: One must build at least two FROUDE models with differing scales; with these, identical experiments are performed each associated with its own REYNOLDS number. With linear interpolation one finds in this way a lineal functional dependence of any measured quantity upon the REYNOLDS number.

This principle, naturally, can also be applied to a case in which several Π-products arise, but is most likely no longer economically feasible. If, for instance, the drag force of a floating body depends also on the surface tension σ, then one has

$$C_D = C_D(\mathbb{F}, \mathbb{R}, \mathbb{W}) , \quad \mathbb{W} := \frac{\rho V^2 L}{\sigma} . \tag{9.2.18}$$

Usually, the WEBER number cannot be controlled in an experiment. In such cases it may be advantageous to seek also a solution with computational means.

9.3 Applications

Example 9.1 Explosion at a Point Source in a Fluid An explosion in an infinite compressible fluid generates a pressure wave; it is radially symmetric if the explosion is confined to a point. The pressure p, of this wave comprises a spherical shock which depends on the distance R of the front from the centre, the initial pressure in the gas p_0, the mass m of the explosive matter and the density ρ and compression modulus E of the fluid,

$$p = f(p_0, R, \rho, E, m) \tag{9.3.1}$$

and therefore via dimensional analysis ($r = 3$)

$$p = p_0 f\left(\frac{p_0}{E}, \frac{m}{\rho R^3}\right) . \tag{9.3.2}$$

This equation implies, if one requires invariance of the Π-products, that in a model the following scale rules must be observed:

$$K_{p_0} = K_p = K_E , \quad K_m = K_\rho K_R^3 . \tag{9.3.3}$$

If in the prototype and the model the explosion is performed in water, then $K_{p_0} = K_p = K_E = K_\rho = 1$, and one has

$$K_m = K_R^3 . \tag{9.3.4}$$

The mass of the explosive must be reduced by the third power of the geometric scale.

Example 9.2 Abrasion at the Glacier Bed It is known that on time scales of centuries and millenia mountain glaciers erode the bedrock; this is achieved by grinding rock powder via the sliding motion of the ice over the bedrock. This process of abrasion is responsible for the formation of the U-shaped mountain valleys which are today ice free but were covered with ice during the last ice age. Common models determining the ice flow in glaciers ignore the abrasion at the glacier bed. To estimate how abrasion could be accounted for, it is now asked how a functional relation between abrasion and input parameters (these are the variables that influence the abrasive process) might look like, and how laboratory experiments could shed further light on the problem.

In the laboratory experiment the situation in nature is put upside down. A piece of rock with plane but rough lower boundary is pulled forth and back over a smooth hard surface (see Fig. 9.1) which we imagine to be quiescent ice; however, its material is not so important, it could be replaced by a metal plate. Physical quantities, which may play a role in this abrasion problem are

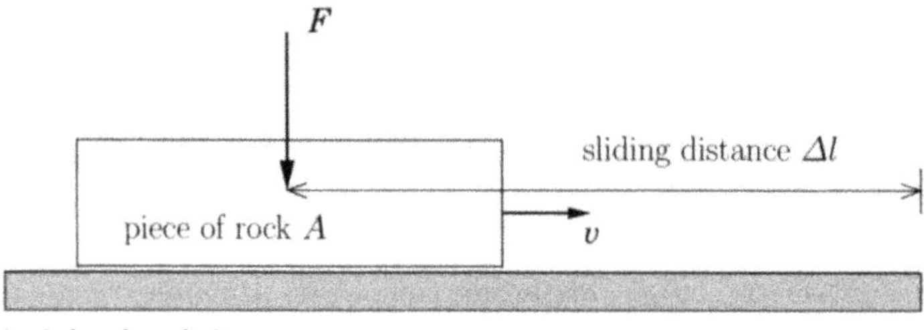

Fig. 9.1. Explaining the determination of the rate of abrasion of rock powder from the glacier bed by the sliding motion of the ice.

F	Weight + dead load, with which the rock is pressed against its support,
A	Contact area between rock and support,
σ_{DF}	Yield pressure (pressure strength) of the rock material,
Δl	Sliding distance,
ΔV	Volume of abrasion per sliding distance Δl.

The motion is assumed to be slow, so that thermal and acceleration effects of the process of abrasion can be ignored. With these quantities the dimensional matrix

	ΔV	Δl	F	A	σ_{DF}
M	0	0	1	0	1
L	3	1	1	2	−1
T	0	0	−2	0	−2

can be formed of which the rank is 2. There are therefore three independent Π-products, namely

$$\Pi_0 = \frac{\Delta V}{(\Delta l)^3} , \quad \Pi_1 = \frac{F/A}{\sigma_{DF}} , \quad \Pi_2 = \frac{F}{(\Delta l)^2 \sigma_{DF}} = \frac{A}{(\Delta l)^2} \frac{F/A}{\sigma_{DF}} , \tag{9.3.5}$$

so that

$$\Delta V = (\Delta l)^3 f(\Pi_1, \Pi_2) . \tag{9.3.6}$$

Further inferences cannot be deduced with methods of dimensional analysis alone. However, it is intuitively clear that $|F/A| \ll \sigma_{DF}$, so that $f(\Pi_1, \Pi_2)$ may be expanded in a TAYLOR series of Π_1 and Π_2 which can be terminated after the linear terms. This yields

$$\Delta V = (\Delta l)^3 \{f(0,0) + \alpha_1 \Pi_1 + \alpha_2 \Pi_2 + \ldots\} . \tag{9.3.7}$$

Alternatively, it is plausible to assume that ΔV grows linearly with Δl. If this is indeed so, which will be assumed, then $f(0,0) = 0$ and $\alpha_1 = 0$. Thus, there only remains a dependence on Π_2:

$$\Delta V \approx \alpha_2 \frac{F \Delta l}{\sigma_{DF}} = \alpha_2 A \left(\frac{F/A}{\sigma_{DF}} \right) \Delta l \, . \tag{9.3.8}$$

If both sides of this equation are divided by $A\Delta t$, and if one defines

$$a_{\mathrm{abr}} := \frac{\Delta V}{A \Delta t} \qquad \text{abrasion rate } [\mathrm{m\,s^{-1}}],$$
$$v := \frac{\Delta l}{\Delta t} \qquad \text{mean velocity,}$$

then (9.3.8) may also be written in the form

$$\frac{a_{\mathrm{abr}}}{v} = \alpha_2 \left(\frac{F/A}{\sigma_{DF}} \right) . \tag{9.3.9}$$

This result is extraordinary. It reduces the determination of the abrasion rate to the determination of a single coefficient α_2, whence to a single experiment. Equation (9.3.9) yields a linear connection between the dimensionless velocity a_{abr}/v and the dimensionless stress $(F/A)/\sigma_{DF}$. If one performs several experiments with several dead loads then one can test the correctness of the TAYLOR series expansion of f and its truncation after the linear terms. In a diagram with abscissa $(F/A)/\sigma_{DF}$ and ordinate a_{abr}/v all "measured" points must then be on a straight line with inclination angle α_2.

If α_2 is determined, then (9.3.9) provides the law of abrasion. F/A namely is the basal pressure of the ice and $p^{\perp}$ and v correspond to the sliding velocity $v_{\parallel}$, so that one obtains

$$a_{\mathrm{abr}} = \frac{\alpha_2}{\sigma_{DF}} p^{\perp} v_{\parallel} \, , \tag{9.3.10}$$

a surprisingly simple relationship! In an experiment, $p^{\perp}$ and $v_{\parallel}$ will be monitored and the obraded material, corresponding to a_{abr}, will be measured for a particular rock. This then determines the coefficient α_2/σ_{DF}.

Example 9.3 Heat Transfer by Forced Convection Consider a sphere with diameter D at rest immersed in a moving fluid. Its far field homogeneous velocity is v, its constant density ρ, the dynamic viscosity η, the specific heat c and heat conductivity κ, all assumed to be constant. The temperature of the sphere is larger than that of the fluid by ΔT degrees. We are interested in the heat transfer (its power N with dimension $[M\, L^2\, T^{-3}]$) from the sphere to the fluid. It is customary to write this transfer to be proportional to the surface of the sphere and proportional to the temperature difference ΔT; thus,

$$N = \alpha \pi D^2 \Delta T \, , \tag{9.3.11}$$

in which the heat transfer coefficient α has dimension $[M\, T^{-3}\, \Theta^{-1}]$. Instead of seeking a relationship for N, one may now determine such a relation for α. The dimensional matrix

$$\begin{array}{c|ccc:cccc:c} & \alpha & v & \Delta T & D & \rho & \eta & c & \kappa \\ \hline M & 1 & 0 & 0 & 1 & 1 & 1 & 0 & 1 \\ L & 0 & 1 & 0 & 0 & -3 & -1 & 2 & 1 \\ T & -3 & -1 & 0 & 0 & 0 & -1 & -2 & -3 \\ \Theta & -1 & 0 & 1 & 0 & 0 & 0 & -1 & -1 \end{array} \tag{9.3.12}$$

has rank 4, since the determinant of the indicated submatrix does not vanish; therefore there are four independent Π-products. Such products are for instance

$$\begin{aligned} \mathbb{N} &:= \frac{\alpha D}{\kappa} && \text{NUSSELT number,} \\ \mathbb{R} &:= \frac{vD}{\eta/\rho} && \text{REYNOLDS number,} \\ \mathbb{E}_c &:= \frac{\Delta T c}{v^2} && \text{ECKERT number,} \\ \mathbb{P}_r &:= \frac{c\eta}{\kappa} && \text{PRANDTL number.} \end{aligned} \tag{9.3.13}$$

The heat transfer in forced convection is therefore characterised by an equation of the form $f(\mathbb{N}, \mathbb{R}, \mathbb{E}_c, \mathbb{P}_r) = 0$. Since the ECKERT and PRANDTL numbers can be combined to form the BRINKMANN number

$$\mathbb{B}_r := \frac{\mathbb{P}_r}{\mathbb{E}_c} = \frac{\eta v^2}{\kappa \Delta T} \qquad \text{BRINKMANN-number,} \tag{9.3.14}$$

the heat transfer can also be characterised by the relation

$$\mathbb{N} = f(\mathbb{R}, \mathbb{B}_r, \mathbb{P}_r) \,. \tag{9.3.15}$$

In a model experiment one must necessarily keep the REYNOLDS number invariant, if one attempts to achieve a similar mapping of the flow in the model to that of the prototype. If in the model the same material is used as in the prototype the invariance requirement of the PRANDTL number is automatically fulfilled. Invariance of the BRINKMANN number, finally, requires

$$\frac{\overline{\Delta T}}{\Delta T} = \frac{\overline{v}^2}{v^2} = \lambda^{-2} \,, \tag{9.3.16}$$

in which λ denotes the length scale. Temperature differences $\overline{\Delta T}$ in the model must therefore be enlarged relative to the prototype by the square of the length scale.

9.4 Model Theory and Differential Equations

In this section the position is taken that for a physical problem the governing equations – e.g. as differential equations and boundary conditions –

are known and that their appropriateness in describing the physical process under consideration is without any doubt. Under such prerequisites the equations must first be dimensionally homogeneous, and second, it must be possible to isolate the dimensionless Π-products, which describe the physical context of the equations. If so, the scale effects contained in the equations must equally be identifiable in the equations.

a) NAVIER–STOKES–FOURIER **Equations**

The best known example, important in applications of technology as well as the natural sciences, are the NAVIER–STOKES–FOURIER (NSF) equations. The balance laws of mass, momentum and energy in this case have the following form:

$$\begin{aligned} &\frac{\partial \rho}{\partial t} + \operatorname{div} \rho \boldsymbol{v} = 0 \,, \\ &\rho \left\{ \frac{\partial \boldsymbol{v}}{\partial t} + (\operatorname{grad} \boldsymbol{v}) \boldsymbol{v} \right\} = -\operatorname{grad} p + \rho \boldsymbol{g} + \operatorname{grad}(\zeta \operatorname{div} \boldsymbol{v}) + 2 \operatorname{div}(\eta \boldsymbol{E}) \,, \\ &\rho T \frac{\mathrm{d}s}{\mathrm{d}t} = \operatorname{div}(\kappa \operatorname{grad} T) + \Phi + \rho \mathfrak{r} \,, \end{aligned} \qquad (9.4.1)$$

in which

$$\boldsymbol{E} = \boldsymbol{D} - \tfrac{1}{3}(\operatorname{div} \boldsymbol{v})\boldsymbol{I} \,, \quad \boldsymbol{D} := \operatorname{sym} \operatorname{grad} \boldsymbol{v} \,, \qquad (9.4.2)$$

$$s = -\frac{\partial \Psi}{\partial T} \,, \quad p = \rho^2 \frac{\partial \Psi}{\partial \rho}$$

and

$$\Phi = \zeta (\operatorname{div} \boldsymbol{v})^2 + 2\eta \operatorname{tr} \boldsymbol{E}^2 \,.$$

Here, $\rho, \boldsymbol{v}, p, \boldsymbol{g}, T, \mathfrak{r}$ are, in turn, the density, velocity, pressure, the Earth's acceleration, the temperature and the specific energy supply rate (radiation). $\boldsymbol{D}$ is the stretching (strain rate) tensor, $\boldsymbol{E}$ its deviator, s the entropy and Ψ the HELMHOLTZ free energy; ζ, η and κ are, finally, the volume and shear viscosities and the heat conductivity, and it is assumed that

$$(\Psi, \zeta, \eta, \kappa) = fcts(T, \rho) \,. \qquad (9.4.3)$$

Equations (9.4.1) are the field equations of a linearly viscous heat conducting fluid with viscous stress tensor

$$\boldsymbol{t}^R = \zeta (\operatorname{div} \boldsymbol{v})\boldsymbol{I} + 2\eta \boldsymbol{E} \,, \qquad (9.4.4)$$

FOURIER law of heat conduction

$$\boldsymbol{q} = -\kappa \operatorname{grad} T \qquad (9.4.5)$$

and caloric equation of state

$$\Psi = \hat{\Psi}(T, \rho)$$

for the HELMHOLTZ free energy. We shall perform the ensuing analysis under the simplifying assumption that $T\mathrm{d}s/\mathrm{d}t$ is given by $c_p(T)\dot{T}$

$$T\frac{\mathrm{d}s}{\mathrm{d}t} \simeq c_p(T)\frac{\mathrm{d}T}{\mathrm{d}t} \,. \tag{9.4.6}$$

Equations (9.4.1)–(9.4.6) describe the thermodynamic behaviour of many fluids and gases, so that scientists and engineers often assign to their validity the absolute truth.

By appropriate non-dimensionalisation, i.e., by scaling the variables, all those dimensionless quantities can be isolated, which are accounted for in the NSF equations. The procedure of the non-dimensionalisation is not unique, first because variables are differently non-dimensionalised for different processes, but, second, also since every scientist introduces some subjective elements which may influence the procedure. In principle every variable f is split into a product of quantities,

$$f = [f]\tilde{f} \,, \tag{9.4.7}$$

in which $[f]$ possesses the same dimension as f, is constant and should have a numerical value so large that the dimensionless variable $\tilde{f}$ assumes values which are of order unity. It is exactly this requirement, which reflects a considerable degree of individuality of the non-dimensionalisation of equations, because a priori knowledge of possible solutions of initial boundary value problems or physical insight enters estimates such as (9.4.7). To let the reader see this a bit more clearly, we presently only look at the mass balance equation $(9.4.1)_1$, and introduce the following two different scalings for ρ, but only one scaling for the velocity and the time, respectively,

$$\rho = [\rho]\tilde{\rho} \quad \text{rsp.} \quad \rho = [\rho](1 + [\sigma]\tilde{\sigma}) \,, \quad \boldsymbol{v} = [v]\tilde{\boldsymbol{v}} \,, \quad t = [\tau]\tilde{t} \,. \tag{9.4.8}$$

In the first scaling, for which $\rho = [\rho]\tilde{\rho}$, one indirectly supposes that $\tilde{\rho}$ may assume all values in the interval $[0, a)$, where $a \sim 1$, in the second case the density apparently deviates form its typical value $[\rho]$ only by a small amount, which is expressed by the dimensionless scale $[\sigma]$. If the representations (9.4.8) are substituted in $(9.4.1)_1$, then one obtains in the first case

$$\mathbb{S}_t\frac{\partial\tilde{\rho}}{\partial\tilde{t}} + \operatorname{div}(\tilde{\rho}\tilde{\boldsymbol{v}}) = 0 \tag{9.4.9}$$

and in the second case

$$[\sigma]\left\{\mathbb{S}_t\frac{\partial\tilde{\sigma}}{\partial\tilde{t}} + \operatorname{div}(\tilde{\sigma}\tilde{\boldsymbol{v}})\right\} + \operatorname{div}\tilde{\boldsymbol{v}} = 0 \,, \tag{9.4.10}$$

with the STROUHAL number

$$\mathbb{S}_t := \frac{[L]}{[v][\tau]} \,. \tag{9.4.11}$$

The above are two completely different representations of the non-dimensionalised mass balance equation. In the first representation only one Π-product, the STROUHAL number arises; its value indicates whether the temporal variation of the density function is significant. In the second case two dimensionless Π-products arise, $\mathbb{S}_t$ und $[\sigma]$. In this form of scaling the mass balance is significant if the density varies around a mean value, but deviations from this mean value are small. Indeed, if the asymptotic limits $[\sigma] \to 0$, $[\sigma]\mathbb{S}_t \to 0$ are justified, (9.4.10) reduces to $\operatorname{div} \tilde{\boldsymbol{v}} = 0$, which is the same equation as for a density preserving fluid, even though here the variation of the density is accounted for. This distinguished limit is called the BOUSSINESQ *approximation.* Of course, one may also consider the limit $[\sigma] \to 0$, $[\sigma]\mathbb{S}_t \neq 0$, this would lead to the approximate equation

$$[\sigma]\mathbb{S}_t \frac{\partial \tilde{\sigma}}{\partial \tilde{t}} + \operatorname{div} \tilde{\boldsymbol{v}} = 0 \tag{9.4.12}$$

which is hardly different from the complete mass balance equation. In cases for which the product $[\sigma]\mathbb{S}_t$ remains finite, it is thus recommended to non-dimensionalise according to (9.4.9). On the other hand, it is plausible that by eliminating certain terms in the governing equations, certain classes of processes may be eliminated. We leave it as an exercise to the reader to convince himself that the BOUSSINESQ assumption eliminates all acoustic phenomena from the NSF equations (see Exercise 1).

If one chooses the scales

$$\begin{array}{lll}
\boldsymbol{x} = [L]\tilde{\boldsymbol{x}} & t = [\tau]\tilde{t} \,, & \rho = [\rho]\tilde{\rho} \,, \\
\zeta = [\zeta]\tilde{\zeta} \,, & c_p = [c_p]\tilde{c}_p \,, & \boldsymbol{v} = [v]\tilde{\boldsymbol{v}} \,, \\
\eta = [\eta]\tilde{\eta} \,, & \mathfrak{r} = [\mathfrak{r}]\tilde{\mathfrak{r}} \,, & p = [p]\tilde{p} \,, \\
\kappa = [\kappa]\tilde{\kappa} \,, & \boldsymbol{g} = [g]\tilde{\boldsymbol{g}} \,, & T = T_0 + [\Delta T]\theta \,, \\
\boldsymbol{D} = [\mathrm{D}]\tilde{\boldsymbol{D}} \,, & \boldsymbol{E} = [\mathrm{E}]\tilde{\boldsymbol{E}} \,, & \omega = [\omega]\tilde{\omega} \,,
\end{array} \tag{9.4.13}$$

then the NAVIER–STOKES–FOURIER equations may be written in the following form

$$
\begin{aligned}
&\left\{\mathbb{S}_t \frac{\partial \tilde{\rho}}{\partial \tilde{t}} + \operatorname{div}(\tilde{\rho}\tilde{\boldsymbol{v}})\right\} = 0 \,, \\
&\tilde{\rho}\left\{\mathbb{S}_t \frac{\partial \tilde{\boldsymbol{v}}}{\partial \tilde{t}} + (\operatorname{grad} \tilde{\boldsymbol{v}})\tilde{\boldsymbol{v}}\right\} = -\mathbb{E}_u \operatorname{grad} \tilde{p} \\
&\qquad\qquad + \frac{1}{\mathbb{R}}\left\{\frac{[\zeta]}{[\eta]} \operatorname{grad}(\tilde{\zeta} \operatorname{div} \boldsymbol{v}) + 2 \operatorname{div}(\tilde{\eta}\tilde{\boldsymbol{E}})\right\} + \frac{1}{\mathbb{F}_r}\tilde{\rho}\tilde{\boldsymbol{g}} \,, \qquad (9.4.14) \\
&\tilde{\rho}\,\tilde{c}_p\left\{\mathbb{S}_t \frac{\partial \theta}{\partial \tilde{t}} + (\operatorname{grad}\theta)\cdot\tilde{\boldsymbol{v}}\right\} = \frac{1}{\mathbb{P}_e} \operatorname{div}(\tilde{\kappa} \operatorname{grad}\theta) + \\
&\qquad\qquad + \frac{1}{\mathbb{E}_d}\left\{\frac{[\zeta]}{[\eta]}\tilde{\zeta}(\operatorname{div}\tilde{\boldsymbol{v}})^2 + 2\tilde{\eta} \operatorname{tr} \tilde{\boldsymbol{E}}^2\right\} + \frac{1}{\mathbb{R}_a}\tilde{\rho}\tilde{\mathfrak{r}} \,,
\end{aligned}
$$

in which the hallow quantities represent the characteristic dimensionless Π-products arising in this non-dimensionalisation

$$
\begin{aligned}
\mathbb{S}_t &:= \frac{[L]}{[v][\tau]} && \text{STROUHAL number ,} \\
\mathbb{E}_u &:= \frac{[p]}{[\rho][v]^2} && \text{EULER number, pressure coefficient,} \\
\mathbb{F}_r &:= \frac{[v]^2}{[g][L]} && \text{FROUDE number (inverse RICHARDSON number),} \\
\mathbb{R} &:= \frac{[v][L]}{[\eta]/[\rho]} && \text{REYNOLDS number,} \\
\mathbb{P}_e &:= \frac{[\rho][c_p][v][L]}{[\kappa]} && \text{PECLET number,} \\
\mathbb{E}_d &:= \frac{[c_p][\Delta T]}{[v]^2}\frac{[v][L]}{[\eta]/[\rho]} && \text{Dissipation number,} \\
\mathbb{R}_a &:= \frac{[c_p][\Delta T][v]}{[L][\mathfrak{r}]} && \text{Radiation number.}
\end{aligned}
\qquad (9.4.15)
$$

To all of these, depending on their position in (9.4.14), a physical meaning can be assigned. The STROUHAL number measures the non-steadiness of the processes, the EULER number those of the pressure gradient, the REYNOLDS number stands for the significance of the internal friction and the FROUDE number for that of the body forces. In the energy equation the PECLET number describes the significance of the heat conduction or heat diffusion, the dissipation number stands for the importance of the heat production due to internal friction and the radiation number measures the significance of the external radiation.

The dimensionless products (9.4.15) are the Π-products usually employed in hydrodynamics; often, however, the following alternatives are also introduced:

$$\mathbb{P}_e = \mathbb{R}\,\mathbb{P}_r\,, \quad \mathbb{E}_d = 2\,\mathbb{R}\,\mathbb{T}_h\,, \tag{9.4.16}$$

in which

$$\begin{aligned} \mathbb{P}_r &:= \frac{[\eta]/[\rho]}{[\kappa]/([\rho][c_p])} = \frac{[\nu]}{[d_{th}]} \qquad \text{PRANDTL number,} \\ \mathbb{T}_h &:= \frac{[c_p][\Delta T]}{[v]^2} \qquad \text{Temperature number,} \end{aligned} \tag{9.4.17}$$

and where d_{th} is the thermal diffusivity. The PRANDTL number is the ratio of the diffusion of momentum (kinematic viscosity) to that of heat, the temperature number expresses the ratio of the thermally stored energy to the kinetic energy.

In the scalings (9.4.13) the typical time $[\tau]$ was chosen independently of the characteristic length $[L]$ and characteristic velocity $[v]$. Analogously, also for the pressure an independent scale $[p]$ was chosen. If considerations of the physical problem formulation should suggest that $[\tau]$ may be interpretable as a time which a material particle needs to propagate a distance $[L]$ with velocity $[v]$, then one may choose in this case $[\tau] = [L]/[v]$ i.e., $\mathbb{S}_t = 1$. Such a scaling is appropriate for all those processes for which instationary and convective features are of similar significance. If, for instance, in a flow around a wing the stagnation pressure, $\frac{1}{2}\rho v^2$ is important, then one may set $[p] = \rho[v]^2$ which corresponds to the choice $\mathbb{E}_u = 1$. For the equations that are non-dimensionalised in this way, only the remaining five Π-products are significant.

If a physical process may be described by the NSF equations, then in a mapping to smaller scale from the prototype to the model all Π-products (9.4.15) must remain invariant, if one wishes to reach complete similarity. This is, as we already know not possible, since in dynamical similitude one cannot simultaneously keep the REYNOLDS as well as the FROUDE number invariant. FROUDE and REYNOLDS similitude, respectively, prevail if

1) $\mathbb{R} \longrightarrow \infty\,,\ \mathbb{F}_r$ finite (FROUDE) ,
2) $\mathbb{R}$ finite, $\mathbb{F}_r \longrightarrow \infty$ (REYNOLDS) .

In the first case the momentum equations reduce to the EULER equations, in the second the body forces play no role. As one can see from the energy equations $(9.4.13)_3$ heat condition and dissipation can only be kept in conformity with the similarity requirement for REYNOLDS models; indeed, PECLET and dissipation numbers are related to the REYNOLDS number via the PRANDTL and the temperature number. The latter are materially dependent and only freely assignable in a restricted sense. We thus conclude:

Theorem 9.2 *As opposed to* REYNOLDS *models, measuring temperatures in* FROUDE *models gives no guarantee of appropriate transfer to corresponding quantities in Nature, or: in* FROUDE *models no homologous temperature field is generated.*

If the body in which the fluid is kept and of which the motion is to be studied, is rotating, a new characteristic time $[\tau]$ is introduced that is given by the time of revolution of the non-inertial frame. Since in such a case the equations are referred to a non-inertial frame, the momentum equation (of the NSF equations) must be complemented by the acceleration terms due to the relative motion. The absolute acceleration may then be written as

$$\left(\frac{\mathrm{d}\boldsymbol{v}}{\mathrm{d}t}\right)_{\text{abs}} = \left(\frac{\mathrm{d}\boldsymbol{v}}{\mathrm{d}t}\right)_{\text{rel}} + 2\boldsymbol{\omega}\times\boldsymbol{v} + \boldsymbol{\omega}\times(\boldsymbol{\omega}\times\boldsymbol{x}) + \dot{\boldsymbol{\omega}}\times\boldsymbol{x}\,, \tag{9.4.18}$$

in which $\boldsymbol{\omega}$ is the angular velocity of the non-inertial frame and a translating acceleration of this frame is thought to be absent. With the non-dimensionalisation (9.4.13) one now obtains for the curly bracket of the left-hand side of $(9.4.14)_2$

$$\mathbb{S}_t\frac{\partial\tilde{\boldsymbol{v}}}{\partial t} + (\operatorname{grad}\tilde{\boldsymbol{v}})\tilde{\boldsymbol{v}} + \mathbb{S}_t(2\tilde{\boldsymbol{\omega}}\times\tilde{\boldsymbol{v}}) + (\mathbb{S}_t)^2\left\{\tilde{\boldsymbol{\omega}}\times(\tilde{\boldsymbol{\omega}}\times\tilde{\boldsymbol{x}}) + \frac{[\dot{\omega}]}{[\omega]^2}\dot{\tilde{\boldsymbol{\omega}}}\times\tilde{\boldsymbol{x}}\right\}, \tag{9.4.19}$$

where we have set $[\omega] = 1/[\tau]$ and $\boldsymbol{\omega} = [\omega]\tilde{\boldsymbol{\omega}}$. The inverse of the STROUHAL number that is formed with $[\omega]$ instead of $[\tau]$ is called the ROSSBY number

$$\mathbb{R}_o := \left.\frac{1}{\mathbb{S}_t}\right|_{[\tau]=[\omega]^{-1}} = \frac{[v]}{[\omega][L]} \qquad \text{(ROSSBY number)}. \tag{9.4.20}$$

Thus, if (9.4.14) is divided by $\mathbb{S}_t$ and if (9.4.19) is used, one obtains

$$\begin{aligned}
&\frac{\partial\tilde{\rho}}{\partial\tilde{t}} + \mathbb{R}_o\operatorname{div}(\tilde{\rho}\tilde{\boldsymbol{v}}) = 0\,,\\
&\tilde{\rho}\left\{\frac{\partial\tilde{\boldsymbol{v}}}{\partial\tilde{t}} + \mathbb{R}_o(\operatorname{grad}\tilde{\boldsymbol{v}})\tilde{\boldsymbol{v}} + 2\tilde{\boldsymbol{\omega}}\times\tilde{\boldsymbol{v}} + \frac{1}{\mathbb{R}_o}\left\{\tilde{\boldsymbol{\omega}}\times(\tilde{\boldsymbol{\omega}}\times\tilde{\boldsymbol{x}}) + \frac{[\dot{\omega}]}{[\omega]^2}\dot{\tilde{\boldsymbol{\omega}}}\times\tilde{\boldsymbol{x}}\right\}\right\}\\
&\qquad = -\mathbb{P}_{[\omega]}\operatorname{grad}\tilde{p} + \mathbb{E}_k\left\{\frac{[\zeta]}{[\eta]}\operatorname{grad}(\tilde{\zeta}\operatorname{div}\tilde{\boldsymbol{v}}) + 2\operatorname{div}(\tilde{\eta}\tilde{\boldsymbol{E}})\right\} + \frac{1}{\mathbb{F}_{r[\omega]}}\tilde{\rho}\tilde{\boldsymbol{g}}\,,\\
&\tilde{\rho}\,\tilde{c}_p\left[\frac{\partial\theta}{\partial\tilde{t}} + \mathbb{R}_o(\operatorname{grad}\theta)\cdot\tilde{\boldsymbol{v}}\right] = \frac{\mathbb{E}_k}{\mathbb{P}_r}\operatorname{div}(\tilde{\kappa}\operatorname{grad}\theta)\\
&\qquad + \frac{\mathbb{E}_k}{\mathbb{T}_h}\left\{\frac{[\zeta]}{[\eta]}\tilde{\zeta}(\operatorname{div}\tilde{\boldsymbol{v}})^2 + 2\tilde{\eta}\operatorname{tr}(\tilde{\boldsymbol{E}}^2)\right\} + \frac{1}{\mathbb{R}_{a[\omega]}}\tilde{\rho}\tilde{\mathfrak{r}}\,,
\end{aligned} \tag{9.4.21}$$

in which

$$
\begin{aligned}
\mathbb{P}_{[\omega]} &:= \frac{[p]}{[\rho][L][\omega][v]} && \text{pressure coefficient,} \\
\mathbb{E}_k &:= \frac{[\eta]/[\rho]}{[\omega][L]^2} && \text{EKMAN number,} \\
\mathbb{F}_{\mathrm{r}\,[\omega]} &:= \frac{[g]}{[\omega][v]} && \text{FROUDE number,} \\
\mathbb{R}_{a\,[\omega]} &:= \frac{[c_p][\Delta T][\omega]}{[\mathfrak{r}]} && \text{radiation number.}
\end{aligned}
\tag{9.4.22}
$$

Usually, the rotation of the frame is steady, $[\dot{\omega}] \equiv 0$; likewise the bulk viscosity ζ is ignored. In (9.4.21) the ROSSBY number measures the significance of the convective terms in the balance laws of mass, momentum and energy. The importance of the effects of internal friction is now governed by the magnitude of the EKMAN (and PRANDTL) numbers. Otherwise (9.4.21) hold for a gas or a fluid.

If the density variation is small, then $\tilde{\rho}$ is replaced in (9.4.21) by $\tilde{\rho} = 1 + [\sigma]\tilde{\sigma}$. If, at the same time the pressure is additively decomposed in a static and a dynamic contribution,

$$
\tilde{p} = \tilde{p}_{\mathrm{stat}} + \tilde{p}_{\mathrm{dyn}} \ , \tag{9.4.23}
$$

in which

$$
\tilde{p}_{\mathrm{stat}} = \frac{\mathbb{P}_{[\omega]}}{\mathbb{F}_{\mathrm{r}\,[\omega]}} \tilde{\boldsymbol{g}} \cdot \tilde{\boldsymbol{x}} = \frac{[p]}{[\rho][g][L]} \tilde{\boldsymbol{g}} \cdot \tilde{\boldsymbol{x}} \ , \tag{9.4.24}
$$

then (9.4.21) take the forms

$$
\begin{aligned}
&[\sigma] \left\{ \frac{\partial \tilde{\rho}}{\partial \tilde{t}} + \mathbb{R}_o \operatorname{div}(\tilde{\rho} \tilde{\boldsymbol{v}}) \right\} = 0 \ , \\
&(1 + [\sigma]\tilde{\sigma}) \left\{ \frac{\partial \tilde{\boldsymbol{v}}}{\partial \tilde{t}} + \mathbb{R}_o (\operatorname{grad} \tilde{\boldsymbol{v}}) \tilde{\boldsymbol{v}} + 2 \tilde{\boldsymbol{\omega}} \times \tilde{\boldsymbol{v}} \right. \\
&\qquad \left. + \frac{1}{\mathbb{R}_o} \left(\tilde{\boldsymbol{\omega}} \times (\tilde{\boldsymbol{\omega}} \times \tilde{\boldsymbol{x}}) + \frac{[\dot{\omega}]}{[\omega]^2} \dot{\tilde{\boldsymbol{\omega}}} \times \tilde{\boldsymbol{x}} \right) \right\} \\
&\quad = -\mathbb{P}_{[\omega]} \operatorname{grad} \tilde{p}_{\mathrm{dyn}} + \mathbb{E}_k \left\{ \frac{[\zeta]}{[\eta]} \operatorname{grad}(\tilde{\zeta} \operatorname{div} \tilde{\boldsymbol{v}}) + 2 \operatorname{div}(\tilde{\eta} \tilde{\boldsymbol{E}}) \right\} \\
&\quad + \frac{[\sigma]}{\mathbb{F}_{\mathrm{r}\,[\omega]}} \tilde{\sigma} \tilde{\boldsymbol{g}} \ , \\
&(1 + [\sigma]\tilde{\sigma})\, \tilde{c}_p \left\{ \frac{\partial \theta}{\partial \tilde{t}} + \mathbb{R}_o (\operatorname{grad} \theta) \cdot \tilde{\boldsymbol{v}} \right\} = \frac{\mathbb{E}_k}{\mathbb{P}_r} \operatorname{div}(\tilde{\kappa} \operatorname{grad} \theta) \\
&\quad + \frac{\mathbb{E}_k}{\mathbb{T}_h} \left\{ \frac{[\zeta]}{[\eta]} \tilde{\zeta} (\operatorname{div} \tilde{\boldsymbol{v}})^2 + 2 \tilde{\eta} \operatorname{tr}(\tilde{\boldsymbol{E}}^2) \right\} + \frac{1}{\mathbb{R}_{a\,[\omega]}} \tilde{\rho} \tilde{\mathfrak{r}} \ .
\end{aligned}
\tag{9.4.25}
$$

From these equations one may deduce the NSF equations of a BOUSSINESQ fluid, if one considers the limit $[\sigma] \to 0$, but keeps all other Π-products, including the BOUSSINESQ number

$$\mathbb{B}_{[\omega]} = \frac{[\sigma][\omega][v]}{[g]} \qquad \text{(BOUSSINESQ number)} \tag{9.4.26}$$

finite. It is then seen from (9.4.25) that the variation of the density is ignored in all terms involving time derivatives, but is kept in the expression for the body force. Moreover, the balance law of mass is reduced to that of density preserving materials, i.e., the statement $\operatorname{div} \boldsymbol{v} = 0$. If, furthermore $[\dot{\omega}] = 0$ and $[\zeta] = 0$ the radiation vanishes and the centrifugal accelerations are incorporated in the gravity term, the BOUSSINESQ approximated dynamical equations take the form

$$\begin{aligned}
&\operatorname{div} \tilde{\boldsymbol{v}} = 0\,, \\
&\frac{\partial \tilde{\boldsymbol{v}}}{\partial \tilde{t}} + \mathbb{R}_o(\operatorname{grad} \tilde{\boldsymbol{v}})\tilde{\boldsymbol{v}} + 2\tilde{\boldsymbol{\omega}} \times \tilde{\boldsymbol{v}} \\
&\qquad = -\mathbb{P}_{[\omega]} \operatorname{grad} \tilde{p}_{\mathrm{dyn}} + 2\mathbb{E}_k \operatorname{div}(\tilde{\eta}\tilde{\boldsymbol{D}}) + \mathbb{B}_{[\omega]}\tilde{\sigma}\tilde{\boldsymbol{g}}\,, \\
&\tilde{c}_p \left(\frac{\partial \theta}{\partial \tilde{t}} + \mathbb{R}_o(\operatorname{grad} \theta)\cdot\tilde{\boldsymbol{v}} \right) = \frac{\mathbb{E}_k}{\mathbb{P}_r} \operatorname{div}(\tilde{\kappa} \operatorname{grad} \theta) + \frac{\mathbb{E}_k}{\mathbb{T}_h} 2\tilde{\eta} \operatorname{tr}(\tilde{\boldsymbol{D}}^2)\,,
\end{aligned} \tag{9.4.27}$$

in which $\tilde{\boldsymbol{E}}$ has been replaced by $\tilde{\boldsymbol{D}}$, since $\operatorname{div} \tilde{\boldsymbol{v}} = 0$. These are probably the most important equations in geophysical fluid mechanics.

b) Diffusion Processes

Models of a mixture of a fluid with a number of tracers do often have merely diffusive structure, i.e., they are described by the balance laws for the mixture as a whole and the tracer masses that arise in dilute form, plus the momentum and energy equations for the mixture as a whole. Under such circumstances one may use the NSF equations for the mixture as a whole; ρ is now the mixture density and $\boldsymbol{v}$ the barycentric velocity. These equations must be complemented by the mass balances for the tracers. If $c^\alpha := \rho^\alpha/\rho$ denotes the concentration of the component α (the ratio of the density of component α to that of the mixture), $\boldsymbol{j}^\alpha$ the flux of mass of constituent α and π^α its specific production, then the mass balance for the tracer α takes the form

$$\rho \frac{\mathrm{d}c^\alpha}{\mathrm{d}t} = -\operatorname{div} \boldsymbol{j}^\alpha + \rho\pi^\alpha\,, \qquad (\alpha = 1, \dots, \nu)\,. \tag{9.4.28}$$

This equation is called FICK's *second law*, according to FICK, who derived it first[2]. For the production terms π^α and the mass fluxes $\boldsymbol{j}^\alpha$ constitutive relations must be formulated such that

[2] ADOLF FICK (1829 –1901), Professor of physics at the Universities Zürich and Würzburg.

$$\sum_{\alpha=1}^{\nu} \pi^\alpha = 0 \,. \tag{9.4.29}$$

In what follows we shall not be concerned any longer with the production terms and shall assume $\pi^\alpha = 0$, for all α.

A gradient type constitutive proposal $\boldsymbol{j}^\alpha$ is known as the first FICK law; it reads

$$\boldsymbol{j}^\alpha = -\rho \sum_{\beta=1}^{\nu} D^{\alpha\beta} \operatorname{grad} c^\beta, \quad \text{with} \quad D^{\alpha\beta} = D^{\beta\alpha} \,. \tag{9.4.30}$$

$D^{\alpha\beta}$ are called *coefficients of diffusion* or *diffusivities*; they have dimension $[\mathrm{m}^2\,\mathrm{s}^{-1}]$, and $D^{\alpha\beta}$, $\alpha \neq \beta$, describe the interaction in the diffusive flux from the constituent β to the constituent α (and vice versa). Often $D^{\alpha\beta}$, $\alpha \neq \beta$, is small in comparison to $D^{\alpha\alpha}$ (no sum over α); then, (9.4.30) reads

$$\boldsymbol{j}^\alpha = -\rho D^{\alpha\alpha} \operatorname{grad} c^\alpha, \quad \text{(no summation over } \alpha) \,. \tag{9.4.31}$$

Substitution of (9.4.30) in (9.4.28) yields the *classical diffusion equation*

$$\rho \frac{\mathrm{d}c^\alpha}{\mathrm{d}t} = \sum_{\beta=1}^{\nu} \operatorname{div}(\rho D^{\alpha\beta} \operatorname{grad} c^\beta) \,. \tag{9.4.32}$$

If one introduces the scaling (9.4.13) together with

$$D^{\alpha\beta} = [D^{\alpha\beta}] \tilde{D}^{\alpha\beta} \,, \tag{9.4.33}$$

then the dimensionless version of (9.4.32) becomes

$$\tilde{\rho} \left\{ \mathbb{S}_t \frac{\partial c^\alpha}{\partial \tilde{t}} + (\operatorname{grad} c^\alpha) \tilde{\boldsymbol{v}} \right\} = \frac{1}{\mathbb{R}} \sum_{\beta=1}^{\nu} \frac{1}{\mathbb{S}^{\alpha\beta}} \operatorname{div}(\tilde{\rho} \tilde{D}^{\alpha\beta} \operatorname{grad} c^\beta) \,, \tag{9.4.34}$$

in which $\mathbb{S}_t$ and $\mathbb{R}$ are the STROUHAL and REYNOLDS numbers and

$$\mathbb{S}^{\alpha\beta} := \frac{[\eta]/[\rho]}{[D^{\alpha\beta}]} \tag{9.4.35}$$

are known as SCHMIDT numbers; they represent the ratios of the kinematic viscosity to the diffusivities of mass $[D^{\alpha\beta}]$, and are the analogues to the PRANDTL number in the energy equation.

If in a physical model diffusion processes are to be mapped to the model size the REYNOLDS and SCHMIDT numbers must be invariants of the mapping. This requirement implies:

Theorem 9.3 *Diffusion processes can only be similarly mapped from the prototype to the model if* REYNOLDS *similitude applies; otherwise stated: measurements of tracer concentrations in* FROUDE *models do not permit a transfer to the scale in Nature.*

9.5 Exercises

1. Show that the hydrodynamic equations in the BOUSSINESQ approximation cannot describe any acoustic phenomena.
2. Prove that equations (9.4.14) are correct.
3. In many problems of technology and physics the geometries of bodies are long in distinct directions and small or shallow in the direction perpendicular to them. Indeed, e.g. the ocean and the atmosphere are rather extent in the horizontal directions but confined to "small" distances in the vertical direction. In such cases it is often so that the motion is characterised by large horizontal velocities but small vertical velocities compared to these.
   Conduct a scaling of geometry and velocities which accounts for this shallowness, i.e., choose
   $$\begin{aligned} x &= [L]\tilde{x}\ ,\ y = [L]\tilde{y}\ ,\ z = [H]\tilde{\tilde{z}}\ , \\ u &= [V]\tilde{u}\ ,\ v = [V]\tilde{v}\ ,\ w = [W]\tilde{\tilde{w}}\ , \end{aligned}$$
   for the lengths and velocities, respectively. How do the non-dimensionalised NSF equations in the BOUSSINESQ approximation look like, if the above distorted scales are introduced?

9.6 Solutions

1. The proof that in a BOUSSINESQ fluid no acoustic waves can propagate, can be conducted in various different ways:

   1) The first proof starts by stating that acoustic waves are longitudinal. With the wave ansatz
   $$\boldsymbol{v} = \boldsymbol{V}_0 \exp(\mathbf{k}\cdot\boldsymbol{x} - \omega t) \tag{9.6.1}$$
   the continuity equation $\operatorname{div}\boldsymbol{v} = 0$ implies
   $$\boldsymbol{V}_0\cdot\mathbf{k} = 0\ , \tag{9.6.2}$$
   in other words, $\boldsymbol{V}_0$ and $\mathbf{k}$ are orthogonal, and since $\mathbf{k}$ marks the direction of propagation the wave is transversal. Longitudinal waves are always directed in the direction of the wave propagation but not perpendicular to it. Thus, there are no acoustic waves in a BOUSSINESQ fluid.

   2) The equations of motion of a barotropic fluid are given by
   $$\begin{aligned} &\frac{\partial\rho}{\partial t} + \operatorname{div}(\rho\boldsymbol{v}) = 0\ , \\ &\rho\frac{\mathrm{d}\boldsymbol{v}}{\mathrm{d}t} = -\operatorname{grad} p\ ,\quad p = p(\rho)\ . \end{aligned} \tag{9.6.3}$$

For small perturbations from a state of rest these equations may be linearised; they then read

$$\begin{aligned} &\frac{\partial \rho}{\partial t} + \rho_0 \operatorname{div} \boldsymbol{v} = 0 \,, \\ &\rho_0 \frac{\partial \boldsymbol{v}}{\partial t} + c_0^2 \operatorname{grad} \rho = 0 \,, \end{aligned} \tag{9.6.4}$$

in which ρ_0, $c_0^2 := \partial p / \partial \rho \,|_{\rho=\rho_0}$ are constants. Elimination of $\boldsymbol{v}$ by taking the divergence of $(9.6.4)_2$ and then substituting $(9.6.4)_1$ leads to the wave equation

$$\frac{\partial^2 \rho}{\partial t^2} - c_0^2 \operatorname{div} \operatorname{grad} \rho = 0 \,. \tag{9.6.5}$$

In the BOUSSINESQ approximation the first term in $(9.6.4)_1$ is absent. In that case one obtains instead of (9.6.5) the equation $\operatorname{div}\operatorname{grad}\rho = 0$. This is an elliptic and not a hyperbolic wave equation.

2. Substituting the scales (9.4.13) into (9.4.1) yields directly the non-dimensional NAVIER–STOKES–FOURIER equations (9.4.14).
3. For the solution of this equation one starts best from the scaled NSF equations in the BOUSSINESQ approximation, (9.4.27). These were non-dimensionlised without geometric distortion. Therefore, all that is to do, is to perform this last step of geometric distortion. To this end we introduce new dimensionless coordinates and velocities according to

$$\begin{aligned} \tilde{\boldsymbol{x}} &= (\tilde{x},\, \tilde{y},\, \tilde{z}) = (\tilde{x},\, \tilde{y},\, \mathbb{A}_L \tilde{\tilde{z}}) \,, \\ \tilde{\boldsymbol{v}} &= (\tilde{u},\, \tilde{v},\, \tilde{w}) = (\tilde{u},\, \tilde{v},\, \mathbb{A}_V \tilde{\tilde{w}}) \end{aligned} \tag{9.6.6}$$

with the aspect ratios

$$\mathbb{A}_L = \frac{[H]}{[L]} \,, \quad \mathbb{A}_V = \frac{[W]}{[V]} \,, \tag{9.6.7}$$

which describe the distortions of the scales of the coordinates and velocities, respectively. With (9.6.7) one obtains

$$\operatorname{div} \tilde{\boldsymbol{v}} = \frac{\partial \tilde{u}}{\partial \tilde{x}} + \frac{\partial \tilde{v}}{\partial \tilde{y}} + \frac{\partial \tilde{w}}{\partial \tilde{z}} = \frac{\partial \tilde{u}}{\partial \tilde{x}} + \frac{\partial \tilde{v}}{\partial \tilde{y}} + \frac{\mathbb{A}_V}{\mathbb{A}_L} \frac{\partial \tilde{\tilde{w}}}{\partial \tilde{\tilde{z}}} \,. \tag{9.6.8}$$

It is a prerequisite of all scalings, which are characterised by different scales in the horizontal and vertical directions and thus introduce explicitly the notion of *shallowness* into the governing equations that the mass balance is treated with equal rigour in all spatial directions; this necessarily suggests the selection

$$\mathbb{A}_L = \mathbb{A}_V = \mathbb{A} \ll 1 \tag{9.6.9}$$

implying the same scales for the coordinates and the velocities. Instead of (9.6.8) one then has

$$\frac{\partial \tilde{u}}{\partial \tilde{x}} + \frac{\partial \tilde{v}}{\partial \tilde{y}} + \frac{\partial \tilde{\tilde{w}}}{\partial \tilde{\tilde{z}}} = 0 \,. \tag{9.6.10}$$

Thus, the divergence operator remains unchanged in the stretched coordinates. Physically, this invariance means that a large weight is assigned to mass balance as one wishes to fulfill it exactly.
If the transformations (9.6.6) and the scalings (9.6.9) are employed, the left-hand sides of $(9.4.27)_2$ take the forms

$$\begin{aligned}
&\frac{\partial \tilde{u}}{\partial \tilde{t}} + \mathbb{R}_o \left(\frac{\partial \tilde{u}}{\partial \tilde{x}} \tilde{u} + \frac{\partial \tilde{u}}{\partial \tilde{y}} \tilde{v} + \frac{\partial \tilde{u}}{\partial \tilde{\tilde{z}}} \tilde{\tilde{w}} \right) + \mathbb{A}\, \tilde{\omega}_y \tilde{\tilde{w}} - \tilde{\omega}_z \tilde{v} \,, \\
&\frac{\partial \tilde{v}}{\partial \tilde{t}} + \mathbb{R}_o \left(\frac{\partial \tilde{v}}{\partial \tilde{x}} \tilde{u} + \frac{\partial \tilde{v}}{\partial \tilde{y}} \tilde{v} + \frac{\partial \tilde{v}}{\partial \tilde{\tilde{z}}} \tilde{\tilde{w}} \right) + \tilde{\omega}_z \tilde{u} - \mathbb{A}\, \tilde{\omega}_x \tilde{\tilde{w}} \,, \\
&\mathbb{A} \left\{ \frac{\partial \tilde{\tilde{w}}}{\partial \tilde{t}} + \mathbb{R}_o \left(\frac{\partial \tilde{\tilde{w}}}{\partial \tilde{x}} \tilde{u} + \frac{\partial \tilde{\tilde{w}}}{\partial \tilde{y}} \tilde{v} + \frac{\partial \tilde{\tilde{w}}}{\partial \tilde{\tilde{z}}} \tilde{\tilde{w}} \right) \right\} + \tilde{\omega}_x \tilde{v} - \tilde{\omega}_y \tilde{u} \,.
\end{aligned} \tag{9.6.11}$$

The pressure term on the right-hand side of $(9.4.27)_2$ may be written as

$$-\mathbb{P}_{[\omega]}\, \mathrm{grad}\, \tilde{p}_{\mathrm{dyn}} = -\mathbb{P}_{[\omega]} \begin{pmatrix} \partial \tilde{p}_{\mathrm{dyn}} / \partial \tilde{x} \\ \partial \tilde{p}_{\mathrm{dyn}} / \partial \tilde{y} \\ \frac{1}{\mathbb{A}} \partial \tilde{p}_{\mathrm{dyn}} / \partial \tilde{\tilde{z}} \end{pmatrix} \,. \tag{9.6.12}$$

Using for $\tilde{\boldsymbol{D}}$ the representation

$$\tilde{\boldsymbol{D}} = \begin{pmatrix} \frac{\partial \tilde{u}}{\partial \tilde{x}} & \frac{1}{2}\left(\frac{\partial \tilde{u}}{\partial \tilde{y}} + \frac{\partial \tilde{v}}{\partial \tilde{x}} \right) & \frac{1}{2}\left(\frac{1}{\mathbb{A}} \frac{\partial \tilde{u}}{\partial \tilde{\tilde{z}}} + \mathbb{A} \frac{\partial \tilde{\tilde{w}}}{\partial \tilde{x}} \right) \\ & \frac{\partial \tilde{v}}{\partial \tilde{y}} & \frac{1}{2}\left(\frac{1}{\mathbb{A}} \frac{\partial \tilde{v}}{\partial \tilde{\tilde{z}}} + \mathbb{A} \frac{\partial \tilde{\tilde{w}}}{\partial \tilde{y}} \right) \\ \mathrm{sym} & & \frac{\partial \tilde{\tilde{w}}}{\partial \tilde{\tilde{z}}} \end{pmatrix} , \tag{9.6.13}$$

one obtains

$$\mathbb{E}_k \tilde{\eta} \cdot 2\, \mathrm{div}\, \tilde{\boldsymbol{D}} = \mathbb{E}_k^v \tilde{\eta} \begin{pmatrix} \frac{\partial^2 \tilde{u}}{\partial \tilde{\tilde{z}}^2} + \mathbb{A}^2 \left(\frac{\partial^2 \tilde{u}}{\partial \tilde{x}^2} + \frac{\partial^2 \tilde{u}}{\partial \tilde{y}^2} \right) \\ \frac{\partial^2 \tilde{v}}{\partial \tilde{y}^2} + \mathbb{A}^2 \left(\frac{\partial^2 \tilde{v}}{\partial \tilde{x}^2} + \frac{\partial^2 \tilde{v}}{\partial \tilde{y}^2} \right) \\ \mathbb{A} \left(\frac{\partial^2 \tilde{\tilde{w}}}{\partial \tilde{\tilde{z}}^2} + \mathbb{A}^2 \left(\frac{\partial^2 \tilde{\tilde{w}}}{\partial \tilde{x}^2} + \frac{\partial^2 \tilde{\tilde{w}}}{\partial \tilde{y}^2} \right) \right) \end{pmatrix} \tag{9.6.14}$$

as well as

$$
\mathbb{E}_k \operatorname{grad} \tilde{\eta} \cdot \tilde{\boldsymbol{D}} =
$$

$$
\mathbb{E}_k^v \left(\begin{array}{c}
\dfrac{\partial \tilde{\eta}}{\partial \tilde{\tilde{z}}} \dfrac{\partial \tilde{u}}{\partial \tilde{\tilde{z}}} + \mathbb{A}^2 \left[2 \dfrac{\partial \tilde{\eta}}{\partial \tilde{x}} \dfrac{\partial \tilde{u}}{\partial \tilde{x}} + \dfrac{\partial \tilde{\eta}}{\partial \tilde{y}} \left(\dfrac{\partial \tilde{u}}{\partial \tilde{y}} + \dfrac{\partial \tilde{v}}{\partial \tilde{x}} \right) + \dfrac{\partial \tilde{\eta}}{\partial \tilde{\tilde{z}}} \dfrac{\partial \tilde{\tilde{w}}}{\partial \tilde{x}} \right] \\
\dfrac{\partial \tilde{\eta}}{\partial \tilde{\tilde{z}}} \dfrac{\partial \tilde{v}}{\partial \tilde{\tilde{z}}} + \mathbb{A}^2 \left[2 \dfrac{\partial \tilde{\eta}}{\partial \tilde{y}} \dfrac{\partial \tilde{v}}{\partial \tilde{y}} + \dfrac{\partial \tilde{\eta}}{\partial \tilde{x}} \left(\dfrac{\partial \tilde{u}}{\partial \tilde{y}} + \dfrac{\partial \tilde{v}}{\partial \tilde{x}} \right) + \dfrac{\partial \tilde{\eta}}{\partial \tilde{\tilde{z}}} \dfrac{\partial \tilde{\tilde{w}}}{\partial \tilde{y}} \right] \\
\mathbb{A} \left\{ 2 \dfrac{\partial \tilde{\eta}}{\partial \tilde{\tilde{z}}} \dfrac{\partial \tilde{\tilde{w}}}{\partial \tilde{\tilde{z}}} + \dfrac{\partial \tilde{\eta}}{\partial \tilde{x}} \dfrac{\partial \tilde{u}}{\partial \tilde{\tilde{z}}} + \dfrac{\partial \tilde{\eta}}{\partial \tilde{y}} \dfrac{\partial \tilde{v}}{\partial \tilde{\tilde{z}}} + \mathbb{A}^2 \left[\dfrac{\partial \tilde{\eta}}{\partial \tilde{x}} \dfrac{\partial \tilde{\tilde{w}}}{\partial \tilde{x}} + \dfrac{\partial \tilde{\eta}}{\partial \tilde{y}} \dfrac{\partial \tilde{\tilde{w}}}{\partial \tilde{y}} \right] \right\}
\end{array} \right) . \tag{9.6.15}
$$

Finally,

$$
\mathbb{B}_{[\omega]} \tilde{\sigma} \tilde{\boldsymbol{g}} = \begin{pmatrix} 0 \\ 0 \\ -\mathbb{B}_{[\omega]}^v \mathbb{A} \tilde{\sigma} g \end{pmatrix} , \tag{9.6.16}
$$

in which

$$
\mathbb{E}_k^v = \frac{\mathbb{E}_k}{\mathbb{A}^2} = \frac{[\eta]/[\rho]}{[\omega][H]^2} , \quad \mathbb{B}_{[\omega]}^v = \frac{\mathbb{B}_{[\omega]}}{\mathbb{A}} = \frac{[\sigma][\omega][W]}{[g]} . \tag{9.6.17}
$$

If one substitutes these expressions into the momentum equation $(9.4.27)_2$, then the following equations are obtained:

In the *horizontal* direction

$$
\begin{aligned}
&\frac{\partial \tilde{u}}{\partial \tilde{t}} + \mathbb{R}_o (\operatorname{grad} \tilde{u}) \cdot \tilde{\boldsymbol{v}} + \mathbb{A}\, \tilde{\omega}_y \tilde{\tilde{w}} - \tilde{\omega}_z \tilde{v} \\
&= -\mathbb{P}_{[\omega]} \frac{\partial \tilde{p}_{\text{dyn}}}{\partial \tilde{x}} + \left\{ \mathbb{E}_k^v \left(\tilde{\eta} \frac{\partial^2 \tilde{u}}{\partial \tilde{\tilde{z}}^2} + \frac{\partial \tilde{\eta}}{\partial \tilde{\tilde{z}}} \frac{\partial \tilde{u}}{\partial \tilde{\tilde{z}}} \right) \right. \\
&\left. + 2\mathbb{A}^2 \left\{ \left(\frac{\partial^2 \tilde{u}}{\partial \tilde{x}^2} + \frac{\partial^2 \tilde{u}}{\partial \tilde{y}^2} \right) + \left[\frac{\partial \tilde{\eta}}{\partial \tilde{x}} \frac{\partial \tilde{u}}{\partial \tilde{x}} + \frac{1}{2} \frac{\partial \tilde{\eta}}{\partial \tilde{y}} \left(\frac{\partial \tilde{u}}{\partial \tilde{y}} + \frac{\partial \tilde{v}}{\partial \tilde{x}} \right) + \frac{1}{2} \frac{\partial \tilde{\eta}}{\partial \tilde{\tilde{z}}} \frac{\partial \tilde{\tilde{w}}}{\partial \tilde{x}} \right] \right\} \right\} , \\
&\frac{\partial \tilde{v}}{\partial \tilde{t}} + \mathbb{R}_o (\operatorname{grad} \tilde{u}) \cdot \tilde{\boldsymbol{v}} + \tilde{\omega}_z \tilde{u} - \mathbb{A}\, \tilde{\omega}_x \tilde{\tilde{w}} \\
&= -\mathbb{P}_{[\omega]} \frac{\partial \tilde{p}_{\text{dyn}}}{\partial \tilde{y}} + \left\{ \mathbb{E}_k^v \left(\tilde{\eta} \frac{\partial^2 \tilde{v}}{\partial \tilde{\tilde{z}}^2} + \frac{\partial \tilde{\eta}}{\partial \tilde{\tilde{z}}} \frac{\partial \tilde{u}}{\partial \tilde{\tilde{z}}} \right) \right. \\
&\left. + 2\mathbb{A}^2 \left\{ \left(\frac{\partial^2 \tilde{v}}{\partial \tilde{x}^2} + \frac{\partial^2 \tilde{v}}{\partial \tilde{y}^2} \right) + \left[\frac{\partial \tilde{\eta}}{\partial \tilde{y}} \frac{\partial \tilde{v}}{\partial \tilde{y}} + \frac{1}{2} \frac{\partial \tilde{\eta}}{\partial \tilde{x}} \left(\frac{\partial \tilde{u}}{\partial \tilde{y}} + \frac{\partial \tilde{v}}{\partial \tilde{x}} \right) + \frac{1}{2} \frac{\partial \tilde{\eta}}{\partial \tilde{\tilde{z}}} \frac{\partial \tilde{\tilde{w}}}{\partial \tilde{y}} \right] \right\} \right\} ,
\end{aligned} \tag{9.6.18}
$$

and in the *vertical* direction

$$
\begin{aligned}
&\mathbb{A}^2\left\{\frac{\partial\tilde{\tilde{w}}}{\partial\tilde{t}}+\mathbb{R}_o(\operatorname{grad}\tilde{\tilde{w}})\cdot\tilde{\boldsymbol{v}}\right\}+\mathbb{A}(\tilde{\omega}_x\tilde{u}-\tilde{\omega}_y\tilde{u})= \\
&\quad-\mathbb{P}_{[\omega]}\frac{\partial\tilde{p}_{\rm dyn}}{\partial\tilde{\tilde{z}}}+\mathbb{A}^2\left\{\mathbb{E}_k^v\left[\tilde{\eta}\frac{\partial^2\tilde{\tilde{w}}}{\partial\tilde{\tilde{z}}^2}+2\mathbb{A}^2\left(\frac{\partial^2\tilde{\tilde{w}}}{\partial\tilde{x}^2}+\frac{\partial^2\tilde{\tilde{w}}}{\partial\tilde{y}^2}\right)\right]\right. \\
&\qquad\left.+\mathbb{E}_k^v\left\{\left[2\frac{\partial\tilde{\eta}}{\partial\tilde{\tilde{z}}}\frac{\partial\tilde{\tilde{w}}}{\partial\tilde{\tilde{z}}}+\frac{\partial\tilde{\eta}}{\partial\tilde{x}}\frac{\partial\tilde{u}}{\partial\tilde{\tilde{z}}}+\frac{\partial\tilde{\eta}}{\partial\tilde{y}}\frac{\partial\tilde{v}}{\partial\tilde{\tilde{z}}}\right]+\mathbb{A}^2\left[\frac{\partial\tilde{\eta}}{\partial\tilde{x}}\frac{\partial\tilde{\tilde{w}}}{\partial\tilde{x}}+\frac{\partial\tilde{\eta}}{\partial\tilde{y}}\frac{\partial\tilde{\tilde{w}}}{\partial\tilde{y}}\right]\right\}\right\} \\
&\quad-\mathbb{B}_{[\omega]}^v\tilde{\sigma}g\,.
\end{aligned}
\tag{9.6.19}
$$

These equations are the basis for the *shallow water equations.* Taking the limit $\mathbb{A}\to 0$, equations $(9.6.18)_{1,2}$ and (9.6.19) reduce to

$$
\begin{aligned}
\frac{\partial\tilde{u}}{\partial\tilde{t}}+\mathbb{R}_o(\operatorname{grad}\tilde{u})\cdot\tilde{\boldsymbol{v}}-\tilde{\omega}_z\tilde{v}&=-\mathbb{P}_{[\omega]}\frac{\partial\tilde{p}_{\rm dyn}}{\partial\tilde{x}}+\mathbb{E}_k^v\frac{\partial}{\partial\tilde{\tilde{z}}}\left(\tilde{\eta}\frac{\partial\tilde{u}}{\partial\tilde{\tilde{z}}}\right), \\
\frac{\partial\tilde{v}}{\partial\tilde{t}}+\mathbb{R}_o(\operatorname{grad}\tilde{v})\cdot\tilde{\boldsymbol{v}}+\tilde{\omega}_z\tilde{u}&=-\mathbb{P}_{[\omega]}\frac{\partial\tilde{p}_{\rm dyn}}{\partial\tilde{y}}+\mathbb{E}_k^v\frac{\partial}{\partial\tilde{\tilde{z}}}\left(\tilde{\eta}\frac{\partial\tilde{v}}{\partial\tilde{\tilde{z}}}\right), \\
0&=-\mathbb{P}_{[\omega]}\frac{\partial\tilde{p}_{\rm dyn}}{\partial\tilde{\tilde{z}}}-\mathbb{B}_{[\omega]}^v\tilde{\sigma}g\,.
\end{aligned}
\tag{9.6.20}
$$

These shallow water equations enjoy the following special properties:

- The vertical momentum balance reduces to a hydrostatic pressure balance between the vertical gradient of the pressure and the buoyancy force. It is only here that density variations play an explicit role.
- It is only the z-component of the angular velocity of the rotating frame of reference that effects the dynamics of the shallow water equations.
- Diffusion of momentum is only operating in the vertical direction and is governed by the vertical EKMAN number.

Finally, we complement the above analysis by also subjecting the energy equation $(9.4.27)_3$ to the distorted scalings (9.6.6) mit (9.6.9). After a short calculation we obtain

$$
\begin{aligned}
\tilde{c}_p\left(\frac{\partial\theta}{\partial\tilde{t}}+\mathbb{R}_o(\operatorname{grad}\theta)\cdot\tilde{\boldsymbol{v}}\right)&=\frac{\mathbb{E}_k^v}{\mathbb{P}_r}\left\{\mathbb{A}^2\left[\frac{\partial}{\partial\tilde{x}}\left(\tilde{\kappa}\frac{\partial\theta}{\partial\tilde{x}}\right)+\frac{\partial}{\partial\tilde{y}}\left(\tilde{\kappa}\frac{\partial\theta}{\partial\tilde{y}}\right)\right]\right. \\
&\quad\left.+\frac{\partial}{\partial\tilde{\tilde{z}}}\left(\tilde{\kappa}\frac{\partial\theta}{\partial\tilde{\tilde{z}}}\right)\right\}+\frac{3}{4}\frac{\mathbb{E}_k}{\mathbb{T}_h}\left\{\left[\left(\frac{\partial\tilde{u}}{\partial\tilde{\tilde{z}}}\right)^2+\left(\frac{\partial\tilde{v}}{\partial\tilde{\tilde{z}}}\right)^2\right]+\mathcal{O}(\mathbb{A}^2)\right\},
\end{aligned}
\tag{9.6.21}
$$

which reduces in the shallow water approximation to

$$
\tilde{c}_p\left(\frac{\partial\theta}{\partial\tilde{t}}+\mathbb{R}_o(\operatorname{grad}\theta)\cdot\tilde{\boldsymbol{v}}\right)=\frac{\mathbb{E}_k^v}{\mathbb{P}_r}\frac{\partial}{\partial\tilde{\tilde{z}}}\left(\tilde{\kappa}\frac{\partial\theta}{\partial\tilde{\tilde{z}}}\right)+\frac{\mathbb{E}_k}{\mathbb{T}_h}\frac{3}{4}\left[\left(\frac{\partial\tilde{u}}{\partial\tilde{\tilde{z}}}\right)^2+\left(\frac{\partial\tilde{v}}{\partial\tilde{\tilde{z}}}\right)^2\right].
\tag{9.6.22}
$$

As in the momentum balance equations all convective terms are accounted for, however of the diffusive and dissipative terms, only those contributions are accounted for which can be traced to the vertical gradients of the motion and temperature distributions.

Part III

Turbulence

10. Fundamental Concepts of Turbulence

All subsequent chapters deal with turbulence; in fact they aim to provide a small introductory insight into the theory of turbulent flows. The question that immediately arises in this connection is, however, why turbulence is dealt with at all in a book of which large parts are devoted to concepts of continuum mechanics and thermodynamics. The obvious answer is that turbulence theory is as much a topic of continuum physics as is continuum thermodynamics, but more importantly, it turned out during the last few decades of intensive research in turbulence modelling that methods of constructing turbulent closure schemes show strong similarities with techniques of establishing constitutive relations of continuum mechanics. There are differences between the two, and these differences are of fundamental nature, but the similarities are so strong that accessing the topic from a structural concept of continuum thermodynamics facilitates the learning of the subject tremendously – apart from providing new insight. To give hints as to these similarities and differences – they will be outlined in the ensuing chapters – we mention that establishing turbulent closure conditions follows similar lines as postulating constitutive equations for a certain material. The complexity of the material behaviour corresponds to the detailed reproduction of the fluctuating quantities by so-called higher order closure schemes. As for rules, valid in continuum thermomechanics for the constitutive relations, these closure relations must be functional relationships between objective tensor fields (of various ranks). However a principle, such as the rule of material objectivity cannot as strictly be imposed as it is commonly done in continuum thermodynamics. Furthermore, equations describing the evolution of the mean fields intend to describe physical observables and are therefore as much subjected to thermodynamic irreversibility requirements as are such fields in continuum thermodynamics. In other words, postulated turbulent closure conditions, which are the analoga to the constitutive relations in continuum thermodynamics, must be such that the second law of thermodynamics is fulfilled. This requirement will constrain the closure conditions in a similar way as the entropy principle constrains the constitutive functionals of continuum mechanics. A complete satisfaction of the second law of thermodynamics is commonly, however, not demonstrated in turbulence modeling – there are exceptions which we will be dealing with in Chap. 12 – but at least some rules are generally obeyed which often may be interpreted as fulfilling certain dissipation principles. These rules are called *realizability conditions.* It follows that turbulent fields that are in conformity with the second law of thermodynamics are automatically physically realizable.

Since the theory of turbulence is a vast field that has extensively been developed in the twentieth century, we can only pick out a few essentials in the hope to give the reader a basis for further study[1]. To this end we shall first provide an insight into the beginnings of the theory of turbulence as it was given by PRANDTL more than 70 years ago in a paper of roughly two pages in length. Then the equations of mean value hydrodynamics will be derived, and it will be explained what modellers mean by turbulence models of zeroth, first and second order. This then leads to the concept of eddy viscosity and turbulent diffusivity, to the k–ε model for incompressible and BOUSSINESQ fluids, and finally to the REYNOLDS stress models (**RSM**). The latter are subsequently reduced to such an extent that instead of partial differential equations only algebraic equations are obtained to describe the turbulent closure except for two differential equations for scalar quantities, usually the *turbulent kinetic energy* and the *turbulent rate of dissipation.* This three dimensional formulation is still rather complicated; for basically one-dimensional problems a further reduction of such a model to one dimension then generates the special one-dimensional k–ε model which is particularly suitable to describe the turbulent mixing in lakes and the ocean for the description of the seasonal development of the water temperature.

In order to be able to describe turbulent as well as other flow phenomena, it would, in principle, suffice to solve the NAVIER–STOKES(–FOURIER) equations, which we have met several times in parts I and II of this book. Indeed, these equations describe *all* hydrodynamic phenomena, which for some processes can be formulated in terms of these equations at all; this is today's firm conviction. However, the problem lies in the fact that different physical processes must simultaneously be viewed at different length scales: for example, large scale motions, which generate small scale eddies at an obstruction. The ranges of length scales, which must be analysed, extend from several meters or even kilometers down to millimeters. If one would attempt to solve the problem on a computer, the mesh size of the discretization would have to orient itself on the smallest entities to be resolved. For this reason, a gigantic computational expenditure would have to be made, both in computational

[1] The literature on turbulence is a vast field. To our knowledge, none of the references known to us treats the subject in the way as outlined in this book. Emphasis in most books is perhaps more on the side of physics, however, the mathematical background needed to understand turbulence profoundly is fairly substantial. The reader is advised to consult such "more physically inclined" books as side reading. A possible selection is as follows: BATCHELOR [21]; FRISCH [80], HINZE [101]; LESIEUR [133]; MCCOMB [153]; MONIN & YAGLOM [156], [157]; LUMLEY [140], [141], [142]; PIQUET [182]; ROTTA [198]; TENNEKES & LUMLEY [235]; TOWNSEND [237] and others that are certainly on the market.

time and storage availability. For many problems such a procedure is uneconomical if not simply unrealistic. Nevertheless, the significance of such high resolution **D**irect **N**umerical **S**imulations (**DNS**) should neither be underestimated, because only such computations at last permit us to examine, whether a different, simpler algorithm leads to similar results for a given problem for which corroborating measurements are available[2].

One could analyse the processes of motion, which are operative at different length and time scales by a dimensional analysis, and thereby separate them via a filtering of the equations due to the scaling. Unfortunately, this spectral separation of turbulent processes is not automatically possible, because the turbulent motions arise on all scales from the mean motion down to very small dimensions, whereby the processes occurring at the different scales are in general strongly coupled. However, the distribution of such *eddies* or vortices can be studied by using statistical considerations; this is also exactly the general procedure one takes in turbulence theory. In principle, one postulates and observes the statistical properties of the turbulent flow, which is described by *mean values* and an infinite, but practically limited number of *correlations* (e.g. the mean velocity and its variance). These quantities are modelled by balance equations or by simple empirical equations, which implies that empirical coefficients arise, which must be known, i.e., a priori determined or determinable by experiments and comparative computations.

10.1 Notation

List of Used Symbols

$\boldsymbol{A}$	....	Arbitrary tensor of second rank, matrix
$\mathcal{A}$	....	Parameter related to the turbulent free energy, definition, see (12.4.18)
$\mathfrak{a}$	....	Anisotropy tensor, derived from the REYNOLDS stress deviator
$\boldsymbol{a}$	....	Arbitrary vector
a	....	Coefficient in the REINER–RIVLIN constitutive relation for stress (10.7.22)
a_{Chl}	.....	Special absorption cross section
B	....	BOWEN ratio
$\mathcal{B}$	....	Parameter related to the turbulent free energy, definition, see (12.4.18)

[2] Because the finding of analytical solutions of the governing equations is practically impossible even for the simplest geometries, such solutions are constructed by numerical simulations. The latter has methodologically been moved so much to the centre that the acronym DNS was created to denote the construction of numerical solutions to the NAVIER–STOKES equations by Direct Numerical Simulation with a resolution in space and time as fine as needed to describe the turbulent flow field without saying anything explicit about the turbulent closure.

$b_1, \dots, b_7$.	Dimensionless scalar coefficients in the isotropic representation for $\mathfrak{Q}$, see (12.4.41)
C_1, C_2, c_ε	Dimensionless coefficient in the parameterization of π^ε, see (12.5.8)
C^T, C_0^T ..	Scale factor relating the turbulent coldness with the turbulent kinetic energy, see (12.4.23)
$c_1, \dots, c_8$.	Dimensionless scalar coefficients in the isotropic representation for the anisotropy tensor $\mathfrak{a}$, see (12.4.39)
c^α	Specific mass concentration of tracer α
c_D	Drag coefficient
c_b	Constant in the law describing the dependence of the radiation intensity upon the degree of cloud cover (13.2.7)
$c_\mu = 0.09$, $c_1 = 0.126$, $c_2 = 1.92$, $c_\varepsilon = 0.07$.	Constants in the k–ε model
c_v, c_p	Specific heat at constant volume, at constant pressure
D	Pipe diameter
$\boldsymbol{D}$	Stretching tensor, strain rate tensor ($\boldsymbol{D} := \operatorname{sym}\operatorname{grad}\boldsymbol{v}$)
$\boldsymbol{D}^0$	JAUMANN derivative of $\boldsymbol{D}$
$\langle\boldsymbol{D}\rangle^{(\varepsilon)}$ ...	Second RIVLIN–ERICKSEN tensor
d	Day
$d_1, \dots, d_7$	Scalar invariants in the constitutive law (10.7.34)
$E(\mathbf{k})$	Spectral density of the kinetic energy
$\mathcal{E}(\mathrm{k})$	Isotropic (radial) spectral density of the kinetic energy
E	Enstrophy
e	Specific enstrophy
f, $\tilde{f}$	Wind function (13.2.21)
$\mathcal{F}\boldsymbol{v}$	FOURIER transform of $\boldsymbol{v}$
$\boldsymbol{g}$, g	Acceleration due to gravity, gravity constant
H	Helicity, water depth
h	PLANCK constant
$\boldsymbol{I}$	Pointing vector
I, I_0	Radiation intensity
I_{max}	Maximum radiation intensity
$\boldsymbol{I}_0$	Velocity-independent part of the system dependent part of the inertial force
$\boldsymbol{j}^\alpha$, $\boldsymbol{j}$	Mass flux vector (of constituents α)
$\boldsymbol{j}_t$	Turbulent mass flux vector
k	Specific turbulent kinetic energy, attenuation coefficient, BOLTZMANN constant
$\boldsymbol{K}$	Flux of turbulent kinetic energy
K_{Chl}	Concentration of chlorophyll
$\boldsymbol{\kappa}^\varepsilon$	Flux of the specific turbulent dissipation

$\boldsymbol{\kappa}^M$	Material part of extra entropy flux
$\boldsymbol{\kappa}^T$	Turbulent part of extra entropy flux
k_{Chl}	Algae–specific attenuation coefficient
k_{w}	Clear–water attenuation coefficient
$\mathbf{k}$	Wavenumber vector
k	Modulus of the wavenumber vector ($\mathrm{k} := \lvert\mathbf{k}\rvert$)
$\boldsymbol{L}$	Spatial velocity gradient ($\boldsymbol{L} := \operatorname{grad} \boldsymbol{v}$)
$L_0, \ldots, L_3$	Operators arising in the balances of mass and momentum (see (10.6.1))
L_K	KOLMOGOROV length scale ($= \eta^{3/4}\varepsilon^{-1/4}$)
L_{ij}^E	EULERian correlation lengths
L_v	Latent heat of evaporation
l	PRANDTL mixing length
N	Length of day in hours
N_{b}	Degree of cloudiness
$\boldsymbol{O}$	Orthogonal transformation ($\boldsymbol{O}\boldsymbol{O}^T = \boldsymbol{O}^T\boldsymbol{O} = \boldsymbol{I}$)
p	Pressure
$\langle p\rangle$, p' ...	Mean value and fluctuation of the pressure p
p_{r}	Reduced pressure ($p_{\mathrm{r}} := p/\rho$)
p_{s}	Saturation pressure
$\wp$	Probability of an event
$\boldsymbol{Q}$	Turbulent heat flux
$\boldsymbol{Q}^M$	Material part of extra heat flux
$\boldsymbol{Q}^T$	Extra turbulent kinetic energy flux
$\left.\begin{array}{l}\boldsymbol{Q}(\boldsymbol{x},t)\\ \boldsymbol{Q}^{(3)}(\boldsymbol{x},t)\\ \boldsymbol{Q}^{(n)}(\boldsymbol{x},t)\end{array}\right\}$.	Statistical moment of the second, third and n-th order of the velocity correlations; two, three and n-fold correlations
$\boldsymbol{\mathfrak{Q}}$	Dimensionless turbulent heat flux vector
Q_{l}	Latent heat flow
Q_{s}	Sensible heat flow
$Q_{\mathrm{ir}}^{\mathrm{a}}, Q_{\mathrm{ir}}^{\mathrm{w}}$ ..	Radiative fluxes of air and water, respectively
$\boldsymbol{q}$, $\boldsymbol{q}_t$	Heat flux vector, turbulent heat flux vector
$\boldsymbol{q}_\varepsilon$	Flux of the dissipative power
R	Gas constant ($= 8.31451\,\mathrm{J\,K^{-1}\,mol^{-1}}$)
$\boldsymbol{R}$	REYNOLDS stress tensor
$\boldsymbol{R}'$	Deviator of $\boldsymbol{R}$
$\boldsymbol{R}^D$	REYNOLDS stress deviator
$\mathbb{R}$	REYNOLDS number
$\mathbb{R}_i$	RICHARDSON number
r	Reflection coefficient (in the SNELLIUS law)
$\boldsymbol{r}$	Difference of two position vectors
$\bar{\boldsymbol{r}}$	Arithmetic mean of two position vectors
$\mathfrak{r}$	Specific radiation of heat

s^M	Material entropy density
s^T	Turbulent entropy density
S, s	Salt concentration
s	Specific entropy density
T^E	EULERian correlation time
T_K	KOLMOGOROV time scale ($=\nu^{1/2}\varepsilon^{-1/2}$)
t	Time
$\bar{t}$	Arithmetic mean of two times
$\boldsymbol{t}$	CAUCHY stress tensor
U, $U_{(10\,\mathrm{m})}$	Wind speed, - 10 m above the water surface
$\bar{u}$	Depth averaged velocity
V	Characteristic velocity
V_K	KOLMOGOROV velocity scale
$\boldsymbol{V}(\mathbf{k}, t)$ ..	FOURIER transform of $\boldsymbol{v}(\boldsymbol{x}, t)$, $\mathbf{k}$ is the wavenumber vector
v_*	Friction velocity, shear stress velocity
$\boldsymbol{v}$	Velocity vector
$\langle\boldsymbol{v}\rangle_T$	Temporal mean value of $\boldsymbol{v}$
$\langle\boldsymbol{v}\rangle_R$	Spatial mean value of $\boldsymbol{v}$
$\langle\boldsymbol{v}\rangle_S$	Statistical mean value of $\boldsymbol{v}$
$\langle\boldsymbol{v}\rangle$, $\boldsymbol{v}'$ ...	Mean value and fluctuation of the velocity $\boldsymbol{v}$
$\boldsymbol{W}$	Spin tensor, vorticity tensor ($\boldsymbol{W} := \mathrm{skw}(\mathrm{grad}\,\boldsymbol{v})$)
$\mathcal{W}$	Absolute vorticity tensor
x, y, z ...	Spatial Cartesian coordinates
$\mathfrak{X}^M, \mathfrak{X}^T$.	Independent material and turbulent constitutive variables
$\mathcal{Y}^M, \mathcal{Y}^T$.	Dependent material and turbulent constitutive variables
α_Θ	Thermal expansion coefficient
$\alpha_1^\kappa, ..., \alpha_6^\kappa$	Scalar parameters in the isotropic representation for $\boldsymbol{K}$, see (12.4.36)
$\alpha_1^\varepsilon, ..., \alpha_6^\varepsilon$	Scalar parameters in the isotropic representation for $\boldsymbol{k}^\varepsilon$, see (12.4.36)
α, β	Parameters arising in the parameterization of Ψ^T, see (12.4.31)
β	Zenith angle of the sun
$\beta_1, ..., \beta_8$.	Scalar parameters in the isotropic representation for $\boldsymbol{R}^D$, see (12.4.33)
$\gamma_1, ..., \gamma_6$.	Scalar parameters in the representation for π^ε, see (12.4.36)
$\Delta, \Delta^{\mathcal{W}}, \Delta^{Z_\alpha}$, $\Delta^{D z_{\alpha\beta}}, \Delta^{\mathcal{W} z_{\alpha\beta}}$	Scalar invariants of turbulent field quantities, see (12.4.8)
δ	Declination
δ_{ij}	Kronecker Delta ($=1$ if $i=j$, $=0$ if $i \neq j$)
ϵ	Specific internal energy

ε	Eddy viscosity, specific turbulent dissipation
ε_{a}	Emissivity of air
ε_{w}	Emissivity of water
ξ_t	Scalar coefficient in the constitutive law (10.7.22) for the REYNOLDS stress tensor
ζ	Displacement of the free water surface
Θ	Absolute temperature
$\langle\Theta\rangle$, Θ'	Mean value, fluctuation of the absolute temperature Θ
$\Theta^M := \langle\frac{1}{\vartheta}\rangle$	Mean of the inverse material coldness
Θ^T	Turbulent temperature
ϑ^M	Mean of the material coldness
ϑ^T	Turbulent coldness
κ	Dynamic bulk viscosity, conductivity of water
κ_{20}	Conductivity of water at 20℃
$\kappa_1, \ldots, \kappa_4$	Scalar parameters in the isotropic representation for $\boldsymbol{Q}$, see (12.4.36)
Λ^ρ	LAGRANGE parameter
λ	Scalar heat conductivity, geographical latitude
μ, μ_t	Dynamic shear viscosity, turbulent dynamic shear viscosity
ν, ν_t	Kinematic viscosity, turbulent kinematic viscosity
ν	Frequency
ν_{ijkl}	Tensor of the kinematic viscosity
ρ, ρ_0	Density, reference density
$\rho(\nu, \Theta)$	Spectral energy density in the PLANCK law of black body radiation
ρ_{a}	Density of air
ϱ_{ij}	Correlation coefficient (10.4.13)
σ_c	Turbulent SCHMIDT number
σ_Θ	Turbulent PRANDTL number
σ_{SB}	STEFAN–BOLTZMANN constant
σ^ε	Supply of turbulent dissipation
τ	Time, difference of two times
$\boldsymbol{\tau}$	Shear stress, wind shear stress
π^ε	Production of turbulent dissipation
$\langle\bar{\pi}\rangle^{\mathrm{ent}}$	Production of turbulent entropy
π^M	Material part of production of turbulent entropy
π^T	Turbulent part of production of turbulent entropy
$\phi^{(c_\alpha)}$	Specific production of tracer c_α
φ	Specific dissipation rate
$\boldsymbol{\phi}$	Entropy flux vector
$\boldsymbol{\phi}^{\mathrm{ent}}_{(T)}$	Turbulent flux of entropy
$\chi^{(c)}$, $\chi^{(\Theta)}$	Mass–, heat diffusivity

$\chi^{\alpha\beta}$	(Cross) mass diffusivities of the tracers α and β
$\chi_{ij}^{(c)}, \chi_{ij}^{(\Theta)}$	Tensor of the turbulent kinematic mass and thermal diffusivities
ψ	HELMHOLTZ free energy
Ψ	Relative humidity ($\Psi = p_\mathrm{a}/p_\mathrm{s}$)
ψ^M	Material HELMHOLTZ free energy
ψ^T	Turbulent HELMHOLTZ free energy
$\boldsymbol{\Omega}$	Earth's angular velocity vector
$\boldsymbol{\Omega} := \dot{\boldsymbol{O}}\boldsymbol{O}^T$	Rotation tensor of the moving observer frame

Special symbols:

$I_{\boldsymbol{A}}, II_{\boldsymbol{A}}, III_{\boldsymbol{A}}$	First, second and third invariants of $\boldsymbol{A}$
$\langle a \rangle$	Mean value of a, either spatial, temporal or statistical
$\{a\}$	FAVRE mean value of a, density weighted mean value of a ($\{a\} := \langle \rho a \rangle / \langle \rho \rangle$)
$\mathcal{F}\boldsymbol{v}$	Spatial, three dimensional FOURIER transform of $\boldsymbol{v}$

10.2 Early Concepts of Turbulence Theory

10.2.1 Experiments of Reynolds

The first basic thoughts and experiments regarding the phenomenon turbulence are due to O. REYNOLDS [193] who studied the flow of a fluid through pipes with circular cross sections. He recognized (by adding dye through a pipette to the fluid that, basically, two flow regimes exist. In one case, the so-called *laminar flow*, the dye forms a coherent thin filament, of which the width grows only little with the distance from the entrance cross section, Fig. 10.1a (top). In the second case, known as *turbulent flow*, the dye filament is torn very quickly after it left the nozzle of the pipette and spread over the entire cross section of the pipe, Fig. 10.1a (bottom) . If one slowly increases the discharge in the pipe, starting from laminar conditions, one observes a sudden change of laminar to turbulent flow. The critical quantity that characterizes this change, can easily be determined with dimensional analysis. The phenomenon "pipe flow" is obviously dictated by the diameter of the pipe D, the kinematic viscosity ν and the axial velocity V (i.e., the average over the cross section). The only dimensionless product that can be formed with these quantities is the REYNOLDS number

$$\mathbb{R} = \frac{V D}{\nu}, \quad \mathbb{R} > 2000 \quad \text{turbulent}, \tag{10.2.1}$$

and follows from a standard analysis of the dimensional matrix

	V	D	ν
L	1	1	2
T	-1	0	-1

.

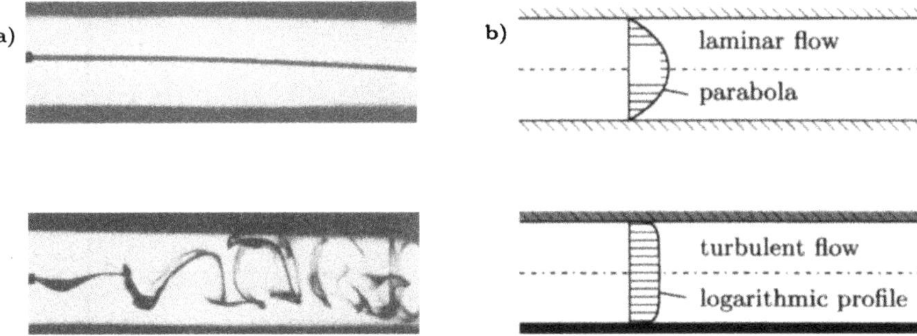

Fig. 10.1. a) Laminar and turbulent flow in a cylindrical pipe. To visualize the flow dye is added to the water through a capillary pipette. The nozzle of this pipette is visible at the left end of the photographs on the left. b) Mean velocity profiles in a circular pipe under steady laminar and turbulent conditions, respectively.

The velocity profiles which are measured in such flows look qualitatively as shown in Fig. 10.1b. The transition from the laminar to the turbulent flow regime takes place in regions of the REYNOLDS number between 500 and 2000. At REYNOLDS numbers larger than 2000 the flow is always turbulent, see Fig. 10.2.

10.2.2 Temporal Averaging

In Fig. 10.1 the velocity profile for the turbulent flow was drawn for the (temporally) *mean* velocity, because the actual velocity profile possesses statistical character and is fraught with many strong fluctuations. This statistical appearance of the flow is, however, a supposition and still not an established fact; indeed the phenomenon "turbulence" may in many details also be described by a cascade of instabilities, from a single first bifurcation to complete deterministic chaos. Despite this fact the statistical nature is a convenient hypothesis of its description. Accordingly, the true (and perhaps measured) velocity profile is a superposition of the mean velocity and a fluctuation velocity. At a fixed time a "snapshot" of a velocity distribution could be as in Fig. 10.3a, and in a fixed point in space, the temporal evolution of the turbulent velocity may look as shown in Fig. 10.3b. Evidently, the velocity oscillates about a certain mean value, and the fluctuations appear to be random. As a rule, the fluctuations are short periodic and have short wave lengths, whilst the evolution or spatial variations of the mean values are much smoother and/or slower, i.e., they have longer periods and larger wave lengths. This fact suggests to additively decompose the velocity into two contributions, the mean velocity $\langle \boldsymbol{v} \rangle$ and the fluctuations $\boldsymbol{v}'$,

$$\boldsymbol{v} = \langle \boldsymbol{v} \rangle + \boldsymbol{v}' \ . \tag{10.2.2}$$

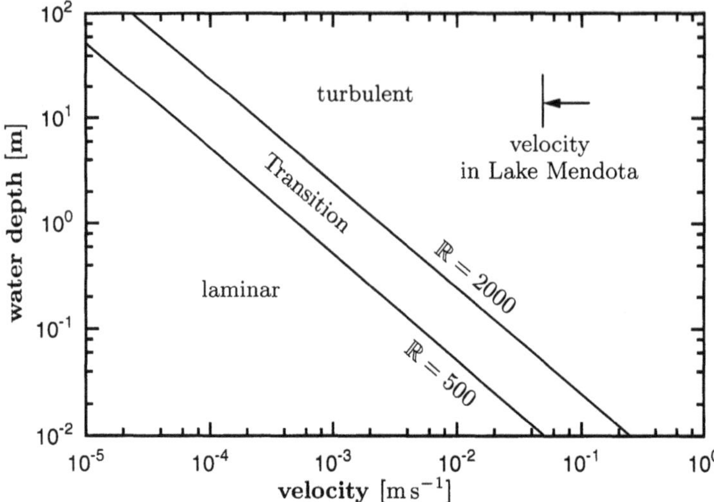

Fig. 10.2. Flow conditions and occurrence of turbulent flows in a lake or in the ocean as they depend on the water depth and the water velocity. As an example, the conditions in Lake Mendota are shown. Turbulence is always possible there, since the REYNOLDS number, defined as $\mathbb{R} = VH/\nu$, where H is the water depth, or the depth of the upper layer, V a typical velocity and ν the kinematic viscosity, is always considerably larger than 2000.

Momentarily, we wish to interpret the mean value $\langle \boldsymbol{v} \rangle$ as a temporal average, defined by

$$\langle \boldsymbol{v} \rangle_T(\boldsymbol{x}, t) := \frac{1}{T} \int_{t-T/2}^{t+T/2} \boldsymbol{v}(\boldsymbol{x}, \tau)\, \mathrm{d}\tau \,, \tag{10.2.3}$$

in which the index $\langle\rangle_T$ denotes this time averaging. In computing such an average from a given time series, it must be observed that the interval T over which the averaging operation is performed is not too small, so that the turbulent contributions are indeed "eliminated". Conversely, this interval should neither be too large, because in so doing important non-turbulent time dependent phenomena may thereby be lost. It transpires that the decomposition of $\boldsymbol{v}$ into $\langle \boldsymbol{v} \rangle$ and $\boldsymbol{v}'$ depends firstly, on the selection of T, and secondly, that this selection may be an important quantity in the modeling.

In the decomposition of the velocity into a mean value and a fluctuating quantity, it is commonly assumed or supposed that the mean value of a fluctuating quantity vanishes, viz.,

$$\langle \boldsymbol{v}' \rangle = \boldsymbol{0}, \quad \text{respectively} \quad \langle \boldsymbol{v} \rangle = \langle \langle \boldsymbol{v} \rangle \rangle \,. \tag{10.2.4}$$

This supposition is, however, not obvious, as one can easily convince oneself; for if one smoothes a curve (by any smoothing operation) that was already

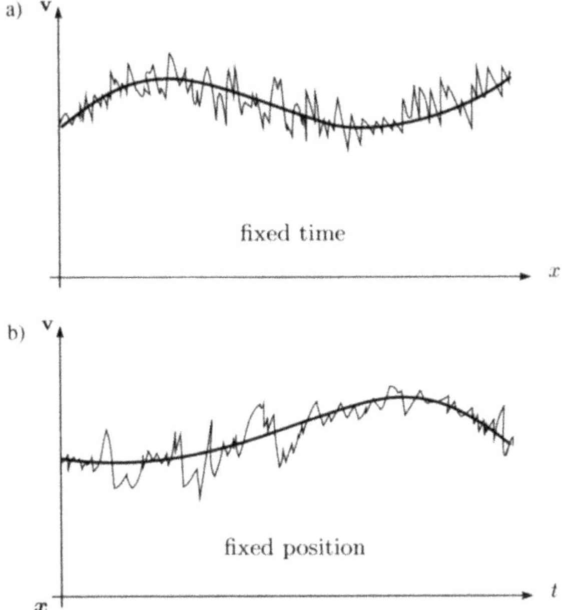

Fig. 10.3. Fluctuations of the turbulent velocities at a fixed time but seen as spatial variations a) and at a fixed spatial point and seen as temporal variations b) [schematic].

smoothed before, then a different result is generally obtained with the second smoothing. Below we shall be further concerned with the properties of such operators that are called *filters*.

10.2.3 Eddy Spectrum and Kolmogorov Scales

It was already mentioned that turbulent flows are accompanied by formations of vortices called "eddies". The generated gyres may assume practically every possible extent that is below the dimension of the region in which the flow arises; an example of such vortices is shown in Fig. 10.4. The kinetic energy that is contained in such a turbulent flow is distributed among all these eddies. An intriguing question is therefore, how this energy is distributed and by which mechanisms it is annihilated and transferred to heat. According to common thinking one supposes that the eddies of a certain size furnish their energy to the eddies of smaller size or may even create these smaller eddies. The process of the energy cascade ends at those scales, where the molecular dissipation due to internal friction commences to play a decisive role. At this lower end of the length scales of the eddies the turbulent kinetic energy is transformed into heat; it is lost for the motion of the fluid. KOLMOGOROV [121] has theoretically studied such *energy cascades*[3].

On the basis that the turbulent eddies are distributed over all length scales, it is more advantageous to study which energy might be stored in which class of dimensions of eddies, rather than to regard the turbulent kinetic energy as a spatially and temporally distributed function[4]. Thus, one studies the *wavenumber* dependence of the energy, expressed as the *energy density* $E(\mathbf{k})$, in wavenumber space, which is connected with the FOURIER transform of the velocity correlation function. Basically, the velocity is decomposed into its contributions with the various wavenumbers $\mathrm{k} := 2\pi/\lambda$,

[3] In this cascade model one imagines that the energy is transferred from large eddies to smaller eddies. This transfer takes place in form of nonlinear interactions or simply a breakup of large eddies into smaller ones. Eventually, the eddies are so small that a further formation of even smaller eddies is no longer possible and transfer of energy will be accomplished as generation of heat. The reverse process, in which large eddies grow at the cost of smaller ones, occasionally occurs as well and is connected with the term *inverse energy cascade.*

[4] We suppose here that the reader possesses some knowledge of the theory of waves in linear continuous media. A typical linear wave equation for a field quantity f is

$$\frac{\partial^2 f}{\partial t^2} = c^2 \Delta f \,, \tag{10.2.5}$$

where $c =$ constant is a velocity, the celerity with which the wave propagates. Δ denotes for LAPLACEian operator and we suppose the spatial dimension to be 1, 2 or 3. Solutions to the above equation are usually sought in terms of plane waves $f = f_0 \exp(i(\mathbf{k}\cdot\boldsymbol{x} - \omega t))$ in which $\mathbf{k}$ is called the wavenumber vector (with dimension 1/length), ω is the circular frequency, and the wave propagates in the direction of $\mathbf{k}$. The wavelength λ is $\lambda = 2\pi/|\mathbf{k}|$. Substituting the plane wave ansatz into (10.2.5) shows that c, $\mathbf{k}$ and ω are not independent, but related to one another by the so-called dispersion relation

$$c = f(\omega, |\mathbf{k}|), \qquad \text{here} \quad c = \frac{\omega}{|\mathbf{k}|} \,. \tag{10.2.6}$$

This velocity is also called phase velocity. For different values of the circular frequency ω_j (the set called the spectrum) there is a corresponding set of wavenumbers $|\mathbf{k}|_j$. There is another celerity arising in wave theory, the so-called group velocity, defined by

$$c_{\mathrm{gr}} := \frac{\partial \omega}{\partial |\mathbf{k}|}, \qquad \text{here} \quad c_{\mathrm{gr}} = c \,. \tag{10.2.7}$$

If $c_{\mathrm{gr}} = c$ holds, then the wave system is called non-dispersive, otherwise it is dispersive.

Solving the wave equation in a certain domain $\mathcal{D} \in \mathcal{R}^3$ subject to certain boundary conditions can be done by superposition, since the operator (10.2.5) and boundary conditions are linear. This superposition can be executed by superposing plane waves of all frequencies. This leads to the method of FOURIER transforms. Solving the formulated boundary value problem then also means selection of the frequencies ω_j such that the superposition of all plane waves having these frequencies solves the boundary value problem. This set of ω_j's defines the spectrum. It can be discrete, countably infinite or uncountably infinite.

For references on wave theory see e.g. WHITHAM [257]

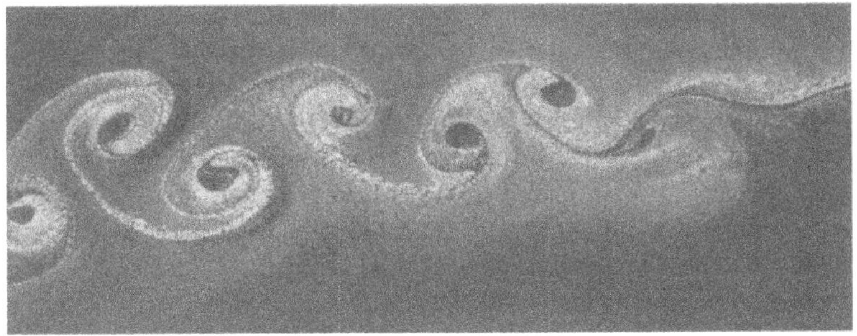

Fig. 10.4. Photograph of vortices (eddies) in form of a vortex street (Photo courtesy of R. PRIEM).

and one now studies the FOURIER transforms of the equations of motion. If $\mathbf{k}$ is the wavenumber vector with $|\mathbf{k}| = \mathrm{k}$ then the three dimensional FOURIER transform is given by

$$\mathcal{F}\boldsymbol{v} \equiv \boldsymbol{V}(\mathbf{k}, t) := \frac{1}{(2\pi)^3} \iiint_{-\infty}^{\infty} \boldsymbol{v}(\boldsymbol{x}, t) \mathrm{e}^{-\mathrm{i}\mathbf{k}\cdot\boldsymbol{x}} \, \mathrm{d}x \mathrm{d}y \mathrm{d}z \, . \tag{10.2.8}$$

The spectral energy density $E(\mathbf{k})$ is now interpreted as that part of the turbulent kinetic energy which lies in the interval $\mathbf{k}$ and $\mathbf{k} + \mathrm{d}\mathbf{k}$, so that one can write for the turbulent kinetic energy k

$$k = \int_{\mathcal{V}(\mathbf{k})} E(\mathbf{k}) \, (\mathrm{d}\mathbf{k})^3 \, , \tag{10.2.9}$$

where $(\mathrm{d}\mathbf{k})^3$ is the three-dimensional volume element in the momentum space $\mathcal{V}(\mathbf{k})$ of wavenumbers. For isotropic turbulence the spectral density can only depend on the modulus $|\mathbf{k}| = \mathrm{k}$ of the wavenumber vector $\mathbf{k}$, so that one may write (10.2.9) also as

$$k = \int_0^{\infty} \underbrace{4\,\pi\, E\,(|\mathbf{k}|)\; |\mathbf{k}|^2}_{\mathcal{E}(\mathrm{k})} \mathrm{dk} = \int_0^{\infty} \mathcal{E}(\mathrm{k}) \, \mathrm{dk} \, , \tag{10.2.10}$$

in which $\mathcal{E}(\mathrm{k})$ is known as isotropic, "radial" spectral energy density; it has the dimension $[L^3 T^{-2}]$. We shall no longer dwell upon these theoretical connections but only deduce qualitative inferences. If one considers the energy flow between such modes, i.e., the energy transfer from eddies of a certain size to the next smaller eddies, it is clear that this transfer is the result of a number of interactions between such eddies. If one follows it over a number of "stations", then as one progresses through this energy cascade, the memory how the turbulence might have been generated will be lost: In other words, the energy spectrum for large wavenumbers (small wave lengths) must be independent of its generation and therefore must assume an universal form as $k \to \infty$. This universal law is the law of KOLMOGOROV; it states that the spectral energy density $\mathcal{E}$ falls for large k as $k^{-5/3}$ ($k \mathrel{\hat{=}}$ wavenumber); this can easily be corroborated by methods of dimensional analysis (Exercise 3). This behaviour can experimentally be verified, see Fig. 10.5).

If the smallest eddies are to be independent of the generation of turbulence then their typical scales can only depend upon the properties of the fluid. Quantities which characterize the latter, are the molecular diffusivity $\nu\,[\mathrm{m^2s^{-1}}]$ and the turbulent dissipation rate $\varepsilon\,[\mathrm{m^2s^{-3}}]$ by which the smallest eddies are dissolved and transformed into heat. With these quantities simple considerations of dimensional analysis allow the definitions of length, time and velocity scales according to

$$L_K = \nu^{3/4}\varepsilon^{-1/4}\,, \quad T_K = \nu^{1/2}\varepsilon^{-1/2}\,, \quad V_K = \nu^{1/4}\varepsilon^{1/4}\,. \tag{10.2.11}$$

They are called the KOLMOGOROV *scales* and represent orders of magnitude for the spatial extent L_K of the smallest eddies, for the characteristic time T_K needed to dissolve the smallest eddies into heat and for the smallest velocities, that give rise to the occurrence of turbulence. In the ocean the above scales can be estimated as approximate values $\nu \approx 10^{-6}\,\mathrm{m^2\,s^{-1}}$ and $\varepsilon \approx 10^{-6}\,\mathrm{m^2\,s^{-3}}$, This yields for the KOLMOGOROV scales $L_K \approx 1\,\mathrm{mm}$, $T_K \approx 1\,\mathrm{s}$ and $V_K \approx 1\,\mathrm{mm\,s^{-1}}$.

The description of turbulence by its wave number content as introduced in (10.2.8) shall not be pursued here any further[5]. Instead, we shall consider in the ensuing analysis the decomposition of the fields into mean and fluctuating quantities, as it was previously introduced already. To this end a clear definition of an averaged field quantity and the averaging operations are needed.

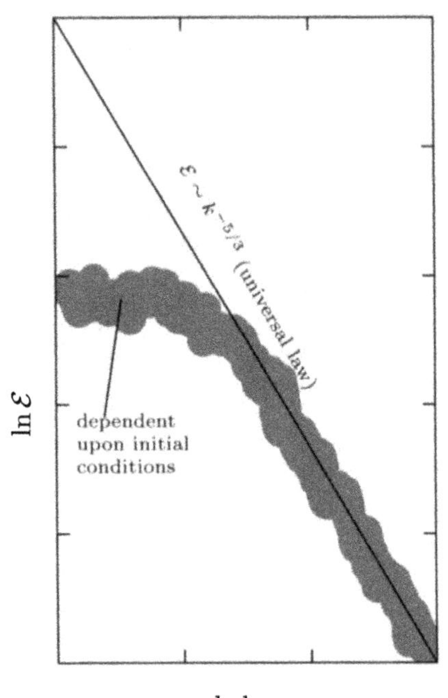

Fig. 10.5. Typical spectrum of the energy density (showing $\ln\mathcal{E}$ plotted against $\ln k$) for a turbulent flow in homogeneous turbulence. The band underlain with gray embraces the observations.

[5] The reader should be aware that this restriction corresponds to the omission of a large part of the physical description of turbulence. We constrain ourselves to this restriction simply because of reasons of space.

10.3 Filters

10.3.1 Definition of Filters

The most important operation in turbulence theory is the application of a filter in order to separate global flow processes from the turbulent contributions. Such filters are subject to their own restrictions, which we now proceed to present.

It was shown before, using the velocity field as an example, viz.,

$$\boldsymbol{v} = \langle \boldsymbol{v} \rangle + \boldsymbol{v}' , \tag{10.3.1}$$

that the velocity field $\langle \boldsymbol{v} \rangle$ can be additively decomposed into a mean contribution of the velocity $\langle \boldsymbol{v} \rangle$ and the superimposed fluctuating contribution $\boldsymbol{v}'$, which is indicated by a prime. The averaging operator or the filter $\langle \rangle$ has not yet been specified, and indeed a representation such as (10.3.1) is always possible with different filters. The most common ones that are of relevance in turbulence theory are the *spatial*, *temporal* and *statistical* averages. The latter says, that a quantity arises with a certain probability; the integral or summation over all probabilities then yields the statistical mean value.

For the *spatial filter* one integrates over a spherical (or nearly spherical) volume $V(\boldsymbol{x}, r)$ with center $\boldsymbol{x}$ and radius r and defines the mean value[6]

$$\langle u \rangle_R := \frac{1}{V} \int_{V(\boldsymbol{x},r)} u(\boldsymbol{x}', t) \,\mathrm{d}V' . \tag{10.3.2}$$

Here, all vortices with wave lengths smaller than $O(V^{1/3})$ are filtered away. The quantity $\langle u \rangle_R$ by itself is a function of position $\boldsymbol{x}$ and time t, but is equally dependent on the region of integration $V(\boldsymbol{x}, r)$, so that in general one has $\langle u \rangle_R = \langle u \rangle_R(\boldsymbol{x}, t; V)$.

For a *time filter* the averaging operation is performed over a time interval T with center t – the actual time –, so that

$$\langle u \rangle_T := \frac{1}{T} \int\limits_{t-T/2}^{t+T/2} u(\boldsymbol{x}, t') \,dt' . \tag{10.3.3}$$

Here, besides the dependence of the filtered quantity on $\boldsymbol{x}$ and t, this quantity will also depend on the interval T: $\langle u \rangle_T = \langle u \rangle_T(\boldsymbol{x}, t; T)$. In this filter, fluctuations with periods smaller than T are eliminated by the filter operation.

The conceptually most complex filter is the *statistical average*. For its definition it is assumed that a variable $u(\boldsymbol{x}, t)$ assumes a certain value with

[6] The volume under consideration is a compact volume, defining the region of influence, that is not exactly spherical but which is sufficiently compact that a typical diameter can be defined for a sphere that may replace the actual volume.

a given probability. If $\wp$ is this probability, then $u(\boldsymbol{x}, t, \wp)$ denotes the probability density with which the function $u(\boldsymbol{x}, t)$ assumes the value assigned to $\wp$. The statistically most probable value of $u(\boldsymbol{x}, t)$ is obtained by summation (here integration) over all probabilities, namely,

$$\langle u \rangle_S := \int u(\boldsymbol{x}, t, \wp)\, \mathrm{d}\wp \,. \tag{10.3.4}$$

Clearly,

$$\int_0^\infty \mathrm{d}\wp = 1 \tag{10.3.5}$$

must hold. One may interpret the statistical mean as the average over an infinite number of realizations of identically conducted experiments from which $u(x, t)$ is obtained.

Since one ordinarily assumes that the various filters are equivalent, which however is only approximately true, one does not differentiate between spatial, temporal or statistical filters; the corresponding indices $\langle\cdot\rangle_R$, $\langle\cdot\rangle_S$ and $\langle\cdot\rangle_T$ will henceforth be omitted. Only when discussing differences of the various filters shall we make the distinction visible.

10.3.2 Properties of Filters

The above defined filters must or ought to satisfy the following properties, if they are meaningfully applied to physical processes:

1. **Linearity**
   Filters are linear operations, i. e., for two quantities, u and w of a turbulent field and a real number a, one has

   $$\langle u + aw \rangle = \langle u \rangle + a\langle w \rangle \,. \tag{10.3.6}$$

   All three filters introduced above satisfy this property because integration is an additive operation.
2. **Commutability with differentiations**
   The filter operation must commute with any temporal or spatial differentiation, viz.,

   $$\langle \partial u \rangle = \partial \langle u \rangle \,, \quad \text{where} \quad \partial \in \{\frac{\partial}{\partial x}, \frac{\partial}{\partial y}, \frac{\partial}{\partial z}, \frac{\partial}{\partial t}\} \,. \tag{10.3.7}$$

   We shall now show that applying the temporal filter to a physical quantity commutes with the spatial or temporal differentiation. Indeed,

$$\begin{aligned}
\langle\frac{\partial u}{\partial x}\rangle_T &= \frac{1}{T}\int_{t-T/2}^{t+T/2}\frac{\partial u(\boldsymbol{x},t')}{\partial x}\,dt' \\
&= \frac{\partial}{\partial x}\left(\frac{1}{T}\int_{t-T/2}^{t+T/2} u(\boldsymbol{x},t')\,dt'\right) \\
&= \frac{\partial\langle u\rangle_T}{\partial x}\,, \\
\langle\frac{\partial u}{\partial t}\rangle_T &= \frac{1}{T}\int_{t-T/2}^{t+T/2}\frac{\partial u(\boldsymbol{x},t')}{\partial t'}\,dt' \\
&= \frac{1}{T}\left[u(\boldsymbol{x},t+T/2)-u(\boldsymbol{x},t-T/2)\right] \\
&= \frac{\partial}{\partial t}\left(\frac{1}{T}\int_{t-T/2}^{t+T/2} u(\boldsymbol{x},t')\,dt'\right) \\
&= \frac{\partial\langle u\rangle_T}{\partial t}\,.
\end{aligned} \tag{10.3.8}$$

For the spatial filter and the statistical filter the corroboration of the commutation rules is similar.

3. **Invariance under multifold averaging**
   An ideal filter should not affect a filtered quantity, i.e.,

$$\langle\langle u\rangle\rangle = \langle u\rangle\,, \tag{10.3.9}$$

or more generally and as a consequence of this

$$\langle\langle\ldots\langle u\rangle\ldots\rangle\rangle = \langle u\rangle\,. \tag{10.3.10}$$

This property is exactly satisfied only by the statistical averaging; indeed,

$$\langle\langle u\rangle\rangle_S = \iint u(x,t,\wp)\,\mathrm{d}\wp\,\mathrm{d}\wp' = \langle u\rangle_S\int \mathrm{d}\wp' = \langle u\rangle_S\,, \tag{10.3.11}$$

in which the integral over all probabilities in (10.3.5) has been normalized to be unity.

In general, neither the spatial nor the temporal averaging satisfy this property. One recognizes this e.g. by the fact that the quantities that are averaged in this way usually depend upon the choice of the time and length interval over which the averaging operation is performed. Only with the restricting condition that the filter period T agrees with the period of a periodic physical quantity does the temporal average satisfy this condition; in this case the averaging results in a constant value of the averaged quantity (Exercise 6).

The request that the multifold average is invariant implies drastic consequences. A filter, which does not satisfy this property, does not imply the vanishing of the averaged fluctuations,

$$\langle\langle u\rangle\rangle \neq \langle u\rangle \quad \not\Rightarrow \quad \langle u'\rangle = 0\,. \tag{10.3.12}$$

Conversely, if $\langle u'\rangle = 0$, then necessarily $\langle\langle u\rangle\rangle = \langle u\rangle$.

4. **Mean value of a product**
   The above property of the invariance of a multifold average implies the following formula for the mean value of a product

$$\langle w\langle u\rangle\rangle = \langle w\rangle\langle u\rangle \ , \tag{10.3.13}$$

   and is again exactly satisfied only by the statistical mean value.

For the following we shall now assume that the chosen filter satisfies all three conditions. Under special assumptions – these are the stationary homogeneous conditions of turbulence – the temporal and spatial averages do also satisfy all three properties. Under such circumstances the three averages are arbitrarily interchangeable. The hypothesis is called **ergodic hypothesis**. It is often used in experiments, if results are being exploited, even in cases when no stationary and homogeneous turbulence prevails. Temporal and spatial averages are then identified with ensemble averages, which is not exact but at best an approximation.

We wish to close this discussion with some remarks. In modern turbulence theory models are being developed which request the invariance of the multifold filtering as well as those which negate it. Into the class of the first belong the so-called **R**EYNOLDS–**A**VERAGED–**N**AVIER–**S**TOKES (**RANS**) models, which we shall discuss in detail. They suppose that the physics of turbulence has statistical nature. To the second class belong all those models, for which $\langle u'\rangle$ may not vanish. In these latter models the large vortices are resolved by the averaged equations, whilst the small eddies are parameterised. These more general models are summarised under the name **L**arge **E**ddy **S**imulation (**LES**) and are, as far as their accuracy is concerned, between **RANS** and **DNS** (Direct Numerical Simulation).

10.3.3 Computation Rules

If a filter satisfies all three properties, then for two statistically varying field quantities u, v, one may prove the following computation rules:

$$\begin{aligned}
&\langle\langle u\rangle\rangle = \langle u\rangle \ , \\
&\langle u'\rangle = \langle u\rangle' = 0 \ , \\
&\langle\langle u\rangle v\rangle = \langle u\rangle\langle v\rangle \ , \\
&\langle\langle u\rangle v'\rangle = 0 \ , \\
&\langle uv\rangle = \langle u\rangle\langle v\rangle + \langle u'v'\rangle \ .
\end{aligned} \tag{10.3.14}$$

Notice that $\langle u'\rangle$ vanishes but $\langle u'v'\rangle$ does not: $\langle u'v'\rangle \neq 0$. This quantity is sometimes called *correlation* or *moment* of the fluctuating quantities u' and v'; in turbulence theory these moments play a significant role.

10.4 Correlations

In the following some statistical facts for correlations of the fluctuating motion are analyzed. To this end the definition of the stationary homogeneous turbulence is needed. Later in this chapter also some symmetry properties will be assigned to the turbulence. Here, we shall only be concerned with *isotropic* turbulence, i.e., turbulence for which quantities are invariant with respect to arbitrary rotations of the frame of reference.

The introduction of such special properties of turbulence exerts a direct influence upon the correlation functions, which below will equally be briefly discussed.

First let us give a definition of homogeneous turbulence. The turbulence of a flow is called *homogeneous*, if all mean values and momenta of the velocity and its gradient are independent of position. If, in addition, there is equally no time dependence for these quantities, then one speaks of *stationary homogeneous turbulence.*

10.4.1 Correlations for Homogeneous Isotropic Turbulence

In turbulence theory moments of second and higher order for the fluctuation velocities are often arising mathematical objects. Let us briefly study these for the case of isotropic turbulence. To this end, consider a spatial point $\boldsymbol{x}$ at the fixed time t and determine the mean value of two fold, three fold etc. dyadic products of the velocity fluctuations at $\boldsymbol{x}$. One then obtains the moments

$$
\begin{aligned}
\boldsymbol{Q}(\boldsymbol{x},t) &:= \langle \boldsymbol{v}'(\boldsymbol{x},t) \otimes \boldsymbol{v}'(\boldsymbol{x},t) \rangle \ , \\
\boldsymbol{Q}^{(3)}(\boldsymbol{x},t) &:= \langle \boldsymbol{v}'(\boldsymbol{x},t) \otimes \boldsymbol{v}'(\boldsymbol{x},t) \otimes \boldsymbol{v}'(\boldsymbol{x},t) \rangle \ , \\
\vdots \quad & \qquad \vdots \\
\boldsymbol{Q}^{(n)}(\boldsymbol{x},t) &:= \underbrace{\langle \boldsymbol{v}'(\boldsymbol{x},t) \otimes \boldsymbol{v}'(\boldsymbol{x},t) \otimes \ldots \otimes \boldsymbol{v}'(\boldsymbol{x},t) \rangle}_{\text{n-fold}} \ .
\end{aligned}
\tag{10.4.1}
$$

$\boldsymbol{Q}$, $\boldsymbol{Q}^{(3)}$, $\boldsymbol{Q}^{(n)}$ are called *second, third* and *n-th statistical moments* of the velocity fluctuations. Alternatively, they are called *two fold, three fold* and *n-fold correlations.* Since $\boldsymbol{x}$ and t are both the same for all fluctuating quantities in (10.4.1), the quantities $\boldsymbol{Q}(\boldsymbol{x},t)$, $\boldsymbol{Q}^{(3)}(\boldsymbol{x},t)$, etc. are also called *single point correlations.*

Generally, one is dealing with two, three etc. different points at two, three, etc., different times and then speaks of *multi-point correlations.* The expressions analogous to (10.4.1) are then given by

$$
\begin{aligned}
&\boldsymbol{Q}(\boldsymbol{x}^1,\boldsymbol{x}^2,t^1,t^2) := \langle \boldsymbol{v}'(\boldsymbol{x}^1,t^1)\otimes \boldsymbol{v}'(\boldsymbol{x}^2,t^2)\rangle\,,\\
&\boldsymbol{Q}^{(3)}(\boldsymbol{x}^1,\boldsymbol{x}^2,\boldsymbol{x}^3,t^1,t^2,t^3)\\
&\qquad := \langle \boldsymbol{v}'(\boldsymbol{x}^1,t^1)\otimes \boldsymbol{v}'(\boldsymbol{x}^2,t^2)\otimes \boldsymbol{v}'(\boldsymbol{x}^3,t^3)\rangle\,,\\
&\boldsymbol{Q}^{(n)}(\boldsymbol{x}^1,\boldsymbol{x}^2,\ldots,\boldsymbol{x}^n,t^1,t^2,\ldots,t^n)\\
&\qquad := \underbrace{\langle \boldsymbol{v}'(\boldsymbol{x}^1,t^1)\otimes \boldsymbol{v}'(\boldsymbol{x}^2,t^2)\otimes\ldots\otimes \boldsymbol{v}'(\boldsymbol{x}^n,t^n)\rangle}_{n-\text{fold}}\,.
\end{aligned}
\tag{10.4.2}
$$

Generally, we shall encounter moments up to order $n = 4$.

Let us now demonstrate, how the second moment $\boldsymbol{Q}(\boldsymbol{x},t)$ simplifies for isotropic turbulence. To this end, the components of the tensor of the single point correlations of the velocity must be invariant under arbitrary rotations and mirror reflections of the coordinate axes. Then it easily follows that $\boldsymbol{Q}$ possesses only a single independent component, for one has

$$
\begin{aligned}
&\langle v'_1 v'_2\rangle = \langle v'_1 v'_3\rangle = \langle v'_2 v'_3\rangle = 0\,,\\
&\langle u^2\rangle := \langle {v'_1}^2\rangle = \langle {v'_2}^2\rangle = \langle {v'_3}^2\rangle\,.
\end{aligned}
\tag{10.4.3}
$$

Here $\langle u^2\rangle$ is the average of the square of the fluctuation–velocity component in an arbitrary direction.

The isotropy conditions just established will now be proven. Starting point for this proof is the fact that for isotropy the components of the correlation tensor do not change, if one performs an arbitrary rotation of the system. If $\boldsymbol{O}$ is a (temporally constant) orthogonal transformation, then, in the rotated system, which will be designated with an asterisk, we must have

$$
\langle {v'_i}^* {v'_j}^*\rangle = \langle O_{ik} v'_k\, O_{jl} v'_l\rangle \overset{!}{=} \langle v'_i v'_j\rangle\,. \tag{10.4.4}
$$

If we consider as an example a rotation about the x_3-axis by an angle $\pi/2$, see Fig. 10.6, then the new and old coordinates are related by

$$
x_1^* = x_2\,,\quad x_2^* = -x_1\,,\quad x_3^* = x_3\,, \tag{10.4.5}
$$

and the new and old velocity components are connected by

$$
{v'_1}^* = v'_2\,,\quad {v'_2}^* = -v'_1\,,\quad {v'_3}^* = v'_3\,. \tag{10.4.6}
$$

With the prerequisite that all single point correlations are invariant under arbitrary rotations and mirror reflections, (10.4.4) implies

$$
\begin{aligned}
&\langle {v'_1}^* {v'_1}^*\rangle \overset{(1)}{=} \langle v'_2 v'_2\rangle \overset{(2)}{=} \langle v'_1 v'_1\rangle\,,\\
&\langle {v'_1}^* {v'_2}^*\rangle \overset{(1)}{=} -\langle v'_1 v'_2\rangle \overset{(2)}{=} \langle v'_1 v'_2\rangle = 0\,.
\end{aligned}
\tag{10.4.7}
$$

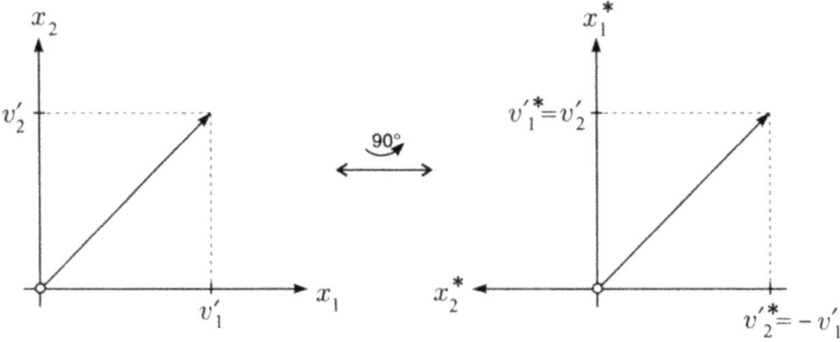

Fig. 10.6. A rotation of a coordinate system about the x_3–axis by an angle $\pi/2$ changes the velocity components in the directions of x_1 and x_2.

The symbols $\overset{(i)}{=}$ indicate which of (10.4.4) is used. (10.4.7) thus implies in particular that $\langle v'_1 v'_2 \rangle = 0$. An analogous rotation about the x_2-axis, also about an angle $\pi/2$, now leads to the result

$$\begin{aligned} \langle {v'_1}^* {v'_1}^* \rangle &\overset{(1)}{=} \langle v'_3 v'_3 \rangle \overset{(2)}{=} \langle v'_1 v'_1 \rangle \;, \\ \langle {v'_1}^* {v'_3}^* \rangle &\overset{(1)}{=} -\langle v'_1 v'_3 \rangle \overset{(2)}{=} \langle v'_1 v'_3 \rangle = 0 \;, \\ \langle {v'_1}^* {v'_2}^* \rangle &\overset{(2)}{=} \langle v'_2 v'_3 \rangle = 0 \;. \end{aligned} \tag{10.4.8}$$

Therefore,

$$\begin{aligned} \langle v'_1 v'_1 \rangle &= \langle v'_2 v'_2 \rangle = \langle v'_3 v'_3 \rangle \;, \\ \langle v'_1 v'_2 \rangle &= \langle v'_1 v'_3 \rangle = \langle v'_2 v'_3 \rangle = 0 \;. \end{aligned} \tag{10.4.9}$$

With the isotropy conditions just derived, the tensor $\boldsymbol{Q}$ may be expressed as

$$\boldsymbol{Q} = \tfrac{2}{3} k \boldsymbol{I} \;, \tag{10.4.10}$$

in which k is the *turbulent kinetic energy*

$$k := \tfrac{1}{2} \langle \boldsymbol{v'}^2 \rangle = \tfrac{1}{2} \sum_{i=1}^{3} \langle {v'_i}^2 \rangle = \tfrac{3}{2} \langle u^2 \rangle \;. \tag{10.4.11}$$

For pure isotropy the tensor of the velocity correlations, $\boldsymbol{Q}$ can only depend on a single scalar quantity, that can be interpreted as the turbulent kinetic energy.

In a similar way, one may also prove that for isotropic turbulence the higher moments $\boldsymbol{Q}^{(3)}$ and $\boldsymbol{Q}^{(4)}$ take the forms (Exercise 7)

$$\begin{aligned} &\boldsymbol{Q}^{(3)} = \boldsymbol{0} \,, \\ &(\boldsymbol{Q}^{(4)})_{ijkl} = \langle u'^2 v'^2 \rangle \delta_{ij}\delta_{kl} + \tfrac{1}{2}(\langle u'^4 \rangle - \langle u'^2 v'^2 \rangle)(\delta_{ik}\delta_{jl} + \delta_{il}\delta_{jk}) \,; \end{aligned} \tag{10.4.12}$$

u' and v' are fluctuations of any two velocity components, which are perpendicular to one another. The twofold correlations or the moments of order two are those that arise more often than the others.

10.4.2 EULERian Length and Time Scales

With the aid of the correlation functions defined above several scales can be constructed, on which turbulence is active. To this end, we consider two different spatial points at two different times (EULER representation) and compute the correlation of the fluctuation velocity according to (10.4.2). The *correlation coefficient* is defined as the following correlation quantity, normalized by the mean values of the square of the velocity components, viz.,

$$\begin{aligned} \varrho_{ij}(\boldsymbol{x}^1, \boldsymbol{x}^2, t^1, t^2) &:= \frac{\langle v'_i(\boldsymbol{x}^1, t^1) v'_j(\boldsymbol{x}^2, t^2) \rangle}{\sqrt{v'_k(\boldsymbol{x}^1, t^1)\, v'_k(\boldsymbol{x}^1, t^1)}\sqrt{v'_l(\boldsymbol{x}^2, t^2)\, v'_l(\boldsymbol{x}^2, t^2)}} \\ &= \frac{Q_{ij}(\boldsymbol{x}^1, \boldsymbol{x}^2, t^1, t^2)}{\sqrt{v'_k(\boldsymbol{x}^1, t^1)\, v'_k(\boldsymbol{x}^1, t^1)}\sqrt{v'_l(\boldsymbol{x}^2, t^2)\, v'_l(\boldsymbol{x}^2, t^2)}} \end{aligned} \tag{10.4.13}$$

This quantity possesses, as $\boldsymbol{Q}$ by itself, objective tensor properties. In other words $\boldsymbol{Q}^{(n)}$ and ϱ_{ij} transform under EUCLIDian transformations as objective tensors; this will further be scrutinized below.

With growing distance between the spatial points $\boldsymbol{x}^1$ and $\boldsymbol{x}^2$ and between the times t^1 and t^2, respectively, one would expect that the signals $\boldsymbol{v}'(\boldsymbol{x}^1, t^1)$ and $\boldsymbol{v}'(\boldsymbol{x}^2, t^2)$ become more and more uncorrelated, so that the correlation coefficients tend to zero with growing distance or time interval; in other words,

$$|\boldsymbol{x}^1 - \boldsymbol{x}^2| \to \infty \Rightarrow \varrho_{ij} \to 0 \,, \quad |t^1 - t^2| \to \infty \Rightarrow \varrho_{ij} \to 0 \,. \tag{10.4.14}$$

Conversely, if $\boldsymbol{x}^1$ and $\boldsymbol{x}^2$ and the times t^1 and t^2 come closer, $\boldsymbol{x}^1 - \boldsymbol{x}^2 \to \boldsymbol{0}$ $t^1 - t^2 \to 0$ then one expects the correlation to approach a maximum. If one supposes the fluctuations of velocity components in directions orthogonal to one another to be uncorrelated, there follows $\varrho_{ij}(\boldsymbol{x}, \boldsymbol{x}, t, t) = \delta_{ij}$. More likely is however a complete correlation so that $\varrho_{ij}(\boldsymbol{x}, \boldsymbol{x}, t, t)$ is a matrix of which all entries are occupied by unity.

Consider now the position of two points in space (in time, analogous relations apply) as displayed in Fig. 10.7. For these points, one has, owing to Fig. 10.7, the representations

$$\bar{\boldsymbol{r}} := \tfrac{1}{2}(\boldsymbol{x}^1 + \boldsymbol{x}^2) \,, \quad \boldsymbol{r} := \boldsymbol{x}^1 - \boldsymbol{x}^2 \,. \tag{10.4.15}$$

Similar formulas hold for the times

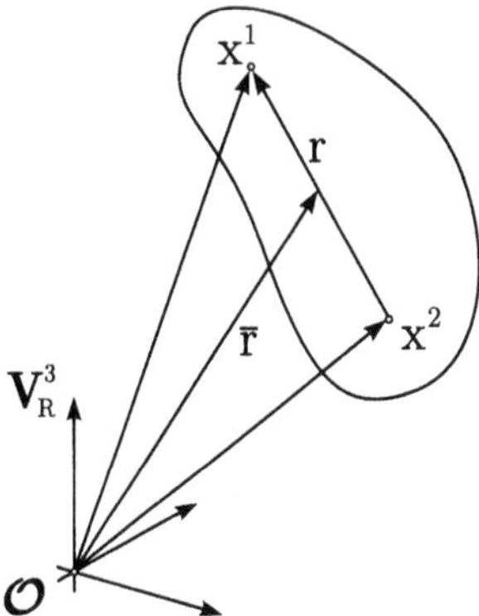

Fig. 10.7. Relative position of two points, expressed also by their distance vector $\boldsymbol{r}$ and the vector of their mean position, $\bar{\boldsymbol{r}}$.

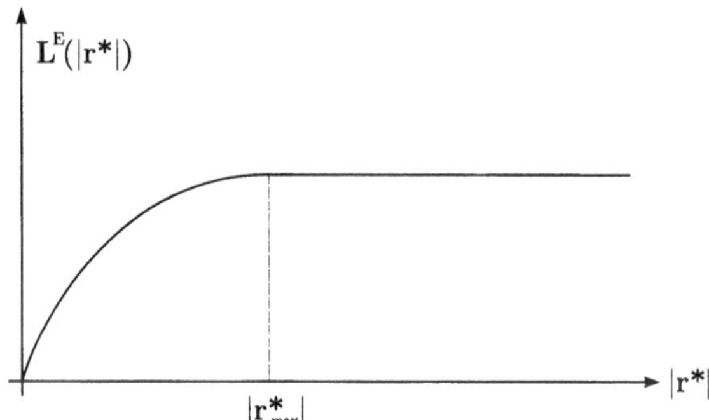

Fig. 10.8. Qualitative dependence of the EULERian length scale as a function of the "correlation lengths" $\boldsymbol{r}^*$.

$$\bar{t} := \tfrac{1}{2}(t^1 + t^2)\,, \quad \tau := t^1 - t^2\,. \tag{10.4.16}$$

If one now considers the correlation coefficient at equal times (e.g. $t^1 - t^2 = 0$) centered at the origin $\bar{\boldsymbol{r}} = \boldsymbol{0}$ – this means that we only account for the distance between the two points – one may define a (tensorial) length scale

$$L^E_{ij}(|\boldsymbol{r}^*|) := \int\limits_{0<|\boldsymbol{r}|<|\boldsymbol{r}^*|} \varrho_{ij}(\boldsymbol{r})\mathrm{d}|\boldsymbol{r}|\,. \tag{10.4.17}$$

For each fixed i and j this length, interpreted as a function of $|\boldsymbol{r}^*|$, behaves qualitatively as shown in Fig. 10.8. One would expect that the fluctuations of the velocity are practically uncorrelated for $|\boldsymbol{r}^*| > |\boldsymbol{r}^*_{\max}|$ so that $L^E_{ij}(|\boldsymbol{r}^*| > |\boldsymbol{r}^*_{\max}|)$ assumes with sufficient accuracy the asymptotic value $L^E_{ij} := L^E_{ij}(\infty)$. An example may explain the significance of this length.

Let a flow in the x_1-direction be given (Fig. 10.9); select two points which are so positioned that their distance vector points in the x_2-direction. Since

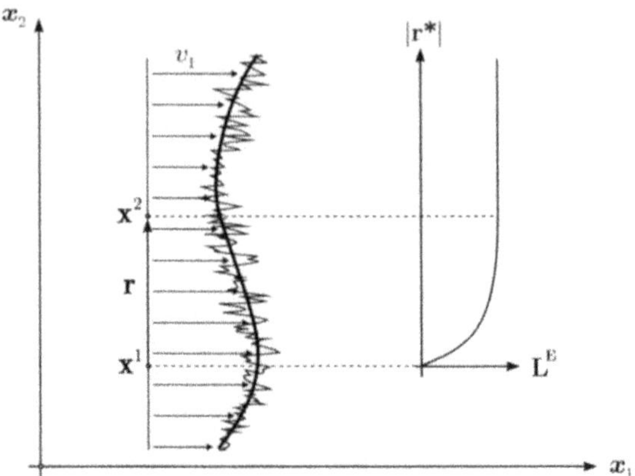

Fig. 10.9. Example explaining EULERian length scales: Two points in a flow field in the x_1-direction, which have the same x_1 coordinate but are separated in the x_2-direction. The main figure shows a "snap shot" of the velocity field and the mean velocity for fixed x_1. The inset figure gives the EULERian length scale as a function of $|\boldsymbol{r}|$.

the flow is only in a single direction, only the correlation built with the fluctuation velocity v'_1 can be different from zero. Thus, the EULERian length scale of this problem is given by

$$L_{11}^E = \int_0^\infty \varrho_{11}(r_2)\mathrm{d}r_2 \,. \tag{10.4.18}$$

This length is a measure for the fact, how much two points must be apart in the x_2-direction, so that fluctuations of their velocities are no longer correlated; for instance, L_{11}^E can be interpreted as a measure for the extent of a vortex or eddy.

Analogously, one can introduce a time scale. To that end, one considers the correlation for the same spatial point, $\boldsymbol{x}^1 = \boldsymbol{x}^2$ and asks for the integral correlation for a temporal distance (i.e., for $\bar{t} = 0$). This leads to the EULERian time scale

$$T^E := \int_0^\infty \varrho(\tau)\mathrm{d}\tau \,, \tag{10.4.19}$$

in which $\varrho(\cdot)$ is any one component of the correlation coefficient. This time measure indicates how much time must elapse, so that the fluctuation velocities at the two times at the same point are uncorrelated. This scale defines the "life time" of the eddies. This time scale expresses how much time must pass until a velocity fluctuation in the past has been forgotten.

The tensor of the twofold velocity correlations plays an important role in turbulence theory. It corresponds to the REYNOLDS *stresses*, which couple the momentum balance (of the mean fields) with the turbulence structure.

Naturally, one can hardly assume that this quantity is isotropic in the above sense. Only two statements can be made in an a priori fashion. First, the tensor must be symmetric by definition and, second, it ought to satisfy the rule of objectivity[7]. However, before we present the concept of the REYNOLDS stress models, we wish to list the field equations, on the basis of which the averaged equations of turbulent motion are derived.

10.5 Equations of Motion

It is the goal of the study of turbulent flows in the geophysical context to comprehend the effect of the small scale eddies on the mean motion. In the relatively short-lived wind-induced currents in lakes a central question is thereby the description of the exchange processes of momentum and energy, whence the parameterization of the corresponding diffusivities. Within time scales of seasonal variations one expects from a turbulence model in oceanography and physical limnology that computational models for the temporal evolution of the thermal stratification are obtained. The derivation of such models is generally very much involved; alternatively, for an adequate understanding of the derivation of turbulence models this general description is not necessary. We shall therefore restrict ourselves first to the balance laws of a fluid with constant density and shall only afterwards account for the density stratification as it naturally arises in the ocean, in a lake or in the atmosphere. We shall also briefly touch upon the formulation of REYNOLDS stress models for compressible fluids.

10.5.1 Material Equations

The material properties of water (and also air) can be assumed as those of a NEWTON*ian fluid with* FOURIER *heat conduction.* Stress tensor and heat flux vector are therefore given by

$$\boldsymbol{t} = \Big(-p + \big(\kappa - \tfrac{2}{3}\mu\big)\,\mathrm{tr}\,\boldsymbol{D}\Big)\boldsymbol{I} + 2\mu\boldsymbol{D}\,, \tag{10.5.1}$$

$$\boldsymbol{D} = \tfrac{1}{2}(\boldsymbol{L} + \boldsymbol{L}^T) \quad \text{with} \quad \boldsymbol{L} := \mathrm{grad}\,\boldsymbol{v} \tag{10.5.2}$$

and

$$\boldsymbol{q} = -\lambda\,\mathrm{grad}\,\Theta\,. \tag{10.5.3}$$

If, moreover, density preserving is supposed, then the stress tensor reduces to

$$\boldsymbol{t} = -p\boldsymbol{I} + \boldsymbol{t}^E = -p\boldsymbol{I} + 2\mu\boldsymbol{D}\,. \tag{10.5.4}$$

[7] What is meant here are the transformation properties under the EUCLIDian transformation group and not the rule of material objectivity. The latter is not appropriate any how, since turbulent behaviour is not based on material properties.

In these formulas, κ, μ, λ are the volume viscosity, the shear viscosity and the heat conduction coefficient, which all may be functions of density and temperature or pressure and temperature; often, however, such dependences are ignored. With these material equations the balance equations are closed.

10.5.2 Balances of Mass and Momentum

NAVIER–STOKES **Equations for a Fluid of Constant Density** The field equations of a NAVIER–STOKES fluid are obtained by substituting into the balance laws of mass and momentum, derived in Chap. 2, the above material equations. We shall momentarily ignore volume forces and the additional forces due to the non-inertial effects of the frame of reference. These effects will be incorporated lateron when the k–ε model is introduced and the density variations are accounted for.

If one defines with

$$p_{\mathrm{r}} = \frac{p}{\rho} \tag{10.5.5}$$

the reduced pressure and replaces the dynamic viscosity μ by the kinematic viscosity $\nu = \mu/\rho$ (which will be assumed to be constant), then the equations of motion are obtained with the aid of which the construction of turbulence models for density preserving fluids will be demonstrated. The equations are

$$\begin{aligned} &\operatorname{div} \boldsymbol{v} = 0 \,, \\ &\frac{\partial \boldsymbol{v}}{\partial t} + (\operatorname{grad} \boldsymbol{v})\boldsymbol{v} = -\operatorname{grad} p_{\mathrm{r}} + \nu \operatorname{div} \operatorname{grad} \boldsymbol{v} \,. \end{aligned} \tag{10.5.6}$$

For later derivations these equations are further changed. Since $\operatorname{div} \boldsymbol{v} = 0$, the identity

$$(\operatorname{grad} \boldsymbol{v})\boldsymbol{v} = \operatorname{div} \left(\boldsymbol{v} \otimes \boldsymbol{v}\right) - \boldsymbol{v} \operatorname{div} \boldsymbol{v} = \operatorname{div} \left(\boldsymbol{v} \otimes \boldsymbol{v}\right) \tag{10.5.7}$$

may be used in the momentum equation to obtain

$$\frac{\partial \boldsymbol{v}}{\partial t} + \operatorname{div} \left(\boldsymbol{v} \otimes \boldsymbol{v}\right) + \operatorname{grad} p_{\mathrm{r}} - \nu \operatorname{div} \operatorname{grad} \boldsymbol{v} = \boldsymbol{0} \,, \tag{10.5.8}$$

which has the form of a local balance equation in conservative form.

10.5.3 Energy Balance

An additional field equation at the disposal is the first law of thermodynamics, which may be written as a field equation for the temperature. For a density preserving viscous fluid as it was considered in Sect. 5.7, the internal energy is only a function of temperature. The equation then takes the form

$$\frac{\partial \Theta}{\partial t} + \operatorname{div} \left(\Theta \boldsymbol{v}\right) = \chi^{(\Theta)} \operatorname{div} \operatorname{grad} \Theta + \frac{1}{\rho c_v} \varphi + \frac{\mathfrak{r}}{c_v} \,, \tag{10.5.9}$$

in which the dissipation function φ is given by

$$\varphi = \mathrm{tr}(\boldsymbol{t}^E \boldsymbol{D}) = 2\mu \boldsymbol{D} \cdot \boldsymbol{D} = \frac{\mu}{2} |\boldsymbol{L} + \boldsymbol{L}^T|^2 \; ; \tag{10.5.10}$$

$\mathfrak{r}$ denotes the specific radiation and

$$\chi^{(\Theta)} := \frac{\lambda}{\rho c_v} \tag{10.5.11}$$

is the thermal diffusivity with dimension $[\mathrm{m}^2\mathrm{s}^{-1}]$ and c_v is the specific heat (at constant volume), both of which were supposed to be constant in the derivation (10.5.9).

10.5.4 Mixtures

If in addition to the velocity and temperature fields, one also wishes to study the dispersion of a passive tracer, i.e., a quantity suspended in the fluid in very small concentration – for instance a pollutant in water or air – the above system of field equations must be extended by an equation modelling the dispersion of the tracer mass. If only a single tracer is present the resulting system is a two component mixture, a so-called binary mixture. The balance laws, written down above for mass, momentum and energy (heat), must then be interpreted as balance laws for the mixture as a whole. The density of the mixture, ρ is obtained in this case as the sum of the partial densities of the individual constituents $\rho = \rho_{\mathrm{Fluid}} + \rho_{\mathrm{Tracer}}$, where the partial densities are the densities measured within the mixture volume. Analogously, $\boldsymbol{v}$ is the so-called *barycentric velocity*, defined by $\rho \boldsymbol{v} = \rho_{\mathrm{Fluid}} \boldsymbol{v}_{\mathrm{Fluid}} + \rho_{\mathrm{Tracer}} \boldsymbol{v}_{\mathrm{Tracer}}$. The different motions of the two constituents – fluid and tracer – is accounted for by a diffusion equation for the concentration c of the tracer component, where $c = \rho_{\mathrm{Tracer}}/\rho$ is the mass ratio of the tracer mass to the mixture mass. This equation has the form of the heat equation and is called FICK's *second law*

$$\frac{\partial c}{\partial t} + \mathrm{div}\,(c\boldsymbol{v}) = -\,\mathrm{div}\,\boldsymbol{j} + \phi^{(c)} \; , \tag{10.5.12}$$

in which

$$\boldsymbol{j} = c(\boldsymbol{v}_{\mathrm{Tracer}} - \boldsymbol{v})$$

is the diffusive flux , and the source $\phi^{(c)}$ describes possible interactions but will be set equal to zero here. The diffusive mass flux (of the tracer relative to the motion of the mixture) is written as a gradient law just as the FOURIER law for the heat flux. More specifically, it is proportional to the concentration gradient and directed from high values of c to small values. This constitutive relation is called FICK's *first law* and reads

$$\boldsymbol{j} = -\chi^{(c)}\,\mathrm{grad}\,c \; , \tag{10.5.13}$$

in which $\chi^{(c)}$ is the mass diffusivity having dimension $[\mathrm{m}^2\mathrm{s}^{-1}]$.

The diffusion equation (10.5.12) has been derived from a mass balance of a tracer substance; it can be written down in that form for an arbitrary number of tracers, see Chap. 7. Instead of (10.5.12) one then arrives at the equation

$$\frac{\partial c_\alpha}{\partial t} + \operatorname{div}\left(c^\alpha \boldsymbol{v}\right) = -\operatorname{div}\boldsymbol{j}^\alpha + \phi^{(c_\alpha)} \tag{10.5.14}$$

with

$$\boldsymbol{j}^\alpha = -\sum_\beta \chi^{\alpha\beta}\operatorname{grad} c_\beta \tag{10.5.15}$$

in which $\chi^{\alpha\beta} = \chi^{\beta\alpha}$. This symmetry condition is a consequence of thermodynamics (Onsager relations), and the matrix is generally diagonally dominant.

10.5.5 Summary of Field Equations

Here, at this place, we shall briefly list the equations which are used in the derivation or formulation of the various different turbulence models. For a fluid with constant density these equations were given above; they read for a

NEWTONian Fluid of Constant Density [8]

- Continuity equation

$$\operatorname{div}\boldsymbol{v} = 0\,,$$

- Momentum balance equation

$$\frac{\partial \boldsymbol{v}}{\partial t} + \operatorname{div}\left(\boldsymbol{v}\otimes\boldsymbol{v}\right) = -\operatorname{grad} p_{\mathrm{r}} + \nu\operatorname{div}\operatorname{grad}\boldsymbol{v} + \boldsymbol{g}\,,$$

- Heat equation (10.5.16)

$$\frac{\partial \Theta}{\partial t} + \operatorname{div}\left(\Theta\boldsymbol{v}\right) = \chi^{(\Theta)}\operatorname{div}\operatorname{grad}\Theta + \frac{1}{\rho_0 c_v}\varphi + \frac{1}{c_v}\mathfrak{r}\,,$$

- Diffusion equation

$$\frac{\partial c}{\partial t} + \operatorname{div}\left(c\boldsymbol{v}\right) = \chi^{(c)}\operatorname{div}\operatorname{grad} c + \phi^{(c)}$$

and constitute equations for the field quantities $\boldsymbol{v}$, p, Θ and c.

For the k–ε model, that will be introduced later on and holds for a fluid enjoying weak density variations, the above description is not sufficient. In

[8] Equations (10.5.16) are written down as referred to an inertial frame of reference. When they are referred to a steadily rotating system then the acceleration is given by

$$\frac{\partial \boldsymbol{v}}{\partial t} + \operatorname{div}\left(\boldsymbol{v}\otimes\boldsymbol{v}\right) - 2\boldsymbol{\omega}\times\boldsymbol{v}$$

and the centripetal acceleration is thought to be absorbed in $\boldsymbol{g}$.

that case one must employ a fluid model for which the density variations are described by a thermal equation of state; this is an equation in which pressure, temperature, density and possibly salt concentration are functionally related to one another. In addition, the currents in the geophysical context (ocean, lakes, atmosphere) may largely be influenced by the CORIOLIS acceleration due to the rotation of the frame of reference. The NAVIER–STOKES equations are under these conditions given by the following equations:

BOUSSINESQ **Fluid**

- Continuity equation

$$\operatorname{div} \boldsymbol{v} = 0 \,,$$

- Momentum equation

$$\frac{\partial \boldsymbol{v}}{\partial t} + \operatorname{div}(\boldsymbol{v} \otimes \boldsymbol{v}) + 2\boldsymbol{\omega} \times \boldsymbol{v} = -\frac{1}{\rho_0} \operatorname{grad} p + \nu \operatorname{div} \operatorname{grad} \boldsymbol{v} + \frac{\rho}{\rho_0} \boldsymbol{g} \,,$$

- Heat equation (10.5.17)

$$\frac{\partial \Theta}{\partial t} + \operatorname{div}\left(\Theta \boldsymbol{v}\right) = \chi^{(\Theta)} \operatorname{div} \operatorname{grad} \Theta + \frac{1}{\rho c_v} \varphi + \frac{1}{c_v} \mathfrak{r} \,,$$

- Thermal equation of state

$$\rho = \hat{\rho}(\Theta) \quad \text{or} \quad \rho = \hat{\rho}(\Theta, p) \quad \text{or} \quad \rho = \hat{\rho}(\Theta, s, p) \,,$$

- Diffusion equation

$$\frac{\partial c}{\partial t} + \operatorname{div}\left(c \boldsymbol{v}\right) = \chi^{(c)} \operatorname{div} \operatorname{grad} c + \phi^{(c)} \,.$$

Besides the CORIOLIS acceleration[9] the momentum equation contains here a buoyancy term. Moreover, the density is now a field variable and must be determined via the thermal equation of state for the density, which for many fluids is a linear function of temperature but in lake hydrodynamics is approximately given as a quadratic equation of the temperature. For deep lakes as e.g. Lake Baikal the pressure dependence of the density must equally be accounted for, and in the ocean an additional dependence is that on the salinity s for which a diffusion equation must hold. In all these cases, the reference density ρ_0 may be position dependent, and need not be constant.

[9] The EULER acceleration is generally ignored if the rotation of the coordinate system is (nearly) constant.

10.6 General Concept of Turbulence Modeling

Now, that the governing equations, upon which the turbulence models to be presented are based are known, and the filter operations in Sect. 10.3 have been introduced, we shall now begin with the mathematical description of turbulent processes in fluids. To this end the existence of a filter is supposed, which fulfills all required properties (linearity, commutativity with spatial and temporal differentiations, vanishing of the mean value of the fluctuating quantity).

Let us first illustrate how one generally proceeds in the derivation of the differential equations for turbulence models; this will isolate the conceptual difficulties that arise. *We emphasize, however, that what we present is only a concept and not the presentation of a complete mathematical theory.* The ensuing argumentation is facilitated if the equations of motion $(10.5.6)_1$ and (10.5.8) are slightly rewritten. To this end, we introduce the operators

$$\begin{aligned} L_0 v &:= \operatorname{div} \boldsymbol{v} \,, \qquad & L_1 v &:= \left(\frac{\partial}{\partial t} - \nu \operatorname{div} \operatorname{grad}\right) \boldsymbol{v} \,, \\ L_2(vv) &:= \operatorname{div}(\boldsymbol{v} \otimes \boldsymbol{v}) \,, \qquad & L_3 p &:= \operatorname{grad} p \,. \end{aligned} \tag{10.6.1}$$

With them the balances of mass and momentum may be rewritten as

$$L_0 v = 0 \,, \qquad L_1 v + L_2(vv) + L_3 p = \mathbf{0} \,. \tag{10.6.2}$$

If one averages these operator equations, evolution equations for the mean velocity and the mean pressure are obtained (notice that averaging and differentiation commute) which read

$$L_0 \langle v \rangle = 0 \,, \qquad L_1 \langle v \rangle + L_2 \langle vv \rangle + L_3 \langle p \rangle = \mathbf{0} \,. \tag{10.6.3}$$

These equations may be regarded as a system of differential equations for the mean velocity $\langle v \rangle$ and the mean pressure $\langle p \rangle$. However, the above symbolic representation also makes clear that a *higher moment* $\langle vv \rangle$ of the velocity must be known, if the mean velocity and mean pressure are to be determined by (10.6.3). In order to procure this moment, $(10.6.2)_2$ is (dyadically) multiplied with the velocity $\boldsymbol{v}$ and the averaging of the resulting equation is performed afterwards. If this calculation is performed, one obtains an evolution equation for the second moment of the velocity (which essentially is the variance of the velocity – the first moment is the mean value)

$$\tilde{L}_1 \langle vv \rangle + \tilde{L}_2 \langle vvv \rangle + \tilde{L}_3 \langle vp \rangle + \ldots = \mathbf{0} \,, \tag{10.6.4}$$

with, in general, new operators $\tilde{L}_1$, $\tilde{L}_2$, $\tilde{L}_3$, where $\tilde{L}_2 \langle vvv \rangle = \operatorname{div}(\langle \boldsymbol{v} \otimes \boldsymbol{v} \otimes \boldsymbol{v} \rangle)$. With the intention to close the system of equations (10.6.3), indeed an equation for the second moment $\langle \boldsymbol{v} \otimes \boldsymbol{v} \rangle$ has been deduced, however this equation contains a dependence on the next higher moment $\langle \boldsymbol{v} \otimes \boldsymbol{v} \otimes \boldsymbol{v} \rangle$ and, additionally, the pressure–velocity correlation $\langle \boldsymbol{v} p \rangle$ arises, for which an evolution equation must also be found to close the system of equations. The scheme appears to be clear; for the computation of the n-th moment knowledge of the next higher moment and even other moments are required, so that one has n equations for at least $n+1$ moments. Furthermore, the pressure–velocity correlations must be determined (and other unknown correlation products of higher moments must be considered). This property is called a *forward coupling* and is a typical symptom of all statistical continuum theories. The problem of turbulence modelling is to cut this (ideally) infinite system and to apply phenomenological parameterizations for the highest unknown moments. These parameterizations comprise the *turbulent closure conditions.* In the most simple case one restricts oneself to the system (10.6.3) and simultaneously postulates an empirical equation for the second moment of the velocities. This procedure defines the *zeroth order closure scheme.* The phenomenological law contains a scalar parameter, the turbulent viscosity which in a first attempt one simply chooses to be constant. This coefficient, however, is not a constant, as has been demonstrated by experiments already very early, and it changes with the degree of turbulence. This fact led to an extension of the model by establishing an evolution equation for a scalar quantity, usually the turbulent kinetic energy and relating this quantity to the turbulent viscosity. This however is not compelling, but the procedure led to the *first order closure models.* The next step has been to formulate or postulate two differential equations for two scalar quantities that determine the turbulent viscosity, e.g. the turbulent kinetic energy and dissipation rate (leading to the k–ε model) or a pair of other quantities. The next level are the so-called REYNOLDS *stress models*, in which differential equations are formulated for all components of the REYNOLDS stress tensor – this is the second moment of the velocities – i.e., one cuts the sequence of evolution equations for higher order elements at the level of (10.6.4). These are then called *turbulence closure models of second order.* This approach suggests a hierarchy of possible turbulence models, which are in principle consecutively constructed in this way. Today's research of turbulence modeling goes as far as to third order with closure conditions for the fourth order moments, see SANDER [208].

10.7 REYNOLDS Stresses

10.7.1 Equations for the Averaged Fields

Let us scrutinize the above discussed concept. For this purpose the field quantities, velocity and pressure, are decomposed, viz.,

$$\boldsymbol{v} = \langle \boldsymbol{v} \rangle + \boldsymbol{v}' \ , \quad p_{\mathrm{r}} = \langle p_{\mathrm{r}} \rangle + p_{\mathrm{r}}' \tag{10.7.1}$$

into mean values and fluctuations, are substituted into the field equations and the latter are filtered using a filter that obeys all requirements of a statistical filter. The balance laws of mass and momentum then take the forms

$$\begin{aligned}
&\operatorname{div}\langle \boldsymbol{v}\rangle = 0\,,\\
&\frac{\partial\langle \boldsymbol{v}\rangle}{\partial t} + \operatorname{div}\left(\langle \boldsymbol{v}\rangle\otimes\langle \boldsymbol{v}\rangle\right) + \operatorname{div}\langle \boldsymbol{v}'\otimes \boldsymbol{v}'\rangle \\
&\qquad + \operatorname{grad}\langle p_{\mathrm{r}}\rangle - \nu \operatorname{div}\operatorname{grad}\langle \boldsymbol{v}\rangle = \mathbf{0}\,,
\end{aligned} \tag{10.7.2}$$

in which use has been made of the fact that

$$\begin{aligned}
\langle \boldsymbol{v}\otimes \boldsymbol{v}\rangle &= \langle ((\langle \boldsymbol{v}\rangle + \boldsymbol{v}')\otimes(\langle \boldsymbol{v}\rangle + \boldsymbol{v}')\rangle\\
&= \langle\langle \boldsymbol{v}\rangle\otimes\langle \boldsymbol{v}\rangle\rangle + \langle\langle \boldsymbol{v}\rangle\otimes \boldsymbol{v}'\rangle + \langle \boldsymbol{v}'\otimes\langle \boldsymbol{v}\rangle\rangle + \langle \boldsymbol{v}'\otimes \boldsymbol{v}'\rangle\\
&= \langle \boldsymbol{v}\rangle\otimes\langle \boldsymbol{v}\rangle + \langle \boldsymbol{v}'\otimes \boldsymbol{v}'\rangle\,,
\end{aligned} \tag{10.7.3}$$

which assumes that $\langle \boldsymbol{v}'\rangle = \mathbf{0}$. Here we now observe the particularities discussed in the last section. Evolution equations are at our disposal for the mean velocity vector $\langle \boldsymbol{v}\rangle$ and the mean pressure $\langle p_{\mathrm{r}}\rangle$, however, a second order moment $\langle \boldsymbol{v}'\otimes \boldsymbol{v}'\rangle$ arises in the unknown fluctuation velocities. This term can be interpreted as a stress tensor or momentum flux; with it the exchange of momentum between the large–scale mean quantities and the small–scale turbulent fluctuations is described. If the flow is not only driven by the velocity gradients of the mean motion but equally also by the fluctuating motion, then the latter must generate stresses. The specific momentum therefore changes on the basis of the transport by these fluctuations. The quantity

$$\boldsymbol{R} := -\rho\langle \boldsymbol{v}'\otimes \boldsymbol{v}'\rangle \tag{10.7.4}$$

is called the REYNOLDS *stress* tensor. It was first introduced by REYNOLDS in [194]. This tensor must be known or determinable by algebraic or differential equations, in order that the system of differential equations (10.7.2) can be closed.

Before addressing the explicit formulation of such closure conditions, let us also write down the governing equations for the fluctuations. To this end the equations describing the mean motion, (10.7.2), are subtracted from the original non-averaged balances of mass and momentum; what obtains reads

$$\begin{aligned}
&\operatorname{div} \boldsymbol{v}' = 0\,,\\
&\frac{\partial \boldsymbol{v}'}{\partial t} + \operatorname{div}(\boldsymbol{v}'\otimes \boldsymbol{v}') + \operatorname{div}\left(\langle \boldsymbol{v}\rangle\otimes \boldsymbol{v}' + \boldsymbol{v}'\otimes\langle \boldsymbol{v}\rangle - \langle \boldsymbol{v}'\otimes \boldsymbol{v}'\rangle\right)\\
&\qquad + \operatorname{grad} p_{\mathrm{r}}' - \nu\operatorname{div}\operatorname{grad} \boldsymbol{v}' = \mathbf{0}\,.
\end{aligned} \tag{10.7.5}$$

If the REYNOLDS stresses are known, this is a system of differential equations for the fluctuations $\boldsymbol{v}'$ and p'. The momentum equation for the fluctuating quantities $(10.7.5)_2$ contains also an additional stress contribution, namely $\langle \boldsymbol{v}\rangle\otimes \boldsymbol{v}' + \boldsymbol{v}'\otimes\langle \boldsymbol{v}\rangle - \langle \boldsymbol{v}'\otimes \boldsymbol{v}'\rangle$, which determines the influence of the mean motion to the fluctuating motion. If one multiplies equation $(10.7.5)_2$ with the fluctuation velocity at the same position (single point correlation), $\boldsymbol{v}'(\boldsymbol{x}, t)$ and subsequently performs an averaging process, then a differential equation is obtained for the unknown correlation $\langle \boldsymbol{v}' \otimes \boldsymbol{v}' \rangle$; however this equation contains a higher order moment of the fluctuation quantities, as was already discussed before.

Notice that the above algebraic procedure involves only single point correlations, i.e., correlations of field quantities defined at the same spatial point and at the same time. The formulation becomes more general if one considers twofold or multifold correlations with different points at different times. If this concept is used, then the equations involving the fluctuation quantities at point $\boldsymbol{x}^1$ and time t^1 are multiplied with the velocity fluctuations at another point and another time, $\boldsymbol{v}'(\boldsymbol{x}^2, t^2)$ and averaging operations are performed afterwards. We shall not be dealt with this modelling approach in this book[10].

For completeness the equation governing the mean temperature will also be stated here. It can be derived by averaging the heat conduction equation (10.5.9) and has the form

$$\begin{aligned} &\frac{\partial \langle \Theta \rangle}{\partial t} + \operatorname{div}\left(\langle \Theta \rangle \langle \boldsymbol{v} \rangle\right) \\ &\quad = \chi^{(\Theta)} \operatorname{div}\operatorname{grad}\langle \Theta \rangle - \operatorname{div}\langle \boldsymbol{v}'\Theta' \rangle + \frac{1}{\rho c_v}\langle \varphi \rangle + \frac{1}{c_v}\mathfrak{r}\,. \end{aligned} \tag{10.7.6}$$

The specific radiation $\mathfrak{r}$ as an external source term does not have to be averaged in this equation, since it is not influenced by the turbulence. To close this equation of the mean temperature, the correlation between the temperature and the velocity fluctuations must be known. Just as for the momentum equation, this is again a flux term, called the turbulent heat flux

$$\boldsymbol{q}_t := \rho c_v \langle \boldsymbol{v}'\Theta' \rangle\,. \tag{10.7.7}$$

For its determination one can proceed in an analogous fashion as before, i.e., one applies an empirically determined equation for this correlation or one establishes also an evolution equation for the turbulent heat flux.

Next, let us compute the mean value for the dissipation, which is a quadratic function of the velocity gradient. If one substitutes the decomposition of the velocity, one obtains

$$\begin{aligned} \frac{1}{\rho}\langle \varphi \rangle &= 2\nu \langle \boldsymbol{D} \cdot \boldsymbol{D} \rangle \\ &= 2\nu \langle \boldsymbol{D} \rangle \cdot \langle \boldsymbol{D} \rangle + 2\nu \langle \boldsymbol{D}' \cdot \boldsymbol{D}' \rangle \\ &= 4\,\nu II_{\langle \boldsymbol{D} \rangle} + \underbrace{4\,\nu \langle II_{\boldsymbol{D}'} \rangle}_{:=\varepsilon}\,, \end{aligned} \tag{10.7.8}$$

[10] The reader interested in such alternative approaches may wish to consult ROTTA [198], BATCHELOR [21], MONIN & YAGLOM [156] and others.

in which the second invariant of a symmetric tensor of rank two, $\boldsymbol{A}$ has been defined as

$$II_{\boldsymbol{A}} := \tfrac{1}{2}\boldsymbol{A}\cdot\boldsymbol{A} = \tfrac{1}{2}\,\mathrm{tr}\,\boldsymbol{A}^2\,, \tag{10.7.9}$$

a definition we shall maintain for the subsequent analysis. In the turbulence literature (10.7.8) is usually written as

$$\frac{1}{\rho}\langle\varphi\rangle = \underbrace{\frac{\nu}{2}\big|\,\mathrm{grad}\,\langle\boldsymbol{v}\rangle + (\mathrm{grad}\,\langle\boldsymbol{v}\rangle)^T\big|^2}_{\text{dissipation due to the mean velocity}} + \underbrace{\frac{\nu}{2}\big\langle\big|\,\mathrm{grad}\,\boldsymbol{v}' + (\mathrm{grad}\,\boldsymbol{v}')^T\big|^2\big\rangle}_{=:\varepsilon\ \text{turbulent dissipation}}\,. \tag{10.7.10}$$

This representation is equivalent to (10.7.8), but, perhaps, not as elegant. The second term in (10.7.8) is called the *turbulent dissipation* or the *turbulent dissipation rate*[11]. It plays a very important role in the formulation of turbulence modeling. Physically, it represents the turbulent energy dissipated per unit time, i.e., the energy of the eddies which, at the end of the energy cascade, is transformed by molecular diffusion into heat per unit time. In applications these dissipation terms can often be neglected, since they are much smaller than the turbulent diffusion of heat that is described by $\mathrm{div}\,\langle\boldsymbol{v}'\Theta'\rangle$.

The mean concentration of a tracer substance follows analogously from the balance equation for the concentration, (10.5.12). By application of the filter, this yields

$$\frac{\partial\langle c\rangle}{\partial t} + \mathrm{div}\,\big(\langle c\rangle\langle\boldsymbol{v}\rangle\big) = \chi^{(c)}\,\mathrm{div}\,\mathrm{grad}\,\langle c\rangle - \mathrm{div}\,\langle\boldsymbol{v}'c'\rangle + \langle\phi^{(c)}\rangle\,. \tag{10.7.11}$$

Here too, the turbulent mass flux,

$$\boldsymbol{j}_t := \langle\boldsymbol{v}'c'\rangle \tag{10.7.12}$$

is a new correlation term that must be determined. As with the heat conduction equation, this term plays an important role in the description of turbulent diffusion. The source term will not be specified any further.

There are many more equations for averaged fields that could be derived and discussed here – and some of them will be dwelled upon in further chapters; we will not do so now, but will rather analyse some transformation properties enjoyed by averaged field quantities.

Before we turn to that, however, let us emphasise once more that the filter function, $\langle\cdot\rangle$, applied here is equivalent to a statistical filter for which $\langle\langle\cdots\langle\cdot\rangle\cdots\rangle\rangle = \langle\cdot\rangle$, or $\langle(\cdot)'\rangle = 0$. The equations that emerge are the REYNOLDS averaged NAVIER–STOKES equations (RANS). These equations are obviously based on a very "restricting" filter function. If the filter properties are relaxed

[11] Notice that the internal energy was denoted by the same symbol as the turbulent dissipation rate is now; this is unfortunate, however this is the classical symbol used in turbulence theory which is maintained here. For internal energy we shall use ϵ, not ε, instead.

such that $\langle(\cdot)'\rangle$ does not necessarily need to vanish, then the form of the averaged balance equations of mass, momentum and energy are more complicated, because they then also involve the linear correlations, $\langle(\cdot)'\rangle$. The emerging equations are no longer REYNOLDS averaged, but more general and used in the so-called "Large Eddy Simulation" (LES) procedures.

10.7.2 Transformation Properties of Turbulent Field Quantities

In the first, continuum mechanical, part of this book the transformation properties of the physical quantities, $\rho, \boldsymbol{v}, \Theta$ etc., as well as those of the balance laws of mass, momentum, energy and entropy under EUCLID*ian transformations* were studied. In particular, it was stated that[12]

$$\begin{array}{ll} \rho, \Theta, \epsilon, s & \text{transform as objective scalars,} \\ \boldsymbol{q}, \boldsymbol{\phi}, \boldsymbol{j} & \text{transform as objective vectors ,} \\ \boldsymbol{t} & \text{transforms as an objective tensor.} \end{array} \tag{10.7.13}$$

Furthermore, it was proved that the balance equations mentioned above possess in each frame of reference the same form, i.e., these equations are formally invariant but may contain terms which are system dependent. The material functionals for ϵ, $\boldsymbol{q}$ and $\boldsymbol{t}$ must therefore not necessarily be independent of the frame of reference; system independence of the material equations is not compelling but a convenience. If it is required, then one says that the constitutive equations obey the *rule of material frame indifference*[13].

Under EUCLIDian transformations

$$\boldsymbol{x}^* = \boldsymbol{O}(t)\boldsymbol{x} + \boldsymbol{b}^*(t) \tag{10.7.14}$$

the velocity vector $\boldsymbol{v}$ transforms according to

$$\boldsymbol{v}^* = \boldsymbol{O}(t)\boldsymbol{v} + \dot{\boldsymbol{O}}(t)\boldsymbol{x} + \dot{\boldsymbol{b}}^*(t) \; ; \tag{10.7.15}$$

so $\boldsymbol{v}$ is not an objective vector, as we long know. Since the frames of reference are not subject to fluctuations, the mean value of (10.7.15) takes the form

$$\langle\boldsymbol{v}^*\rangle = \boldsymbol{O}(t)\langle\boldsymbol{v}\rangle + \dot{\boldsymbol{O}}(t)\boldsymbol{x} + \dot{\boldsymbol{b}}^*(t) \; , \tag{10.7.16}$$

in which we have set $\langle\boldsymbol{x}\rangle = \boldsymbol{x}$. Hence neither the mean velocity $\langle\boldsymbol{v}\rangle$ is an objective vector. On the other hand, by taking the difference of (10.7.15) and (10.7.16) one obtains

[12] Here, ρ, Θ, ϵ, s, $\boldsymbol{q}$, $\boldsymbol{\phi}$, $\boldsymbol{j}$, $\boldsymbol{t}$ denote the mass density, absolute temperature, specific internal energy, entropy, heat flux, entropy flux, mass flux and CAUCHY stress tensor.

[13] In Chap. 4 these properties were called frame "invariance" and "frame indifference": The balance laws are frame invariant but not indifferent and the constitutive relations were assumed to be indifferent.

$$v'^* = v^* - \langle v^* \rangle = O(t)(v - \langle v \rangle) = O(t)v' . \tag{10.7.17}$$

Under EUCLID*ian transformations the fluctuation velocity v' transforms as an objective vector.* This is a very important result.

It follows, since $\langle \Theta \rangle, \Theta'$ and $\langle c \rangle, c', \langle \varphi \rangle$ transform as objective scalars that[14]

$$\begin{array}{lll} \boldsymbol{R} := -\rho \langle \boldsymbol{v}' \otimes \boldsymbol{v}' \rangle & \Rightarrow & \boldsymbol{R}^* = \boldsymbol{O}\boldsymbol{R}\boldsymbol{O}^T , \\ \boldsymbol{q}_t := \rho c_v \langle \Theta' \boldsymbol{v}' \rangle & \Rightarrow & \boldsymbol{q}_t{}^* = \boldsymbol{O}\boldsymbol{q}_t , \\ \boldsymbol{j}_t := \langle c' \boldsymbol{v}' \rangle & \Rightarrow & \boldsymbol{j}_t{}^* = \boldsymbol{O}\boldsymbol{j}_t , \\ \frac{1}{\rho}\langle \varphi \rangle := 4\nu II_{\langle \boldsymbol{D} \rangle} + \varepsilon & \Rightarrow & \varepsilon^* = \varepsilon . \end{array} \tag{10.7.18}$$

$\boldsymbol{R}$ transforms under EUCLIDian transformations as an objective symmetric second rank tensor, $\boldsymbol{q}_t$ and $\boldsymbol{j}_t$ as two objective vectors, $\langle \varphi \rangle$ as an objective scalar and ε as an objective scalar. *Thus, the* REYNOLDS *stress tensor and the turbulent heat and mass fluxes as well as the turbulent dissipation rate are objective quantities.*

With this knowledge we ask next, whether the balance laws for the mean motion (10.7.2), temperature (10.7.6) and tracer concentration (10.7.11) behave under EUCLIDian transformations as the corresponding balance equations in continuum mechanics, whence that they are invariant under such transformations. The answer to this question is easy insofar as (10.7.2), (10.7.6) and (10.7.11) all contain averaged variables which possess the same transformation properties as their original (non-averaged) counterparts that arise in the original equations; additional variables are only those listed in (10.7.18), which have been seen above to be objective quantities. It therefore follows that *equations (10.7.2), (10.7.6) and (10.7.11), because they are balance laws, have in every frame of reference the same form, whereby the momentum equation may contain system dependent terms as is the case for the momentum balance in continuum mechanics.*

If closure equations must be postulated for the turbulent fluxes (10.7.18), then these closure conditions have – at least formally – a very similar meaning as the material equations have in continuum mechanics. Therefore, the question whether a *rule* or a *principle of objectivity* may be an adequate postulate for turbulent closure equations, is well justified. Such a requirement would say that the functionals for the fluxes (10.7.18) could not contain system dependent terms. Such a rule, if appropriate, could be called *rule of turbulent objectivity* or *rule of turbulent frame indifference.* It is known from the literature that, depending upon the author and application, such a rule

[14] This list does not contain the variable $\langle \Theta'^2 \rangle$. This variable, however, does arise in a BOUSSINESQ fluid for which the density may nonlinearly depend on the temperature. (This is so for water).

is conjectured to hold or negated. In what follows we shall come back to this point at several instances; in particular we shall show in the next section that early postulates for turbulent closure conditions were obeying the rule of turbulent frame indifference.

10.7.3 REYNOLDS Hypothesis

A very important concept of the turbulence theory is the REYNOLDS hypothesis and the introduction of the *eddy viscosity*. It may be deduced from the idea that the state of turbulence of the velocity field is connected with the mean field through its gradient; the larger the gradient of the mean velocity is, the larger will be the generated turbulent activities. Certainly, this cannot be considered to be a rule of general validity. As a counter example we may mention the flow over an airplane wing; this flow has large velocity gradients at the nose, close to the stagnation point, but it is laminar there and becomes turbulent at the rear portion of the wing.

The fundamental idea goes back to BOUSSINESQ [34], who described the connection between the turbulent shear stress τ and the velocity gradient $\partial v_1/\partial x_2$ perpendicular to the main flow with an exchange coefficient in the form $\tau = \rho \nu_t \, \partial v_1/\partial x_2$. The form of this exchange coefficient and its dependence upon quantities such as the turbulent mixing length or the turbulent kinetic energy is an important basis for turbulence modelling, but was not spelled out by BOUSSINESQ. As an example we shall in the next section demonstrate this for PRANDTL's mixing length. Here, let us first treat the connection between the REYNOLDS stress tensor and the mean velocity gradient.

According to the above mentioned desire of observing turbulent objectivity one may, in an attempt of generalising the BOUSSINESQ hypothesis, write the REYNOLDS stress tensor as a function of the mean stretching,

$$\langle \boldsymbol{D} \rangle = \mathrm{sym}\big(\,\mathrm{grad}\,\langle \boldsymbol{v} \rangle\big) = \tfrac{1}{2}\big(\,\mathrm{grad}\,\langle \boldsymbol{v} \rangle + (\mathrm{grad}\,\langle \boldsymbol{v} \rangle)^T\big) \,, \qquad (10.7.19)$$

since only the symmetric part of the velocity gradient is an objective quantity. This yields

$$\boldsymbol{R} = \hat{\boldsymbol{R}}(\langle \boldsymbol{D} \rangle) \,. \qquad (10.7.20)$$

Since both $\boldsymbol{R}$ as well as $\langle \boldsymbol{D} \rangle$ transform under EUCLIDian transformations as objective symmetric tensors of rank two, $\boldsymbol{R}^* = \boldsymbol{O}\boldsymbol{R}\boldsymbol{O}^T$, $\langle \boldsymbol{D}^* \rangle = \boldsymbol{O}\langle \boldsymbol{D} \rangle \boldsymbol{O}^T$, in which $\boldsymbol{O} \in \mathbb{O}$ is a time-dependent orthogonal transformation, the ansatz (10.7.20) yields[15]

[15] This is not a rule similar to the rule of material objectivity since turbulence properties are properties of the motion. Once (10.7.20) is made as a closure assumption, (10.7.21) follows strictly by transformation rules and must be fulfilled for all orthogonal transformations $\boldsymbol{O}$. However, the requirement $\hat{\boldsymbol{R}}^*(\langle \boldsymbol{D} \rangle^*) = \hat{\boldsymbol{R}}(\langle \boldsymbol{D} \rangle^*)$ is a genuine statement, that must newly be motivated. The absence of $\langle \boldsymbol{W} \rangle = \mathrm{skw}(\mathrm{grad}\,\langle \boldsymbol{v} \rangle) = \frac{1}{2}(\mathrm{grad}\,\langle \boldsymbol{v} \rangle - (\mathrm{grad}\,\langle \boldsymbol{v} \rangle)^T)$ as an independent variable in (10.7.20) is, however, the expression of "turbulent" objectivity.

$$\hat{\boldsymbol{R}}(\boldsymbol{O}\langle\boldsymbol{D}\rangle\boldsymbol{O}^T) = \boldsymbol{O}\hat{\boldsymbol{R}}(\langle\boldsymbol{D}\rangle)\boldsymbol{O}^T \; . \tag{10.7.21}$$

According to these requirements the REYNOLDS stress tensor is an isotropic tensor function of $\langle\boldsymbol{D}\rangle$ and may therefore be written as

$$\boldsymbol{R} = a\boldsymbol{I} + 2\mu_t\langle\boldsymbol{D}\rangle + \xi_t\langle\boldsymbol{D}\rangle^2 \; , \tag{10.7.22}$$

in which the coefficients are functions of the invariants of $\langle\boldsymbol{D}\rangle$. The isotropic contribution $a\boldsymbol{I}$ can still be absorbed into the pressure (if one so desires) since the latter is an independent field that is determined by solving the field equations, viz.,

$$p_{\mathrm{r}}^{\circ} = p_{\mathrm{r}} - \frac{a}{\rho} \quad \Rightarrow \quad \operatorname{grad} p_{\mathrm{r}} - \frac{1}{\rho}\operatorname{div}(a\boldsymbol{I}) = \operatorname{grad}\left(p_{\mathrm{r}} - \frac{a}{\rho}\right) = \operatorname{grad} p_{\mathrm{r}}^{\circ} \; . \tag{10.7.23}$$

Consider, in particular the "linear" case for which $\xi_t = 0$; for this case (10.7.22) allows a physical interpretation of the coefficient a, if one simultaneously recalls the definition of the turbulent kinetic energy. The latter is obtained from the REYNOLDS stress tensor by forming its trace,

$$k := \tfrac{1}{2}\langle v'_i v'_i\rangle = \tfrac{1}{2}\operatorname{tr}\langle\boldsymbol{v}'\otimes\boldsymbol{v}'\rangle = -\frac{1}{2\rho}\operatorname{tr}\boldsymbol{R} = -\frac{3a}{2\rho} \; . \tag{10.7.24}$$

Therefore, the constant a is given by the turbulent kinetic energy and the density (see §10.4.1), and in the "linear" case the formula for the REYNOLDS stresses and the correlation of the velocity fluctuations may be written as

$$-\langle\boldsymbol{v}'\otimes\boldsymbol{v}'\rangle \;=\; \frac{1}{\rho}\boldsymbol{R} \;=\; -\tfrac{2}{3}k\boldsymbol{I} + 2\frac{\mu_t}{\rho}\langle\boldsymbol{D}\rangle \; , \tag{10.7.25}$$

which now contains two scalar coefficients, k and μ_t. The term

$$\nu_t = \mu_t/\rho \tag{10.7.26}$$

is called *kinematic turbulent viscosity* or *eddy viscosity*. The representation (10.7.25) shows that, in the "linear" and density preserving case it is sufficient to work only with the deviator $\boldsymbol{R}'$ of $\boldsymbol{R}$ and to assume $\boldsymbol{R}' = 2\mu_t\langle\boldsymbol{D}\rangle$.

In much the same way, one may parameterize the turbulent heat flux vector. Here we may assume that this vector "orients itself" on the gradient of the mean temperature field. The objective vector of the turbulent heat flux is therefore an isotropic function of a vector, the mean temperature gradient, $\boldsymbol{q}_t = \hat{\boldsymbol{q}}_t(\operatorname{grad}\langle\Theta\rangle)$. In analogy to the FOURIER law of heat conduction one may thus write

$$\frac{1}{\rho c_v}\boldsymbol{q}_t = \langle\boldsymbol{v}'\Theta'\rangle = -\chi_t^{(\Theta)}\operatorname{grad}\langle\Theta\rangle \; . \tag{10.7.27}$$

For the diffusive mass flux of a tracer substance one may proceed analogously. Indeed, in much the same way as for FOURIER's law one may assume for the turbulent diffusive mass flux a closure equation of the form $\boldsymbol{j}_t = \hat{\boldsymbol{j}}_t(\text{grad}\,\langle c\rangle)$ and then finds by imposing the rules of EUCLIDian transformations the analogue to FICK's first law as

$$\boldsymbol{j}_t = \langle \boldsymbol{v}'c'\rangle = -\chi_t^{(c)}\,\text{grad}\,\langle c\rangle\,. \tag{10.7.28}$$

The newly introduced scalar quantities $\chi_t^{(\Theta)}$ and $\chi_t^{(c)}$ are called *eddy diffusivities* of heat and mass. On the basis of their construction these may depend upon the scalar quantities k, ε, $\langle\Theta\rangle$ and the moduli $|\,\text{grad}\,\langle\Theta\rangle|$ and $|\,\text{grad}\,\langle c\rangle|$, respectively[16]. This latter dependence is, however, usually, ignored.

10.7.4 Eddy Viscosity and Diffusivity

If one interprets the parameterizations for $\boldsymbol{R}$, $\boldsymbol{q}_t$ and $\boldsymbol{j}_t$ just as constitutive relations of a material theory and applies for their representation the rule of equipresences, then as a set of independent variables one may start from the list of variables

$$\mathfrak{X} := \left(\langle\boldsymbol{D}\rangle,\ \langle\Theta\rangle,\ \langle c\rangle,\ \text{grad}\,\langle\Theta\rangle,\ \text{grad}\,\langle c\rangle,\ k,\ \varepsilon\ \langle{\Theta'}^2\rangle,\ldots\right) \tag{10.7.29}$$

and apply these to the closure quantities $\Psi := (\boldsymbol{R}',\ \boldsymbol{q}_t,\ \boldsymbol{j}_t)$ in the form

$$\boldsymbol{\Psi} = \hat{\boldsymbol{\Psi}}(\mathfrak{X})\,. \tag{10.7.30}$$

Here in (10.7.29) we have assumed that, besides $\langle\boldsymbol{D}\rangle$, $\langle\Theta\rangle$ and $\langle c\rangle$, also the gradients grad $\langle\Theta\rangle$ and grad $\langle c\rangle$ arise as variables as do the turbulent kinetic energy, k the turbulent dissipation rate, ε as well as the mean value of the square of the temperature fluctuations[17], $\langle{\Theta'}^2\rangle$. The points indicate further possible dependences.

Under EUCLIDian transformations the following transformation rules apply for $\mathfrak{X}$ and $\boldsymbol{\Psi}$

$$\begin{aligned} \mathfrak{X}^* := \big(&\boldsymbol{O}\langle\boldsymbol{D}\rangle\boldsymbol{O}^t,\ \langle\Theta\rangle,\ \langle c\rangle,\ \boldsymbol{O}\,\text{grad}\,\langle\Theta\rangle,\\ &\boldsymbol{O}\,\text{grad}\,\langle c\rangle,\ k,\ \varepsilon,\ \langle{\Theta'}^2\rangle,\ldots\big)\,,\\ \boldsymbol{\Psi}^* = \big(&\boldsymbol{O}\boldsymbol{R}'\boldsymbol{O}^T,\ \boldsymbol{O}\boldsymbol{q}_t,\ \boldsymbol{O}\boldsymbol{j}_t\big) =: \boldsymbol{O}^*\hat{\boldsymbol{\Psi}}(\mathfrak{X})\,, \end{aligned} \tag{10.7.31}$$

so that by adopting the rule of turbulent frame indifference one must necessarily fulfil the requirement

$$\hat{\boldsymbol{\Psi}}(\mathfrak{X}^*) = \boldsymbol{O}^*\hat{\boldsymbol{\Psi}}(\mathfrak{X})\,. \tag{10.7.32}$$

[16] A dependence on the turbulent kinetic energy k and its dissipation rate ε is not evident at this stage of the computations and, indeed, one often assumes only a dependence on $\langle\Theta\rangle$. We will, however, maintain a $(k,\,\varepsilon)$-dependence.

[17] We list here $\langle{\Theta'}^2\rangle$ as an independent constitutive variable, because it is assumed that we may also have an evolution equation for $\langle{\Theta'}^2\rangle$ at our disposal. We shall show later how such an equation can be derived.

$\boldsymbol{O}^*$ in (10.7.31) and (10.7.32) represents the action of the orthogonal transformation $\boldsymbol{O}$ on $\boldsymbol{\Psi}$ as defined in relation $(10.7.31)_2$, and $\boldsymbol{O}$ may be any orthogonal transformation. Applied to $\boldsymbol{R}'$ and $\boldsymbol{q}_t$, (10.7.32) means

$$\hat{\boldsymbol{R}}'(\mathfrak{X}^*) = \boldsymbol{O}\hat{\boldsymbol{R}}'(\mathfrak{X})\boldsymbol{O}^T \,, \quad \hat{\boldsymbol{q}}(\mathfrak{X}^*) = \boldsymbol{O}\hat{\boldsymbol{q}}(\mathfrak{X}) \,. \tag{10.7.33}$$

Thus $\hat{\boldsymbol{R}}'$ and $\hat{\boldsymbol{q}}_t$ are an isotropic tensor function and an isotropic vector function of a tensor valued variable, two vector valued variables and five scalar variables. One could write down for these the most general representation; its usefulness would, however, be rather doubtful, since it would be so complicated that hardly any meaningful inferences could be drawn from it. If one assumes instead (in explicit violation of the rule of equipresence) that $\boldsymbol{R}'$ is affine to $\langle \boldsymbol{D} \rangle$, $\boldsymbol{q}_t$, is collinear to grad $\langle \Theta \rangle$ and $\boldsymbol{j}_t$ is affine to grad $\langle c \rangle$, then the representations (10.7.25), (10.7.27) and (10.7.28) hold, in which the eddy viscosity ν_t and the eddy diffusivities $\chi_t^{(\Theta)}$ and $\chi_t^{(c)}$ are functions of the form

$$\begin{aligned}
& f = \hat{f}(k, \varepsilon, \langle \Theta'^2 \rangle, II_{\langle \boldsymbol{D} \rangle}, III_{\langle \boldsymbol{D} \rangle}, |\operatorname{grad} \langle \Theta \rangle|, |\operatorname{grad} \langle c \rangle|, d_j) \; (j = 1, \ldots, 7) \,, \\
& d_1 := \langle \operatorname{grad} \Theta \rangle \cdot \langle \boldsymbol{D} \rangle \langle \operatorname{grad} \Theta \rangle \,, \quad d_2 := \langle \operatorname{grad} \Theta \rangle \cdot \langle \boldsymbol{D}^2 \rangle \langle \operatorname{grad} \Theta \rangle \,, \\
& d_3 := \langle \operatorname{grad} c \rangle \cdot \langle \boldsymbol{D} \rangle \langle \operatorname{grad} c \rangle \,, \quad d_4 := \langle \operatorname{grad} c \rangle \cdot \langle \boldsymbol{D}^2 \rangle \langle \operatorname{grad} c \rangle \,, \\
& d_5 := \langle \operatorname{grad} \Theta \rangle \cdot \langle \boldsymbol{D} \rangle \langle \operatorname{grad} c \rangle \,, \quad d_6 := \langle \operatorname{grad} \Theta \rangle \cdot \langle \boldsymbol{D}^2 \rangle \langle \operatorname{grad} c \rangle \,, \\
& d_7 := \langle \operatorname{grad} \Theta \rangle \langle \operatorname{grad} c \rangle \,,
\end{aligned} \tag{10.7.34}$$

where $f \in \{\nu_t, \chi_t^{(\Theta)}, \chi_t^{(c)}\}$. Naturally, in individual cases not all the listed dependencies must arise. It is also customary to introduce the ratios between the eddy viscosity and the diffusivities of heat and mass

$$\sigma_\Theta := \frac{\nu_t}{\chi_t^{(\Theta)}} \,, \quad \sigma_c := \frac{\nu_t}{\chi_t^{(c)}} \,, \tag{10.7.35}$$

respectively, and to call σ_Θ *turbulent* PRANDTL *number* and σ_c *turbulent* SCHMIDT *number*. The turbulent heat flux and turbulent mass flux can then be written as

$$\langle \boldsymbol{v}'\Theta' \rangle = -\frac{\nu_t}{\sigma_\Theta} \operatorname{grad} \langle \Theta \rangle \,, \quad \langle \boldsymbol{v}'c' \rangle = -\frac{\nu_t}{\sigma_c} \operatorname{grad} \langle c \rangle \,. \tag{10.7.36}$$

As long as one assumes for ν_t, σ_Θ and σ_c the dependencies (10.7.34), no restriction is introduced by the representation (10.7.36). If, however σ_Θ and σ_c are assumed to be constant, then the laws (10.7.36) are restricted insofar as the functional dependencies of the turbulent heat and mass diffusivities, $\chi_t^{(\Theta)}$ and $\chi_t^{(c)}$ are following the functional dependence of the momentum viscosity ν_t. This is a kind of similarity rule. Such a restriction is often implied, but cannot experimentally be corroborated. In Chap. 12, it will be seen how the dependencies can be described more generally but still approximately by an algebraic stress model.

By introducing the above "gradient laws", which are only adequate for flows with isotropic turbulence, one has achieved a tremendous reduction in the necessary parameterizations. Instead of the six unknown velocity correlations (the REYNOLDS stresses are necessarily symmetric) and the twice three unknown correlations between the temperature, the concentration and velocity fluctuations, only three scalar parameters – ν_t, σ_Θ and σ_c – must be identified. Notice, however, that these coefficients depend via (10.7.34) upon the turbulent processes themselves. Indeed, among other variables k, ε and $\langle\Theta'^2\rangle$ arise for which no field equations are (yet) at our disposal. For very simple turbulent closure conditions one usually dispenses oneself with a parameterization of these quantities; in turbulent closure schemes of the first order differential equations are postulated for k and ε, and $\langle\Theta'^2\rangle$ is simply ignored as a variable.

A possible generalization of the diffusivities just introduced is obtained if one assumes that the turbulence is not isotropic; this means that the turbulent diffusivities are not scalar quantities, but tensors of fourth (turbulent viscosities) and second rank (mass and thermal diffusivities) according to

$$(R)_{ij} = \nu_{ijkl}\langle D\rangle_{kl}\,, \quad (q_t)_i = \chi_{ik}^{(\Theta)}\langle\Theta\rangle,_k\,, \quad (j_t)_i = \chi_{ik}^{(c)}\langle c\rangle,_k\,. \qquad (10.7.37)$$

For the diffusion of a pollutant or nutrient in a lake or the ocean and atmosphere such generalizations are significant, since the *turbulent exchange coefficients* in the horizontal and vertical directions do not have the same values. The concept of this notion of *anisotropic turbulence* does not satisfy objectivity requirements and must be treated with caution. We shall return to anisotropic closure conditions lateron.

10.8 General Definition of Turbulence Models

10.8.1 Turbulence Models of Various Orders

In order to make the balance laws for the mean velocity, temperature and tracer concentration, derived in the previous sections, integrable, closure conditions must be postulated. This can be done in various different ways, of which the most simple possibilities lead to the following turbulence models:

- **Turbulence Model of Zeroth Order** For the second moments $\langle\boldsymbol{v}'\otimes\boldsymbol{v}'\rangle$, $\langle\boldsymbol{v}'\Theta'\rangle$ and $\langle\boldsymbol{v}'c'\rangle$, i.e., for the second order correlations of the velocity, temperature and concentration fluctuations phenomenological assumptions of the form (10.7.25), (10.7.27) and (10.7.28) or even more general assumptions are made. The coefficients (turbulent viscosity and diffusivities) are assumed to be known algebraic functions of the variables k, ε, $\langle\Theta'^2\rangle$, $II_{\langle\boldsymbol{D}\rangle}$,

$III_{\langle \boldsymbol{D} \rangle}$, $|\operatorname{grad}\langle \Theta \rangle|$, $|\operatorname{grad}\langle c \rangle|$; they are further simplified or simply set equal to constants. Because k and $\langle \Theta'^2 \rangle$ do not arise in the equations and ε only occurs in the averaged energy balance, further phenomenological statements are needed for these variables. The common procedure is to simply ignore them at this level of closure.

- **Turbulence Model of First Order** For the specific turbulent kinetic energy or another scalar quantity related to it a transport equation is established, and the turbulent viscosity and diffusivities are algebraically connected with this quantity that is evolving in time and space.

  In the somewhat more general variants of the first order models transport equations are established for *two scalar quantities*; these variables are generally the specific turbulent kinetic energy k and the specific turbulent dissipation rate ε, however, also other combinations of scalar quantities are possible. The turbulent viscosity and diffusivities are again connected with these variables; these latter relations are often motivated by means of dimensional analysis[18].
- **Turbulence Models of Second Order** In these closure schemes transport equations are established for the second moments $\langle \boldsymbol{v}' \otimes \boldsymbol{v}' \rangle$, $\langle \boldsymbol{v}'\Theta' \rangle$, $\langle \boldsymbol{v}'c' \rangle$ and $\langle \Theta'^2 \rangle$; the transport equations for these quantities contain higher, third order moments and possibly other new correlation terms which are parameterized by closure conditions of the gradient type or other parameterizations, which are often motivated by methods of dimensional analysis. These closure schemes are called REYNOLDS **Stress Models (RSM)**. In most cases only transport equations of momentum and heat play a role, whilst a tracer mass balance only enters in special cases. A reduction of these full models are the so-called **Algebraic** REYNOLDS **Stress Models (ARSM)** in which the time derivatives of the stresses are ignored.

Besides these categories of models, mixed types are equally possible and often also applied; for instance a closure scheme of second order may be applied for the REYNOLDS stresses but the turbulent heat flux may be parameterized by a model of the zeroth order.

[18] We have seen above that the turbulent kinetic energy k is a variable which arises quite naturally in the definition of the REYNOLDS stress tensor (10.7.25). That two scalar quantities are needed follows alone from the fact that one wishes to express dimensionally the turbulent kinematic viscosity ν_t in terms of k and another scalar quantity. The turbulent dissipation rate ε with dimension $[\mathrm{m}^2\mathrm{s}^{-3}]$, a frequency ω $[\mathrm{s}^{-1}]$ and a length l $[\mathrm{m}]$ are natural choices. They yield

$$\nu_t = c_\mu^\varepsilon \frac{k^2}{\varepsilon}, \qquad \nu_t = c_\mu^l \sqrt{k} l, \qquad \nu_t = c_\mu^\omega k \omega$$

to express ν_t in terms of k and ε, k and l or k and ω, respectively. The coefficients c_μ^ε, c_μ^l and c_μ^ω are dimensionless, but not necessarily constants. Because the choice of the second scalar is not unanimously the same among different modellers, one occasionally speaks of k-z models, leaving the second scalar, z, unspecified.

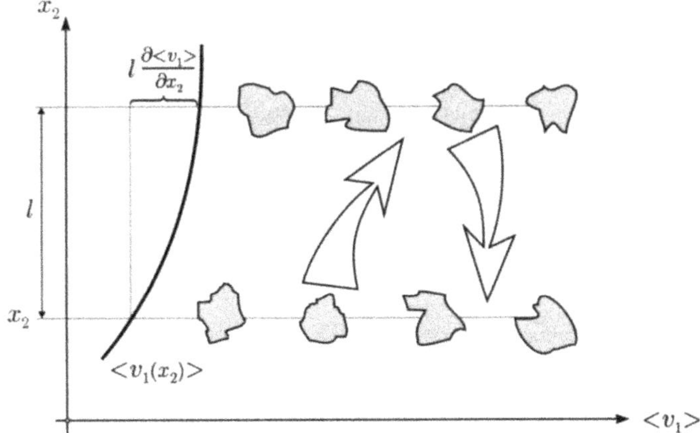

Fig. 10.10. Explaining the concept of the PRANDTL mixing length.

In the subsequent chapters both first and second order models – REYNOLDS stress models – will be motioned or derived. Before we turn to these, we wish to present two early models of turbulence closure of zeroth and first order.

10.8.2 PRANDTL's Mixing Length

PRANDTL [186], [187] developed a simple model which in principle describes the eddy viscosity as a function of the (mean) velocity gradient. To present this model, let us imagine a flow in the x_1-direction (Fig. 10.10) and consider the correlation $\langle v'_1 v'_2 \rangle$. This term describes the turbulent flux (exchange) of momentum from the x_1-direction in the x_2-direction. PRANDTL describes this exchange with the aid of lumps of fluid which propagate perpendicularly to the direction of the flow with a fluctuation velocity v'_2. This transverse motion is thought to take place over a certain length l without any exchange of momentum. Thus, the PRANDTL *mixing length* l is a "mean free path" for turbulent momentum exchange between fluid lumps. Such a lump possesses at position x_2 the mean momentum $\langle v_1(x_2) \rangle$. At a new position $x_2 + l$ after traversing the mean free path the fluid packet takes on the momentum $\langle v_1(x_2 + l) \rangle$. The increase of momentum results in a fluctuation velocity in the x_1-direction, which to first order is proportional to this change. Formulated as an equation, this assumption reads

$$v'_1 \mathrm{d}m = \big(\langle v_1(x_2 + l) \rangle - \langle v_1(x_2) \rangle\big) \mathrm{d}m \approx l \, \frac{\partial \langle v_1 \rangle}{\partial x_2} \mathrm{d}m \, . \tag{10.8.1}$$

Moreover, if one further assumes that the fluctuations in the x_1- and x_2-directions are of the same order of magnitude (a fact that is compelling due to continuity arguments) the correlation of the fluctuation velocities or specific momentum exchanges may be given as

$$\langle v'_1 v'_2 \rangle = l^2 \left(\frac{\partial \langle v_1 \rangle}{\partial x_2} \right)^2 . \tag{10.8.2}$$

Except for a factor ρ the turbulent shear stress is equal to the negative of this expression,

$$\frac{1}{\rho} (\boldsymbol{R})_{12} = -l^2 \left(\frac{\partial \langle v_1 \rangle}{\partial x_2} \right)^2 . \tag{10.8.3}$$

The turbulent viscosity may therefore be written as (note this is always positive!)

$$\nu_t = l^2 \left| \frac{\partial \langle v_1 \rangle}{\partial x_2} \right| , \tag{10.8.4}$$

or more generally

$$\nu_t \propto \left| \operatorname{grad} \langle \boldsymbol{v} \rangle + (\operatorname{grad} \langle \boldsymbol{v} \rangle)^T \right| \quad \Rightarrow \quad \nu_t \propto \sqrt{II_{\langle \boldsymbol{D} \rangle}} . \tag{10.8.5}$$

If one also uses the definition of the fluctuation velocity, one obtains

$$\nu_t = v'_2 l \quad \text{or dimensionally} \quad \nu_t = wl , \tag{10.8.6}$$

where w is a typical turbulent velocity.

This discussion of PRANDTL's ansatz of turbulent mixing lengths will now be complemented by the original text [186].

– Bericht über Untersuchungen zur ausgebildeten Turbulenz

'...

II. Weiter möchte ich von einem Ansatz berichten, der dazu dienen sollte, die Verteilung der Grundströmung einer turbulenten Bewegung unter den verschiedensten Bedingungen hydrodynamisch zu berechnen. Nach verschiedenen vergeblichen Versuchen konnte hier ein erfreulicher Erfolg erzielt werden, und es zeigte sich überdies, daß der Ansatz für die durch den Impulsaustausch hervorgebrachte scheinbare Schubspannung τ, um die es sich hier handelt, auch einer recht anschaulichen Begründung fähig ist. In der BOUSSINESQschen Formel

– Report about investigations regarding fully developed turbulences

'...

II. Furthermore, I wish to report about an ansatz which should serve as a means to hydrodynamically compute under various conditions the distribution of the mean flow of a turbulent motion. After several fruitless attempts a gratifying success could be reached in this regard, and it turned out, in addition, that the formula of concern for the fictitious shear stress τ that is generated by the momentum exchange, can also clearly be motivated. In the BOUSSINESQ formula

$$\tau = \rho \varepsilon \frac{\partial u}{\partial y} \tag{10.8.7}$$

ist ε ein Maß für den turbulenten 'Austausch' und ist seiner Dimension nach, die gleich derjenigen von ν ist, das Produkt einer Länge und einer Geschwindigkeit. Diese Länge und die Geschwindigkeit lassen sich nun vorstellungsmäßig fassen. Die letztere ist die

ε is a measure for the turbulent 'exchange' and, according to its dimension, which is the same as that of ν, is the product of a length and a velocity. This length and the velocity can now conceptually be understood. The latter is the

Quergeschwindigkeit w, mit der im Mittel die von beiden Seiten herankommenden Flüssigkeitsballen durch die Schicht mit dem zeitlichen Mittelwert der Qergeschwindigkeit u hindurchtreten.
Die von der Seite der größeren Geschwindigkeiten kommenden Flüssigkeitsballen bringen auch größere Werte der Geschwindigkeit u mit, die von der Seite der kleineren Geschwindigkeiten dagegen kleinere, so daß immer mehr Impuls in der einen Richtung transportiert wird als in der entgegengesetzten (abgesehen von der Stelle von u_{max}). Die gesuchte Länge l ist nun dadurch charakterisiert, daß sie die Entfernung von der betrachteten Schicht angibt, in der die durchschnittlichen u–Geschwindigkeiten, die die Flüssigkeitsballen bei ihrem Durchtritt haben, als zeitlicher Mittelwert der Strömungsgeschwindigkeit angetroffen werden. Genähert sind diese Geschwindigkeiten also $u + l\partial u/\partial y$ und $u - l\partial u/\partial y$. Daß l der Größenordnung nach mit dem Durchmesser der Flüssigkeitsballen übereinstimmt, sei nebenher erwähnt (genauer ist es der 'Bremsweg' des Flüssigkeitsballens in der übrigen Flüssigkeit, der aber dem Durchmesser proportional ist). Ueber die Länge l kann einstweilen nur ausgesagt werden, daß sie an der Wand gegen Null gehen muß, da hier nur noch Ballen, deren Durchmesser kleiner als der Wandabstand ist, sich wie besprochen bewegen können. Im übrigen soll l einen möglichst regelmäßigen Verlauf haben. Ist β der durchschnittliche verhältnismäßige Anteil der Fläche, der von den von der einen Seite durchtretenden Flüssigkeitsballen eingenommen wird, so tritt an dieser Seite sekundlich ein Impuls $\beta\rho w \cdot l\partial u/\partial y$ durch die Flächeneinheit, von der anderen Seite ungefähr der gleiche Betrag, so daß wir den BOUSSINESQschen Ansatz also bestätigen und $\varepsilon = 2\beta wl$ setzen können.
Es handelt sich jetzt noch darum, für die Mischgeschwindigkeit w einen brauchbaren Ansatz zu machen. Diese Mischgeschwindigkeit wird immer rasch abgebremst und muß immer wieder neu

transverse velocity w by which, on average, the fluid packages enter from both sides the fluid layer that moves with a temporal mean of the transverse velocity u.
The fluid packages coming from the side with the larger velocities also carry with them larger values of the velocity u, those from the side with the smaller velocities, however, smaller ones, so that always more momentum is transported in one direction than in the other (except at a position where u_{max}). The sought length l is now characterised by the fact that it provides the distance from the considered layer in which the average u-velocities, which the fluid packages have on their passage, are encountered as a temporal mean value. Approximations of those velocities are therefore $u + l\partial u/\partial y$ and $u - l\partial u/\partial y$. That l agrees in order of magnitude with the diameter of the fluid packages is only remarked here parenthetically (more accurately, it is the "stopping distance" of the fluid package in the remaining fluid, which, however, is proportional to the diameter). About the length l one can presently only say, that it must go to zero at the wall, since only packages, of which the diameter is smaller than the distance from the wall, can move as discussed. Besides, l should have a behaviour as regular as possible. If β is the averaged relative fraction of area that is encountered by the fluid packages passing from one side, then the momentum per second entering from this side is $\beta\rho w \cdot l\partial u/\partial y$, from the opposite side about the same amount, so that we can corroborate BOUSSINESQ's ansatz and set $\varepsilon = 2\beta wl$.
The remainder now consists in making a useful hypothesis for the mixing velocity w. This mixing velocity is always very quickly attenuated and must continuously be

geschaffen werden. Wir nehmen daher an, daß sie beim Zusammentreffen von zwei Ballen mit verschiedener Geschwindigkeit u erzeugt wird und darum dem Geschwindigkeitsunterschied, also dem Betrage von $l\partial u/\partial y$ proportional ist. Damit wird aber, falls wir alle unbekannten Zahlenfaktoren auf die nicht genauer bekannte Länge l werfen, die scheinbare Schubspannung σ

newly created. We therefore suppose that it is generated in an encounter of two packages with different velocity u and thus is proportional to the velocity difference, whence the modulus of $l\partial u/\partial y$. With this, and provided we throw all unknown factors on this not exactly known length l, the fictitious shear stress σ becomes

$$\tau = \rho l^2 \left| \frac{\partial u}{\partial y} \right| \cdot \frac{\partial u}{\partial y} .$$

Dieser Ansatz bedarf noch einer Berichtigung für den Fall, daß $\partial u/\partial y = 0$ wird. Für die Erzeugung der Geschwindigkeit w wirkt die Nachbarschaft in einer gewissen Breite zusammen; sie wird nicht Null, wenn $\partial u/\partial y = 0$ ist, wird vielmehr einem statistischen Mittelwert von $|\partial u/\partial y|$ proportional gesetzt werden können, also proportional $|\partial u/\partial y|$; verändert sich das Geschwindigkeitsprofil in der Strömungsrichtung, wie bei verengten und erweiterten Kanälen, so wird die Stelle, über die gemittelt wird, auch um einen gewissen Betrag stromauf gelegt werden müssen, da der Vorgang der Ausbildung der Geschwindigkeit w Zeit beansprucht.
...'

This formula still needs to be amended for the case that $\partial u/\partial y = 0$. For the creation of velocity w the neighbourhood of a certain width is active; it does not become zero, if $\partial u/\partial y = 0$, it may rather be set proportional to an average value of $|\partial u/\partial y|$, thus proportional to $|\partial u/\partial y|$ itself; if the velocity profile changes in the direction of the flow, as is the case in contracting and diverging channels, then the position about which the average is taken will have to be moved somewhat upstream, because the process of the creation of the velocity w will take some time.
...'

This is PRANDTL's text both in its original German version and our translation into the English language. PRANDTL's formula constitutes the most simple turbulent closure condition. All quantities in the formula for the mean velocity (10.7.2) are now known. However, the mixing length must be prescribed; depending upon the experiment it can assume different values, and so it is not a constant.

10.8.3 Turbulence Model of First Order

Turbulence closure models of the first order go one step further. If one applies PRANDTL's model of mixing length, then a length l arises for which a numerical value must be found. The extension of merely assigning a single value has been proposed by PRANDTL himself in [188] by postulating a differential equation of the form

$$\frac{\partial l}{\partial t} + \operatorname{div}(l\langle \boldsymbol{v} \rangle) + l \left| \operatorname{grad} \langle \boldsymbol{v} \rangle + (\operatorname{grad} \langle \boldsymbol{v} \rangle)^T \right| + \ldots = 0 , \tag{10.8.8}$$

or

$$\frac{\partial l}{\partial t} + \text{div}\,\big(l\langle \boldsymbol{v}\rangle\big) + 2\,l\sqrt{2II_{\langle \boldsymbol{D}\rangle}} + \ldots = 0\,, \tag{10.8.9}$$

which is a balance for the mixing length.

Other models foresee a differential equation for the turbulent kinetic energy k, whereby this quantity is dimensionally related to the turbulent viscosity via

$$\nu_t \propto \sqrt{k}l\,. \tag{10.8.10}$$

Here too, the length scale must be experimentally determined, or it must a priori be prescribed. The balance law of turbulent kinetic energy is obtained if the equation for the fluctuation velocities $(10.7.5)_2$ is (scalarly) multiplied with the fluctuation velocity and the resulting equation is filtered. In Chap. 11 when treating the k–ε model this equation will be derived in detail.

On the other hand, if one considers the expression (10.8.10) which establishes a connection between the turbulent viscosity, and the turbulent kinetic energy and a characteristic length, then one of the scalar quantities l and k remains unknown in the above procedure and must be set equal to a constant or else parameterized. In extending the above procedure it appears to be more adequate to formulate balance equations for both quantities or for a parameter that can be connected with them. Such quantities are for instance

- the turbulent dissipation rate $\varepsilon \propto k^{3/2}l^{-1}$,
- the turbulent kinetic energy k,
- the turbulent vorticity k/l^2.

For all these models a direct connection with the turbulent viscosity must still be established. This is obtained with the aid of dimensional analysis by appropriately constructing the physical dimension of the quantity under question with the turbulent length scale and turbulent kinetic energy. In this process unknown proportionality factors always arise to which numerical values must be assigned in each particular case. These parameters should have values which enjoy universal properties if the balance laws indeed should reflect some notion of universality. In the next chapter we shall demonstrate the procedure of derivation of a turbulence model of first order, the so-called k–ε model.

10.9 Exercises

1. As an example of illustration of a repetitive application of an averaging operation, compute the temporal average of the function

$$f(t) = \cos\left(\frac{1}{10}\omega t\right) + \cos(\omega t) + \cos\left(\frac{5}{2}\omega t\right)\,.$$

   How does the average of $f(t)$ depend on the choice of the averaging interval T?

2. With the aid of dimensional analysis, show that the decay rate of the turbulent radial energy density for large k, but outside the region where molecular diffusion plays a role, is $k^{-5/3}$.
3. Show that temporal, spatial and statistical averagings commute with temporal and spatial derivatives.
4. Prove the rules (10.3.14) for the filter operations.
5. Prove for isotropic turbulence that the third moment of the velocity fluctuations vanishes. Show, moreover, that the fourth moment can be expresses as stated in formula (10.4.12) .
6. If one wishes to model the anisotropic character of the Reynolds stress tensor, then a possibility is to apply the law $\boldsymbol{R} = \boldsymbol{c}\langle \boldsymbol{D}\rangle$, where $\boldsymbol{c}$ is a fourth order tensor. Prove that this law does not obey the objective transformation rules if the diffusivities c_{ijkl} do not form an isotropic tensor.
7. For a compressible viscous fluid the dependence of the internal energy on the temperature was studied in Chap. 5. Using the balance of energy, derive the heat conduction equation.
8. Using FOURIER transforms, demonstrate that the temporal averaging operation defined in (10.3.3) can never satisfy the condition $\langle\langle u\rangle\rangle = \langle u\rangle$ for multifold averaging. Construct the spectrum of the filter function.
9. Derive the evolution (balance) equation for the mean temperature and the fluctuations of the temperature.
10. Derive the balance equations for the turbulent heat flux $\langle ftemp\boldsymbol{v}'\rangle$ for the turbulent heat flux $\langle \Theta'\boldsymbol{v}'\rangle$ and the temperature variance $\langle {\Theta'}^2\rangle$.

10.10 Solutions

1. We start from

$$\begin{aligned}\langle f\rangle_T &= \frac{1}{T}\int_{t-T/2}^{t+T/2} f(t')\,\mathrm{d}t' \\ &= \frac{1}{T}\int_{t-1/2}^{t+T/2}\left(\cos(\frac{\omega}{10}t') + \cos(\omega t') + \cos(\frac{\omega}{5}t')\right)\mathrm{d}t'\end{aligned} \tag{10.10.1}$$

and employ the formula

$$\int \cos(a\omega t)\,\mathrm{d}t = \frac{1}{a\omega}\sin(a\omega t)\,. \tag{10.10.2}$$

Then we obtain

$$
\begin{aligned}
\langle f\rangle_T &= \frac{1}{T}\left[\frac{10}{\omega}\sin\left(\frac{1}{10}\omega t\right)+\frac{1}{\omega}\sin(\omega t)+\frac{2}{5\omega}\sin\left(\frac{5}{2}\omega t\right)\right]_{t-T/2}^{t+T/2} \\
&= \frac{1}{T}\left(\frac{10}{\omega}\sin\left(\frac{1}{10}\omega\left(t+\frac{T}{2}\right)\right)+\frac{1}{\omega}\sin\left(\omega\left(t+\frac{T}{2}\right)\right)\right. \\
&\quad +\frac{2}{5\omega}\sin\left(\frac{5}{2}\omega\left(t+\frac{T}{2}\right)\right)-\frac{10}{\omega}\sin\left(\frac{1}{10}\omega\left(t-\frac{T}{2}\right)\right) \\
&\quad \left.-\frac{1}{\omega}\sin\left(\omega\left(t-\frac{T}{2}\right)\right)-\frac{2}{5\omega}\sin\left(\frac{5}{2}\omega\left(t-\frac{T}{2}\right)\right)\right) \\
&= \frac{1}{T}\left(\frac{10}{\omega}\sin\left(\frac{1}{10}\omega t+\frac{\omega T}{20}\right)+\frac{1}{\omega}\sin\left(\omega t+\frac{\omega T}{2}\right)\right. \\
&\quad +\frac{2}{5\omega}\sin\left(\frac{5\omega t}{2}+\frac{5\omega T}{4}\right)-\frac{10}{\omega}\sin\left(\frac{\omega t}{10}-\frac{\omega T}{20}\right) \\
&\quad \left.-\frac{1}{\omega}\sin\left(\omega t-\frac{\omega T}{2}\right)-\frac{2}{5\omega}\sin\left(\frac{5\omega t}{2}-\frac{5\omega T}{4}\right)\right)\,. \qquad (10.10.3)
\end{aligned}
$$

Application of the trigonometric addition theorems for sin() und cos() leads to

$$
\begin{aligned}
\langle f\rangle_T &= \frac{1}{T}\left[\frac{20}{\omega}\cos\left(\frac{\omega t}{10}\right)\sin\left(\frac{\omega T}{20}\right)+\frac{2}{\omega}\cos(\omega t)\sin\left(\frac{\omega T}{2}\right)\right. \\
&\quad \left.+\frac{4}{5\omega}\cos\left(\frac{5\omega}{2}t\right)\sin\left(\frac{5\omega}{4}T\right)\right]\,. \qquad (10.10.4)
\end{aligned}
$$

Averaging once more yields now

$$
\begin{aligned}
\langle\langle f\rangle\rangle_T &= \int_{t-T/2}^{t+T/2}\langle f\rangle_T\,\mathrm{d}t = \frac{1}{T^2}\int_{t-T/2}^{t+T/2}\left(\frac{20}{\omega}\sin\left(\frac{\omega T}{20}\right)\cos\left(\frac{\omega t}{10}\right)\right. \\
&\quad \left.+\frac{2}{\omega}\sin\left(\frac{\omega T}{2}\right)\cos(\omega t)+\frac{4}{5\omega}\sin\left(\frac{5\omega T}{4}\right)\cos\left(\frac{5\omega t}{2}\right)\right)\,\mathrm{d}t \\
&= \frac{1}{T^2}\left[\frac{200}{\omega^2}\sin\left(\frac{\omega T}{20}\right)\sin\left(\frac{\omega t}{10}\right)+\frac{2}{\omega^2}\sin\left(\frac{\omega T}{2}\right)\sin(\omega t)\right. \\
&\quad \left.+\frac{8}{25\omega^2}\sin\left(\frac{5\omega T}{4}\right)\sin\left(\frac{5\omega t}{2}\right)\right]_{t-T/2}^{t+T/2} \\
&= \frac{1}{T^2}\left[\frac{200}{\omega^2}\sin\left(\frac{\omega T}{20}\right)\left(\sin\left(\frac{\omega t}{10}+\frac{\omega T}{20}\right)-\sin\left(\frac{\omega t}{10}-\frac{\omega T}{20}\right)\right)\right. \\
&\quad +\frac{2}{\omega^2}\sin\left(\frac{\omega T}{2}\right)\left(\sin\left(\omega t+\frac{\omega T}{2}\right)-\sin\left(\omega t-\frac{\omega T}{2}\right)\right) \\
&\quad \left.+\frac{8}{25\omega^2}\sin\left(\frac{5\omega T}{4}\right)\left(\sin\left(\frac{5\omega t}{2}+\frac{5\omega T}{4}\right)-\sin\left(\frac{5\omega t}{2}-\frac{5\omega T}{4}\right)\right)\right]
\end{aligned}
$$

$$= \frac{1}{T^2}\left[\frac{400}{\omega^2}\sin\left(\frac{\omega T}{20}\right)\cos\left(\frac{\omega t}{10}\right)\sin\left(\frac{\omega T}{20}\right) + \frac{4}{\omega^2}\sin\left(\frac{\omega T}{2}\right)\sin(\omega t)\sin\left(\frac{\omega T}{2}\right) + \frac{16}{25\omega^2}\sin\left(\frac{5\omega T}{4}\right)\sin\left(\frac{5\omega t}{2}\sin\left(\frac{5\omega T}{4}\right)\right)\right] . \quad (10.10.5)$$

Direct comparison of (10.10.4) and (10.10.5) immediately shows that

$$\langle f\rangle_T \neq \langle\langle f\rangle\rangle_T .$$

2. Variables which occur are the spectral density $\mathcal{E}$ (for its definition, see (10.2.7)), the kinematic molecular viscosity ν, the wave number k and the turbulent dissipation rate ε. These have the dimensions shown in the following dimensional matrix

	$\mathcal{E}$	ν	k	ε
L	3	2	−1	2
T	−2	−1	0	−3

(10.10.6)

Since its rank is 2, there are two dimensionless products. If one chooses

$$\pi = \mathcal{E}^{k_1}\nu^{k_2}\mathrm{k}^{k_3}\varepsilon^{k_4} , \quad (10.10.7)$$

one obtains the system of linear equations

$$\begin{aligned} 3k_1 + 2k_2 - k_3 + 2k_4 &= 0 , \\ -2k_1 - k_2 \qquad - 3k_4 &= 0 , \end{aligned} \quad (10.10.8)$$

with two independent solutions

$$\begin{aligned} k_1 = 1 , \quad k_2 = 0 , \quad k_3 = \tfrac{5}{3} , \quad k_4 = -\tfrac{2}{3} , \\ k_1 = 0 , \quad k_2 = 1 , \quad k_3 = \tfrac{4}{3} , \quad k_4 = -\tfrac{1}{3} , \end{aligned} \quad (10.10.9)$$

so that

$$\pi_1 = \mathcal{E}\mathrm{k}^{5/3}\varepsilon^{-2/3} , \quad \pi_2 = \nu\mathrm{k}^{4/3}\varepsilon^{-1/3} \quad (10.10.10)$$

are two independent solutions. Therefore one has

$$\pi_1 = f(\pi_2) = const. = c , \quad (10.10.11)$$

in which the last step follows, because the molecular viscosity is supposed not to play a role. Together with definition $(10.10.10)_1$ equation (10.10.11) therefore implies

$$\mathcal{E} = c\varepsilon^{2/3}\mathrm{k}^{-5/3}$$

For large wavenumbers, i. e., small length scales, the radial energy spectrum falls as $\mathrm{k}^{-5/3}$.

3. (i) The proof $\frac{\partial}{\partial x}\langle f\rangle_T = \langle\frac{\partial f}{\partial x}\rangle_T$ and $\frac{\partial}{\partial t}\langle f\rangle_T = \langle\frac{\partial f}{\partial t}\rangle_T$ has already been given in (10.3.8) and will not be repeated here.

(ii) For the corresponding analogous proof for the spatial filter we start with the integral

$$\frac{1}{V}\int_V f(\boldsymbol{x}',t)\;\mathrm{d}\boldsymbol{x}'^3 = \frac{1}{V}\int_V f(\boldsymbol{x}+\boldsymbol{\xi},t)\;\mathrm{d}\boldsymbol{\xi}^3 \tag{10.10.12}$$

and decompose the integration variable $\boldsymbol{x}' = \boldsymbol{x}+\boldsymbol{\xi}$ into a contribution $\boldsymbol{x}$ that is independent of the integration (it marks the center of the volume V) and a deviation $\boldsymbol{\xi}$ from this center. Then, we obviously have

$$\begin{aligned}\frac{1}{V}\int_V \mathrm{grad}_{\boldsymbol{x}'} f\;\mathrm{d}\boldsymbol{x}'^3 &= \underbrace{\frac{1}{V}\int_V \mathrm{grad}_{\boldsymbol{x}} f\;\mathrm{d}\boldsymbol{\xi}^3}_{(I)}\\ &= \underbrace{\frac{1}{V}\int_V \mathrm{grad}_{\boldsymbol{\xi}} f\;\mathrm{d}\boldsymbol{\xi}^3}_{(II)} = \langle\mathrm{grad}\, f\rangle_R\,.\end{aligned} \tag{10.10.13}$$

In the integral (I) the operator $\mathrm{grad}_{\boldsymbol{x}}$ can be pulled in front of the integration, since $\boldsymbol{x}$ is, relative to the $\boldsymbol{\xi}$-integration variable a constant. Hence, one has

$$\begin{aligned}\mathrm{grad}_{\boldsymbol{x}}\,\langle f\rangle_R &= \mathrm{grad}_{\boldsymbol{x}}\,\frac{1}{V}\int_V f\;\mathrm{d}\boldsymbol{\xi}^3\\ &= \frac{1}{V}\int_V \mathrm{grad}_{\boldsymbol{\xi}} f\;\mathrm{d}\boldsymbol{\xi}^3 = \langle\mathrm{grad}_{\boldsymbol{\xi}}\, f\rangle_R\,.\end{aligned} \tag{10.10.14}$$

The analogous proof for the time derivative is easier, since the time is kept fixed in the integral (10.10.12). Hence, the differentiation with respect to time can directly be pulled out of the integral, so that

$$\begin{aligned}&\frac{1}{V}\int_V \frac{\partial}{\partial t} f(\boldsymbol{x}'+\boldsymbol{\xi},t)\;\mathrm{d}\boldsymbol{\xi}^3 = \frac{\partial}{\partial t}\frac{1}{V}\int_V f(\boldsymbol{x}+\boldsymbol{\xi})\;\mathrm{d}\boldsymbol{\xi}^3\\ \text{or}\quad &\langle\frac{\partial f}{\partial t}\rangle_R = \frac{\partial}{\partial t}\langle f\rangle_R\,,\quad \textit{qed.}\end{aligned} \tag{10.10.15}$$

(iii) For the statistical filter the proof follows the lines of formula (10.10.15)

$$\begin{aligned}\langle\mathrm{grad}_x\, f\rangle_S &= \int \mathrm{grad}_x\, f(\boldsymbol{x},t;\wp)\;\mathrm{d}\wp\\ &= \mathrm{grad}_x \int f(\boldsymbol{x},t;\wp)\;\mathrm{d}\wp\\ &= \mathrm{grad}\,\langle f\rangle_S\,,\quad \textit{qed.}\end{aligned} \tag{10.10.16}$$

Analogously,

$$\langle \frac{\partial}{\partial t} f \rangle_S = \int \frac{\partial}{\partial t} f(\boldsymbol{x}, t; \wp)\, \mathrm{d}\wp = \frac{\partial}{\partial t} \int f(\boldsymbol{x}, t; \wp)\, \mathrm{d}\wp = \frac{\partial}{\partial t} \langle f \rangle_S ,$$

completing the proof.

4. For the solution of this problem it is assumed that all three properties of a filter are fulfilled. With $u = \langle u \rangle + u'$, we have

(i)
$$\begin{aligned} \langle u \rangle &= \langle \langle u \rangle \rangle + \langle u' \rangle \quad (\text{ Linearity }) \\ &= \langle u \rangle + \langle u' \rangle \quad (\langle \langle\ \rangle \rangle = \langle\ \rangle) \\ \Rightarrow \quad & \langle u' \rangle = 0 . \end{aligned}$$

(ii) If one forms the fluctuating quantity $\langle u \rangle = u - u'$ one gets

$$\langle u \rangle' = u' - u'' = 0 , \quad \text{since} \quad u'' = u' .$$

This proof can also be conducted as follows. Decompose $\langle u \rangle$ into mean value and fluctuation; then one obtains

$$\begin{aligned} \langle u \rangle &= \langle \langle u \rangle \rangle + \langle u \rangle' \\ &= \langle u \rangle + \langle u \rangle' \quad \Rightarrow \quad \langle u \rangle' = 0 . \end{aligned}$$

(iii) Because in view of (10.3.13) we have $\langle w \langle u \rangle \rangle = \langle w \rangle \langle u \rangle$, and thus one may deduce

$$\begin{aligned} \langle \langle u \rangle w \rangle &= \langle \langle u \rangle (\langle w \rangle + w') \rangle \\ &= \langle \langle u \rangle \langle w \rangle + \langle u \rangle w' \rangle \\ &= \langle u \rangle \langle w \rangle + \langle u \rangle \langle w' \rangle \end{aligned}$$

and hence

$$\langle \langle u \rangle w' \rangle = 0 .$$

(iv)
$$\begin{aligned} \langle uv \rangle &= \langle (\langle u \rangle + u')(\langle v \rangle + v') \rangle \\ &= \langle \langle u \rangle \langle v \rangle \rangle + \underbrace{\langle u' \rangle \langle v \rangle}_{=0} + \underbrace{\langle u \rangle \langle v' \rangle}_{=0} + \langle u'v' \rangle \\ &= \langle u \rangle \langle v \rangle + \langle u'v' \rangle \quad \textit{qed.} \end{aligned}$$

5. (i) The *third moment* of $\boldsymbol{v}'$ is given by

$$Q^{(3)}_{ijk} = \langle v'_i v'_j v'_k \rangle \tag{10.10.17}$$

and transforms under EUCLIDian transformations as

$$\langle v^{*\prime}_i v^{*\prime}_j v^{*\prime}_k \rangle = \langle O_{il} v'_l O_{jm} v'_m O_{kn} v'_n \rangle \stackrel{!}{=} \langle v'_i v'_j v'_k \rangle , \tag{10.10.18}$$

in which the last step follows from the condition of isotropy. We only present a sketch of the proof.

For a rotation about the x_3- and x_1-axes, respectively, by an angle $\pi/2$, the following relations must follow

$$\left.\begin{array}{lll} x_1^* = x_2, & x_2^* = -x_1, & x_3^* = x_3, \\ v_1^{*\prime} = v_2', & v_2^{*\prime} = -v_1', & v_3^{*\prime} = v_3' \end{array}\right\} \text{rotation about } x_3\text{–axis,} \tag{10.10.19}$$

$$\left.\begin{array}{lll} x_1^{**} = x_1, & x_2^{**} = -x_3, & x_3^{**} = x_2, \\ v_1^{**\prime} = v_1', & v_2^{**\prime} = -v_3', & v_3^{**\prime} = v_2' \end{array}\right\} \text{rotation about } x_1\text{–axis.} \tag{10.10.20}$$

Therefore, (10.10.18) implies

1)
$$\left.\begin{array}{l} \langle v_1^{*\prime 3}\rangle = \langle {v_2'}^3\rangle = \langle {v_1'}^3\rangle, \\ \langle v_2^{*\prime 3}\rangle = -\langle {v_1'}^3\rangle = \langle {v_2'}^3\rangle, \end{array}\right\} \Rightarrow \langle {v_1'}^3\rangle = 0\,. \tag{10.10.21}$$

2)
$$\left.\begin{array}{l} \langle v_2^{**\prime 3}\rangle = -\langle {v_3'}^3\rangle = \langle {v_2'}^3\rangle, \\ \langle v_3^{**\prime 3}\rangle = \langle {v_2'}^3\rangle = \langle {v_3'}^3\rangle, \end{array}\right\} \Rightarrow \langle {v_3'}^3\rangle = \langle {v_2'}^3\rangle = 0\,. \tag{10.10.22}$$

3)
$$\begin{array}{l} \langle v_1^{*\prime} v_2^{*\prime} v_3^{*\prime}\rangle = -\langle v_1' v_2' v_3'\rangle = \langle v_1' v_2' v_3'\rangle\,, \\ \Rightarrow \quad \langle v_1' v_2' v_3'\rangle = 0\,. \end{array} \tag{10.10.23}$$

4)
$$\begin{array}{l} \left.\begin{array}{l} \langle v_1^{*\prime 2} v_2^{*\prime}\rangle = -\langle v_1' {v_2'}^2\rangle = \langle {v_1'}^2 v_2'\rangle, \\ \langle v_1^{*\prime} v_2^{*\prime 2}\rangle = \langle {v_1'}^2 v_2'\rangle - \langle v_1' {v_2'}^2\rangle, \end{array}\right\} \\ \Rightarrow \langle v_1' {v_2'}^2\rangle = \langle {v_1'}^2 v_2'\rangle = 0\,. \end{array} \tag{10.10.24}$$

5)
$$\begin{array}{l} \langle v_2^{*\prime 2} v_3^{*\prime}\rangle = \langle {v_1'}^2 v_3\rangle = \langle {v_2'}^2 v_3'\rangle, \\ \langle v_2^{*\prime} v_3^{*\prime 2}\rangle = -\langle {v_1'}^2 v_3'\rangle = \langle v_2' {v_3'}^2\rangle, \\ \langle v_2^{**\prime} v_3^{**\prime 2}\rangle = -\langle v_3' {v_2'}^2\rangle = \langle v_2' {v_3'}^2\rangle, \\ \langle v_1^{**\prime 2} v_2^{**\prime}\rangle = -\langle {v_3'}^2 v_1'\rangle = \langle {v_1'}^2 v_2'\rangle, \\ \langle v_1^{*\prime 2} v_2^{*\prime}\rangle = -\langle {v_2'}^2 v_1'\rangle = \langle {v_1'}^2 v_2'\rangle, \\ \langle v_1^{**\prime 2} v_3^{**\prime}\rangle = \langle {v_1'}^2 v_2'\rangle = \langle {v_1'}^2 v_3'\rangle, \\ \langle v_1^{**\prime} v_2^{**\prime 2}\rangle = \langle v_1' {v_3'}^2\rangle = \langle v_1' {v_2'}^2\rangle. \end{array} \tag{10.10.25}$$

(10.10.25) and the preceding results now imply that $\langle v_i'^2 v_j' \rangle = 0$ for all $i, j = 1, 2, 3$. Indeed, $(10.10.25)_{6,7}$ and (10.10.24) yields

$$\boldsymbol{Q}^{(3)} \equiv \boldsymbol{0}\,, \quad qed.$$

(ii) The proof for $\boldsymbol{Q}^{(4)}$ is differently conducted. From continuum mechanics we know that an isotropic tensor of rank four has the form

$$Q^4_{ijkl} = \lambda \delta_{ij}\delta_{kl} + 2\mu(\delta_{ik}\delta_{jl} + \delta_{il}\delta_{jk}) \tag{10.10.26}$$

with arbitrary λ, μ. Thus, one now must determine λ and μ. With the representation (10.10.26) we may now show that

$$\begin{aligned} &\langle v_1'^4 \rangle = \langle v_2'^4 \rangle = \langle v_3'^4 \rangle = \lambda + 4\mu\,, \\ &\langle v_i'^3 v_j' \rangle = 0\,, \quad \forall\, i \neq j\ (= 1,\ 2,\ 3)\,, \\ &\langle v_i'^2 v_j'^2 \rangle = \lambda\,, \quad \forall\, i \neq j\ (= 1,\ 2,\ 3)\,, \\ &\langle v_i'^2 v_j' v_k' \rangle = 0\,, \quad \forall\, i \neq j \neq k \neq i\ (= 1,\ 2,\ 3)\,. \end{aligned} \tag{10.10.27}$$

Therefore, one has

$$\begin{aligned} \lambda &= \langle v_i'^2 v_j'^2 \rangle\,, \quad \forall\, i \neq j\ (= 1,\ 2,\ 3)\,, \\ \mu &= \tfrac{1}{4}(\langle v_i'^4 \rangle - \langle v_i'^2 v_j'^2 \rangle)\,, \quad \forall\, i \neq j\ (= 1,\ 2,\ 3)\,, \end{aligned} \tag{10.10.28}$$

qed.

6. In (10.7.21) it was shown that under EUCLIDian transformations the REYNOLDS stress tensor transforms as

$$\hat{\boldsymbol{R}}(\boldsymbol{O}\langle\boldsymbol{D}\rangle\boldsymbol{O}^T) = \boldsymbol{O}\hat{\boldsymbol{R}}(\langle\boldsymbol{D}\rangle)\boldsymbol{O}^T\,. \tag{10.10.29}$$

In other words $\hat{\boldsymbol{R}}(\cdot)$ is an isotropic tensor function of the symmetric tensor $\langle\boldsymbol{D}\rangle$ and possesses the representation

$$\boldsymbol{R} = \alpha\boldsymbol{I} + \beta\langle\boldsymbol{D}\rangle + \gamma\langle\boldsymbol{D}\rangle^2\,,$$

in which α, β, γ depend upon the invariants of $\langle\boldsymbol{D}\rangle$. A representation

$$\boldsymbol{R} = \boldsymbol{c}\langle\boldsymbol{D}\rangle\,, \tag{10.10.30}$$

in which $\boldsymbol{c}$ is a general tensor of rank four cannot exist. Nevertheless such a form is used in meteorology and oceanography, it is of course an inobjective representation.
If one intends to formulate an objective anisotropic law, then $\boldsymbol{R}$ must depend on $\langle\boldsymbol{D}\rangle$ as well as on a structure tensor $\boldsymbol{M}$: $\boldsymbol{R} = \hat{\boldsymbol{R}}(\langle\boldsymbol{D}\rangle, \boldsymbol{M})$. See more on this in Chap. 12.

7. The local form of the energy equation is given by

$$\rho\dot{\varepsilon} = -\operatorname{div}\boldsymbol{q} + \operatorname{tr}(\boldsymbol{t}\boldsymbol{D}) + \rho\mathfrak{r} \tag{10.10.31}$$

with the same notation as in Chap. 5. One decomposes the stress tensor

$$\boldsymbol{t} = -p\boldsymbol{I} - \boldsymbol{t}^R \tag{10.10.32}$$

into a pressure and a frictional tensor and then obtains

$$\begin{aligned} \rho(\dot{\epsilon} + p(1/\rho)^{\cdot}) &= -\operatorname{div}\boldsymbol{q} + \Phi + \rho\mathfrak{r} \,, \\ \Phi &:= \operatorname{tr}(\boldsymbol{t}^R\boldsymbol{D}) \,, \end{aligned} \tag{10.10.33}$$

where use has also been made of the mass balance $\operatorname{div}\boldsymbol{v} = -\dot{\rho}/\rho = \rho(1/\rho)^{\cdot}$. Φ is the dissipation function. If, on the left-hand side also the GIBBS relation (5.6.24)

$$\dot{\epsilon} = \Theta\dot{s} - p\left(\frac{1}{\rho}\right)^{\cdot} \,, \tag{10.10.34}$$

is used, then (10.10.33) takes the form

$$\rho\Theta\frac{\mathrm{d}s}{\mathrm{d}t} = -\operatorname{div}\boldsymbol{q} + \Phi + \rho\mathfrak{r} \,. \tag{10.10.35}$$

For a NAVIER–STOKES–FOURIER fluid one has

$$\boldsymbol{q} = -\lambda(\Theta,\rho)\operatorname{grad}\Theta \,, \quad \boldsymbol{t}^R = 2\mu(\Theta,\rho)\underbrace{\left(\boldsymbol{D} - \tfrac{1}{3}I_{\boldsymbol{D}}\boldsymbol{I}\right)}_{\boldsymbol{E}} \,, \tag{10.10.36}$$

if one assumes that the bulk viscosity vanishes. One then writes for the dissipation

$$\Phi = 2\mu(\Theta,\rho)\operatorname{tr}(\boldsymbol{E}^2) \,. \tag{10.10.37}$$

With all this, the energy equation takes the form

$$\begin{aligned} \rho(\dot{\epsilon} + p(1/\rho)^{\cdot}) &= \operatorname{div}(\lambda(\Theta,\rho)\operatorname{grad}\Theta) \\ &\quad + 2\mu(\Theta,\rho)\operatorname{tr}(\boldsymbol{E}^2) + \rho\mathfrak{r} \,. \end{aligned} \tag{10.10.38}$$

There remains to find an expression for the internal energy (or the entropy). For an *ideal gas* $\epsilon = \hat{\epsilon}(\Theta)$ and

$$\dot{\epsilon} = \hat{c}_v(\Theta)\dot{\Theta} \,, \quad \hat{c}_v = \frac{\partial\hat{\epsilon}}{\partial\Theta} \,. \tag{10.10.39}$$

Substitution of this expression into (10.10.38) then yields the heat conduction equation for an ideal gas.

For a general heat-conducting viscous fluid the functions $\hat{c}_v(\Theta, 1/\rho)$ and $\hat{p}(\Theta, 1/\rho)$ are determined by experiment and $\hat{\epsilon}(\Theta, 1/\rho)$ is obtained from the GIBBS relation (10.10.34) by integration. This integration leads to

$$\begin{aligned}\hat{\epsilon}(\Theta, 1/\rho) = &\int_{\Theta_0}^{\Theta} \hat{c}_v(T, 1/\rho)\, \mathrm{d}T \\ &+ \int_{1/\rho_0}^{1/\rho} \hat{p}(\Theta_0, \boldsymbol{\xi})\Big(\Theta\hat{\beta}(\Theta_0, \boldsymbol{\xi}) - 1\Big)\mathrm{d}\boldsymbol{\xi} \\ &+ \hat{\epsilon}(\Theta_0, 1/\rho_0)\,,\end{aligned} \tag{10.10.40}$$

in which

$$\beta := \frac{1}{p}\frac{\partial p}{\partial \Theta}\bigg|_{\rho} \tag{10.10.41}$$

denotes the so-called *isochoric* stress coefficient, which is known if $\hat{p}(\Theta, 1/\rho)$ is known. Equations (10.10.38) and (10.10.40) together form the heat conduction equation. In many practical situations the second term in (10.10.40) is ignored. If so, (10.10.38) reads

$$\begin{aligned}\rho\left(\hat{c}_v(\Theta, \rho)\frac{\mathrm{d}\Theta}{\mathrm{d}t} + \hat{p}(\Theta, \rho)\frac{\mathrm{d}}{\mathrm{d}t}\left(\frac{1}{\rho}\right)\right) &= \operatorname{div}(\hat{\lambda}(\Theta, \rho)\operatorname{grad}\Theta) \\ &\quad + 2\hat{\mu}(\Theta, \rho)\operatorname{tr}(\boldsymbol{E}^2) + \rho\mathfrak{r}\,,\end{aligned} \tag{10.10.42}$$

and if the second term on the left-hand side is ignored and c_v, λ and μ are set equal to constants

$$\begin{aligned}\frac{\mathrm{d}\Theta}{\mathrm{d}t} &= \chi^{(\Theta)} \,\triangle\, \Theta + \underbrace{\frac{2\nu}{c_v}\operatorname{tr}\boldsymbol{E}^2}_{\frac{1}{\rho c_v}\phi} + \frac{\mathfrak{r}}{c_v}\,, \\ \chi^{(\Theta)} &:= \frac{\lambda}{\rho c_v} \quad \text{Diffusivity.}\end{aligned} \tag{10.10.43}$$

This is the form of the heat conduction equation as it is normally used in applications.

8. The temporal mean value of $f(\boldsymbol{x}, t)$,

$$\langle f\rangle_T := F(t) = \frac{1}{T}\int_{t-T/2}^{t+T/2} f(\boldsymbol{x}, \tau)\, \mathrm{d}\tau \tag{10.10.44}$$

can be interpreted as a *convolution* of the function $f(t)$ with a *window function*

$$w(t) := \begin{cases} \frac{1}{T}\,, & t \in [-\frac{T}{2}, \frac{T}{2}], \\ 0\,, & \text{otherwise.}\end{cases} \tag{10.10.45}$$

Indeed,

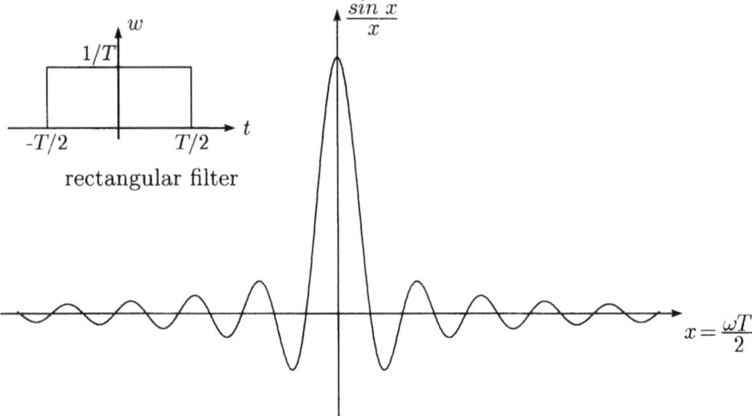

Fig. 10.11. The function $\sin x/x$ is the FOURIER transform of the rectangular filter.

$$F(t) = \frac{1}{T}\int_{-\infty}^{\infty} f(\tau)\ (\tau - t)\ \mathrm{d}\tau =: f \circ w\ , \tag{10.10.46}$$

this is a FOURIER convolution integral in which the FOURIER transform and its inverse are defined by

$$\tilde{f}(\omega) = \int_{-\infty}^{\infty} f(t)\ e^{-i\omega t}\ \mathrm{d}t \tag{10.10.47}$$

and

$$f(t) = \frac{1}{2\pi}\int_{\infty}^{\infty} \tilde{f}(t)\ e^{i\omega t}\ \mathrm{d}\omega\ , \tag{10.10.48}$$

respectively. For these the following convolution theorem holds:

$$\widetilde{f \circ w} = \tilde{f} \cdot \tilde{w}\ . \tag{10.10.49}$$

A multiplication of two FOURIER-transformed functions (one says: multiplication in frequency space) therefore corresponds to a convolution in physical space. The filtered signal of a time series $f(t)$ is obtained as a multiplication of the spectrum of f with the spectrum of the filter.
In order to separate various regions of the periodogram one should apply in frequency space an "abrupt" filter, such as e.g. a rectangular filter (see Fig. 10.11). Now, to the temporal average, there corresponds a rectangular filter in the physical domain. The FOURIER transform of this filter is the function $\sin x/x$, which oscillates and embraces the entire frequency space. Since this is not a rectangular filter, there can never hold an equation such as $\langle\langle f\rangle\rangle_T = \langle f\rangle_T$.

9. We start with the heat conduction equation in the form

$$\frac{\partial \Theta}{\partial t} + \operatorname{div}(\Theta \boldsymbol{v}) = \operatorname{div}\left(\chi^{(\Theta)} \operatorname{grad} \Theta\right) + \frac{1}{\rho c_v}\Phi + \frac{1}{c_v}\mathfrak{r}\,, \qquad (10.10.50)$$

where use has been made of the fact that $\operatorname{div} \boldsymbol{v} = 0$. Because

$$\langle \Theta \boldsymbol{v} \rangle = \langle \Theta \rangle \langle \boldsymbol{v} \rangle + \langle \Theta' \boldsymbol{v}' \rangle\,, \qquad (10.10.51)$$

there follows from (10.10.50) by averaging

$$\frac{\partial \langle \Theta \rangle}{\partial t} + \operatorname{div}(\langle \Theta \rangle \langle \boldsymbol{v} \rangle) = -\operatorname{div}(\boldsymbol{q}_t - \chi^{(\Theta)} \operatorname{grad} \langle \Theta \rangle) + \frac{\langle \Phi \rangle}{\rho c_v} + \frac{1}{c_v}\mathfrak{r}\,. \quad (10.10.52)$$

Here,

$$\boldsymbol{q}_t := \langle \Theta' \boldsymbol{v}' \rangle$$

denotes the turbulent heat flux. If one subtracts (10.10.52) from (10.10.50) and defines

$$\Theta' = \Theta - \langle \Theta \rangle\,, \qquad (10.10.53)$$

then the equation for the temperature fluctuations is obtained,

$$\begin{aligned} &\frac{\partial \Theta'}{\partial t} + \operatorname{div}(\boldsymbol{v}'\langle \Theta \rangle + \langle \boldsymbol{v} \rangle \Theta' + \boldsymbol{v}'\Theta' - \langle \boldsymbol{v}'\Theta' \rangle) \\ &\qquad = \chi^{(\Theta)} \operatorname{div} \operatorname{grad} \Theta' + \frac{1}{\rho c_v}\Phi'\,. \end{aligned} \qquad (10.10.54)$$

This equation is needed, if balance laws for $\langle \Theta' \boldsymbol{v}' \rangle$ and $\langle {\Theta'}^2 \rangle$ are sought.

10. For the determination of a balance equation for the variance of the temperature fluctuation we start from (10.10.54), multiply this equation with Θ', and subsequently average the resulting equation. This yields

$$\begin{aligned} &\underbrace{\left\langle \frac{\partial \Theta'}{\partial t}\Theta' \right\rangle}_{(1)} + \underbrace{\langle \operatorname{div}(\boldsymbol{v}'\langle \Theta \rangle)\,\Theta' \rangle}_{(2)} + \underbrace{\langle \operatorname{div}(\langle \boldsymbol{v} \rangle \Theta')\,\Theta' \rangle}_{(3)} \\ &\quad + \underbrace{\langle \operatorname{div}(\boldsymbol{v}'\Theta')\,\Theta' \rangle}_{(4)} - \underbrace{\langle \operatorname{div}(\langle \boldsymbol{v}'\Theta' \rangle)\,\Theta' \rangle}_{(5)} \\ &\quad = \underbrace{\left\langle \operatorname{div}\left(\chi^{(\Theta)} \operatorname{grad} \Theta'\right)\Theta' \right\rangle}_{(6)} - \underbrace{\left\langle \frac{\Phi'\Theta'}{\rho c_v} \right\rangle}_{(7)}\,. \end{aligned} \qquad (10.10.55)$$

Reverting to Cartesian tensor notation the seven terms can be transformed as follows:

$$(1) = \frac{\partial}{\partial t}\left(\frac{\left\langle \Theta'^2 \right\rangle}{2}\right) ,$$

$$(2) = \left\langle (v_i'\langle\Theta\rangle)_{,i}\,\Theta' \right\rangle = \left\langle v_i'\langle\Theta\rangle_{,i}\Theta' \right\rangle = \langle\Theta\rangle_{,i}\langle v_i'\Theta'\rangle ,$$

$$(3) = \left\langle (\langle v_i\rangle\Theta')_{,i}\,\Theta' \right\rangle = \langle\langle v\rangle_i\Theta'_{,i}\Theta'\rangle$$

$$= \langle v\rangle_i\left\langle\frac{\Theta'^2}{2}\right\rangle_{,i} + \underbrace{\langle v\rangle_{i,i}}_{0}\left\langle\frac{\Theta'^2}{2}\right\rangle = \left(\langle v\rangle_i\left\langle\frac{\Theta'^2}{2}\right\rangle\right)_{,i} ,$$

$(4) = 0\,,$ since it is linear in the fluctuation,

$$(5) = \left\langle (v_i'\Theta')_{,i}\,\Theta' \right\rangle = \left\langle v_i'\left(\frac{\Theta'^2}{2}\right)_{,i} + \underbrace{v_{i,i}'}_{0}\frac{\Theta'^2}{2} \right\rangle = \left\langle v_i'\frac{\Theta'^2}{2}\right\rangle_{,i} ,$$

$$(6) = \left\langle \left(\chi^{(\Theta)}\Theta'_{,i}\right)_{,i}\Theta' \right\rangle = \left\langle \chi_{,i}^{(\Theta)}\Theta'_{,i}\Theta' + \chi^{(\Theta)}\Theta'_{,ii}\Theta' \right\rangle$$

$$= \left\langle \chi_{,i}^{(\Theta)}\left(\frac{\Theta'^2}{2}\right)_{,i} + \chi^{(\Theta)}\left[\left(\frac{\Theta'^2}{2}\right)_{,i}\right]_{,i} - \chi^{(\Theta)}\Theta'_{,i}\Theta'_{,i} \right\rangle$$

$$= \left\langle \chi^{(\Theta)}\left(\frac{\Theta'^2}{2}\right)_{,i} \right\rangle_{,i} - \left\langle \chi^{(\Theta)}\Theta'_{,i}\Theta'_{,i} \right\rangle ,$$

$$(7) = \left\langle \frac{\Phi'\Theta'}{\rho c_v} \right\rangle .$$

Substituting these relations in the above equation and rearranging yields

$$\underbrace{\frac{\partial\left\langle\Theta'^2\right\rangle}{\partial t}}_{\text{local change}} + \underbrace{\operatorname{div}\left(\langle \boldsymbol{v}\rangle\left\langle\Theta'^2\right\rangle\right)}_{\text{convective change}} = -\underbrace{\operatorname{div}\left\langle \boldsymbol{v}'\Theta'^2\right\rangle}_{\text{diffusive flux contrib.}} \underbrace{- 2\operatorname{grad}\langle\Theta\rangle\underbrace{\langle \boldsymbol{v}'\Theta'\rangle}_{\propto \boldsymbol{q}_t}}$$

$$+ \underbrace{\operatorname{div}\left\langle \chi^{(\Theta)}\Theta'^2\right\rangle} - \underbrace{2\left\langle \chi^{(\Theta)}\left(\operatorname{grad}\Theta'\right)^2\right\rangle} - \underbrace{\left\langle\frac{\Phi'\Theta'}{\rho c_v}\right\rangle} , \quad (10.10.56)$$

in which the last four terms are production and annihilation terms, respectively. If it is assumed that the molecular thermal diffusivity $\chi^{(\Theta)}$ is constant, and, moreover, if $\langle\Phi'\Theta'/\rho c_v\rangle$ is ignored, then the above equation simplifies to

$$\frac{\partial\left\langle\Theta'^2\right\rangle}{\partial t}+\operatorname{div}\left(\langle\boldsymbol{v}\rangle\left\langle\Theta'^2\right\rangle\right)=-\operatorname{div}\left\langle\boldsymbol{v}'\Theta'^2\right\rangle-2\operatorname{grad}\langle\Theta\rangle\boldsymbol{q}_t$$
$$+\chi^{(\Theta)}\operatorname{div}\left\langle\Theta'^2\right\rangle-2\chi^{(\Theta)}\langle|\operatorname{grad}\Theta'|^2\rangle\,. \qquad (10.10.57)$$

Because $\boldsymbol{q}_t$ is given by its own evolution equation, there are still closure conditions needed. Possibilities are

$$\left\langle\boldsymbol{v}'\Theta'^2\right\rangle=-\frac{\nu_t}{\sigma_{\Theta^2}}\operatorname{grad}\left\langle\Theta'^2\right\rangle=-\frac{c_\mu}{\sigma_{\Theta^2}}\frac{k^2}{\varepsilon}\operatorname{grad}\left\langle\Theta'^2\right\rangle$$
$$=-c_{\Theta^2}\frac{k^2}{\varepsilon}\operatorname{grad}\left\langle\Theta'^2\right\rangle, \qquad (10.10.58)$$
$$\left\langle 2\chi^{(\Theta)}\left(\operatorname{grad}\Theta'\right)^2\right\rangle=c_{\operatorname{grad}\Theta^2}\frac{\varepsilon^2}{k^3}\left\langle\Theta'^2\right\rangle$$

with dimensionless parameters c_{Θ^2} and $c_{\operatorname{grad}\Theta^2}$.

11. k–ε Model for Density Preserving and BOUSSINESQ Fluids

In this chapter a simple turbulence model of first order – called the k–ε model – will be developed, which merely uses evolution equations for the specific turbulent kinetic energy, k and the specific turbulent dissipation rate, ε. In the derivation concepts will become known, which illustrate how also higher order models might be handled. This so-called k–ε model shall in the subsequent paragraphs be developed for a density preserving fluid and is then extended for a BOUSSINESQ fluid i.e., a fluid in which density variations are only accounted for in the body (gravity) force. In particular the CORIOLIS acceleration and the buoyancy effects play often a role in combination in such geophysical applications, e.g. in a stratified atmosphere or in the ocean which are typical configurations in meteorology, oceanography and limnology.

This chapter serves as a transition to the more general models such as the higher order REYNOLDS *stress models* and the *algebraic* REYNOLDS *stress models*, etc. Whereas in the REYNOLDS stress models transport equations are developed for the six components of the REYNOLDS stress tensor and the three components of the heat flux vector, this is not necessary for the algebraic REYNOLDS stress models, as one might imply simply from the name *algebraic* REYNOLDS *stress model.* Such a model may be obtained, on the one hand, by a generalization of the classical k–ε model by simply postulating a more general closure condition for the REYNOLDS stress tensor and the heat flux vector. On the other hand, an algebraic REYNOLDS stress model may be obtained from the REYNOLDS stress model by functionally relating certain transport quantities with the turbulent kinetic energy. This also yields a k–ε model, however, with more complicated closure expressions than in the simple k–ε model. The closure procedures are often done without consideration of thermodynamic irreversibility, but, of course, conformity with the second law of thermodynamics is actually compulsory. In this chapter such considerations will be disregarded.

Before presenting the model equations, an important relation between the turbulent viscosity and the turbulent kinetic energy and dissipation rate should be mentioned. Basis of a k–ε model is the hypothesis of REYNOLDS (or BOUSSINESQ) that the REYNOLDS stress tensor is proportional to the mean stretching tensor with a factor of proportionality ν_t, called turbulent kinematic viscosity. This scalar parameter of the k–ε model is the only place where the averaged equations of motion are connected with the turbulence equations. The evolution equations for k and ε therefore only serve to compute the turbulent viscosity. The balance laws of the averaged fields are therefore only coupled with the k–ε equations via the turbulent kinematic viscosity; however the turbulent kinetic energy, k and its dissipation, ε are themselves coupled with the physical transport equations of mass, momentum and energy via the gradient of the mean velocity field and the mean temperature gradient and the parameterisations of the production terms. A relation between the turbulent kinetic energy, its dissipation rate and the turbulent kinematic viscosity can be deduced with the aid of arguments of dimensional analysis. A simple evaluation of the (only) dimensionless product that exists on the basis of the dimensional matrix

	ν_t	k	ε
L	2	2	2
T	-1	-2	-3

yields, according to BUCKINGHAM's theorem, the formula $\Pi := k^2/(\nu_t \varepsilon) =$ const. so that

$$\nu_t = c_\mu \frac{k^2}{\varepsilon} , \tag{11.0.1}$$

in which c_μ is a coefficient of proportionality of which the value was identified as 0.09. It is not a universal constant, despite the fact that this is sometimes claimed; all the more, this coefficient may also depend in principle upon the invariants of the REYNOLDS stress tensor and the temperature gradient and possible other quantities. The above mentioned value has been obtained with the aid of experiments and inverse adjustment by numerical modelling. In the ensuing derivation of the evolution equations for k and ε many other terms will arise, which may equally be parameterized with considerations of dimensional analysis.

Historically, the k–ε model has originally been developed by HANJALIC & LAUNDER [94], JONES & LAUNDER [114] and LAUNDER & SPALDING [126]. It has recently attracted great attention in the engineering and geophysical modelling community. RODI [196], [197], [197] describes its applicability in geophysics and hydraulic engineering and UMLAUF [246] and WEIS [254] put it into the proper perspective with other two-equation models as well as certain algebraic REYNOLDS stress models. We shall come back to those connections later on.

11.1 Model Equations

11.1.1 Definitions and Balance Laws

Because the derivation of the model equations and the parameterizations of the various terms are rather complex so that the overview may be lost in the many derivations, we shall first state the k–ε model and discuss its significance and derive it only afterwards. Let a filter be given which satisfies all requirements of Sect. 10.3.2 of Chap. 10. Such a filter is e.g. the statistical filter. The mean velocities, the pressure and the REYNOLDS stress tensor then obey the balance laws

$$\begin{aligned} &\operatorname{div}\langle \boldsymbol{v}\rangle = 0 , \\ &\frac{\partial \langle \boldsymbol{v}\rangle}{\partial t} + \operatorname{div}(\langle \boldsymbol{v}\rangle \otimes \langle \boldsymbol{v}\rangle) - \frac{1}{\rho}\operatorname{div}\boldsymbol{R} + \operatorname{grad}\langle p_{\mathrm{r}}\rangle - \nu \operatorname{div}\langle \boldsymbol{D}\rangle = \boldsymbol{0} . \end{aligned} \tag{11.1.1}$$

The turbulent kinetic energy, k and the turbulent dissipation rate, ε are equally obtained from equations in balance form, viz.,

$$\begin{aligned} \frac{\partial k}{\partial t} + \operatorname{div}(k\langle \boldsymbol{v}\rangle) &= \operatorname{div}\left(\left[\frac{\nu_t}{\sigma_k} + \nu\right]\operatorname{grad} k\right) + 4\nu_t II_{\langle \boldsymbol{D}\rangle} - \varepsilon , \\ \frac{\partial \varepsilon}{\partial t} + \operatorname{div}(\varepsilon\langle \boldsymbol{v}\rangle) &= \operatorname{div}\left(\left[\frac{\nu_t}{\sigma_\varepsilon} + \nu\right]\operatorname{grad} \varepsilon\right) + 4c_1 k II_{\langle \boldsymbol{D}\rangle} - c_2\frac{\varepsilon^2}{k} . \end{aligned} \tag{11.1.2}$$

Here we could also have replaced the PRANDTL number σ_ε by c_μ / c_ε. In the above equations the REYNOLDS stress tensor is given by (10.7.25), i.e.,

$$\frac{1}{\rho}\boldsymbol{R} = 2\nu_t\langle \boldsymbol{D}\rangle - \tfrac{2}{3}k\boldsymbol{I} , \tag{11.1.3}$$

and the turbulent kinetic energy and its dissipation rate are defined by

$$k := \tfrac{1}{2}\langle \boldsymbol{v}' \cdot \boldsymbol{v}'\rangle , \quad \varepsilon := 4\nu\langle II_{\boldsymbol{D}'}\rangle . \tag{11.1.4}$$

Finally, the turbulent viscosity is postulated in the form (11.0.1) or

$$\nu_t = c_\mu \frac{k^2}{\varepsilon} . \tag{11.1.5}$$

All coefficients arising in the above equations are usually taken as the following constants

$$c_\mu = 0.09 , \quad c_1 = 0.126 , \quad c_2 = 1.92 , \quad c_\varepsilon = 0.07 . \tag{11.1.6}$$

The PRANDTL number σ_k has a value of about unity, $\sigma_k = 1$. Moreover, the second invariant of the symmetric tensor $\boldsymbol{A}$ is defined by[1]

$$II_{\boldsymbol{A}} := \tfrac{1}{2}\operatorname{tr}(\boldsymbol{A}^2) = \tfrac{1}{2}\boldsymbol{A}\cdot\boldsymbol{A} = \tfrac{1}{2}I_{\boldsymbol{A}^2} \tag{11.1.7}$$

as is customary when $I_{\boldsymbol{A}} = 0$. The identity $\langle |\operatorname{grad}\boldsymbol{v}'|^2\rangle = 4\langle II_{\boldsymbol{D}'}\rangle$ is proven in Exercise 5. The above equations describe the k–ε model for a density

[1] Observe that for density preserving media the definition of the second invariant of the stretching tensor differs from that given in (1.5.26). This has the advantage that the production terms arising in (11.1.2) are positive.

preserving fluid; we shall motivate and derive it in the following paragraphs and indicate the necessary hypotheses and approximations.

Equations (11.1.2) show that the fluxes of the turbulent kinetic energy and its dissipation rate are modelled by a *gradient law*; in other words these fluxes are proportional to the gradient of the respective field quantity with negative sign, so that the flow of turbulent kinetic energy and its dissipation is from a high level to a lower level of the variable, viz.,

$$\boldsymbol{\phi}^{(k)} = -\left(\frac{\nu_t}{\sigma_k} + \nu\right) \operatorname{grad} k \,, \quad \boldsymbol{\phi}^{(\varepsilon)} = -\left(\frac{\nu_t}{\sigma_\varepsilon} + \nu\right) \operatorname{grad} \varepsilon \,. \tag{11.1.8}$$

Consequently, the parameterization of these fluxes is structurally of the same form as the BOUSSINESQ type ansatz for the REYNOLDS stress tensor and the heat flux vector. However, turbulent kinetic energy and dissipation are also produced by the production terms

$$\pi_k^+ = 4\nu_t II_{\langle \boldsymbol{D} \rangle} \,, \quad \pi_\varepsilon^+ = 4c_1 k II_{\langle \boldsymbol{D} \rangle} \tag{11.1.9}$$

and destroyed by the annihilation rates

$$\pi_k^- = -\varepsilon \,, \quad \pi_\varepsilon^- = -c_2 \frac{\varepsilon^2}{k} \,, \tag{11.1.10}$$

but there is no turbulent kinetic energy and dissipation that is supplied from the outside, because in density preserving fluids the mass densities are not subject to any fluctuations and thus no fluctuations of the gravitational forces can arise. A supply occurs in those cases in which the buoyancy in the stratified fluid is taken into account; in such cases a production of turbulent kinetic energy (and corresponding dissipation) as a result of density fluctuations must be accounted for. These processes will later be analysed.

The production rates are therefore given by

$$\pi^{(k)} = 4\nu_t II_{\langle \boldsymbol{D} \rangle} - \varepsilon \,, \quad \pi^{(\varepsilon)} = 4c_1 k II_{\langle \boldsymbol{D} \rangle} - c_2 \frac{\varepsilon}{k} \varepsilon \,. \tag{11.1.11}$$

The first says that the stretchings of the mean motion – these are primarily the shearing rates – produce turbulent kinetic energy, whilst the dissipation rate destroys the latter. Compared to this, the production of the dissipation rate is so parameterized that it is proportional to the production of the turbulent kinetic energy; (the larger the production of the turbulent kinetic energy is, the larger will equally also be the production of its dissipation rate), whereby the factors k and (ε/k) in (11.1.11) have been inserted in view of arguments of mere dimensional analysis.

With this simple description the terms arising in the k–ε equations can be explained. Whereas the equation for the turbulent kinetic energy can quite easily be analytically derived, the corresponding derivation or motivation of the balance equation of turbulent dissipation requires considerably more assumptions; in this case an acceptable alternative could equally be to regard the proposed balance law $(11.1.2)_2$ as an ab-initio postulate.

11.2 k–ε Model

11.2.1 Turbulent Kinetic Energy

The fundamental idea for the derivation of additional equations of the turbulent motion is the recognition that multiplication of the local momentum balance equation with the velocity field (or with any other quantity) generates an identity, but that this identity, after averaging, becomes independent of the averaged momentum balance.

We commence with the derivation of the equation of the turbulent kinetic energy. To this end one multiplies the (momentum) equation for the fluctuation velocities (10.7.5), which we write down here once more,

$$\begin{aligned}\frac{\partial \boldsymbol{v}'}{\partial t} &+ \operatorname{div}\left(\langle\boldsymbol{v}\rangle \otimes \boldsymbol{v}' + \boldsymbol{v}' \otimes \langle\boldsymbol{v}\rangle\right) + \operatorname{div}\left(\boldsymbol{v}' \otimes \boldsymbol{v}' - \langle\boldsymbol{v}' \otimes \boldsymbol{v}'\rangle\right) \\ &+ \frac{1}{\rho_*}\operatorname{grad} p' - \nu \operatorname{div}\operatorname{grad}\boldsymbol{v}' = \mathbf{0}\,,\end{aligned} \tag{11.2.1}$$

scalarly with the fluctuation velocity and then applies the filter $\langle\rangle$ to the emerging equation. If use is made of the continuity equations

$$\operatorname{div}\langle\boldsymbol{v}\rangle = 0 \qquad \text{and} \qquad \operatorname{div}\boldsymbol{v}' = 0\,, \tag{11.2.2}$$

one obtains

$$\begin{aligned}&\underbrace{\left\langle\frac{\partial \boldsymbol{v}'}{\partial t}\cdot\boldsymbol{v}'\right\rangle}_{(1)} + \underbrace{\langle\operatorname{div}\left(\boldsymbol{v}'\otimes\langle\boldsymbol{v}\rangle\right)\cdot\boldsymbol{v}'\rangle}_{(2)} + \underbrace{\langle\operatorname{div}\left(\langle\boldsymbol{v}\rangle\otimes\boldsymbol{v}'\right)\cdot\boldsymbol{v}'\rangle}_{(3)} \\ &+ \underbrace{\langle\operatorname{div}\left(\boldsymbol{v}'\otimes\boldsymbol{v}'\right)\cdot\boldsymbol{v}'\rangle}_{(4)} + \frac{1}{\rho_*}\operatorname{div}\langle p'\boldsymbol{v}'\rangle - \nu\langle(\operatorname{div}\operatorname{grad}\boldsymbol{v}')\cdot\boldsymbol{v}'\rangle = 0\,,\end{aligned} \tag{11.2.3}$$

in which a term linear in $\boldsymbol{v}'$ has been omitted since $\langle\boldsymbol{v}'\rangle = \mathbf{0}$. For the individual terms one obtains in due order

$$\begin{aligned}(1) &= \left\langle\frac{\partial}{\partial t}\left(\tfrac{1}{2}\boldsymbol{v}'\cdot\boldsymbol{v}'\right)\right\rangle = \frac{\partial k}{\partial t}\,, \\ (2) &= \langle(v'_i\langle v_j\rangle)_{,j}\,v'_i\rangle = \langle v'_{i,j}\,v'_i\langle v_j\rangle + 0\rangle = \langle\left(\tfrac{1}{2}v'_iv'_i\right)_{,j}\langle v_j\rangle\rangle \\ &= \left(\operatorname{grad}\langle\tfrac{1}{2}\boldsymbol{v}'\cdot\boldsymbol{v}'\rangle\right)\cdot\langle\boldsymbol{v}\rangle = (\operatorname{grad}k)\cdot\langle\boldsymbol{v}\rangle + \underbrace{(k\operatorname{div}\langle\boldsymbol{v}\rangle)}_{=0} \\ &= \operatorname{div}(k\langle\boldsymbol{v}\rangle)\,, \\ (3) &= \langle((\langle v_i\rangle v'_j)_{,j}\,v'_i\rangle = \langle\langle v_i\rangle_{,j}\,v'_jv'_i\rangle = \langle v_i\rangle_{,j}\,\langle v'_jv'_i\rangle \\ &= -\frac{1}{\rho}\boldsymbol{R}\cdot\operatorname{grad}\langle\boldsymbol{v}\rangle = -\frac{1}{\rho}\boldsymbol{R}\cdot\langle\boldsymbol{D}\rangle\,, \\ (4) &= \langle(v'_iv'_j)_{,j}\,v'_i\rangle = \langle v'_{i,j}\,v'_jv'_i\rangle = \langle\tfrac{1}{2}(v'_iv'_i)_{,j}\,v'_j\rangle \\ &= \left\langle\operatorname{grad}\left(\tfrac{1}{2}\boldsymbol{v}'\cdot\boldsymbol{v}'\right)\cdot\boldsymbol{v}' + \left(\tfrac{1}{2}\boldsymbol{v}'\cdot\boldsymbol{v}'\right)\operatorname{div}\boldsymbol{v}'\right\rangle = \operatorname{div}\left\langle\tfrac{1}{2}(\boldsymbol{v}'\cdot\boldsymbol{v}')\boldsymbol{v}'\right\rangle\,,\end{aligned} \tag{11.2.4}$$

so that (11.2.3) takes the form

$$\frac{\partial k}{\partial t} + \operatorname{div}(k\langle \boldsymbol{v}\rangle) + \operatorname{div}\langle(\tfrac{1}{2}\boldsymbol{v}'\cdot\boldsymbol{v}')\cdot\boldsymbol{v}'\rangle$$
$$\underbrace{-\frac{1}{\rho}\boldsymbol{R}\cdot\langle\boldsymbol{D}\rangle}_{\text{(i)}} + \frac{1}{\rho_*}\underbrace{\operatorname{div}\langle p'\boldsymbol{v}'\rangle}_{\text{(ii)}} - \underbrace{\nu\langle(\operatorname{div}\operatorname{grad}\boldsymbol{v}')\cdot\boldsymbol{v}'\rangle}_{\text{(iii)}} = 0\,, \tag{11.2.5}$$

which indeed does enjoy the form of a local balance law, in which we subsequently will transform the labelled terms in the second line.

Using the representation (11.1.3) for the REYNOLDS stress tensor one obtains

$$\text{(i)} = \frac{1}{\rho}\boldsymbol{R}\cdot\langle\boldsymbol{D}\rangle = 2\nu_t\langle\boldsymbol{D}\rangle\cdot\langle\boldsymbol{D}\rangle - \frac{2}{3}k\underbrace{\operatorname{tr}\langle\boldsymbol{D}\rangle}_{0} = 4\nu_t II_{\langle\boldsymbol{D}\rangle} = 4c_\mu\frac{k^2}{\varepsilon}II_{\langle\boldsymbol{D}\rangle}\,, \tag{11.2.6}$$

where use has also been made of (11.1.5).

To simplify the term (ii) one supposes ergodicity and replaces the statistical mean by a spatial mean, so that

$$\text{(ii)} = \frac{1}{V}\int_V \operatorname{div}(p'\boldsymbol{v}')\mathrm{d}v = \frac{1}{V}\int_{\partial V}(p'\boldsymbol{v}'\cdot\boldsymbol{n})\,\mathrm{d}a\,, \tag{11.2.7}$$

in which differentiability of the fields was assumed and the divergence theorem was employed; V is a small representative volume which contains the considered point $\boldsymbol{x}$. If one additionally supposes the turbulence to be homogeneous and isotropic, which is an approximation, then $\boldsymbol{v}'$ is an even function relative to the centre $\boldsymbol{x}$, whilst p' must be *odd* (because the momentum balance forces $\operatorname{grad} p'$ to be even if $\boldsymbol{v}'$ is even). All this implies that the pressure term in (11.2.5) must approximately vanish, viz.,

$$\text{(ii)} = \frac{1}{V}\int_{\partial V} p'(\boldsymbol{v}'\cdot\boldsymbol{n})\mathrm{d}a \approx 0\,. \tag{11.2.8}$$

The last term (iii) in (11.2.5) may, via a simple mathematical transformation, be written as follows:

$$\begin{aligned}
\frac{1}{\nu}\text{(iii)} &= \langle(\operatorname{div}\operatorname{grad}\boldsymbol{v}')\cdot\boldsymbol{v}'\rangle = \langle v'_{i,kk}\,v'_i\rangle \\
&= \langle(v'_{i,k}\,v'_i)_{,k} - v'_{i,k}\,v'_{i,k}\rangle \\
&= \langle(\tfrac{1}{2}v'_i\cdot v'_i)\rangle_{,kk} - \langle v'_{i,k}\,v'_{i,k}\rangle \\
&= \operatorname{div}\operatorname{grad} k - \langle|\operatorname{grad}\boldsymbol{v}'|^2\rangle \\
&\overset{(11.1.4)}{=} \operatorname{div}\operatorname{grad} k - \frac{\varepsilon}{\nu}\,.
\end{aligned} \tag{11.2.9}$$

Thus we achieved to express all correlation terms in (11.2.3) in terms of known averaged field variables as well as k and ε. There remains the triple correlation

$$\boldsymbol{q}^{(k)} := \langle \tfrac{1}{2}(\boldsymbol{v}' \cdot \boldsymbol{v}')\boldsymbol{v}' \rangle \,, \tag{11.2.10}$$

a vectorial flux quantity, which one may parameterize with the "gradient law"

$$\boldsymbol{q}^{(k)} = -\chi_t^{(k)} \operatorname{grad} k = -\frac{\nu_t}{\sigma_k} \operatorname{grad} k = -c_\mu \frac{k^2}{\sigma_k \varepsilon} \operatorname{grad} k \,. \tag{11.2.11}$$

One thus requires that, firstly, the flux of the turbulent kinetic energy points in the direction of steepest descent of k (just as the heat flux vector in FOURIER's heat law points in the direction of the strongest temperature drop). Secondly, one supposes that the associated diffusion coefficient $\chi_t^{(k)}$ is affine to the turbulent viscosity ν_t with a proportionality factor $1/\sigma_k$; σ_k is called PRANDTL number of the turbulent kinetic energy and is a constant, otherwise the affinity postulate expressed above would not hold.

If the intermediate results (11.2.6)–(11.2.11) are substituted in (11.2.5) one obtains the balance law of turbulent kinetic energy in its final form, viz.,

$$\frac{\partial k}{\partial t} + \operatorname{div}(k\langle \boldsymbol{v} \rangle) = \operatorname{div}\Big(\big(\frac{c_\mu k^2}{\sigma_k \varepsilon} + \nu\big) \operatorname{grad} k\Big) + 4c_\mu \frac{k^2}{\varepsilon} II_{\langle \boldsymbol{D} \rangle} - \varepsilon \,. \tag{11.2.12}$$

In this equation $\operatorname{div}(k\langle \boldsymbol{v} \rangle)$ could also be replaced by $(\operatorname{grad} k) \cdot \langle \boldsymbol{v} \rangle$, so that its left-hand side could also be written as $\mathrm{d}^{\langle\rangle} k/\mathrm{d}t$.

The individual terms in this equation for the turbulent kinetic energy represent the following subprocesses:

1. the temporal change of the turbulent kinetic energy,
2. its convective transport with the mean velocity,
3. the diffusion of the turbulent kinetic energy based on the fluctuating motion,
4. its production due to the REYNOLDS stresses by the gradient of the mean velocity, and
5. its molecular dissipation, i.e., the annihilation of the turbulent kinetic energy (i.e., change into heat)

In this equation for the turbulent kinetic energy there arise, apart from the mean velocity $\langle \boldsymbol{v} \rangle$ that is governed by the mean momentum balance, only the constant c_μ and the turbulent dissipation rate ε , for which we shall now deduce a further differential equation. Before turning to this we briefly wish to derive a connection between this quantity and other field quantities that are important in turbulence theory.

11.2.2 Vorticity and Enstrophy

At this place we briefly introduce two new physical quantities which will be encountered in the derivation of the balance law for the turbulent dissipation rate.

The *vorticity* (or vectorial vortex strength) is defined as the curl of the velocity field

$$\boldsymbol{\omega} := \operatorname{curl} \boldsymbol{v} \, . \tag{11.2.13}$$

Starting from the additive decomposition of the spatial velocity gradient $\boldsymbol{L} = \boldsymbol{D} + \boldsymbol{W}$ the meaning of the vorticity becomes immediately clear; it is the dual vector of the skew symmetric part of the spatial velocity gradient and thus defined by the formula

$$\boldsymbol{W a} = \tfrac{1}{2} \boldsymbol{\omega} \times \boldsymbol{a}, \quad \forall\, \boldsymbol{a} \in \mathbb{R}^3 \, . \tag{11.2.14}$$

In exactly the same way as one associates with the velocity $\boldsymbol{v}$ a scalar quantity via the inner product of the velocity by itself, revealing the kinetic energy (multiplied by 2), one may do this with the vorticity vector $\boldsymbol{\omega}$ and so introduce a new quantity, called *enstrophy*. For a material volume both are defined according to

$$\boldsymbol{v} \longrightarrow \frac{1}{\rho} K = \frac{1}{V} \int_V \frac{\boldsymbol{v} \cdot \boldsymbol{v}}{2} \mathrm{d}v \quad \text{(kinetic energy)}, \tag{11.2.15}$$

$$\boldsymbol{\omega} \longrightarrow \frac{1}{\rho} E = \frac{1}{V} \int_V \frac{\boldsymbol{\omega} \cdot \boldsymbol{\omega}}{2} \mathrm{d}v \quad \text{(enstrophy)} \tag{11.2.16}$$

and can, if defined in terms of turbulent fluctuations, be associated with the specific turbulent kinetic energy, k and the specific *turbulent enstrophy per unit volume*[2], defined as e,

$$\boldsymbol{v}' \quad \longrightarrow \quad k = \tfrac{1}{2} \langle \boldsymbol{v}' \cdot \boldsymbol{v}' \rangle, \qquad \boldsymbol{\omega}' \quad \longrightarrow \quad e = \tfrac{1}{2} \langle \boldsymbol{\omega}' \cdot \boldsymbol{\omega}' \rangle \, . \tag{11.2.17}$$

There is yet a third quantity which can be built with $\boldsymbol{v}$ and $\boldsymbol{\omega}$; it is the *helicity* and defined by

$$\frac{1}{\rho} H = \frac{1}{V} \int_V \boldsymbol{\omega} \cdot \boldsymbol{v} \, \mathrm{d}v \, , \tag{11.2.18}$$

in the subsequent derivations on turbulence, it, however, plays no role.

The significance of kinetic energy, enstrophy and helicity becomes more clear, if one considers these quantities in a material volume of an EULER fluid. The following theorem holds

Proposition *Let a force free* EULER *fluid fill the three-dimensional physical space. Select a material volume with boundary ∂V and assume that the boundary ∂V of V is free of vorticity $\boldsymbol{\omega}$. Then, kinetic energy, enstrophy and helicity are conserved in the sense that*

$$\frac{\mathrm{d}K}{\mathrm{d}t} = 0 \, , \quad \frac{\mathrm{d}E}{\mathrm{d}t} = 0 \, , \quad \frac{\mathrm{d}H}{\mathrm{d}t} = 0 \, , \tag{11.2.19}$$

hold. ■

[2] We define here the specific turbulent enstrophy e according to $\frac{1}{2}\langle \boldsymbol{\omega}' \cdot \boldsymbol{\omega}' \rangle$, which is different from the customary definition by a factor $\frac{1}{2}$.

The proof is given in Exercise 6. The functional relation between the turbulent enstrophy e and the dissipation rate ε is obtained by the following identities,

$$\begin{aligned}\varepsilon &= 4\nu\langle II_{\boldsymbol{D}'}\rangle = 2\nu\langle \boldsymbol{D}'\cdot\boldsymbol{D}'\rangle = 2\nu\langle \boldsymbol{L}'\cdot\boldsymbol{D}'\rangle \\ &= \nu\langle \boldsymbol{L}'\cdot\boldsymbol{L}'\rangle + \nu\langle \boldsymbol{L}'\cdot\boldsymbol{L}'^{T}\rangle \simeq \frac{\nu}{2}\langle|\operatorname{grad}\boldsymbol{v}' + \operatorname{grad}^T\boldsymbol{v}'|^2\rangle \\ &= \nu\langle|\operatorname{curl}\boldsymbol{v}'|^2\rangle = \nu\langle\boldsymbol{\omega}'\cdot\boldsymbol{\omega}'\rangle = 2\nu e\ ,\end{aligned} \tag{11.2.20}$$

which holds true for a density preserving fluid. In other words, the turbulent dissipation ε for a density preserving fluid equals the product of the molecular viscosity with the enstrophy (multiplied by 2). In what follows it will be assumed that the molecular kinematic viscosity is a constant.

At this stage of the development of the theory the reader might ask "why do we wish to have a balance law for the turbulent dissipation or the enstrophy (which is proportional to it)?" The answer is that this quantity is likely to have a significant influence in the evolution of the turbulent processes. Firstly, this is evidenced by the fact that ε enters the production–annihilation term in the balance law of turbulent kinetic energy, (11.2.12). Secondly, we know intuitively, that shearing has a strong influence in generating turbulent intensity; simple shearing is the most simple process with nontrivial vorticity. So by requesting a balance law to be obeyed for the dissipation of turbulent kinetic energy or for the enstrophy assigns a high weight to the significance of this variable. Of course this argument is more heuristic than sharp and serves here simply as a motivation. With this we now proceed in motivating the respective balance law.

11.2.3 Turbulent Dissipation

The above connection between turbulent dissipation and enstrophy is now used to derive a balance law for the turbulent dissipation ε. To this end one first needs a differential equation for the fluctuations of the vorticity. This balance is obtained by forming the curl of the balance equation for the fluctuation velocities (11.2.1); it reads

$$\begin{aligned}\operatorname{curl}\Big(&\frac{\partial\boldsymbol{v}'}{\partial t} + \operatorname{div}(\boldsymbol{v}'\otimes\boldsymbol{v}') + \operatorname{div}(\langle\boldsymbol{v}\rangle\otimes\boldsymbol{v}' + \boldsymbol{v}'\otimes\langle\boldsymbol{v}\rangle - \langle\boldsymbol{v}'\otimes\boldsymbol{v}'\rangle) \\ &+ \frac{1}{\rho_*}\operatorname{grad}p' - \nu\operatorname{div}\operatorname{grad}\boldsymbol{v}'\Big) = 0\end{aligned} \tag{11.2.21}$$

or (see Exercise 7)

$$\begin{aligned}&\frac{\partial\boldsymbol{\omega}'}{\partial t} + \operatorname{div}\Big(\boldsymbol{\omega}'\otimes(\langle\boldsymbol{v}\rangle + \boldsymbol{v}') - (\langle\boldsymbol{v}\rangle + \boldsymbol{v}')\otimes\boldsymbol{\omega}' + \langle\boldsymbol{\omega}\rangle\otimes\boldsymbol{v}' - \boldsymbol{v}'\times\langle\boldsymbol{\omega}\rangle\Big) \\ &\qquad - \nu\operatorname{div}\operatorname{grad}\boldsymbol{\omega}' - \operatorname{div}\big(\langle\boldsymbol{\omega}'\otimes\boldsymbol{v}'\rangle - \langle\boldsymbol{v}'\otimes\boldsymbol{\omega}'\rangle\big) = 0\ .\end{aligned} \tag{11.2.22}$$

This equation incidentally also follows, if in a first step one takes the vorticity transport equation and subtracts in a second step the averaged vorticity transport equation from it.

As a next step we now must form from (11.2.22) a balance equation for the turbulent enstrophy $\frac{1}{2}\langle|\boldsymbol{\omega}'|^2\rangle$ or the turbulent dissipation $\varepsilon = \nu\langle|\boldsymbol{\omega}'|^2\rangle = 2\nu e$. To this end, one multiplies (11.2.22) scalarly with $2\nu\boldsymbol{\omega}'$ and applies to the resulting equation the filter operation $\langle\ \rangle$. Because of the requested properties of the filter operator this yields the relation

$$\begin{aligned}
&\text{(i)} && \frac{\partial \varepsilon}{\partial t} + \langle 2\nu\boldsymbol{\omega}' \cdot \operatorname{div}\left[\boldsymbol{\omega}' \otimes (\langle \boldsymbol{v}\rangle + \boldsymbol{v}')\right]\rangle \\
&\text{(ii)} && - \langle 2\nu\boldsymbol{\omega}' \cdot \operatorname{div}\left[(\langle \boldsymbol{v}\rangle + \boldsymbol{v}') \otimes \boldsymbol{\omega}'\right]\rangle \\
&\text{(iii)} && + \langle 2\nu\boldsymbol{\omega}' \cdot \operatorname{div}\left[\langle\boldsymbol{\omega}\rangle \otimes \boldsymbol{v}' - \boldsymbol{v}' \otimes \langle\boldsymbol{\omega}\rangle\right]\rangle \\
&\text{(iv)} && - \langle 2\nu^2\boldsymbol{\omega}' \cdot \operatorname{div}\operatorname{grad}\boldsymbol{\omega}'\rangle \\
&\text{(v)} && - \langle 2\nu\boldsymbol{\omega}' \cdot \operatorname{div}\big(\langle\boldsymbol{\omega}' \otimes \boldsymbol{v}'\rangle - \langle\boldsymbol{v}' \otimes \boldsymbol{\omega}'\rangle\big)\rangle \\
& && = 0\,.
\end{aligned} \tag{11.2.23}$$

We shall subsequently scrutinise and transform the above five terms in a similar way as we have done this also for the balance law of turbulent kinetic energy, see (11.2.4) and (11.2.5).

(i) By exploiting the condition of density preserving, $\operatorname{div}\boldsymbol{v} = 0$, there follows

$$\begin{aligned}
2\omega'_i(\omega'_i v_j)_{,j} &= 2\omega'_i\,\omega'_{i,j} v_j \\
&= (\omega'_i\,\omega'_i)_{,j}\, v_j \\
&= (\omega'_i\,\omega'_i)_{,j}\, v_j + (\omega'_i\,\omega'_i)\, v_{j,j} \\
&= (\omega'_i\,\omega'_i\, v_j)_{,j} \\
&= \operatorname{div}\left((\boldsymbol{\omega}' \cdot \boldsymbol{\omega}')\,\boldsymbol{v}\right) \\
&= \operatorname{div}\left((\boldsymbol{\omega}' \cdot \boldsymbol{\omega}')\,\langle\boldsymbol{v}\rangle\right) + \operatorname{div}\left((\boldsymbol{\omega}' \cdot \boldsymbol{\omega}')\,\boldsymbol{v}'\right)\,.
\end{aligned} \tag{11.2.24}$$

With the definition of the dissipation rate $\varepsilon = \nu\langle(\boldsymbol{\omega}' \cdot \boldsymbol{\omega}')\rangle$ we thus have for constant ν

$$\text{(i)} = \frac{\partial \varepsilon}{\partial t} + \operatorname{div}\left(\varepsilon\langle\boldsymbol{v}\rangle\right) + \operatorname{div}\boldsymbol{q}^{(\varepsilon)}\,, \tag{11.2.25}$$

where

$$\boldsymbol{q}^{(\varepsilon)} := \langle\nu\,(\boldsymbol{\omega}' \cdot \boldsymbol{\omega}')\,\boldsymbol{v}'\rangle \tag{11.2.26}$$

is the flux of dissipation rate which, as a triple correlation, will be parameterized. This parameterization is described as the gradient law

$$\boldsymbol{q}^{(\varepsilon)} = -\chi_t^{(\varepsilon)} \operatorname{grad} \varepsilon \, , \tag{11.2.27}$$

in which the diffusivity $\chi_t^{(\varepsilon)}$ is written as

$$\chi_t^{(\varepsilon)} = \frac{\nu_t}{\sigma_\varepsilon} \tag{11.2.28}$$

and is thus connected with the turbulent viscosity ν_t. Possible deviations from the latter are corrected by the PRANDTL number σ_ε, which is assumed to be constant. This allows, since $\nu_t = c_\mu k^2/\varepsilon$, to write

$$\sigma_\varepsilon = \frac{c_\mu}{c_\varepsilon} \, , \quad c_\varepsilon = 0.07 \, , \tag{11.2.29}$$

so that (11.2.27) transforms into

$$\boldsymbol{q}^{(\varepsilon)} = -c_\varepsilon \frac{k^2}{\varepsilon} \operatorname{grad} \varepsilon \, . \tag{11.2.30}$$

With this, (11.2.25) may be written in the form

$$(\mathrm{i}) = \frac{\partial \varepsilon}{\partial t} + \operatorname{div}\left(\varepsilon \langle \boldsymbol{v} \rangle\right) - \operatorname{div}\left(c_\varepsilon \frac{k^2}{\varepsilon} \operatorname{grad} \varepsilon \right) \, . \tag{11.2.31}$$

(ii) Because div curl $(\cdot) = 0$ one may conclude here that

$$\begin{aligned} (\mathrm{ii}) &= -\langle 2\nu \boldsymbol{\omega}' \cdot ((\operatorname{grad} \langle \boldsymbol{v} \rangle) \, \boldsymbol{\omega}') \rangle - \langle 2\nu \boldsymbol{\omega}' \cdot ((\operatorname{grad} \boldsymbol{v}') \, \boldsymbol{\omega}') \rangle \\ &= -2 \operatorname{tr}\left[\nu \langle \boldsymbol{\omega}' \otimes \boldsymbol{\omega}' \rangle \langle \boldsymbol{L} \rangle\right] - 2 \operatorname{tr}\left[\langle \nu \, (\boldsymbol{\omega}' \otimes \boldsymbol{\omega}') \, \boldsymbol{L}' \rangle\right] \\ &= \underbrace{-2 \operatorname{tr}\left[\nu \langle \boldsymbol{\omega}' \otimes \boldsymbol{\omega}' \rangle \langle \boldsymbol{D} \rangle\right]}_{\mathrm{a}} \underbrace{-2 \operatorname{tr}\left[\langle \nu \, (\boldsymbol{\omega}' \otimes \boldsymbol{\omega}') \, \boldsymbol{D}' \rangle\right]}_{\mathrm{b}} \, . \end{aligned} \tag{11.2.32}$$

In the step from the second to the third line we have used the fact that $\boldsymbol{\omega}' \otimes \boldsymbol{\omega}'$ is symmetric, so that the trace of its product with a skew symmetric tensor must vanish. In the terms on the third row of (11.2.32) new correlations arise which must be parameterized.

The term (a) possesses the same form as the dissipation rate of the REYNOLDS stresses does in the balance law of kinetic energy. (One only needs to replace $\nu \langle \boldsymbol{\omega}' \otimes \boldsymbol{\omega}' \rangle$ by $\boldsymbol{R}$). This fact will here be used to motivate a corresponding parameterization. In analogy to the REYNOLDS hypothesis we therefore assume that $\nu \langle \boldsymbol{\omega}' \otimes \boldsymbol{\omega}' \rangle$ only depends upon the gradient of the mean velocity $\langle \boldsymbol{L} \rangle$ as well as upon k and ε, viz.,

$$\nu \langle \boldsymbol{\omega}' \otimes \boldsymbol{\omega}' \rangle = \boldsymbol{f}\left(\langle \boldsymbol{L} \rangle, k, \varepsilon\right) \, . \tag{11.2.33}$$

Now, $\nu \langle \boldsymbol{\omega}' \otimes \boldsymbol{\omega}' \rangle$ transforms under EUCLIDian transformations as an objective symmetric tensor. If one also postulates that turbulent closure conditions are objective, i.e., if one requires that[3]

[3] An asterisk denotes a field quantity in a reference system of an observer which moves, relative to another observer according to $\boldsymbol{x}^* = \boldsymbol{O}(t)\boldsymbol{x} - \boldsymbol{c}^*(t)$, in which $\boldsymbol{O}$ is an arbitrary orthogonal transformation.

$$\boldsymbol{f}^* \left(\langle \boldsymbol{L} \rangle^*, k^*, \varepsilon^* \right) = \boldsymbol{f} \left(\langle \boldsymbol{L} \rangle^*, k^*, \varepsilon^* \right) \,, \tag{11.2.34}$$

then

$$\boldsymbol{f} \left(\langle \boldsymbol{L} \rangle^*, k^*, \varepsilon^* \right) = \boldsymbol{O}^* \, \boldsymbol{f} \left(\langle \boldsymbol{L} \rangle, k, \varepsilon \right) \, \boldsymbol{O}^{*T} \tag{11.2.35}$$

must hold, an equation which can only be fulfilled, provided $\boldsymbol{f}$ depends upon $\langle \boldsymbol{L} \rangle$ only through its symmetric part $\langle \boldsymbol{D} \rangle$ of $\langle \boldsymbol{L} \rangle$. With these prerequisites we have

$$\boldsymbol{f} \left(\boldsymbol{O}^* \langle \boldsymbol{D} \rangle \, \boldsymbol{O}^{*T}, k^*, \varepsilon^* \right) = \boldsymbol{O}^* \, \boldsymbol{f} \left(\langle \boldsymbol{D} \rangle, k, \varepsilon \right) \, \boldsymbol{O}^{*T} \,. \tag{11.2.36}$$

The tensor $\nu \langle \boldsymbol{\omega}' \otimes \boldsymbol{\omega}' \rangle$ is therefore expressible as an isotropic tensor function of the variables $\langle \boldsymbol{D} \rangle, k, \varepsilon$ and may be represented in the form

$$\nu \langle \boldsymbol{\omega}' \otimes \boldsymbol{\omega}' \rangle = a \boldsymbol{I} + b \langle \boldsymbol{D} \rangle + c \langle \boldsymbol{D} \rangle^2 \,, \tag{11.2.37}$$

in which a, b, c are functions of the invariants of $\langle \boldsymbol{D} \rangle$ and of k, ε. If one requires linearity of this relation in $\langle \boldsymbol{D} \rangle$, then necessarily $c = 0$; if one further accounts for the fact that for a density preserving fluid the isotropic contribution, $a\boldsymbol{I}$ may be dropped (only the contraction is used in the term (a)), then one may finally write

$$\nu \langle \boldsymbol{\omega}' \otimes \boldsymbol{\omega}' \rangle = b \left(I\!I_{\langle \boldsymbol{D} \rangle}, I\!I\!I_{\langle \boldsymbol{D} \rangle}, k, \varepsilon \right) \langle \boldsymbol{D} \rangle \,. \tag{11.2.38}$$

To quantify the scalar function b, the recognition that the dimension of b is $[\mathrm{m}^2\,\mathrm{s}^{-2}]$ may be helpful. This gives rise to the simple parameterization by dimensional analysis

$$b = c_1 k \,, \quad c_1 = 0.126 \,. \tag{11.2.39}$$

The value of c_1 has been determined by numerical experiment; so c_1 is not an universal constant. In summary the term (a) in (11.2.32) takes the form

$$(\mathrm{a}) = -2 c_1 k \, \mathrm{tr}(\langle \boldsymbol{D} \rangle^2) = -4 c_1 k I\!I_{\langle \boldsymbol{D} \rangle} \,. \tag{11.2.40}$$

The term (b) on the right-hand side of (11.2.32) is neglected in the parameterization of the k–ε model, thus

$$(\mathrm{b}) = -\langle 2\nu \left(\boldsymbol{\omega}' \otimes \boldsymbol{\omega}' \right) \cdot \boldsymbol{D}' \rangle \approx 0 \,. \tag{11.2.41}$$

A convincing argument for this assumption can not be given. However, one may justify the choice (11.2.41) by the following plausibility argument: If one namely supposes that the turbulence at the point $\boldsymbol{x}$ is homogeneous and isotropic, then $\boldsymbol{v}'$ is an even function relative to the center $\boldsymbol{x}$; consequently, $\boldsymbol{\omega}'$ and $\boldsymbol{D}'$ must be odd functions. This implies that $\boldsymbol{\omega}' \otimes \boldsymbol{\omega}' \cdot \boldsymbol{D}'$ is equally an odd function. The mean value $\langle \boldsymbol{\omega}' \otimes \boldsymbol{\omega}' \cdot \boldsymbol{D}' \rangle$, now interpreted as a spatial mean value, must therefore vanish. In summary we thus have

$$(\mathrm{ii}) = -4 c_1 k I\!I_{\langle \boldsymbol{D} \rangle} \,. \tag{11.2.42}$$

(iii) This term in the balance relation (11.2.23) is transformed according to

$$
\begin{aligned}
(\mathrm{iii}) &= \langle 2\nu\boldsymbol{\omega}' \cdot \operatorname{div}\left[\langle\boldsymbol{\omega}\rangle \otimes \boldsymbol{v}' - \boldsymbol{v}' \otimes \langle\boldsymbol{\omega}\rangle\right]\rangle \\
&= \langle 2\nu\boldsymbol{\omega}' \cdot \left[(\operatorname{grad}\langle\boldsymbol{\omega}\rangle)\,\boldsymbol{v}' - (\operatorname{grad}\boldsymbol{v}')\,\langle\boldsymbol{\omega}\rangle\right]\rangle \\
&= \langle 2\nu\boldsymbol{\omega}' \cdot \operatorname{curl}\,(\boldsymbol{v}' \times \langle\boldsymbol{\omega}\rangle)\rangle\ ,
\end{aligned}
\tag{11.2.43}
$$

where, in the second line, use was made of the density preserving assumption and the identity $\operatorname{div}\boldsymbol{\omega}' = 0$ was observed. The step from the second to the third line can easily be verified (Exercise). If $\langle\ \rangle$ is now interpreted as a spatial average (ergodic hypothesis), then from the third line of (11.2.43) we deduce

$$
\begin{aligned}
\frac{1}{2\nu}(\mathrm{iii}) &= \frac{1}{V}\int_V \varepsilon_{ijk}(\boldsymbol{v}' \times \langle\boldsymbol{\omega}\rangle)_{k,j}\,\omega'_i\,\mathrm{d}v \\
&= \frac{1}{V}\int_V \left[\varepsilon_{ijk}(\boldsymbol{v}' \times \langle\boldsymbol{\omega}\rangle)_k\,\omega'_i\right]_{,j}\,\mathrm{d}v \\
&\quad -\frac{1}{V}\int_V \varepsilon_{ijk}\omega'_{i,j}\,(\boldsymbol{v}' \times \langle\boldsymbol{\omega}\rangle)_k\,\mathrm{d}v \\
&= \frac{1}{V}\oint_{\partial V} \varepsilon_{ijk}(\boldsymbol{v}' \times \langle\boldsymbol{\omega}\rangle)_k\omega'_i\,n_j\,\mathrm{d}a \\
&\quad -\frac{1}{V}\int_V (\operatorname{curl}\boldsymbol{\omega}') \cdot (\boldsymbol{v}' \times \langle\boldsymbol{\omega}\rangle)\,\mathrm{d}v\ .
\end{aligned}
\tag{11.2.44}
$$

The second integral, because of

$$
\operatorname{curl}\boldsymbol{\omega}' = \operatorname{curl}\operatorname{curl}\boldsymbol{v}' = -\operatorname{div}\operatorname{grad}\boldsymbol{v}'\ ,
\tag{11.2.45}
$$

can be written as

$$
\begin{aligned}
&-\frac{1}{V}\int_V v'_{i,ll}\ \varepsilon_{ijk}v'_k\langle\omega_j\rangle\,\mathrm{d}v \\
&= -\langle\omega_j\rangle\left(\frac{1}{V}\int_V \varepsilon_{ijk}\,v'_{i,ll}\ v'_k\,\mathrm{d}v\right) \\
&= -\langle\omega_j\rangle\frac{1}{V}\int_V \left(\varepsilon_{ijk}(v'_{i,l}\,v'_k)_{,l} - \underbrace{\varepsilon_{ijk}v'_{i,l}\,v'_{k,l}}_{=0}\right)\mathrm{d}v \\
&= -\langle\omega_j\rangle\frac{1}{V}\oint_{\partial V} \varepsilon_{ijk}v'_{i,l}\ v'_k\,n_l\,\mathrm{d}a\ .
\end{aligned}
\tag{11.2.46}
$$

If we assume again homogeneous isotropic turbulence, then the surface integrals in (11.2.44) and (11.2.46) vanish because $\boldsymbol{v}' \times \langle\boldsymbol{\omega}\rangle$ and $\boldsymbol{\omega}'$ are odd functions, but $\boldsymbol{v}'$ is even, whilst $\operatorname{grad}\boldsymbol{v}'$ is again odd relative to the center $\boldsymbol{x}$. Therefore, the entire expression (11.2.44) vanishes,

$$
(\mathrm{iii}) = 0\ .
\tag{11.2.47}
$$

(iv) The fourth term in (11.2.23) may be transformed as follows:

$$\begin{aligned}
(\text{iv}) &= -\langle 2\nu^2 {\omega'}_i {\omega'}_{i,jj} \rangle \\
&= -\langle 2\nu^2 ({\omega'}_i {\omega'}_{i,j})_{,j} \rangle + \langle 2\nu^2 {\omega'}_{i,j} {\omega'}_{i,j} \rangle \\
&= -\langle 2\nu^2 \tfrac{1}{2} ({\omega'}_i {\omega'}_i)_{,ij} \rangle + 2\nu^2 \langle |\operatorname{grad} \boldsymbol{\omega}'|^2 \rangle \\
&= -\nu \operatorname{div} \operatorname{grad} \underbrace{\langle \nu\, \boldsymbol{\omega}' \cdot \boldsymbol{\omega}' \rangle}_{\varepsilon} + 2\nu^2 \langle |\operatorname{grad} \boldsymbol{\omega}'|^2 \rangle \\
&= -\operatorname{div} (\nu \operatorname{grad} \varepsilon) + 2\nu^2 \langle |\operatorname{grad} \boldsymbol{\omega}'|^2 \rangle \, .
\end{aligned} \tag{11.2.48}$$

The first of these two terms is collected as $\boldsymbol{q}^{(\varepsilon)}$ together with the corresponding term in (i); the resulting term is then the combined molecular and turbulent diffusive flux (see (11.2.30)). The second term must be parameterized and may be assumed in the form

$$2\nu^2 \langle |\operatorname{grad} \boldsymbol{\omega}'|^2 \rangle = f \left(II_{\langle \boldsymbol{D} \rangle}, III_{\langle \boldsymbol{D} \rangle}, k, \varepsilon \right) \tag{11.2.49}$$

($I_{\langle \boldsymbol{D} \rangle} = 0$ does not arise as a variable). The most simple parameterization can again be found by methods of dimensional analysis. Since the dimension of f is $[\mathrm{m}^2\ \mathrm{s}^{-4}]$, one immediately is led to the relation

$$2\nu^2 \langle |\operatorname{grad} \boldsymbol{\omega}'|^2 \rangle = c_2 \frac{\varepsilon^2}{k} \, , \qquad c_2 = 1.92 \, . \tag{11.2.50}$$

The entire term (iv) is now given as

$$(\text{iv}) = -\operatorname{div} (\nu \operatorname{grad} \varepsilon) + c_2 \frac{\varepsilon^2}{k} \, . \tag{11.2.51}$$

There remains the term (v); however, this term must necessarily vanish, since it involves fluctuating quantities only in a linear form,

$$(\text{v}) = 0 \, . \tag{11.2.52}$$

If the results (11.2.31), (11.2.40), (11.2.41), (11.2.47), (11.2.51), (11.2.52) are substituted into (11.2.23) one obtains the balance law of turbulent dissipation in its final form

$$\begin{aligned}
&\frac{\partial \varepsilon}{\partial t} + \operatorname{div} (\varepsilon \langle \boldsymbol{v} \rangle) \\
&= \operatorname{div} \left(\left(c_\varepsilon \frac{k^2}{\varepsilon} + \nu \right) \operatorname{grad} \varepsilon \right) + 4 c_1 k II_{\langle \boldsymbol{D} \rangle} - c_2 \frac{\varepsilon^2}{k} \, ,
\end{aligned} \tag{11.2.53}$$

in which instead of $\operatorname{div} (\varepsilon \langle \boldsymbol{v} \rangle)$ one also could write $(\operatorname{grad} \varepsilon) \cdot \langle \boldsymbol{v} \rangle$. This equation has the standard form of a balance relation. The local and convective changes of the turbulent dissipation are balanced by the divergence of its flux as well as its rate of production and annihilation. The total diffusivity has a turbulent, $c_\varepsilon k^2/\varepsilon$, and a laminar, ν, contribution, and the production rate is caused by friction on the mean motion.

Together with the continuity equation and the momentum equations for the mean motion (11.1.1) the equations for the turbulent kinetic energy (11.2.12) and its dissipation (11.2.53) form a closed system of differential equations which, upon the prescription of boundary conditions, can be solved. Analytical solutions can hardly be found, so that one needs to rely upon numerical schemes. The prescription of physically motivated boundary conditions will be dealt with in Sect. 11.7.

As the above derivation has shown, the balance law of turbulent dissipation can hardly be described to be derived; there are simply too many ad-hoc assumptions involved that the product – equation (11.2.53) – qualifies to be derived. A derivation deserves more rigour and so we may justly say that the computations leading to (11.2.53) serve as a *motivation* for it. One may, alternatively, have taken the position to simply *postulate* a balance law of turbulent dissipation without recourse to any physical motivation. Such a postulated approach would directly yield

$$\frac{\partial \varepsilon}{\partial t} + \operatorname{div}(\varepsilon \langle \boldsymbol{v} \rangle) = - \operatorname{div} \boldsymbol{\phi}^\varepsilon + \pi^\varepsilon \tag{11.2.54}$$

and a counter gradient assumption for its flux would lead to

$$\boldsymbol{\phi}^\varepsilon = -\nu_\varepsilon \operatorname{grad} \varepsilon = -c_\varepsilon \frac{k^2}{\varepsilon} \operatorname{grad} \varepsilon \,, \tag{11.2.55}$$

in which ν_ε has been parameterized with k and ε (as previously done with arguments of dimensional analysis). This approach would bring us rather close to (11.2.53), however with no explicit motivation for the production term π^ε, which in (11.2.53) comprises two contributions.

11.3 Summary of the Underlying Assumptions

In closing we will now once more collect in one place all assumptions which led to the differential equations for ε and k.

1. REYNOLDS hypothesis for $\langle \boldsymbol{v}' \otimes \boldsymbol{v}' \rangle$,

$$-\rho \langle \boldsymbol{v}' \otimes \boldsymbol{v}' \rangle = \boldsymbol{R}(\langle \boldsymbol{D} \rangle, k, \varepsilon) \,.$$

   One assumes for the velocity correlation $\langle \boldsymbol{v}' \otimes \boldsymbol{v}' \rangle$ a functional relation which satisfies the rules of turbulence objectivity and in so doing is led for $\boldsymbol{R}(\langle \boldsymbol{D} \rangle, k, \varepsilon)$ to an isotropic tensorial function

$$\frac{1}{\rho} \boldsymbol{R} = -\langle \boldsymbol{v}' \otimes \boldsymbol{v}' \rangle = a\boldsymbol{I} + b\langle \boldsymbol{D} \rangle + c\langle \boldsymbol{D} \rangle^2 \,.$$

- Neglecting of the quadratic dependence on $\langle \boldsymbol{D} \rangle^2$ and requesting that $\mathrm{tr}(\boldsymbol{R}/\rho) = -2k$ leads to

$$\frac{1}{\rho}\boldsymbol{R} = 2\nu_t \langle \boldsymbol{D} \rangle - \frac{2}{3} k \boldsymbol{I} \,, \quad \nu_t = \nu_t(k, \varepsilon) \,.$$

- The explicit form of ν_t is obtained by arguments of dimensional analysis; this suggests to set ν_t proportional to k^2/ε

$$\nu_t = c_\mu \frac{k^2}{\varepsilon} \,,$$

- whereby c_μ is generally regarded as an (universal) constant and given by

$$c_\mu = 0.09 \,.$$

2. An analogous hypothesis is made for the "enstrophy stress"

$$\langle \boldsymbol{\omega}' \otimes \boldsymbol{\omega}' \rangle = \mathfrak{f}(\langle \boldsymbol{D} \rangle, k, \varepsilon)$$

for which similar objectivity requirements yield

$$-\nu \langle \boldsymbol{\omega}' \otimes \boldsymbol{\omega}' \rangle = a \boldsymbol{I} + b \langle \boldsymbol{D} \rangle + c \langle \boldsymbol{D} \rangle^2$$

with coefficients a, b, c different from those above.
- Linearization in $\langle \boldsymbol{D} \rangle$ and omission of the isotropic contribution ($a = 0$, this contribution does physically not become visible) leads to

$$-\nu \langle \boldsymbol{\omega}' \otimes \boldsymbol{\omega}' \rangle = b \langle \boldsymbol{D} \rangle \,, \quad b = \hat{b}(k, \varepsilon) \,.$$

- Arguments of dimensional analysis show that b has the dimension of a kinetic energy, so that one may write

$$b = c_1 k$$

- with the (universal) constant

$$c_1 = 1.26 \,.$$

3. All terms which are expressible as divergences of certain convective quantities and cannot directly be expressed in terms of averaged fields, are described by a gradient law. For the flux of turbulent kinetic energy one obtains in this way e.g.

$$\begin{aligned} \boldsymbol{q}^{(k)} &= \left\langle \left(\tfrac{1}{2} \boldsymbol{v}' \cdot \boldsymbol{v}'\right) \boldsymbol{v}' \right\rangle = \hat{\boldsymbol{q}}^{(k)}(\langle \boldsymbol{D} \rangle, k, \operatorname{grad} k, \varepsilon) \,, \\ &= -\chi_t^{(k)} \operatorname{grad} k \,, \end{aligned}$$

and for the flux of turbulent dissipation

$$\boldsymbol{q}^{(\varepsilon)} = \langle \nu\, (\boldsymbol{\omega}' \cdot \boldsymbol{\omega}')\, \boldsymbol{v}' \rangle = \hat{\boldsymbol{q}}^{(\varepsilon)}(\langle \boldsymbol{D} \rangle, k, \varepsilon, \operatorname{grad} \varepsilon)$$

$$= -\chi_t^{(\varepsilon)} \operatorname{grad} \varepsilon \, .$$

Strictly, the isotropic function representations of these parameterizations would in addition allow for the dependence on $\langle \boldsymbol{D} \rangle \operatorname{grad} \varepsilon$ and $\langle \boldsymbol{D} \rangle^2 \operatorname{grad} \varepsilon$. With the intention of a linear representation these terms were left out of consideration in the above laws. A dependence on $\langle \boldsymbol{D} \rangle$ is therefore only possible via the invariants, but would then again be nonlinear.

- The diffusivities are only functions of k and ε, namely

$$\chi_t^{(k)} = c_k \frac{k^2}{\varepsilon} = \frac{\nu_t}{\sigma_k} \, , \quad \chi_t^{(\varepsilon)} = c_\varepsilon \frac{k^2}{\varepsilon} = \frac{\nu_t}{\sigma_\varepsilon} \, ,$$

where these representations are obtained via pure considerations of dimensional analysis and the coefficients c_k and c_ε (or σ_k and σ_ε, respectively) are given as (universal) constants,

$$\sigma_k = 1.0 \, , \quad c_\varepsilon = 0.07 \, .$$

4. Wherever necessary use is made of the ergodic hypothesis, i.e., one interprets the mean $\langle \Psi \rangle$ as a spatial (and not as a statistical) average

$$\langle \Psi \rangle = \langle \Psi \rangle_V = \frac{1}{V} \int_V \Psi \, \mathrm{d}V \, .$$

5. Often use is made of the assumption of a symmetric statistical distribution of the velocity fluctuations about every spatial point $\boldsymbol{x}$. The assumption of homogeneous isotropic turbulence allows us to ignore surface integrals of odd functions, e.g.

$$\int_{\partial \omega} \boldsymbol{v}' \cdot (\operatorname{grad} \boldsymbol{v}' \cdot \boldsymbol{n}) \, \mathrm{d}a \approx 0 \, .$$

(This assumption can in some points be relaxed).
6. The introduction of the closure condition for the second moment

$$\nu^2 \langle | \operatorname{grad} \boldsymbol{\omega}' |^2 \rangle = c_2 \frac{\varepsilon^2}{k}$$

is made on the basis of considerations of dimensional analysis. (This second moment is assumed to be solely a function of k and ε). The coefficient c_2 is assumed to be a constant with value

$$c_2 = 1.92 \, .$$

Besides the statistical assumptions 4 and 5 and the assumption of the gradient laws for the diffusion terms 3 the k–ε model is based chiefly on the parameterizations involving the turbulent kinetic energy and its dissipation (as well as the mean stretchings). Therefore the diffusivities (of a physical field Φ), which possess the dimension $[\mathrm{m}^2\,\mathrm{s}^{-1}]$ must always be proportional to k^2/ε, so that

$$\chi_t^{(\Phi)} = c_\Phi \frac{k^2}{\varepsilon}, \qquad \text{respectively}, \qquad \chi_t^{(\Phi)} = \frac{\nu_t}{\sigma_\Phi},$$

with the corresponding PRANDTL number which connects the turbulent momentum diffusivity with the diffusivity of the "Φ-variable".

11.4 Determination of the Coefficients of the k–ε Model

The numerical determination of the phenomenological parameters arising through closure of the above k–ε model – these are four parameters, see e.g. (11.1.6) – has been a major undertaking in the development of the two-equation models. Several approaches were used. Generally simple flow configurations were studied. For these a DNS computation may be used and results compared with corresponding results obtained by using the k–ε model. Alternatively, experimental measurements may be available for certain simple flow configurations for which k–ε computations may be performed and the parameters be identified by comparison with the experimental findings. Sometimes analytical solutions may be found, more often, however, numerical computations are needed.

The constants to be determined are c_1, c_2, c_μ and c_ε. c_2 is determined by studying the behaviour of attenuation of the turbulence (c_2 is the factor of proportionality of the annihilation rate of ε). c_1 and c_μ can be determined via measurements in shear flows; however c_1 also depends on the values of c_2. The coefficient is the factor of proportionality in the production of turbulent kinetic energy by shearing and c_1 has the same meaning for the turbulent dissipation. The fourth coefficient c_ε is obtained via analysis of a turbulent boundary layer (in a turbulent flow above a plate at rest). It determines the strength of the diffusion of ε. More on this will be said in Chap. 12.

11.5 Compressible Fluids

An extension of the above derived k–ε equations to the compressible case is not possible with the representation shown above, since one must now work with the mass balance equation in the form

$$\frac{\partial \rho}{\partial t} + \operatorname{div}(\rho \boldsymbol{v}) = 0\,.$$

All simplifications that were used because the velocity field was solenoidal, $\operatorname{div} \boldsymbol{v} = 0$, are not applicable. Moreover, additional equations are needed to close the system of equations. For instance, the density ρ is given by a thermal equation of state, which via the second law is related to the specific heat and the enthalpy (see Exercise 8). For a simple fluid this equation of state is assumed in the form $\rho = \hat{\rho}(p, \Theta)$. Furthermore, equations must be deduced for the temperature and its fluctuations. Besides, it now becomes difficult to separate in the balance laws density and velocity fluctuations. When averaging the mass balance equation, second moments of the form $\langle \rho' \boldsymbol{v}' \rangle$ are generated, yet one wishes to avoid these, firstly, to avoid a mass flux in the averaged mass balance equation, but, secondly, also to obtain averaged balance laws which are formally similar to those of an unaveraged compressible fluid. However, without additional thoughts, this is not possible. To recognize the problem, let us write down the averaged equation of conservation of mass

$$\begin{aligned} &\frac{\partial \langle \rho \rangle}{\partial t} + \operatorname{div}(\langle \rho \boldsymbol{v} \rangle) = 0 , \qquad \text{resp.} \\ &\frac{\partial \langle \rho \rangle}{\partial t} + \operatorname{div}(\langle \rho \rangle \langle \boldsymbol{v} \rangle) + \operatorname{div}(\langle \rho' \boldsymbol{v}' \rangle) = 0 . \end{aligned} \tag{11.5.1}$$

Here, besides the mean fields $\langle \rho \rangle$ and $\langle \boldsymbol{v} \rangle$ there appears an additional second moment $\langle \rho' \boldsymbol{v}' \rangle$, of which the parameterization one wishes to avoid. This can be achieved by introducing a *density weighted average of the velocity vector.* Interpreted in a somewhat more general way, the second moments that are formed with the density fluctuations are transformed to a newly defined mean value.

Definition *The density weighted average* $\{a\}$ of a field quantity a is defined as

$$\{a\} := \frac{\langle \rho a \rangle}{\langle \rho \rangle} , \tag{11.5.2}$$

so that

$$a = \{a\} + a'' \tag{11.5.3}$$

is decomposed into the so-called FAVRE*–average* $\{a\}$ *and fluctuating deviations* a'' *from it.* ■

Note that applied to the velocity, the FAVRE–average of the velocity may be interpreted as a barycentric velocity. Furthermore, a'' is not the fluctuation of a'. It is simply a notation to separate this new fluctuating quantity from its old definition. The FAVRE–average possesses all filter properties which were also presumed for the statistical filter $\langle \rangle$ (see Exercise 9). Besides, one may easily show that

$$\{a\} = \langle a \rangle + \frac{\langle \rho' a' \rangle}{\langle \rho \rangle} , \tag{11.5.4}$$

$$a'' = a' - (\{a\} - \langle a \rangle) = a' - \frac{\langle \rho' a' \rangle}{\langle \rho \rangle} . \tag{11.5.5}$$

By employing the FAVRE averaging to the velocity but the ordinary filter to the density field, the averaged balance of mass takes the form

$$\frac{\partial\langle\rho\rangle}{\partial t} + \operatorname{div}(\langle\rho\rangle\{\boldsymbol{v}\}) = 0 \,. \tag{11.5.6}$$

This equation again possesses the usual form of a mass balance equation without a mass production and a mass flux, as one would expect. If the momentum equation is similarly averaged by using FAVRE–averages for the velocity, field equations are obtained that involve $\{\boldsymbol{v}\}$ as a variable rather than $\langle\boldsymbol{v}\rangle$. The derivation of the k–ε equations must then similarly be conducted by using FAVRE–averages for the velocities. In this process, k and ε, and possibly other quantities must obviously be redefined in terms of FAVRE fluctuations. This general formulation shall not be pursued here. Fortunately, it need not be done for most questions addressed in meteorology, oceanography and limnology, because density variations are often rather small. Typical variations of the density on the basis of variations of the temperature, salinity and pressure fields are generally far less than 1% of the densities of a given state of rest[4]. This fact justifies it usually to ignore the pressure variations in the densities of the thermal equation of state and to ignore the density variations in the balances of mass, momentum and energy everywhere except in the gravity force. Thus one neglects in the mass balance equation the term $d\rho/dt$ in comparison to $\rho \operatorname{div}\boldsymbol{v}$, one replaces in the momentum equation $\rho\dot{\boldsymbol{v}}$ by $\rho^*\dot{\boldsymbol{v}}$ and in the energy equation $\rho c\dot{\Theta}$ by $\rho^* c\dot{\Theta}$, where ρ^* is a constant density (e.g. at normal pressure and at 4°C for water). These assumptions, which are due to BOUSSINESQ, lead to the field equations of a BOUSSINESQ *fluid* and were justified already in Chap. 7.

This simplification is somewhat relaxed e.g. for very deep lakes – as for instance Lake Baikal – and the deep ocean for which in the heat transport equation the power of the thermodynamic pressure can not be neglected. Therefore, in this case the energy equation possesses the form

$$\rho\dot{\epsilon} = -p \operatorname{div}\boldsymbol{v} - \operatorname{div}\boldsymbol{q} + \operatorname{tr}(\boldsymbol{t}^E\boldsymbol{D}) + \rho\mathfrak{r} \tag{11.5.7}$$

with the decisive pressure term on the right-hand side[5]. With the help of the mass balance equation, this form of the energy balance may also be written as

$$\rho\left(\dot{\epsilon} + p\left(\frac{1}{\rho}\right)^{\cdot}\right) = \rho\left(\dot{h} - \frac{\dot{p}}{\rho}\right) = \rho\Theta\dot{s} = -\operatorname{div}\boldsymbol{q} + \operatorname{tr}(\boldsymbol{t}^E\boldsymbol{D}) + \rho\mathfrak{r}\,, \tag{11.5.8}$$

[4] For water this is certainly correct in practically all situations of limnology and oceanography, for air or a gas a "thumb rule" is that the local MACH number ought to be smaller than 1/3. Of course, whenever sonic conditions arise, then density variations are larger.

[5] Recall that we use ϵ to denote the specific internal energy and ε for the specific turbulent dissipation.

where $h = \epsilon + p/\rho$ is the enthalpy and s the entropy, the latter being connected with the HELMHOLTZ free energy via $\Psi_\epsilon(\Theta, \rho) = \epsilon - \Theta s$, respectively the free enthalpy $\psi_h(\Theta, p) = \epsilon - \Theta s + p/\rho$, such that

$$s = -\frac{\partial \Psi_\epsilon(\Theta, \rho)}{\partial \Theta} = -\frac{\partial \psi_h(\Theta, p)}{\partial \Theta} \ .$$

For the determination of the specific entropy the reader may consult Exercise 8.

11.6 k–ε Equations for a BOUSSINESQ Fluid

Because for a BOUSSINESQ fluid the balance of mass reduces to the continuity equation, $\operatorname{div} \boldsymbol{v} = 0$, the mean velocities $\langle \boldsymbol{v} \rangle$ and the fluctuation velocities $\boldsymbol{v}'$ are equally solenoidal. Moreover, the density fluctuations ρ' are negligibly small; this implies that no distinction exists between FAVRE and classical averages. Despite this fact there result some changes relative to the equations of a density preserving fluid. We illustrate the computational procedure by using a pure fluid (water, no salinity nor impurity) and will in the accelerations also account for the CORIOLIS accelerations.

11.6.1 Heat Transport Equation

Apart from the "old" field equations – mean continuity, averaged momentum balance, k and ε equations – one needs in addition the heat conduction equation for the averaged temperature (10.7.6). An equation for the temperature fluctuations is only needed when a second order closure scheme is applied; in that situation evolution equations for the averages of the square of the temperature fluctuations $\langle \Theta'^2 \rangle$ and the turbulent heat flux $\langle \Theta' \boldsymbol{v}' \rangle$ will be needed. In the first order closure approach, used here, a gradient type law will be employed – the analogon to the REYNOLDS stress hypothesis – so that the turbulent heat flux vector or the temperature velocity correlation takes the form

$$\langle \boldsymbol{v}' \Theta' \rangle = -\chi_t^{(\Theta)} \operatorname{grad} \langle \Theta \rangle \ . \tag{11.6.1}$$

The turbulent diffusivity is parameterized as the ratio of the eddy viscosity (of the momentum diffusivity) and the PRANDTL number σ_Θ

$$\chi_t^{(\Theta)} = \frac{\nu_t}{\sigma_\Theta} \ . \tag{11.6.2}$$

With such a closure approach all terms arising in the energy equation are known or expressed in terms of the turbulent kinetic energy and its dissipation and mean velocity field. With it the mean temperature distribution can be computed provided the radiation and mean velocity fields are prescribed. For the radiation term (in the ocean or in a lake) we shall use the most simple representation[6], which later shall more explicitly be justified,

$$\rho\mathfrak{r} = \frac{\partial I}{\partial z} \; . \tag{11.6.3}$$

The dissipation in the heat conduction equation (mean temperature equation) is generally ignored in comparison to the much more dominant term of the turbulent diffusion of heat. Thus one obtains the averaged heat conduction equation in the form

$$\frac{\partial\langle\Theta\rangle}{\partial t} + \operatorname{div}(\langle\Theta\rangle\langle\boldsymbol{v}\rangle) = \operatorname{div}\left[\left(\chi^{(\Theta)} + \frac{\nu_t}{\sigma_\Theta}\right)\operatorname{grad}\langle\Theta\rangle\right] + \frac{1}{\rho c_v}\frac{\partial I}{\partial z} \, , \tag{11.6.4}$$

in which the convective term could again be replaced by $\operatorname{grad}\langle\Theta\rangle \cdot \langle\boldsymbol{v}\rangle$. The above differential equation accounts for the variations of the temperature by a convective heat transport, molecular and turbulent diffusion as well as radiation.

Together with the k–ε model, i.e., the equation of averaged mass and momentum balance (11.1.1) and the differential equations for the turbulent kinetic energy and its dissipation (11.1.2) this equation can be integrated, if appropriate external boundary conditions such as wind stress, heat flux etc. are prescribed. They then yield the temporal and spatial evolution of the density variation as a result of temperature variations.

Before we simplify these equations the influence of the density stratification and the external forces to these equations need be described.

11.6.2 Thermal Equation of State

The thermal equation of state is generally a statement of the form $F(\Theta, \rho, p, s) = 0$, involving temperature, density, thermodynamic pressure and salinity (or impurities). For lake and ocean water detailed explicit formulas are known for these. Here we content ourselves with a functional relation e.g. for fresh water with $p \leq 2 \times 10^6 Pa$ in the temperature range 0°C to 30°C and write either generally

$$\rho = \hat{\rho}(\Theta) \, , \tag{11.6.5}$$

or in quadratic approximation

[6] The radiation is actually defined via the *pointing vector* $\boldsymbol{I}$, the energy flux density, so that

$$\rho\mathfrak{r} = \operatorname{div}\boldsymbol{I} \, .$$

This formulation has the geophysical application in mind where the radiation is essentially due to solar irradiation. If the z-coordinate is perpendicular to the Earth's surface then the dominant variation of the pointing vector is vertical so that $\operatorname{div}\boldsymbol{I}$ reduces to $\partial I/\partial z$ where I is the z-component of $\boldsymbol{I}$.

$$\rho = \rho_* \left(1 - \tilde{\alpha}(\Theta - \Theta_*)^2\right) , \quad \Theta_* = 4°[\mathrm{C}] , \quad \rho_* = \rho(\Theta_*) , \tag{11.6.6}$$

which for computational purposes in fresh water lakes is sufficiently accurate. An important parameter is the *coefficient of thermal expansion*

$$\alpha_\Theta \quad := \quad -\frac{1}{\rho}\frac{\partial \hat{\rho}(\Theta)}{\partial \Theta} \tag{11.6.7}$$

$$\stackrel{(11.6.6)}{=} \frac{2\tilde{\alpha}(\Theta - \Theta_*)}{1 - \tilde{\alpha}(\Theta - \Theta_*)^2} \approx 2\tilde{\alpha}(\Theta - \Theta_*) . \tag{11.6.8}$$

For a fluid obeying the simple thermal equation of state (11.6.6) it is, in the mentioned temperature range, a linear function of the temperature.

In a BOUSSINESQ fluid the density variation only arises in the buoyancy term $\rho \boldsymbol{g}$, where $\boldsymbol{g}$ is the vector of the acceleration due to gravity. If we also account for the fact that the Earth as a system of reference moves with (constant) angular velocity $\boldsymbol{\Omega}$, then the momentum balance equation takes the form

$$\frac{\partial \boldsymbol{v}}{\partial t} + (\operatorname{grad} \boldsymbol{v})\boldsymbol{v} + 2\boldsymbol{\Omega} \times \boldsymbol{v} + \frac{1}{\rho_*}\operatorname{grad} p - \nu \operatorname{div} \operatorname{grad} \boldsymbol{v} - \frac{\rho \boldsymbol{g}}{\rho_*} = \boldsymbol{0} , \tag{11.6.9}$$

in which ρ_* is a constant reference density $\rho_* = \rho(4°\,\mathrm{C})$ and ρ is variable and only arises in the buoyancy term. The equations for the turbulent kinetic energy and its dissipation rate must be altered according to this change in the momentum equation.

Fluctuations of the temperature cause, via the thermal equation of state, also corresponding fluctuations of the density. For a quadratic equation of state one obtains with $\Theta = \langle\Theta\rangle + \Theta'$

$$\begin{aligned} \rho(\Theta) &= \rho(\langle\Theta\rangle + \Theta') = \rho_*\left[1 - \tilde{\alpha}(\langle\Theta\rangle + \Theta' - \Theta_*)^2\right] \\ &= \rho_*\left[1 - \tilde{\alpha}(\langle\Theta\rangle - \Theta_*)^2\right] - \rho_*\tilde{\alpha}\left({\Theta'}^2 + 2(\langle\Theta\rangle - \Theta_*)\Theta'\right) \end{aligned} \tag{11.6.10}$$

and after filtering

$$\begin{aligned} \langle\rho\rangle &= \rho_*\left[1 - \tilde{\alpha}(\langle\Theta\rangle - \Theta_*)^2 - \tilde{\alpha}\langle{\Theta'}^2\rangle\right] , \\ \rho' &= -\rho_*\tilde{\alpha}\left[({\Theta'}^2 - \langle{\Theta'}^2\rangle) + 2(\langle\Theta\rangle - \Theta_*)\Theta'\right] . \end{aligned} \tag{11.6.11}$$

These equations contain the quadratic correlation term $\langle{\Theta'}^2\rangle$, also known as variance of the temperature fluctuation, which in most closure schemes of oceanographic, limnological and meteorological applications is assumed to be small[7]. If this term is ignored one is led to the approximations

[7] The reader may consult Exercise 12 of Chap. 10 to see how an evolution equation for $\langle{\Theta'}^2\rangle$ is derived.

$$\begin{aligned} \langle \rho \rangle &= \rho(\langle \Theta \rangle) = \rho_* \left[1 - \tilde{\alpha}(\langle \Theta \rangle - \Theta_*)^2\right] , \\ \rho' &= -2\rho_* \tilde{\alpha}(\langle \Theta \rangle - \Theta_*)\Theta' . \end{aligned} \tag{11.6.12}$$

When compared with the exact expressions (11.6.11), these formulas bear the advantage that $\langle \Theta'^2 \rangle$ must not be parameterized; as a result the temperature fluctuations are linearly transferred into density fluctuations. Thereby, the density fluctuations arising in the turbulence equations may directly be replaced by temperature fluctuations.

For a general density function of the form $\rho = \hat{\rho}(\Theta)$ a TAYLOR series expansion yields

$$\begin{aligned} \rho &= \hat{\rho}(\Theta) = \hat{\rho}(\langle \Theta \rangle + \Theta') \\ &\approx \hat{\rho}(\langle \Theta \rangle) + \Theta' \frac{\partial \rho}{\partial \Theta}(\langle \Theta \rangle) + O(\Theta'^2) . \end{aligned} \tag{11.6.13}$$

In the linear approximation we thus have for an arbitrary equation of state

$$\begin{aligned} \langle \rho \rangle &= \hat{\rho}(\langle \Theta \rangle) , \\ \rho' &= \Theta' \frac{\partial \rho}{\partial \Theta}(\langle \Theta \rangle) = -\langle \rho \rangle \langle \alpha_\Theta \rangle \Theta' , \end{aligned} \tag{11.6.14}$$

in which

$$\langle \alpha_\Theta \rangle = \hat{\alpha}_\Theta(\langle \Theta \rangle) = -\frac{1}{\hat{\rho}(\langle \Theta \rangle)} \frac{\partial \rho}{\partial \Theta}(\langle \Theta \rangle) \tag{11.6.15}$$

is the averaged coefficient of thermal expansion; it is the volumetric coefficient of thermal expansion evaluated for the mean temperature field.

11.6.3 Changes in the Turbulence Equations

The equations for k and ε originated from the balance law of the fluctuation velocities, which itself was obtained by averaging the momentum balance equation. If one performs the averaging with $\boldsymbol{v} = \langle \boldsymbol{v} \rangle + \boldsymbol{v}'$, $\rho = \langle \rho \rangle + \rho'$ then by closing the REYNOLDS stresses with $\boldsymbol{R}/\rho_* = 2\nu_t \langle \boldsymbol{D} \rangle$, one obtains

$$\begin{aligned} &\frac{\partial \langle \boldsymbol{v} \rangle}{\partial t} + \operatorname{div}(\langle \boldsymbol{v} \rangle \otimes \langle \boldsymbol{v} \rangle) - \operatorname{div}\left(2(\nu + \nu_t)\langle \boldsymbol{D} \rangle\right) \\ &\qquad + \frac{1}{\rho_*} \operatorname{grad} \langle p \rangle + 2\boldsymbol{\Omega} \times \langle \boldsymbol{v} \rangle - \frac{\langle \rho \rangle \boldsymbol{g}}{\rho_*} = 0 , \end{aligned} \tag{11.6.16}$$

in which the last two terms are new as compared with (11.1.1). When deriving the turbulence equations for a BOUSSINESQ fluid it therefore suffices to focus attention to the implications of those terms.

If one subtracts the averaged momentum balance from its non-averaged counterpart, one obtains the equation of the fluctuation velocities which is now equally complemented by the two terms mentioned above; it reads

$$\frac{\partial \boldsymbol{v}'}{\partial t} + \operatorname{div}(\boldsymbol{v}' \otimes \boldsymbol{v}') + \operatorname{div}\left(\langle \boldsymbol{v} \rangle \otimes \boldsymbol{v}' + \boldsymbol{v}' \otimes \langle \boldsymbol{v} \rangle\right) - \langle \boldsymbol{v}' \otimes \boldsymbol{v}' \rangle\big)$$
$$+ \frac{1}{\rho_*} \operatorname{grad} p' - \nu \operatorname{div} \operatorname{grad} \boldsymbol{v}' + 2\boldsymbol{\Omega} \times \boldsymbol{v}' - \frac{\boldsymbol{g}\rho'}{\rho_*} = \boldsymbol{0} .$$

This equation is the starting point for the derivation of the balance laws of turbulent kinetic energy and its dissipation.

Taking the inner product of this equation with the fluctuation velocity and averaging the emerging equation yields the differential equation for the turbulent kinetic energy in the form

$$\begin{aligned} &\frac{\partial k}{\partial t} + \operatorname{div}(k\langle \boldsymbol{v} \rangle) - \operatorname{div}\left(\left(\frac{c_\mu k^2}{\sigma_k \varepsilon} + \nu\right) \operatorname{grad} k\right) \\ &\qquad - 4c_\mu \frac{k^2}{\varepsilon} I\!I_{\langle \boldsymbol{D} \rangle} + \varepsilon - \frac{\boldsymbol{g}}{\rho_*} \cdot \langle \rho' \boldsymbol{v}' \rangle = 0 . \end{aligned} \tag{11.6.17}$$

Since the CORIOLIS force is gyroscopic it does not perform any work and therefore does not arise in this equation ($(2\boldsymbol{\Omega} \times \boldsymbol{v}') \cdot \boldsymbol{v}' \equiv 0$).

The balance law for the turbulent dissipation is obtained, if the curl of the equation for the velocity fluctuations is formed. If this equation is scalarly multiplied by the vorticity $\boldsymbol{\omega}' = \operatorname{curl} \boldsymbol{v}'$ and the resulting equation is subsequently filtered, one obtains

$$\begin{aligned} &\frac{\partial \varepsilon}{\partial t} + \operatorname{div}\left(\varepsilon \langle \boldsymbol{v} \rangle\right) - \operatorname{div}\left(\left(c_\varepsilon \frac{k^2}{\varepsilon} + \nu\right) \operatorname{grad} \varepsilon\right) - 4c_1 k I\!I_{\langle \boldsymbol{D} \rangle} + c_2 \frac{\varepsilon^2}{k} \\ &-2\nu \left\langle \operatorname{curl}\left(\frac{\boldsymbol{g}\rho'}{\rho_*}\right) \cdot \operatorname{curl} \boldsymbol{v}' \right\rangle + 2\nu \langle \operatorname{curl}\left(2\boldsymbol{\Omega} \times \boldsymbol{v}'\right) \cdot \operatorname{curl} \boldsymbol{v}' \rangle = 0 . \end{aligned} \tag{11.6.18}$$

In (11.6.17) the last term on the left-hand side is new, and in (11.6.18) the two last terms on the left-hand side are new. They are, respectively, due to the small density variations in the buoyancy term and the CORIOLIS force. We will now address the parameterizations of these terms.

Turbulent Kinetic Energy We now wish to simplify the newly added terms and apply to this end a thermal equation of state with quadratic temperature dependence as shown in (11.6.12). We transform thereby the density fluctuations and employ a gradient type law for the mean turbulent kinetic energy flux; the term

$$\begin{aligned} -\frac{\boldsymbol{g}}{\rho_*} \cdot \langle \rho' \boldsymbol{v}' \rangle &= 2\tilde{\alpha} \boldsymbol{g}(\langle \Theta \rangle - \Theta_*) \cdot \langle \Theta' \boldsymbol{v}' \rangle \\ &\cong -2\tilde{\alpha} \boldsymbol{g}(\langle \Theta \rangle - \Theta_*) \frac{\nu_t}{\sigma_\Theta} \cdot \operatorname{grad} \langle \Theta \rangle \end{aligned} \tag{11.6.19}$$

describes the production (respectively annihilation) of the turbulent kinetic energy due to the buoyancy. This ignores the terms involving $\langle \Theta'^2 \rangle$. The flux $\langle \rho' \boldsymbol{v}' \rangle$ is called the *turbulent buoyancy flux.* This term may be positive or negative. If the temperature decreases with depth[8], i.e., if $\partial\langle\Theta\rangle/\partial z > 0$, then $\boldsymbol{g}\cdot\operatorname{grad}\langle\Theta\rangle < 0$; this corresponds to a stable stratification and the expression on the right-hand side of (11.6.19) is positive. For the turbulent kinetic energy k, this corresponds to a decrease. This result may be expressed as follows: If a stable density stratification is changed by a turbulent motion, turbulent kinetic energy is consumed. The stratification is thereby weakened. Thus, if one wishes to transport an amount of air or water from a deeper (i.e., heavier) layer to higher levels, then energy must be supplied. Alternatively, if a stratification is unstable, then the expression on the right-hand side of (11.6.19) is negative and in the balance law (11.6.17) a growth of turbulent kinetic energy is accompanied with it.

With the additional buoyancy term the equation of the turbulent kinetic energy now reads

$$\begin{aligned}\frac{\partial k}{\partial t} &+ \operatorname{div}(k\langle\boldsymbol{v}\rangle) - \operatorname{div}\left(\left(\frac{c_\mu k^2}{\sigma_k\varepsilon}+\nu\right)\operatorname{grad}k\right)\\ &- 4c_\mu\frac{k^2}{\varepsilon}II_{\langle\boldsymbol{D}\rangle} + \varepsilon - 2\tilde{\alpha}\frac{\nu_t}{\sigma_\Theta}(\langle\Theta\rangle-\Theta_*)\boldsymbol{g}\cdot\operatorname{grad}\langle\Theta\rangle = 0\,.\end{aligned} \tag{11.6.20}$$

For an arbitrary thermal equation of state of the kind $\rho = \hat{\rho}(\Theta)$, it takes the form

$$\begin{aligned}\frac{\partial k}{\partial t} &+ \operatorname{div}(k\langle\boldsymbol{v}\rangle) - \operatorname{div}\left(\left(\frac{c_\mu k^2}{\sigma_k\varepsilon}+\nu\right)\operatorname{grad}k\right)\\ &- 4c_\mu\frac{k^2}{\varepsilon}II_{\langle\boldsymbol{D}\rangle} + \varepsilon - \frac{\langle\rho\rangle\langle\alpha_\Theta\rangle}{\rho_*}\frac{\nu_t}{\sigma_\Theta}\boldsymbol{g}\cdot\operatorname{grad}\langle\Theta\rangle = 0\,.\end{aligned} \tag{11.6.21}$$

This is the balance law of turbulent kinetic energy for a BOUSSINESQ fluid.

Turbulent Dissipation The simplifications for the dissipation are somewhat more complicated.

1. For the first of the two last terms in (11.6.18), which arises here because of the buoyancy effect, the following estimation can be made[9]

[8] The z-coordinate points upwards, opposite to the direction of gravity.

[9] The simplification is achieved in two steps. First, one transforms the term

$$\operatorname{curl}\left(\frac{\boldsymbol{g}\rho'}{\rho_*}\right)\cdot\operatorname{curl}\boldsymbol{v}' = \frac{g_j}{\rho_*}\rho'_{,i}\left(v'_{j,i} - v'_{i,j}\right)\,.$$

With the aid of the ergodic hypothesis and after truncation of surface-integral terms (isotropic turbulence), the above relation becomes

$$\left\langle\operatorname{curl}\left(\frac{\boldsymbol{g}\rho'}{\rho_*}\right)\cdot\operatorname{curl}\boldsymbol{v}'\right\rangle \approx \left\langle\frac{g_j}{\rho_*}\rho'_{,i}\,v'_{j,i}\right\rangle = \frac{\boldsymbol{g}}{\rho_*}\cdot\langle(\operatorname{grad}\boldsymbol{v}')\operatorname{grad}\rho'\rangle\,.$$

This formula is obtained because

$$\frac{1}{V}\int\limits_V \rho'_{,i}\,v'_{i,j}\,\mathrm{d}v = -\frac{1}{V}\int\limits_V \rho' v'_{i,ij}\,\mathrm{d}v + \frac{1}{V}\int\limits_{\partial V}\rho' v'_{i,j}n_i\,\mathrm{d}a = 0\,,$$

$$-2\nu\left\langle \operatorname{curl}\left(\frac{\boldsymbol{g}\rho'}{\rho_*}\right)\cdot \operatorname{curl}\boldsymbol{v}'\right\rangle \approx -2\nu\frac{\boldsymbol{g}}{\rho_*}\cdot\langle(\operatorname{grad}\boldsymbol{v}')\operatorname{grad}\rho'\rangle\,, \qquad (11.6.22)$$

which needs to be parameterized by a closure condition. If one assumes that the term $\langle(\operatorname{grad}\boldsymbol{v}')\operatorname{grad}\rho'\rangle$ depends on the turbulent buoyancy flux $\langle\rho'\boldsymbol{v}'\rangle$ and k and ε, then dimensional analysis implies

$$2\nu\langle(\operatorname{grad}\boldsymbol{v}')\operatorname{grad}\rho'\rangle = c_3\frac{\varepsilon}{k}\langle\rho'\boldsymbol{v}'\rangle \qquad (11.6.23)$$

with dimensionless c_3. Therefore, the buoyancy term in the balance of turbulent dissipation may be approximated as[10]

$$-2\nu\left\langle \operatorname{curl}\left(\frac{\boldsymbol{g}\rho'}{\rho_*}\right)\cdot \operatorname{curl}\boldsymbol{v}'\right\rangle \approx -\frac{\boldsymbol{g}}{\rho_*}\cdot c_3\frac{\varepsilon}{k}\langle\rho'\boldsymbol{v}'\rangle\,. \qquad (11.6.24)$$

With (11.6.14) and the corresponding gradient-type assumption (11.6.1) the buoyancy flux can now be transformed so that

$$-2\nu\left\langle \operatorname{curl}\left(\frac{\boldsymbol{g}\rho'}{\rho_*}\right)\cdot \operatorname{curl}\boldsymbol{v}'\right\rangle \approx -\frac{c_3 c_\mu}{\sigma_\Theta}\frac{\langle\rho\rangle\langle\alpha_\Theta\rangle}{\rho_*}k\boldsymbol{g}\cdot\operatorname{grad}\langle\Theta\rangle \qquad (11.6.25)$$

is obtained. SVENSSON [231] sets the constant c_3 equal to zero. Such an assumption may be justified on the basis of additional simplifications regarding the distribution of fluctuations: Consider to this end the simplified situation of turbulent processes with density fluctuations which cannot depend on the horizontal coordinates, so that

$$\rho' = \hat{\rho}'(z)\,. \qquad (11.6.26)$$

Therefore, since $\boldsymbol{g} = -g_3\hat{\boldsymbol{e}}_z$, where $\hat{\boldsymbol{e}}_z$ is the unit vector in the z-direction,

$$\begin{aligned}\operatorname{curl}\left(\frac{\boldsymbol{g}\rho'}{\rho_*}\right)\cdot\operatorname{curl}\boldsymbol{v}' &= \frac{g_j}{\rho_*}\rho'_{,i}\left({v'}_{j,i} - {v'}_{i,j}\right)\\ &= -\frac{g_3}{\rho_*}\rho'_{,i}\left({v'}_{3,i} - {v'}_{i,3}\right).\end{aligned} \qquad (11.6.27)$$

Because it was assumed that the density fluctuations depend only on the z-coordinate, the index i must be identified with $i = 3$; but for this case the right-hand side of (11.6.27) vanishes identically.

since ${v'}_{i,i} = 0$, and since ${v'}_{i,j}$ in the surface integral is an odd function.

[10] One may criticize this transformation as being superfluous, because it does actually hardly motivate the parametrization of eq. (11.6.23).The following motivation does not seem to be less convincing: on the left–hand side of (11.6.22) ρ' and $\boldsymbol{v}'$ are the only arising fluctuation variables which as averages also arise in the buoyancy flux $\langle\rho'\boldsymbol{v}\rangle$. The only scalar combination of $k, \varepsilon, \nu, \rho_*, \boldsymbol{g}, \langle\rho\boldsymbol{v}'\rangle$ with the same dimension as the left–hand side of (11.6.22) is then the right–hand side of (11.6.24).

2. The term which in (11.6.18) may be traced back to the CORIOLIS acceleration may be written in the following form

$$\begin{aligned}\langle \operatorname{curl}(2\boldsymbol{\Omega} \times \boldsymbol{v}') \cdot \operatorname{curl} \boldsymbol{v}' \rangle &= \langle (\varepsilon_{ijk} 2\Omega_i v'_j)_{,l}\, \varepsilon_{lkm} \varepsilon_{pqm} v'_{q,p} \rangle \\ &= \langle 2\Omega_i v'_{q,p} (v'_{j,j}\, \varepsilon_{pqi} - v'_{j,i}\, \varepsilon_{pqj}) \rangle \qquad (11.6.28) \\ &= -2\Omega_i \langle v'_{j,i}\, v'_{q,p}\, \varepsilon_{pqj} \rangle \, .\end{aligned}$$

SVENSSON [231] ignores this in comparison to the production term in (11.6.18) which is multiplied with c_1. One may show that this term equally vanishes, if similarly to the above procedure, the velocity fluctuations are merely z-dependent: $\boldsymbol{v}' = \hat{\boldsymbol{v}}'(z)$. Since in this case non-vanishing derivatives of field variables are only those with respect to z (or x_3), we have

$$v'_{j,i}\, v'_{q,p}\, \varepsilon_{pqj} = v'_{j,3}\, v'_{q,3}\, \varepsilon_{3qj} = v'_{2,3}\, v'_{1,3} - v'_{1,3}\, v'_{2,3} = 0 \, . \qquad (11.6.29)$$

The omission of this term can also be justified with the analogy (already mentioned above) between the production of turbulent kinetic energy and the production of turbulent dissipation. Because the CORIOLIS acceleration does not produce any turbulent kinetic energy, the corresponding production in the dissipation equation should equally vanish. It has thus been shown that *under weak simplifying assumptions the differential equation for the turbulent dissipation for a* BOUSSINESQ *fluid is the same as the corresponding equation for a density preserving fluid.* Equation (11.2.53) can be taken over without any changes. CORIOLIS acceleration and buoyancy practically do not have an influence on the turbulent dissipation.

11.7 Summary of the k–ε Model for Turbulence of a BOUSSINESQ Fluid

We wish to collect here all equations of the complete k–ε model in three-dimensional form.

1. The balances of mass and momentum for the averaged field quantities and the averaged heat transport equation take the forms

$$\begin{aligned}&\operatorname{div} \langle \boldsymbol{v} \rangle = 0 \, , \\ &\frac{\partial \langle \boldsymbol{v} \rangle}{\partial t} + \operatorname{div}(\langle \boldsymbol{v} \rangle \otimes \langle \boldsymbol{v} \rangle) + 2\boldsymbol{\Omega} \times \langle \boldsymbol{v} \rangle \qquad (11.7.1) \\ &\quad = -\frac{1}{\rho_*} \operatorname{grad} \langle p \rangle + \operatorname{div}\Big(2(\nu + \nu_t) \langle \boldsymbol{D} \rangle \Big) + \frac{\langle \rho \rangle}{\rho_*} \boldsymbol{g} \, ,\end{aligned}$$

$$\begin{aligned}\frac{\partial\langle\Theta\rangle}{\partial t} &+ \operatorname{div}(\langle\boldsymbol{v}\rangle\langle\Theta\rangle)\\ &= \operatorname{div}\left(\left(\chi^{(\Theta)}+\frac{\nu_t}{\sigma_\Theta}\right)\operatorname{grad}\langle\Theta\rangle\right)+\frac{1}{\rho_* c_v}\operatorname{div}\boldsymbol{I}\,.\end{aligned} \tag{11.7.2}$$

These equations have the same form as the non-averaged balances of mass, momentum and energy. The only difference lies in the fact that in place of the molecular viscosity there arises now the sum of the molecular and turbulent viscosities. In the hydrodynamic literature it is customary to call this sum "viscosity", without referring to the turbulence theory and without explicitly assigning to this diffusivity a certain value or a certain known functional relation.

2. The coupling of the mean fields with the fluctuating quantities is accomplished solely via the eddy viscosity, i.e. the turbulent momentum diffusivity for which the relation

$$\nu_t = c_\mu \frac{k^2}{\varepsilon} \tag{11.7.3}$$

is postulated. As already mentioned one can in some simple hydrodynamic models prescribe this eddy viscosity and then does not need in these cases an additional turbulence model.

3. The k–ε equations, however, reflect the turbulent processes dynamically and are given by

$$\begin{aligned}\frac{\partial k}{\partial t} &+ \operatorname{div}(k\langle\boldsymbol{v}\rangle) = \operatorname{div}\left(\left(\nu+\frac{\nu_t}{\sigma_k}\right)\operatorname{grad} k\right)\\ &+4\nu_t II_{\langle\boldsymbol{D}\rangle} - \varepsilon + \frac{\langle\rho\rangle\langle\alpha_\Theta\rangle}{\rho_*}\frac{\nu_t}{\sigma_\Theta}\boldsymbol{g}\cdot\operatorname{grad}\langle\Theta\rangle\,,\\ \frac{\partial\varepsilon}{\partial t} &+ \operatorname{div}(\varepsilon\langle\boldsymbol{v}\rangle) = \operatorname{div}\left(\left(\nu+\frac{\nu_t}{\sigma_\varepsilon}\right)\operatorname{grad}\varepsilon\right)\\ &+4c_1 k II_{\langle\boldsymbol{D}\rangle} - c_2\frac{\varepsilon^2}{k} + \frac{c_3 c_\mu}{\sigma_\Theta}\frac{\langle\rho\rangle\langle\alpha_\Theta\rangle}{\rho_*}k\boldsymbol{g}\cdot\operatorname{grad}\langle\Theta\rangle\,.\end{aligned} \tag{11.7.4}$$

These equations express the temporal changes plus the convective transports of the turbulent kinetic energy and its dissipation (left-hand sides of the equations) as consequences of the diffusive transports – molecular as well as turbulent – the production by shearing, the dissipative loss and, in case of turbulent kinetic energy, also the buoyancy flux.

4. To the many empirical constants, which arise here, numerical values must be assigned. In general, the numerical values are

$$\begin{aligned} &c_\mu = 0.09, \quad c_1 = 0.126, \quad c_2 = 1.92\,,\\ &c_3 \simeq 0, \qquad \sigma_k = 1.4, \qquad \sigma_\varepsilon = 1.3\end{aligned} \tag{11.7.5}$$

and have been thus obtained by simple model calculations (see e.g. RODI [197]. In the model by SVENSSON [231], which we shall present later in

Chap. 13, only the PRANDTL number for the temperature diffusivity is presented by a functional of the form

$$\sigma_\Theta = \mathfrak{s}(k, \varepsilon, \langle\Theta\rangle, \operatorname{grad}\langle\Theta\rangle) . \tag{11.7.6}$$

Because the goal of the model is also the determination of the temperature distribution, the parameterization of this quantity requires special attention. This relation is obtained in the context of an algebraic stress model.

5. As further material parameters one additionally needs the thermal equation of state for the density e.g. $\rho = \hat{\rho}(\Theta)$, the coefficient of thermal expansion α_Θ, the specific heat at constant volume, c_v and the molecular diffusivities, ν and $\chi^{(\Theta)}$, here for water

$$\begin{aligned} &\rho = \hat{\rho}(\Theta) , && \text{see (11.6.6)}, && c_v = 4190\,\mathrm{m}^2\,\mathrm{K}^{-1}\,\mathrm{s}^{-2} , \\ &\alpha_\Theta = -\frac{1}{\rho}\frac{\partial\hat{\rho}}{\partial\Theta} , && \text{see (11.6.8)}, && \nu = 1.3\cdot 10^{-6}\,\mathrm{m}^2\,\mathrm{s}^{-1} , \\ & && && \chi^{(\Theta)} = 1.3\cdot 10^{-7}\,\mathrm{m}^2\,\mathrm{s}^{-1} . \end{aligned} \tag{11.7.7}$$

6. Apart from the boundary conditions the outside world manifests itself by the radiation, which represents an energy supply within the entire body. In lakes and in the ocean this supply varies primarily in the vertical direction from the surface, so that one has approximately

$$\begin{aligned} &\operatorname{div}\boldsymbol{I} \approx \frac{\partial I}{\partial z} \quad \text{with} \quad I = I_0 \exp\left(\textstyle\int_0^z (k_\mathrm{w} + \ldots)\mathrm{d}z \right) , \\ &k_\mathrm{w} = 0.3 , \end{aligned} \tag{11.7.8}$$

where $z < 0$, and I is the z-component of the POINTING vector; k_w is the *extinction coefficient* of clear water that varies from position to position in the ocean or in a lake and the dots indicate additional possible influences of absorption by a phytoplankton population.

7. The fluxes on the free surface of the ocean or lake concern the prediction of wind stress and heat fluxes

$$\boldsymbol{\tau} = \rho_a c_D |\boldsymbol{U}_w| \boldsymbol{U}_w , \tag{11.7.9}$$

$$\boldsymbol{q}\cdot\boldsymbol{n} = Q_\mathrm{ir}^\mathrm{a} - Q_\mathrm{ir}^\mathrm{w} + Q_\mathrm{l} + Q_\mathrm{s} . \tag{11.7.10}$$

Here, $\rho_a, c_D, \boldsymbol{U}_w$ are the density of air, the drag coefficient $c_D \sim 1.2 \times 10^{-3}$ and the wind speed (ordinarily 10 m above the water), Q_ir^a and Q_ir^w are the (gray body) radiation of air and water above the surface, Q_l and Q_s are the latent and sensible heat flows in air. These thermal fluxes are treated in greater detail in Chap. 13.

8. For the averaged equations, dynamical and kinematic boundary conditions are needed. To this end one prescribes stress and heat flux at the

free surface and the base. The boundary conditions for the turbulent equations, on the other hand, are rather difficult to postulate. Ordinarily, one will prescribe values for k and ε and/or their normal derivatives on the boundary. Such values or formulas can often only be obtained by boundary layer consideration.

In particular, the following boundary conditions are used: The traction vector at the lake surface is continuous; it is given by the wind shear and the atmospheric pressure. Furthermore, a kinematic boundary condition is needed here, which describes the evolution of the surface displacement. The heat flow across the lake surface is also continuous and is determined by the normal component of the heat flux vector. At the base, dynamic boundary conditions for the traction and continuity of the heat flow perpendicular to the surface must hold; additionally, a kinematic boundary condition guarantees for the tangentiality of the current field at the bottom, or one requires the no-slip condition.
For the turbulent quantities one may assume, at least for a deep lake or the deep ocean, that turbulence has died out. In such a case it is justified to request that[11]

$$k = 0\,, \quad \varepsilon = 0 \quad \text{at the bottom.} \tag{11.7.11}$$

Such conditions apply in general, if one supposes that the turbulent zone is separated from the bottom surface by a laminar sublayer. At the free surface a physically appropriate formulation of boundary conditions is more complicated and also more critical. One may for instance assume that the normal flows of the turbulent kinetic energy and its dissipation may vanish. The simplest possibility for a horizontal surface is thus to set

$$\frac{\partial k}{\partial z} = 0\,, \quad \frac{\partial \varepsilon}{\partial z} = 0 \quad \text{at the free surface.} \tag{11.7.12}$$

One may, on the other hand, also find equations which relate k and ε with the wind shear stress and the heat flow at the surface plus a boundary layer thickness, see SVENSSON [231].

The equations presented above in all detail will in Chap. 13 be used to derive the one-dimensional k–ε model. In that chapter, which treats a *simple* application of turbulence theory to lake hydrodynamics, the surface flows that are only treated in a gross fashion here, will be specified more explicitly. Before turning to those equations, the turbulent closure schemes should be further scrutinized, in particular with regard to second order closure – the so-called

[11] Close to solid walls the $k - \varepsilon$ model requires the introduction of wall functions to properly describe the turbulent boundary layer. This means that the boundary condition (11.7.10) is too simple and must be regarded as a gross simplification of the correct behaviour.

REYNOLDS stress models and their algebraic reductions – and the thermodynamic foundation of such models and their embedding into an irreversibility principle.

11.8 Exercises

1. For a density preserving fluid, prove the identities

$$\text{(i)}\qquad \langle \operatorname{div}(\langle \boldsymbol{v}\rangle \otimes \boldsymbol{v}')\cdot \boldsymbol{v}'\rangle = -\frac{1}{\rho}\boldsymbol{R}\cdot\langle \boldsymbol{D}\rangle\,,$$

$$\text{(ii)}\qquad \langle \operatorname{div}(\boldsymbol{v}'\otimes(\boldsymbol{v}'+\langle \boldsymbol{v}\rangle))\cdot \boldsymbol{v}'\rangle = \left\langle \operatorname{grad}\left(\frac{\boldsymbol{v}'\cdot\boldsymbol{v}'}{2}\right)\cdot(\boldsymbol{v}'+\langle \boldsymbol{v}\rangle)\right\rangle .$$

2. For every odd one-dimensional function $f(x)$ one has identically

$$\int_{-a}^{a} f(x)\, f'(x)\,\mathrm{d}x = 0\,.$$

3. For a density preserving fluid, using the ergodic assumption and the assumption of isotropic turbulence, prove that

$$\langle \boldsymbol{v}'\times\Delta\boldsymbol{v}'\rangle = 0\,. \tag{11.8.1}$$

4. Under the prerequisites of ergodicity and isotropic turbulence, prove for a density preserving fluid that

$$\begin{aligned}\langle \boldsymbol{v}'\cdot\operatorname{curl}(\operatorname{curl}\boldsymbol{v}')\rangle &= -\langle \Delta\boldsymbol{v}'\cdot\boldsymbol{v}'\rangle\\ &= \langle |\operatorname{grad}\boldsymbol{v}'|^2\rangle\\ &= \langle |\operatorname{curl}\boldsymbol{v}'|^2\rangle\,.\end{aligned} \tag{11.8.2}$$

5. Continuing with the above problem 4, prove the statements

$$\langle \boldsymbol{L}'\cdot\boldsymbol{L}'^{T}\rangle \cong 2\langle \boldsymbol{D}'\cdot\boldsymbol{D}'^{T}\rangle \cong 2\langle \boldsymbol{W}'\cdot\boldsymbol{W}'^{T}\rangle\,, \tag{11.8.3}$$

   in which "$\cong$" means that ergodicity and isotropic turbulence are assumed to hold.
6. Prove relation (11.2.19).
7. On using the vorticity transport equation, derive equation (11.2.22) for the fluctuations of the vorticity vector.
8. Given a heat conducting, compressible viscous fluid, derive the relation that must hold between the thermal equation of state, the specific enthalpy and the specific entropy. To this end, one may start by assuming that the specific heat at constant pressure, $c_p = \check{c}_p(\Theta, p)$ and the coefficient of thermal expansion $\alpha = \check{\alpha}(\Theta, p)$ are known functions determined by experiment. Does knowledge about these functions suffice for the determination of the thermal equation of state?
9. Prove that the FAVRE–average $\{a\}$ of a fluctuating quantity a satisfies the same filter rules that are also supposed by $\langle a\rangle$.

11.9 Solutions

1. Because $\langle \boldsymbol{v} \rangle$ and $\boldsymbol{v}'$ are both solenoidal, $\operatorname{div} \langle \boldsymbol{v} \rangle = \operatorname{div} \boldsymbol{v}' = 0$ one may show that

$$\text{(i)} \qquad \langle ((\langle v_i \rangle v_j')_{,j}\, v_i' \rangle = \langle \langle v_i \rangle_{,j}\, v_j' v_i' \rangle$$

$$= \langle v_i' v_j' \rangle \langle v_i \rangle_{,j} = \langle v_i' v_j' \rangle \langle v_{i,j} \rangle = -\frac{1}{\rho} R_{ij} \langle L_{ij} \rangle = -\frac{1}{\rho} \boldsymbol{R} \cdot \langle \boldsymbol{D} \rangle ,$$

$$\text{(ii)} \qquad \left\langle (v_i'(v_j' + \langle v \rangle_j))_{,j}\, v_i' \right\rangle = \left\langle v_{i,j}' v_i' (v_j' + \langle v \rangle_j) \right\rangle$$

$$= \left\langle \left(\frac{v_i' v_i'}{2} \right)_{,j} (v_j' + \langle v \rangle_j) \right\rangle = \left\langle \operatorname{grad} \left(\frac{\boldsymbol{v}' \cdot \boldsymbol{v}'}{2} \right) \cdot (\boldsymbol{v}' + \langle \boldsymbol{v} \rangle) \right\rangle .$$

2. Since $f(x) f'(x) = \frac{1}{2} (f^2(x))'$, one may write

$$\int_{-a}^{a} f(x)\, f'(x)\, \mathrm{d}x = \frac{1}{2} \int_{-a}^{a} [f^2(x)]'\, \mathrm{d}x = \frac{1}{2} f^2(x) \Big|_{-a}^{a}$$
$$= \frac{1}{2} [f^2(a) - f^2(-a)] = 0 ,$$

which must vanish, since $f(a) = -f(-a)$ and therefore $f^2(a) = f^2(-a)$, *qed.*

3. Written in Cartesian index notation we have

$$\begin{aligned} \langle \varepsilon_{ijk} v_j' v_{k,ll}' \rangle &= \langle (\varepsilon_{ijk} v_j' v_{k,l}')_{,l} - \varepsilon_{ijk} v_{j,l}' v_{k,l}' \rangle \\ &= \langle (\varepsilon_{ijk} v_j' v_{k,l}')_{,l} \rangle . \end{aligned} \qquad (11.9.1)$$

The second term vanishes, because it involves a contracting operation over the indices j, k of a skewsymmetric tensor with a symmetric one. The mean value (11.8.1) is interpreted as the spatial mean value:

$$(11.9.1) = \frac{1}{V} \int_V (\varepsilon_{ijk} v_j' v_{k,l}')_{,l}\, \mathrm{d}v \overset{\text{Gauß}}{=} \frac{1}{V} \int_{\partial V} \varepsilon_{ijk} v_j' v_{k,l}' n_l\, \mathrm{d}a = 0 ,$$

where the surface integral vanishes because $\boldsymbol{v}'$ is an even function relative to the center $\boldsymbol{x}$ of V, whereas $\operatorname{grad} \boldsymbol{v}$ is odd. With the ergodic assumption and the assumption of isotropic turbulence (11.8.1) thus holds. *qed.*

4. In Cartesian index notation we have

$$\begin{aligned} \langle \varepsilon_{ijk} \varepsilon_{klm} v_{m,lj}' v_i' \rangle &= \langle (\delta_{il} \delta_{jm} - \delta_{im} \delta_{jl}) v_{m,lj}' v_i' \rangle \\ &= \langle \underbrace{v_{j,ij}'}_{0}\, v_i' - v_{i,jj}' v_i' \rangle = -\langle \triangle \boldsymbol{v}' \cdot \boldsymbol{v} \rangle \\ &= -\langle (v_{i,j}' v_i')_{,j} - v_{i,j}' v_{i,j}' \rangle \\ &= \langle v_{i,j}' v_{i,j}' \rangle - \frac{1}{V} \oint_{\partial V} v_{i,j}' v_i' n_j\, \mathrm{d}a \\ &= \langle |\operatorname{grad} \boldsymbol{v}'|^2 \rangle . \end{aligned} \qquad (11.9.2)$$

(Here, $\langle\rangle$ has been interpreted as a spatial average. The divergence term has been transformed with the divergence theorem into a surface integral. However, this surface integral vanishes, since for isotropic turbulence $\boldsymbol{v}'$ is an even function relative to the point $\boldsymbol{x}$ (the center of V), so that $\operatorname{grad}\boldsymbol{v}'$ must be odd. Hence, the result $(\operatorname{grad}^T \boldsymbol{v}')\boldsymbol{v}'$ must also be odd, so that the surface integral vanishes, since positive and negative contributions of the integral function arise in pairs. Equation $(11.9.2)_2$ proves $(11.8.2)_1$ and $(11.9.2)_4$ proves $(11.8.2)_2$. Furthermore, with $\omega_i' = \varepsilon_{ijk}v_{k,j}'$, we have

$$
\begin{aligned}
\langle \boldsymbol{v}' \cdot \operatorname{curl}(\operatorname{curl}\boldsymbol{v}')\rangle &= \langle \varepsilon_{ijk}\omega_{k,j}' v_i' \rangle \\
&= \langle \underbrace{\varepsilon_{ijk}(\omega_k' v_i')_{,j}}_{\sim 0} - \underbrace{\varepsilon_{ijk}v_{i,j}'}_{-\omega_k'}\omega_k' \rangle \qquad (11.9.3)\\
&= \langle \boldsymbol{\omega}' \cdot \boldsymbol{\omega}' \rangle = \langle |\operatorname{curl}\boldsymbol{v}'|^2 \rangle \,,
\end{aligned}
$$

which proves $(11.8.2)_3$

5. In Exercise 4 it was proved that $\langle |\operatorname{grad}\boldsymbol{v}'|^2\rangle \cong \langle |\operatorname{curl}\boldsymbol{v}'|^2\rangle$. If one compares

$$
\begin{aligned}
\langle |\operatorname{grad}\boldsymbol{v}'|^2\rangle &= \langle \boldsymbol{L}' \cdot \boldsymbol{L}'^T \rangle \\
&= \langle (\boldsymbol{D}' + \boldsymbol{W}') \cdot (\boldsymbol{D}'^T + \boldsymbol{W}'^T)\rangle \qquad (11.9.4)\\
&= \langle \boldsymbol{D}' \cdot \boldsymbol{D}'^T \rangle + \langle \boldsymbol{W}' \cdot \boldsymbol{W}'^T \rangle
\end{aligned}
$$

with

$$
\begin{aligned}
\langle |\operatorname{curl}\boldsymbol{v}'|^2\rangle &= \langle \boldsymbol{L}' \cdot \boldsymbol{L}'^T - \boldsymbol{L}' \cdot \boldsymbol{L}' \rangle = \langle \boldsymbol{L}' \cdot (\boldsymbol{L}'^T - \boldsymbol{L}')\rangle \\
&= \langle \boldsymbol{L}' \cdot 2\boldsymbol{W}^T \rangle = 2\underbrace{\langle \boldsymbol{D}' \cdot \boldsymbol{W}'^T \rangle}_{0} + 2\langle \boldsymbol{W}' \cdot \boldsymbol{W}'^T \rangle \qquad (11.9.5)\\
&= 2\langle \boldsymbol{W}' \cdot \boldsymbol{W}'^T \rangle \,,
\end{aligned}
$$

then by equating these two expressions one obtains the result (11.8.3), *qed*.

6. Consider a density preserving EULER fluid of which the momentum equation is given by

$$
\rho \frac{\mathrm{d}\boldsymbol{v}}{\mathrm{d}t} = -\operatorname{grad} p \qquad (11.9.6)
$$

(specific body forces are ignored).
(i) Scalar multiplication of (11.9.6) with $\boldsymbol{v}$ yields

$$
\rho \frac{\mathrm{d}(\boldsymbol{v}^2/2)}{\mathrm{d}t} = -(\operatorname{grad} p) \cdot \boldsymbol{v} = -\operatorname{div}(p\boldsymbol{v}) \qquad (11.9.7)
$$

and after integration over a material volume

$$
\int_V \rho \frac{\mathrm{d}(v^2/2)}{\mathrm{d}t} \mathrm{d}V = -\int_V \operatorname{div}(p\boldsymbol{v}) = -\int_{\partial V} p\boldsymbol{v} \cdot \boldsymbol{n} \mathrm{d}a \,. \qquad (11.9.8)
$$

If in addition the REYNOLDS transport theorem is used, there follows

$$\underbrace{\frac{\mathrm{d}}{\mathrm{d}t}\int_V \left(\rho\frac{v^2}{2}\right)\mathrm{d}V}_{K} = \underbrace{-\int_{\partial V} p(\boldsymbol{v}\cdot\boldsymbol{n})\mathrm{d}a}_{\substack{=\ 0,\ \text{if}\ p\ =\ 0\ \text{or}\\ \boldsymbol{v}\cdot\boldsymbol{n}=0\ \text{on}\ \partial V}}\ . \tag{11.9.9}$$

If on ∂V the surface forces vanish, then

$$\frac{\mathrm{d}K}{\mathrm{d}t} = 0\,, \quad K := \rho\int_V \frac{v^2}{2}\mathrm{d}V\,, \tag{11.9.10}$$

which proves the first conjecture.
(ii) In a density preserving fluid the curl of the momentum equation yields the HELMHOLTZ vorticity equation,

$$\frac{\mathrm{d}\boldsymbol{\omega}}{\mathrm{d}t} = (\operatorname{grad}\boldsymbol{v})\boldsymbol{\omega}\,. \tag{11.9.11}$$

Scalar multiplication of (11.9.11) with $\boldsymbol{v}$ leads to

$$\frac{\mathrm{d}\boldsymbol{\omega}}{\mathrm{d}t}\cdot\boldsymbol{v} = ((\operatorname{grad}\boldsymbol{v})\boldsymbol{\omega})\cdot\boldsymbol{v}\,. \tag{11.9.12}$$

If one adds to this the scalar product of the momentum equation with $\boldsymbol{\omega}$,

$$\boldsymbol{\omega}\cdot\frac{\mathrm{d}\boldsymbol{v}}{\mathrm{d}t} = -\operatorname{grad}\left(\frac{p}{\rho}\right)\cdot\boldsymbol{\omega}\,, \tag{11.9.13}$$

then one obtains

$$\frac{\mathrm{d}}{\mathrm{d}t}(\boldsymbol{\omega}\cdot\boldsymbol{v}) = ((\operatorname{grad}\boldsymbol{v})\boldsymbol{\omega})\cdot\boldsymbol{v} - \operatorname{div}(\frac{p}{\rho}\boldsymbol{\omega})\,. \tag{11.9.14}$$

Here use was made of the fact that $\operatorname{div}\boldsymbol{\omega} = \operatorname{div}\operatorname{curl}\boldsymbol{v} \equiv 0$. Multiplying both sides of (11.9.14) with ρ and then integrating over a material volume yields

$$\begin{aligned}
\int_V \rho\frac{\mathrm{d}}{\mathrm{d}t}(\boldsymbol{\omega}\cdot\boldsymbol{v})\mathrm{d}V &= \frac{\mathrm{d}}{\mathrm{d}t}\int_V \rho\boldsymbol{\omega}\cdot\boldsymbol{v}\mathrm{d}V \\
&= \int_V \rho\left[(v_{i,j}\omega_j)v_i - \left(\frac{p}{\rho}\omega_j\right)_{,j}\right]\mathrm{d}V \\
&= \int_V \rho\left[\left(\frac{v_i v_i}{2}\right)_{,j}\omega_j - \left(\frac{p}{\rho}\omega_j\right)_{,j}\right]\mathrm{d}V \\
&= \int_V \left[\rho\left(\frac{v_i v_i}{2}\omega_j - \frac{p}{\rho}\omega_j\right)\right]_{,j}\mathrm{d}V \\
&= \int_{\partial V} \left[\rho\left(\frac{v_i v_i}{2}\omega_j - \frac{p}{\rho}\omega_j\right)\right] n_j\mathrm{d}a\,,
\end{aligned}$$

where again use was made of the fact that $\operatorname{div}\operatorname{curl}\boldsymbol{v} = \operatorname{div}\boldsymbol{\omega} = 0$. The material volume must be so chosen that on ∂V p and $\boldsymbol{\omega}$ vanish. Therefore, since $\rho = constant$,

$$\frac{\mathrm{d}}{\mathrm{d}t}\int \boldsymbol{\omega}\cdot\boldsymbol{v}\mathrm{d}V = \frac{\mathrm{d}H}{\mathrm{d}t} = 0\,, \tag{11.9.15}$$

proving the second conjecture.
(iii) For the enstrophy one starts with (11.9.11) and obtains, after a multiplication with $\boldsymbol{\omega}$

$$\frac{\mathrm{d}}{\mathrm{d}t}\left(\frac{\boldsymbol{\omega}\cdot\boldsymbol{\omega}}{2}\right) = \operatorname{div}(\boldsymbol{v}\otimes\boldsymbol{\omega})\cdot\boldsymbol{\omega}\,,$$

and after multiplication with ρ and integration over the material volume

$$\int_V \rho\frac{\mathrm{d}}{\mathrm{d}t}\left(\frac{\boldsymbol{\omega}\cdot\boldsymbol{\omega}}{2}\right)\mathrm{d}V = \frac{\mathrm{d}}{\mathrm{d}t}\int_V \rho\frac{\boldsymbol{\omega}\cdot\boldsymbol{\omega}}{2}\mathrm{d}V = \int_V \rho\operatorname{div}(\boldsymbol{v}\otimes\boldsymbol{\omega})\cdot\boldsymbol{\omega}\mathrm{d}V = 0\,.$$

which shows that $\mathrm{d}E/\mathrm{d}t = 0$.

7. We first form the curl of the momentum equation

$$\operatorname{curl}\left(\frac{\partial\boldsymbol{v}}{\partial t} + \operatorname{grad}\left(\frac{v^2}{2}\right) - \boldsymbol{v}\times\operatorname{curl}\boldsymbol{v} = -\operatorname{grad}\frac{p}{\rho_*} + \nu\operatorname{div}\operatorname{grad}\boldsymbol{v} - \operatorname{grad}U\right),$$

in which U is the gravitational potential. From this, there follows

$$\frac{\partial\boldsymbol{\omega}}{\partial t} + \operatorname{div}(\boldsymbol{\omega}\otimes\boldsymbol{v}) - \operatorname{div}(\boldsymbol{v}\otimes\boldsymbol{\omega}) - \nu\operatorname{div}\operatorname{grad}\boldsymbol{\omega} = 0\,. \tag{11.9.16}$$

If this equation is averaged, one obtains

$$\begin{aligned}\frac{\partial\langle\boldsymbol{\omega}\rangle}{\partial t} + \operatorname{div}\Big(\langle\boldsymbol{\omega}\rangle\otimes\langle\boldsymbol{v}\rangle - \langle\boldsymbol{v}\rangle\otimes\langle\boldsymbol{\omega}\rangle + \langle\boldsymbol{\omega}'\otimes\boldsymbol{v}'\rangle \\ - \langle\boldsymbol{v}'\otimes\boldsymbol{\omega}'\rangle\Big) - \nu\operatorname{div}\operatorname{grad}\langle\boldsymbol{\omega}\rangle = 0\,.\end{aligned} \tag{11.9.17}$$

Subtracting next (11.9.17) from (11.9.16) and defining

$$\boldsymbol{\omega}' = \boldsymbol{\omega} - \langle\boldsymbol{\omega}\rangle\,,$$

yields

$$\begin{aligned}\frac{\partial\boldsymbol{\omega}'}{\partial t} + \operatorname{div}\Big(\boldsymbol{\omega}\otimes\boldsymbol{v} - \boldsymbol{v}\otimes\boldsymbol{\omega} - \langle\boldsymbol{\omega}\rangle\otimes\langle\boldsymbol{v}\rangle + \langle\boldsymbol{v}\rangle\otimes\langle\boldsymbol{\omega}\rangle \\ - \langle\boldsymbol{\omega}'\otimes\boldsymbol{v}'\rangle + \langle\boldsymbol{v}'\otimes\boldsymbol{\omega}'\rangle\Big) - \nu\operatorname{div}\operatorname{grad}\boldsymbol{\omega}' = 0\,.\end{aligned} \tag{11.9.18}$$

The critical term is the expression with the divergence operator. This term can be transformed as follows:

$$\begin{aligned}\operatorname{div}() &= \operatorname{div}\Big((\boldsymbol{\omega} - \langle\boldsymbol{\omega}\rangle) \otimes \langle\boldsymbol{v}\rangle + \boldsymbol{\omega} \otimes \boldsymbol{v}' - (\boldsymbol{v} - \langle\boldsymbol{v}\rangle) \otimes \langle\boldsymbol{\omega}\rangle - \boldsymbol{v} \otimes \boldsymbol{\omega}' \\ &\qquad - \langle\boldsymbol{\omega}' \otimes \boldsymbol{v}'\rangle + \langle\boldsymbol{v}' \otimes \boldsymbol{\omega}'\rangle\Big) \\ &= \operatorname{div}\Big(\boldsymbol{\omega}' \otimes (\langle\boldsymbol{v}\rangle + \boldsymbol{v}') + \langle\boldsymbol{\omega}\rangle \otimes \boldsymbol{v}' - \boldsymbol{v}' \otimes \langle\boldsymbol{\omega}\rangle - (\langle\boldsymbol{v}\rangle + \boldsymbol{v}') \otimes \boldsymbol{\omega}' \\ &\qquad - \langle\boldsymbol{\omega}' \otimes \boldsymbol{v}'\rangle + \langle\boldsymbol{v}' \otimes \boldsymbol{\omega}'\rangle\Big) .\end{aligned}$$

Substitution in (11.9.18) yields the equation listed as (11.2.22), qed.

8. In Chapter 5 the GIBBS equation for a compressible, heat conducting viscous fluid was shown to take the following forms:

$$\begin{aligned} \mathrm{d}s &= \frac{1}{\Theta}\left\{\frac{\partial h}{\partial s}\mathrm{d}s + \left(\frac{\partial h}{\partial p} - v\right)\mathrm{d}p\right\} \\ &= \frac{1}{\Theta}\left\{\frac{\partial \breve{g}}{\partial \Theta}\mathrm{d}\Theta + \left(\frac{\partial \breve{g}}{\partial p} - v\right)\mathrm{d}p\right\} , \\ \mathrm{d}g &= -s\mathrm{d}\Theta + v\mathrm{d}p ; \end{aligned} \tag{11.9.19}$$

$h = h(s,p)$ is the specific enthalpy, $g = \breve{g}(\Theta,p)$ the free enthalpy or the GIBBS free energy and $v := 1/\rho$ the specific volume. From $(11.9.19)_{2,3}$ one immediately deduces

$$\Theta = \frac{\partial h}{\partial s} , \quad v = \frac{\partial h}{\partial p} , \quad s = -\frac{\partial \breve{g}}{\partial \Theta} , \quad v = \frac{\partial \breve{g}}{\partial p} , \tag{11.9.20}$$

which imply the MAXWELL relations

$$\frac{\partial \Theta}{\partial p} = \frac{\partial v}{\partial \Theta} , \quad \frac{\partial \breve{s}}{\partial p} = -\frac{\partial \breve{v}}{\partial \Theta} . \tag{11.9.21}$$

Consequently, $(11.9.19)_2$ and $(11.9.21)_2$ imply

$$\Theta\frac{\partial \breve{s}}{\partial p} = -\Theta\frac{\partial \breve{v}}{\partial \Theta} = \frac{\partial \breve{h}}{\partial p} - \breve{v} \tag{11.9.22}$$

or after an integration with respect to p

$$\begin{aligned} \breve{h}(\Theta,p) &= -\int_{p_0}^{p}\left(\Theta\frac{\partial \breve{v}}{\partial \Theta} - \breve{v}\right)(\Theta,\bar{p})\mathrm{d}\bar{p} + F(\Theta) \\ &= -\int_{p_0}^{p}(\breve{v}(\Theta\breve{\alpha} - 1))(\Theta,\bar{p})\mathrm{d}\bar{p} + F(\Theta) , \end{aligned} \tag{11.9.23}$$

in which $\alpha := \frac{1}{v}\partial\breve{v}/\partial\Theta$ is the coefficient of thermal expansion. $F(\Theta)$ is a "constant of integration". Differentiation of (11.9.23) with respect to Θ leads to

$$\frac{\mathrm{d}F}{\mathrm{d}\Theta} = \underbrace{\frac{\partial \breve{h}}{\partial \Theta}}_{\breve{c}_p(\Theta,p)} + \frac{\partial}{\partial \Theta}\int_{p_0}^{p}(\breve{v}\Theta(\breve{\alpha} - 1))(\Theta,\bar{p})\mathrm{d}\bar{p} \tag{11.9.24}$$

and, after a further integration with respect to Θ,

$$F(\Theta) = \int_{\Theta_0}^{\Theta} \check{c}_p(\bar{\Theta}, p)\mathrm{d}\bar{\Theta} + \int_{p_0}^{p} (\check{v}(\Theta\check{\alpha} - 1))(\Theta, \bar{p})\mathrm{d}\bar{p}\Big|_{\Theta_0}^{\Theta} + h(\Theta_0, p_0) \,. \tag{11.9.25}$$

Substitution into (11.9.23) yields

$$\begin{aligned}\check{h}(\Theta, p) &= \int_{\Theta_0}^{\Theta} (\check{c}_p)(\bar{\Theta}, p)\mathrm{d}\bar{\Theta} \\ &\quad - \int_{p_0}^{p} \check{v}(\Theta_0, \bar{p})(\Theta_0\check{\alpha}(\Theta_0, \bar{p}) - 1)\mathrm{d}\bar{p} + \check{h}(\Theta_0, p_0) \,.\end{aligned} \tag{11.9.26}$$

The entropy follows from $(11.9.21)_2$ by integration with respect to p,

$$\begin{aligned} s &= -\int_{p_0}^{p} \frac{\partial\check{v}(\Theta, \bar{p})}{\partial\Theta}\mathrm{d}\bar{p} + H(\Theta) \\ &= -\int_{p_0}^{p} \check{v}(\Theta, \bar{p})\check{\alpha}(\Theta, \bar{p})\mathrm{d}\bar{p} + H(\Theta) \,.\end{aligned} \tag{11.9.27}$$

The 'constant' of integration is obtained by differentiating (11.9.27) with respect to Θ, yielding

$$\frac{\partial\check{s}}{\partial\Theta} = \frac{\mathrm{d}H}{\mathrm{d}\Theta} - \frac{\partial}{\partial\Theta}\int_{p_0}^{p} \check{v}(\Theta, \bar{p})\check{\alpha}(\Theta, \bar{p})\mathrm{d}\bar{p} \,, \tag{11.9.28}$$

From this as well as equation (11.9.19) and the definition of $c_p = \check{c}_p(\Theta, p)$ one obtains

$$\frac{\partial\check{s}}{\partial\Theta} = \frac{1}{\Theta}\frac{\partial\check{h}}{\partial\Theta} = \frac{\check{c}_p(\Theta, p)}{\Theta} \,. \tag{11.9.29}$$

Therefore, (11.9.28) implies, after another integration with respect to Θ,

$$H = \int_{\Theta_0}^{\Theta} \frac{\check{c}_p(\bar{\Theta}, p)}{\bar{\Theta}}\mathrm{d}\bar{\Theta} + \int_{p_0}^{p} \check{v}(\Theta, \bar{p})\check{\alpha}(\Theta, \bar{p})\mathrm{d}\bar{p}\Big|_{\Theta_0}^{\Theta} + \check{s}(\Theta_0, p_0) \,. \tag{11.9.30}$$

Substitution of this result in (11.9.27) yields for the entropy the expression

$$\check{s}(\Theta, p) = \int_{\Theta_0}^{\Theta} \frac{\check{c}_p(\bar{\Theta}, p)}{\bar{\Theta}}\mathrm{d}\bar{\Theta} - \int_{p_0}^{p} \check{v}(\Theta_0, \bar{p})\check{\alpha}(\Theta_0, \bar{p})\mathrm{d}\bar{p} + \check{s}(\Theta_0, p_0) \,. \tag{11.9.31}$$

The important formulas are (11.9.26) and (11.9.31); from them it is manifest that besides the functions $\check{c}_p(\Theta, p)$ and $\check{\alpha}(\Theta_0, p)$ also the constants of integration $\check{h}(\Theta_0, p_0)$, $\check{s}(\Theta_0, p_0)$ as well as the function $\check{v}(\Theta_0, p)$ must be known. The latter cannot be determined from $\alpha := \partial(\ln\check{v}(\Theta, p))/\partial\Theta$ via

$$\check{v}(\Theta,p) = v(\Theta_0,p)\exp\left\{\int_{\Theta_0}^{\Theta} \check{\alpha}(\bar{\Theta},p)\mathrm{d}\bar{\Theta}\right\}, \tag{11.9.32}$$

because in (11.9.26) and (11.9.31) one needs $v(\Theta_0, p)$, a function that cannot be determined from α. Therefore, c_p and $v = 1/\rho$ must be determined experimentally as functions of both Θ and p, if the enthalpy and entropy are to be determined.

One may also easily verify that for an *ideal gas*, $pv = R\Theta$, the following results hold.

$$\begin{aligned} \check{\alpha} &= \frac{1}{\Theta}, \quad \check{s} = \int_{\Theta_0}^{\Theta} \frac{\check{c}_p(\bar{\Theta})}{\bar{\Theta}}\mathrm{d}\bar{\Theta} - R\ln\frac{p}{p_0}, \\ \check{h} &= \int_{\Theta_0}^{\Theta} \check{c}_p(\bar{\Theta})\mathrm{d}\bar{\Theta} \end{aligned} \tag{11.9.33}$$

as one would expect.

For a nearly density preserving material, i. e., a fluid for which $|\Theta_0\alpha(\Theta_0,p)| \ll 1$ one obtains

$$\begin{aligned} \check{s} &\approx \int_{\Theta_0}^{\Theta} \frac{\check{c}_p(\Theta,p)}{\bar{\Theta}}\mathrm{d}\bar{\Theta} + \check{s}(\Theta_0,p_0), \\ \check{h} &\approx \int_{\Theta_0}^{\Theta} \check{c}_p(\bar{\Theta},p)\mathrm{d}\bar{\Theta} + \int_{p_0}^{p} \check{v}(\Theta_0,\bar{p})\mathrm{d}\bar{p} + \check{h}(\Theta_0,p_0). \end{aligned} \tag{11.9.34}$$

Only if the second term is small compared to the first, $\check{h}$ will be given simply by c_p.

9. (i) Linearity: Let a_1 and a_2 be two variables which are subject to fluctuations and let c be a constant. Then, according to (11.5.4) we have

$$\begin{aligned} \{a_1 + ca_2\} &= \langle a_1 + ca_2\rangle + \frac{\langle\rho'(a_1' + ca_2')\rangle}{\langle\rho\rangle} \\ &= \langle a_1\rangle + c\langle a_2\rangle + \frac{\langle\rho' a_1'\rangle}{\langle\rho\rangle} + c\frac{\langle\rho' a_2'\rangle}{\langle\rho\rangle} \\ &= (\langle a_1\rangle + \frac{\langle\rho' a_1'\rangle}{\langle\rho\rangle}) + c(\langle a_2\rangle + \frac{\langle\rho' a_2'\rangle}{\langle\rho\rangle}) \\ &= \{a_1\} + c\{a_2\}, \end{aligned}$$

which corroborates linearity.

(ii) $\langle\langle a\rangle\rangle = \langle a\rangle \;\Rightarrow\; \{\{a\}\} = \{a\}$.
The proof follows with the aid of the definition of $\{a\}$ as follows:

$$
\begin{aligned}
\{\{a\}\} &= \left\{ \langle a \rangle + \frac{\langle \rho' a' \rangle}{\langle \rho \rangle} \right\} \\
&= \left\langle \langle a \rangle + \frac{\langle \rho' a' \rangle}{\langle \rho \rangle} \right\rangle + \left\langle \frac{\rho' (\langle a \rangle + \frac{\langle \rho' a' \rangle}{\langle \rho \rangle})'}{\langle \rho \rangle} \right\rangle \\
&= \langle \langle a \rangle \rangle + \left\langle \frac{\langle \rho' a' \rangle}{\langle \rho \rangle} \right\rangle + \underbrace{\frac{\langle \rho' \rangle (\langle a \rangle' + (\frac{\langle \rho' a' \rangle}{\langle \rho \rangle})')}{\langle \rho \rangle}}_{=0,\ \text{since}\ \langle \rho' \rangle = 0} \\
&= \langle a \rangle + \frac{\langle \rho' a' \rangle}{\langle \rho \rangle} , \qquad \text{qed.}
\end{aligned}
$$

(iii) $\langle a' \rangle = 0 \Rightarrow \{a''\} = 0$.
The proof follows with the definition of a'' by using (11.5.5):

$$
\begin{aligned}
\{a''\} &= \left\{ a' - \frac{\langle \rho' a' \rangle}{\langle \rho \rangle} \right\} = \{a'\} - \left\{ \frac{\langle \rho' a' \rangle}{\langle \rho \rangle} \right\} \\
&= \langle a' \rangle + \frac{\langle \rho' (a')' \rangle}{\langle \rho \rangle} - \left(\left\langle \frac{\langle \rho' a' \rangle}{\langle \rho \rangle} + \frac{\langle \rho' \langle \rho' a' \rangle \rangle}{\langle \rho \rangle^2} \right\rangle \right) .
\end{aligned}
$$

Since $(a')' = a'$ (because the mean value of a' vanishes) one has

$$
\{a''\} = \langle a' \rangle + \frac{\langle \rho' a' \rangle}{\langle \rho \rangle} - \frac{\langle \rho' a' \rangle}{\langle \rho \rangle} - 0 = \langle a' \rangle = 0 .
$$

(iv) $\langle a \rangle' = 0 \Rightarrow \{a\}'' = 0$. For the proof one forms

$$
\begin{aligned}
\{a\}'' &= \left(\langle a \rangle + \frac{\langle \rho' a' \rangle}{\langle \rho \rangle} \right)'' \\
&= \underbrace{\langle a \rangle'}_{0} + \underbrace{\left(\frac{\langle \rho' a' \rangle}{\langle \rho \rangle} \right)'}_{0} - \frac{1}{\langle \rho \rangle} \left\langle \rho' \underbrace{\left(\langle a \rangle + \frac{\langle \rho' a' \rangle}{\langle \rho \rangle} \right)'}_{0} \right\rangle = 0 .
\end{aligned}
$$

Here, in the first row use was made of (11.5.4), and in the second row $\langle\rangle' = 0$ was employed. This completes the proof.

12. Thermodynamic Formulation of Turbulent Closure Relations of First Order Level – Algebraic REYNOLDS Stress Models[1]

12.1 Background

In the last two Chaps. 10 and 11, some of the elements of the foundations of turbulence theory and turbulent closure procedures were laid down. It was shown, how the REYNOLDS averaged field equations could be derived from the full balance laws of physics by applying to them a filter operation $\langle\cdot\rangle$. The properties of the filter were so assigned to include the ergodic hypothesis, which, in the averaging process, means that multiple averaging is an invariant operation[2], i.e., does not lead to new results, viz. $\langle f\rangle = \langle\langle\cdots\langle f\rangle\cdots\rangle\rangle$ for any physical quantity f. As for the physical balance laws of mass, momenta and energy this procedure led to the so-called REYNOLDS averaged balance equations (of NAVIER–STOKES and FOURIER). They are the foundation for the so-called REYNOLDS Averaged Numerical Simulation (RANS), in which the REYNOLDS stresses and the turbulent heat flux vector are parameterized by writing

$$
\begin{aligned}
\frac{1}{\rho}\boldsymbol{R} &= -\langle \boldsymbol{v}'\otimes\boldsymbol{v}'\rangle = -\tfrac{2}{3}k\boldsymbol{I} + 2\,\nu_t\langle\boldsymbol{D}\rangle,\\
\boldsymbol{Q} &= \langle\rho\rangle\langle\epsilon'\boldsymbol{v}'\rangle = -\lambda_t\,\mathrm{grad}\,\langle\Theta\rangle,
\end{aligned}
\tag{12.1.1}
$$

in which ν_t is the turbulent kinematic viscosity and $\boldsymbol{D} = \mathrm{sym}\,\mathrm{grad}\,\boldsymbol{v}$ is the stretching tensor. The zeroth order closure scheme simply assigns to ν_t and λ_t numerical values; the first order closure approach, presented in Chap. 11, consisted in postulating the relation

$$
\nu_t = c_\mu\frac{k^2}{\varepsilon}\,,\quad c_\mu = 0.09\,,\quad \lambda_t = \frac{\nu_t}{\sigma}\,,
\tag{12.1.2}
$$

in which k is the turbulent kinetic energy and ε its dissipation rate, for which balance relations were postulated to hold. They were motivated or derived in Chap. 11, and a one-dimensional application is given in Chap. 13; however, it is equally long recognized that the above closure procedure is limited, because computational results are often at variance with observations of the physics

[1] The text for this chapter was thorougly read and criticised by AMSINI SADIKI.

[2] In mathematics operators with such properties are also called projections.

of many turbulent flow classes. In those cases, the above closure procedure is obviously too simple.

An obvious generalization could be to employ a second order closure scheme, i.e., to formulate a transport equation for the REYNOLDS stress tensor, but such schemes require five[3] additional transport equations for the REYNOLDS stresses (and three for the turbulent heat flux vector and additional three for the mass flux vector of each tracer substance if heat and concentration transport are considered). These must then be solved along with the other balance laws. This is computationally very expensive. More economical are so-called *nonlinear* or *algebraic* REYNOLDS *stress models* (and corresponding models for the heat flux and the tracer mass fluxes). Such models establish a general closure condition for $\boldsymbol{R}^D$, a functional relation, which may make $\boldsymbol{R}^D = \boldsymbol{R} + (2/3)\rho k \boldsymbol{I}$ depending not only on $\langle \boldsymbol{D} \rangle$, but also on $\langle \boldsymbol{W} \rangle := \langle \operatorname{skew} \operatorname{grad} \boldsymbol{v} \rangle$, $\langle \Theta \rangle$ and $\operatorname{grad} \langle \Theta \rangle$, where $\langle \Theta \rangle$ is the temperature, and possibly further variables. Such models are called *nonlinear* REYNOLDS *stress models.* If they are derived from the transport equation of $\boldsymbol{R}^D$, they are called *algebraic* REYNOLDS *stress models.* Since the two formulations are mathematically the same, we shall use either denotation[4]. In these models $\boldsymbol{R}^D$ and $\langle \boldsymbol{D} \rangle$ may not be collinear (or affine) to one another. Such non-linearities and anisotropies[5] (or the corresponding non-collinearity of the heat flux vector with the temperature gradient and of the tracer mass flux with the concentration gradient) may account – as we shall see – for apparent nonlinear dependences of the REYNOLDS stress on the averaged stretching, streamline curvature and frame rotation, etc. Such a closure approach is still of first order, if balance equations are formulated for the turbulent kinetic energy and its dissipation rate so that k and ε still play the role of independent fields and no evolution equation for $\boldsymbol{R}^D$ is considered. However, contrary to the form which the REYNOLDS stress tensor has in the k–ε model its closure statements are now very involved, similar to second and third grade fluids in

[3] The REYNOLDS stress deviator has only five independent components.

[4] The motivation of the nonlinear REYNOLDS stress models is very similar to that used in phenomenological continuum mechanics: The REYNOLDS stress tensor and other closure variables are assumed to depend on a certain set of variables, and then thermodynamic arguments (second law) are used to reduce these dependences. In the other approach the balance law for the REYNOLDS stresses is simplified; it leads to an implicit relation for the REYNOLDS stress tensor which may equally be viewed as a phenomenological relation of a certain constitutive class. They simply express the dependencies more directly.

[5] The term "anisotropy" in connection with the closure functional for the REYNOLDS stress tensor is differently used in the turbulence literature than in the continuum mechanics literature. When turbulence modellers speak of the anisotropic REYNOLDS stress, then $\boldsymbol{R}^D$, the deviator of the REYNOLDS stress tensor, is meant. Any deviation of $\boldsymbol{R}^D$ from $\boldsymbol{R}^D = \rho \nu_t \langle \boldsymbol{D} \rangle$ is then called to contribute to the anisotropy of $\boldsymbol{R}^D$. This is different from the classical notion of anisotropy which is a property usually expressed in the reference configuration. The reader should be aware of the difference of the jargons.

rheology, and, as stated above, often they are an algebraic reduction of the balance law for the REYNOLDS stress tensor in second order closure models. If properly parameterized it should not be a surprise that these models perform much better than the standard k–ε model when they are compared with experiments.

Turbulent processes are irreversible processes. Therefore, they must fulfill the irreversibility requirements of the second law of thermodynamics as all physical processes must do. In the theory of material behaviour (Chap. 5) the requirements of irreversibility operate as constraints to the material equations. Since the motion within a fluid body can arise in both forms, laminar and turbulent, it is only natural to require that the second law of thermodynamics takes a form such that it may – for laminar processes – constrain the material functionals, and once this is achieved, they may – for the turbulent flows – reduce the functionals postulated for the turbulence description. Thus, if the postulate of irreversibility is requested to hold for the turbulent flows of a fluid obeying a certain material behaviour, then the second law must necessarily constrain the constitutive relations as phenomenological relations for the material behaviour as well as closure relations for the parameterization of the turbulence.

The above outline indicates that a possible entropy principle must cover both cases: on the *microscale* level it accounts for the loss of degrees of freedom in the molecular structure of the fluid, on the *mesoscale* level, in which the turbulent eddies form the subscale structure of the turbulent motion, it constrains the way in which the subscale eddies affect the *macroscale* closure relations.

It is intuitively clear that the processes at the microscale level may affect those in the mesoscale but not vice versa. It also seems to be plausible that the possible interaction of the two levels may complicate the thermodynamics. In particular, if one interprets temperature to be a measure of the kinetic energy of the fluctuating motion of the atoms and molecules, then the turbulent kinetic energy likely plays the analogous role for the turbulent motion. Heat flux and turbulent kinetic-energy flux are the corresponding flux quantities arising in the balance laws of internal energy and turbulent kinetic energy, respectively, but it is not clear at all at the present stage of development how the entropy flux may be related to these or, more generally, how it should be postulated. Thus, it is inescapable here to regard the entropy flux as a phenomenological quantity to be determined as part of the entropy principle and not to assume it in the classical form "heat flux divided by absolute temperature". That such an a priori assignment must most likely be wrong follows simply from the above rules played by the temperature and turbulent kinetic energy[6].

[6] We shall give further reasons based on significant physical results, as we proceed in our developments, when they are better understood than here.

Reduction of nonlinear algebraic REYNOLDS stress relations by an entropy principle ascertains that the closure conditions satisfy the irreversibility conditions. In turbulence theory turbulent closure conditions are required to be in conformity with what is called *realizability conditions.* Authors do not unanimously agree on what these conditions represent[7]. These conditions seem to be inequality requirements for variables that must be satistified for obvious mathematical reasons. Such relations are e.g. either the SCHWARZ inequalities for the REYNOLDS stress components or conditions necessary to be fulfilled for obvious reasons (e.g. $k > 0$ by definition). In our, more complete, formulation turbulent processes are simply said to be *realizable*, if they obey the second law of thermodynamics. Indeed, any other process is physically not realizable by mere violation of a fundamental physical law. This more general understanding of the notion "realizability" is always embracing the more restrictive definitions used in turbulence theory, but it may also be regarded as more adequate because it is based on a physical law (the second law of thermodynamics) rather than on ad-hoc assumptions. In particular, an ad-hoc condition of realizability may simply follow as a subcondition of the entropy inequality (e.g. $\varepsilon > 0$ as dissipation is thought to be positive).

12.2 Basic Equations

Consider a BOUSSINESQ fluid, in which density variations are only accounted for in the buoyancy force, and the mass balance requires the velocity field to be solenoidal. Assume this fluid to obey the NAVIER–STOKES–FOURIER equations as presented e.g. in Chap. 7. Apply an averaging operator to these equations which satisfies the ergodic property. Then, the REYNOLDS averaged balance laws of mass, linear momentum, internal energy, turbulent kinetic energy and entropy in a non-inertial frame take the following forms

$$\operatorname{div}\langle \boldsymbol{v}\rangle = 0\,, \tag{12.2.1}$$

$$\frac{\mathrm{d}\langle \boldsymbol{v}\rangle}{\mathrm{d}t} = \frac{1}{\langle\rho\rangle}\operatorname{div}\langle \boldsymbol{t}\rangle + \frac{1}{\rho_*}\operatorname{div}\boldsymbol{R} + \langle \boldsymbol{f}\rangle + (\boldsymbol{I}_0 + 2\boldsymbol{\Omega}\times\langle \boldsymbol{v}\rangle)\,, \tag{12.2.2}$$

$$\langle\rho\rangle\frac{\mathrm{d}\langle\epsilon\rangle}{\mathrm{d}t} = -\operatorname{div}\langle \boldsymbol{q}\rangle - \operatorname{div}\boldsymbol{Q} + \operatorname{tr}(\langle \boldsymbol{t}\rangle\operatorname{grad}\langle \boldsymbol{v}\rangle) + \langle\rho\rangle\varepsilon + \langle\rho\rangle\mathfrak{r}\,, \tag{12.2.3}$$

$$\langle\rho\rangle\frac{\mathrm{d}k}{\mathrm{d}t} = \operatorname{tr}(\boldsymbol{R}\operatorname{grad}\langle \boldsymbol{v}\rangle) - \operatorname{div}\boldsymbol{K} - \langle\rho\rangle\varepsilon\,, \tag{12.2.4}$$

$$\langle\rho\rangle\frac{\mathrm{d}\langle s\rangle}{\mathrm{d}t} + \operatorname{div}\langle\boldsymbol{\phi}^{\mathrm{ent}}\rangle + \operatorname{div}\boldsymbol{\phi}^{\mathrm{ent}}_{(T)} - \langle\eta^{\mathrm{ent}}\rangle \geq 0\,. \tag{12.2.5}$$

In these equations, $\langle\cdot\rangle$ denotes the averaging operation and

$$\frac{\mathrm{d}f}{\mathrm{d}t} = \dot{f} = \frac{\partial f}{\partial t} + (\operatorname{grad} f)\langle \boldsymbol{v}\rangle \tag{12.2.6}$$

[7] Papers dealing with such realizability conditions are e.g. by DU VACHAT [247], RUNG et al. [200], SHIH et al. [214].

is the time derivative of f following the mean motion. Moreover, $\langle\rho\rangle, \langle\epsilon\rangle, k, \varepsilon$ are the mean density, the mean internal energy, the turbulent kinetic energy and the specific turbulent dissipation rate[8]; $\langle\boldsymbol{f}\rangle$ is the averaged imposed body force (of e.g. gravitational origin), $\boldsymbol{I}_0$ the velocity independent part of the system dependent inertial force (sum of the EULER force, translatory and centrifugal force), whilst $2\boldsymbol{\Omega} \times \langle\boldsymbol{v}\rangle$ is its velocity dependent part (known as CORIOLIS force). Furthermore, as already demonstrated in Chap. 11,

$$
\begin{aligned}
&\frac{1}{\rho}\boldsymbol{R} := -\langle\boldsymbol{v}' \otimes \boldsymbol{v}'\rangle\,, \\
&k := \frac{1}{2\rho}\operatorname{tr}\boldsymbol{R} = \tfrac{1}{2}\langle\boldsymbol{v}' \cdot \boldsymbol{v}'\rangle\,, \\
&\boldsymbol{Q} := \langle\rho\rangle\langle\epsilon'\boldsymbol{v}'\rangle\,, \\
&\boldsymbol{K} := \langle\boldsymbol{t}'\boldsymbol{v}'\rangle - \tfrac{1}{2}\langle\rho\rangle\langle\boldsymbol{v}'(\boldsymbol{v}' \cdot \boldsymbol{v}')\rangle\,, \\
&\varepsilon := \frac{1}{\langle\rho\rangle}\langle\operatorname{tr}(\boldsymbol{t}^D \operatorname{grad}\boldsymbol{v}')\rangle\,, \\
&\boldsymbol{\phi}^{\mathrm{ent}}_{(T)} := -\langle\rho\boldsymbol{v}'s'\rangle
\end{aligned}
\tag{12.2.7}
$$

define the REYNOLDS stress tensor, the turbulent kinetic energy, the turbulent heat flux vector, the flux vector of the turbulent kinetic energy, the turbulent dissipation rate and the turbulent entropy flux vector. The external source terms of momentum, energy and entropy are given by $\langle\boldsymbol{f}\rangle, \mathfrak{r}$ and $\langle\eta^{\mathrm{ent}}\rangle$, respectively[9].

As mentioned in the background information, we are deducing here a model for the turbulent description of a NAVIER–STOKES–FOURIER fluid subject to the BOUSSINESQ approximation. For such a fluid, the CAUCHY stress deviator, the internal energy and the heat flux vector satisfy the material relations

$$
\boldsymbol{t}^D = 2\mu\boldsymbol{D}\,, \quad \epsilon = \epsilon(\Theta)\,, \quad \boldsymbol{q} = -\kappa\operatorname{grad}\Theta \tag{12.2.8}
$$

valid for the non-averaged fields. Thermodynamics for a viscous heat conducting fluid with constitutive relations of the form $\mathcal{C} = \mathcal{C}(\boldsymbol{D}, \Theta, \operatorname{grad}\Theta)$ has shown (see Chap. 5) that the entropy flux vector and the entropy supply density are given by

$$
\boldsymbol{\phi}^{\mathrm{ent}} = \frac{\boldsymbol{q}}{\Theta}\,, \quad \eta^{\mathrm{ent}} = \rho\frac{\mathfrak{r}}{\Theta}\,, \tag{12.2.9}
$$

[8] The reader should be aware of the difference in notation: ϵ is used for internal energy, ε for the turbulent dissipation rate.

[9] The radiative source is assumed not to have any fluctuation. This makes it a true source.

where Θ may be identified with the absolute temperature. Moreover, from thermodynamic equilibrium, in which the entropy production must both vanish and reach a minimum, it followed that the molecular shear viscosity and heat conductivity are non-negative functions of the temperature

$$\mu(\Theta) \geq 0\,, \quad \kappa(\Theta) \geq 0\,. \tag{12.2.10}$$

Here in this chapter μ and κ will be treated as constants for which case the filtered variants of (12.2.8) yield

$$\langle \boldsymbol{t}^D \rangle = 2\mu \langle \boldsymbol{D} \rangle\,, \quad \langle \epsilon \rangle = \epsilon(\langle \Theta \rangle)\,, \quad \langle \boldsymbol{q} \rangle = -\kappa \langle \operatorname{grad} \Theta \rangle\,. \tag{12.2.11}$$

The second of these statements requires precision. We may start with

$$\begin{aligned} \epsilon(\Theta) &= \epsilon(\langle \Theta \rangle + \Theta') \\ &\simeq \epsilon(\langle \Theta \rangle) + \frac{d\epsilon}{d\Theta}|_{\Theta=\langle\Theta\rangle} \Theta' + \frac{1}{2}\frac{d^2\epsilon}{d\Theta'}|_{\Theta=\langle\Theta\rangle} \Theta'^2 + \cdots \end{aligned}$$

and average this to obtain

$$\begin{aligned} \langle \epsilon(\Theta) \rangle &= \langle \epsilon(\langle \Theta \rangle) \rangle + \tfrac{1}{2} \left\langle \frac{d^2\epsilon}{d\Theta^2}|_{\Theta=\langle\Theta\rangle} \Theta'^2 \right\rangle + \cdots \\ &= \epsilon(\langle \Theta \rangle) + \tfrac{1}{2} \frac{d^2\epsilon}{d\Theta^2}|_{\Theta=\langle\Theta\rangle} \langle \Theta'^2 \rangle + \cdots \end{aligned}$$

The second term on the right contains a new correlation, $\langle \Theta'^2 \rangle$, but is generally ignored since it is regarded as small and its coefficient vanishes for constant specific heat. So we will set

$$\langle \epsilon(\Theta) \rangle = \epsilon(\langle \Theta \rangle).$$

This is exact for constant specific heat and approximate otherwise with an error proportional to $\langle \Theta'^2 \rangle$. As is easily seen, relations (12.2.11) do not close eqs. (12.2.1)–(12.2.5), because new variables arise, which are defined in (12.2.7) and are all expressed as averages of fluctuation products. These variables must be functionally related to the independent mean fields just as material equations are expressed in terms of independent constitutive variables. These closure relations must be formulated in a thermodynamically consistent way, i.e., they must conform with the second law of thermodynamics. Thereby, the expected equations will take account of the material behaviour of the fluid by means of thermodynamically consistent equations (as stated above) as well as the turbulence behaviour by prescribing the corresponding turbulent closure relations.

The above list of equations does involve the specific turbulent dissipation rate ε as a variable, but not (yet) a corresponding balance law. This additional balance law will now be added as a postulate in the form

$$\langle \rho \rangle \frac{\mathrm{d}\varepsilon}{\mathrm{d}t} = -\operatorname{div} \boldsymbol{k}^\varepsilon + \pi^\varepsilon + \sigma^\varepsilon\,, \tag{12.2.12}$$

in which $\boldsymbol{k}^{\varepsilon}, \pi^{\varepsilon}, \sigma^{\varepsilon}$ are the flux vector of turbulent dissipation and its production and supply rate densities. The spirit in introducing (12.2.12) is to arrive at a theory as closely as possible to that of Chap. 11, but extending it with a more detailed REYNOLDS stress expression. Alternatively, one may think of ε as an internal variable, which satisfies a balance law with flux and production terms. The introduction of the supply term σ^{ε} is here rather formal; an equation describing the evolution of the dissipative structure of the turbulence (an internal property of the flow!) can not be governed by outside source terms; so we will set $\sigma^{\varepsilon} = 0$.

The above equations must be complemented by constitutive/closure conditions. As far as material behaviour is concerned we have seen already in Chap. 5 that for a heat conducting fluid the dependent constitutive variables are a thermodynamic energy, the entropy, the heat flux vector and the CAUCHY stress, and these variables are functionally related to the stretching tensor $\boldsymbol{D}$, the temperature Θ and temperature gradient $\operatorname{grad}\Theta$. For the turbulent closure such relations are needed for all variables stated in (12.2.7) (except the turbulent kinetic energy which we treat as an independent field) plus $\boldsymbol{k}^{\varepsilon}$ and π^{ε} in (12.2.12), but not ε which again is treated as an independent field. To these variables additional thermodynamic variables will be added whose meaning is presently not clear, and all these quantities will depend on $k, \varepsilon, \operatorname{grad} k, \operatorname{grad}\varepsilon, \langle\boldsymbol{D}\rangle, \ldots$ and the material variables mentioned above. Here the dots stand for possibly other dependences such as objective time derivatives of $\langle\boldsymbol{D}\rangle$ and perhaps frame dependent fields such as the vorticity tensor, etc. The closure conditions must be such that the balance laws, material equations and closure conditions together form a well posed set of functional differential equations that allow (at least in principle) the unique determination of the independent fields for suitably prescribed boundary and initial conditions. The criterion is often such that the number of equations must be the same as that of the unknowns or if not, (as is for instance the case for the MAXWELL equations of electrodynamics), that appropriate integrability or compatibility conditions are fulfilled. A mathematical existence proof is generally not conducted. One rather conjectures that with such conditions being fulfilled, the system of functional equations is well posed and admits solutions. *All such solutions are called thermodynamic turbulent processes.*

The way how to proceed becomes much clearer if the entropy principle is introduced first and detailed closure conditions applied afterwards. This will be addressed now.

12.3 Entropy Principle for Turbulent Processes

To express the irreversibility requirement of the physical processes, the second law of thermodynamics is now postulated as the statement that for all processes satisfying the physical balance laws (12.2.1)–(12.2.4) as well as necessary closure conditions the entropy production must be non-negative in every point of the region where the flow takes place at all times. Thus, the entropy inequality (12.2.5) must hold for all thermodynamic processes. This inequality contains the entropy density $\langle s \rangle$ and the entropy fluxes $\langle \boldsymbol{\phi}^{\text{ent}} \rangle$ and $\boldsymbol{\phi}^{\text{ent}}_{(T)}$ which must be prescribed by constitutive relations. The exact entropy postulate is as follows:

Entropy Principle for Turbulent Processes

1. *In every fluid material which possesses the potential of forming laminar and turbulent motions there exists an extensive quantity, called entropy with non-negative production, such that (12.2.5) holds. All quantities in that law, i.e., entropy density, entropy supply and entropy fluxes, respectively, are objective scalar and vector valued quantities.*
2. *To each field variable in the entropy balance there exists a microscopic (molecular) and a mesoscopic (turbulent) contribution which are additive and given by*

$$\langle s \rangle = s^M + s^T \ , \ \boldsymbol{\phi}^{\text{tot}} = \langle \boldsymbol{\phi}^{\text{ent}} \rangle + (\boldsymbol{\phi}^{\text{ent}})^T \ , \ \langle \eta^{\text{ent}} \rangle = \eta^M + 0. \quad (12.3.1)$$

In these expressions the indexed quantities $(\cdot)^M$ and $(\cdot)^T$ describe the material and turbulent parts, respectively. The constitutive quantities of the material part describe exclusively the material behaviour of the body; those of the turbulent part are, however, given in terms of material as well as turbulent field quantities.
3. *The turbulent parts of the quantities in (12.3.1) vanish when the flow is purely laminar.*
4. *Entropy supply possesses no turbulent contribution, and the material contribution is a linear combination,*

$$\eta^M = \boldsymbol{\Lambda}^{\boldsymbol{v}} \cdot \langle \boldsymbol{f} \rangle + \Lambda^{\varepsilon} \langle \rho \rangle \mathfrak{r}$$

of the momentum and energy supply terms. The vector $\boldsymbol{\Lambda}^{\boldsymbol{v}}$ and the scalar Λ^{ε} are such that neither the material constitutive quantities nor the turbulent closure quantities depend on any supply terms.
5. *There exist two (empirical) temperatures Θ^M and Θ^T representing measures for the intensity of the molecular and turbulent fluctuating motions, respectively. Associated with these are also variables, ϑ^M and ϑ^T called material and turbulent coldness variables*[10].
6. *The entropy production is non-negative for all thermodynamic processes, i.e.,*

[10] In rational thermodynamics in which the entropy principle is formulated with the CLAUSIUS–DUHEM inequality, entropy flux equals heat flux divided by absolute temperature. Here no distinction is made between empirical and absolute temperature and coldness is set equal to the inverse temperature $\vartheta = 1/\Theta$ by definition. Generally, the coldness is not necessarily regarded to be equal to the inverse temperature, but as a derived quantity that could possibly satisfy such a relationship, particularly in equilibrium.

$$\langle\bar{\pi}\rangle^{ent} = \pi^M + \pi^T \geq 0 \tag{12.3.2}$$

for all solutions of the field equations.

We repeat that the field equations are the union of (i) all balance laws and (ii) constitutive and closure relations describing the material and turbulence behaviour in such a way that the field variables can be determined by integration, at least in principle.

The above entropy principle is an outgrowth of the corresponding entropy principle applied in Chaps. 5 and 7. Point 1 is based on the earlier assumption that entropy is an additive quantity. Furthermore, objectivity properties are a natural consequence that the non-averaged quantities are objective, that averaging does not destroy the objectivity property of an objective quantity and that the correlation products defined in (12.2.7) only contain objective quantities.

Item 2 is a natural postulate as the material quantities $(\cdot)^M$ must exist by themselves, if laminar processes are assumed to exist, which is done here. Then, if $(\cdot)^M$ is the material part of $\langle(\cdot)\rangle$, we may simply define the turbulent part by the difference $(\cdot)^T = \langle(\cdot)\rangle - (\cdot)^M$. This additive decomposition requires specification. If the variable under consideration is the CAUCHY stress deviator, then $(\boldsymbol{t}^D)^M$ is given by $(12.2.11)_1$ and not by $(12.2.8)_1$, i.e., $(\boldsymbol{t}^D)^M$ denotes the material part after averaging.

Item 3 of the entropy principle is an obvious requirement to recover purely laminar behaviour. Item 4 requires that the constitutive equations must not depend on the supplies arising in the theory. This property is simply taken over from the corresponding property expressed in the material theory of Chaps. 5 and 7.

The postulate 5 in the above entropy principle is fundamental and may be subject to criticism. The existence of a turbulent temperature is postulated here only on the basis of a suggestive analogy between the meaning of temperature in the kinetic theory of gases or in statistical mechanics and the kinetic energy of the fluctuating motion. That the kinetic energy of the fluctuating motion in the turbulence theory may have a similar meaning as the fluctuation energy for the molecular motion is suggestive but not compelling. In particular the vanishing of the turbulent temperature should correspond to purely laminar processes which is not a state of absolute rest as for the molecular motion[11].

Finally item 6 is obvious, but the attribute "for all thermodynamic processes" must be qualified and made more precise, because there are different views how to interpret it. We take here the position of open systems thermodynamics and therefore assume that the sources, $\langle\boldsymbol{f}\rangle$ in the momentum equations and $\mathfrak{r}$ in the energy equation, can take any value we please. Thus, these two laws will not affect the exploitation of the entropy inequality; how-

[11] Very close to absolute zero on the molecular level, quantum mechanical effects play a role; to these we have no analogue in the turbulent case.

ever, the balance laws of turbulent kinetic energy, its dissipation and the conservation law of mass must be accounted for as constraints in satisfying the entropy inequality, because these equations do not have free source terms. This view is very much in the spirit of COLEMAN & NOLL [50][12].

We mentioned above that the entropy flux should be treated as a constitutive quantity to be determined by the entropy principle. For the material part we may thus set

$$(\boldsymbol{\phi}^{\text{ent}})^M = \frac{\boldsymbol{q}}{\Theta} + \boldsymbol{k} = \vartheta \boldsymbol{q}\,, \quad \vartheta := \frac{1}{\Theta}\,, \quad \boldsymbol{k} = \boldsymbol{0}\,, \tag{12.3.3}$$

in which ϑ is called coldness[13] and $\boldsymbol{k}$ the "extra entropy flux vector" which is simply the difference between the entropy flux vector and the heat flux vector divided by the absolute temperature and, for a NAVIER–STOKES fluid (and a large class of other materials), has been shown to vanish. In fact, for a NAVIER–STOKES fluid we have shown that

$$\boldsymbol{k} = \boldsymbol{0}\,, \quad \eta^{\text{ent}} = \frac{\rho\mathfrak{r}}{\Theta}\,, \quad \boldsymbol{\Lambda}^v = \boldsymbol{0}\,, \quad \Lambda^\epsilon = \frac{1}{\Theta} = \vartheta\,. \tag{12.3.4}$$

Now, if one filters (12.3.3) with $\boldsymbol{k} = \boldsymbol{0}$, then one obtains

$$\langle \boldsymbol{\phi}^{\text{ent}} \rangle = \langle \vartheta \rangle \langle \boldsymbol{q} \rangle + \langle \boldsymbol{q}' \vartheta' \rangle\,, \quad \langle \vartheta \rangle := \langle 1/\Theta \rangle\,, \tag{12.3.5}$$

implying that *even when the entropy flux of the unfiltered processes is collinear to the heat flux, this cannot necessarily be true any longer for the filtered entropy flux. In other words, the averaged entropy flux is not equal to "averaged heat flux divided by averaged temperature"*. There is a turbulent correction to this that is not necessarily collinear to the heat flux.

On the basis of this result, we are now introducing the following postulates for the entropy fluxes arising in (12.2.5),

$$\begin{aligned} \langle \boldsymbol{\phi}^{\text{ent}} \rangle &= \langle \boldsymbol{\phi}^{\text{ent}}_{(M)} \rangle + \boldsymbol{k}^M = \vartheta^M (\langle \boldsymbol{q} \rangle + \boldsymbol{Q}^M)\,, \\ (\boldsymbol{\phi}^{\text{ent}})^T &= -\langle \rho s' \boldsymbol{v}' \rangle = \vartheta^T \boldsymbol{K} + \boldsymbol{k}^T = \vartheta^T (\boldsymbol{K} + \boldsymbol{Q}^T)\,. \end{aligned} \tag{12.3.6}$$

Here, $\boldsymbol{k}^M$ and $\boldsymbol{k}^T$ are extra entropy flux vectors for the material and fluctuating behaviour, respectively, and ϑ^M, ϑ^T are coldness variables, which can be identified with

[12] We are taking this view of open systems thermodynamics here mainly in order to simplify the computations which otherwise become rather complicated. Research in thermodynamics of turbulence is at a rather early stage. With knowledge accumulating it may well become necessary to repeat the analysis for closed systems thermodynamics.

[13] When $\dot{\Theta}$ is a further independent constitutive variable, then $\vartheta = \hat{\vartheta}(\Theta.\dot{\Theta})$ is called coldness rather than $1/\Theta$. In this case $(\boldsymbol{\phi}^{\text{ent}})^M = \hat{\vartheta}(\Theta, \dot{\Theta})\boldsymbol{q}$. We shall here not be dealt with this case.

$$\vartheta^M = \frac{1}{\Theta^M} = \langle 1/\Theta \rangle \,, \quad \vartheta^T = \frac{1}{\Theta^T} \,, \tag{12.3.7}$$

i.e., the mean of the inverses of the absolute temperature $\langle 1/\Theta \rangle$ and a variable Θ^T, which we call *turbulent temperature.* Furthermore, it was tempting to set $(\boldsymbol{\phi}^{\mathrm{ent}})^T$ equal to the sum of a term proportional to the flux of the turbulent kinetic energy, $\boldsymbol{K}$ plus a correction that is not necessarily collinear to $\boldsymbol{K}$, the *extra turbulent entropy flux,* $\boldsymbol{k}^T$. $\boldsymbol{Q}^M$ and $\boldsymbol{Q}^T$ are convenient auxiliary variables

$$\boldsymbol{Q}^M = \Theta^M \boldsymbol{k}^M \,, \quad \boldsymbol{Q}^T = \Theta^T \boldsymbol{k}^T \,. \tag{12.3.8}$$

There is no necessity to introduce both $\boldsymbol{k}^M$ and $\boldsymbol{k}^T$, since they arise additively together, but with $(12.3.8)_1$ and the identification $\boldsymbol{Q}^M = \boldsymbol{Q}$, which is defined in (12.2.7), the exploitation of the entropy inequality is facilitated (see Exercise 1). Finally, it is convenient to introduce the thermodynamic potentials

$$\psi^M := \langle \epsilon \rangle - \frac{s^M}{\vartheta^M} \,, \quad \psi^T := k - \frac{s^T}{\vartheta^T} \,. \tag{12.3.9}$$

These are motivated by the classical counterpart – the HELMHOLTZ free energy – and will analogously be called the HELMHOLTZ *free energies of the mean thermal and turbulent processes.*

With the decompositions (12.3.1) and the representations (12.3.6) and (12.3.9) the original entropy inequality (12.2.5) takes the form (Exercise 1)

$$\begin{aligned}
&\vartheta^M \Bigg\{ - \langle \rho \rangle \left[\frac{\mathrm{d}\psi^M}{\mathrm{d}t} - \frac{s^M}{(\vartheta^M)^2} \frac{\mathrm{d}\vartheta^M}{\mathrm{d}t} \right] + \frac{1}{\vartheta^M} (\langle \boldsymbol{q} \rangle + \boldsymbol{Q}) \cdot \operatorname{grad} \vartheta^M \\
&+ \operatorname{tr} \Big[(\langle \boldsymbol{t} \rangle + \Lambda^\rho \boldsymbol{I}) \operatorname{grad} \langle \boldsymbol{v} \rangle \Big] + \langle \rho \rangle \varepsilon \Bigg\} \\
&+ \vartheta^T \Bigg\{ - \langle \rho \rangle \left[\frac{\mathrm{d}\psi^T}{\mathrm{d}t} - \frac{s^T}{(\vartheta^T)^2} \frac{\mathrm{d}\vartheta^T}{\mathrm{d}t} \right] + \frac{1}{\vartheta^T} \boldsymbol{K} \cdot \operatorname{grad} \vartheta^T + \frac{1}{\vartheta^T} \operatorname{div} \boldsymbol{k}^T \\
&+ \operatorname{tr} \Big[\boldsymbol{R} \operatorname{grad} \langle \boldsymbol{v} \rangle \Big] - \langle \rho \rangle \varepsilon \Bigg\} \;\geq\; 0 \,.
\end{aligned} \tag{12.3.10}$$

This inequality represents the averaged form of the entropy inequality as stated in (12.3.2), $\pi^M + \pi^T \geq 0$, but it accounts via the term $\operatorname{tr}(\Lambda^\rho \boldsymbol{I} \operatorname{grad} \langle \boldsymbol{v} \rangle)$ for the density preserving of the processes. Λ^ρ is the corresponding LAGRANGE multiplier accounting for the constraint of the balance of mass. The balances of turbulent kinetic energy and its dissipation are not yet taken into account in (12.3.10). We shall demonstrate below how this will be done.

Before we turn to a systematic exploitation of the entropy principle, note that an *immediate corollary of item 2 of the entropy principle is that*

$$\pi^M \geq 0 \quad \text{and} \quad \pi^M + \pi^T \geq 0 \tag{12.3.11}$$

must hold for all thermodynamic processes, because the entropy principle must also hold for purely laminar flows. On the other hand, it is trivial to see that the two independent inequalities

$$\pi^M \geq 0\,, \quad \pi^T \geq 0 \tag{12.3.12}$$

are sufficient to satisfy the inequalities (12.3.11), in other words

$$\{\pi^M \geq 0\,, \quad \pi^T \geq 0\} \quad \Longrightarrow \quad \{\pi^M \geq 0\,, \quad \pi^M + \pi^T \geq 0\}\,;$$

however, the converse is not true. Now, for *highly developed* turbulent motions it is well known that the laminar contributions to the total fields are in most cases negligibly small. So, one may well stipulate that $\pi^M \ll \pi^T$ in these cases.

12.4 Closure Conditions in Explicit Form

In the foregoing subsections no explicit form of the turbulent closure conditions were given, but the entropy principle requires these to be formulated to become amenable to explicit computation. In order to be able to "rationally" postulate such relations we shall briefly review certain peculiarities that were already encountered in connection with the postulation of constitutive relations describing the material behaviour. Recall that a constitutive relation was only written down for an objective quantity – internal energy, entropy, heat flux vector, CAUCHY stress tensor – and that the functionals representing these field quantities were postulated to obey the rule of material frame indifference.

These two rules express two distinct *notions of objectivity* which are here repeated once more to clarify the issues.

- First, recall that EUCLIDian transformations

$$\boldsymbol{x}^* = \boldsymbol{O}(t)\boldsymbol{x} - \boldsymbol{c}(t)\,, \quad \boldsymbol{O}\boldsymbol{O}^T = \boldsymbol{1} \tag{12.4.1}$$

are time dependent transformations which express the position of a particle in two different observer frames. The unstarred coordinates $\boldsymbol{x}$ are referred to an inertial frame, the starred position $\boldsymbol{x}^*$ refers the same particle to an observer frame which performs, relative to the inertial frame, a possibly time dependent rigid body motion.
A scalar a, vector $\boldsymbol{a}$, second rank tensor $\boldsymbol{A}$ and tensor $\mathbb{A}$ of n-th rank which transform under EUCLIDian transformations as

$$a^* = a\,, \quad \boldsymbol{a}^* = \boldsymbol{O}\boldsymbol{a}\,, \quad \boldsymbol{A}^* = \boldsymbol{O}\boldsymbol{A}\boldsymbol{O}^T\,,$$

$$\mathbb{A}^*_{\underbrace{ij\ldots n}_{n\text{-times}}} = \underbrace{O_{ip}O_{jq}\ldots O_{nr}}_{n\text{-times}}\,\mathbb{A}_{\underbrace{pq\ldots r}_{n\text{-times}}} \tag{12.4.2}$$

are called an *objective* scalar, *objective* vector, *objective* second rank tensor and *objective* tensor of rank n. This is the first notion of objectivity. Recall that we have shown in Chaps. 10 and 11 that the averaged balance laws of mass, momentum, energy and entropy, as well as the balance laws of turbulent kinetic energy and turbulent dissipation rate assume the same form in all EUCLIDian frames, but that in non-inertial frames they may involve frame dependent terms. Thus, the equations are invariant under EUCLIDian transformations but they are not frame indifferent. Objective scalars, vectors, tensors etc. can also be defined for turbulent field quantities; for instance all quantities defined in (12.2.7) are objective scalars (k, ε), vectors $(\boldsymbol{Q}, \boldsymbol{K}, (\boldsymbol{\phi}^{\mathrm{ent}})^T)$ and an objective second rank tensor $(\boldsymbol{R})$.

- The second notion of objectivity was the *rule of material frame indifference* or *rule of material objectivity*. It has exclusively to do with the postulation of constitutive relations. Let $\mathcal{Y} = \mathcal{C}(\mathfrak{X})$ be such a constitutive relation between independent, $\mathfrak{X}$, and dependent, $\mathcal{Y}$, variables, referred to an inertial frame. It is obvious that $\mathcal{Y} = \mathcal{C}(\mathfrak{X})$ does not involve any frame dependent terms. Alternatively, these same constitutive relations, referred to the moving observer frame, $\mathcal{Y}^* = \mathcal{C}^*(\mathfrak{X}^*)$, may be frame dependent, i.e., the functional $\mathcal{C}^*(\cdot)$ may explicitly depend on $\dot{\boldsymbol{O}}\boldsymbol{O}^T = \boldsymbol{\Omega}$ and/or $\dot{\boldsymbol{c}}$; this dependence is expressed by the $*$ in the functional symbol $\mathcal{C}^*(\cdot)$. The rule of material frame indifference states that constitutive relations should be independent of the observer i.e., frame indifferent. This implies that $\mathcal{C}^*(\cdot) = \mathcal{C}(\cdot)$ or $\mathcal{Y}^* = \mathcal{C}(\mathfrak{X}^*)$.

It is worth mentioning that expressions for the CAUCHY stress tensor and heat flux vector derived by methods of statistical mechanics are not frame indifferent in general when e.g. the frame of reference is rotating. Are the turbulent closure relations functional relationships which should obey a similar *rule of turbulence objectivity*; in other words, if $\mathcal{Y} - \mathcal{C}(\mathfrak{X})$ is a closure relation, does this relation have to be frame independent? No, it does not need to, because the averaged equations of turbulence theory have a structure analogous to the governing equations in statistical mechanics. More importantly, however, observations and/or the following gedanken experiment make it clear that a rule of turbulence objectivity cannot possibly hold in general for turbulent closure relations: rotation of a reference frame attenuates or amplifies the turbulent intensity of a certain flow.

As an imaginative example consider a fluid in the annulus between two concentric cylinders. Let the inner and outer cylinder be able to rotate about their common axis with their own angular velocities. Imagine the outer cylinder to rotate and the inner cylinder to remain stationary. In the frame of the rotating cylinder the centrifugal forces will be directed in the radial direction, and they increase with growing distance from the cylinder axis. This corresponds to a stable force distribution as with a stably stratified layer of fluid that is heated from above. Deviations from the pure azimuthal velocity profile will be attenuated, the flow configuration is stable and so turbulent fluctuations, if they exist, will be attenuated. Now, if the inner cylinder rotates (with the same angular velocity) and the outer cylinder is still, then the centrifugal forces in the same frame of the rotating cylinder are again directed in the outward radial direction but they are large close to the inner cylinder and small close to the outer cylinder. This corresponds to an "unstable" force system, i.e., there will be a threshold of rotation beyond which the pure azimuthal velocity field will necessarily be perturbed. This argument should demonstrate that possible velocity fluctuations are more likely to be amplified in this situation than in the previous one. Thus, depending upon the conditions, turbulent intensity can be amplified by the rotation of the frame of reference or attenuated. We thus summarize

Rules of Turbulent Closure Conditions *Constitutive relations of any turbulence theory need not satisfy the rule of turbulence frame indifference or rule of turbulence objectivity. Such closure relations must be expressed as equations between objective scalar, vector and tensor valued quantities; they are therefore expressions for objective scalar, vector and tensor valued quantities with functional relations which need not be frame indifferent.*[14]

The above principle of approach and item 2 of the entropy principle together direct the method, how closure relations for turbulent constitutive quantities ought to be formulated. On the one hand, there are the material relations, $\mathcal{Y}^m = \hat{\mathcal{C}}^m(\mathfrak{X}^M)$, for the averaged dependent material variables, $\mathcal{Y}^m$ expressed in terms of the averaged independent variables, $\mathfrak{X}^M$. Obviously, for a density preserving, viscous heat conducting material these are given as follows:

$$\left.\begin{aligned} \mathcal{Y}^M &= (\psi^M, s^M, \langle \boldsymbol{q} \rangle, \langle \boldsymbol{t} \rangle) \\ \mathfrak{X}^M &= (\vartheta^M, \operatorname{grad} \vartheta^M, \langle \boldsymbol{D} \rangle) \end{aligned}\right\} \Longrightarrow \quad \mathcal{Y}^M = \hat{\mathcal{C}}^M(\mathfrak{X}^M) \qquad (12.4.3)$$

and $\mathcal{Y}^M = \hat{\mathcal{C}}^M(\mathfrak{X}^M)$ must obey the rule of material frame indifference, because these relations derive from material equations.

On the other hand, the closure conditions for turbulent correlation quantities must be of the form $\mathcal{Y}^T = \hat{\mathcal{C}}^T(\mathfrak{X}^M \cup \mathfrak{X}^T)$ for the averaged dependent turbulent variables $\mathcal{Y}^T$ expressed as functionals of the union of the independent material variables $\mathfrak{X}^M$ and the independent turbulent variables $\mathfrak{X}^T$. We choose as an example

$$\begin{aligned} \mathcal{Y}^T &:= (\boldsymbol{Q}^M, \boldsymbol{K}, \boldsymbol{k}^\varepsilon, \pi^\varepsilon, \boldsymbol{R}, \psi^T, s^T, \boldsymbol{k}^T) , \\ \mathfrak{X}^T &:= (\vartheta^T, \operatorname{grad} \vartheta^T, \langle \boldsymbol{D} \rangle, \langle \boldsymbol{\mathcal{W}} \rangle, \langle \boldsymbol{D} \rangle^\circ, \varepsilon, \operatorname{grad} \varepsilon) , \end{aligned} \qquad (12.4.4)$$

[14] Inspite of this rule, authors proposing higher order turbulent closure conditions have occasionally not followed this rule and nevertheless proposed turbulent closure relations which do obey the rule of material frame indifferance. This is, of course, permissible, but it is not compelling.

so that

$$\mathcal{Y}^T = \hat{\mathcal{C}}^T(\mathfrak{X}^M \cup \mathfrak{X}^T)\,. \tag{12.4.5}$$

The dependent closure variables are the extra entropy flux, $\boldsymbol{Q}^M$, the fluxes of the turbulent kinetic energy, $\boldsymbol{K}$ and its dissipation rate, $\boldsymbol{k}^\varepsilon$, the production rate of the specific turbulent dissipation, π^ε, the REYNOLDS stress tensor, $\boldsymbol{R}$, the turbulent free energy, ψ^T, the turbulent entropy, s^T, and the extra turbulent entropy flux, $\boldsymbol{k}^T$ or $\boldsymbol{Q}^T$. The balance laws of mass, momentum, energy, turbulent kinetic energy and its dissipation rate form together with (12.4.3), (12.4.4) and (12.4.5) a set of seven partial differential equations for the unknown fields

$$\begin{aligned}
&\textit{mean pressure} \quad \langle p \rangle\,,\\
&\textit{mean velocity} \quad \langle \boldsymbol{v} \rangle\,,\\
&\textit{material coldness} \quad \vartheta^M\,,\\
&\textit{turbulent coldness} \quad \vartheta^T\,,\\
&\textit{turbulent dissipation} \quad \varepsilon\,.
\end{aligned} \tag{12.4.6}$$

Thus, there only remains an explanation for the choice of independent variables in $(12.4.4)_2$. The dependences of $\mathcal{Y}^T$ on ϑ^T, $\operatorname{grad}\vartheta^T$ and $\langle \boldsymbol{D} \rangle$ are analogous to the selections (12.4.3). Incorporation of $\operatorname{grad}\vartheta^T$ will yield a diffusive contribution in the equation of the turbulent kinetic energy. Analogously, a dependence on ε and $\operatorname{grad}\varepsilon$ guarantees that also the ε-equation receives a diffusive term[15]. The remaining two variables are

$$\begin{aligned}
&\langle \boldsymbol{D} \rangle^\circ = \frac{\mathrm{d}\langle \boldsymbol{D} \rangle}{\mathrm{d}t} + \langle \boldsymbol{D} \rangle \langle \boldsymbol{\mathcal{W}} \rangle - \langle \boldsymbol{\mathcal{W}} \rangle \langle \boldsymbol{D} \rangle\\
&\langle \boldsymbol{\mathcal{W}} \rangle := \langle \boldsymbol{W} \rangle - \boldsymbol{\Omega}\,, \quad \langle \boldsymbol{W} \rangle := \operatorname{skw}\operatorname{grad}\langle \boldsymbol{v} \rangle\,, \quad \boldsymbol{\Omega} = \dot{\boldsymbol{O}}\boldsymbol{O}^T\,,
\end{aligned} \tag{12.4.7}$$

where $\langle \boldsymbol{D} \rangle^\circ$ is the so-called JAUMANN derivative of $\langle \boldsymbol{D} \rangle$, and $\langle \boldsymbol{\mathcal{W}} \rangle$ is the absolute vorticity tensor; both are objective second rank tensor quantities (Exercise 2). It follows that (12.4.5) connects the objective dependent variables $(12.4.4)_1$, $\mathcal{Y}^T$ to the objective independent variables $(12.4.3)_2$ and $(12.4.4)_2$. Writing relation (12.4.5) down in a moving frame, $\mathcal{Y}^{*T} = \hat{\mathcal{C}}^T(\mathfrak{X}^{M^*} \cup \mathfrak{X}^{T^*})$, and expressing the starred variables in terms of the unstarred variables shows that each of the constitutive functions $\hat{\mathcal{C}}^T(\cdot)$ is an isotropic function of its arguments. In other words, for each relation the most general isotropic function representation could be written down to obtain the explicit form of the closure relations of the class (12.4.5). This indicates that the turbulent fields can be regarded as if they were the result of a laminar motion of a very com-

[15] These arguments anticipate that $\boldsymbol{K} \propto \operatorname{grad}\vartheta^T$ and $\boldsymbol{k}^\varepsilon \propto \operatorname{grad}\varepsilon$ or at least that $\boldsymbol{K}$ and $\boldsymbol{k}^\varepsilon$ will contain such contributions.

plex non-NEWTONian fluid[16]. This fact suggests to use knowledge as much as possible from rheology of such fluids.

Naturally, it would not be helpful to write down the complete isotropic function relation of the closure equation (12.4.5). The formulas would be unwieldy and little if anything could be inferred from such complexity. Some terms ought to be dropped, and the underlying principle we take is to assume that velocity gradients and temperature gradients are small. As for the dependence upon $\langle \boldsymbol{D} \rangle$ caution should be observed. In fact, it is known that the quadratic terms of $\langle \boldsymbol{D} \rangle$ occurring in the closure statement of the REYNOLDS stress tensor give rise to normal stress effects, better known in turbulence theory as the anisotropy of the normal stresses[17], see Exercise 3. However, analyses of curved shear flows[18] have indicated that the quadratic terms do not decisively improve the agreement between computed and measured flows. Because the stress tensor in a REINER–RIVLIN fluid is given by a combination of $\boldsymbol{I}$, $\langle \boldsymbol{D} \rangle$ and $\langle \boldsymbol{D} \rangle^2$, this implies that either the coefficients must depend on invariants of $\langle \boldsymbol{D} \rangle$ or/and further dependences on $\langle \boldsymbol{\mathcal{W}} \rangle$ and $\langle \boldsymbol{D} \rangle^\circ$ must be included.

These are the reasons why we have chosen the variables as shown in (12.4.4). In the ensuing analysis certain combinations of these variables are now introduced to properly define isotropic tensorial functions of rank 0,1 and 2. More specifically, we define the variables

$$
\begin{aligned}
&\boldsymbol{Z} := (\Theta^M, \Theta^T, \varepsilon), \;\; Z_1 = \Theta^M, \;\; Z_2 = \Theta^T, \;\; Z_3 = \varepsilon, \\
&\Delta := \operatorname{tr}(\langle \boldsymbol{D} \rangle^2), \\
&\Delta^{\mathcal{W}} := \operatorname{tr}(\langle \boldsymbol{\mathcal{W}} \rangle^2), \\
&\Delta^{Z_\alpha} := \| \operatorname{grad} Z_\alpha \|^2 \qquad (\alpha = 1,2,3), \\
&\Delta^{D_{Z\alpha\beta}} := \operatorname{grad} Z_\alpha \cdot \langle \boldsymbol{D} \rangle \operatorname{grad} Z_\beta \qquad (\alpha,\beta = 1,2,3), \\
&\Delta^{\mathcal{W}_{Z\alpha\beta}} := \operatorname{grad} Z_\alpha \cdot \langle \boldsymbol{\mathcal{W}} \rangle \operatorname{grad} Z_\beta \qquad (\alpha,\beta = 1,2,3), \\
&\boldsymbol{\Delta} := \left(\Delta, \Delta^{\mathcal{W}}, \Delta^{Z_\alpha}, \Delta^{D_{Z\alpha\beta}}, \Delta^{\mathcal{W}_{Z\alpha\beta}}\right),
\end{aligned}
\qquad (12.4.8)
$$

[16] See LUMLEY [140] and SAFFMAN [206]. This statement is not exactly correct, because the variable list $(12.4.4)_2$ contains $\langle \boldsymbol{\mathcal{W}} \rangle$ which is not usually an independent constitutive variable, because non-NEWTONian behaviour is always postulated to be materially objective, i.e., indifferent. Such non-NEWTONian type closure relations have been proposed by many authors, among others by SPEZIALE [220], [222], [226], SPEZIALE et al. [221], [223], SPEZIALE & GATSKI [225], SPEZIALE & XU [224], WANG [253], CRAFT et al. [55], [54] and POPE [183], [184].

[17] See SPEZIALE [219].

[18] See CRAFT et al. [55].

where $\boldsymbol{Z}$ and $\boldsymbol{\Delta}$ are the indicated ordered arrays. These are not a complete set of scalar invariants of the variable list $(12.4.4)_2$; for instance third order terms in $\langle \boldsymbol{D} \rangle$ and $\langle \boldsymbol{\mathcal{W}} \rangle$ are ignored, and scalar invariants of $\langle \boldsymbol{D} \rangle^\circ$ are equally left aside. We will assume that of the latter only $I_{\langle \boldsymbol{D} \rangle^\circ}$ will be important. If dependences on these variables suffice for an adequate turbulence closure of a certain flow class to be described, a fact which can only be a posteriori tested, then a scalar valued closure variable a is a function of the form

$$a = \hat{a}(\boldsymbol{Z}, \boldsymbol{\Delta}, I_{\langle \boldsymbol{D} \rangle^\circ}) , \tag{12.4.9}$$

whilst vector and tensor valued closure variables $\boldsymbol{a}$ or $\boldsymbol{A}$ take the forms

$$\begin{aligned} \mathfrak{a} &= \hat{\mathfrak{a}}(\boldsymbol{Z}, \operatorname{grad} \boldsymbol{Z}, \langle \boldsymbol{D} \rangle, \langle \boldsymbol{\mathcal{W}} \rangle, \langle \boldsymbol{D} \rangle^\circ) , \\ \boldsymbol{A} &= \hat{\boldsymbol{A}}(\boldsymbol{Z}, \operatorname{grad} \boldsymbol{Z}, \langle \boldsymbol{D} \rangle, \langle \boldsymbol{\mathcal{W}} \rangle, \langle \boldsymbol{D} \rangle^\circ) , \end{aligned} \tag{12.4.10}$$

of which isotropic representations will be given below. Before turning to that, we substitute the constitutive relations[19]

$$\begin{aligned} \psi^M &= \psi^M(\vartheta^M, \| \operatorname{grad} \vartheta^M \|^2, \Delta) , \\ \psi^T &= \psi^T(\vartheta^M, \vartheta^T, \varepsilon, \operatorname{grad} \varepsilon, \Delta, \Delta \boldsymbol{Z}_1, \Delta^{\mathcal{W}}, \Delta^{D z_{\alpha\beta}}, \Delta^{\mathcal{W} z_{\alpha\beta}}, I_{\langle \boldsymbol{D}^\circ \rangle}) , \end{aligned} \tag{12.4.11}$$

where the argument list in $(12.4.11)_2$ is that of (12.4.9) and

$$\Delta \boldsymbol{Z}_1 = (\| \operatorname{grad} \vartheta^M \|^2, \| \operatorname{grad} \vartheta^T \|^2),$$

into the inequality (12.3.10) and then obtain, after some calculation and reordering, the following inequality[20]:

[19] The independent variables in $(12.4.11)_2$ are the same so those listed in (12.4.9), but are written differently. The form listed in $(12.4.11)_2$ facilitates the explicit computations.

[20] We use, for instance, the fact that $\operatorname{tr}(\langle \boldsymbol{D}^\circ \rangle \langle \boldsymbol{D} \rangle) = \operatorname{tr}(\langle \boldsymbol{D} \rangle^\circ \langle \boldsymbol{D} \rangle)$ which can easily be proven with the aid of the definitions $(12.4.7)_1$.

$$
\begin{aligned}
\vartheta^M \Bigg\{ &- \langle\rho\rangle \left[\left(\frac{\partial \psi^M}{\partial \vartheta^M} - \frac{s^M}{(\vartheta^M)^2} \right) \dot{\vartheta}^M + \frac{\partial \psi^M}{\partial \Delta} \dot{\Delta} \right. \\
&\left. + \frac{\partial \psi^M}{\partial \parallel \operatorname{grad} \vartheta^M \parallel^2} 2 (\operatorname{grad} \vartheta^M) \cdot (\operatorname{grad} \vartheta^M)^{\cdot} \right] \\
&+ \frac{1}{\vartheta^M} \Big(\langle \boldsymbol{q} \rangle + \boldsymbol{Q} \Big) \cdot \operatorname{grad} \vartheta^M + \operatorname{tr} \Big((\langle \boldsymbol{t} \rangle + \Lambda^\rho \mathbf{1}) \operatorname{grad} \langle \boldsymbol{v} \rangle \Big) + \langle\rho\rangle \varepsilon \Bigg\} \\
+ \vartheta^T \Bigg\{ &- \langle\rho\rangle \left[\frac{\partial \psi^T}{\partial \vartheta^M} \dot{\vartheta}^M + \left(\frac{\partial \psi^T}{\partial \vartheta^T} - \frac{s^T}{(\vartheta^T)^2} \right) \dot{\vartheta}^T + \frac{\partial \psi^T}{\partial \varepsilon} \dot{\varepsilon} \right. \\
&+ \frac{\partial \psi^T}{\partial \operatorname{grad} \varepsilon} \cdot (\operatorname{grad} \varepsilon)^{\cdot} + \sum_{\alpha=1}^{2} \frac{\partial \psi^T}{\partial \Delta Z_1^\alpha} (\Delta Z_1^\alpha)^{\cdot} + \frac{\partial \psi^T}{\partial \Delta^{\mathcal{W}}} (\Delta^{\mathcal{W}})^{\cdot} \\
&+ \sum_{\alpha,\beta=1}^{3} \frac{\partial \psi^T}{\partial \Delta^{\boldsymbol{D} z_{\alpha\beta}}} (\Delta^{D z_{\alpha\beta}})^{\cdot} + \sum_{\alpha,\beta=1}^{3} \frac{\partial \psi^T}{\partial \Delta^{\mathcal{W}}_{Z_{\alpha\beta}}} \left(\Delta^{\mathcal{W}}_{Z_{\alpha\beta}} \right)^{\cdot} \\
&\left. + \frac{\partial \psi^T}{\partial I_{\langle \boldsymbol{D} \rangle^\circ}} \left(I_{\langle \boldsymbol{D} \rangle^\circ} \right)^{\cdot} \right] + \frac{1}{\vartheta^T} \boldsymbol{K} \cdot \operatorname{grad} \vartheta^T + \frac{1}{\vartheta^T} \operatorname{div} \boldsymbol{k}^T \\
&+ \operatorname{tr} \left((\boldsymbol{R} - \langle\rho\rangle \frac{\partial \psi^T}{\partial \Delta} \langle \boldsymbol{D} \rangle^\circ) \cdot \langle \boldsymbol{D} \rangle \right) - \langle\rho\rangle \varepsilon \Bigg\} \geq 0 \,. \qquad (12.4.12)
\end{aligned}
$$

This inequality must hold for any thermodynamic process, i.e., solutions of the field equations. The balance law of mass has been accounted for by adding $\Lambda^\rho \operatorname{div} \langle \boldsymbol{v} \rangle$ to it, where Λ^ρ is a LAGRANGE parameter. Balances of momentum and internal energy need not be considered because they have external source terms and balance of turbulent kinetic energy has already been substituted in (12.4.12). Only the balance of turbulent dissipation has not yet been substituted. This could also be done with the LAGRANGE multiplier method, but we prefer a direct substitution as will shortly be shown. Thus, except for the terms involving $\dot{\varepsilon}$ and $\operatorname{grad} \dot{\varepsilon}$, inequality (12.4.12) must hold for any values of the thermodynamic variables $\mathfrak{X}^M \cup \mathfrak{X}^T$, irrespective of how they vary with space and time. Symbolically it can be written as

$$
\vartheta^M (\boldsymbol{\alpha}^M \cdot \dot{\boldsymbol{y}}^M + \beta^M) + \vartheta^T (\boldsymbol{\alpha}^T \cdot \dot{\boldsymbol{y}}^T + \beta^T) \geq 0 \,, \qquad (12.4.13)
$$

in which $\boldsymbol{y}^M$ and $\boldsymbol{y}^T$ are given by the set $\mathfrak{X}^M$ and $\mathfrak{X}^M \cup \mathfrak{X}^T$, respectively, and $\boldsymbol{\alpha}^M$ and $\boldsymbol{\alpha}^T$ are functions of these variables but not of $\dot{\boldsymbol{y}}^M$ and $\dot{\boldsymbol{y}}^T$. Likewise, β^M and β^T are functions of $\boldsymbol{y}^M$ and $\boldsymbol{y}^T$, respectively, but not $\dot{\boldsymbol{y}}^M$ and $\dot{\boldsymbol{y}}^T$. Therefore, (12.4.13) is linear in $\dot{\boldsymbol{y}}^M$ and $\dot{\boldsymbol{y}}^T$ and these variables may have any arbitrary values. It follows that[21]

[21] Notice that β^T contains as elements the terms involving $\dot{\varepsilon}$ and $\operatorname{grad} \dot{\varepsilon}$.

$$\boldsymbol{\alpha}^M = \mathbf{0}\,, \quad \boldsymbol{\alpha}^T = \mathbf{0}\,, \quad \text{and} \quad \vartheta^M \beta^M + \vartheta^T \beta^T \geq 0\,, \tag{12.4.14}$$

for otherwise the inequality could be violated. These arguments yield,

(i) from the first bracket $\vartheta^M\{\cdot\}$:

$$\frac{\partial \psi^M}{\partial \Delta} = 0\,, \quad \frac{\partial \psi^M}{\partial \parallel \operatorname{grad} \vartheta^M \parallel^2} = 0\,,$$

$$\Longrightarrow \psi^M = \psi^M(\vartheta^M)\,, \qquad s^M = (\vartheta^M)^2 \frac{\partial \psi^M}{\partial \vartheta^M}\,. \tag{12.4.15}$$

(ii) from the second bracket $\vartheta^T\{\cdot\}$:

$$\frac{\partial \psi^T}{\partial \vartheta^M} = \frac{\partial \psi^T}{\partial \Delta Z_1^\alpha}\Big|_{\alpha=1,2} = \frac{\partial \psi^T}{\partial \Delta^{\mathcal{W}}} = \frac{\partial \psi^T}{\partial I_{\langle \boldsymbol{D}\rangle^\circ}} = 0\,,$$

$$\frac{\partial \psi^T}{\partial \Delta^{D z_{\alpha\beta}}}\Big|_{\alpha,\beta=1,2,3} = \frac{\partial \psi^T}{\partial \Delta^{\mathcal{W} z_{\alpha\beta}}} = 0\,,$$

$$\Longrightarrow \psi^T = \psi^T(\vartheta^T, \varepsilon, \parallel \operatorname{grad} \varepsilon \parallel^2, \Delta)\,, \qquad s^T = (\vartheta^T)^2 \frac{\partial \psi^T}{\partial \vartheta^T}\,. \tag{12.4.16}$$

The results (12.4.15) recover the thermodynamic implications of a materially viscous heat conducting fluid, whilst (12.4.16) provide new information for the turbulence, namely that the turbulent free energy depends only on a considerably reduced number of variables. Note also that the "material" and "turbulent" entropies are given by analogous formulas. With these results, the inequality (12.4.12) reduces to

$$\begin{aligned}
&\vartheta^M \left\{ \frac{1}{\vartheta^M} \Big(\langle \boldsymbol{q} \rangle + \boldsymbol{Q} \Big) \cdot \operatorname{grad} \vartheta^M + \operatorname{tr} \Big((\langle \boldsymbol{t} \rangle + \Lambda^\rho \mathbf{1}) \operatorname{grad} \langle \boldsymbol{v} \rangle \Big) + \langle \rho \rangle \varepsilon \right\} \\
&+ \vartheta^T \Bigg\{ \underbrace{- \langle \rho \rangle \frac{\partial \psi^T}{\partial \varepsilon} \dot{\varepsilon} - \langle \rho \rangle \frac{\partial \psi^T}{\partial \operatorname{grad} \varepsilon} \cdot (\operatorname{grad} \varepsilon)^{\cdot} + \frac{1}{\vartheta^T} \operatorname{div} \boldsymbol{k}^T}_{(1)} \\
&+ \underbrace{\frac{1}{\vartheta^T} \boldsymbol{K} \cdot \operatorname{grad} \vartheta^T}_{(2)} + \operatorname{tr} \Big(\Big(\boldsymbol{R} - \langle \rho \rangle \frac{\partial \psi^T}{\partial \Delta} \langle \boldsymbol{D} \rangle^\circ \Big) \langle \boldsymbol{D} \rangle \Big) - \langle \rho \rangle \varepsilon \Bigg\} \geq 0\,.
\end{aligned} \tag{12.4.17}$$

This inequality is now transformed as follows: With the definitions

$$\mathcal{A} := - \frac{\partial \psi^T}{\partial \varepsilon} - \operatorname{div} \boldsymbol{\mathcal{B}}\,, \quad \boldsymbol{\mathcal{B}} := - \frac{\partial \psi^T}{\partial \operatorname{grad} \varepsilon} \tag{12.4.18}$$

the terms summarized by the curly bracket as (1) may be written in the form

$$(1) = \langle\rho\rangle\mathcal{A}\dot{\varepsilon} + \langle\rho\rangle\boldsymbol{\mathcal{B}}\cdot(\operatorname{grad}\varepsilon)^{\boldsymbol{\cdot}} + \langle\rho\rangle(\operatorname{div}\boldsymbol{\mathcal{B}})\dot{\varepsilon} + \operatorname{div}\left(\frac{1}{\vartheta^T}\boldsymbol{k}^T\right) + \frac{1}{(\vartheta^T)^2}\boldsymbol{k}^T\cdot\operatorname{grad}\vartheta^T\,. \qquad (12.4.19)$$

This can still further be transformed by recognizing that

$$\begin{aligned}&\boldsymbol{\mathcal{B}}\cdot(\operatorname{grad}\varepsilon)^{\boldsymbol{\cdot}} = \operatorname{div}(\boldsymbol{\mathcal{B}}\dot{\varepsilon}) - (\operatorname{div}\boldsymbol{\mathcal{B}})\dot{\varepsilon} + \operatorname{grad}\varepsilon\cdot\langle\boldsymbol{L}\rangle\boldsymbol{\mathcal{B}}\,,\\ &\operatorname{grad}\varepsilon\cdot\langle\boldsymbol{L}\rangle\boldsymbol{\mathcal{B}} = \operatorname{grad}\boldsymbol{\varepsilon}\cdot\langle\boldsymbol{D}\rangle\boldsymbol{\mathcal{B}}\,.\end{aligned} \qquad (12.4.20)$$

Here, we have used the fact that

$$\begin{aligned}\operatorname{grad}\varepsilon\cdot\langle\boldsymbol{\mathcal{W}}\rangle\boldsymbol{\mathcal{B}} &= \operatorname{grad}\varepsilon\cdot\langle\boldsymbol{\mathcal{W}}\rangle\frac{\partial\psi^T}{\partial\parallel\operatorname{grad}\varepsilon\parallel^2}2\operatorname{grad}\varepsilon\\ &= 2\frac{\partial\psi^T}{\partial\parallel\operatorname{grad}\varepsilon\parallel^2}\operatorname{tr}\Big[\langle\boldsymbol{\mathcal{W}}\rangle(\operatorname{grad}\varepsilon\otimes\operatorname{grad}\varepsilon)\Big] = 0\,,\end{aligned} \qquad (12.4.21)$$

which vanishes, since $\langle\boldsymbol{\mathcal{W}}\rangle$ is skew symmetric, whilst $\operatorname{grad}\varepsilon\otimes\operatorname{grad}\varepsilon$ is symmetric; in obtaining this result use was also made of the reduced functional relation stated in (12.4.16) . With the above findings we have, instead of (12.4.19),

$$(1) = \langle\rho\rangle\mathcal{A}\dot{\varepsilon} + \operatorname{div}\left(\langle\rho\rangle\boldsymbol{\mathcal{B}}\dot{\varepsilon} + \frac{1}{\vartheta^T}\boldsymbol{k}^T\right) + \frac{1}{(\vartheta^T)^2}\boldsymbol{k}^T\cdot\operatorname{grad}\vartheta^T + \langle\rho\rangle\operatorname{tr}\Big[(\operatorname{sym}(\operatorname{grad}\varepsilon\otimes\boldsymbol{\mathcal{B}}))\langle\boldsymbol{D}\rangle\Big]\,. \qquad (12.4.22)$$

In a second step we replace the material, ϑ^M, and turbulent, ϑ^T, coldnesses, respectively, by the temperature and the turbulent kinetic energy as follows:

$$\vartheta^M = \langle\frac{1}{\Theta}\rangle = \frac{1}{\Theta^M}\,,\quad \vartheta^T = \frac{C^T}{k} = \frac{C_0^T\varepsilon}{k}\,,\quad C_0^T = \text{const}\,. \qquad (12.4.23)$$

The first introduces the mean temperature as the mean of the inverse coldness, the second recognizes that the turbulent kinetic energy plays the same role in turbulence as the temperature does in the kinetic theory of gases. C^T is a scale factor analogous to the specific heat and it was set proportional to the specific turbulent dissipation rate. Such representations were already suggested by others[22]. With $(12.4.23)_2$ we may write for the subbraced term (2) in (12.4.18)

$$(2) = \frac{1}{\vartheta^T}\boldsymbol{K}\cdot\operatorname{grad}\vartheta^T = \frac{1}{k}\boldsymbol{K}\cdot\left(\frac{k}{\varepsilon}\operatorname{grad}\varepsilon - \operatorname{grad}k\right)\,. \qquad (12.4.24)$$

[22] See for instance CHOWDHURY & AHMADI [48], CABON & SCOTT [41], CRAFT, LAUNDER & SUGA [53], [55], [54].

If the results (12.4.22) – (12.4.24) are substituted into inequality (12.4.18), and simultaneously the definition for $\mathcal{A}$ is used, then the residual inequality (12.4.17) takes the form

$$
\begin{aligned}
&\frac{1}{\Theta^M}\left\{-\frac{1}{\Theta^M}(\langle\boldsymbol{q}\rangle+\boldsymbol{Q})\cdot\operatorname{grad}\Theta^M+\operatorname{tr}\left((\langle\boldsymbol{t}\rangle^E+(\Lambda^\rho-p)\mathbf{1})\langle\boldsymbol{D}\rangle\right)+\langle\rho\rangle\varepsilon\right\}\\
&+\frac{C^T}{k}\left\{\left(-\frac{\partial\psi^T}{\partial\varepsilon}-\operatorname{div}\boldsymbol{\mathcal{B}}\right)\underbrace{(-\operatorname{div}\boldsymbol{k}^\varepsilon+\pi^\varepsilon)}_{\langle\rho\rangle\dot{\varepsilon}}+\operatorname{div}\Big(\underbrace{\langle\rho\rangle\boldsymbol{\mathcal{B}}\dot{\varepsilon}+\frac{1}{\vartheta^T}\boldsymbol{k}^T}_{*}\Big)\right.\\
&-\frac{1}{k}\boldsymbol{K}\cdot\left(\operatorname{grad}k-\frac{k}{\varepsilon}\operatorname{grad}\varepsilon\right)+\frac{1}{(\vartheta^T)^2}\boldsymbol{k}^T\cdot\operatorname{grad}\vartheta^T\\
&\left.+\operatorname{tr}\left(\left(\boldsymbol{R}-\langle\rho\rangle\frac{\partial\psi^T}{\partial\Delta}\langle\boldsymbol{D}\rangle^\circ+\langle\rho\rangle\operatorname{sym}(\operatorname{grad}\varepsilon\otimes\boldsymbol{\mathcal{B}})\right)\langle\boldsymbol{D}\rangle\right)-\langle\rho\rangle\varepsilon\right\}\geq 0\,,
\end{aligned}
\tag{12.4.25}
$$

in which the balance law of turbulent dissipation has also been substituted, and $\langle\boldsymbol{t}\rangle$ has been split into the pressure term $-p\mathbf{1}$ and the deviator $\langle\boldsymbol{t}\rangle^E$, viz., $\langle\boldsymbol{t}\rangle=\langle\boldsymbol{t}\rangle^E-p\mathbf{1}$, where p is the constraint pressure.

Inequality (12.4.25) must hold for any value of the constraint pressure. This is only possible if this constraint pressure does not occur in it at all. Thus, of necessity we have

$$
\Lambda^\rho = p\,. \tag{12.4.26}
$$

The LAGRANGE parameter of the continuity equation equals the pressure that maintains the density preserving constraint. This result is no surprise and known from earlier formulations of volume preserving.

It is plausible to assume that the extra turbulent entropy flux $\boldsymbol{k}^T$ differs from zero provided the turbulent dissipation rate ε is not materially constant. This, together with the term * in inequality (12.4.25), suggests to set

$$
\boldsymbol{k}^T = -\langle\rho\rangle\vartheta^T\boldsymbol{\mathcal{B}}\dot{\varepsilon} = C_0^T\frac{\langle\rho\rangle\varepsilon}{k}\dot{\varepsilon}\frac{\partial\psi^T}{\partial\operatorname{grad}\varepsilon}\,, \tag{12.4.27}
$$

so that the extra turbulent entropy flux is constitutively fixed once the HELMHOLTZ free energy ψ^T is prescribed as a function of grad ε. This result requires the following remark: It appears from (12.4.27) as if $\boldsymbol{k}^T$ would depend on $\dot{\varepsilon}$ and therefore be a function of a variable not originally contained in the variable set (12.4.4), (12.4.5). However, this is not so because $\dot{\varepsilon}$ in (12.4.27) can be replaced by its expression from the balance law for ε.

Next, let us determine the functional relationship for the turbulent HELMHOLTZ free energy. To this end, we start from $(12.3.9)_2$ and $(12.4.23)_2$, which, when being combined, yield

$$
k = \psi^T + \frac{s^T}{\vartheta^T} = \frac{C_0^T\varepsilon}{\vartheta^T}\,. \tag{12.4.28}
$$

If the last of equations $(12.4.16)_1$ is used, which expresses the turbulent entropy s^T in terms of ψ^T, then the following differential equation for the HELMHOLTZ free energy is obtained:

$$\frac{\partial}{\partial \vartheta^T}(\vartheta^T \psi^T) = \frac{C_0^T \varepsilon}{\vartheta^T}; \tag{12.4.29}$$

it can easily be integrated to yield

$$\begin{aligned} \psi^T &= \frac{C_0^T \varepsilon}{\vartheta^T}\Big\{ \int_{\vartheta_0^T}^{\vartheta^T} \frac{d\bar{\vartheta}^T}{\bar{\vartheta}^T} + \bar{f}(\varepsilon, \| \operatorname{grad}\varepsilon \|^2, \Delta)\Big\} \\ &= k\left\{ \ln\left(\frac{C_0^T \varepsilon}{k}\right) + f(\varepsilon, \| \operatorname{grad}\varepsilon \|^2, \Delta)\right\}, \end{aligned} \tag{12.4.30}$$

in which $f(\cdot)$ is a differentiable function. To fix it, we now choose the most simple nontrivial representation, namely

$$f = \alpha\varepsilon^2 \Delta + \beta \| \operatorname{grad}\varepsilon \|^2 + A_\circ$$

with constant coefficients α, β and $A_\circ$, so that

$$\psi^T = k\left\{ \ln \frac{C_0^T \varepsilon}{k} + \alpha\varepsilon^2 \Delta + \beta \| \operatorname{grad}\varepsilon \|^2 + A_\circ \right\}. \tag{12.4.31}$$

This choice of ψ^T guarantees that ψ^T varies with Δ only provided ε differs from zero. Moreover, when $k \to 0$, then equally $\psi^T \to 0$ and the quadratic dependence on ε makes ψ^T stationary with regard to ε when ε is zero. This is also reasonable. These results generalize those of AHMADI [7]. Finally, the value of $A_\circ$ is not significant, as only differentiated forms of ψ^T enter previous formulas.

With the representation (12.4.31) explicit expressions can also be given for $\boldsymbol{\mathcal{B}}, \operatorname{div}\boldsymbol{\mathcal{B}}$ and $\boldsymbol{k}^T$, namely,

$$\begin{aligned} \boldsymbol{\mathcal{B}} &= -\frac{\partial \psi^T}{\partial \operatorname{grad}\varepsilon} = -2\beta k \operatorname{grad}\varepsilon, \\ \operatorname{div}\boldsymbol{\mathcal{B}} &= -2\beta\,(k \operatorname{div}\operatorname{grad}\varepsilon + \operatorname{grad}k \cdot \operatorname{grad}\varepsilon), \\ \boldsymbol{k}^T &= 2\, C_0^T\, \beta\, \langle\rho\rangle\, \varepsilon\, \dot{\varepsilon} \operatorname{grad}\varepsilon. \end{aligned} \tag{12.4.32}$$

To further explore the residual inequality (12.4.25), explicit representations for the dependent closure variables must be written down. To this end we take the view, already well accepted in turbulence theory that the turbulent flow of a NAVIER–STOKES fluid may be viewed as a laminar flow of a certain non–NEWTONian fluid. Therefore we let ourselves be motivated by the constitutive theory used in thermodynamics of non–NEWTONian fluid mechanics. In so doing we restrict considerations to postulates of so-called non-NEWTONian fluids of third grade that are complemented by additional dependences on $\boldsymbol{\mathcal{W}}$, $\operatorname{grad}\Theta$ and $\operatorname{grad}\varepsilon$, which appear to play a significant role in turbulence modeling. For the closure relations of the REYNOLDS stress deviator we make the following assumptions:

- Dependences on $\langle\boldsymbol{L}\rangle$ arise in the forms

$$\text{linear}: \langle\boldsymbol{D}\rangle$$

$$\begin{aligned}\text{quadratic}: &\langle\boldsymbol{D}\rangle^2 - \tfrac{1}{3}\operatorname{tr}(\langle\boldsymbol{D}\rangle^2)\boldsymbol{I}\,,\\ &\langle\boldsymbol{\mathcal{W}}\rangle^2 - \tfrac{1}{3}\operatorname{tr}(\langle\boldsymbol{\mathcal{W}}\rangle^2)\boldsymbol{I}\,,\\ &\langle\boldsymbol{D}\rangle\langle\boldsymbol{\mathcal{W}}\rangle - \langle\boldsymbol{\mathcal{W}}\rangle\langle\boldsymbol{D}\rangle\,,\end{aligned}$$

$$\begin{aligned}\text{cubic}: &\operatorname{tr}(\langle\boldsymbol{D}\rangle^2)\langle\boldsymbol{D}\rangle\,,\quad \operatorname{tr}(\langle\boldsymbol{\mathcal{W}}\rangle^2)\langle\boldsymbol{D}\rangle\,,\quad \langle\boldsymbol{D}\rangle^3 - \tfrac{1}{3}III_{\langle D\rangle}\boldsymbol{I}\,,\\ &\langle\boldsymbol{D}\rangle^2\langle\boldsymbol{\mathcal{W}}\rangle - \langle\boldsymbol{\mathcal{W}}\rangle\langle\boldsymbol{D}\rangle^2\,,\\ &\langle\boldsymbol{\mathcal{W}}\rangle^2\langle\boldsymbol{D}\rangle + \langle\boldsymbol{D}\rangle\langle\boldsymbol{\mathcal{W}}\rangle^2 - \tfrac{2}{3}(\operatorname{tr}\langle\boldsymbol{D}\rangle\langle\boldsymbol{\mathcal{W}}\rangle^2)\boldsymbol{I}\,.\end{aligned}$$

- Dependences on $\langle\boldsymbol{D}\rangle^\circ$ arise only linearly
- Dependences on $\operatorname{grad} Z_\alpha$, arises in the combination

$$\begin{aligned}\boldsymbol{Z}^{\alpha\beta} := \tfrac{1}{2}(\operatorname{grad} Z_\alpha \otimes \operatorname{grad} Z_\beta + \operatorname{grad} Z_\beta \otimes \operatorname{grad} Z_\alpha)&\\ - \tfrac{1}{3}(\operatorname{grad} Z_\alpha \cdot \operatorname{grad} Z_\beta)\boldsymbol{I}\,,&\end{aligned}$$

in which $\alpha, \beta = 2, 3$. Notice that, since $\langle\boldsymbol{D}\rangle$ and $\langle\boldsymbol{\mathcal{W}}\rangle$ are traceless, each of the tensors in the above list is symmetric and traceless, so that a linear combination of all of them forms a permissible representation of $\boldsymbol{R}^D$:

$$\begin{aligned}\boldsymbol{R}^D = &\beta_1\Big(\langle\boldsymbol{D}\rangle^3 - \tfrac{1}{3}III_{\langle D\rangle}\boldsymbol{I}\Big) + \left(\beta_2^0 + \beta_2^1\Delta + \beta_2^2\Delta^{\mathcal{W}}\right)\langle\boldsymbol{D}\rangle\\ &-\beta_3\Big(\langle\boldsymbol{D}\rangle^2 - \tfrac{1}{3}\Delta\boldsymbol{I}\Big) - \beta_4\Big(\langle\boldsymbol{\mathcal{W}}\rangle^2 - \tfrac{1}{3}\Delta^{\mathcal{W}}\boldsymbol{I}\Big)\\ &-\beta_4^1\left(\langle\boldsymbol{D}\rangle^\circ - \tfrac{1}{3}\operatorname{tr}(\langle\boldsymbol{D}\rangle^\circ)\boldsymbol{I}\right)\\ &+\sum_{\alpha,\beta}^{3}\beta_5^{\alpha\beta}\boldsymbol{Z}^{\alpha\beta} + \beta_6\Big(\langle\boldsymbol{D}\rangle\langle\boldsymbol{\mathcal{W}}\rangle - \langle\boldsymbol{\mathcal{W}}\rangle\langle\boldsymbol{D}\rangle\Big)\\ &+\beta_7\Big(\langle\boldsymbol{D}\rangle^2\langle\boldsymbol{\mathcal{W}}\rangle - \langle\boldsymbol{\mathcal{W}}\rangle\langle\boldsymbol{D}\rangle^2\Big)\\ &+\beta_8\Big(\langle\boldsymbol{\mathcal{W}}\rangle^2\langle\boldsymbol{D}\rangle + \langle\boldsymbol{D}\rangle\langle\boldsymbol{\mathcal{W}}\rangle^2 - \tfrac{2}{3}\operatorname{tr}(\langle\boldsymbol{D}\rangle\langle\boldsymbol{\mathcal{W}}^2\rangle)\boldsymbol{I}\Big)\,.\end{aligned}\tag{12.4.33}$$

In this expression, the various coefficients β may be thought to be functions of $\boldsymbol{Z}$ and $\boldsymbol{\Delta}$ as indicated in (12.4.8). They involve variables which describe the thermal and turbulent flow states of the fluid under consideration and thus couple in a natural way the mechanical and thermodynamic effects arising in the flow. The ansatz (12.4.33) is more general than any expression we have found in the literature for an algebraic closure relation of the REYNOLDS stress deviator, and none of them has been subjected to thermodynamic scrutiny[23]. So, the proposed relation (12.4.33) is likely the most general form of an algebraic REYNOLDS stress model that has been proposed so far, and it couples in a natural way mechanical and thermodynamic effects occuring in the flow.

Apart from the REYNOLDS stress tensor, isotropic representations are also needed for the vectors $\boldsymbol{Q}$ (turbulent heat flux vector), $\boldsymbol{K}$ (flux of turbulent kinetic energy), $\boldsymbol{k}^{\varepsilon}$ (flux of turbulent dissipation) and the scalar π^{ε} (production of turbulent dissipation). If gradient type relationships are assumed, the structure of these closure relations for $\boldsymbol{Q}, \boldsymbol{K}$ and $\boldsymbol{k}^{\varepsilon}$ is as follows:

$$\left.\begin{aligned}
\boldsymbol{Q} &= \boldsymbol{K}_{\Theta^M\Theta^M} \operatorname{grad}\Theta^M + \boldsymbol{K}_{\Theta^M\vartheta^T} \operatorname{grad}\left(\frac{1}{\vartheta^T}\right) + \boldsymbol{K}_{\Theta^M\varepsilon} \operatorname{grad}\varepsilon\,, \\
\boldsymbol{K} &= \boldsymbol{K}_{\vartheta^T\Theta^M} \operatorname{grad}\Theta^M + \boldsymbol{K}_{\vartheta^T\vartheta^T} \operatorname{grad}\left(\frac{1}{\vartheta^T}\right) + \boldsymbol{K}_{\vartheta^T\varepsilon} \operatorname{grad}\varepsilon\,, \\
\boldsymbol{k}^{\varepsilon} &= \boldsymbol{K}_{\varepsilon\Theta^M} \operatorname{grad}\Theta^M + \boldsymbol{K}_{\varepsilon\vartheta^T} \operatorname{grad}\left(\frac{1}{\vartheta^T}\right) + \boldsymbol{K}_{\varepsilon\varepsilon} \operatorname{grad}\varepsilon
\end{aligned}\right\} \quad (12.4.34)$$

with tensorial coefficients $\boldsymbol{K}_{\Theta^M\Theta^M}, \ldots, \boldsymbol{K}_{\varepsilon\varepsilon}$ that depend on $\langle\boldsymbol{D}\rangle$ and $\langle\boldsymbol{\mathcal{W}}\rangle$ and other variables of the list (12.4.4). Our intention is not to include all dependences but only those which are thought to be significant. It will be conjectured that in (12.4.34) only

$$\boldsymbol{K}_{\Theta^M\Theta^M},\ \boldsymbol{K}_{\vartheta^T\Theta^M},\ \boldsymbol{K}_{\vartheta^T\vartheta^T},\ \boldsymbol{K}_{\varepsilon\Theta^M},\ \boldsymbol{K}_{\varepsilon\varepsilon} \qquad (12.4.35)$$

will be different from zero. Moreover, it will be assumed that the tensors (12.4.35) depend tensorially only on $\langle\boldsymbol{D}\rangle$ and $\langle\boldsymbol{\mathcal{W}}\rangle$ but not on $\langle\boldsymbol{D}\rangle^{\circ}$ or any of the $\boldsymbol{Z}$. All this implies the representations

[23] The representation (12.4.33) is a slight generalisation of that presented by SADIKI and HUTTER [204]. It differs from that presented in [204] by the term involving β_1 which is stated as $\beta_1\langle\boldsymbol{D}\rangle$. Both are possible forms because in a density preserving material the difference can be absorbed in the pressure.

$$
\begin{aligned}
\boldsymbol{Q} = \Bigg\{ & -\kappa \boldsymbol{I} + \Big(\kappa_2^0 + \kappa_2^1 \Delta + \kappa_2^2 \Delta^{\mathcal{W}} \langle \boldsymbol{D} \rangle \\
& + \kappa_3 \langle \boldsymbol{D} \rangle^2 + \Big(\kappa_4^0 + \kappa_4^1 \Delta + \kappa_4^2 \Delta^{\mathcal{W}} \langle \boldsymbol{\mathcal{W}} \rangle \\
& + \kappa_5 \langle \boldsymbol{\mathcal{W}} \rangle^2 + \kappa_6 \langle \boldsymbol{D} \rangle \langle \boldsymbol{\mathcal{W}} \rangle + \kappa_7 \langle \boldsymbol{\mathcal{W}} \rangle \langle \boldsymbol{D} \rangle \Bigg\} \operatorname{grad} \Theta^M , \\
\boldsymbol{K} = \Bigg\{ & \alpha_1^k \boldsymbol{I} + \Big(\alpha_{20}^k + \alpha_{21}^k \Delta + \alpha_{22}^k \Delta^{\mathcal{W}}\Big) \langle \boldsymbol{D} \rangle \\
& + \alpha_3^k \langle \boldsymbol{D} \rangle^2 + \alpha_4^k \langle \boldsymbol{\mathcal{W}} \rangle^2 \Bigg\} \Big(\operatorname{grad} k - \frac{k}{\varepsilon} \operatorname{grad} \varepsilon \Big) \\
& + \Bigg\{ \alpha_5^k \langle \boldsymbol{D} \rangle + \alpha_6^k \langle \boldsymbol{\mathcal{W}} \rangle \Bigg\} \operatorname{grad} \Theta^M , \\
\boldsymbol{k}^\varepsilon = \Bigg\{ & \alpha_1^\varepsilon \boldsymbol{I} + (\alpha_{20}^\varepsilon + \alpha_{21}^\varepsilon \Delta + \alpha_{22}^\varepsilon \Delta^{\mathcal{W}}) \langle \boldsymbol{D} \rangle + \alpha_3^\varepsilon \langle \boldsymbol{D} \rangle^2 + \alpha_4^\varepsilon \langle \boldsymbol{\mathcal{W}} \rangle^2 \Bigg\} \operatorname{grad} \varepsilon \\
& + \Big\{ \alpha_5^\varepsilon \langle \boldsymbol{D} \rangle + \alpha_6^\varepsilon \langle \boldsymbol{\mathcal{W}} \rangle \Big\} \operatorname{grad} \Theta^M , \\
\pi^\varepsilon = {} & \gamma_1 k + \gamma_2 \varepsilon + \gamma_3 \Delta + \gamma_4 \Delta^{\mathcal{W}} + \gamma_5 \parallel \operatorname{grad} k \parallel^2 + \gamma_6 \parallel \operatorname{grad} \varepsilon \parallel^2 ,
\end{aligned}
\tag{12.4.36}
$$

in which a dependence of $\boldsymbol{Q}$ on $\operatorname{grad} \varepsilon$ has been omitted as indicated after (12.4.34). In fully turbulent flows this seems to be reasonable. Furthermore, instead of $\operatorname{grad}(1/\vartheta^T)$ the combination $(\operatorname{grad} k - (k/\varepsilon) \operatorname{grad} \varepsilon)$ was used (see (12.4.24)) and the scalar coefficients can be considered to be functions of $\boldsymbol{Z}$ and $\boldsymbol{\Delta}$. Often they are, however, regarded to be constant. In a complete turbulent model these coefficients must be numerically known. This identification of the parameters is in turbulence theory one of the most important problems to solve; it is often done by (semi)-inverse techniques. Special flow configurations, for which the mean velocity and temperature fields (and any other field variables) have been measured, are computationally reproduced with preselected values of the closure parameters, and their numerical values are changed until optimal agreement is obtained. In general, this is a very elaborous endeavour; indeed, a substantial part of the literature on turbulence modelling is concerned with such identifications of closure parameters. The models for which this parameter identification is done are generally much simpler than the above model since coupling effects are often ignored in them. We suggest in the following to derive the restricting implications from the above model via the exploitation of the entropy principle. This will allow us to derive inferences from the second law for a model that embraces a large number of turbulence models that already exist in the literature.

The representations (12.4.33) and (12.4.36) are objective isotropic tensorial relations of rank 2, 1 and 0, but they have yet been demonstrated to be thermodynamically consistent. If conditions on the closure parameters are derived which will guarantee thermodynamical consistency, then the deduced model can be compared with existing turbulence models and demonstrate their agreement or disagreement with the second law of thermodynamics.

It so happens that in theoretical work on turbulence the expressions for the REYNOLDS stress tensor and the turbulent heat flux vector are generally written in dimensionless form. It is customary to introduce dimensionless versions of $\boldsymbol{R}^D$ and $\boldsymbol{Q}$ as follows:

$$\mathfrak{a} := -\frac{1}{\rho k}\boldsymbol{R}^D = -\frac{1}{\rho k}\boldsymbol{R} + \tfrac{2}{3}\boldsymbol{I}\,, \qquad \mathfrak{Q} := \frac{k^2}{\varepsilon}\boldsymbol{Q}\,. \tag{12.4.37}$$

$\mathfrak{Q}$ is referred to as dimensionless turbulent heat flux vector; alternatively, $\mathfrak{a}$ (a second rank tensor, despite its notation as a lower case letter) is called **anisotropy tensor**. We caution the reader not to interpret this as a fact that $\mathfrak{a}$ would not be represented as an isotropic tensor function. It is simply customary in the turbulence literature to interpret any deviation of the REYNOLDS stress tensor from the "isotropic" relation $\frac{2}{3}k\boldsymbol{I}$ as its anisotropic part. In the jargon of continuum mechanics $\mathfrak{a}$ is a dimensionless deviator proportional to the negative of the REYNOLDS stress deviator[24]. When written in terms of the variables (12.4.37), then $\mathfrak{a}$ and $\mathfrak{Q}$ prove to be extensions of a number of models that have already been proposed in the literature[25]. This is comforting for, if we succeed in constraining the coefficient functions by the exploitation of the entropy principle the emerging model will automatically satisfy the realizability conditions.

[24] To be more precise, the classical BOUSSINESQ type closure $\boldsymbol{R}^D = \nu_t\langle\boldsymbol{D}\rangle$ generates normal stresses $\boldsymbol{R}^D_{11} = \boldsymbol{R}^D_{22} = \boldsymbol{R}^D_{33}$, which are all equal. A closure scheme that is more general than this produces, in general, diagonal components of $\boldsymbol{R}^D$ of which not all are equal to one another. This fact alludes to "anisotropy". In the continuum mechanical literature differences in the normal stresses are referred to as normal stress effects.

[25] For instance, if

$$\beta_4' = \beta_5^{\alpha\beta} = \beta_7 = \beta_8 = \beta_9^{\alpha\beta} = \beta_{10}^{\alpha\beta} = 0\,,$$

then the emerging expression for the REYNOLDS stress deviator corresponds to the algebraic REYNOLDS stress model as presented by YOSHIZAWA [260], [261] and RUBINSTEIN & BARTON [199], but derived by them in an entirely different way. YOSHIZAWA used the two-scale Direct Interaction Approximation (DIA) method and RUBINSTEIN & BARTON employed the ReNormalization Group (RNG) method. SHIH [214] arrived at the representation (12.4.33) and $(12.4.36)_1$ by using the "invariant modelling" method. Applied to the quadratic cases (third order terms are ignored) this model reduces to the quadratic model of SPEZIALE [220] if the "realizability constraints" of SCHUMANN [213], LUMLEY [141], [142] and the "rapid distortion constraints" (REYNOLDS [195]; MANSOUR et al. [144]) are observed, however without the gradient terms in $\langle\boldsymbol{D}\rangle$. A cubic model in curved shear flows was developed by CRAFT et al. [53], [54], [55] and LAUNDER [128].

Comparison of explicit representations for $\mathfrak{a}$ and $\mathcal{Q}$ is facilitated, if these representations are written down in a form that scalar coefficients are dimensionless quantities; such dimensionless forms can be obtained by premultiplying each term arising in the representations of $\mathfrak{a}$ and $\mathcal{Q}$ with the appropriate power product of k and ε. Defining

$$\nu_t = \frac{c_\mu k^2}{\varepsilon}, \quad (c_\mu = 0.09) \tag{12.4.38}$$

which is the kinematic turbulent viscosity, it may be straightforward to corroborate that $\mathfrak{a}$ and $\mathcal{Q}$ may be represented as

$$\begin{aligned}
\mathfrak{a} = {} & \frac{\nu_t}{\varepsilon^2} c_1 \left(\langle \boldsymbol{D} \rangle^3 - I\!I\!I_{\langle \boldsymbol{D} \rangle} \boldsymbol{I} \right) - \frac{\nu_t}{k} \left(c_2^0 + c_2^1 \frac{k^2}{\varepsilon^2} \Delta + c_2^2 \frac{k^2}{\varepsilon^2} \Delta^{\mathcal{W}} \right) \langle \boldsymbol{D} \rangle \\
& + \frac{\nu_t}{\varepsilon} c_3 \left(\langle \boldsymbol{D} \rangle^2 - \tfrac{1}{3} \Delta \boldsymbol{I} \right) + \frac{\nu_t}{\varepsilon} c_4 \left(\langle \boldsymbol{\mathcal{W}} \rangle^2 - \tfrac{1}{3} \Delta^{\mathcal{W}} \boldsymbol{I} \right) \\
& + \frac{\nu_t}{\varepsilon} c_4' \left(\langle \boldsymbol{D} \rangle^{o} - \tfrac{1}{3} \left(\operatorname{tr} \langle \boldsymbol{D} \rangle^{o} \right) \boldsymbol{I} \right) \\
& + \sum_{\alpha,\beta=1}^{3} c_5^{\alpha\beta} \boldsymbol{Z}^{\alpha\beta} + \frac{\nu_t}{\varepsilon} c_6 \left(\langle \boldsymbol{D} \rangle \langle \boldsymbol{\mathcal{W}} \rangle - \langle \boldsymbol{\mathcal{W}} \rangle \langle \boldsymbol{D} \rangle \right) \\
& + \frac{\nu_t k}{\varepsilon^2} c_7 \left(\langle \boldsymbol{D} \rangle^2 \langle \boldsymbol{\mathcal{W}} \rangle - \langle \boldsymbol{\mathcal{W}} \rangle \langle \boldsymbol{D} \rangle^2 \right) \\
& + \frac{\nu_t k}{\varepsilon^2} c_8 \left(\langle \boldsymbol{\mathcal{W}} \rangle^2 \langle \boldsymbol{D} \rangle + \langle \boldsymbol{D} \rangle \langle \boldsymbol{\mathcal{W}} \rangle^2 - \tfrac{2}{3} \operatorname{tr} \left(\langle \boldsymbol{D} \rangle \langle \boldsymbol{\mathcal{W}} \rangle^2 \right) \boldsymbol{I} \right),
\end{aligned} \tag{12.4.39}$$

where

$$\boldsymbol{c}_5 = \begin{pmatrix} \dfrac{k^4}{\varepsilon^2} c_5^{10} & 0 & \dfrac{1}{2} \dfrac{k^3}{\varepsilon} c_5^{13} \\ 0 & 0 & 0 \\ \dfrac{1}{2} \dfrac{k^3}{\varepsilon} c_5^{13} & 0 & \dfrac{k^3}{\varepsilon^4} c_5^{33} \end{pmatrix}, \tag{12.4.40}$$

and

$$\begin{aligned}
\mathcal{Q} = \frac{k^2}{\varepsilon} \Bigg\{ & -b_1 \boldsymbol{I} + \frac{k}{\varepsilon} \left(b_2^0 + \frac{k^2}{\varepsilon^2} \left(b_2^1 \Delta + b_2^2 \Delta^{\mathcal{W}} \right) \right) \langle \boldsymbol{D} \rangle \\
& + \frac{k^2}{\varepsilon^2} b_3 \langle \boldsymbol{D} \rangle^2 + \frac{k}{\varepsilon} \left(b_4^0 + \frac{k^2}{\varepsilon^2} \left(b_4^1 \Delta + b_4^2 \Delta^{\mathcal{W}} \right) \right) \langle \boldsymbol{\mathcal{W}} \rangle \\
& + \frac{k^2}{\varepsilon^2} \left(b_5 \langle \boldsymbol{\mathcal{W}} \rangle^2 + b_6 \langle \boldsymbol{D} \rangle \langle \boldsymbol{\mathcal{W}} \rangle + b_7 \langle \boldsymbol{\mathcal{W}} \rangle \langle \boldsymbol{D} \rangle \right) \Bigg\} \operatorname{grad} \Theta^M
\end{aligned} \tag{12.4.41}$$

in which all coefficients $b_1, \ldots, b_7$ and $c_1, \ldots, c_8$ are dimensionless, with the following correspondences to the β's and κ's:

$$\begin{aligned}
&\beta_1 = -\frac{\rho\nu_t k}{\varepsilon^2}c_1, \qquad (\beta_2^0, \beta_2^1, \beta_2^2) = \frac{\rho\nu_t}{k}(c_2^0, c_2^1, c_2^2),\\
&(\beta_3, \beta_4, \beta_4^1) = \frac{\rho\nu_t}{\varepsilon}(c_3, c_4, c_4^1), \qquad \beta_5^{\alpha\beta} = -\rho k c_5^{\alpha\beta},\\
&\beta_6 = -\frac{\rho\nu_t k}{\varepsilon}c_6, \qquad (\beta_7, \beta_8) = -\frac{\rho\nu_t k^2}{\varepsilon^2}(c_7, c_8);
\end{aligned} \tag{12.4.42}$$

$$\begin{aligned}
&\kappa = b_1, \quad \kappa_2^0 = \frac{k}{\varepsilon}b_2^0, \quad (\kappa_2^1, \kappa_2^2) = \frac{k^3}{\varepsilon^3}(b_2^1, b_2^2),\\
&\kappa_3 = \frac{k^2}{\varepsilon^2}b_3, \quad \kappa_4^0 = \frac{k}{\varepsilon}b_4^0, \quad (\kappa_4^1, \kappa_4^2) = \frac{k^3}{\varepsilon^3}(b_4^1, b_4^2),\\
&(\kappa_5, \kappa_6, \kappa_7) = \frac{k^2}{\varepsilon^2}(b_5, b_6, b_7).
\end{aligned} \tag{12.4.43}$$

Specialisations of these representations have extensively been proposed in the literature. For instance, SPEZIALE et al. [219], [220], [222], SHIH et al. [215], CRAFT et al. [53], [54], [55], YOSHIZAWA [260], [261] and RUBINSTEIN & BARTON [199] have all presented models, which can be regarded as special cases of (12.4.33) or (12.4.39) in which

$$\beta_4^1 = \beta_5^{\alpha\beta} = 0 \;\text{ or }\; c_4^1 = c_5^{\alpha\beta} = 0 \;\; (\alpha, \beta = 1, 2, 3). \tag{12.4.44}$$

The model coefficients of these earlier closure relations are based on eq. (12.4.33) and, with (12.4.44) implemented, will be collected in Table 12.1 for comparison purposes.

12.5 Thermodynamic Compatibility

We shall now indicate how the restrictions that are imposed by the irreversibility requirements constrain the coefficient functions in (12.4.33)–(12.4.44). This is done by substituting the (incomplete) isotropic functions for $\boldsymbol{R}^D$ (or $\mathfrak{a}$), $\boldsymbol{Q}$ (or $\mathfrak{Q}$), $\boldsymbol{K}$ and $\boldsymbol{k}^\varepsilon$ and π^ε into the entropy inequality (12.4.25) (that is reduced by accounting for the relations (12.4.26) and (12.4.27)). The domains of validity of the model parameters that are deduced thereby will then define the *thermodynamic realizability conditions* of the flow, better called the *thermodynamic consistency conditions*.

The residual inequality that emerges by this substitution process must be non-negative for all values of the variables $\langle\boldsymbol{D}\rangle$, $\langle\boldsymbol{\mathcal{W}}\rangle$, $\langle\boldsymbol{D}^0\rangle$, $\boldsymbol{Z}^{\alpha\beta}$ etc. The exploitation of this requirement is computationally rather cumbersome and will not be presented in all details here; we confine ourselves to presenting only the result. It states that the model coefficients arising in (12.4.33)–(12.4.41) must be restricted as follows:

- $\mu \geq 0, \;\; \kappa + \kappa^T \geq 0,$

$$\begin{aligned}
&\text{with } \kappa \geq 0, \; \kappa_2^0 \geq -\beta_5^{13},\\
&\kappa_2^2 = \kappa_3 = \kappa_4^0 = \kappa_4^1 = \kappa_4^2 = 0,
\end{aligned} \tag{12.5.1}$$

- $\alpha_1^k \geq 0,\ \alpha_2^k \neq 0,$

$$\text{with } \alpha_5^k = \frac{C^T}{k\varepsilon}\beta_5^{13}, \qquad (12.5.2)$$
$$\alpha_3^k = \alpha_4^k = \alpha_2^{1k} = \alpha_2^{2k} = \alpha_6^k = 0,$$

- $\varepsilon \geq 0,\ \alpha_1^\varepsilon \neq 0,$

$$\alpha_3^\varepsilon = \alpha_4^\varepsilon = \alpha_{21}^\varepsilon = \alpha_{22}^\varepsilon = 0, \qquad (12.5.3)$$

- $\gamma_4 = 0, \qquad (12.5.4)$

- $$\beta_4^1 = \rho\frac{\partial\psi^T}{\partial\Delta},\ \text{with } \beta_2^0 \geq 0,\ \beta_2^1\Delta \geq -\beta_2^0,$$
$$\beta_1 = \beta_2^2 = \beta_4 = \beta_7 = \beta_8 = \beta_5^{11} = 0, \qquad (12.5.5)$$

- $(\mathcal{Q}_i)^2 \geq 0, \qquad (12.5.6)$

- $$0 \leq R_{ij}R_{ij} \leq R_{ii}R_{jj} \Longrightarrow R_{ii}R_{jj} - R_{ij}R_{ij} \geq 0,\ R_{ii} \geq 0 \qquad (12.5.7)$$
(no summation)

Furthermore, all coefficients depend at most on Θ^M, ε, $||\operatorname{grad}\varepsilon||$ and Δ, as was already shown for the free emerges ψ^T in (12.4.16).

Let us detail the inferences that follow from these results:

a) The relations (12.5.7) are known in the literature as "realizability conditions", see [213], [141], [247]. They state that the Cartesian components of the REYNOLDS stress tensor must satisfy the SCHWARZ inequality and that all "components" forming the turbulent kinetic energy must be non-negative. The constraints (12.5.7) may also be stated as follows: the REYNOLDS stress tensor assures to conform with the second law of thermodynamics or guarantees realizability of the thermodynamic processes, if it constitutes a positive definite matrix with non-negative real eigenvalues. Any turbulence model violating these constraints is then said to be unrealizable. There is a considerable number of models which indeed violate conditions (12.5.7), see RUNG et al. [200]. From the above conditions (12.5.1)–(12.5.7) it is obvious that the realizability constraints (12.5.7) due to SCHUMANN et al represent only a subset of all conditions that must be obeyed to guarantee thermodynamic compatibility.

b) The coefficient β_2^2 vanishes because otherwise the scalar coefficient of $\langle \boldsymbol{D} \rangle$ in the expression for the REYNOLDS stress tensor (12.4.33) would depend on $\Delta^{\mathcal{W}}$ which is excluded by the statement immediately following (12.5.7). All in all, the conditions (12.5.5) state that the prefactor of the linear term in (12.4.33) is positive and that the explicitly nonlinear terms which are cubic in and quadratic or cubic in the products of $\langle \boldsymbol{D} \rangle$ and $\langle \boldsymbol{\mathcal{W}} \rangle$ must vanish. This inference can also be justified by methods of *extended* thermodynamics, but will not further be analysed here, see however SADIKI [201].

c) Besides the relations (12.5.1)–(12.5.7) listed above additional restrictions can be derived, if special representations for the coefficient functions are selected. For instance, if the coefficient functions for the production π^ε in $(12.4.36)_4$ are chosen according to

$$
\begin{aligned}
&(i) \quad \gamma_1 = \rho C_1 \frac{\varepsilon^2}{k^2}, \quad \gamma_2 = \rho C_2 \frac{\varepsilon}{k}, \quad C_1 + C_2 =: c_\varepsilon, \quad c_\varepsilon \geq 0 \\
&(ii) \quad \gamma_5 ||\operatorname{grad} k||^2 + \gamma_6 ||\operatorname{grad} \varepsilon||^2 = c_\varepsilon^1 \frac{\varepsilon}{k} ||\operatorname{grad} k + \frac{k}{\varepsilon} \operatorname{grad} \varepsilon||^2, \qquad (12.5.8) \\
&(iii) \quad \gamma_6 = -\frac{k^2}{\varepsilon^2} \gamma_5
\end{aligned}
$$

with dimensionless C_1, C_2, c_ε and c_ε^1 and these are substituted into $(12.4.36)_4$ and then into (12.4.25), one finds that the coefficients in the ε-equation cannot be calibrated independently of the coefficients of the k-equation. Indeed, one finds the inequalities

$$
-\left(\frac{C_0^T}{\varepsilon} + 2k\varepsilon\alpha\Delta \right) \frac{\varepsilon}{k^2} c_\varepsilon^1 + \frac{\alpha_1^k}{k} \geq 0, \quad c_\varepsilon^1 \geq 0 \qquad (12.5.9)
$$

and can demonstrate that

$$
\beta_5^{33} \text{ is proportional to } 2\left(\beta + \frac{k}{\varepsilon^2} \alpha_{20}^k \right). \qquad (12.5.10)
$$

This last relation states that, when identifying the parameters, the coefficient α_{20}^k in the k-equation must be determined together with β_5^{33} such that the ratio of these parameters remains constant.

One further thermodynamic condition that must be fulfilled is the fact that the turbulent free energy ψ^T, (12.4.31), assumes its minimum in thermodynamic equilibrium; this implies

$$
\alpha \geq 0 \Longrightarrow \beta_4^1 \leq 0 \text{ and } \beta \geq 0. \qquad (12.5.11)
$$

These results closely mimic those obtained for non-NEWTONian fluids, see MÜLLER & WILMANSKI [166].

d) In the above the thermodynamic compatibility statements for the coefficient functions of the REYNOLDS stress tensor were stated in terms of the coefficients β. They can, of course, also be stated in terms of the dimensionless parameters $c_1, \ldots, c_7$. When (12.5.5) is combined with (12.5.10), (12.5.11) using (12.4.42), the following results are obtained:

$$
\begin{aligned}
&\nu_t \geq 0,\ c_2^1 \Delta \geq -\nu_t,\ \ (c_2^0 > 0) \\
&\qquad c_1 = c_2^2 = c_4 = c_7 = c_8 = c_5^{11} = 0, \\
&\qquad c_{11} \text{ is proportional to } \alpha_{20}^k, \\
&\qquad c_4^1 \leq 0, \\
&\qquad c_5^{33} \text{ is proportional to } 2\left(\beta + \frac{k}{\varepsilon^2}\alpha_{20}^k\right), \\
&\qquad b_1 \geq 0,\ b_2^0 = b_2^1 = b_2^2 = b_3 = b_4^0 = b_4^1 = b_4^2 = b_5 = b_6 = b_7 = 0.
\end{aligned}
\tag{12.5.12}
$$

These relations constrain the anisotropy tensor considerably.

e) In the representation (12.4.33) for the REYNOLDS stress tensor one term consisted of a deviatoric contribution of the JAUMANN derivative $\langle \boldsymbol{D} \rangle^o$ of $\langle \boldsymbol{D} \rangle$ (see in (12.4.33) the term with the coefficient β_4^1). In principle, it is irrelevant which time derivative of $\langle \boldsymbol{D} \rangle$ is used to parameterize $\boldsymbol{R}^D$ (or $\boldsymbol{a}$) as long as this time derivative is objective. It turns out that, if $\langle \boldsymbol{D} \rangle^o$ is replaced by the second RIVLIN–ERICKSEN tensor $\langle \boldsymbol{D} \rangle^{(2)} = \langle \boldsymbol{D} \rangle^o + \langle \boldsymbol{L} \rangle \langle \boldsymbol{D} \rangle + \langle \boldsymbol{D} \rangle \langle \boldsymbol{L} \rangle^T$, the following inequalities can be derived (for derivation see JOU et al. [116]):

$$
\begin{aligned}
&\left(\frac{C_0^T}{\varepsilon} + \frac{2\beta}{\varepsilon}||\operatorname{grad}\varepsilon||^2\right)\left(c_\varepsilon \frac{\varepsilon^2}{k} - c_\varepsilon^1 \frac{\varepsilon}{k}\right)\left(\frac{\mu}{k} - c_6 \frac{\mu k}{\varepsilon^2}\Delta\right) \\
&+\frac{2\alpha\varepsilon\alpha_1^\varepsilon}{2\beta}\left(2\beta + 2\alpha\Delta - \frac{C_0^T}{\varepsilon}\right) \\
&+2\alpha\varepsilon c_\varepsilon^1 \left(\operatorname{grad} k - \frac{k}{\varepsilon}\operatorname{grad}\varepsilon\right)\cdot\left(\operatorname{grad} k - \frac{k}{\varepsilon}\operatorname{grad}\varepsilon\right) \\
&-(c_1 + 2c_4^1)(\operatorname{tr}\langle \boldsymbol{D} \rangle^3)/\Delta \geq 0, \qquad\qquad (12.5.13) \\
&\alpha_1^\varepsilon \geq 0.
\end{aligned}
$$

These inequalities may look complicated, however, they show together with (12.5.13) that the model parameters α_1^k and α_1^ε cannot be estimated independently of c_ε, c_ε^1, c_1, c_4^1 and c_6; in parameter identifications such restrictions must be observed. These constraints are new and have first been given by SADIKI & HUTTER [205]. In the existing k–ε models they have so far not been implemented.

f) The closure expression (12.4.33) for the REYNOLDS stress deviator possesses the structure of the stress tensor of a RIVLIN–ERICKSEN *fluid of grade 3*. For a density preserving fluid the stress deviator of such a fluid is given by

$$
\frac{1}{\rho}\boldsymbol{t}^D = \nu \boldsymbol{A}_1 + \alpha_1 \boldsymbol{A}_2 + \alpha_2 \boldsymbol{A}_1^2 + \beta_1 \boldsymbol{A}_3 + \beta_2(\boldsymbol{A}_1\boldsymbol{A}_2 + \boldsymbol{A}_2\boldsymbol{A}_1) + \beta_3 \operatorname{tr}(\boldsymbol{A}_2)\boldsymbol{A}_1
\tag{12.5.14}
$$

with coefficients that are constant. In the above

$$\boldsymbol{A}_{n+1} = \dot{\boldsymbol{A}}_n + \boldsymbol{A}_n \boldsymbol{L} + \boldsymbol{L}^T \boldsymbol{A}_n \, , \quad \boldsymbol{A}_0 = \boldsymbol{I}$$

are the RIVLIN–ERICKSEN tensors (of order $n+1$). ν is a kinematic viscosity, α_1 and α_2 are called normal stress moduli and β_i $(i = 1,2,3)$ are third grade fluid moduli. The analogy of (12.4.33) with (12.5.14) is perhaps somewhat stretching the similarity. However, what can be learnt from it, are the following facts which we state without proof: RAJAGOPAL [190] studies the thermodynamics and stability of incompressible fluids of third grade using the CLAUSIUS–DUHEM inequality and the COLEMAN–NOLL approach as its underlying entropy principle. The same was done by FOSDICK & RAJAGOPAL [78] for a fluid of second grade. On the other hand, MÜLLER & WILMANSKI [166] studied the thermodynamics and stability of fluids of the second grade using the entropy principle as introduced by MÜLLER. All these studies showed that the stability of the second grade fluid depended upon the sign of the normal stress moduli. In important points some inferences using the two approaches are significantly different. One conclusion, derived in [78] and [190], which concerns the last term on the left-hand side of $(12.5.13)_1$ and carries over to this situation implies the inequalities

$$-\frac{|c_3 + 2c_4^1|}{\sqrt{6}} ||\boldsymbol{D}||^3 \leq (c_3 + 2c_4^1)\,\mathrm{tr}(\boldsymbol{D}^3) \leq \frac{|c_3 + 2c_4^1|}{\sqrt{6}} ||\boldsymbol{D}||^3, \quad \text{for all } \boldsymbol{D}.$$

Cumbersome, but straightforward transformations allow to derive an inequality, which is quadratic in $||\boldsymbol{D}||$ from which the statements

$$c_3 + 2c_4^1 \neq 0 \ \text{ and } \ c_4^1 \leq 0 \tag{12.5.15}$$

may be derived.

g) Note that the coefficients α_1^k, α_1^ε and b_1 are in the literature often given by expressions of the form

$$\alpha_1^k = \frac{\rho \nu_t}{\sigma^k}, \quad \alpha_1^\varepsilon = \frac{\rho \nu_t}{\sigma^\varepsilon}, \quad k_t = \frac{k^2}{\varepsilon} b_1 = \frac{\rho \nu_t}{\sigma^T} \tag{12.5.16}$$

in which σ^k, σ^ε and σ^T are called the PRANDTL–SCHMIDT numbers of the turbulent kinetic energy, turbulent dissipation and heat. These are ordinarily determined by computer optimization and in most cases treated as constants. If this is so, the parameterizations (12.5.16) imply that the diffusivities of the turbulent kinetic energy, its dissipation and of heat all vary as ν_t; this is an assumption that is occasionally violated. It would therefore be advantageous to directly identify the parameters α_1^k, α_1^ε and k_t.

In summary, the statements (12.5.9)–(12.5.15) define the thermodynamic consistency conditions. They prescribe the validity domains of the model coefficients. At the first level of closure, any turbulence model of which the coefficients obey these conditions is thermodynamically consistent, and solutions constructed with the field equations with such consistent values of the turbulence parameters are physically realizable processes. For ease of reference Table 12.1 summarizes these restrictions.

Table 12.1. Restrictive conditions imposed on the turbulent model coefficients.

Constitutive quantities	Restrictions imposed on the model coefficients
$\boldsymbol{R}^D$	$\nu_t \geq 0$; $c_2^1 \Delta \geq -\nu_t$; $c_1 = c_2^2 = c_4 = c_7 = c_8 = c_5^{11} = 0$, $c_3 + 2c_4^1 \neq 0$; $c_4^1 \leq 0$; $R_{ii}R_{jj} - R_{ij}R_{ij} \geq 0$, $R_{ii} \geq 0$ (no summation)
$\boldsymbol{\mathcal{Q}}$	$b_1 \geq 0$; $b_2^0 = b_2^1 = b_2^2 = b_3 = b_4^0 = b_4^1 = b_4^2 = b_5 = b_6 = b_7 = 0$; $(\mathcal{Q}_i)^2 \geq 0$;
$\boldsymbol{K}$	$\alpha_1^k \geq 0$; $\alpha_2^k \neq 0$; $\alpha_3^k = \alpha_4^k = \alpha_4^{1k} = \alpha_4^{2k} = \alpha_6^k = 0$;
$\boldsymbol{k}^\varepsilon$	$\alpha_1^\varepsilon \neq 0$; $\alpha_3^\varepsilon = \alpha_4^\varepsilon = \alpha_{21}^\varepsilon = \alpha_{22}^\varepsilon = \alpha_5^\varepsilon = \alpha_6^\varepsilon = 0$;
π^ε	$\gamma_4 = 0$; $\varepsilon \geq 0$; (12.5.8)
Coupling relation	$k_2 = -\beta_5^{13}$; $\alpha_5^k = \frac{c^T}{k\varepsilon}\beta_5^{11}$; (12.5.9), (12.5.10), (12.5.11), (12.5.13)

12.6 Critical Evaluation of Existing Models

12.6.1 The Algebraic Reynolds Stress Model of Ahmadi et al. [5], [7], [48]

Ahmadi & Chowdhury were among the first to construct an algebraic Reynolds stress model such that this model was obeying requirements of a second law of thermodynamics. Their Reynolds stress tensor conforms with (12.4.33) but with $\beta_2^2 = \beta_4 = \beta_5^{\alpha\beta}$ $(\alpha, \beta = 1, 2, 3) = \beta_8 = 0$. The emerging model corresponds formally to a Rivlin–Ericksen model of third grade. As mentioned above, such fluids were studied by Rajagopal [190] and Rajagopal & Fosdick [78]. These authors were relying on an entropy principle in the form of the Clausius–Duhem inequality and so were Ahmadi and Chowdhury. By analogy they simply took over the results of Rajagopal and Fosdick. In our notation the decisive results are the inequalities

$$\beta_4^1 \geq 0, \quad \beta_2^1 \geq 0, \quad |\beta_3 + \beta_4^1| \leq \sqrt{24\mu\beta_2^1}. \tag{12.6.1}$$

These results recover corresponding results derived by Dunn & Fosdick [58] for the thermodynamic stability of *fluids of the second grade* by taking

$$\beta_2^1 = 0 \quad \Longrightarrow \quad \beta_3 + \beta_4^1 = 0, \quad \beta_4^1 \geq 0. \tag{12.6.2}$$

These results imply that β_3 is negative, $\beta_3 \leq 0$. However, in experiments on several non-Newtonian fluids, these theoretical results were not corroborated, see Eringen [67]; such experiments seem to indicate that the sign of

β_3 does not play a great role and that β_3 and β_4^1 ought to satisfy

$$\beta_3 + \beta_4^1 \neq 0, \qquad \beta_4^1 \leq 0, \tag{12.6.3}$$

in contrast to (12.6.2). It transpires that the results (12.6.1) and (12.6.2) contradict observations. AHMADI et al and FOSDICK & RAJAGOPAL based their results on the CLAUSIUS–DUHEM inequality and for their turbulence model assumed that the production of the turbulent kinetic energy must always be positive, a fact which is violated in some cases. SADIKI & HUTTER [201], [205], whose results are reported here, used a more general entropy principle than the CLAUSIUS–DUHEM inequality – MÜLLER's entropy principle – and derived the result (12.5.15), or equivalently

$$c_1 + 2c_4^1 \neq 0, \quad c_4^1 \leq 0 \ \ (\beta_4^1 \leq 0). \tag{12.6.4}$$

This result constitutes two simple conditions required to hold in order that they fulfil the entropy inequality; it has been obtained by coupling the thermodynamic inequality with the basic assumption that the (turbulent) free energy is a minimum in thermodynamic equilibrium. The experimental, (12.6.3), and our thermodynamically implied conditions (12.6.4) can both be satisfied by setting

$$c_1 \geq 0, \qquad \beta_3 \geq 0 \ \text{ or } \ \beta_3 \leq 0$$

without violating the physics of non-NEWTONian fluids of grade 2 for which $c_2^1 = \beta_2^1 = 0$. These are results which are in conformity with analogous results, obtained for turbulence modelling in extended thermodynamics as obtained by SADIKI [205] following MÜLLER & WILMANSKI [166] and JOU et al. [116].

12.6.2 Thermodynamic Consistency and Other Existing First Order Closure Models

Nonlinear closure schemes for the REYNOLDS stress tensor and heat flux vector going beyond the simple gradient representation and therefore modelling anisotropy phenomena have been proposed by many scientists, among others by POPE [183], GATSKI & SPEZIALE [82] and LAUNDER et al. [127], but not by using an entropy principle. As far as the REYNOLDS stress tensor is concerned the nonlinear part is quadratic, [48], [220], or cubic [7], [214], [215], [53], [54], [55] in the stretching and vorticity tensors. The parameterization influences the convergence properties of the numerical methods. Experience has also shown, that the calibration of the model coefficients plays an important role in this process and determines, among other things, the stability performance of the models, see SADIKI et al. [204].

The constraint conditions in Table 12.1 can be used to judge whether a particular model is thermodynamically consistent or not, once the unknown coefficients they contain are determined. Table 12.2 shows values chosen for

Table 12.2. Different coefficients used in the literature and their values.

Author	c_2^1	c_2^2	c_3	c_4	c_6	c_7	c_8	c_μ
CARFT et al. [55]	$-5c_\mu^2$	$5c_\mu^2$	−0.1	0.26	0.6	$-10c_\mu^2$	0	$f(S,W)$
SHIH [214]	–	–	0.7	4.8	0.8	–	–	$f(S,\Omega)$
MYONG & KASAGI [170]	–	–	0.025	0.0045	0.022	–	–	0.09
RUBINSTEIN & BARTON [199]	–	–	0.057	−0.047	0.012	–	–	0.0845
NISIZIMA & YOSHIZAWA [175]	–	–	−0.068	0.094	0.016	–	–	0.09
GATSKI & SPEIZIALE [82]	–	–	−0.17475	–	0.1856	–	–	$f(S,W)$

the coefficients in some particular models in the literature. As mentioned above, there is a remarkable disparity among the values recommended for the model coefficients although each model originator had developed his scheme by reference to simple shear and one other class of flow.

Based on the thermodynamic consistency conditions (12.5.12) the following inferences for the anisotropy tensor $\mathfrak{a}$ are evident:
Remark:

1. The cubic model by CRAFT et al. [53]-[55], with $c_4 \neq 0$ and $c_2^2 \neq 0$ as well as $c_7 \neq 0$ and $c_8 \neq 0$ is thermodynamically inconsistent. CRAFT et al used cubic terms of the mean velocity gradient to properly account for the effects of streamline curvature and frame rotation and fixed their model coefficients by reference to curved shear flows where the linear eddy viscosity schemes do badly and where quadratic versions do no better.
2. The quadratic models proposed by RUBINSTEIN & BARTON [199], NISIZIMA & YOSHIZAWA [175] do not fulfil the consistency condition (12.5.12). The same is true for the quadratic models by SHIH [214], and SHIH et al. [215] and MYONG & KASAGI [170]. REYNOLDS stress-strain relationship at second and third power terms, respectively, and calibrated the coefficients by means of rapid distortion theory and realizability condition arguments. First, two extreme cases were considered: a pure strain flow and a pure shear flow in which a rapid distortion theory analysis and realizability constraints [141], [213] on the REYNOLDS stresses are carried out to ensure positive energy components and SCHWARZ's inequality is applied. This allows to determine the model coefficients c_μ, c_3, c_4. To determine the free model coefficients c_7 and c_8 a study of a fully developed rotating pipe flow is carried out by using the experimental data. The

transport equations for the turbulent kinetic energy and the dissipation rate have been used in the usual known standard form [161].

3. The quadratic model by GATSKI & SPEZIALE (12.5.9) can be thermodynamically consistent if condition (12.5.9) between c_ε^1 and α_1^k is also satisfied. These authors derived from the REYNOLDS stress transport equation containing the equilibrium turbulent models for the unclosed terms (see SPEZIALE [222] and SPEZIALE et al. [223] an explicit algebraic REYNOLDS stress expression which is formally comparable to the quadratic eddy viscosity models of SPEZIALE [219]. No additional calibration of the coefficients is necessary here as the calibration was performed on the level of second order closure.
4. The standard k–ε model is not thermodynamically consistent; it obviously returns a negative kinetic turbulence energy in flows with a large positive strain rate.
5. Even though the "realizability–contraints" have been used by SHIH [214] to determine the model coefficients, and such restrictions have not been taken into account by CRAFT et al. [53]-[55] and GATSKI & SPEZIALE [82], the results of the present investigation clearly show that all these models are thermodynamically inadmissible. Thus, the "classical realizability constraints" do not alone guarantee realizability of a turbulent thermodynamic process, which is inherent to fulfil the second law of thermodynamics.
6. For the turbulent heat-flux vector in a non-inertial frame, a constitutive relation independent of $\langle \boldsymbol{\mathcal{W}} \rangle$ naturally emerges. At the first order closure level a constitutive expression depending on $\langle \boldsymbol{\mathcal{W}} \rangle$ can be expected if $(\operatorname{grad} \Theta^M)^\bullet$ is considered as an independent constitutive variable. All existing models based on $(12.4.36)_1$ or (12.4.41) with this vorticity dependence therefore appear to be thermodynamically inconsistent (e.g. SHIH [214]). This inference incidentally also applies to the turbulent flux vectors $\boldsymbol{K}$ and $\boldsymbol{k}^\varepsilon$. In the framework of second order closure, such a dependence may appear in the context of extended thermodynamics, as pointed out by SADIKI [201]. Furthermore, the flux vector of the turbulent kinetic energy depends on the temperature gradient only if the REYNOLDS stress tensor depends on $\operatorname{grad} \varepsilon \otimes \operatorname{grad} \Theta^M$.

It must be emphasized that FOSDICK & RAJAGOPAL [78] and MÜLLER & WILMANSKI [166] as well as JOU et al. [116] showed in their studies on non-NEWTONian fluids that the thermodynamic consistency is very important for determining the thermodynamic stability properties of the flow. This fact has been numerically investigated by SADIK et al. [204] for some nonlinear existing turbulence models through the behaviour of the model coefficients determined by different calibration strategies. They found that thermodynamically consistent models predict the stability behaviour of the flow well and remain in good agreement with the results of the linear hydrodynamic instability analysis. Furthermore, they investigated the importance of the different terms in nonlinear anisotropic models. It transpired that the quadratic terms are vital and allow to better capture the anisotropy of the normal stress components and to make possible the description of streamline curvature effects. Their inaccurate prediction of body force effects was clearly demonstrated. These last (frame rotation) effects could be, however, captured by the cubic terms involving the coefficient c_1 and the second invariants of the mean stretching tensor.

12.7 Summary of Governing Equations

Since the implications of the entropy inequality are so large it is advantageous to list the governing filed equations of balance of mass, momentum, internal energy, turbulent kinetic energy and turbulent dissipation at one place with the constitutive and closure relations as reduced by the second law of thermodynamics substituted. The relevant equations are (12.2.1)–(12.2.4), (12.2.12), (12.4.33), (12.4.36), (12.4.39)–(12.4.41) with the restrictions of Table 12.1 implemented. This process yields the following field equations, valid for a density preserving fluid in a rotating frame of reference:

$$\operatorname{div}\langle \boldsymbol{v}\rangle = 0\,, \tag{12.7.1}$$

$$\begin{aligned}\frac{\mathrm{d}\langle \boldsymbol{v}\rangle}{\mathrm{d}t} &== -\operatorname{grad}\left(\frac{\langle p\rangle}{\rho} + \frac{2}{3}k\right)\\ &\quad + \operatorname{div}\left(2\mu\langle \boldsymbol{D}\rangle - \boldsymbol{a}\right) + \langle \boldsymbol{f}\rangle + \boldsymbol{I}_0 + 2\boldsymbol{\Omega}\times\langle \boldsymbol{v}\rangle\,,\end{aligned} \tag{12.7.2}$$

$$\begin{aligned}\rho\frac{\mathrm{d}(c\Theta^M)}{\mathrm{d}t} &= \operatorname{div}\left\{\left[\left(\kappa + b_1\frac{k^2}{\varepsilon}\right) + b_2\frac{k^3}{\varepsilon^2}\langle \boldsymbol{D}\rangle\right]\operatorname{grad}\Theta^M\right\}\\ &\quad +2\rho\nu\left(\operatorname{tr}\langle \boldsymbol{D}\rangle^2\right) + \rho\varepsilon + \rho\mathfrak{r}\,,\end{aligned} \tag{12.7.3}$$

$$\begin{aligned}\rho\frac{\mathrm{d}k}{\mathrm{d}t}t &= P^k + c_4'\frac{\rho\nu_t}{\varepsilon}\operatorname{tr}\left(\langle \boldsymbol{D}\rangle^{o}\langle \boldsymbol{D}\rangle\right)\\ &\quad +\frac{c_5^{33}}{k}\operatorname{tr}\left[(\operatorname{grad}\varepsilon\otimes\operatorname{grad}\varepsilon)\langle \boldsymbol{D}\rangle\right]\\ &\quad +\frac{c_5^{13}}{\varepsilon}\operatorname{tr}\left[(\operatorname{grad}\varepsilon\otimes\operatorname{grad}\Theta)\langle \boldsymbol{D}\rangle\right]\\ &\quad +\operatorname{div}\left\{\left(\alpha_1^k\boldsymbol{I} + \alpha_{20}^k\langle \boldsymbol{D}\rangle\right)\left(\operatorname{grad}k - \frac{k}{\varepsilon}\operatorname{grad}\varepsilon\right)\right\}\\ &\quad +\operatorname{div}\left\{\alpha_5^k\frac{k^3}{\varepsilon}\langle \boldsymbol{D}\rangle\operatorname{grad}\Theta^M\right\} - \rho\varepsilon\,,\end{aligned} \tag{12.7.4}$$

$$
\begin{aligned}
\rho \frac{\mathrm{d}\varepsilon}{\mathrm{d}t} &= -c_\varepsilon^1 \frac{\varepsilon}{k} P^{k*} + c_\varepsilon^1 \left\{ c_3 \frac{\rho \nu_t}{k} \operatorname{tr}\left(\langle \boldsymbol{D} \rangle^3 \right) \right\} \\
&+ \frac{c_5^{33}}{k} \operatorname{tr}\left[(\operatorname{grad} \varepsilon \otimes \operatorname{grad} \varepsilon) \langle \boldsymbol{D} \rangle \right] \\
&+ c_4^1 \frac{\rho \nu_t}{\varepsilon} \left[\operatorname{tr}(\langle \boldsymbol{D} \rangle^o \langle \boldsymbol{D} \rangle) + c_5^{13} \operatorname{tr}\left((\operatorname{grad} \varepsilon \otimes \operatorname{grad} \varepsilon) \langle \boldsymbol{D} \rangle \right) \right] \\
&+ c_\varepsilon^3 \frac{\varepsilon}{k^2} \| \operatorname{grad} k - \frac{k}{\varepsilon} \operatorname{grad} \varepsilon \|^2 - \rho c_\varepsilon \frac{\varepsilon^2}{k} + \operatorname{div}\left(\alpha_1^\varepsilon \operatorname{grad} \varepsilon \right) . \quad (12.7.5)
\end{aligned}
$$

In these equations

$$
c_\varepsilon^1 := \frac{\gamma_3}{\frac{k}{\nu_t}\left(1 + c_2^1 k \Delta\right)} , \tag{12.7.6}
$$

$$
P^k := -\frac{\rho \nu_t}{k} \Delta + c_2^2 \frac{\rho \nu_t k}{\varepsilon^2} \Delta^2 - c_3 \frac{\rho \nu_t}{\varepsilon} \operatorname{tr}\left(\langle \boldsymbol{D} \rangle^3 \right) , \tag{12.7.7}
$$

$$
\begin{aligned}
-P^{k*} := &-P^k - c_4^1 \frac{\rho \nu_t}{\varepsilon} \operatorname{tr}\left(\langle \boldsymbol{D} \rangle^o \langle \boldsymbol{D} \rangle \right) \\
&- c_5^{13} \frac{\rho \nu_t k^3}{\varepsilon^2} \operatorname{tr}\left(\left(\operatorname{grad} \varepsilon \otimes \operatorname{grad} \Theta^M \right) \langle \boldsymbol{D} \rangle \right) \\
&- c_5^{33} \frac{1}{k} \operatorname{tr}\left((\operatorname{grad} \varepsilon \otimes \operatorname{grad} \varepsilon) \langle \boldsymbol{D} \rangle \right) . \quad (12.7.8)
\end{aligned}
$$

It is evident that the production P^{k*} can depend on the gradients of Θ^M and ε as well as on dissipation rates of the strain rate and strain acceleration. The model coefficients must obey the constraints listed in Tables 12.1 and 12.2 and they are, in general, calibrated by inverse techniques and comparison with experiments.

Equations (12.7.1)–(12.7.5) form seven (parabolic) partial differential equations for the seven unknowns $\langle \boldsymbol{v} \rangle$, $\langle p \rangle$, k, Θ^M and ε. They constitute a thermodynamically consistent nonlinear $k - \varepsilon$ model. In (12.7.4) the time rate of change (following the mean motion) is balanced by its production P^k (12.7.7) and its diffusive transport due to heat, turbulent kinetic energy and its dissipation (see the last two terms in (12.7.4)). In (12.7.5) the material time rate of change of the turbulent dissipation rate, ε, is balanced by the net effect of the generation of ε by vortex stretching and vortex tilting of turbulent filaments and its destruction by viscous action, represented collectively by the first four terms on the right-hand side of (12.7.5) and the diffusive transport in the last term.

The equations (12.7.1)–(12.7.8) with the anisotropy stress given by (12.4.39) and constrained by the conditions of Table 12.1 constitute a well posed set of evolution equations for the unknowns $\langle \boldsymbol{v} \rangle$, $\langle p \rangle$, k, Θ^M and ε, which is thermodynamically consistent and incorporates by the satisfaction of the entropy and thermodynamic stability conditions the "realizability conditions" of SCHUMANN. In this respect the model is to be preferred over earlier models which are not based on an entropy principle or use the CLAUSIUS-DUHEM inequality as its basis.

12.8 Exercises

1. By using the decompositions (12.3.1) and the representations (12.3.6) and (12.3.9) prove that the entropy inequality (12.2.5) takes the form (12.3.10).
2. Prove that the JAUMANN derivative $\boldsymbol{A}^{\circ}$ of an objective tensor $\boldsymbol{A}$ and the absolute vorticity tensor are objective tensor quantities.
3. Let the REYNOLDS stress tensor be given by

$$\boldsymbol{R} = \tfrac{1}{3}\kappa\boldsymbol{I} + \nu_t\langle\boldsymbol{D}\rangle + a_2\langle\boldsymbol{D}\rangle^2$$

and consider simple shear of a density preserving fluid. Show that $\boldsymbol{R}^D$ has normal stress components not all of which are the same.

12.9 Solutions

1. With (12.3.1), $(12.3.4)_2$ and (12.3.6) the entropy inequality (12.2.5) takes the form

$$\begin{aligned}&\langle\rho\rangle\frac{\mathrm{d}s^M}{\mathrm{d}t} + \langle\rho\rangle\frac{\mathrm{d}s^T}{\mathrm{d}t} + \operatorname{div}(\vartheta^M\langle\boldsymbol{q}\rangle + \boldsymbol{Q}^M) + \operatorname{div}(\vartheta^T\boldsymbol{K})\\ &+\operatorname{div}\boldsymbol{k}^T - \vartheta^M\langle\rho\rangle\mathfrak{r} \quad\geq\quad 0\,,\end{aligned} \tag{12.9.1}$$

where the identification

$$\Lambda^{\varepsilon} = \langle\frac{1}{\theta}\rangle = \vartheta^M \tag{12.9.2}$$

has been substituted. Eliminating the radiation $\mathfrak{r}$ and $\operatorname{div}\boldsymbol{K}$ with the aid of the balance relations (12.3.3) and (12.3.4) yields

$$\begin{aligned}&\vartheta^M\Bigg\{-\langle\rho\rangle\left(\frac{\mathrm{d}\langle\epsilon\rangle}{\mathrm{d}t} - \frac{1}{\vartheta^M}\frac{\mathrm{d}s^M}{\mathrm{d}t}\right) + \operatorname{div}\left(\vartheta^M(\boldsymbol{Q}^M - \boldsymbol{Q})\right)\\ &+\frac{1}{\vartheta^M}\left(\langle\boldsymbol{q}\rangle + \boldsymbol{Q}^M\right)\cdot\operatorname{grad}\vartheta^M + \operatorname{tr}\left[\langle\boldsymbol{t}\rangle\operatorname{grad}\langle\boldsymbol{v}\rangle\right] + \langle\rho\rangle\varepsilon\Bigg\}\\ &\vartheta^T\Bigg\{-\langle\rho\rangle\left(\frac{\mathrm{d}\boldsymbol{k}}{\mathrm{d}t} - \frac{1}{\vartheta^T}\frac{\mathrm{d}s^T}{\mathrm{d}t}\right) + \frac{1}{\theta^T}\boldsymbol{K}\cdot\operatorname{grad}\vartheta^T + \frac{1}{\vartheta^T}\operatorname{div}\boldsymbol{k}^T\\ &+\operatorname{tr}\left[\boldsymbol{R}\operatorname{grad}\langle\boldsymbol{v}\rangle\right] - \langle\rho\rangle\varepsilon\Bigg\} \;\geq\; 0\,.\end{aligned} \tag{12.9.3}$$

Because the choice of $\boldsymbol{Q}^M$ is arbitrary, we now select

$$\boldsymbol{Q}^M = \boldsymbol{Q} \tag{12.9.4}$$

to simplify (12.9.3). This choice is also physically reasonable since for laminar flow $\boldsymbol{Q} = \boldsymbol{0}$, as it must be. If next, the definitions (12.3.9) are substituted and if

$$\vartheta^M \Lambda^\rho \operatorname{div}\langle \boldsymbol{v}\rangle = \vartheta^M \Lambda^\rho \operatorname{tr}[\boldsymbol{I} \operatorname{grad}\langle \boldsymbol{v}\rangle]$$

is added to (12.9.3) to account for the balance of mass, then inequality (12.3.10) is obtained. Λ^ρ is the LAGRANGE multiplier of the continuity equation (12.2.1).

2. (i) The JAUMANN derivative of $\boldsymbol{A}^*$ is given by

$$\begin{aligned}
\boldsymbol{A}^{*\circ} &= \frac{\mathrm{d}\boldsymbol{A}^*}{\mathrm{d}t} + \boldsymbol{A}^*\boldsymbol{W}^* - \boldsymbol{W}^*\boldsymbol{A}^* \\
&= \frac{\mathrm{d}(\boldsymbol{O}\boldsymbol{A}\boldsymbol{O}^T)}{\mathrm{d}t} + (\boldsymbol{O}\boldsymbol{A}\boldsymbol{O}^T)(\boldsymbol{O}\boldsymbol{W}\boldsymbol{O}^T + \dot{\boldsymbol{O}}\boldsymbol{O}^T) \\
&\quad -(\boldsymbol{O}\boldsymbol{W}\boldsymbol{O}^T + \dot{\boldsymbol{O}}\boldsymbol{O}^T)(\boldsymbol{O}\boldsymbol{A}\boldsymbol{O}^T) \\
&= \boldsymbol{O}\dot{\boldsymbol{A}}\boldsymbol{O}^T + \dot{\boldsymbol{O}}\boldsymbol{A}\boldsymbol{O}^T + \boldsymbol{O}\boldsymbol{A}\dot{\boldsymbol{O}}^T + \boldsymbol{O}\boldsymbol{A}\boldsymbol{W}\boldsymbol{O}^T + \boldsymbol{O}\boldsymbol{A}\boldsymbol{O}^T\dot{\boldsymbol{O}}\boldsymbol{O}^T \\
&\quad -\boldsymbol{O}\boldsymbol{W}\boldsymbol{A}\boldsymbol{O}^T - \dot{\boldsymbol{O}}\boldsymbol{A}\boldsymbol{O}^T \\
&= \boldsymbol{O}\left\{\dot{\boldsymbol{A}} + \boldsymbol{A}\boldsymbol{W} - \boldsymbol{W}\boldsymbol{A}\right\}\boldsymbol{O}^T, \quad \text{qed.}
\end{aligned}$$

(ii) In the non-starred fixed system the absolute voticity tensor is given by $\langle \boldsymbol{\mathcal{W}}\rangle = \langle \boldsymbol{W}\rangle$ according to $(12.4.7)_2$ with $\boldsymbol{O} = \boldsymbol{1}$. In the starred moving system the absolute voticity tensor $\langle \boldsymbol{\mathcal{W}}\rangle^*$ can be written as

$$\begin{aligned}
\langle \boldsymbol{\mathcal{W}}\rangle^* &= \langle \boldsymbol{W}\rangle^* - \boldsymbol{\Omega} = \boldsymbol{O}\langle \boldsymbol{W}\rangle\boldsymbol{O}^T + \dot{\boldsymbol{O}}\boldsymbol{O}^T - \dot{\boldsymbol{O}}\boldsymbol{O}^T = \boldsymbol{O}\langle \boldsymbol{W}\rangle\boldsymbol{O}^T \\
&= \boldsymbol{O}\langle \boldsymbol{\mathcal{W}}\rangle\boldsymbol{O}^T, \quad \text{qed.}
\end{aligned}$$

3.

$$\langle \boldsymbol{D}\rangle = \begin{pmatrix} 0 & \dot{\gamma}/2 & 0 \\ \dot{\gamma}/2 & 0 & 0 \\ 0 & 0 & 0 \end{pmatrix}, \tag{12.9.5}$$

which corresponds to simple shear. Thus

$$\langle \boldsymbol{D}\rangle^2 = \begin{pmatrix} \dot{\gamma}^2/4 & 0 & 0 \\ 0 & \dot{\gamma}^2/4 & 0 \\ 0 & 0 & 0 \end{pmatrix}. \tag{12.9.6}$$

It follows that a parameterization

$$\boldsymbol{R} = \tfrac{2}{3}\kappa \boldsymbol{I} + \nu_t\langle \boldsymbol{D}\rangle + a_2\langle \boldsymbol{D}\rangle^2 \tag{12.9.7}$$

for the REYNOLDS stress tensor yields normal stress effects, if $a_2 \neq 0$.

13. Application of the k–ε Model to the Description of the Diurnal and Seasonal Temperature Variation in Lakes

13.1 Introduction

13.1.1 Motivation

In the last chapters a great mathematical stride was taken to deduce several different models for the description of turbulent processes in fluid flow situations. These models find their application in many hydrodynamical problems of mechanical and civil engineering, but equally also meteorology, oceanography and physical limnology (lake hydrodynamics). In the latter applications the intention may be to compute the evolution of the vertical temperature profiles in a large body of water. One can easily imagine that the temperature distribution in a lake is, apart from light, one of the most important quantities that affects the seasonal development of the biological processes. Phyto– and zooplankta grow according to the temperature and light environment they encounter. Let us touch upon some of the phenomena related to this fascinating biological–physical coupling problem.

Firstly, the intrusion of turbulent intensity in a lake or in the ocean by the wind is chiefly dependent upon the density stratification. If the latter consists of a light upper layer and a heavy lower layer, then the intrusion of turbulence into the heavy lower layer is hampered, a fact that also slows down the transport of oxygen to depth. Conversely, turbulent activity in a homogeneous lake can much easier be transported into large depths. Second, an increase of the temperature in natural waters reduces the solubility of the oxygen necessary for life, whilst simultaneously the oxygen dependent metabolism of living organs accelerates. By these mutually amplifying effects a depletion of oxygen may arise, a fact which negatively affects the growth and reproduction of the living organisms. If the concentration of oxygen is drastically reduced, its content in the water may fall to zero and thus annihilate all aerobic life. One says then that the conditions are anaerobic in the lake and that the lake looses its balance.

In the course of a year a lake (situated at medium geographical latitude) changes its temperature distribution drastically. This happens on the basis of seasonal, but also daily (diurnal) variations of the external activities of wind, solar irradiation, etc. Furthermore, it is very important for the mostly–passively moving algal species how they are transported in the lake, whether they are advected by the currents or essentially mixed by the turbulent processes. All these processes in their entirety lead to the fact that the plankton concentration nearly completely disappears in winter, whilst in spring the living climate for the plankton species is substantially improved, because of the formation of a light upper layer and the associated warming; the algal spring blossoming is initiated. In addition, algae now find new nutrients in the waters, which during the complete winter mixing were distributed from the lower layers over the entire lake. A further point which favours the rapid reproduction of algae is the increase of the radiation intensity in spring, which the algal populations need for their photo synthesis. Because of the intensed algal blossoming, the zoo plankton, in turn, also increases its reproduction, because it now finds sufficient nutrients for itself. Due to the accompanied *predation–pressure* the algal population very often suddenly collapses. What results is the so-called *clear water phase*. In the consecutive processes during the year, depending upon weather conditions and availability of nutrients (eutrophe–nutritient rich, mesotrophe and oligotrophe–nutritient poor lake), new algal blossomings may arise. In autumn the solar irradiation into the lake is reduced, and the heat losses due to reduced radiation and increased evaporation exceed the heat input; in addition, storms enhance the turbulent mixing, the upper layer deepens and simultaneously cools down until, in winter, a uniform temperature distribution of 4°C is reached.

This seasonal behaviour does not hold for all lakes, but is adequate for most deep lakes in moderate climate zones, e.g. Alpine lakes such as e.g. Lake Constance and Lake Ammer (both in Southern Germany), to name two among many. In Fig. 13.1 temperature profiles for Lake Ammer are shown through the year which allow us to identify and corroborate the above explained behaviour. The figure displays for certain dates during the year 1996 and for the indicated position (of largest depth) the measured temperature distribution with depth from the free surface to 30 m depth. One recognizes in these curves the transition from the (more or less well) mixed upper layer (*epilimnion*) via an intermediate layer with strong vertical temperature gradient (*metalimnion*) to a lower layer above the lake bottom with nearly constant temperature (*hypolimnion*). Usually, the wind causes with its strong induced turbulent intensity a fairly homogeneous mixed upper layer which is bounded at its lower end by a strong vertical temperature gradient. The thickness of this upper layer is not uniquely defined; its depth may be identified with the depth of the (absolutely) largest temperature gradient – this position is called the *thermocline*. However, it is also possible to define the lower end of the upper layer by the turbulent kinetic energy, which is largely governed by the free surface fluxes (wind, heat flux) and by radiation; this boundary is called the *turbocline*. In Fig. 13.2 the seasonal variation of the upper-layer depth and the thermocline, respectively, are drawn together with the buoyancy fre-

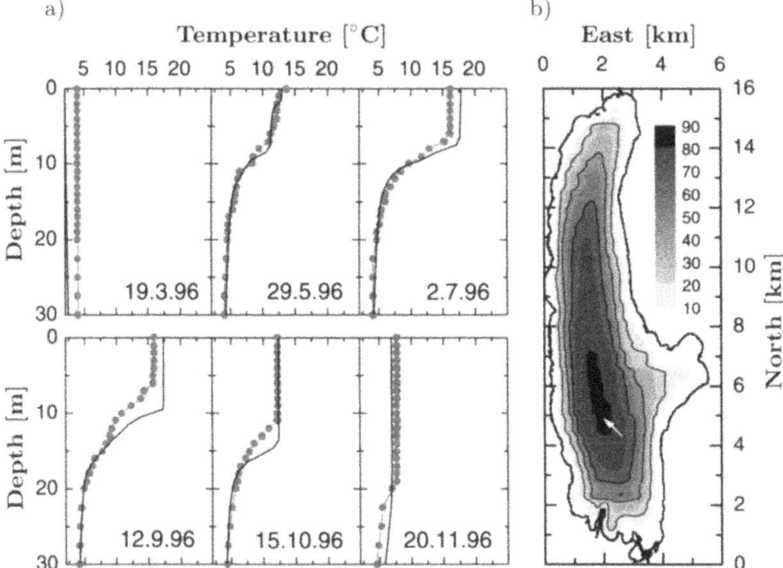

Fig. 13.1.
a) Temperature profile for lake Ammer in its seasonal variation (1996). The panels show the measurements at the indicated dates (symbols) and the result from simulations (solid lines).
b) Bathymetric chart of Lake Ammer with 20 m equidistance of the isobaths. The arrow marks the deepest position of the lake (82 m), where the measurements were taken.

quency[1] ($N^2 = (g/\rho)(d\rho/dz)$); the results follow from simulations based on the measurements shown in Fig. 13.1.

The stratification is not only relevant for the biological processes; the inflow of pollutants and the spreading of the water from tributaries is strongly coupled to the temperature distribution in the lake. Depending upon the density (which is given by the distributions of temperature, mineral composition and possible tracer concentrations, etc.) of the incoming water, this water will deposit itself at a certain depth of the lake. And finally, also the diffusive spreading of tracers, such as pollutants and nutrients depends strongly on the stratification.

It will now be our goal, to derive or sketch a mathematical model for the determination of the temperature distribution, through time when external wind shear and solar irradiation are prescribed as functions of time (and possibly, position). In so doing we will restrict ourselves to the derivation of a vertical temperature profile at a fixed position of the lake (one-dimensional

[1] The buoyancy frequency is an indicator of the local "stability" of a water mass. If $N^2 \geq 0$ ($N^2 < 0$), then the stratification is stable (unstable), provided z points in the direction opposite to gravity, see Exercise 1.

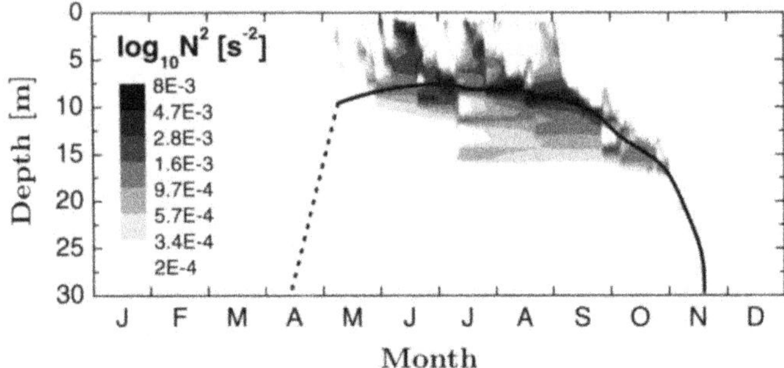

Fig. 13.2. Buoyancy frequency (stability) N computed with a simulation for Lake Ammer with meteorological data from the year 1996. The solid line gives the approximate position of the upper layer depth at the thermocline depth. Note that in spring, several local maxima of the gradient of the temperature occur. The increase of the upper layer depth (shown as dashed line), is thought to be an indicative trend here.

model). This will certainly not be representative for the entire lake, and in particular not for the shallow coastal region (called litoral), but it can be regarded as a representative mean for the free–water zone (pelagial). Therefore, the model cannot cope with large–scale–flow processes such as global circulation, seiches, etc., processes which can also lead to horizontal variations of the temperature.

13.1.2 Water Circulation in a Lake

In the last subsection, a typical cycle of temperature distribution through the year was described. Here, the same cycle will be explained once more, this time in more detail and with emphasis to physical processes; moreover, we shall give a typification of lakes according to their circulation pattern.

The phenomena to be discussed are based upon the unique property of pure water that has a coefficient of thermal expansion which does not monotonically depend upon the temperature. More specifically, water has its largest mass density a 4°C [2]. In winter the entire water column often has a temperature of approximately 4°C; very close to the surface the temperature can, however, be smaller. During the *winter–stagnation* the heavy water therefore lies at the bottom or in the lower most layer. In spring, the water close to the surface is slowly warmed up; the water becomes thereby heavier

[2] In general, the thermal equation of state is an equation between density, pressure, temperature and salinity, which can e.g. be written in the form $\rho = \hat{\rho}(T, p, S)$. For fresh water the dependence upon the salinity is omitted; however for many lakes such an assumption does not seem to be reasonable. The dependence upon the pressure is only significant for very deep waters (lake Baikal, deep ocean). For sea water with salinity $S = 34\ ^0\!/_{00}$ the mentioned non-monotonicity is also lost. Fresh water lakes, therefore, are special in this regard.

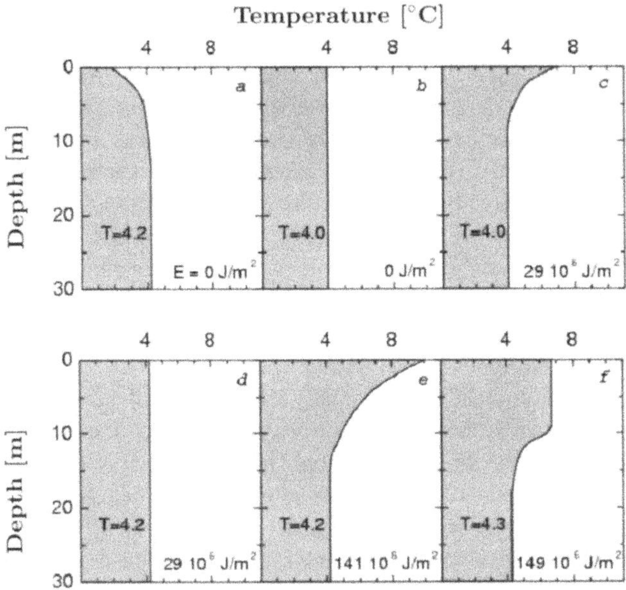

Fig. 13.3. Formation of the thermal stratification in spring. *(a)* Winter stagnation, *(b)* Heating and mixing during the spring circulation *(c)* Bright weather phase. *(d)* Mixing after a storm. *(e)* Renewed bright weather phase, *(f)*. As long as there is no net energy exchange with the atmosphere the heat content (relative to a homogeneous layer of 4°C) is constant. The mixing by wind, $a \to b$ and $c \to d$, respectively, does not enlarge the heat content. The build-up of the stratification from accomplished by both mixing by wind as well as heating and cooling.

and has the tendency to fall to larger depth and to mix with the underlying water. In this unstable situation every energy input by wind leads to a mixing of the entire lake (*spring circulation*), a process which also transports nutrients from lower layers upwards. The entire water body finally reaches again a uniform temperature of 4°C. During a few days with calm and warm weather, a stable temperature profile will now be built (see Fig. 13.3). In this situation a small storm suffices to mix again the entire water body (through its depth). The subsequent further warming in spring after such a storm yields again a stable stratification but with temperatures at the top above 4°C. Since the density does not change linearly with the temperature, more energy is now needed than would otherwise be to relocate the heavy water masses. By a storm (which intrudes turbulence energy) it is now likely that an insufficient amount of energy is brought into the lake, in order to completely mix the entire water column; this energy may mainly be sufficient to achieve a homogeneous upper layer of a certain depth. So, an upper layer with larger, practically uniform, temperature is obtained. The daily variation of the incoming and outgoing radiation then yields a diurnal variation of the upper layer thickness. In the course of the summer the upper layer will continue to warm; it will thereby become more stable, implying that the daily variations of the density and upper layer thickness are now small. In autumn, the upper layer is again slowly cooled, it starts again to loose its stability and consequently grows in thickness, because the necessary energy for mixing a water packet becomes smaller with a decrease of the temperature and density, respectively. This process, favoured by the ongoing cooling at the end of the year, and the now frequent storms continue until the entire water column is mixed; the autumn circulation with unstable stratification is now reached. If the water at the lake surface is cooled below 4°C, then the situation corresponds again to the winter stagnation.

This description indicates, and Figs. 13.1 and 13.3 support this, that in the course of the year there prevails always a more or less distinct double layering of the lake. This double layering of the lake, which is imprinted by the thermocline, may be constructively used in the dynamical description of the lake, in particular its wave dynamics, which employs it as an approximation.

The term *thermocline* means *thermal inclination* and designates that depth of the temperature profile at which the vertical temperature gradient is largest. This water level separates the warmer *epilimnion* from the cooler *hypolimnion* (greek: *epi*: above; *hypo*: below; *limnion*: lake). The transition region, which is characterised by a strong temperature gradient, is called *metalimnion* (meta: in–between), see Fig. 13.4.

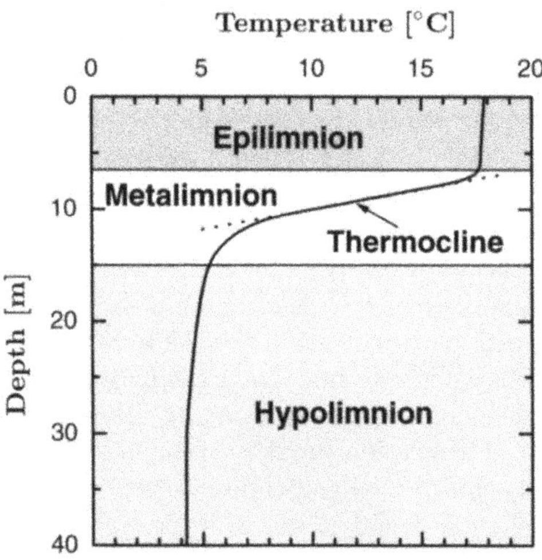

Fig. 13.4. Typical temperature profile in a lake for a summer situation. The lake is divided into the warm upper layer, called epilimnion, a transition layer with strong temperature gradient (thermocline), called metalimnion, and the underlying hypolimnion.

Lakes are classified according to their scheme of circulation. If the mixing reaches at any one time during the year the lake bottom, then it is called *holomictic* (*holo*, entire; *mictic*, mixing). In deep lakes or in lakes with a lower layer containing increased salinity, the heavy water may, on occasion, not participate in the mixing processes; in those cases the lake is called *meromictic* (*mero*, partly). In such waters, in which virtually no deep water exchange occurs, a shortage of available oxygen usually arises. Other circulation processes may, however, also occur, so that a slow water exchange may still take place without any observable vertical turbulent mixing. Such an *exchange of deep water* occurs for instance in Lake Baikal.

The above described mixing and heating processes are typical of *holomictic* lakes, especially for *dimictic* ones (*di*, two, double). The classification of lakes according to their mixing behaviour goes back to works by FOREL [76] and was extended and updated by FINDELEGG [73], [74], [75] and HUTCHINSON & LÖFFLER [103] to its present used form. In general, one may differentiate three classes: *holomictic* lakes, which are mixed once or several times during a year down to the bottom; *meromictic* lakes only mix to a certain depth, which chiefly depends on how much a lake is exposed to wind; they never (or only very seldom) mix to the ground; and, finally, *amictic* lakes which never mix. *Amictic* lakes for instance arise in Antarctica where a whole-year ice cover attenuates the energy fluxes into the water body, and thus generation of turbulent intensity is blocked.

Holomictic lakes – they are the rule – are further differentiated according to how much they mix during a year. At moderate latitudes one often encounters *dimictic* lakes, which experience in spring and autumn a complete circulation. The scheme shown in Fig. 13.3 applies to this lake type; in spring the inverse (but stable) temperature profile goes through a complete circulation and then establishes a stable (and normal) stratification; in fall the opposite phenomenon arises, namely the transition from an upper layer structure to an inverse temperature profile. If one of these transitions is missing, and if the density maximum (i.e. the 4°C temperature) is never crossed then this lake is *monomictic*, it mixes only once per year. If the temperature remains always below the temperature corresponding to the density maximum, then one speaks of a *cold monomictic* lake, in the reverse case, however, of a *warm monomictic* one.

If in the course of a year several mixing events take place – e.g. because frequent wind events trigger complete mixing in a shallow lake, or because only a weak stratification has formed and the air temperature is subject to strong variations (which often occurs in the tropics) – then the lake is called *polymictic* (*poly*, many).

Essential conditions which make the various states likely are the geographical position (latitude, altitude), exposition to wind in connection with the lake surface and depth, and a possible chemical stratification. Some characterizations of such lake types are illustrated in Fig. 13.5.

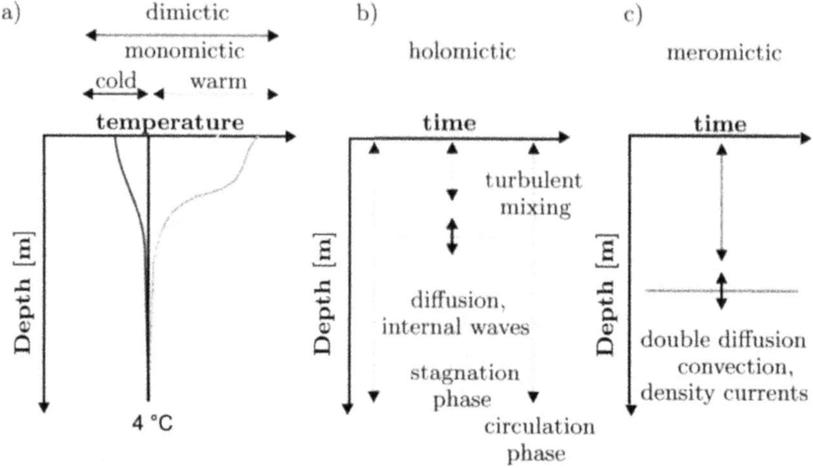

Fig. 13.5. Lake types characterised according to how they mix. a) Mono– and dimictic lakes are characterized by the vertical temperature profile and how it varies near the surface; b) c) holomictic and meromictic lakes are characterized by the fact whether the mixed layer may reach the bottom or not.

13.1.3 General Processes Arising in Lakes

In order to be able to derive the diurnal and seasonal development of the stratification within a lake (or within the ocean), one must describe, which processes are significant in the build-up of a stratification: The two most significant ones are the wind-input and the heating due to the solar irradiation. Figure 13.6 sketches these processes, which in the following are of interest. Let us commence with those effects of turbulence which were already dealt with in the last chapters. Depending upon the strength of the existing velocity gradient, turbulent flows may prevail or be minor. At the intermediate layer between the epilimnion and hypolimnion shear instabilities may arise, which may lead to intensified mixing of the water masses that are involved. These instabilities are called KELVIN–HELMHOLTZ instabilities and the RICHARDSON number[3] is the parameter characterizing them. On the basis of the cooling of the upper most water layer, an unstable stratification

[3] It is defined as

$$\mathbb{R}_i = \frac{-\dfrac{g}{\rho}\dfrac{\mathrm{d}\rho}{\mathrm{d}z}}{\left\|\dfrac{\mathrm{d}\boldsymbol{v}}{\mathrm{d}z}\right\|^2},$$

and gives the ratio of the stabilizing buoyancy forces to the destabilising shear forces. As a thumb rule one has $\mathbb{R}_i < 1/4$ for unstable situations and $\mathbb{R}_i > 1/4$ for stable ones. Compare also Example 8.10 in Chap. 8.

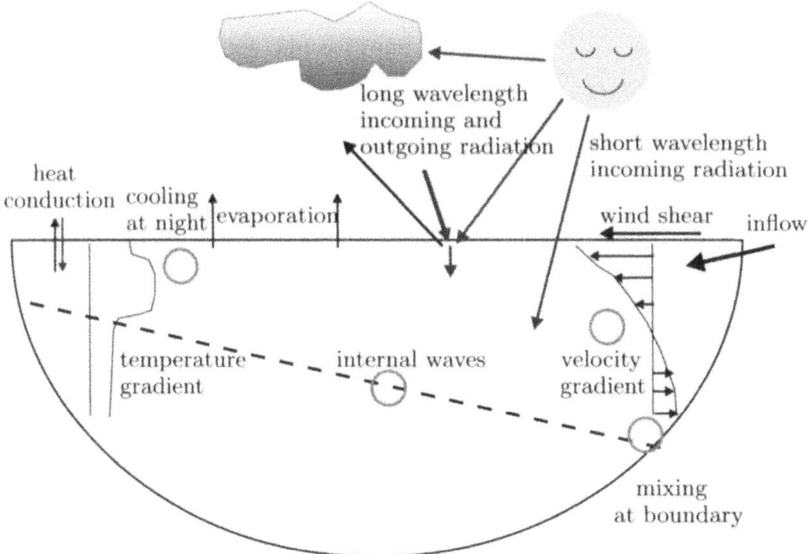

Fig. 13.6. Physically relevant processes in the water column which influence the structure of the stratification and vertical temperature distribution in a lake.

is caused, which subsequently leads to convective mixing. We shall ignore the mixing at the side boundaries where the thermocline touches the base, as we do neglect the mixing processes which are caused by the inflow and (less) by the outflow of water by the tributaries. Moreover, the geothermal heat flow and the flow of heat into or from the sediment can be ignored, but both could be accounted for without difficulty, and must be, at locations of unusually high geothermal heat as e.g. in lakes within and the ocean near Iceland. Far more important is the wind shear i.e., the traction at the free surface due to the action of the wind. It generates, via its momentum flux within the upper layer of the lake, turbulent kinetic energy and therefore plays the role of a boundary condition.

The heat content of a lake is influenced primarily by the *radiation* from above; besides this it also is affected by the redistribution of the water by turbulent and diffusive processes. To describe it, let us assume that the part of the total solar irradiation that is directly reflected (short wavelength radiation) or *rescattered* by the clouds is known; we thus may start from a known net radiation at the lake surface. In determining this net radiation the various *radiative* contributions with different wavelengths would have to be differentiated, in order to arrive at an appropriate value of the sum of all these contributions. Instead, to account for the total spectrum of the incoming radiation we divide it into merely two contributions. For one, this is the long wavelength radiation, which only penetrates the upper most few millimeters to centimeters of the water and leads to a warming of this very thin surface interface. This part shall be lumped together with the back radiation by the water surface to form the heat flow above the lake surface. The other, short wavelength contribution (the visible light) penetrates deeper into the water and produces heat within the water as a result of *radiation absorption* by the water molecules, tracers and algae. Strictly, this absorption is wavelength dependent, which in some models is accounted for. Short wave radiative absorption is energetically a supply term in the (internal) energy balance equation.

Because the surface temperature of the water often does not agree with the air temperature in the lower most atmospheric layer, there is generally a heat transfer taking place from the (usually) cold free surface to the atmosphere which is called *sensible heat flow.* A further heat sink is produced at the free surface by evaporation processes. This contribution is known as *latent heat flow.* These individual contributions together with the long wave heat radiation shall be encountered again when the boundary conditions for the heat conduction equation will be derived.

In the next section the above discussed effects will be scrutinized in detail and thereby also quantified.

13.2 Physical Processes

13.2.1 Solar Irradiation

We now wish to detail all influences described in the last section, which affect the thermal stratification of a lake; the goal is to arrive at explicit statements and to present formulas that quantify the various partial effects. We begin with the solar radiation, which is shown in Fig. 13.7. In this figure the radiative intensity [$\mathrm{Wm^{-2}nm^{-1}}$] is plotted against the wavelength in the wavelength range from 100 nm to 3200 nm (nanometer). The dashed curve shows the radiation distribution of a black body with a temperature of 6000 K. The solar irradiation outside the Earth's atmosphere nearly exactly follows this distribution. Within the atmosphere a part of this radiation is, however, absorbed, so that only a part of the incoming solar radiation reaches the Earth surface. This absorption is due to the greenhouse gases (H_2O, O_3, CO_2, ...) and strongly depends on wavelength. This external net irradiation close to the Earth's surface is also shown in Fig. 13.7 as the garlandic curve and its total amount is given by the black area. If one integrates the total irradiation, the solar constant is obtained, which at the Earth's surface has the value $1360\,\mathrm{Wm^{-2}}$. Its reduction by absorption, scattering and reflection in the Earth's atmosphere depends, among other things, on the geographical latitude and the season throughout the year. Without the influence of the clouds, one obtains about $800 - 1000\ \mathrm{W\,m^{-2}}$. One may divide the solar irradiation that falls on the Earth's surface roughly into three categories:

- **300 – 380** nm: Ultraviolet light; it harms the organisms, if it occurs with too great intensity.
- **380 – 750** nm: This is the visible light. It is particularly important in the range 400–700 nm for the life in a lake, i.e., for the photosynthesis. This contribution of the radiation is called PAR (Photosynthetically Active Radiation)
- **750 – 3000** nm: Infrared radiation forming the heat radiation.

The incoming radiation from the sun is already partly absorbed in the atmosphere and, depending upon the weather conditions, cloud cover and fog, etc., it experiences a further attenuation. We shall not dwell any further upon the difference between this direct and diffuse radiation, but will rather assume that with the degree of cloud cover the light intensity will decrease.

For a cloudless day the irradiation onto the Earth's surface can be computed as a function of its latitude and time of the day. Since the spectrum of this incoming energy is not homogeneous and since the radiation for different wavelengths is differently attenuated or absorbed, it would actually be indispensable to account for the proper spectral distribution. We shall not do this and only assume that the short wave radiation at the Earth's surface generates a radiative flux of magnitude $I_0(t)$, independent of wavelength. The seasonal and diurnal variation of this irradiation is qualitatively displayed in Fig. 13.8. The computation on which the graph is based was done for a situation pertinent to Lake Ammer, Bavaria, and the observed cloud cover was accounted for. In a realistic model for the determination of the diurnal and seasonal temperature distributions the quantity I_0 will be prescribed as a measured quantity. For the simulation of certain typical situations or an idealised annual variation, it is, however important that this quantity can be computed. For an ideal case we shall motivate formulas, however without deriving them in detail.[4] We ignore in our description of the solar irradiation the distinction between direct and diffuse radiation and thus shall neither describe the scattering effects by the atmosphere and the clouds. Furthermore, the differences in the reflectivity of the water surface, which depend upon the angle of the incoming radiation, the surface roughness of the water (wave forms and height) and possible foam formation, will not be accounted for. The latter two can, in principle, be taken into account by an empirical formula which contains the air temperature and the wind speed as parameters, see PREISENDORF & MOBLEY [189]. In order to explicitly compute the solar irradiation, it is necessary to explain certain astronomical facts in particular with regard to the motion of the Earth around the Sun. The formulas that will be presented below (KIRK, 1983; FORSYTHE, 1995) are approximations to the parameters which can be determined very accurately. The irradiation to the Earth depends upon the position of the sun at a certain time of the year and the day. Depending upon the incoming angle of the rays the radiation travels a longer or shorter distance through the atmosphere and thus

[4] See in particular KIRK [118].

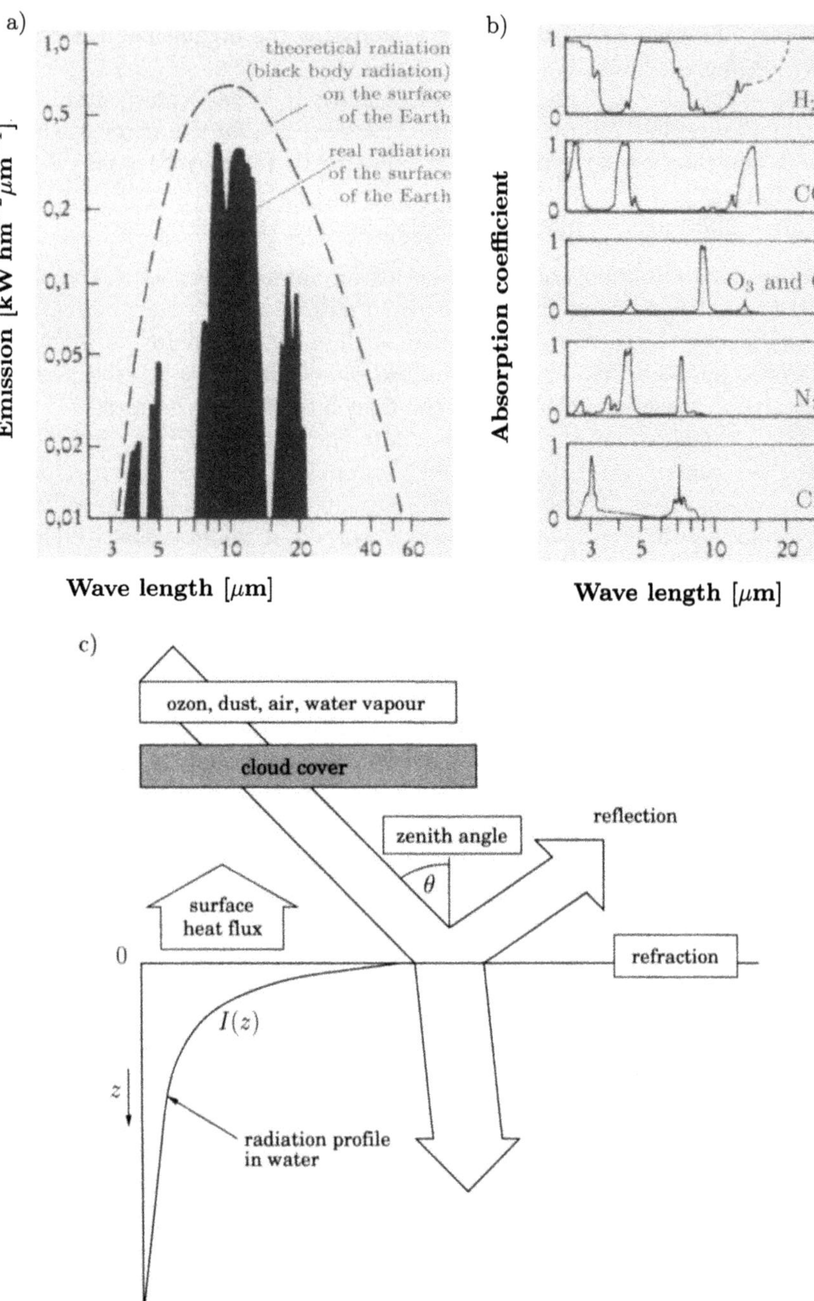

Fig. 13.7. a) Spectrum of the solar radiation. As dashed line the black body radiation of a body at 6000° K (the theoretical radiation) is shown. The black areas show the radiation on the surface of the Earth. b) Absorption coefficients caused by absorption of Ozone, Oxygen, Hydrogen, etc. c) Schematic sketch of the solar radiation on a water body.

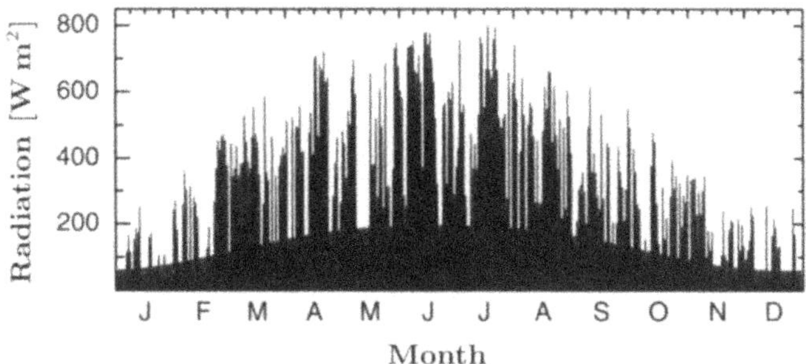

Fig. 13.8. Diurnal (daily) and seasonal (yearly) variation of the solar irradiation at a geographical latitude of 48° (appropriate for Lake Ammer, Bavaria).

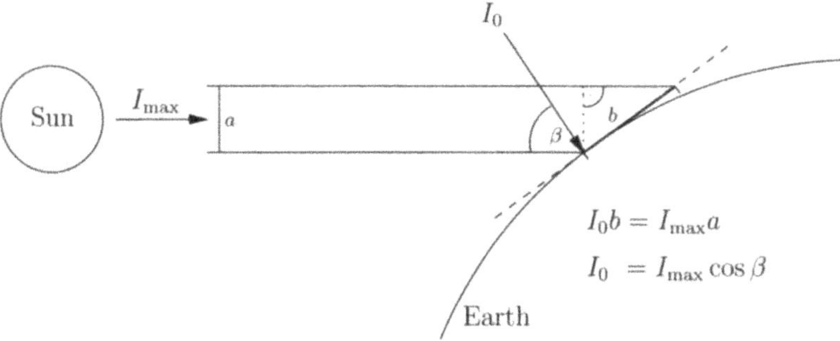

Fig. 13.9. The rays from the Sun falling on the Earth's surface define the incoming irradiation.

also experiences different absorptions by the water molecules, ozone and other greenhouse gases. By the non-normality to the Earth's surface the radiation flux per unit area is equally reduced according to

$$I_0 \propto \cos\beta \,, \tag{13.2.1}$$

see Fig. 13.9, in which β denotes the zenith angle of the Sun. Since the axis of rotation of the Earth relative to the normal to the orbital plane of the elliptical trajectory of the Earth is inclined by approximately 23.27°, an observer at the equator will recognise that in the course of a year the Sun will not always stay perpendicularly above him, but that its position varies. The angle between the connecting line *Earth–Sun* and the intersection line between the plane of the equator and the plane that is perpendicular to the orbital plane of the Earth passing through this connecting line, is called declination δ. Figure 13.10 illustrates the definition of this angle. The angle between the plane of the Earth's trajectory and the plane of the equator defines its extreme values of ±23.27°, which are reached on 22 June and 27 December. The minimum absolute value of δ, namely zero, is reached on 21

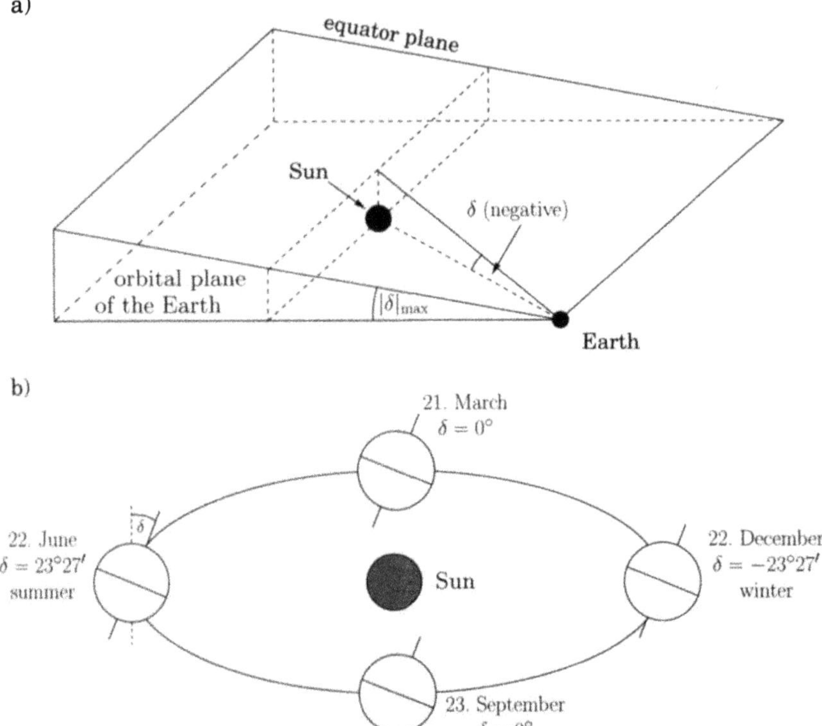

Fig. 13.10. Angle of declination: a) Definition of δ for an arbitrary position of the Earth relative to the Sun b), extreme values of δ during the year.

March and 23 September. The variation of δ through the year as a function of the day d of the year can approximately be computed from the formula

$$\begin{aligned}\delta = 0.39637 - 22.9133\cos(\psi) + 4.02543\sin(\psi)\,, \\ - 0.3872\cos(2\psi) + 0.052\sin(2\psi)\,,\end{aligned} \qquad \psi = \frac{360°\, d}{365}\,, \tag{13.2.2}$$

in which $d = 0$ corresponds to 1. January (KIRK [118]). The position of the Sun for every time of the day throughout the year at any position on Earth, i.e., at any geographical latitude is obtained from the angle between the directions to the Sun and the zenith, β. If t_d is the time at a day in hours from midnight and if λ is the geographical latitude of the position on Earth, then the zenith angle, i.e., the angle between the horizon and the Sun is given by

$$\cos\beta = \sin\lambda\sin\delta - \cos\lambda\cos\delta\cos\tau\,, \quad \tau = \frac{360°\, t_d}{24}\,. \tag{13.2.3}$$

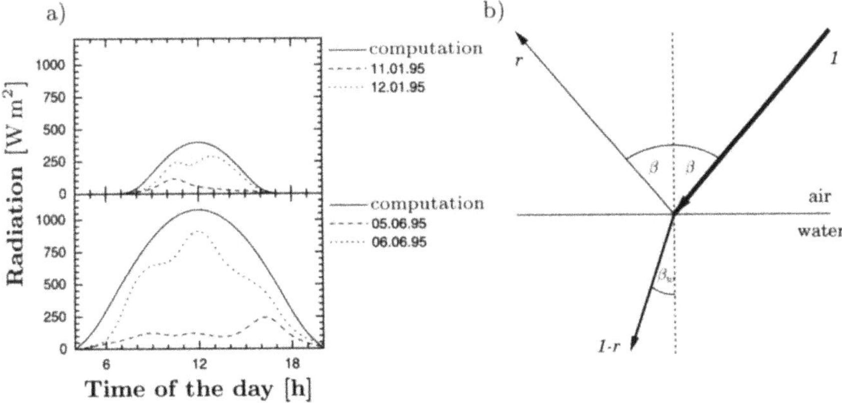

Fig. 13.11. a) Solar irradiation for $\lambda = 48°$ as computed (solid lines) and as measured for 4 January and 5 June and the following day. The large differences are due to effects of the cloud cover.
b) Incoming (1), reflected (r) and transmitted (1-r) radiation in the law of SNELLIUS.

If I_{max} is the maximum irradiation, i.e., the irradiation parallel to the orbital plane of the Earth at sea level and without any cloud cover, then one may now compute at every position on Earth (λ), at every day (d), at every hour (t_d) the irradiation according to

$$I_0 = I_{\text{max}} \cos \beta \,. \tag{13.2.4}$$

If one wishes here to also account for a certain cloud cover, then the right-hand side of this equation is simply multiplied with a corresponding reduction factor (< 1), which accounts for this fact.

The length of a day can also be calculated from the above formula (13.2.3); to this end one only needs to recognize that sun rise begins with $\cos \beta = 0$. The time of the sun rise may therefore be found from

$$\cos \tau_s = \tan \lambda \tan \delta \,, \quad \tau_s = \frac{360° \, t_s}{24} \,. \tag{13.2.5}$$

The length of the day, N is then $24h - 2t_s$, because of symmetry relative to the zenith; expressed in terms of latitude and declination, one thus obtains

$$N = 0.133 \arccos(-\tan \lambda \tan \delta) \,. \tag{13.2.6}$$

In Fig. 13.11a the solar irradiation is shown for a geographical latitude $\lambda = 48°$ – corresponding to the position of lake Ammer or lake Constance – for a day in summer and a day in winter. The figure also displays results from measurements which were taken at the same days. The large discrepancies between measured and computed values are due to the fact that computations were performed without taking the cloud cover into account. With a simple linear model it may for instance be parameterized by

$$I_0 = I_{\max} \cos\beta(1 - c_b N_{\mathrm{b}}(t)) , \quad c_b \approx 0.7 . \tag{13.2.7}$$

Here, c_b is an empirical constant and $N_{\mathrm{b}}(t)$ denotes the degree of cloudiness, i.e., the areal part of the cloud covered sky at a certain time. $N_{\mathrm{b}}(t) = 1$ for a completely covered sky and $N_{\mathrm{b}}(t) = 0$ for a free sky without any clouds. With the aid of the SNELLIUS law between air and water one has

$$\frac{\sin\beta}{\sin\beta_w} = 1.33 , \tag{13.2.8}$$

in which β_w is the angle between the vertical and the direction of the impinging radiation, the FRESNEL formula allows computation of the coefficient of reflectivity[5], viz.,

$$r = \frac{1}{2}\left[\left(\frac{\tan(\beta-\beta_w)}{\tan(\beta+\beta_w)}\right)^2 + \left(\frac{\sin(\beta-\beta_w)}{\sin(\beta+\beta_w)}\right)^2\right] . \tag{13.2.9}$$

The radiation entering the water is therefore given by

$$I = I_{\max} \cos\beta(1 - c_b N_{\mathrm{b}}(t))(1 - r) . \tag{13.2.10}$$

13.2.2 Short Wave Radiative Input Into the Water

The short wave radiation in the range from about 400 to 700 nm enters the water and is attenuated there by absorption. The underwater light field is thereby influenced by several factors such as

- clear water attenuation
- absorption by chlorophyll pigments in the phytoplankton
- absorption by tracers in solution and in particulate form.

The clear water attenuation, which is quantified by an attenuation coefficient, is strongly wavelength dependent (see Table 13.1). Red light is much more attenuated than blue light. This is the reason why bodies become more and more blue with diving depth. If one does not wish to resolve the spectral distribution in a model, then one needs to concentrate at least to its most important aspects. For the attenuation coefficient this means that a few spectral regions are specially accounted for. Here we confine ourselves to the assumption of a mean representative value. Since the radiation occurs chiefly in the wavelength range of about 600 nm (at larger wavelengths the absorption by the atmosphere is small, see Fig. 13.1) and other effects as e.g. the absorption by matter in solution only can enlarge the attenuation, one takes a value of about $k_{\mathrm{w}} \approx 0.3\mathrm{m}^{-1}$, depending upon the turbidity of the lake. This value is, of course, not just the clear water attenuation coefficient.

[5] See any book on elementary physics or optics.

Table 13.1. Attenuation coefficient of water, k for a number of wavelengths λ (modified after KIRK, [118]). Note that red light (720nm) is attenuated much more than blue light (475nm) and that k does not monotonically change with λ.

λ (nm)	k (m^{-1})	λ (nm)	k (m^{-1})	λ (nm)	k (m^{-1})
310	0.105	480	0.0176	650	0.349
320	0.0844	490	0.0196	660	0.400
330	0.0678	500	0.0257	670	0.430
340	0.0561	510	0.0357	680	0.450
350	0.0463	520	0.0477	690	0.500
360	0.0379	530	0.0507	700	0.650
370	0.0300	540	0.0558	710	0.839
380	0.0220	550	0.0638	720	1.169
390	0.0191	560	0.0708	730	1.799
400	0.0171	570	0.0799	740	2.38
410	0.0162	580	0.108	750	2.47
420	0.0153	590	0.157	760	2.55
430	0.0144	600	0.244	770	2.51
440	0.0145	610	0.289	780	2.36
450	0.0145	620	0.309	790	2.16
460	0.0156	630	0.319	800	2.07
470	0.0156	640	0.329		

The absorption by the algae is also wavelength dependent, a dependence we shall here, however ignore. If one wishes to establish a model in which biological and physical processes are coupled, it is often sufficient to regard the attenuation as independent of wavelength and to account for the algal concentration in a gross fashion. Because regularly a large number of different algal species occupy a lake, such a procedure appears to be a reasonable compromise. The attenuation coefficient due to the algal population is then set proportional to its concentration or to the concentration of the chlorophyll, $K_{\mathrm{Chl}}[\mathrm{mgChl\ m^{-3}}]$; thus one writes

$$k_{\mathrm{Chl}}(\boldsymbol{x},\,t) = a_{\mathrm{Chl}} K_{\mathrm{Chl}}(\boldsymbol{x},\,t)\;, \tag{13.2.11}$$

in which the dependence on space and time accounts for the large variation on these variables. Occasionally the spatial dependence is restricted to a dependence on the depth alone. The coefficient a_{Chl} with the dimension $[\mathrm{m^2(mgChl\,m^{-3})^{-1}}]$ is called the *specific absorption cross section* ; it constitutes the sum of all absorbing cross sections per unit volume and unit concentration, and it possesses a value of about $a_{\mathrm{Chl}} \approx 0.015\ \mathrm{m^2(mgChl\,m^{-3})^{-1}}$.

The decrease of the light intensity with the water depth z by absorption and scattering is mathematically described by the so-called *law of* LAMBERT–BEER. It supposes that *the reduction of radiation per unit depth is proportional to the radiation itself*, so that with a dependence on z alone

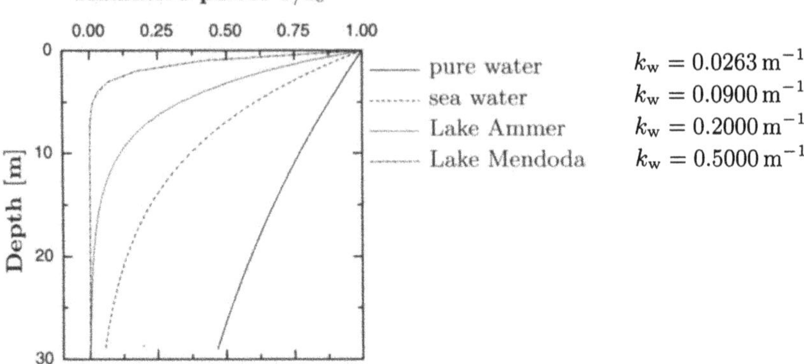

Fig. 13.12. Distribution with depth of the radiation of light for different values of the attenuation coefficient k_w.

$$\frac{\partial I}{\partial z} = -k(z,\, t)I \qquad \Rightarrow \qquad I(z) = I_0 \exp\Big(-\int_0^z k(z,\, t)\,\mathrm{d}z\Big) \tag{13.2.12}$$

emerges, in which the above introduced distinction between clear water attenuation and absorption by the algae may be accounted for by writing

$$I(z) = I_0(t) \exp\Big(-k_w z - a_{\mathrm{Chl}} \int_0^z K_{\mathrm{Chl}}(z,\, t)\,\mathrm{d}z\Big)\,. \tag{13.2.13}$$

In this formula the radiation intensity at the free surfaces, I_0 and concentration of algae, K_{Chl} are time dependent. The former can be computed with the aid of (13.2.10). For pure water the exponential decay with depth as shown in Fig. 13.12 is obtained. For an attenuation coefficient of $k_w = 0.3$ the radiation at 15m depth below the surface is already reduced to 1% of its surface value. In order to determine the photosynthetic production in the water column, one must account in (13.2.10) only for the photosynthetically active part; usually, one uses $I_{\mathrm{PAR}} = 0.46\, I$.

13.2.3 Long Wave Radiation

The long wave infrared radiation of heat is modelled just as a black body radiation; this is an idealized body which radiates a certain spectrum of heat that depends on its temperature. The spectral energy density as a function of the temperature Θ and frequency ν is thereby given by PLANCK's *radiation law*

$$\rho(\nu, \Theta) \propto \frac{8\pi h \nu^3}{\mathrm{e}^{h\nu/k\Theta} - 1}\,, \tag{13.2.14}$$

in which $\rho(\nu,\, \Theta)\,\mathrm{d}\nu$ denotes the energy density, which at a certain given temperature of the radiating body is emitted between the frequencies ν and

$\nu + d\nu$; h is PLANCK's constant and k is BOLTZMANN's constant. The energy flux or heat flux generated by this emission is obtained by integrating the PLANCK law over all frequencies. One obtains (see Exercises)

$$Q_{\mathrm{ir}} \propto \sigma_{\mathrm{SB}} \Theta^4 \tag{13.2.15}$$

with the STEFAN–BOLTZMANN constant $\sigma_{\mathrm{SB}} = 5.671 \times 10^{-8} \mathrm{W\,m^{-2}K^{-4}}$.

Every body radiates according to this law, but depending on its composition with different intensity, which explains the proportionality sign in (13.2.15). In the present situation there are two different bodies, the water and the atmosphere and both contribute to corresponding radiation flows, which are given by the STEFAN–BOLTZMANN law with different prefactors. It is customary to account for these differences by different prefactors, ε_{a} and ε_{w} and to write

$$Q_{\mathrm{ir}}^{\mathrm{a}} = \varepsilon_{\mathrm{a}} \sigma_{\mathrm{SB}} \Theta_{\mathrm{a}}^4 \quad \text{and} \quad Q_{\mathrm{ir}}^{\mathrm{w}} = \varepsilon_{\mathrm{w}} \sigma_{\mathrm{SB}} \Theta_{\mathrm{w}}^4 \,. \tag{13.2.16}$$

ε_{a} and ε_{w} are called *emissivities* and the indices $(.)_{\mathrm{a}}$ and $(.)_{\mathrm{w}}$ stand for atmosphere and water, respectively. The temperatures, correspondingly, refer to the air temperature and surface temperature of the water body, respectively. The emissivity of water is

$$\varepsilon_{\mathrm{w}} = 0.97 \,, \tag{13.2.17}$$

that of the atmosphere depends on the temperature, whereby one would have to take into account that the radiation depends on the temperature of the clouds and aerosols at various heights. This dependence is usually only accounted for by taking a corresponding mean value for the temperature. Several empirical formulas have been deduced from measurements.

For a completely clear sky a quadratic dependence of the emissivity upon the temperature (immediately above the water body) is assumed[6]. If one also wishes to account for the cloud cover, this is achieved by a multiplicative factor, which is larger than unity (as it is an additional radiative source). The emissivity of the atmosphere is in this case given by the simple formula

$$\varepsilon_{\mathrm{a}} = \alpha_{\mathrm{a}} \Theta_{\mathrm{a}}^2 (1 + \beta_{\mathrm{a}} N_{\mathrm{b}}^2) \tag{13.2.18}$$

with the empirical dimensional coefficient[7])

$$\alpha_{\mathrm{a}} = 0.937 \times 10^{-5} \left[\mathrm{K}^{-2}\right] \quad \text{and} \quad \beta_{\mathrm{a}} = 0.17 \,, \tag{13.2.19}$$

in which N_{b} as in §13.2.1 is the degree of cloudiness.

The balance of heat radiation can now be formulated as

$$\begin{aligned} Q_{\mathrm{ir}} &= Q_{\mathrm{ir}}^{\mathrm{a}} - Q_{\mathrm{ir}}^{\mathrm{w}} \\ &= 0.937 \times 10^{-5} \left[\mathrm{K}^{-2}\right] \sigma_{\mathrm{SB}} \Theta_{\mathrm{a}}^6 (1 + 0.17 N_{\mathrm{b}}^2) - 0.97 \sigma_{\mathrm{SB}} \Theta_{\mathrm{w}}^4 \left[\mathrm{W\,m^{-2}}\right] . \end{aligned} \tag{13.2.20}$$

[6] SWINBAECK [232]

[7] see STRASKRABA & GRANK, [228]

A positive value on the right-hand side corresponds here to a heat input into the lake. Such a formulation has the advantage that only three empirical constants arise. These are $\varepsilon_{\rm w}$, $\alpha_{\rm a}$ and $\beta_{\rm a}$. Other authors account also for additional effects as e.g. the dependence on the vapour pressure above the lake, but then additional constants must be determined by measurements.

13.2.4 Latent and Sensible Heat Flux

Additional heat fluxes through the surface are activated when (1) water evaporates and (2) temperature differences exist between the near surface water and the air close to the water.

The *latent heat flux* is that heat flux which extracts energy from the near-water air for evaporation. It depends upon the humidity of the air and the wind speed above the lake surface and the vertical gradient of the horizontal wind velocity above the lake, which is responsible for the turbulent mixing of the humid air. Nonetheless, water can also evaporate when no wind prevails. This suggests to parameterize the latent heat flux with the humidity gradient above the lake surface and to set it proportional to this gradient as well as the gradient of the wind (or the wind speed v_*), at some height above the lake, and to add to this parameterization a constant term. Instead of a dependence upon the humidity one may also parameterize the latent heat flow with the differences of the vapour pressure at the lake surface (this is its saturation value at the corresponding air temperature, i.e., the saturation pressure), $p_{\rm w}$ and the vapour pressure in the air, $p_{\rm a}$. Then, the latent heat flow is given by

$$Q_{\rm l} = f(\Theta_{\rm w}, v_*)(p_{\rm w} - p_{\rm a}) \,, \qquad (13.2.21)$$

where the wind function f is usually taken to be linear in v_*.

The *sensible heat flux* is analogously parameterized with the temperature difference between air and lake water at the surface

$$Q_{\rm s} = \tilde{f}(\Theta_{\rm w}, v_*)(\Theta_{\rm w} - \Theta_{\rm a}) \,. \qquad (13.2.22)$$

The conductive heat flux is determined via the BOWEN ratio

$$B := \frac{Q_{\rm s}}{Q_{\rm l}} = \frac{\tilde{f}}{f}\frac{\Theta_{\rm w} - \Theta_{\rm a}}{p_{\rm w} - p_{\rm a}} \,, \qquad (13.2.23)$$

where the ratio of the wind functions takes the value

$$\frac{\tilde{f}}{f} \approx 62 \,\left[\mathrm{Pa\,K^{-1}}\right] \,. \qquad (13.2.24)$$

The wind function of the latent heat flux is directly proportional to the *heat of evaporation* $L_v(\Theta_{\rm w})$ and is parameterised as[8]

[8] MARTI & IMBODEN [150]

$$f(\Theta_\mathrm{w}, v_*) = \rho_\mathrm{w} L_v(\Theta_\mathrm{w}) \times a(1 + b\,U_{(10m)}) \tag{13.2.25}$$

with the empirical constants

$$a = 1.36 \times 10^{-11} \left[\mathrm{m\,s^{-1}\,Pa^{-1}}\right] , \tag{13.2.26}$$

$$b = 0.59 \left[(\mathrm{m\,s^{-1}})^{-1}\right] . \tag{13.2.27}$$

Here the wind speed must be taken as that at the 10 m height above the lake surface, $U_{(10m)}$.

The model, that describes the latent and sensible heat fluxes also requires knowledge of the vapour pressure[9]. Directly above the water (i.e., in the millimeter range of the surface) the pressure will be the saturation vapour pressure $p_\mathrm{w} = p_\mathrm{s}(\Theta_\mathrm{w})$; it depends upon the temperature Θ_w of the water surface. It can be computed with the aid of the CLAUSIUS–CLAPEYRON equation (see Chap. 6, formula (6.2.11)); for water vapour, treated as an ideal gas, it takes the form

$$\frac{1}{p_\mathrm{s}}\frac{\mathrm{d}p_\mathrm{s}}{\mathrm{d}\Theta} = \frac{L_v M_v}{R\Theta^2} . \tag{13.2.28}$$

The index s in p_s stands here for the saturation value of the vapour pressure. L_v is the latent heat of evaporation of water, which is weakly temperature dependent, viz.,

$$L_v = 2500900 - 2365(\Theta_w - 273.15\,[\mathrm{K}]) \quad [\mathrm{J\ kg^{-1}}] . \tag{13.2.29}$$

Here, $R = 8.31451\,[\mathrm{J\ K^{-1}\ mol^{-1}}]$ as well as $M_v = 0.018\,[\mathrm{kg\ mol^{-1}}]$ are the gas constant and the mol weight of water vapour. In addition, the vapour pressure in the atmosphere (above the water surface) must be measured, or alternatively, the relative humidity

$$\Psi = \frac{p_\mathrm{a}}{p_\mathrm{s}(\Theta_\mathrm{air})} \times 100 \quad [\%] \tag{13.2.30}$$

must be known. The latter is 100%, if the atmosphere is saturated with water vapour, and it is 0% if no water vapour is present in the atmosphere. Notice, that the saturation vapour pressure in the atmosphere, i.e., at the air temperature Θ_air must be employed here.

With the deductions described above, the latent heat is given by

$$Q_\mathrm{l} = 1.36 \times 10^{-11}\,\rho_\mathrm{w} L_v(1 + 0.59\,U_{(10m)}) \left(p_\mathrm{s}(\Theta_w) - p_\mathrm{s}(\Theta_\mathrm{air})\frac{\Psi}{100}\right) .$$

[9] The vapour pressure can be visualized as follows: In a vacuated vessel, that is partly filled with a fluid, part of the fluid will evaporate. The so established equilibrium is characterized by a pressure, the so-called *saturation vapour pressure*. In the atmosphere, this equilibrium will most likely never be established because of the ongoing temperature and air–pressure variations. The ratio between the actual vapour pressure and the saturation vapour pressure is called the relative humidity.

The saturation vapour pressure at a certain temperature is obtained by integrating the CLAUSIUS–CLAPEYRON equation; if the latent heat is taken to be constant, $L(\Theta_w = 10° \text{ C}) \approx 2477250\,[\text{J kg}^{-1}]$, then one has

$$p_s = p_s(\Theta_{s0}) \exp\left(-\frac{L_v M_v}{R}\left(\frac{1}{\Theta} - \frac{1}{\Theta_{s0}}\right)\right) . \tag{13.2.31}$$

If the reference pressure is taken as the pressure of the saturated water vapour at the same temperature, $\Theta_{s0} = 10°$ C, $p_s(10° \text{ C}) = 1230$ Pa, then the above equation may also be written as

$$p_s = 1230 \times \exp\left(5362.97\left(0.0035317 - \frac{1}{\Theta}\right)\right) [\text{Pa}] , \quad \Theta\,[\text{K}] . \tag{13.2.32}$$

In these equations the relative humidity, the temperature of the atmosphere and the wind speed are to be measured 10 m above the lake surface, or they must be computed there from an atmospheric model. Once the latent heat is determined, one may, with the aid of the BOWEN ratio, compute the sensible heat. This functional relationship reflects the fact that the sensible heat flux vanishes, if there is no temperature difference between the surface water and the near-water air layer; yet this does not necessarily imply a vanishing latent heat flux, as it often arises when a constant BOWEN ratio ($Q_s/Q_l \approx 0.2$) is used.

The annual variation of the above discussed heat fluxes and the radiation, respectively, is shown in Fig. 13.13a. The radiation fluxes are hereby of the order of magnitude of 3×10^3 [W m^{-2}]; the sensible and the latent heat flux, on the other hand, are only about 50 [W m^{-2}]. During day light, as would be expected, the energy supplied by radiation is dominant, whilst at night the smaller heat fluxes are of larger influence. In Fig. 13.13b the sum of the energy fluxes – except for that caused by wind – are displayed. One clearly recognises, how in spring the energy balance turns positive; this is the beginning of the formation of the upper layer. This layering can, however, again be destroyed by the vigour of the wind energy, and such a potential exists until the wind energy does no longer suffice to distribute the heat over the entire water column.

13.2.5 Wind Shear

On the water surface, due to the wind shear stresses, besides the heat fluxes, also the momentum fluxes are active. As was the case for the latent heat flux, these tractions depend upon the wind speed. The turbulent kinetic energy generated by such a momentum transfer can cause an intensive mixing in the water body below the water surface. This is the case in particular, when the stratification is not yet strongly established. This effect caused by wind shear enters the lake system as a boundary condition at the lake surface for the differential equations of the turbulence model that still are to be derived.

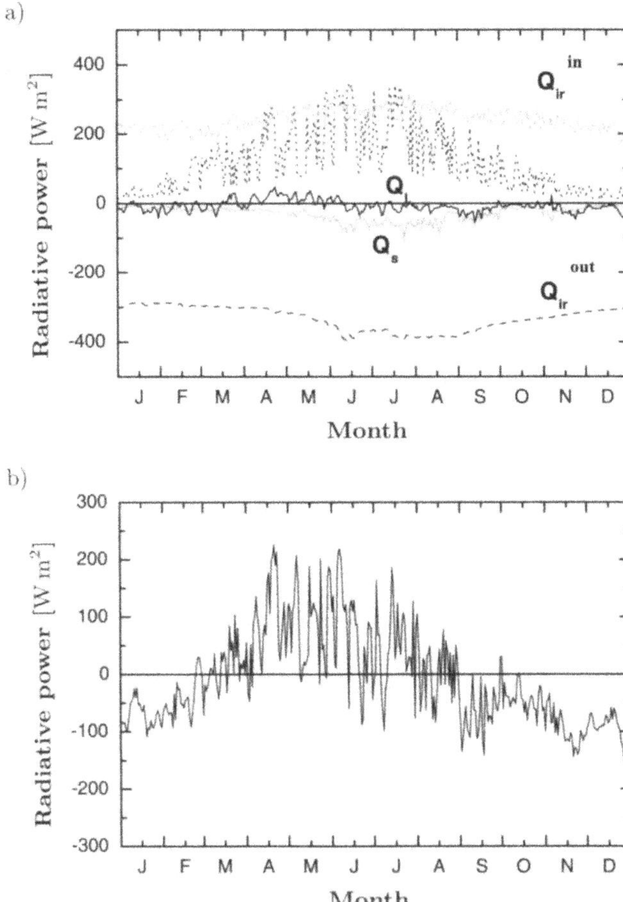

Fig. 13.13. a) Mean daily heat fluxes throughout the year at the lake surface. Q_l is the latent, Q_s the sensible heat flux, Q_{ir} the heat due to (incoming and outgoing) heat radiation of the short wave radiation. b) Sum of all energy fluxes above the lake surface.

The formulation of the wind shear stress is mostly expressed as a quadratic dependence upon the *frictional velocity* v_* according to

$$\tau = \rho_w (v_*)^2 \ . \tag{13.2.33}$$

The shear stresses, caused by the wind are equally assumed to depend quadratically upon a velocity, viz.,

$$\tau = \rho_a c_D U^2 \ , \tag{13.2.34}$$

where c_D is a drag coefficient, $\rho_a = 1.2\,\text{kg m}^{-3}$ is the density of air and U is the wind speed at a certain height above the lake. However, the drag coefficient can itself also depend upon the velocity. In this regard, there exist a large number of possible parameterizations ($c_D \propto U^{\mathrm{const}}$ or $c_D = c_1 + c_2 U$), of which we shall choose one,

$$c_{D(10m)} = (0.75 + 0.067 U_{(10m)}) \times 10^{-3} \,, \quad U_{(10m)} \text{ in } [\mathrm{m\ s^{-1}}] \,. \tag{13.2.35}$$

This empirical equation holds for the wind speed at the 10m height above the lake. Generally, the c_D-value has the order of magnitude of 10^{-3}, in fact its value is about

$$c_D \approx 1.3 \times 10^{-3} \,. \tag{13.2.36}$$

The wind shear stress, therefore, enters the surface boundary condition approximately as

$$\tau \approx 1.3 \times 10^{-3} \rho_{\mathrm{a}} U_{(10m)}^2 \,. \tag{13.2.37}$$

13.3 Material Behaviour of Water

13.3.1 Density

To completely establish the turbulence model, a number of material properties of water are needed which we shall now describe. The most important one of these is the density. Via the thermal equation of state it is a function of the temperature, the pressure, the salinity and also other possibly present tracer substances. In lakes, which often possess a small, nearly constant mineral content, which does not affect the density function, also the pressure dependence may be ignored down to 200–300m. In this case, there only remains the temperature that affects the value of the density.

In the temperature range of 0℃ $\leq \Theta \leq$ 100℃ the density is not a monotonous function of the temperature. For almost all other substances an increase of density is measured when the temperature decreases. This normal behaviour does only hold for water above the temperature of 4°C; below it, water shows an abnormal behaviour. Pure water has a density maximum at approximately 4°C, i.e., the density decreases when the temperature falls. This property has important consequences for all waters and their living populations. If this density maximum would not exist above the freezing point a lake would freeze in winter starting from the bottom with its lowest temperature and eventually freeze through its entire depth. The solar irradiation would be too small to melt the ice (Exercise). The living creatures would be without nutrients and would mostly die. Instead, the lake freezes because of its non-monotonic density–temperature relationship from its surface; the ice floats on the surface because of an additional density decrease at freezing from water to ice. This ice forms an insulator to the cold from above for the entire lake below, which in winter assumes a temperature between 0°C – 4°C. The density variation of pure water as a function of temperature in the interval 0°C $\leq \Theta \leq$ 40°C is shown in Fig. 13.14. Despite the fact

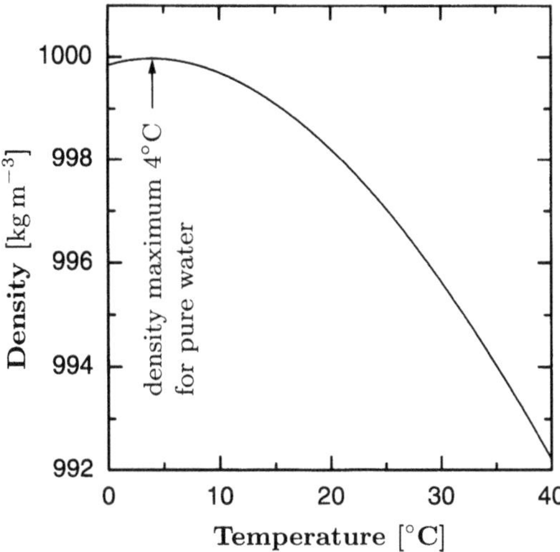

Fig. 13.14. Density variation of pure water as a function of the temperature.

that this non-monotonicity of the density function as a function of temperature is obviously very small (the relative density change in the temperature range of 20°C is of the order of 10^{-3}) it is of immense significance. The temperature–salinity dependence of the density of water has very accurately been measured up to six significant figures; the relationship is expressed by several mathematical formulas with different accuracy. Today the standard formula in use is that of CHEN & MILLERO [47], viz.,

$$\begin{aligned}\rho(\Theta, s) = 999.8395 &+ 6.7914 \times 10^{-2}\,\Theta - 9.0894 \times 10^{-3}\,\Theta^2 \\ &+ 1.0171 \times 10^{-4}\,\Theta^3 - 1.2846 \times 10^{-6}\,\Theta^4 \\ &+ 1.1592 \times 10^{-8}\,\Theta^5 - 5.0125 \times 10^{-11}\,\Theta^6 \\ + \; s \cdot (8.181 \times 10^{-1} &- 3.85 \times 10^{-3}\,\Theta + 4.96 \times 10^{-5}\,\Theta^2) \\ &[\mathrm{kg\,m^{-3}}],\end{aligned} \tag{13.3.1}$$

in which the temperature must be given in centigrades[10] and the salinity in grams of solved substance per kilogram of water. The values of the density are listed in Table 13.2. A somewhat simpler formula, in which the temperature is expressed as a polynomial of third degree is

$$\begin{aligned}\rho(\Theta, s) = 999.8395 &+ 6.7914 \cdot 10^{-2}\,\Theta \\ &- 9.0894 \cdot 10^{-3}\,\Theta^2 \\ &+ 1.0171 \cdot 10^{-4}\,\Theta^3 \,.\end{aligned} \tag{13.3.2}$$

This formula was used by BÜHRER & AMBÜHL [40]; it describes the density of water in the interval $0°\mathrm{C} < \Theta < 30°\mathrm{C}$ accurate to six significant figures.

[10] In all subsequent formulas it is understood that the temperature is given in centigrades CELSIUS.

Table 13.2. Properties of pure water in the temperature range of 0°C to 35°C.

°C temp [°C]	ρ density [kg m^{-3}]	c_p spec. heat [J kg^{-1} K^{-1}]	α_Θ coeff. of thermal expansion [K^{-1}]	ν kinematic viscosity [m^2 s^{-1}]
0	999.8395	4217.4	$-6.8 \cdot 10^{-5}$	$1.78029 \cdot 10^{-6}$
1	999.89843	4213.8683	$-5.00943 \cdot 10^{-5}$	$1.72178 \cdot 10^{-6}$
2	999.93976	4210.58612	$-3.27746 \cdot 10^{-5}$	$1.66633 \cdot 10^{-6}$
3	999.96408	4207.54076	$-1.60112 \cdot 10^{-5}$	$1.61372 \cdot 10^{-6}$
4	999.97192	4204.71986	$2.24665 \cdot 10^{-5}$	$1.56375 \cdot 10^{-6}$
5	999.96378	4202.11143	$1.59601 \cdot 10^{-5}$	$1.51624 \cdot 10^{-6}$
6	999.94016	4199.70382	$3.1221 \cdot 10^{-5}$	$1.47102 \cdot 10^{-6}$
7	999.90151	4197.48579	$4.6032 \cdot 10^{-5}$	$1.42793 \cdot 10^{-6}$
8	999.84827	4195.44641	$6.04165 \cdot 10^{-5}$	$1.38685 \cdot 10^{-6}$
9	999.78086	4193.57514	$7.43968 \cdot 10^{-5}$	$1.34764 \cdot 10^{-6}$
10	999.69967	4191.8618	$8.79937 \cdot 10^{-5}$	$1.31018 \cdot 10^{-6}$
11	999.60508	4190.29657	$1.01227 \cdot 10^{-4}$	$1.27437 \cdot 10^{-6}$
12	999.49745	4188.86998	$1.14117 \cdot 10^{-4}$	$1.24011 \cdot 10^{-6}$
13	999.3771	4187.57294	$1.26679 \cdot 10^{-4}$	$1.2073 \cdot 10^{-6}$
14	999.24437	4186.39671	$1.38932 \cdot 10^{-4}$	$1.17587 \cdot 10^{-6}$
15	999.09957	4185.33292	$1.5089 \cdot 10^{-4}$	$1.14573 \cdot 10^{-6}$
16	998.94297	4184.37356	$1.6257 \cdot 10^{-4}$	$1.1168 \cdot 10^{-6}$
17	998.77486	4183.51098	$1.73984 \cdot 10^{-4}$	$1.08903 \cdot 10^{-6}$
18	998.59551	4182.73788	$1.85147 \cdot 10^{-4}$	$1.06235 \cdot 10^{-6}$
19	998.40516	4182.04734	$1.9607 \cdot 10^{-4}$	$1.03671 \cdot 10^{-6}$
20	998.20405	4181.4328	$2.06765 \cdot 10^{-4}$	$1.01203 \cdot 10^{-6}$
21	997.99242	4180.88805	$2.17242 \cdot 10^{-4}$	$9.88287 \cdot 10^{-7}$
22	997.77048	4180.40726	$2.27513 \cdot 10^{-4}$	$9.65418 \cdot 10^{-7}$
23	997.53844	4179.98494	$2.37586 \cdot 10^{-4}$	$9.43384 \cdot 10^{-7}$
24	997.2965	4179.61598	$2.4747 \cdot 10^{-4}$	$9.22143 \cdot 10^{-7}$
25	997.04486	4179.29562	$2.57174 \cdot 10^{-4}$	$9.01654 \cdot 10^{-7}$
26	996.7837	4179.01948	$2.66705 \cdot 10^{-4}$	$8.81883 \cdot 10^{-7}$
27	996.51319	4178.78351	$2.7607 \cdot 10^{-4}$	$8.62795 \cdot 10^{-7}$
28	996.2335	4178.58405	$2.85276 \cdot 10^{-4}$	$8.44358 \cdot 10^{-7}$
29	995.9448	4178.4178	$2.9433 \cdot 10^{-4}$	$8.26542 \cdot 10^{-7}$
30	995.64725	4178.2818	$3.03237 \cdot 10^{-4}$	$8.09317 \cdot 10^{-7}$
31	995.34099	4178.17348	$3.12002 \cdot 10^{-4}$	$7.92658 \cdot 10^{-7}$
32	995.02618	4178.09061	$3.20631 \cdot 10^{-4}$	$7.76539 \cdot 10^{-7}$
33	994.70295	4178.03133	$3.29129 \cdot 10^{-4}$	$7.60936 \cdot 10^{-7}$
34	994.37144	4177.99415	$3.375 \cdot 10^{-4}$	$7.45828 \cdot 10^{-7}$
35	994.03178	4177.97792	$3.45749 \cdot 10^{-4}$	$7.31191 \cdot 10^{-7}$

The ambient pressure is equally exercising an influence on the density; this can approximately be accounted for by a multiplicative factor according to

$$\rho(\Theta,p) = \rho(\Theta,p=1\text{bar}) \times \frac{1}{1-\Phi\cdot p}\,, \quad p \quad \text{in [bar]}\,, \tag{13.3.3}$$

in which the pressure must be expressed in bar, and the function Φ is given by

$$\begin{aligned}\Phi(\Theta) = 19652.17 + 148.133\,\Theta \;-\; 2.293\,\Theta^2 \\ + 1.256\cdot 10^{-2}\,\Theta^3 \;-\; 4.180\cdot 10^{-5}\,\Theta^4\,.\end{aligned} \tag{13.3.4}$$

A somewhat simpler description, which is adequate for lakes uses instead of salinity the electrical conductivity of water, κ; in terms of this one has

$$\rho(\Theta,\kappa) = \rho(\Theta)\left(1+\beta_\kappa\cdot\kappa_{20}(\Theta)\right)\,, \tag{13.3.5}$$

whereby the conductivity of water is given in $\mu\,\text{S m}^{-1}$ and referred to a standard value of 20°C,

$$\kappa_{20} = \kappa(1.72228 - 0.0541369\Theta + 0.001148420\Theta^2 - 0.0000222651\Theta^3)\,. \tag{13.3.6}$$

In this formula κ is the value of the conductivity measured at the temperature Θ [°C]. IMBODEN & WÜEST [112] give for β_κ the value

$$\beta_\kappa = \frac{1}{\rho}\frac{\partial\rho}{\partial\kappa_{20}} = 0.705\cdot 10^{-4}\mu\,\text{S m}^{-1}\,, \tag{13.3.7}$$

a value appropriate for calcium carbonate at 20°C.

In limnology, the temperature range is between 0°C – 30°C and larger temperatures virtually never arise. For computational purposes the rather complicated formulas given above can then also be replaced by a quadratic equation

$$\rho(\Theta) = \rho_0\left(1-\tilde{\alpha}(\Theta-\Theta_0)^2\right)\,. \tag{13.3.8}$$

The reference temperature is chosen as that of the density maximum, $\Theta_0 =$ 4°C with the corresponding density $\rho_0 = \rho(4°\text{C})$; $\tilde{\alpha}$ has the value

$$\tilde{\alpha} = 6.493\times 10^{-6}\,[\text{K}^{-1}]\,. \tag{13.3.9}$$

If one also needs the coefficient of thermal expansion,

$$\alpha_\Theta = -\frac{1}{\rho}\frac{\partial\rho}{\partial\Theta}\,, \tag{13.3.10}$$

then for (13.3.8) it is given by

$$\alpha_\Theta = \frac{2\tilde{\alpha}(\Theta-\Theta_0)}{1-\tilde{\alpha}(\Theta-\Theta_0)^2} \approx 2\tilde{\alpha}(\Theta-\Theta_0)\,. \tag{13.3.11}$$

Table 13.2 and Fig. 13.15 display the variation of the coefficient of thermal expansion as a function of temperature. One recognizes that the temperature dependence is nearly linear, so that the approximation (13.3.11) is justified.

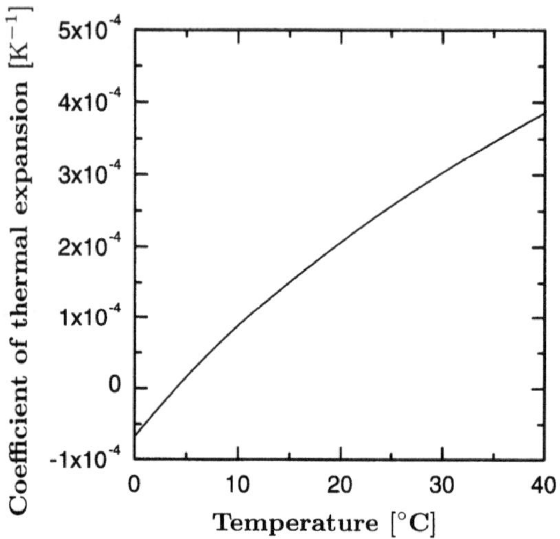

Fig. 13.15. Coefficient of thermal expansion for pure water.

13.3.2 Molecular Viscosity

As a compressible fluid water possesses two viscosities, a shear and a bulk viscosity. The latter is very difficult to determine experimentally and is not reliably known. We shall therefore set it to zero. For this reason the shear viscosity is simply called *molecular viscosity*; it varies slightly with temperature. A representative value for the *dynamic viscosity* is

$$\mu \approx 1.3 \times 10^{-3} [\mathrm{N\,s\,m^{-2}}] , \tag{13.3.12}$$

which, with a density of $1000\,\mathrm{kg\,m^{-3}}$, yields the *kinematic viscosity*

$$\nu = \frac{\mu}{\rho} \approx 1.3 \times 10^{-6} [\mathrm{m^2\,s^{-1}}] . \tag{13.3.13}$$

If one wishes to take the temperature dependence into account, one uses the formula

$$\nu(\Theta) = \frac{1.78 \times 10^{-3}\ [\mathrm{kg\,m^{-1}\,s^{-1}}]}{1 + 0.0337\,\Theta + 0.00022\,\Theta^2} \frac{1}{\rho} , \quad \Theta[^\circ C],\ \rho[\mathrm{kg\,m^{-3}}] . \tag{13.3.14}$$

Table 13.2 and Fig. 13.16 show the dependence of the molecular viscosity with temperature. A dependence on salinity may also be incorporated as follows:

$$\nu(\Theta, S) = (1 - \alpha S)\nu(\Theta) , \quad \alpha = 2.5 . \tag{13.3.15}$$

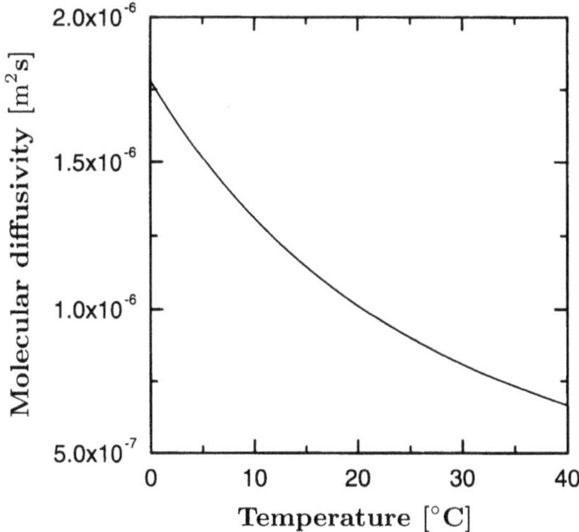

Fig. 13.16. Molecular kinematic viscosity of pure water as a function of temperature.

13.3.3 Specific Heat and Thermal Diffusivity

Further parameters arising in the turbulence model proposed below are the heat capacity and molecular thermal diffusivity. The specific heat at constant pressure, c_p is again weakly temperature dependent (see HENDERSON–SELLERS [100]),

$$c_p = 4190.0 + \exp(46.40 - 0.156\Theta)\ [\mathrm{m}^2\,\mathrm{K}^{-1}\,\mathrm{s}^{-2}]\,,\ \Theta\,[^\circ\,\mathrm{C}]\,, \tag{13.3.16}$$

or when pressure p (depth) [bar] and salinity s [g kg^{-1}] are taken into account

$$\begin{aligned} c_p = {} & 4217.4 - 3.6608\,\Theta + 0.13129\,\Theta^2 \\ & -2.210 \times 10^{-3}\,\Theta^3 + 1.508 \times 10^{-5}\,\Theta^4 \\ & +s\,(-6.616 \times 10^{-2} + 9.28 \times 10^{-3}\,\Theta - 2.39 \times 10^{-5}\,\Theta^2) \\ & +p\,(-4.917 \times 10^{-1} + 1.335 \times 10^{-2}\,\Theta - 2.177 \times 10^{-4}\,\Theta^2) \\ & +s\,p\,3.441 \times 10^{-3} \\ & +p^2\,1.50 \times 10^{-4}\,, \end{aligned} \tag{13.3.17}$$

see CHEN & MILLERO [47]. Table 13.2 lists numerical values and Fig. 13.17 displays the function (13.3.16) graphically. The thermal diffusivity is an order of magnitude smaller than the molecular viscosity; its value is

$$\chi^{(\Theta)} = \frac{\lambda}{\rho c_p} \approx 1.3 \times 10^{-7}\ [\mathrm{m}^2\,\mathrm{s}^{-1}]\,. \tag{13.3.18}$$

In turbulence modelling the molecular diffusivities are taken as constant; so no temperature dependence is accounted for; the reason is that molecular diffusivities are much smaller than their turbulent counterparts.

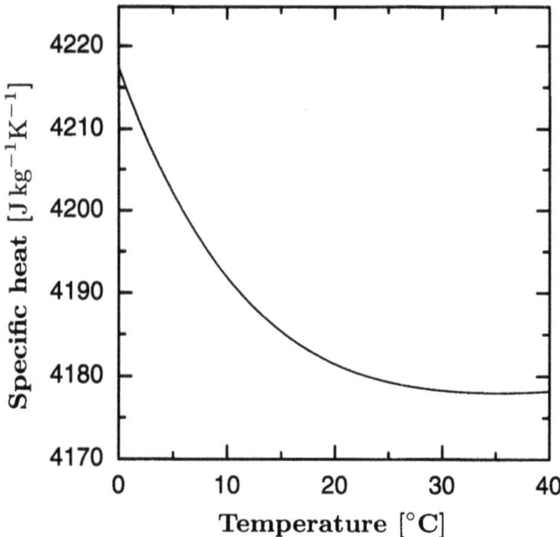

Fig. 13.17. Specific heat at constant pressure for pure water as a function of temperature.

13.4 One-Dimensional Turbulence Models

Now that all input quantities are known and all parameterizations have been given, the k–ε model presented in Chap. 11 can be applied to any lake. Since our intention is the presentation of a vertically one-dimensional k–ε model, special thoughts are necessary about how such a model is being driven. Even though the turbulence equations are one-dimensional, the three-dimensionality of the balances of mass and momentum need to be accounted for to compute the currents in the lake and, consequently, to be able to derive at all a representative form for the vertical profile of the vertical gradient of the horizontal velocity field. This will be more closely analyzed in the next subsection, before we turn to representative computations for this model for Lake Ammer.

13.4.1 Governing Equations and Parameterization of the Pressure Gradient

In order to keep the model simple and practical, we will restrict considerations to processes which are not influenced by boundaries. In other words we will compute a temperature profile that varies only with the depth coordinate, whereby one must assume that the currents are essentially horizontal and the velocities equally only vary with depth. If the horizontal component of the angular velocity of the Earth is then ignored and the x, y, z coordinates are chosen to point towards East, North and vertically upwards, i.e., if one chooses

$$
\begin{aligned}
& u := \langle v_1 \rangle \, , \quad v := \langle v_2 \rangle \, , \quad w := \langle v_3 \rangle \approx 0 \, , \\
& \frac{\partial}{\partial x} = \frac{\partial}{\partial y} \approx 0 \qquad \text{(except for the mean pressure)}, \\
& f := 2|\boldsymbol{\Omega}| \sin\phi \qquad \phi = \text{geogr. latitude},
\end{aligned}
\tag{13.4.1}
$$

then the averaged momentum balances (continuity is automatically satisfied) and the heat equation in the BOUSSINESQ approximation read

$$
\begin{aligned}
\frac{\partial u}{\partial t} - fv &= -\frac{1}{\rho_0}\frac{\partial \langle p\rangle}{\partial x} + \frac{\partial}{\partial z}\left((\nu+\nu_t)\frac{\partial u}{\partial z}\right) , \\
\frac{\partial v}{\partial t} + fu &= -\frac{1}{\rho_0}\frac{\partial \langle p\rangle}{\partial y} + \frac{\partial}{\partial z}\left((\nu+\nu_t)\frac{\partial v}{\partial z}\right) , \\
0 &= \frac{\partial \langle p\rangle}{\partial z} + \langle \rho\rangle g \, , \\
\frac{\partial \Theta}{\partial t} &= \frac{\partial}{\partial z}\left(\left(\chi^{(\Theta)} + \frac{\nu_t}{\sigma_\Theta}\right)\frac{\partial \Theta}{\partial z}\right) + \frac{1}{\rho_0 c_p}\frac{\partial I}{\partial z} ,
\end{aligned}
\tag{13.4.2}
$$

whilst the k–ε equations simplify themselves for the horizontal mean currents as follows:

$$
\begin{aligned}
\frac{\partial k}{\partial t} &= \frac{\partial}{\partial z}\left(\left(\nu + \frac{\nu_t}{\sigma_k}\right)\frac{\partial k}{\partial z}\right) + \frac{c_\mu k^2}{\varepsilon}\left(\left(\frac{\partial u}{\partial z}\right)^2 + \left(\frac{\partial v}{\partial z}\right)^2\right) \\
&\quad - \frac{\langle \rho\rangle \langle \alpha_\Theta\rangle}{\rho_0}\frac{\nu_t}{\sigma_\Theta} g \frac{\partial \Theta}{\partial z} - \varepsilon \, , \\
\frac{\partial \varepsilon}{\partial t} &= \frac{\partial}{\partial z}\left(\left(\nu + \frac{\nu_t}{\sigma_\varepsilon}\right)\frac{\partial \varepsilon}{\partial z}\right) + c_1 k\left(\left(\frac{\partial u}{\partial z}\right)^2 + \left(\frac{\partial v}{\partial z}\right)^2\right) \\
&\quad - c_3 \frac{\langle \rho\rangle \langle \alpha_\Theta\rangle}{\rho_0}\frac{\nu_t}{\sigma_\Theta} g \frac{\partial \Theta}{\partial z}\cdot\frac{\varepsilon}{k} - \frac{c_2 \varepsilon^2}{k} \, .
\end{aligned}
\tag{13.4.3}
$$

It is evident that none of these equations contains any convective terms; these are eliminated because of the assumptions $(13.4.1)_{1-5}$ i.e., that the currents are horizontal and vary only with z. Such terms thus play a role when horizontal gradients of horizontal velocities are important that is in the shore zones and boundary regions of a lake. A two-dimensional model which accounts for these littoral processes is much more elaborate than the procedure followed here both with regard to the equations as well as the numerical solutions.

To solve the above equations, one must determine the horizontal pressure variations and the initiation of the mean currents which are important as driving mechanism of the model. The horizontal derivatives of the velocities must be taken into account only at this place, because without them no circulating current can form at all. This can be corroborated by an exact derivation of the equations followed by a dimensional analysis.

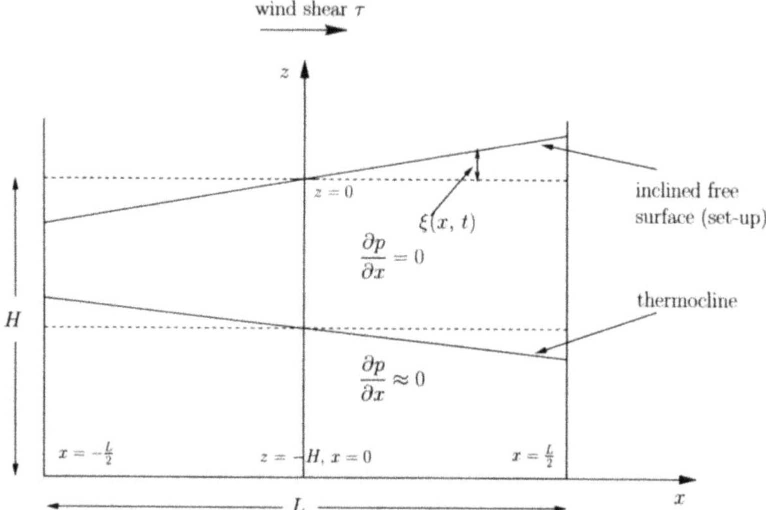

Fig. 13.18. Build-up of a pressure gradient in an enclosed water basin on the basis of wind shear.

The dependence of the pressure upon the horizontal coordinate can, in a lake, easily be envisaged. A wind persistently blowing into the same direction generates in a homogeneous lake a mass transport such that in an enclosed basin the free surface is tilted. The inclination of the free surface then generates a horizontal pressure gradient, see Fig. 13.18. By the mass transport a surface displacement will be initiated. In a simple problem, in which only the hydrostatic pressure is accounted for and the density is constant, we have

$$\langle p \rangle = \rho_0 g \zeta \,. \tag{13.4.4}$$

This set-up will not persist, but an oscillation – a so-called seiche – will be established. With a simple model for a rectangular basin (because of symmetry) the horizontal pressure variation can be derived from it; this pressure variation then enters the momentum balance of the vertically one-dimensional model. Let us now reproduce this computation.

The free oscillations of the free surface of a rectangular basin can quite easily be computed. With the denotations of Fig. 13.18 and the scales indicated there, the displacements of the free surface may be written as

$$\zeta = \zeta_{\max} \cos\left(\frac{2\pi t}{T}\right) \sin\left(\frac{\pi x}{L}\right) , \tag{13.4.5}$$

where T denotes the period of the oscillation (see Exercises). Consider now the position $x = 0$ in the middle of the lake, then the total mass flux is there $\bar{u}H$, where $\bar{u}$ is the velocity averaged over the depth. This mass flux must balance with the corresponding mass change due to the surface displacements, viz.,

$$\bar{u}_{x=0} = \int_0^{L/2} \frac{\partial \zeta}{\partial t}\,\mathrm{d}x = -\frac{2\zeta_{\max} L}{T} \sin\left(\frac{2\pi t}{T}\right). \tag{13.4.6}$$

Thus, the temporal changes of $\partial\langle p\rangle/\partial x$ at $x=0$ can be written as

$$\begin{aligned}\frac{\partial}{\partial t}\left(\frac{\partial\langle p\rangle}{\partial x}\right)_{x=0} &= \frac{\partial}{\partial t}\left(\rho_0 g \frac{\partial \zeta}{\partial x}\right)_{x=0} \\ &= -\rho_0 g \zeta_{\max} \frac{2\pi^2}{LT} \sin\left(\frac{2\pi t}{T}\right) = \rho_0 g \frac{\pi^2 \bar{u} H}{L^2}\,,\end{aligned} \tag{13.4.7}$$

where the expression for the mass flux was used to eliminate the explicit time dependence. For the y-component one proceeds in principle in the same way, or one solves the free oscillation problem in a three dimensional box. This procedure is obviously only a very rough approximation, but such kind of approximations are needed, if one wishes to deduce a one-dimensional model that enjoys any realistic behaviour. However, with it and with a given mass flux (which can be computed from the velocities) and with the mean dimensions of the lake one may compute the horizontal variations and thence solve the equations of the vertically one-dimensional turbulence model.

SVENSSON [231] complements this ansatz by a multiplicative term to account for the fact that, besides the free surface displacement, in particular the thermocline displacement is significant. This factor is parameterized by the temperature difference (density difference) of the upper and the lower layer. Despite the boldness of this parameterization and also the many ad-hoc assumptions of the turbulence equations, this model allows relatively easily the computation of a realistic distribution of the vertical temperature profile; Unfortunately, this is only possible for relatively short times; entire annual cycles can for realistic wind input, etc., not so well be reproduced; this failure, however is not so much the result of the model as such, but rather the empirical parameterization of the wind function in the surface fluxes, which are always only valid in certain ranges of the wind intensity and which certainly change throughout the year.

13.4.2 Example Computation for the One-Dimensional Model

The numerical computation for solutions of the turbulence equations requires, apart form the knowledge of a few constants, which are supposed to be known, also some measured quantities such as wind speed at some altitude above the lake surface, heat fluxes and short wave radiation, humidity and vapour pressure of the atmosphere and pressure and temperature above the lake surface. We now present an example (Lake Ammer) for which the evolution of the temperature is to be demonstrated; we select the period from 1. January 1996 to 31. December 1996. The meteorological input quantities are displayed in Fig. 13.19. Figure 13.20 shows the temperature profiles measured during the

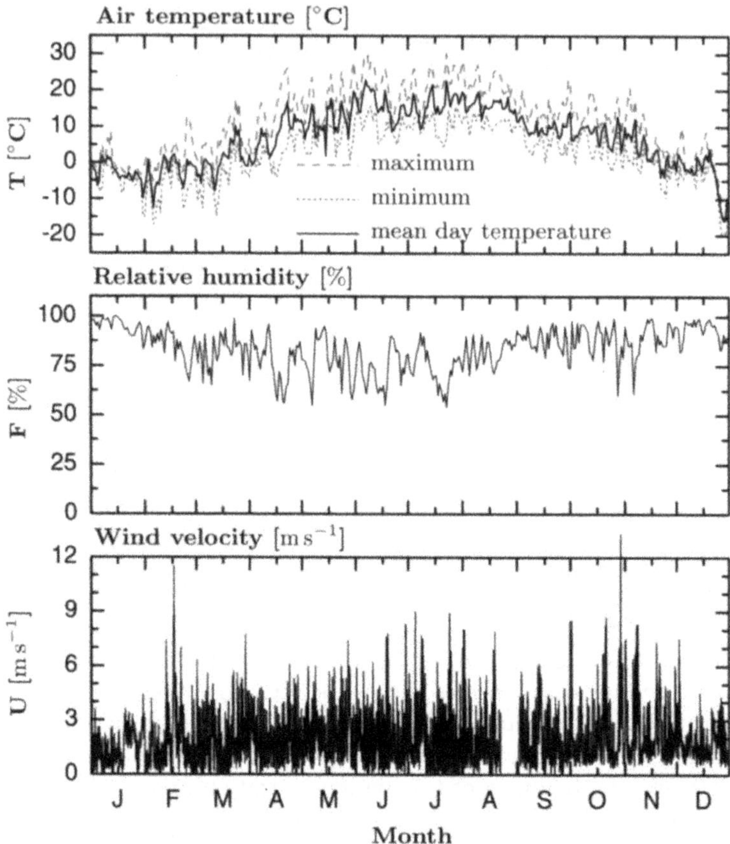

Fig. 13.19. Meteorological input quantities for the simulations in Fig. 13.20 and 13.21. In the panel for the air temperature daily mean values and minima and maxima are shown, the relative humidity is the measured mean value and the wind speed was hourly measured.

year 1996 in comparison with those computed. The corresponding measured profiles have also been used in Fig. 13.1. The simulations, which start on 1. January 1996 with a uniform temperature distribution at 4°C, agree very well with the measured profiles. The seasonal variation of the temperature in the epi- and hypolimnion are well reproduced, only the position of the thermocline is somewhat overestimated.

In order to also visualize the computed turbulence properties we have displayed in Fig. 13.21 the turbulent kinetic energy and the turbulent diffusivity, both in logarithmic representation. It is clearly seen how in spring until mid April the turbulent kinetic energy causes mixing throughout the entire depth. Then this momentum transport stops abruptly and is, during

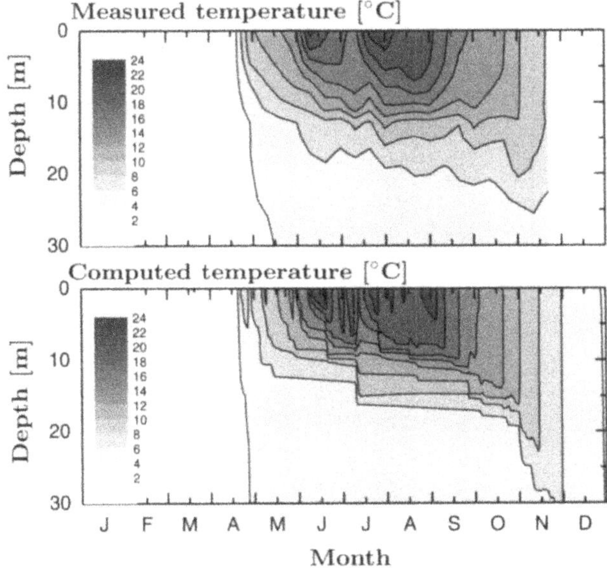

Fig. 13.20. Measured (top) and simulated (bottom) temperature stratification in Lake Ammer for the year 1996.

the stagnation phase, restricted to the epilimnion. Starting in September the intrusion of turbulent energy is initiated again due to the increased turbulence by the stronger winds and the increased cooling of the water; the upper layer depth is growing again, though slower than it was established in the spring formation of the epilimnion.

13.5 Concluding Remarks

The spirit of conceptual approach in this chapter was different than in the previous chapters. Great efforts were devoted to the description of that information which appears to be peripheral to the model: Boundary conditions. To a certain extert this was intentional. Throughout the entire book the focus was the development of model equations, and indeed the content of the book should have made it clear to the reader that rather complicated physical and mathematical thoghts are necessary, so that well justified descriptions of the physical processes emerge.

Here in this chapter the description of the conditions that describe the communication of the system under consideration with the outside world were in focus. It should have made it clear that at least in the geophysical context of the response of a lake to meteorological conditions, the formulation of boundary conditions is of similar complexity as is the derivation of the model

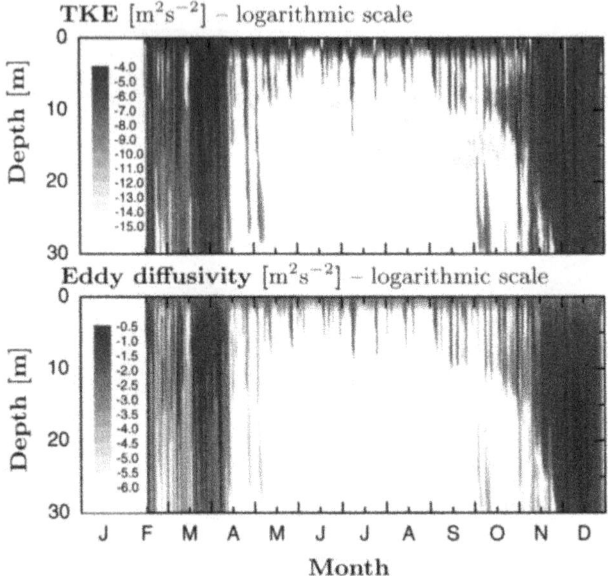

Fig. 13.21. Turbulent kinetic energy (top) and eddy diffusivity (bottom) in logarithmic scale for the simulation in Fig. 13.20.

for which they are used. This is rather typical for most physical systems in a geophysical context and should be borne in mind, not only the field equations describe our continua, conditions at their boundaries are similarly important.

13.6 Exercises

1. Consider a vertically (linearly) stratified fluid layer at rest. Its density is ρ, and $d\rho/dz$ =constant. Isolate a fluid particle at a certain position and displace it in imagination a small distance ζ into the vertical. Assume that there is no interaction of this particle with the fluid surraounding it, except that it is subjected to the gravity force and buoyancy force. Show that when being left free, the particle will either oscillate about its rest position or move away from it indefinitely. Show that the sign of

$$N^2 = -\frac{g}{\rho}\frac{d\rho}{dz} \qquad \left[\mathrm{s}^{-2}\right]$$

determines the particle's stability or instability, and under "stable conditions" the particle oscillates with frequency N.
2. Read EINSTEIN's paper "About electromagnetic radiation".
3. Derive, by using PLANCK's law of radiation, the law of STEFAN–BOLTZMANN on black body radiation.

4. Compute the free oscillations of the surface in a rectangular basin (two-dimensional, x, z coordinates as in Fig. 13.17). Assume that $f = 0$ and that the dissipation can be ignored and the density is constant.

13.7 Solutions

1. Let z be the Cartesian distance, measured against the direction of gravity; let, moreover, ζ be vertical displacement of the fluid particle displaced from its rest position. Because the particle does not interact with the surrounding fluid,it is subjected to the forces

$$\begin{array}{ll} \rho(0)\, g\, V & \text{weight, pointing downwards,} \\ \rho(\zeta)\, g\, V & \text{buoyancy force, pointing upwards.} \end{array} \tag{13.7.1}$$

V is the volume of the particle which does not change while it is displaced, because the particle does not interact with the fluid. Using TAYLOR series expansion,

$$\rho(\zeta) g V = \left(\rho(0) + \frac{d\rho}{dt}(0)\zeta + ...\right) gV \ ,$$

the net force, exected on the particle, is given by

$$\frac{d\rho}{dz}(0)\zeta g V.$$

Applying NEWTON's second law to the particle yields

$$\rho(0) V \ddot{\zeta} = \frac{d\rho}{dt}(0) V g \zeta$$

or

$$\ddot{\zeta} + N^2 \zeta = 0 \ , \tag{13.7.2}$$

where

$$N^2 := -\frac{g}{\rho(0)} \frac{d\rho}{dz}(0) \qquad \left[\mathrm{s}^{-2}\right].$$

N is called the buoyancy frequency or the BRUNT–VIÄSÄLÄ frequency. Equation (13.7.2) is the differential equation of a harmonic oscillation if $N^2 > 0$, but becomes exponential if $N^2 < 0$. In the first case, $(d\rho/dz)(0)$ is negative which is reminiscent of a stable stratification, in the second case, $(d\rho/dz)(0)$ is positive and corresponds to an unstable stratification, because with an exponential solution

$$\zeta = A \exp(Nt) + B \exp(-Nt) \ ,$$

the displacement ζ approaches infinity if only $A \neq 0$.

3. From (13.2.14) one obtains by integration over all frequencies $\nu \in [0, \infty)$

$$Q_{ir} = \int_0^\infty \frac{8\pi h\nu^3}{e^{h\nu/k\Theta} - 1} \mathrm{d}\nu = \frac{8\pi k^4 \Theta^4}{h^3} \underbrace{\int_0^\infty \frac{x^3 \mathrm{d}x}{e^x - 1}}_{I_3} , \tag{13.7.3}$$

where the substitution $x := h\nu/k\Theta$ has been used. Integrals of the sort I_l (here $l = 3$) cannot be solved by elementary methods, but they are expandable as series of RIEMANN ζ-functions,

$$\zeta(l) = \sum_{n=1}^\infty \frac{1}{n^l} . \tag{13.7.4}$$

Transforming I_l further, i.e.,

$$\begin{aligned} I_l &= \int_0^\infty \frac{x^l \mathrm{d}x}{e^x - 1} = \int_0^\infty \sum_{n=1}^\infty x^l e^{-nx} \mathrm{d}x \\ &= \sum_{n=1}^\infty \int_0^\infty x^l e^{-nx} \mathrm{d}x = \sum_{n=1}^\infty \frac{l!}{n^{l+1}} , \end{aligned} \tag{13.7.5}$$

one obtains

$$I_l = l!\zeta(l+1) = \Gamma(l+1)\zeta(l+1) , \tag{13.7.6}$$

where Γ is the *Gamma function* so that (13.7.6) also holds for non-integer l. Therefore, the determination of the BOLTZMANN constant is reduced to the computation of $\Gamma(4) = 3! = 6$ and $\zeta(4) = \pi^4/90$, so that

$$\sigma_{SB} = \frac{8\pi k^4}{h^3} 6 \times \frac{\pi^4}{90} = 5.671 \times 10^{-8} \quad \left[\mathrm{Wm^{-2}K^{-4}}\right] ,$$

and

$$Q_{ir} = \sigma_{SB} \Theta^4 . \tag{13.7.7}$$

The radiative power is therefore proportional to the fourth power of the temperature.

4. Ignoring the CORIOLIS and dissipative effects the governing equations read

$$\begin{aligned} \frac{\partial u}{\partial t} &= -\frac{1}{\rho} \frac{\partial \langle p \rangle}{\partial x} , \\ \frac{\partial \langle p \rangle}{\partial z} &= -\rho g \quad \text{and} \quad \frac{\partial u}{\partial x} + \frac{\partial w}{\partial z} = 0 . \end{aligned} \tag{13.7.8}$$

Integrating the hydrostatic equation yields

$$\langle p \rangle = \int_z^\zeta \rho g \, \mathrm{d}z = -\rho g(\zeta - z) . \tag{13.7.9}$$

The boundary condition at the free surface ($s = z - \zeta(x,t)$) yields

$$\frac{\mathrm{d}s}{\mathrm{d}t} = w - \frac{\partial \zeta}{\partial t} = 0 \tag{13.7.10}$$

and together with the depth integrated continuity equation

$$(H+\zeta)\frac{\partial u}{\partial x} - \underbrace{w(z=-H)}_{=0} + w(z=\zeta) = 0 \tag{13.7.11}$$

one obtains with ($H \gg \zeta$)

$$\frac{\partial \zeta}{\partial t} = -H\frac{\partial u}{\partial x}, \quad \frac{\partial u}{\partial t} = -g\frac{\partial \zeta}{\partial x} . \tag{13.7.12}$$

The surface displacement follows from the equation

$$\frac{\partial^2 \zeta}{\partial t^2} = H\,g\frac{\partial^2 \zeta}{\partial x^2}, \quad \frac{\partial \zeta}{\partial x} = 0 \quad \text{for} \quad x = \pm\frac{L}{2}, \tag{13.7.13}$$

via the separation ansatz

$$\zeta(x,t) = C(x)\cdot \Phi(t) . \tag{13.7.14}$$

With the separation constant λ this becomes

$$\frac{\mathrm{d}^2 \Phi}{\mathrm{d}t^2} = -Hg\lambda^2\Phi \quad \text{and} \quad \frac{\mathrm{d}^2 C}{\mathrm{d}x^2} = -\lambda^2 C . \tag{13.7.15}$$

The boundary conditions are

$$\frac{\mathrm{d}C}{\mathrm{d}x}\left(-\frac{L}{2}\right) = \frac{\mathrm{d}C}{\mathrm{d}x}\left(+\frac{L}{2}\right) = 0 . \tag{13.7.16}$$

So, the horizontal dependence of C has the general form

$$C(x) = a\cos(\lambda x) + b\sin(\lambda x) , \tag{13.7.17}$$

and the boundary conditions lead to

$$a = 0 \quad \text{and} \quad \lambda = \frac{2n-1}{L}\pi, \quad n \in \mathbb{N}^+ . \tag{13.7.18}$$

For $\Phi(t)$ one obtains

$$\Phi(t) = \cos\left(\frac{2\pi}{T_n}t\right) \tag{13.7.19}$$

with the eigenperiod

$$T_n = \frac{2L}{\sqrt{gH}}\frac{1}{2n-1} . \tag{13.7.20}$$

The fundamental oscillation for $n = 1$ has the form of a semi cosine wave and reads

$$\zeta^{(1)} = \zeta_{\max}\cdot\sin\left(\frac{\pi x}{L}\right)\cdot\cos\left(\frac{2L}{\sqrt{gH}}t\right) . \tag{13.7.21}$$

References

1. D.J. Acheson: *Elementary Fluid Dynamics*. Oxford Appl. Math. and Comp. Sci. Series (Clarendon Press, Oxford 1994, 397p.)
2. G. Ahmadi: On the $k-\varepsilon$ model of turbulence. *Int. J. Eng. Sci.*, **23**(8), 847–856 (1985a)
3. G. Ahmadi: Thermodynamics of multi-temperature fluids with applications to turbulence. *Appl. Math. Modelling*, **9**, 271–274 (1985b)
4. G. Ahmadi: On material frame indifference of turbulence closure models. *Geophys. Astrophys. Fluid Dyn.* **38**, 131–144 (1987)
5. G. Ahmadi: Thermodynamically consistent models for compressible turbulent flows. *Appl. Math. Modelling* **12**, 391–398 (1988)
6. G. Ahmadi: A two-equation turbulence model for compressible flows based on the second law of thermodynamics. *J. Non-equil. Thermodyn.* **14**, 45–49 (1989)
7. G. Ahmadi: A thermodynamically consistent rate-dependent model for turbulence. Part I: Formulation. *Int. J. Non-lin. Mech.* **26**, 595–607 (1991a)
8. G. Ahmadi, S. Abu-Zaid: A thermodynamically consistent stress transport model for rotating turbulent flows. *Geophys. Astrophys. Fluid Dyn.* **61**, 109–125 (1991b)
9. G. Ahmadi: On thermodynamics of turbulence. *Iranian J. Sci. Technology*, **15**, 67–84 (1991c)
10. G. Ahmadi, S.J. Chowdhury: A rate dependent algebraic stress model for turbulence. *Appl. Math. Modellling.* **15**, 516–524 (1991d)
11. H.-D. Alber: *Materials with Memory. Initial-Boundary Value Problems for Constitutive Equations with Internal Variables* (Springer, Berlin 1998, 166p.)
12. J. Altenbach, H. Altenbach: *Einführung in die Kontinuumsmechanik* (B. G. Teubncr, Stuttgart 1994, 285p.)
13. T. Alts: *Thermodynamik elastischer Körper mit thermo-kinematischen Zwangsbedingungen – fadenverstärkte Materialien.* Habilitationsschrift, Universitätsbibliothek der TU Berlin, ISBN 37983 6674 5 (1979, 245p.)
14. T. Alts, K. Hutter: Continuum description of the dynamics and thermodynamics of phase boundaries between ice and water, Part I. Surface balance laws and their interpretation in terms of three-dimensional balance laws averaged over the phase change boundary layer. *J. Non-Equilibrium Thermodyn.* **13**, 224–257 (1988a)
15. T. Alts and K. Hutter: Continuum description and thermodynamics of phase boundaries between ice and water. Part II: Thermodynamics. *J. Non-Equilibrium Thermodyn.* **13**, 259–280 (1988b)
16. T. Alts, K. Hutter: Continuum description and thermodynamics of phase boundaries between ice and water. Part III: Thermostatics and its consequences. *J. Non-Equilibrium Thermodyn.* **13**, 301–329 (1988c)

17. T. Alts and K. Hutter: Continuum description and thermodynamics of phase boundaries between ice and water. Part IV: On thermostatic stability and well posedness. *J. Non-Equilibrium Thermodyn.* **14**, 1–22 (1989)
18. S.S. Atatürk and K.B. Katsaros: Wind stress and surface waves observed on Lake Washington. *J. Phys. Oceanogr.* **29**, 633–650 (1999)
19. G.I. Barenblatt: *Dimensional Analysis* (Gordon and Breach Science Publishers, New York etc. 1987, 135p.)
20. G.I. Barenblatt, A.J. Chorin, V.M. Prostokishin: Self-similar intermediate structures in turbulent boundary layers at large Reynolds numbers. *Journal of Fluid Mechanics* **410**, 263–283 (2000)
21. G.K. Batchelor: *The Theory of Homogeneous Turbulence* (Cambridge University Press, Cambridge 1953, 197p.)
22. G. Bauer: *Thermodynamische Betrachtung einer gesättigten Mischung.* Doctoral dissertation, Department of Mechanics, Darmstadt University of Technology (1997, 126p.)
23. J. Bear: *Dynamics of Fluids in Porous Media* (Dover Publications, Inc. New York 1988, 764p.)
24. E. Becker, W. Bürger: *Kontinuumsmechanik* (B. G. Teubner, Stuttgart 1975, 228p.)
25. J. Betten: *Tensorrechnung für Ingenieure* (B. G. Teubner, Stuttgart 1987, 320p.)
26. M.A. Biot: General theory of three dimensional consolidation. *J. Appl. Phys.* **12**, 155–164 (1941)
27. M.A. Biot: Theory of deformation of a porous viscoelastic anisotropic solid. *J. Appl. Phys.* **27**, 459–467 (1956)
28. H.D. Block: *Introduction to Tensor Analysis* (Charles E. Merrill Books Inc. Columbers, Ohio 1962, 68p.)
29. J. Bluhm: *A Consistent Model for Saturated and Empty Porous Media.* Forschungsberichte aus dem Fachbereich Bauwesen, 74, Universität GH Essen (1997, 133p.)
30. J. Bluhm: *Constitutive relations for thermoelastic porous solids within the framework of finite deformations,* Report Mech. 99/1 FB 10/Mechanik, Universität–GH Essen (1999)
31. R. de Boer: *Vektor- und Tensorrechnung für Ingenieure* (Springer-Verlag, Berlin, etc. (1982, 1260p.)
32. R. de Boer: Highlights in the historical development of the porous media theory - toward a consistent macroscopic theory. *Appl. Mech. Rev.* **49**, 201–262 (1996)
33. R. de Boer: *Theory of Porous Media. Highlights in the Historical Development and Current State* (Springer, Berlin etc. 2000, 618p.)
34. J. Boussinesq: *Essai sur la théorie des eaux courantes.* Mémoires, présentés par div. Savants á l'Académie des Sciences de l'Institut de France. 1877, Tome 23 (with Suppl. in Vol.24)
35. R.M. Bowen: Incompressible porous media models by use of the theory of mixtures. *Int. J. Engn. Sci.* **18**, 1129–1148 (1980)
36. R.M. Bowen: Compressible porous media models by use of the theory of mixtures *Int. J. Engn. Sci.* **20**, 697–763 (1982)
37. R.M. Bowen: *Introduction to Continuum Mechanics for Engineers* (Plenum Press 1989, 261p.)
38. R.M. Bowen, C.-C. Wang: *Introduction to Vectors and Tensors, Vol. 1: Linear and MultillinearAlgebra, Vol. 2: Vector and Tensor Analysis* (Plenum Press 1976, 434p.)

39. E. Buckingham: On physically similar systems; Illustrations of the use of dimensional equations. *Phys. Rev.* **4**(4), 345–376 (1914)
40. H. Bührer and H.A. Ambühl: Die Einleitung von gereinigtem Abwasser in Seen. *Schweiz. Z. Hydrol.* **37**, 374–369 (1975)
41. C. Cabon, J.F. Scott: Linear and non-linear models of anisotropic turbulence, *Annual Rev. Fluid Mech.* **31**, 1–53 (1999)
42. C. Carathéodory: Untersuchungen über die Grundlagen der Thermodynamik. *Mathematische Annalen* **67**, 355–386 (1909)
43. D.E. Carlson: On some new results in dimensional analysis. *Arch. Rational Mech. Anal.* **68**, 191–220 (1978)
44. D.E. Carlson: A mathematical theory of physical units, dimensions and measures. *Arch. Rational Mech. Anal.* **70**, 289–305 (1979)
45. S. Chandrasekhar: *Hydrodynamic and Hydromagnetic Stability* (Dover Publications. Inc. New York 1981, 643p.)
46. P. Chadwick: *Continuum Mechanics, Concise Theory and Problems* (George Atten & Unwin Ltd., 1976, 174p.; also Dover, New York 1999, 187p.)
47. C.T. Chen and F.J. Millero: Precise thermodynamic properties for natural waters covering only the limnological range. *Limnol. Oceanogr.* **31**, 657–662 (1986)
48. S.J. Chowdhury, G. Ahmadi: A thermodynamically consistent rate-dependent model for turbulence, Part I. Theory, Part II: Computational results. *Int. J. Non-Linear Mech.*, **27**, 705–718 (1992)
49. C.F. Colebrook: Turbulent flow in pipes with particular reference to the transition region between the smooth and rough pipe laws. *J. Institution of Civ. Engrs.* London **11**, 133–136 (1938–39)
50. B.D. Coleman, W.Noll: The thermodynamics of elastic materials with heat conduction and viscosity, Arch. *Rational Mech. Anal.* **13**, 167–178 (1963)
51. B.D. Coleman: Thermodynamics of materials with memory. *Arch. Rational Mech. Anal.* **17**, 1–46 (1964)
52. O. Coussy: *Mechanics of Porous Continua* (John Wiley & Sons, Chichester, etc. 1995, 455p.)
53. T.L. Craft, B.E. Launder, A. Suga: A non-linear eddy viscosity model including sensitivity to stress anisotropy. *Proc. of 10th Symp. on Turbulent Shear Flow, The Pennsylvania State Univ., USA,* August 14–16 (1995) Paper N.23–4
54. T.L. Craft, B.E. Launder, A. Suga: A development and application of a cubic eddy-viscosity model of turbulence. *Int. J. of Heat and Fluid Flow* **17**(2), 108–115 (1996)
55. T.L. Craft, B.E. Launder, A. Suga: The prediction of turbulence transitional phenomena with a non-linear eddy viscosity model. *Int. J. of Heat and Fluid Flow* **18**, 15–28 (1997)
56. F. dell'Isola, P. Seppecher: Edge contact forces and quasi-balanced power. *Mechanica* **32**, 33–52 (1997)
57. D.A. Drew, S.L. Passman: *Theory of multicomponent fluids.* Applied Mathematical Sciences, Vol. 135l. (Springer, New York etc. 1998, 308p.)
58. J.E. Dunn, R.L. Fosdick: Thermodynamics stability and boundedness of fluids of complexity 2 and fluids of second grade. *Arch. Rat. Mech. Anal.* **56**, 191–252 (1974)
59. P.A. Durbin: Separated flow computations with $k - \varepsilon - \boldsymbol{v}^2$ - model. *AIAA J.* **33**(4), 659–664 (1995a)
60. P.A. Durbin: *Turbulence modelling for non-equilibrium flows.* Annual Research Briefs, Center for Turbulence Research, Stanford Univ., USA (1995b)

61. W. Ehlers: *Poröse Medien – Ein kontinuumsmechanisches Model auf der Basis der Mischungstheorie*. Forschungsberichte aus dem Fachbereich Bauwesen, 47, Universität GH Essen (1989, 332p.)
62. W. Ehlers: Constitutive equations for granular materials in geomechanical context. In: *Continuum Mechanics in Environmental Sciences and Geophysics. CISM Courses and Lectures No. 337*, ed. by K. Hutter (Springer, Vienna, etc. 1993) pp. 313–402.
63. H. Ehrentraut: Kontinuumsmechanik anisotroper Festkörper und Fluide. Habilitationsschrift, FB Mechanik, TU Darmstadt, (2002, 237p.)
64. A. Einstein: *Physikalische Zeitschrift*, **18**, 121 (1917)
65. J.L. Ericksen: Tensor fields, Appendix to "The Classical Field Theories" . In: *Encyclopedia of Physics,* Vol. III/1 ed. by S. Flügge (Springer, Berlin, etc. 1960), pp. 794–858
66. A.C. Eringen: *Nonlinear Theory of Continuous Media* (McGraw-Hill, Book Company, New York 1962, 477p.)
67. A.C. Eringen: *Mechanics of Continua* (John Wiley and Sons 1967, 502p.)
68. H. Ertel: Ein neuer hydrodynamischer Wirbelsatz. *Meteorologische Zeitschrift* **59**(9), 277–281 (1942)
69. L. Euler: Découverte d'un nouveau principe de mécanique. *Mém. Acad. Sci.* Berlin **6**, 185–217 (1750) or *Opera omnia* **5**(2), 81–108 (1752)
70. L. Euler: Theoria Motus Corporum Solidorum seu Rigidorum ex Primis nostrae Cognitionis Principiis Stabilita et ad Omnis Motus, qui in hujusmodi Corpora Cadere Possunt Accomodata. Rostock = Opera omnia (2) **3** and **4**, 3–293 (1765)
71. G. Farkas: A Fourier-file mechanikai etv alkalmazai. *Mathematikai* és Természettundományi, Ertesitö. **12**, 457–472 (1894)
72. R.P. Feynman, R.B. Leighton and M. Sand: *The Feynman Lectures on Physics* Vol. 1 (Addison-Wesley Publishing company 1966)
73. I. Findenegg: Die Schichtungsverhältnisse im Wörthersee. *Arch. Hydrobiol.* **24**, 253–262 (1932)
74. I. Findenegg: Zur Naturgeschichte des Wörthersees. *Carinthia II, Sonderheft* **2**, 1–63 (1933a)
75. I. Findenegg: Alpenseen ohne Vollzirkulation, *Int. Rev. Hydrobiol.* **28**, 295–311 (1933b)
76. F.A. Forel, Le Léman: monographie limnologique, (F. Rouge, Librairée de l'Université de Lausanne Vol. 1, 1892, 539p.; Vol. 2, 1895, 651p.; Vol. 3, 1904, 715p.)
77. W.C. Forsythe, E.J. Rykiel Jr., R.S. Stahl, H. Wu and R.M. Schoolfield: A model comparison for daylength as a function of latitude and day of year. *Ecol. Modelling*, **80**, 87–95 (1995)
78. R.L. Fosdick, K.R. Rajagopal: Thermodynamics and instability of fluids of third grade. *Proc. Roy. Soc. Lond.* **339**, 351–377 (1980)
79. E. Fried: Energy release, friction and supplemental relations at phase interfaces. *Continuum Mech. Thermodyn.* **7**, 111–121 (1995)
80. U. Frisch: *Turbulence, the Legacy of A. N. Kolmogorov* (Cambridge University Press, Cambridge, etc. 1995, 296p.)
81. Y.C. Fung: *A First Course in Continuum Mechanics* (Prentice Hall 1977, 340p.)
82. T.B. Gatski, C.G. Speziale: On explicit algebraic stress models for complex turbulent flows. *J. Fluid Mech.* **254**, 59–78 (1993)
83. H. Giesekus: *Phänomenologische Rheologie* (Springer, Berlin etc. 1994, 659p.)
84. H. Görtler: *Dimensionsanalyse* (Springer, Berlin etc. 1975, 247p.)

85. M.A. Goodman, S.C. Cowin: A continuum theory of granular materials. *Arch. Rational Mech Anal.* **44**, 249–266 (1972)
86. W.G. Gray: Elements of a systematic procedure for the derivation of macroscale conservation equations for multiphase flows in porous media. In: *Kinematic and Continuum Theories of Granular and Porous Media. CISM Courses and Lectures No. 400*, ed. by K. Hutter, (Springer, Berlin 1999) pp. 67–129
87. A.E. Green, P.M. Naghdi and J.A. Trapp: Thermodynamics of a continuum with internal constraints. *Int. J. Engn. Sci.* **8**, 891-901 (1971)
88. R. Greve: Kontinuumsmechanik. Ein Grundkurs (Springer, Berlin etc. 2003, 302p.)
89. M.E. Gurtin: On the thermodynamics of chemically reacting fluid mixtures. *Arch. Rational Mech. Anal.* **43**, 198–212 (1971)
90. M.E. Gurtin and P.P. Gindugli: The thermodynamics of constrained materials. Arch. Rat. Mech. Anal., **51**, 192–208 (1973)
91. M.E. Gurtin: *An Introduction to Continuum Mechanics* (Academic Press 1981, 265p.)
92. M.E. Gurtin: *Thermomechanics of Evolving Phase Boundaries in the Plane* (Oxford University Press, Oxford 1993, 148p.)
93. M. Hallbäck, A.V. Johansson, A.D. Burden: The basics of turbulence modeling. In: *Turbulence and Transition Modeling*, eds. by Hallbäck et al. (Grademic Publisher, Holland 1996)
94. K. Hanjalic and B.-E. Launder: A Reynolds stress model of turbulence and its application to thin shear flows. *J. Fluid Mech.* **52**, 609–638 (1972)
95. S.M. Hassaniazadeh, W.G. Gray: General conservation equations for multiphase systems: 1. Averaging procedure. *Advances in Water Resources* **2**, 25–40 (1979a)
96. M. Hassaniazadeh and W. G. Gray: General conservation equations for multiphase systems: 1. Averaging procedure. *Adv. Water Res.*, **2**, 131–144 (1979a)
97. S.M. Hassaniazadeh, W.G. Gray: General conservation equations for multiphase systems: 2. Mass, momenta, energy and entropy equations. *Advances in Water Resources* **2**, 191–208 (1979b)
98. P. Haupt: *Continuum Mechanics and Theory of Materials* (Springer, Berlin etc. 2000, 583p.)
99. R.A. Hauser, N. Kirchner: A historical note on the entropy principle of Müller and Liu. *Continuum Mech. Thermodyn.* **14**, 223–226 (2002)
100. B. Henderson-Sellers: Engineering Limnology (Pitman Boston, 1–265, 1984, 356p.)
101. J. O. Hinze: *Turbulence* (McGraw Hill, New York 1975, 790p.)
102. L.N. Howard: Note on a Paper of John W. Miles. *J. Fluid Mech.* **10**, 509–512 (1961)
103. G.E. Hutchinson and H. Löffler: The thermal classification of lakes. *Proc. Natl. Acad. Sci.* **42**, 84–86 (1956)
104. K. Hutter: The foundations of thermodynamics, its basic postulates and implications. A review article. *Acta Mech.* **27**, 1–54 (1977)
105. K. Hutter: *Theoretical Glaciology, Material Science of Ice and the Mechanics of Glaciers and Ice Sheets.* (D. Reidel Publ. Comp. Dordrecht, etc. 1983, 510p.)
106. K. Hutter (Hrsg.): *Die Anfänge der Mechanik. Newtons Principia gedeutet aus ihrer Zeit und ihrer Wirkung auf die Physisk* (Springer, Berlin, etc. 1989, 98p.)
107. K. Hutter (ed.): *Continuum Mechanics in Environmental Sciences and Geophysics. CISM Courses and Lectures No. 337* (Springer, Vienna, etc. 1993a, 522p.)

108. K. Hutter: Thermomechanically coupled ice sheet response: Cold, polythermal, temperate. *J. Glaciology* **99** (131), 65–86 (1993b)
109. K. Hutter: *Fluid und Thermodynamik, eine Einführung* (Springer, Berlin, etc. 1995), second ed.(2002, 445p.)
110. K. Hutter, K. Wilmanski (eds.): *Kinetic and Continuum Theories of Granular and Porous Media. CISM Courses and Lectures No. 400* (Springer, Vienna, etc. 1999, 308p.)
111. K. Hutter, Y. Wang: Phenomenological Thermodynamics and entropy principles. In: *Entropy*, eds. by A. Greven, G. Keller & G. Warnecke (Princeton University press, 2003, pp. 57–78)
112. D.M. Imboden and A. Wüest: Mixing mechanisms in lakes. In: A. Lerman, D.M. Imboden, J.R. Gat (Eds.): Physics and Chemistry of Lakes (Springer-Verlag, Berlin, etc., 1995), pp.83–138
113. W. Jaunzemis: *Continuum Mechanics* (The McMillan Company. 1967, 604p.)
114. W. P. Jones and B. E, Launder: The prediction of laminarization with a two-equation model of turbulence. *Int. J. Heat Mass Transfer* **15**, 301–314 (1972)
115. D.D. Joseph: *Stability of fluid motions I and II.* Tracts in Natural Philosophy Vols. 27 & 28. Springer Verlag, New York etc, (1976, Vol. 1 : 282p., Vol. 2 : 274p.)
116. D. Jou, J. Casas-Vasquez, G. Lebon: *Extended Irreversible Thermodynamics.* 2nd. edn. (Springer, Berlin 1996, 383p.)
117. J.P. Joule: On the existence of an equivalent relation between heat and the ordinary forms of mechanical power. *Phil. Mag.* **27**(3) 205–207 (1845)
118. J. Kirk: Light and photosynthesis in aquatic ecosystems (Cambridge University press, Cambridge 1983, 401p.)
119. C. Kittel: Introduction to Solid State Physics (3rd edition) (John Wiley and Sons, Inc. New York, etc. 1968, 744p.)
120. E. Klingbeil: *Tensorrechnung für Ingenieure* (B. I. Wissenschaftsverlag, 1989, 197p.)
121. A.N. Kolmogorov: The local structure of turbulence in incompressible viscous fluid for very large Reynolds number, *Dokl. Akad. Nauk SSSR*, **30**. 9-13 (1971) [reprinted in *Proc. R. Soc. London*, **A434**, 9–13 (1991)]
122. B. Kamb: The thermodynamic theory of non-hydrostatically stressed solids. *J. Geophys. Res.* **66**, 259–271 (1961)
123. W. Kosiński: *Field Singularities and Wave Analysis in Continuum Mechanics.* (Ellis Horwood Ltd. Publ. Chichester 1986, 251p.)
124. W. Kosiński, A.I. Murdoch (eds.): *Modelling Macroscopic Phenomena at Liquid Boundaries. CISM Courses and Lectures 318* (Springer, Vienna etc. 1991, 288p.)
125. H.L. Langhaar: *Dimensional Analysis and Model Theory* (John Wiley and Sons, New York 1964, 166p.)
126. B. E. Launder and D. B. Spalding: The numerical computation of turbulent flow. *Comp. Meth. Appl. Mech. Eng.* **3**, 269–288 (1974)
127. B.E. Launder, G.J. Reece, W. Rodi: Progress in the development of a Reynolds stress turbulence closure. *J. Fluid Mech.* **68**, 537–566 (1975)
128. B.E. Launder: Advanced turbulence models for industrial applications. In: Turbulence and Transition Modelling, (Hallbäck et al. (ed.)) *Gradenic Publ. Holland:* 193–231 (1996)
129. P. Le Blond, L.A. Mysak: *Waves in the Ocean*, (Elsevier Scientific Publ. Company, Amsterdam 1978, 602p.)
130. G. Lebon, M.S. Boukary: Objectivity, kinetic theory and extended irreversible thermodynamics. *Int. J. Engng. Sci.* **26**(5), 471–483 (1988)

131. G. Lebon, P.C. Dauby, A. Palumbo, G. Valenti: Rheological properties of dilute polymer solutions: An extended thermodynamic approach. *Rheologica Acta* **29**, 127–136 (1990)
132. D.C. Leigh: *Nonlinear Continuum Mechanics* (McGraw-Hill, 1968, 240p.)
133. M. Lesieur: *Turbulence in Fluids.* 2nd ed. (Kluwer, Dordrecht 1990, 411p.)
134. F.M. Leslie: Some constitutive equations for liquid crystal. *Arch. Rat. Mech. Anal.* **28**, 265–283 (1968)
135. R.W. Lewis, B.A. Schrefler: *The Finite Element Method in the Consolidation and Deformation of Porous Media* (J. Wiley, Chichester 1998, 492p.)
136. I-Shih Liu: Method of Lagrange multipliers for exploitation of the entropy principle. *Arch. Rational Mech. Anal.* **46**, 131–148 (1972)
137. I-Shih Liu: *Introduction to Continuum Mechanics* (Springer-Verlag, Berlin, etc., 2002, 297p.)
138. I-Shih Liu: *Continuum Mechanics* (Springer, Berlin etc, 2002, 312p.)
139. A.E. H. Love: *A Treatise on the Mathematical Theory of Elasticity* (4th edition) (Dover Publications, New York (1927, 643p.)
140. J.L. Lumley: *Stochastic Tools in Turbulence* (Academic Press, New York, 1970, 194p.)
141. J.L. Lumley: Computational modeling of turbulence flows. *Advances in Applied Mechanics* **18**, 123–176 (1978)
142. J.L. Lumley: Turbulence modeling. *J. Appl. Mech. Trans. ASME* **50**, 1097–1103 (1983)
143. L.E. Malvern: *Introduction to the Mechanics of a Continuous Medium* (Prentice Hall 1969, 713p.)
144. N.M. Mansour, J. Kim, P. Moin: Reynolds stress and dissipation rate budgets in turbulent channel flow. *J. Fluid Mech.* **194**, 15–44 (1988)
145. J.E. Marsden, T.J.R. Hughes: *Mathematical Foundations of Elasticity.* (Prentice Hall 1983, 556p.)
146. J.E. Marsden, M.J. Hoffmann: *Elementary Classical Analysis* (W. H. Freeman & Company 1993, 738p.)
147. J.S. Marshall, P.M. Naghdi: Thermodynamical theory of turbulence: I. Basic developments. *Phil. Trans. R. Soc. Lond.* **A327**, 415–448 (1989)
148. J.S. Marshall, P.M. Naghdi: A thermodynamical theory of turbulence: II. Determination of Constitutive coefficients and illustrative examples. *Phil. Trans. R. Soc. Lond.* **A327**, 449–475 (1989)
149. J.S. Marshall, P.M. Naghdi: Consequences of the second law for a turbulent fluid flow. *Continuum Mech. Thermodyn.* **3**, 65–77 (1991)
150. D.E. Martin and D Imboden: Thermische Energieflüsse an der Wasseroberfläche: Beispiel Sempachersee. *Schweiz. Z. Hydrol.* **48**, 196-228 (1986)
151. G.A. Maugin: Internal variables and dissipative structures. *J. Non-Equilib. Thermodyn.* **5**, 173–192 (1990)
152. J.R. Mayer: Bemerkungen über die Kräfte der unbelebten Natur. *Liebigs Ann. Chem.* **42**, 233–240 (1842)
153. W. D. McComb: *The Physics of Fluid Turbulence* (Clarendon Press, Oxford 1990, 572p.)
154. J.W. Miles: On the stability of heterogeneous shear flows. *J. Fluid Mech.* **10**, 496–508 (1961)
155. H. Minkowski: *Geometrie der Zahlen* (Teubner, Leipzig 1896, 256p.)
156. A. S. Monin and A. M Yaglom: *Statistical Fluid Mechanics*, Vol 1 ed. J. Lumley (MIT Press, Cambridge, Mass. 1971, 769p.)
157. A. S. Monin and A. M Yaglom: *Statistical Fluid Mechanics*, Vol. 2 ed. J. Lumley (MIT Press, Cambridge, Mass. 1975, 874p.)

158. L.F. Moody: Friction factors for pipe flow. *Trans. Amer. Soc. Mech. Engrs.* **66**, 671–684 (1944)
159. I. Müller: *Zur Ausbreitungsgeschwindigkeit von Störungen in kontinuierlichen Medien.* Dissertation, Techn. Hochschule Aachen (1966, 111p.)
160. I. Müller: Zum Paradox der Wärmeleitungstheorie. *Zeitschrift für Physik* **198**, 329–344 (1967)
161. I. Müller: A thermodynamic theory of mixtures of fluids. *Arch. Rational. Mech. Anal.* **28**, 1–39 (1968)
162. I. Müller: On the frame dependence of stress and heat flux. *Arch. Rat. Mech. Anal.* **45**, 241–250 (1972)
163. I. Müller: *Thermodynamik. Die Grundlagen der Materialtheorie* (Bertelsmann Universitätsverlag, 1973, 232p.)
164. I. Müller: On the frame dependence of electric current and heat flux in a metal. *Acta Mech.* **24**, 117–128 (1976)
165. I. Müller: *Thermodynamics* (Pitman 1985, 521p.)
166. I. Müller, K. Wilmansky: Extended thermodynamics of non-Newtonian fluid. *Rheol. Acta* **25**, 335–349 (1986)
167. I. Müller, T. Ruggeri: *Rational Extended Thermodynamics*, 2nd edn. (Springer, Berlin, etc. 1998, 396p.)
168. W. Muschik: Fundamentals of non-equilibrium thermodynamics. In: *Nonequilibrium thermodynamics with applications to solids,* ed. by W. Muschik (Springer, Wien 1993) pp. 1–63
169. W. Muschik: Objectivity and frame indifference, revisited. *Arch. Mech.* **50**, 541–547 (1998)
170. H.K. Myong, N. Kasagi: A new approach to the improvement of $k-\varepsilon$ - turbulence model for wall-bounded shear flows. *JSME Int. J.* **33**(1), 63–72 (1990)
171. I. Newton: *Philosophiae naturalis principia mathematica.* (Jussu Soc. Reg. ac Typis J. Streater, London 1987)
172. J. Nikuradse: *Untersuchungen über die Geschwindigkeitsverteilung in turbulenten Strömungen.* Dissertation, Göttingen (1926), VDI – Forschungsheft **281,** Berlin (1926, 44p.)
173. J. Nikuradse: Gesetzmäßigkeiten der turbulenten Strömung in glatten Rohren. *Forschungsarb. Ing-wesen* **356** (1932, 36p.)
174. J. Nikuradse: Strömungsgesetze in rauhen Rohren. *Forschungsarb. Ing-wesen* **361** (1933b, 22p.)
175. S. Nisizima, A. Yoshizawa: Turbulent channel and Couette flows using an anisotropic $k-\varepsilon$ model, *AIAA Journal,* **25**, 414–420 (1987)
176. W. Noll: A mathematical theory of the mechanical behaviour of continuous media. *Arch. Rational Mech. Anal.* **2**, 197–226 (1958/59)
177. W. Noll: The foundation of classical mechanics in the light of recent advances in continuum mechanics. *Proc. Berkeley Symposium on Aximatic Methods, Amsterdam* 226–228 (1959)
178. W. Noll, E.G. Virga: On edge interactions and surface tension. *Arch. Rational Mech. Anal.* **111**(1), 1–31 (1990)
179. M. Oberlack: Non-isotropic dissipation in non-homogeneous turbulence. *J. Fluid Mech.* **350**, 351–374 (1997)
180. R.W. Ogden: *Nonlinear Elastic Deformations* (John Wiley & Sons 1984, 532p.)
181. A. Okubo: Some Speculations on oceanic diffusion diagrams. *Rapp. P.-v. Run. Const. Int. Explor. Mer.*, **167**, 77–85 (1974)
182. J. Piquet: Turbulent flows, models and physics. (Springer Verlag, Berlin, etc. 1999, 761p.)

183. S.B. Pope: A more generally effective viscosity hypothesis. *J. Fluid Mech.* **72**, 331–340 (1975)
184. S.B. Pope: A perspective on turbulence modelling /CASE/LaRC/AFORSR. In: *Modelling Complex Turbulent Flows* ed. by M.Salar et.al. (Kluwer, Dordrecht 1999) pp.53–67
185. E.P. Popov: Introduction to Mechanics of Solids (Prentice Hall Inc., Englewood Cliffs, N.J. 1968, 571p.)
186. L. Prandtl: Bericht über Untersuchungen zur ausgebildeten Turbulenz. *Zeitschrift für angewandte Mathematik und Mechanik (ZAMM)* **5**(2), 136–139 (1925)
187. L. Prandtl: Neuere Ergebnisse der Turbulenzforschung. *Zeitschr. VDI* **77**, 105–113 (1933)
188. L. Prandtl: Über ein neues Formelsystem für die ausgebildete Turbulenz. *Nachr. Akad. Wiss. Göttingen Math.-Phys. Klasse*, 6–19 (1945)
189. R.W. Preisendorfer and C.D. Mobley: Albedos and glitten patterns of a wind-roughened sea surface. *J. Phys. Oceanogr.* **16**, 1293–1316 (1986)
190. K.R. Rajagopal: On Stability of third grade fluids. *Arch. Mech.* **32**, 867–875 (1980)
191. L. Rayleigh (J.W.S. Strutt): On the viscosity of argon as affected by temperature. *Proc. R. Soc. London* **66**, 68–74 (1899)
192. L. Rayleigh (J.W.S. Strutt): The principle of similitude. *Nature* **95** (66) 591 and 644 (1915)
193. O. Reynolds: An experimental investigation of the circumstances which determine whether the motion of water shall be direct or sinuous, and of the law of resistance in parallel channels *Phil. Trans. R. Soc. London* **A 174**, 935–982 (1883)
194. O. Reynolds: On the dynamical theory of turbulent incompressible viscous fluids and the determination of the criterion. *Phil. Trans. R. Sec. London* **A 186**, 123–164 (1894)
195. W.C. Reynolds: Effects of rotation on homogeneous turbulence. *Proc. 10th Austr. Fluid Mech. Conf. Univ. Melbourne*, 1–6 (1989)
196. W. Rodi: Examples of calculation methods for flow and mixing in stratified fluids. *J. Geophys. Res. (C5)*, **92**, 5305–5328 (1987)
197. W. Rodi: *Turbulence Models and Their Applications in Hydraulics.* IAHR Monograph Series. A. A. 3 ed. (Balkema, Rotterdam/Brookfield 1993, 104p.)
198. J.C. Rotta: *Turbulente Strömungen. Eine Einführung in die Theorie und ihre Anwendung* (Teubner, Stuttgart, 1972, 267p.)
199. R. Rubinstein, J.M. Barton: Non-linear Reynolds stress models and the renormalization group. *Phys. Fluids* **A2**, 1472–1476 (1990)
200. T. Rung, F. Thiele, S. Fu: On the realizability of non-linear stress-strain relationships for Reynolds-stress closures. *Flow, Turbulence and Combustion* **60**, 333–359 (1999)
201. A. Sadiki: *Thermodynamik und Turbulenzmodellierung.* Habilitationsschrift, Fachbereich Mechanik, Technische Universität Darmstadt (1998, 102p.)
202. A. Sadiki, K. Hutter: On the frame dependence and form invariance of the transport equations for the Reynolds stress tensor and the turbulent heat flux vector: its consequences on closure models in turbulence modelling. *Continuum Mech. Thermodyn.* **8**, 341–349 (1996)
203. A. Sadiki, K. Hutter, J. Janicka: Thermodynamically consistent second order turbulence modelling based on extended thermodynamics. In: *Engineering Turbulence Modelling and Experiments* 4, Corsica, eds. by W. Rodi, D. Laurence (Elsevier Sci. Ltd. 1999) pp.99–102

204. A. Sadiki, W. Bauer, K. Hutter: Thermodynamically consistent coefficient calibration in nonlinear and anisotropic closure models for turbulence. *Continuum Mech. Thermodyn.* **12**, 131–149 (2000a)
205. A. Sadiki, K. Hutter: On thermodynamics of turbulence: Development of first order closure models and critical evaluation of existing models. *J. Non-equil. Thermodyn.* **23**, 131–160 (2000b)
206. G.P. Saffman: A model for inhomogenous turbulent flow. *Proc. Roy. Soc. London* **A 317**, 417–433 (1970)
207. Sameer Abu-Zaid and G. Ahmadi: A thermodynamically consistent rate-dependent model for turbulent two-phase flows. *Int. J. Non-lin. Mech.* **30**, 509–529 (1995)
208. J. Sander: Dynamical equations and turbulent closures in Geophysics. *Continuum Mech. Thermodyn.* **10**, 1–28 (1998)
209. H. Schlichting: *Boundary Layer Theory.* Translation by J. Keslin. (Pergamon Press Ltd, London 1955, 535p.)
210. A. Schrijver: Theory of Linear and Integer Programming, Wiley-Interscience Series in Discrete Mathematics (John Wiley & Sons, Chichester, etc. 1996, 471p.)
211. W. Schröder: *Geophysical Hydrodynamics and Ertel's Potential Vorticity,* selected papers of Hans Ertel. (Bremen – Rönnebeck 1991, 218p.)
212. W. Schröder, H.-J. Treder: *Theoretical Concepts and Observational Implications in Meteorology and Geophysics,* (selected papers from the IAGA – symposium to commomerate the 50th anniversary of Ertel's potential vorticity). International Assoc. of Geomagnetism and Aeronomy – Newsletters of the Interdivisional commission on history of the IAGA, 17 (Bremen – Rönnebeck 1993, 206p.)
213. U. Schumann: Realizability of Reynolds-stress turbulence models. *Phys. Fluids* **20**, 721–725 (1997)
214. T.H. Shih: Constitutive relations and realizability of single-point turbulence closures. In: *Turbulence and transition modelling,* eds. by Hallbäck et al (Grademic Publisher, Holland 1996) pp. 155–192
215. T.H. Shih, J. Zhu, J.L. Lumley: A realizable Reynolds stress algebraic equation model. *NASA TM-105993* (1993)
216. G.E. Smith: On isotropic integrity integrity bases, *Arch. Rational Mech. Anal.* **18**, 282–292 (1965)
217. A.J.M. Spencer: *Theory of invariants.* In: A.C. Eringen (Ed.) *Continuum Physics* (Academic Press, New York - London Vol. 1, 1971) pp.240–353
218. A. Spencer: *Continuum Mechanics* (Longman London, New York 1980, 183p.)
219. C.G. Speziale: On non-linear $k - l$ and $k - \varepsilon$ models of turbulence. *J. Fluid Mech.* **178**, 459–475 (1987)
220. C.G. Speziale: Turbulence modeling in non-inertial frames of reference. *Theor. Comput. Fluid Dyn.* **1**, 3–19 (1989)
221. C.G. Speziale, T.B. Gatski, Mac G. Mhuiris: A critical comparison of turbulence models for homogeneous shear flows in a rotating frame. *Phys. Fluids A* **2**(9), 1678–1684 (1990)
222. C.G. Speziale: Analytical methods for the development of Reynolds stress closures in turbulence. *Annu. Rev. Fluid Mech.* **23**, 107–157 (1991)
223. C.G. Speziale, S. Sarkar, T.B. Gatski: Modelling the pressure-strain correlation of turbulence: An invariant dynamical system approach. *J. Fluid Mech.* **227**, 245–272 (1991)
224. C.G. Speziale, Xiang-Hua Xu: Towards the development of second order closure models for non-equilibrium turbulent flows. *Int. J. Heat and Fluid Flows,* **17**, 238–244 (1996)

225. C.G. Speziale, T.B. Gatski: Analysis and modelling of anisotropies in the dissipation rate of turbulence. *J. Fluid Mech.* **344**, 155–180 (1997)
226. C.G. Speziale: Turbulence modelling for time-dependent RANS/VLES, A Review. *AIAA-J.* **36** 173–184 (1998)
227. J.H. Spurk: *Dimensionsanalyse in der Strömungslehre* (Springer, Berlin, etc. 1992, 270p.)
228. M. Straskraba and A. Gnauck: Aquatische Oekosysteme, Modellierung und Simulation (G. Fischer Verlgag 1983, 279p.)
229. B. Straughan: *The Energy Method, Stability and Nonlinear Convection.* Appl. Mathematical Aciences, Vol. 91 (Springer, New York etc. 1992, 242p.)
230. B. Svendsen and K. Hutter: On the thermodynamics of a mixture of isotropic viscous materials with kinematic constraints. *Int.-J. Eng. Sci.* **33**, 2021–2054 (1995)
231. U. Svensson: *A Mathematical Model of the Seasonal Thermocline.* Report 1002, Department of Water Resources Engineering, University of Lund, Sveden (1978, 187p.)
232. W.C. Swinbaeck: Long-wave radiation from clear skies. *Quart. J. R. Meteorol. Soc.* **89**, 339–348 (1963)
233. I. Szabo: *Geschichte der mechanischen Prinzipien.* 3nd edn. (Birkhäuser-Verlag, Basel 1987, 571p.)
234. G.I. Taylor: The formation of a blast wave by a very intense explosion. Part I: Theoretical discussion. Part II: The atomic explosion of 1945. *Proc. R. Soc. London*, **A 201**, 159–186 (1950)
235. H. Tennekes, J.L. Lumley: *A First Course in Turbulence* (M.I.T Press 1972, 300p.)
236. S. Timoshenko: *History of Strength of Materials* (McGraw-Hill Book Company New York 1953, 452p.)
237. A. A. Townsend: *The Structure of Turbulent Shear Flow* (Cambridge University Press, Cambridge 1976, 429p.)
238. C.A. Truesdell, R.A. Toupin: The Classical Field Theories. In: *Encyclopedia of Physics,* Vol. III/1, ed. by S. Flügge (Springer, Berlin etc. 1957) pp.226–793
239. C.A. Truesdell: Solle basi della termomechanica. Rend. Accad. Lincei (8/22m 33-88, 158-166. [Engl. Translation in Rational Mechanics of Materials. Int. Sci. Rev. Ser. New York (Gordon and Breach, 1965)]
240. C.A. Truesdell, W. Noll: The Non-Linear Field Theories of Mechanics. In: *Encyclopedia of Physics,* Vol. III/3, ed. by S. Flügge (Springer, Berlin etc. 1965, 602p.)
241. C.A. Truesdell: *Essays in the History of Mechanics* (Springer, Berlin etc. 1968, 383p.)
242. C.A. Truesdell, R.G. Muncaster: *Fundamentals of Maxwell's Kinetic Theory of a Simple Monotonic Gas treated as a Branch of Rational Mechanics.* (Academic Press, New York etc. 1980, 593p.)
243. C.A. Truesdell: *Rational Thermodynamics.* 2nd edn. (Springer, Berlin etc. 1984, 578p.)
244. C.A. Truesdell: *The Elements of Continuum Mechanics* (Springer, Berlin etc. 1985, 279p.)
245. C.A. Truesdell: *Newtons Einfluss auf die Mechanik des 18. Jahrhunderts.* In: (K. Hutter, ed.) *Die Anfänge der Mechanik. Newtons Principia gedeutet aus ihrer Zeit und ihrer Wirkung auf die Physik* (Springer, Berlin, etc. 1989) pp.47–73
246. L. Umlauf: *Turbulence Parameterization in Hydrobiological Models for Natural Waters.* Ph. D. Dissertation, Department of Mechanics, Darmstadt University of Technolgy, Darmstadt, Germany (2001, 231p.)

247. R. Du Vachat: Realizability inequalities in turbulent flows *Phys. Fluids*, **20**(4), 551–556 (1977)
248. M.G. Velarde, C. Normand: Convection *Scietific American*, **243**(1), 92–& (1980)
249. W. Voigt: *Lehrbuch der Kristallphysik* (Teubner, Leipzig & Berlin, 1910, 946p.)
250. C.C. Wang: A new representation theorem for isotropic functions: Parts I and II. *Arch. Rational Mech. Anal.* **36**, 166–197, 198–223 (1970)
251. C.C. Wang: Corrigendum to my recent paper on "Representations for isotropic functions", *Arch. Rational Mech. Anal.* **43**, 392–395 (1971)
252. C.-C. Wang, C.A. Truesdell: *Introduction to Rational Elasticity* (Noordhoff Int. Publ. 1973, 556p.)
253. L. Wang: Frame-indifferent and positive-definite Reynolds stress-strain relation. *J. Fluid Mech.* **352**, 341–358 (1997)
254. J. Weis: *Ein algebraisches Reynoldsspannungsmodell (An algebraic Reynolds stress model)*. Ph. D. Dissertation, Department of Mechanics, Darmstadt University of Technology, Darmstadt, Germany (2001, 111p.)
255. D.C. Wilcox: *Turbulence modelling for CFD.* DCW Industries (1993, 460p.)
256. K. Wilmanski: *Thermomechanics of Continua* (Springer, Berlin etc. 1998, 273p.)
257. G.B. Whitham: *Linear and Nonlinear Waves* (John Wilex Interscience, Chichester, 1974, 636p.)
258. V. Yakhot, S.A. Orszag: Renormalization group analysis of turbulence, I. Basic theory. *J. Sci. Comput.* **1**, 3–51 (1986)
259. Z. Yang, T.H. Shih: New time scale based $k - \varepsilon$ - model for near-wall turbulence. *AIAA J.* **31**(7), 1191–1198 (1993)
260. A. Yoshizawa: Statistical analysis of the derivation of the Reynolds stress from its eddy viscosity representation. *Phys. Fluids* **27**, 1377–1387 (1984)
261. A. Yoshizawa: Statistically derived system of equations for turbulent shear flows. *Phys. Fluids* **28**, 59–63 (1985)
262. H. Xiao, O.T. Bruhns and A. Meyers: Logarithmic Strain, logarithmic spin and logarithmic rate. *Acta Mech.* **124**, 89–105 (1997)
263. R.K. Zeytounian: *Meteorological Fluid Dynamics Asymptotic Modelling, Stability and Chaotic Atmospheric Motion.* Lecture Notes in Physics, Monograph 5, (Springer, Berlin etc. 1991, 346p.)

Name Index

Index

GPSR Compliance
The European Union's (EU) General Product Safety Regulation (GPSR) is a set of rules that requires consumer products to be safe and our obligations to ensure this.

If you have any concerns about our products, you can contact us on

ProductSafety@springernature.com

In case Publisher is established outside the EU, the EU authorized representative is:

Springer Nature Customer Service Center GmbH
Europaplatz 3
69115 Heidelberg, Germany

www.ingramcontent.com/pod-product-compliance
Ingram Content Group UK Ltd.
Pitfield, Milton Keynes, MK11 3LW, UK
UKHW021902190726
13853UKWH00003B/1386

* 9 7 8 3 6 6 2 0 6 4 0 3 0 *